THE SCIENCE OF ANIMALS THAT SERVE HUMANITY

McGraw-Hill Publications in the Agricultural Sciences

Consulting Editor in Animal Science
John R. Campbell, *University of Illinois*

Consulting Editor in Agricultural Engineering
Carl Hall, *College of Engineering, Washington State University*

Brown: Farm Electrification
Campbell and Lasley: The Science of Animals That Serve Humanity
Campbell and Marshall: The Science of Providing Milk for Man
Christopher: Introductory Horticulture
Edmonds, Senn, Andrews, and Halfacre: Fundamentals of Horticulture
Jones and Aldred: Farm Power and Tractors
Kipps: The Production of Field Crops
Kohnke and Bertrand: Soil Conservation
Krider, Conrad, and Carroll: Swine Production
Lassey: Planning in Rural Environments
Laurie, Kiplinger, and Nelson: Commercial Flower Forcing
Maynard, Loosli, Hintz, and Warner: Animal Nutrition
Metcalf, Flint, and Metcalf: Destructive and Useful Insects
Smith and Wilkes: Farm Machinery and Equipment
Sorenson: Animal Reproduction: Principles and Practices
Thompson and Troeh: Soils and Soil Fertility
Thompson and Kelly: Vegetable Crops
Treshow: Environment and Plant Response
Walker: Plant Pathology
Warwick and Legates: Breeding and Improvement of Farm Animals

THE SCIENCE OF ANIMALS THAT SERVE HUMANITY

THIRD EDITION

John R. Campbell

Dean, College of Agriculture
University of Illinois
Formerly Professor of Dairy Science
University of Missouri

John F. Lasley

Professor Emeritus of Animal Science
University of Missouri

McGRAW-HILL BOOK COMPANY

New York   St. Louis   San Francisco   Auckland   Bogotá
Hamburg   Johannesburg   London   Madrid   Mexico   Montreal   New Delhi
Panama   Paris   São Paulo   Singapore   Sydney   Tokyo   Toronto

This book was set in Times Roman by Black Dot, Inc. (ECU).
The editor was Mary Jane Martin;
the production supervisor was Marietta Breitwieser.
Project supervision was done by The Total Book.
Halliday Lithograph Corporation was printer and binder.

THE SCIENCE OF ANIMALS THAT SERVE HUMANITY

1 2 3 4 5 6 7 8 9 0   HALHAL   8 9 8 7 6 5 4

ISBN 0-07-009700-3

Library of Congress Cataloging in Publication Data

Campbell, John R., date
    The science of animals that serve humanity.

    Rev. ed. of: The science of animals that serve
mankind. 2nd ed. 1975.
    Includes index.
    1. Livestock. I. Lasley, John Foster, date   .
II. Campbell, John R., date   . Science of animals
that serve mankind. 2nd ed. III. Title.
SF61.C219 1985   636   84-7950
ISBN 0-07-009700-3

We respectfully dedicate this book
to our students from whom
we received the inspiration
to prepare the text.

CONTENTS

PREFACE   xv

1 Animal Agriculture and Humanity   1

1.1 HISTORY AND DEVELOPMENT OF ANIMAL AGRICULTURE   1
1.2 DOMESTICATION OF ANIMALS   2
1.3 HISTORY OF AGRICULTURAL EDUCATION AND RESEARCH   5
1.4 ANIMAL AGRICULTURE AND THE WORLD ECONOMY   6
1.5 ANIMAL AGRICULTURE AND THE UNITED STATES ECONOMY   32
1.6 SUMMARY   38

2 Breeds of Livestock and Poultry   43

2.1 INTRODUCTION   43
2.2 DEVELOPMENT OF BREEDS   43
2.3 BREEDS OF LIVESTOCK AND POULTRY   52
2.4 SUMMARY   59

3 Animal Products and Humanity   64

3.1 HISTORY OF AVAILABILITY   64
3.2 COMPOSITION AND COMPARATIVE NUTRITIONAL CONTRIBUTIONS OF SELECTED ANIMAL PRODUCTS   66
3.3 PURCHASING FOOD NUTRIENTS VIA ANIMAL PRODUCTS   76
3.4 TRENDS IN PER CAPITA CONSUMPTION OF ANIMAL PRODUCTS   80
3.5 ATHEROSCLEROSIS   84
3.6 THE ROLE OF DOMESTIC ANIMALS IN THE ACCUMULATION OF RADIONUCLIDES IN FOODS   89
3.7 FORTIFICATION OF ANIMAL PRODUCTS   90

3.8 PRESERVATION OF ANIMAL PRODUCTS   92
3.9 THE FUTURE OF ANIMAL PRODUCTS   101
3.10 SUMMARY   107

4 Fundamental Principles of Genetics   110

4.1 THE CELL THEORY OF INHERITANCE   110
4.2 CHROMOSOMAL ABNORMALITIES   112
4.3 DETERMINATION OF SEX   113
4.4 CELL DIVISION   113
4.5 THE GENE AND HOW IT FUNCTIONS   119
4.6 GENES AND EMBRYOLOGICAL DEVELOPMENT   129
4.7 VIRUSES   130
4.8 GENETIC ENGINEERING   130
4.9 SEGREGATION AND RECOMBINATION OF GENES   135
4.10 LAWS OF PROBABILITY AND ANIMAL BREEDING   136
4.11 MUTATIONS   137
4.12 PHENOTYPIC EXPRESSION OF GENES (NONADDITIVE)   138
4.13 PHENOTYPIC EXPRESSION OF GENES (ADDITIVE)   141
4.14 SEX-LINKED INHERITANCE   142
4.15 SEX-INFLUENCED INHERITANCE   143
4.16 SEX-LIMITED TRAITS   144
4.17 SUMMARY   144

5 Principles of Selecting and Mating Farm Animals   150

5.1 PHENOTYPIC VARIATIONS IN QUANTITATIVE TRAITS   150
5.2 STATISTICAL EVALUATION OF QUANTITATIVE TRAITS   151
5.3 FREQUENCY OF GENES IN A POPULATION   153
5.4 CAUSES OF PHENOTYPIC VARIATION   156
5.5 SELECTION   157
5.6 SELECTION FOR DIFFERENT KINDS OF GENE ACTION   158
5.7 SELECTION OF SUPERIOR BREEDING STOCK   164
5.8 PREDICTING THE AMOUNT OF PROGRESS POSSIBLE
     THROUGH SELECTION   172
5.9 GENETIC CORRELATIONS   174
5.10 MATING SYSTEMS FOR LIVESTOCK IMPROVEMENT   174
5.11 SUMMARY   180

6 Anatomy and Physiology of Farm Animals   184

6.1 INTRODUCTION   184
6.2 EXTERNAL BODY PARTS   185
6.3 THE SKELETAL SYSTEM   190
6.4 THE MUSCULAR SYSTEM   193

6.5   THE CIRCULATORY SYSTEM   195
6.6   THE DIGESTIVE SYSTEM   203
6.7   THE RESPIRATORY SYSTEM   203
6.8   THE NERVOUS SYSTEM   206
6.9   THE URINARY SYSTEM   208
6.10 SUMMARY   211

7   The Application of Endocrinology to Selected
     Animals and Humans   213

7.1   THE SCIENCE OF ENDOCRINOLOGY   213
7.2   ENDOCRINE GLANDS AND THEIR SECRETIONS   215
7.3   THE CHEMICAL NATURE OF HORMONES   218
7.4   FUNCTIONS OF HORMONES   219
7.5   MECHANISM OF HORMONE ACTION   220
7.6   REGULATION OF HORMONE SECRETION   221
7.7   PROSTAGLANDINS (PG)   221
7.8   PRACTICAL USES OF NATURAL AND SYNTHETIC
      HORMONES   222
7.9   SUMMARY   224

8   The Physiology of Growth and Senescence   226

8.1   INTRODUCTION   226
8.2   THE PHENOMENON OF GROWTH   227
8.3   THE CELL IS THE UNIT OF GROWTH   230
8.4   PERIODS OF GROWTH   232
8.5   HORMONAL CONTROL OF GROWTH   238
8.6   NUTRITION AND GROWTH   242
8.7   HEREDITARY MECHANISMS IN GROWTH   244
8.8   SENESCENCE (AGING)   247
8.9   SOME THEORIES OF AGING   251
8.10 SUMMARY   252

9   Anatomy and Physiology of Reproduction in
     Farm Mammals   255

9.1   INTRODUCTION   255
9.2   ANATOMY OF THE MAMMALIAN MALE REPRODUCTIVE
      TRACT   256
9.3   ANATOMY OF THE MAMMALIAN FEMALE
      REPRODUCTIVE TRACT   261
9.4   PHYSIOLOGY OF REPRODUCTION IN FARM MAMMALS   265
9.5   APPLICATION OF RECENT RESEARCH FINDINGS IN
      THE PHYSIOLOGY OF REPRODUCTION   281
9.6   SUMMARY   292

10   Artificial Insemination   **295**

   10.1   NOMENCLATURE AND DEFINITION   295
   10.2   HISTORY AND DEVELOPMENT OF ARTIFICIAL
          INSEMINATION   295
   10.3   IMPORTANCE AND IMPLICATIONS OF ARTIFICIAL
          INSEMINATION   300
   10.4   SEMEN COLLECTION   302
   10.5   EVALUATION OF SEMEN   307
   10.6   EXTENSION OF SEMEN   312
   10.7   SEMEN STORAGE   314
   10.8   REGULATIONS GOVERNING ARTIFICIAL INSEMINATION
          IN CATTLE   318
   10.9   INSEMINATING THE COW   318
   10.10 THE NATIONAL ASSOCIATION OF ANIMAL BREEDERS, INC.   320
   10.11 ARTIFICIAL INSEMINATION IN POULTRY   320
   10.12 ARTIFICIAL INSEMINATION IN BEES   321
   10.13 ARTIFICIAL INSEMINATION IN HUMANS   321
   10.14 THE FUTURE OF ARTIFICIAL INSEMINATION   324
   10.15 SUMMARY   327

11   Physiology of Lactation   **329**

   11.1   INTRODUCTION   329
   11.2   MAMMARY GLAND DEFINED   330
   11.3   ANATOMY AND ARCHITECTURE OF MAMMARY
          GLANDS   330
   11.4   GROWTH AND DEVELOPMENT OF MAMMARY GLANDS   334
   11.5   HORMONAL REGULATION OF LACTATION   340
   11.6   HOW MILK IS MADE   344
   11.7   HOW MILK IS DISCHARGED (SECRETED)   347
   11.8   THE PHENOMENON OF MILK LETDOWN   350
   11.9   REGRESSION (INVOLUTION) OF THE MAMMARY
          GLAND   355
   11.10 FACTORS AFFECTING LACTATION   356
   11.11 FACTORS AFFECTING THE COMPOSITION OF MILK   360
   11.12 IMMUNOLOGICAL ASPECTS OF COLOSTRUM   364
   11.13 SUMMARY   366

12   Physiology of Egg Laying   **369**

   12.1   INTRODUCTION   369
   12.2   EGG COLORS, SHAPES, AND KINDS   370
   12.3   THE STRUCTURE OF AN EGG   372
   12.4   REPRODUCTION AND EGG FORMATION   375
   12.5   HORMONAL REGULATION OF EGG LAYING   379

12.6 HOW AN EGG IS LAID (OVIPOSITION)   382
12.7 FACTORS AFFECTING EGG LAYING   383
12.8 FACTORS AFFECTING THE COMPOSITION AND
      CHARACTERISTICS OF EGGS   390
12.9 FACTORS AFFECTING EGG SIZE   391
12.10 IMMUNOLOGICAL AND MEDICAL ASPECTS OF EGGS   393
12.11 SUMMARY   394

13 Ecology and Environmental Physiology   397

13.1 INTRODUCTION   397
13.2 HEREDITY AND ENVIRONMENT   400
13.3 ADAPTATION TO ENVIRONMENT   403
13.4 STRESS   405
13.5 HOMEOSTASIS AND HOMEOTHERMY   406
13.6 TEMPERATURE REGULATION   409
13.7 NUTRITIONAL CONSIDERATIONS OF ENVIRONMENTAL
      CONDITIONS   412
13.8 FEVER (PYREXIA)   415
13.9 THE THERMONEUTRAL ZONE   416
13.10 HEAT PRODUCTION   418
13.11 HEAT DISSIPATION   420
13.12 EFFECTS OF CLIMATE ON PRODUCTION   428
13.13 SUMMARY   428

14 Principles of Nutrition—Plant and Animal Composition   432

14.1 INTRODUCTION   432
14.2 COMPOSITION OF PLANTS AND ANIMALS   434
14.3 ANALYSIS OF FOODSTUFFS   447
14.4 DETERMINATION OF THE DIGESTIBILITY OF FEEDS   449
14.5 THE ENERGY CONTENT OF FOODS   451
14.6 FEED ADDITIVES   456
14.7 SUMMARY   457

15 The Physiology of Digestion in Nutrition   459

15.1 INTRODUCTION   459
15.2 TYPES AND CAPACITIES OF DIGESTIVE SYSTEMS   460
15.3 THE PROCESS OF DIGESTION   463
15.4 APPETITE   466
15.5 THE PREHENSION OF FOOD   466
15.6 THE MASTICATION OF FOOD   466
15.7 ENZYMES OF THE DIGESTIVE TRACT   467
15.8 AVIAN DIGESTION   469
15.9 ABSORPTION OF FOOD NUTRIENTS   470
15.10 FACTORS AFFECTING THE DIGESTIBILITY OF FEEDS   470

15.11 EFFICIENCY OF FOOD CONVERSION — 473

15.12 FACTORS AFFECTING THE EFFICIENCY OF FOOD CONVERSION — 474

15.13 SUMMARY — 476

16 The Nutritional Application of Vitamins to Human and Animal Health — 478

16.1 INTRODUCTION — 478

16.2 VITAMINS DEFINED — 479

16.3 THE FAT-SOLUBLE VITAMINS — 479

16.4 THE WATER-SOLUBLE VITAMINS AND RELATED COMPOUNDS — 491

16.5 VITAMIN ASSAYS — 509

16.6 EXPRESSING VITAMINS A AND D QUANTITATIVELY — 510

16.7 SUPPLYING VITAMINS TO FARM MAMMALS AND POULTRY — 510

16.8 SUMMARY — 510

17 The Nutritional Contributions of Minerals to Humans and Animals — 512

17.1 INTRODUCTION — 512

17.2 THE MACROELEMENTS — 513

17.3 THE MICROELEMENTS (TRACE ELEMENTS) — 521

17.4 SUMMARY — 534

18 Animal Disease and the Health of Humans — 536

18.1 INTRODUCTION — 536

18.2 DISEASE AND HEALTH — 538

18.3 SELECTED ANIMAL DISEASES TRANSMISSIBLE TO HUMANS — 546

18.4 SELECTED HUMAN DISEASES TRANSMISSIBLE TO ANIMALS — 581

18.5 DISEASES TRANSMISSIBLE BY ANIMALS AS PASSIVE CARRIERS — 582

18.6 TOXIC PLANTS — 583

18.7 GOVERNMENTAL SAFEGUARDS FOR ANIMAL AND HUMAN HEALTH — 586

18.8 PROTECTING UNITED STATES LIVESTOCK FROM FOREIGN DISEASES — 590

18.9 SUMMARY — 593

19 Selected Insects and Parasites of Significance to Humans and Animals — 596

19.1 INTRODUCTION — 596

19.2 TAXONOMY — 598

19.3 CONTRIBUTIONS OF INSECTS TO HUMANITY   **601**
19.4 HARMFUL EFFECTS OF INSECTS   **605**
19.5 SELECTED ARTHROPODS AFFECTING DOMESTIC
      ANIMALS AND/OR HUMANS   **608**
19.6 ARTHROPOD CONTROL—ESSENTIAL FOR HUMANITY   **617**
19.7 SUMMARY   **627**

20 Animal Behavior   **630**

20.1 INTRODUCTION   **630**
20.2 CAUSES OF BEHAVIORIAL RESPONSES IN ANIMALS   **632**
20.3 MOTIVATION   **636**
20.4 METHODS OF ANIMAL COMMUNICATION   **637**
20.5 ORIENTATION (NAVIGATION OR HOMING) BEHAVIOR   **639**
20.6 TYPES OF ANIMAL BEHAVIOR   **639**
20.7 SOCIAL DOMINANCE   **654**
20.8 POPULATION DENSITY AND ANIMAL BEHAVIOR   **655**
20.9 SUMMARY   **656**

21 Horses in the Service of Humanity   **658**

21.1 INTRODUCTION   **658**
21.2 CHARACTERISTICS AND TYPES OF HORSES   **660**
21.3 SELECTION OF HORSES   **665**
21.4 CARE AND MANAGEMENT OF BREEDING ANIMALS   **668**
21.5 NUTRITION OF HORSES   **675**
21.6 TRAINING AND GROOMING HORSES   **678**
21.7 COMMON DEFECTS AND UNSOUNDNESS
      IN HORSES   **679**
21.8 DETERMINING THE AGE OF HORSES   **686**
21.9 DISEASE AND PARASITE CONTROL   **689**
21.10 SUMMARY   **691**

22 Animal Research in Retrospect and Prospect **698**

22.1 INTRODUCTION   **698**
22.2 KINDS OF ANIMAL RESEARCH   **701**
22.3 THE SCIENTIFIC METHOD   **703**
22.4 RESEARCH CONTROL   **706**
22.5 STATISTICS AS A RESEARCH TOOL   **706**
22.6 COMPUTERS: AID TO RESEARCH   **708**
22.7 LITERATURE AND THE LIBRARY   **709**
22.8 RESEARCH ORGANIZATIONS   **710**
22.9 ATOMS IN ANIMAL RESEARCH   **712**
22.10 RESEARCH IN ENDOCRINOLOGY   **726**
22.11 GAS CHROMATOGRAPHY (GC) IN ANIMAL
      RESEARCH   **727**
22.12 HIGH-PRESSURE LIQUID CHROMATOGRAPHY (HPLC)   **729**

22.13 IN PRAISE OF PIGS AS RESEARCH ANIMALS   730
22.14 FISTULATED ANIMALS: AID TO RESEARCH   734
22.15 ANIMAL-DISEASE RESEARCH   735
22.16 FINANCING AGRICULTURAL RESEARCH   741
22.17 TRENDS AFFECTING FUTURE ANIMAL RESEARCH   743
22.18 OPPORTUNITIES IN ANIMAL RESEARCH   756
22.19 SUMMARY   756

APPENDIXES
A Common Terms or Names Applied to Selected Farm Animals   761
B Convenient Conversion Data   762
C Tables of Weights and Measures   764
D Agricultural Colleges and Experiment Stations in the
   United States   766
E Alphabetical List of Elements and Symbols   769

GLOSSARY   771
INDEX   I-1

PREFACE

This book is designed to serve as a text for college students who desire a comprehensive introduction to the fundamental principles of animal science, emphasizing the study of the animals that serve humanity. We hope it communicates our own enthusiasm for this exciting field of science.

The third edition of *The Science of Animals That Serve Humanity* (formerly titled *Mankind*) resembles the previous ones in its organization, literary style, and reading level. New sections on the application of biology and technology to animal science in many areas, such as reproductive physiology, genetics, and animal health, have been added. A new, well-illustrated chapter on breeds of livestock and poultry has been added, and tabular and illustrative materials have been updated throughout.

The initial chapter presents materials related to the economic impact of animal agriculture on the United States and the world. In subsequent chapters, information pertaining to the nutritional contributions of animal products and the principles of animal genetics, anatomy, and physiology are presented. Included in these principles are digestion, growth, senescence, lactation, egg laying, the physiology of reproduction, and the ecology and stress responses of animals. Other materials related to artificial insemination, endocrinology, nutrition, animal disease and public health, parasites, insects and their biological control, and animal behavior are included as well. Because of the increasing interest in horses, a chapter is devoted to their contributions to humans. Problems associated with population and food production and those related to animal-waste disposal have gained public attention, especially as they relate to environmental quality, and so these timely topics are discussed. The final chapter presents an overview of the all-important field of animal science research. After reviewing examples of the countless contributions animal research has made to the betterment of humanity, we hope many curious, science-oriented students will be inspired to pursue a career in this challenging and rewarding field.

To foster a better understanding of the materials presented, a glossary has been included. (Glossary words have been set in **boldface** type the first time they appear in a particular chapter.) Study questions are provided at the end of each chapter so that students can test their knowledge and understanding of the materials presented.

The subject matter has not been treated exhaustively. Rather, the object has been to introduce the field of animal science and the contributions of animals to the progress of humanity. No attempt has been made to fully document the materials presented. However, special references have been cited where data or direct quotations are involved or when fundamental historical contributions are mentioned. It is our belief that an undergraduate textbook on this subject should be designed to develop and portray an integrated and coherent picture rather than to serve as a reference to the literature of animal science. Emphasis has been given to the basic principles of animal science; management and marketing aspects are reserved for advanced courses.

A survey conducted by the publisher of our teaching colleagues at other schools was especially helpful in determining that different teachers assign higher priority to certain chapters than to others. However, the consensus was that although the book contains more information than can be included in a single course, all chapters are being used. For this reason, we have not deleted any chapters that were included in the first two editions.

The sequence followed in presenting the subject matter is the one developed and revised by us and our colleagues in teaching animal science; however, units are independent and may be presented in the order the instructor desires.

To emphasize the importance of studying the supporting disciplines, we have purposely used many anatomical, behavioral, chemical, ecological, genetic, mathematical, microbiological, nutritional, physiological, statistical, and taxonomic terms. (Many are defined in the Glossary.) We trust that such terms will open new vistas to students and thereby increase the scope of their understanding.

Each chapter is preceded by an appropriate quotation. This practice is intended to stimulate students to "think" as they begin reading each chapter:

Think, reader, for thyself, so God allow thee profit from thy reading, think I say!

Dante Alighieri (1265–1321)

Students, teachers, and others throughout the United States, Canada, and other countries who used earlier editions and who kindly made valuable suggestions for improvement were particularly helpful in the revision process. The overall value of their contributions cannot be measured adequately, and we are most grateful to them.

Scientists who contributed to research, teaching, and public service over the years—those who generated the data and information recorded here—are too numerous to name and impossible to repay. But we are especially indebted to the more than 25 animal science and veterinary medicine faculty colleagues at the Universities of Florida, Illinois, and Missouri who reviewed the manuscript and contributed immensely to this edition. They are acknowledged in the respective chapters. In addition, we gratefully acknowledge the excellent editorial assistance, including countless contributions made to clarity and consistency, by several members of the McGraw-Hill editorial and production team: Mary Jane Martin, editor; Annette Bodzin, project supervisor; Karen McDermott, copy editor; and Josephine Satloff, proofreader.

Finally, students continue to provide the primary source of motivation for our keeping the book current. We hope that through their study of the exciting phenomena associated with the biology of farm animals and of the recent research discoveries presented in this revised edition, student pleasure, excitement, and curiosity will run high and result in further exploration of the challenging field of animal science in both the classroom and the library.

John R. Campbell
John F. Lasley

THE SCIENCE OF ANIMALS THAT SERVE HUMANITY

ANIMAL AGRICULTURE AND HUMANITY

A land poor in livestock is never rich, and a land rich in livestock is never poor.

Arab philosopher

1.1 HISTORY AND DEVELOPMENT OF ANIMAL AGRICULTURE

The success of human beings on earth is attributable largely to the animals that have fed, clothed, and carried them and cultivated their fields. Animal agriculture utilizes biological processes to produce animal products useful to humans. Animal science embraces all disciplines in the biological and physical sciences that influence animal life.

The field of animal science developed with the **domestication**[1] of animals in the Neolithic (new stone age) period, when the world population is estimated to have been a meager 5 million. This era also marked the first step toward civilization of the most primitive tribes of humans. It was the beginning of humanity's transformation from the savage to the civilized way of life—from nomads, or wanderers, to, eventually, urban dwellers. The herding of animals became indicative of the superiority of one tribe over another. Historically, the great livestock countries of the world have supported the most advanced civilizations and have been the most powerful.

Throughout most of their existence, humans were nomadic, their numbers small, and their technologies rudimentary. Agricultural research is really very new when viewed in terms of the perhaps millions of years of human existence. The domestication of animals that serve humanity is only about 10 to 12 thousand years old. Although there have been "growing pains" through the years, with each succeeding step in their advancing civilizations people have become more dependent on animals and their products. Supporting successful animal production are years of experience and scientific research. From this research, Texas Longhorns were replaced by meat-type **steers;** Arkansas **Razorbacks** were replaced by improved meat-type hogs; black, brown, and spotted sheep were replaced by improved **mutton-** and wool-type sheep; inefficient **poultry** were replaced by fast-gaining birds with high **feed conversion,** birds that

[1]Words set in boldface type are defined in the Glossary at the back of the book.

1

now convert 2 pounds (lb) of feed into 1 lb of meat; and **cows** originally selected for meat, milk, and **draft** in many countries have been replaced by high-producing dairy cows.

Just as they do today, animals served humanity in early times in many ways other than as food. They provided leather and wool for clothing, bones for tools, and **dung** for fertilizer and fuel. They were a means of transportation and were used for entertainment and in religious offerings. Today portions of various animals are used in the manufacture of certain pharmaceuticals, fuel, fertilizer, oil, gelatin, glue, and other industrial products. Catgut (commonly made from sheep intestines) is used for violin and tennis racket strings and for sutures. Pigskin, the hide of swine, is synonymous with football in the United States. Animals serve humanity as subjects in experiments for medical and scientific research. Painters enjoy brushing farm scenes that include animal life. In short, animals contribute greatly to both the mental and physical health and well-being of humanity.

> He who looks on his cattle merely as meat and milk has lost the art of living. There is beauty in the scene of feeding cattle equalling that of music or theater. Both associate man with the meaningful things of life.
>
> *A. P. Schultze*

> The power bestowed on the horse, the dog, the ox, the sheep, the cat, and many species of domestic fowls, of supporting almost every climate, was given expressly to enable them to follow man throughout all parts of the globe in order that we might obtain their services, and they our protection.
>
> *C. Lyell* (1837)

1.2   DOMESTICATION OF ANIMALS

Domesticated cattle have long been a hallmark of civilization. Where beef and dairy animals were raised, humans enjoyed improved health and prosperity. Cultivation of plants and domestication of animals began at approximately the same place, but at different times. The place was the hills of southwestern Asia in the Zagros, Lebanese, and Palestinian mountains.[2] The first agriculturalists and stock raisers were of the Mediterranean race, similar to the people dwelling in that region at the present time.

The shift from food gathering to food cultivation began about 10,000 to 16,000 years ago. Mortars and pestles used for grinding grain have been found

---

[2] Anthropologists at the University of Massachusetts found evidence recently that animal husbandry may have begun some 15,000 years ago in east Africa. The findings include bones and teeth of cattle at three separate sites in the Kenya highlands, about 25 miles from Nairobi. Using modern radiocarbon dating techniques (see Chapter 22), the Massachusetts scientists were able to identify the animals and to determine approximately when they lived. Archeological examinations indicated that the cattle must have been domesticated, because there probably were never wild cattle in the area. Moreover, other studies have shown that tsetse flies, the primary cause of cattle deaths in Africa (see Chapter 19), would have wiped out any wild animals that were roaming the area freely. The assumption that civilization began in Africa long before the iron age is supported by discoveries of 18,000-year-old domesticated grain crops in Africa.

TABLE 1.1
THE DOMESTICATION OF ANIMALS

Species	When, B.P.,* years	Where	Why	How
Dog	8500–9000	Old and new worlds	Pet, companion	Wolf or jackal
Goat	8500–9000	Old world	Food, milk, and clothing	Wild goat
Pig	8000–9000	Old world	Food and sport	European wild boar
Sheep	6000–7000	Old world	Food, milk, and clothing	European mouflon and Asiatic urial
Cattle	6000–6500	Old world	Religious reasons	Aurochs
Chickens	5000–5500	India, Sumatra, and Java	Cockfights, shows, food, and religion	Jungle fowl
Horse	4000–5000	Old world	Transportation	Wild horse
Ducks	?	Probably China	Food and feathers	Wild duck
Geese	?	Greece and Italy	Food and feathers	Wild goose
Turkeys	?	Mexico or North America	Food and feathers	Wild turkey

*B.P. means *before present*.

that verify this belief. Domestication of animals came somewhat later. The first animals to be domesticated may have been the dog and the goat, probably 8500 to 9000 years ago. The probable time of domestication for selected animals, together with other information, is given in Table 1.1.

Some plants and a few animals were domesticated in the new world, but most were domesticated in the old world. Animals domesticated in the new world include the llama, alpaca, vicuna, Andean guinea pig, and turkey. Dogs and bees and bee products were common to both the new and old worlds.

Several important crops of today were first cultivated in the new world. Among these are the white and sweet potatoes, chili, sunflower, peanut, common bean and other varieties of bean, pumpkin, gourd, squash, tomato, pineapple, tobacco, and Indian corn, or maize. In 1965 in Athens, Ohio, an ear of maize was discovered that yielded radiocarbon dated at 280 B.C. Tobacco, Indian corn, and the white potato probably were the new world's greatest contributions to crops of the world. The impact of introducing the white potato into Ireland was sensational. In the latter part of the seventeenth century, Ireland had a population of nearly 2 million living in hunger. Then the white potato was introduced from the new world. This crop rapidly became popular because the soil and **climate** were ideal for its growth and production. The white potato produced more food per acre in that country than had ever been produced before. By 1835, the population of Ireland had increased from 2 to 8 million persons, largely because of the increased food supply. Then came a potato-crop failure, resulting in a great famine. It is said that 2 million people died of starvation. Another 2 million migrated to other countries. Since that time, the population of Ireland has been nearly stabilized at approximately 4 million, but the white potato is still an important food crop.

Famines are as old as the twelfth chapter of Genesis, when Abraham went down to Egypt "and there was a famine in the land." In 1125 a famine reduced by one-half the population of Germany. Hungary experienced a serious famine in 1505. England records a terrible famine in 1586, and in 1870 to 1872 Persia lost one-fourth of its population to hunger.

Some 10 million Chinese died of starvation in 1877 to 1878. Famines in India claimed some 3 million lives in 1769 to 1770, 1.5 million in 1865 to 1866, and 0.5 million in 1877. In 1891 to 1892 a Russian famine brought severe hardship to an estimated 27 million people.

One of the great historical events in Europe during the twentieth century was the Russian Revolution of 1917. Included on its banner was the inscription "Bread and Peace." These two words are related and have always been important to the welfare and perpetuation of the human race.

Indian corn has become one of the greatest crops in the history of the world, especially in the corn belt area of the United States. Billions of bushels of corn are produced in that region, and it is a chief source of energy for growing and **finishing** millions of **livestock** annually. It is also an important source of food for humans in many parts of the world. Through the development of a high-protein variety having a better balance of amino acids, corn promises to become an even more important food crop for humans. Moreover, the discovery of high-lysine corn has spurred the quest for a similar gene in wheat, rice, and grain sorghum. Perhaps genetic engineering of food crops will result in other research findings of considerable consequence for humanity.

The development of cities began with the cultivation of crops and domestication of animals. The first small cities appeared about 5500 years ago. Growth of cities was rather slow over the next few centuries. Before 1850, no society could be described as predominantly urbanized. By 1900, Great Britain was the only nation that was highly urbanized. Today, all industrial nations are highly urbanized and are moving further in that direction each year. The rapid movement from farms to cities has left the job of food production in the hands of fewer and fewer people and has resulted in an important problem of proper food distribution. This problem points out the immense importance to the urbanized economy of an excellent transportation system.

No important new **species** of animal or crop has been domesticated in recent years. However, some crops and animals that originated in one country have been introduced into others, where they have become quite popular and productive. Crops recently introduced into the United States include Korean lespedeza and soybeans. Examples of livestock that have been introduced into the United States are Brahman, Charolais, Limousin, and Simmental cattle and Landrace swine (see Chapter 2).

The major efforts of both plant and animal breeders in the 1900s have been directed toward the more efficient production of a better-quality product through close attention to breeding, feeding, and management methods. Specific breeding methods used will be discussed in Chapter 5.

1.3 HISTORY OF AGRICULTURAL EDUCATION AND RESEARCH

The signing of the Land-Grant College Act on July 2, 1862, by President Lincoln began a new era in **agriculture.** The sponsor of the bill was Justin Smith Morrill, a senator from Vermont. The Morrill Act proposed that a portion of federally owned land be sold and the proceeds used for the establishment of at least one college in each state, the main goal of which would be to teach branches of learning related to agricultural and mechanical arts. This was to be done without exclusion of other scientific and classical studies. The United States now has a total of 68 land-grant colleges and universities. The name *land grant* comes from the granting of land by the federal government for the establishment of these colleges. Names and addresses of the agricultural colleges and experimental stations in the United States are given in Appendix D.

Teaching agriculture in the colleges in early years was difficult. No textbooks on agricultural subjects were available, and there were no bulletins or circulars that instructors could assign to students or use for lecture material. Little or no research had been done, so that principles of agricultural production based on research could not be taught. This situation was quite different from the present one, in which there usually is a choice of more than one textbook for each course and current articles in agricultural journals come monthly to individuals and libraries.

There was a clear need for research in agriculture in the early years of the land-grant system. This need resulted in the passage of the Hatch Act in 1887.[3] It provided for the establishment of agricultural experiment stations in all states and territories of the United States. From these experiment stations and the U.S. Department of Agriculture (USDA) has come a wealth of information on agricultural subjects; this information has helped to produce animals and crops more efficiently, thereby contributing to making the United States the world's richest agricultural nation.

It was recognized that information gained from agricultural research must be made available to people on farms as well as to students. To facilitate the spreading of research information to the user group, the Smith-Lever Act was passed by Congress in 1914. The act provided for cooperative financing of the present-day county agent system under the direction of land-grant colleges. This system has as its main objective the carrying of new research information to people on farms, where they can put it into actual practice.

---

[3]George Washington, in his 1796 message to Congress, requested a board of agriculture with one of its purposes to be the encouragement of experimentation. However, it was not until 1887 that the land-grant colleges of agriculture had federal funding for agricultural experimentation.

The Smith-Hughes Act (passed by Congress in 1917) made federal funds available, if matched by state funds, for the study of vocational agriculture and vocational home economics and for education in the trades and industry. Land-grant colleges often are designated as the institutions that train teachers in these subjects. This program also has been very successful.

The threefold objective of agricultural education, then, is (1) gaining knowledge through research, (2) teaching established principles to students of agriculture in high schools and colleges, and (3) disseminating new information directly to the farm, to be applied with minimal delay. What has been accomplished in efficient agricultural production in the past has been amazing. What will be accomplished in the future could be phenomenal. Indeed, it will need to be, to feed the increasing world population.

1.4 ANIMAL AGRICULTURE AND THE WORLD ECONOMY

> Let us never forget that the cultivation of the earth is the most important labor of man. When tillage begins, other arts follow. The farmers, therefore, are the founders of human civilization.
>
> *Daniel Webster* (1782–1852)

Agriculture is the world's oldest and largest primary industry. It plays a vital role in the economic life of virtually all nations regardless of their state of development. It employs more than one-half of the world's population. (In the developing countries, more than two-thirds of the people live on farms.) The basic necessities of life, food, clothing, and shelter, are supplied by the people on the land.

An imminent and great challenge to human ingenuity is the problem of the uneven distribution of various populations relative to agricultural resources and national income. As is shown in Figure 1.1, more than 50 percent of the world's people live in Asia, and yet that portion of the earth yields only 27 percent of the world's agricultural production and has only about 12 percent of the world's income, whereas North America, with 6 percent of the world's population, produces 22 percent and has about 40 percent of the world's income. These extremes make it difficult to foresee the fulfillment of the aim of the Universal Declaration of Human Rights adopted by the General Assembly of the United Nations: "A comman standard of achievement for all peoples of all nations."

> When the price of rice goes higher than a common man can pay, Heaven ordains a new ruler.
>
> *Old Chinese proverb*

1.4.1 World Population Trends

It took all the years from humanity's first appearance on the earth until Christ's time to reach an estimated world population of one-quarter billion. The

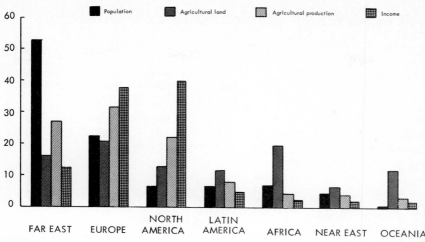

FIGURE 1.1
Regional distribution of the world's population, agricultural land, income, and agricultural production. (Agriculture in the World Economy, *FAO, Rome.*)

population doubled to one-half billion by 1600 and doubled again to 1 billion by 1830. The world's population now exceeds 4.5 billion people and is growing at a rate of about 1.8 percent annually (Table 1.2). A population doubles in only 42 years at this rate (Table 1.3).

The world population doubled during the 100-year period from 1830 to 1930. It doubled again during the 30-year period from 1930 to 1960 (Figure 1.2). It took only 15 years to add the next billion, and the fifth billion will be added by 1990. This is a growth rate that more closely approximates a geometric than an arithmetic increase. Each year, the world population increases by some 85 million people, or enough to populate a new nation larger than West Germany or Great Britain. Considering this accelerating rate of increase, it seems highly probable that by the year 2000 the population of the world will reach, and may well exceed, 6 billion (Figure 1.2).

The current annual growth rate of the United States population is about 0.6 percent. Largely because of legalized abortion, Japan has the lowest population growth rate in Asia, 0.7 percent, and other countries having planned parenthood and/or legalized abortions (e.g., Belgium, Hungary, Sweden, and Great Britain) have rates of 0.5 percent or less. West Germany, East Germany, and Luxembourg have also nearly achieved population stability.

Populations of the world's underdeveloped countries are growing much faster than those of industrialized nations (Figure 1.3 and Table 1.4). Therefore, areas with the *least food available* have the greatest population explosion. The seriousness of "the stork outrunning the plow" in many developing countries is

TABLE 1.2
POPULATION PROJECTIONS FOR THE WORLD, MAJOR REGIONS, AND SELECTED COUNTRIES

	1975	2000	Percent increase by 2000	Average annual percent increase	Percent of world population in 2000
	millions				
World	4090	6351	55	1.8	100
More developed regions	1131	1323	17	0.6	21
Less developed regions	2959	5028	70	2.1	79
Major regions					
Africa	399	814	104	2.9	13
Asia and Oceania	2274	3630	60	1.9	57
Latin America	325	637	96	2.7	10
U.S.S.R. and eastern Europe	384	460	20	0.7	7
North America, western Europe, Japan, Australia, and New Zealand	708	809	14	0.5	13
Selected countries and regions					
People's Republic of China	935	1329	42	1.4	21
India	618	1021	65	2.0	16
Indonesia	135	226	68	2.1	4
Bangladesh	79	159	100	2.8	2
Pakistan	71	149	111	3.0	2
Philippines	43	73	71	2.1	1
Thailand	42	75	77	2.3	1
South Korea	37	57	55	1.7	1
Egypt	37	65	77	2.3	1
Nigeria	63	135	114	3.0	2
Brazil	109	226	108	2.9	4
Mexico	60	131	119	3.1	2
United States	214	248	16	0.6	4
U.S.S.R.	254	309	21	0.8	5
Japan	112	133	19	0.7	2
Eastern Europe	130	152	17	0.6	2
Western Europe	344	378	10	0.4	6

Source: Global 2000 Technical Report, as cited in *Global 2000 Report to the President—Entering the Twenty-First Century,* vol. 1, 1980.

made even worse by those countries' lack of funds to import foods. Moreover, estimates of world population increases by the Food and Agriculture Organization of the United Nations (FAO) project that the situation will become even more serious (Figure 1.4). Developed countries have both the opportunity and the obligation to share their agricultural and industrial technology and expertise in assisting less fortunate countries.

TABLE 1.3
THE RELATION BETWEEN
ANNUAL INCREASE AND
TIME REQUIRED TO
DOUBLE A POPULATION

Annual increase, %	Doubling time, years
0.5	139
0.8	87
1.0	70
2.0	35
3.0	23
4.0	17

FIGURE 1.2
World population, in billions, since 1830. After an estimated 2 million years of human life, world population reached 1 billion in 1830. But since 1830, each successive billion has been added in fewer and fewer years. *(United Nations, Overseas Development Council.)*

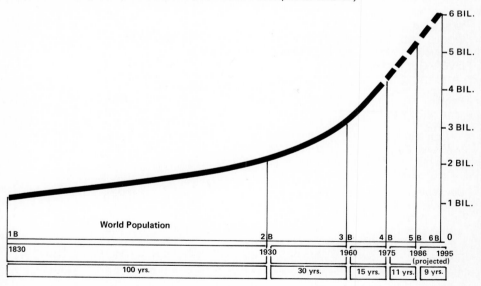

I know of no pursuit in which more real and important service can be rendered any country than by improving its agriculture.

George Washington

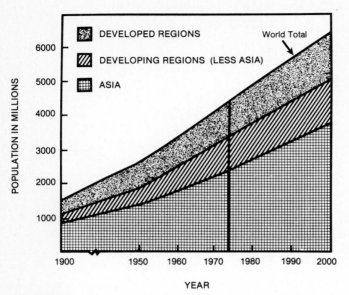

FIGURE 1.3
Past, present, and projected world population, 1900 to 2000. Note that the largest population increases continue to be in Asia and other developing regions of the world. *(FAO, Rome.)*

TABLE 1.4
WORLD AND REGIONAL AVERAGE ANNUAL PERCENT CHANGE IN
HUMAN AND ANIMAL POPULATIONS FROM 1961–1965
(AVERAGE) TO 1980

Region	Human	Cattle and buffalo	Sheep and goats	Pigs
Developed	0.9	0.6	−0.5	1.8
North America	0.9	0.4	−4.1	1.2
Western Europe	0.6	0.6	−0.3	2.1
Oceania	1.5	1.6	−0.2	0.7
Other developed	1.4	0.2	−0.8	4.1
Developing	2.4	1.1	1.2	0.3
Africa	2.7	1.1	2.2	3.5
Latin America	2.8	1.8	−0.7	0.0
Near east	2.4	2.0	1.3	4.6
Far east	2.2	0.6	1.6	1.1
Other	2.3	3.7	1.6	11.8
Centrally planned	0.5	1.0	1.1	2.3
Asia	2.1	0.4	2.0	2.6
Europe and U.S.S.R.	0.9	1.5	0.3	1.7
World	1.8	1.0	0.7	1.8

Source: Developed from *FAO Production Yearbooks,* 1974 and 1980.

We cannot live in a world divided between, on one hand, two-thirds who do not eat properly and, knowing the causes of their hunger, revolt, and, on the other, one-third who eat well—sometimes too much—but who can sleep no longer for fear of revolt on the part of the two-thirds who do not have enough to eat.

David Rockefeller

Effective measures for the control of malaria, yellow fever, smallpox, cholera, and other infectious diseases brought sharp reductions in the death rate in many countries during the twentieth century, resulting in substantial population increases. Thailand is a classic example of how declining death rates increase population growth. The death rate decreased from 30 per thousand after World War II to below 20 per thousand in the 1950s and is now about 10 per thousand. In 1937 life expectancy in Thailand was approximately 35 years, but it currently exceeds 60 years. FAO projections indicate that, because of a 2.3 percent annual population increase in Thailand, its present 42 million people will expand to 75 million by the year 2000. Other countries such as Bangladesh, Brazil, Mexico, Nigeria, and Pakistan are experiencing even larger percentages of annual increases in population. As a major region of the world, Africa has the world's largest annual percent increase (2.9 percent) in population (Table 1.2) and, except for Africa, progress was made during the 1970s to bring food production and population increase into a more favorable balance (Figure 1.4).

Approximately 40 percent of the world's population is under 15 years of age. This means there are far more young people who soon will be reproducing, or adding to the world's population, than there are old people who soon will be dying, or subtracting from it. This fact points to the need for bringing the birthrate and death rate closer together. A high birthrate leads to a high death rate due to hunger. Malthus warned of this in 1798:

I wish to make two postula. First, that food is necessary for the existence of man. Secondly, that passion between the sexes is necessary, and will remain nearly in its present state. I therefore conclude that the power of population is indefinitely greater than the power in the earth to produce subsistence for man.

Many believe that the explosive growth of the human population is the most significant event in history. Almost all agree that next to the pursuit of peace, the greatest challenge to humanity is the race between food production and population increases. People multiply—land does not.

Mankind's future is at stake in a formidable race between population growth and famine.

Arnold Toynbee

1.4.2 World Animal Production Trends

World food production is inadequate to ensure a balanced diet for all people of all lands. Providing food to meet caloric needs is not enough. Adequate protein is also required for normal maintenance of body tissues and functions and

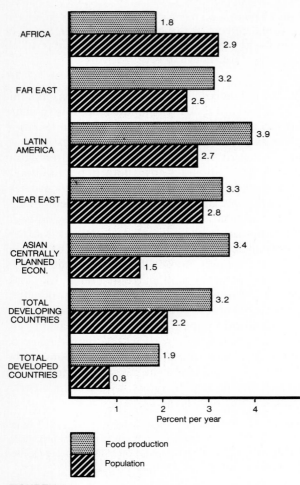

FIGURE 1.4
Increases in population and food production in selected regions of the world from 1971 to 1980. (The State of Food and Agriculture, *FAO Agr. Series No. 12, FAO, Rome, 1981.*)

additionally for growth, maturation, pregnancy, lactation, and recovery from disease. Supplies of protein are particularly scarce and costly[4] for the populations of most developing nations.

Malnutrition is the world's number one health problem. It adversely affects

---

[4]Poverty is perhaps the world's greatest cause of hunger. In 38 nations, the average annual per capita income is $100 or less. People in these countries have little money to purchase food or fertilizer. The underdeveloped countries have approximately 55 percent of the world's land but only 10 percent of the commercial fertilizer needed to increase food production.

mental and physical development, productivity, and the span of working years, all of which significantly influence the economic potential of humanity. However, malnutrition does not arouse the sense of urgency that accompanies an outbreak of a contagious disease such as smallpox.

Recent estimates compiled by the Economic and Social Council of the United Nations indicate that more than 500 million children and perhaps an equal number of adults throughout the world are malnourished. (An estimated 20 million people starve to death annually.[5]) These children's lack of protein, calories, vitamins, and minerals is reflected in retarded physical growth and development, and for many of them mental development, learning, and behavior will also be impaired. (An estimated 80 percent of brain cell growth occurs during the first 2 years of life.) Protein and caloric deficiencies also affect the health and economic productivity of adult populations.

If the nutritional status of the world's hungry masses is to improve, food production and distribution must increase at an unprecedented rate. What are the prospects of this happening? Can animal agriculture be expected to contribute further to the health and well-being of humanity? Foods of animal origin provide high-quality protein, vitamins, minerals, and other dietary essentials. Meat, milk, eggs, and wool (including the feed that goes into their production) represent about two-fifths of the value of the world's agricultural output. Meat and milk constitute about 85 percent of the major livestock products (Figure 1.5). In the following the worldwide base for the production of animal protein will be considered.

Livestock   The livestock base for the world's food production consists of about 1.36 billion cattle and buffalo, 1.16 billion sheep, 710 million pigs, and 410 million goats. Horses number approximately 76 million. The largest number of cattle is in the far east (Figure 1.6), where they serve mainly as draft animals and suppliers of milk. However, the human population is increasing at a much faster rate than the cattle population in that region (Table 1.4).

Although animal numbers and production are usually recorded to show changes in their respective populations, a preferred method of expressing animal production is in meat and milk production per capita. This has been done in Table 1.5. In Africa and Latin America, meat production per capita decreased from 1960 to 1980; only in Europe and the U.S.S.R. did it increase appreciably (from 18.8 to 26.0 kilograms, kg) during this 20-year period (Table 1.5). During the 20-year period 1960 to 1980, significant per capita milk production increases occurred only in Europe and the U.S.S.R. Per capita worldwide production of

---

[5]Although there are varying degrees of starvation, it is accepted that starvation results in adults when daily caloric intake is consistently below 1600 calories. Child starvation is demonstrated when individuals are below 60 percent of standard body weight for their age. Any death that would not have occurred had the individual been properly nourished is considered to be due to starvation regardless of the ultimate cause of death. About one-fourth of the children in the developing countries die before the age of 5 years, mostly from nutrition-related causes. More than 100,000 children go blind annually for lack of vitamin A.

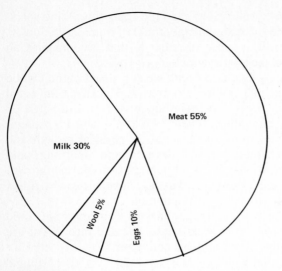

FIGURE 1.5
Worldwide output (percent of dollar value) of major livestock products. Farmers' total estimated income from these products was $156 billion in 1982. *(Compiled from various FAO publications.)*

TABLE 1.5
WORLD AND REGIONAL AVERAGE PER CAPITA PRODUCTION OF MEAT AND MILK, KG, 1961–1965 (AVERAGE) AND 1980*

Region	Total meat 1961–1965		1980		Total milk 1961–1965	1980
Developed	60.9	(28.7)	83.4	(31.8)	295.2	293.8
North America	98.4	(45.4)	117.7	(45.0)	323.5	268.9
Western Europe	49.6	(20.3)	76.7	(25.8)	337.0	383.2
Oceania	190.6	(171.6)	213.4	(182.3)	958.9	702.8
Other	12.2	(6.7)	28.2	(7.9)	47.6	65.0
Developing	11.3	(7.7)	12.5	(7.0)	44.6	44.0
Africa	12.5	(8.9)	12.1	(7.5)	24.0	20.4
Latin America	37.8	(27.9)	39.4	(23.1)	89.5	91.5
Near east	12.3	(9.6)	15.9	(10.6)	75.9	67.4
Far east	3.6	(1.5)	4.0	(1.5)	32.9	34.8
Other	13.0	(2.8)	13.0	(2.5)	15.4	12.3
Centrally planned	26.8	(8.0)	34.6	(8.8)	98.6	98.6
Asian	18.2	(3.0)	23.1	(2.6)	6.7	7.5
Europe and U.S.S.R.	45.3	(18.8)	66.0	(26.0)	297.3	348.7
World	27.2	(12.3)	32.2	(12.0)	116.6	106.4

*Data in parentheses refer to beef, veal, mutton, and goat meat.
Source: Developed from *FAO Production Yearbooks,* 1974 and 1980.

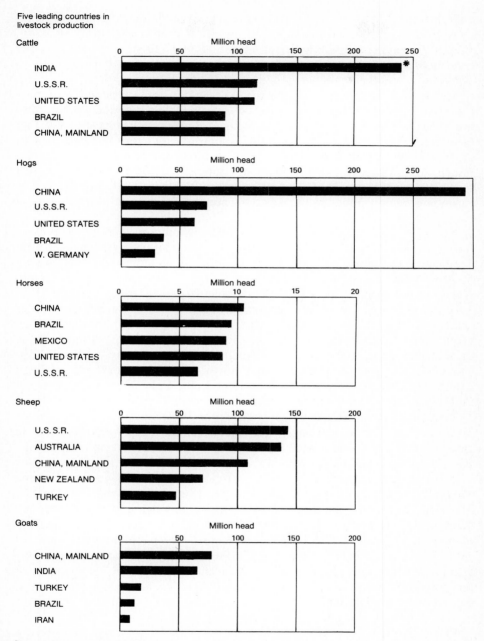

Five leading countries in livestock production

FIGURE 1.6
Countries that lead in the production of selected livestock and animal products (FAO Production Yearbook, *Rome, 1980; Foreign Agriculture Service, USDA, 1983; and ERS, USDA,* China World Agr. Reg. Supp., Review of 1982 and Outlook for 1983, *June 1983.*)

✳inc. an est. 70 million water buffalo

milk decreased from 116 to 106 kg (Table 1.5). In 1980 approximately 11 times more milk was produced per capita in Europe than in the far east. Similarly, Europe (including the U.S.S.R.) produced about 17 times as much meat per capita as did the far east. The five leading countries of the world in major classes of livestock and their products are given in Figure 1.7.

The data in Table 1.5 show that the average per capita production of meat increased in most developed countries during the 20-year period 1960 to 1980 but changed little in the developing regions. Recent studies have shown that present cattle and sheep numbers could be doubled with intensified management of tropical and subtropical pastures, coupled with increased yields of grain and forage crops. Moreover, swine and poultry numbers could be increased greatly if feed were made available. The question arises: Will land, seed, fertilizer, water (including water supplied by the weather), and the ability to cultivate be in large enough supply?

Not all of the earth's surface is suitable for agriculture. If all of the present food-producing areas and all of the areas of potential crop-growing soil were combined into a single band encircling the earth, it would occupy only about 5 percent of the total surface. As is shown in Figure 1.8, the world's land presently is classified as approximately 10 percent arable, 20 percent pasture and meadow, 30 percent forest, and 40 percent nonproductive (desert, rock, etc.). It is estimated that one-third of the forest land has potential for agricultural production. Additionally, the yield per acre may be increased in much of the present productive arable land; however, cropland per person is diminishing. The per capita grain-producing area in North America, for example, declined from 1.7 to 1.0 acres between 1940 and 1980. In the United States, 1.7 acres per capita is in crops, but about 40 percent of the crops is exported.

Virtually all human nutrients come from three sources: crops, animal products (meat, milk, and eggs), and aquatic foods. Crops provide about 90 percent of all calories and about two-thirds of all protein consumed by humans. Animal products and fish provide about one-third of humanity's protein.

If water can be made available for irrigation, large acreages will become available worldwide for crop production. Most human food comes directly or indirectly from grain, and there is over 3 times as much arid land within 300 miles of the sea as the world now employs in grain production. The desert is human beings' greatest land bank: it offers more than 8 million square miles of space for human occupation and use. This wondrously rich bank may someday turn green when humans tap desalinized seawater for irrigation. The majority of this potentially arable land is in Africa, South America, and Asia—continents in serious need of additional food.

Fish   The 20 leading fish-catching countries in the world are given in Figure 1.9. Our marine resources could yield a vastly increased supply of high-quality protein. World fish production and consumption have increased substantially during each of the past two decades. Marine biologists believe an annual global fish yield of 160 million metric tons is attainable. (The current annual fish

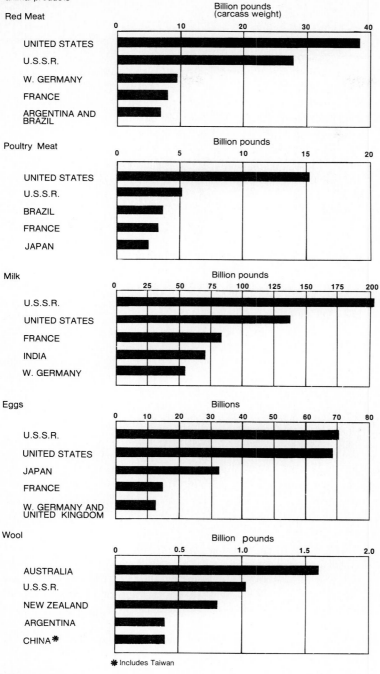

Five leading countries in animal products

Red Meat

Billion pounds
(carcass weight)

UNITED STATES
U.S.S.R.
W. GERMANY
FRANCE
ARGENTINA AND BRAZIL

Poultry Meat

Billion pounds

UNITED STATES
U.S.S.R.
BRAZIL
FRANCE
JAPAN

Milk

Billion pounds

U.S.S.R.
UNITED STATES
FRANCE
INDIA
W. GERMANY

Eggs

Billions

U.S.S.R.
UNITED STATES
JAPAN
FRANCE
W. GERMANY AND UNITED KINGDOM

Wool

Billion pounds

AUSTRALIA
U.S.S.R.
NEW ZEALAND
ARGENTINA
CHINA ✳

✳ Includes Taiwan

FIGURE 1.7
Countries that lead in the production of selected animal products (FAO Production Yearbook, *Rome, 1980; Foreign Agriculture Service, USDA, 1983; and ERS, USDA,* China World Agr. Reg. Supp., Review of 1982 and Outlook for 1983, *June 1983.*)

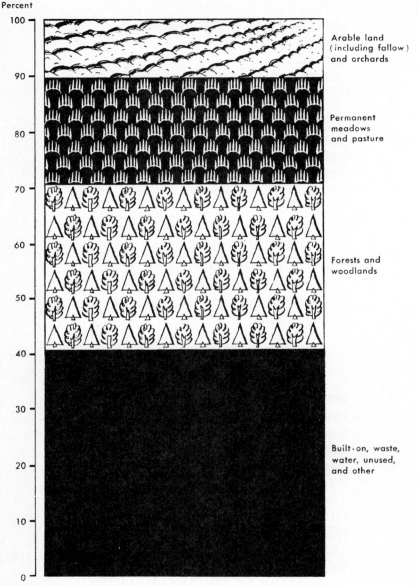

Percent

FIGURE 1.8
Classification of the world's land resources. (Agriculture in the World Economy, *FAO*.)

harvest is about 72 million metric tons.) Fish at present provide about 15 percent of the world's animal protein.

Slightly over half of the world's fish catch is used directly as food for people; the balance is fed to cattle, swine, poultry, and pet animals. Fish constitutes an

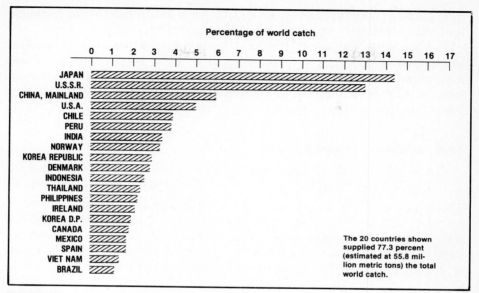

FIGURE 1.9
Contributions of 20 countries to the world catch of fish, crustaceans, and mollusks. These 20 countries supply more than 77 percent of the total world catch. *(FAO,* Yearbook of Fishery Statistics, *Rome, 1981.)*

average of only 1 percent of humanity's total diet throughout the world. Japan, however, depends on the sea for approximately half of its protein for human consumption. There is a fish flour made from waste products of the sea that, once its fat has been removed, can be added to wheat flour to provide a balanced protein diet needed desperately in many parts of the world, especially in the tropics.

Development of **aquaculture** offers another means of supplying humans with fish protein. Fish were domesticated in ponds by wealthy people as early as 3000 B.C., and by 400 B.C. raising fish was commonplace in China and Persia. Recently scientists of Israel have been producing as much as 9000 kg of fish annually per hectare of lake or pond.

Aquaculture Fish is the only major class of foods that is still gathered, in large part, from the wild. But this is changing. American farmers are now producing fish under managed conditions, just as they have produced meat, milk, and eggs. This is called **aquaculture.**

During the decade 1970 to 1980, the harvesting of channel catfish increased from 20 million lb to 125 million lb annually. The channel catfish has great potential for commercial production because it can be grown in all parts of the United States, and it reproduces and grows well under managed conditions. A well-managed catfish farm can yield well over 100,000 pounds of fish per person-year of labor annually. Yields of 5000 lb/acre are commonly harvested from earthen ponds that average about 10 acres in size. It takes about 18 months for a channel catfish to grow to its market size of 1 lb.

Fish are highly efficient in converting feed into edible human proteins. They gain a pound of weight for every 1.5 lb of feed, approximating the **feed conversion** of chickens and exceeding that of beef or swine. Moreover, through research on genetics, disease, and nutrition, the efficiency of feed utilization of fish is expected to increase significantly in the years ahead.

1.4.3 Availability of Animal Protein

A hungry people listens not to reason nor are its demands turned aside by prayers.
Lucius Annaeus Seneca (4 B.C.–A.D. 65)

The foremost reason for maintaining our animal populations is to provide a nutritious and desirable form of food for people. It has been well established that, nutritionally, animal proteins are superior to vegetable proteins for humans. This superiority results largely from the better balance of amino acids in animal products. However, when animals are kept for the purpose of

FIGURE 1.10
Number of days of protein requirement produced by 1 acre via selected foods. *(After L. H. Bean*, Protein Advisory Group News Bull., *6, WHO/FAE/UNICEF, 1966.)*

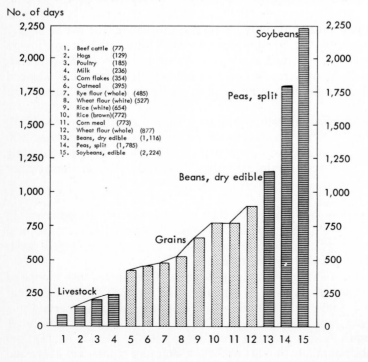

producing meat, milk, and eggs, food cannot be produced as efficiently as when grains are consumed by humans themselves (Figure 1.10). Daily per capita grain usage of developing and developed countries is approximately 1 and 5 lb, respectively. In developed countries, grain is consumed mostly indirectly, as animal products. The dairy cow is the most efficient among farm animals in converting feed into both protein and energy; poultry and swine follow (Table 1.6). Beef cattle are the least efficient of the farm animals in this respect. It should be noted, however, that much of the feed utilized by animals, particularly the **ruminant,** cannot be utilized directly by people. The ruminant, in fact, is actually a creator of food nutrients in that it can synthesize **essential amino acids** and B-complex vitamins. *Ruminant is the name given to a* **herbivorous** *animal that chews its* **cud** *and has split hooves.* Such animals as the ox, sheep, cow, llama, deer, goat, antelope, and giraffe are ruminants (see Chapters 14 and 15). The central role of ruminants in feeding a hungry world is outlined schematically in Figure 1.11.

The people of North America, Oceania, and Europe are blessed with an abundance of animal protein, whereas the balance of the earth's people obtain the larger portion of their dietary protein from other sources (Figure 1.12). Since a diet containing more livestock products requires more land and thus feeds fewer persons per acre, it may be anticipated that, with the projected population boom ahead, animal proteins will be at a premium. However, supplementation of predominantly grain diets with animal products will certainly improve the nutrient balance of the populations consuming these diets. Moreover, people prefer animal foods to plant foods, and as their economic status improves, their consumption of this type of food increases. This, coupled with the accelerating rate of population increase and the awakening of a world

TABLE 1.6
ESTIMATE OF RELATIVE PERCENTAGES OF FEED NUTRIENTS CONVERTED INTO EDIBLE PRODUCTS BY SELECTED FARM ANIMALS

Animal product	Energy conversion, %	Protein conversion, %	Gross edible product output as percent of feed intake
Milk	20	30	90
Chicken (broilers)	10	25	45
Eggs	15	20	33
Pork	15	20	30
Turkey	10	20	29
Beef	8	15	10
Lamb	6	10	7

Source: Adapted from R. E. Hodgson, "Place of Animals in World Agriculture," *J. Dairy Sci.* 54:442–447, 1971.

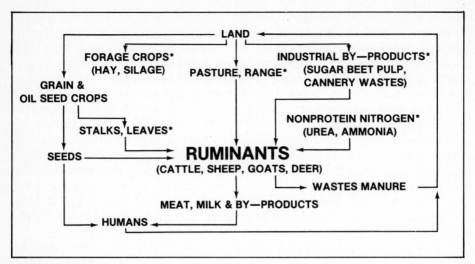

FIGURE 1.11
The central role of ruminant animals in human nutrition is shown. Items marked with an asterisk are utilized by ruminants but not used directly as food by humans. *(From* Ruminants as Food Producers—Now and for the Future, *Council for Agricultural Science & Technology, Special Pub. No. 4, 1975.)*

conscience in human nutrition, offers a tremendous potential to the producers of animal protein throughout the world.

The world is confronted with an increasingly severe protein shortage that threatens much of its present and future population. An estimated 40 percent of the world's people consume inadequate quantities of protein. Worldwide estimates of future food needs are equally disturbing (Figure 1.13). Measures are needed *now* to improve the balance between population and food supply. The material and human resources required to avert the impending protein crisis are indeed large but are within the support capabilities of the presently industrialized nations. However, current import-export trends do not indicate that the situation will improve in the near future. For example, although developed countries annually export approximately 2.5 million tons of protein to developing countries, the latter export about 3.5 million tons of protein (especially fish meal and press cakes of oilseeds) to the developed countries. The developed countries use most of it to feed poultry, livestock, and pets.[6] Over 50

---

[6]People in the United States spent more than $4.5 billion for pet foods in 1982 (approximately 5 times as much as was spent for infant foods). There were an estimated 48,846,000 dogs and 39,000,000 cats in the United States in 1982. The foods consumed annually in the United States by dogs, cats, birds, horses, and other pet and/or companion animals would feed an estimated 40 million people. Yet many advantages accrue from this means of marketing food products. Companion animals contribute greatly to the mental health and well-being of millions of children and adults alike.

countries (e.g., Mexico, Panama, India, and Hong Kong) supply the United States with shrimp, which could otherwise provide valuable nutrition to protein-starved children in these countries. Peru exports to the developed countries large quantities of fish that could greatly relieve protein deficiencies in Latin America.

Approximately 7 million tons of nonfat milk solids are produced annually in the world. However, about 60 percent of this valuable source of nutrition is consumed by animals. Much of the world's dry nonfat milk solids is used in the production of meat in western Europe. The availability of an economical substitute for milk solids in animal feeding would increase the availability of this valuable source of nutrition for needy populations. It is possible that animal feeds, especially protein supplements, will be commercially available from the

FIGURE 1.12
Regional per capita protein supplied from major food groups. *(FAO,* The Fourth World Food Survey, *Rome, 1977.)*

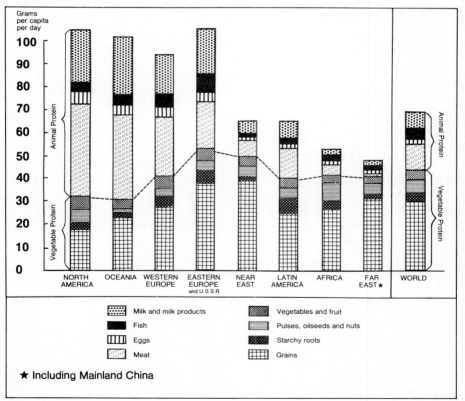

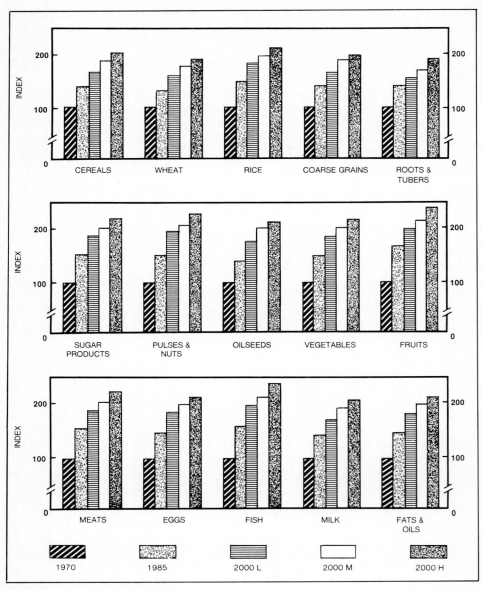

FIGURE 1.13
Projected consumption of selected foods in the world in 1985 and 2000 relative to consumption in 1970 as 100. Three values are given for the year 2000 based on low (L), mean (M), and high (H) FAO estimates of world population. *(From* A Hungry World: The Challenge to Agriculture, *FAO, 1980.)*

laboratory within the next two decades, and this will spare conventional foods for human consumption.

Other potential ways of providing additional protein-rich foods include herding such animals as the South American capybara (a rodent) and the African eland (an antelope), converting water hyacinths and other aquatic weeds into ruminant feed, making cattle feed from wood products, extracting protein from leaves and small fishes, and culturing algae in the fecal slime of sewage-treatment plants.

There is no single or simple solution to the complex problem of providing sufficient quantities of proteins of high **biological value** economically and in an acceptable form. It is readily apparent that the supply must be increased. This means intensifying efforts to increase production of conventional plant, animal, and fish sources of protein; it means improving ways of preserving and storing proteins to minimize wastes. Furthermore, it means that people must develop and produce new and unconventional sources of protein, such as foods incorporating oilseed meals, fish protein concentrate, single-cell protein and fortified cereals, for human consumption. Additionally, with respect to consumption, it is imperative that greater emphasis be placed on improving the distribution and marketing of protein-rich foods.

As is shown in Figure 1.14, FAO studies indicate that poultry meat production, as a major source of meat production in 90 developing countries, is projected to increase from 17 percent in 1975 to 27 percent in the year 2000.

During the 10-year period from 1970 to 1980, the worldwide production of meat from all sources increased from 96 to 129 million metric tons (34 percent). The increase was, as is shown in Figure 1.15, largely accounted for by nonruminant meat (pig and poultry). Concurrently, the world population increased from 3.596 to 4.415 billion, or by 23 percent, suggesting that on the average the world's people received an additional 7 grams (g) of carcass meat daily, which represents an increase of only about 1 g of protein daily. The problem of unequal distribution still prevailed, however, so that most of the world's population received little or no increase in dietary protein from meat. For example, the amount of animal protein in the diet in Indonesia still totals only 1.8 g per person daily.

1.4.4 Animals as Competitors with Humans for Food

The degree to which animals compete with humans for food is of special interest in a hungry world. Ruminants (particularly cattle, sheep, and goats) can consume and utilize large quantities of roughage, pasturage, and forage. This fact is of special significance in the harvesting of such feeds from rough, hilly lands in which crop production is impractical. In semiarid areas, the only practical means of harvest is by livestock.

Although animals are not especially efficient converters of plant nitrogen into protein, an advisory committee of the United Nations stated, "Livestock will nevertheless be an increasingly important source of protein in the developing

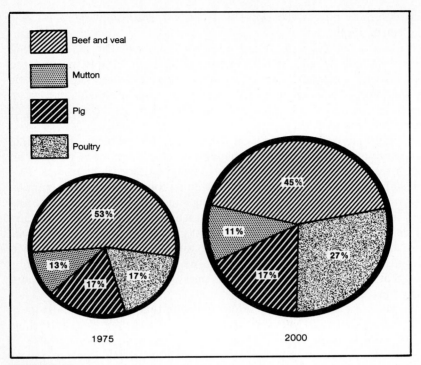

FIGURE 1.14
Major sources of meat production in 90 developing countries in 1975 and projected in the year 2000. *(From* World Animal Review, *no. 40, October–December 1981.)*

countries."[7] This statement suggests that efforts to increase the efficiency of producing milk and meat from cattle, meat from sheep and swine, and eggs and meat from poultry should be expanded throughout the developing world. This means the developed countries must share their knowledge related to genetic selection for efficient feed converters, increased productivity by use of certain feed additives and balanced rations, and greater use of protective measures against livestock diseases and parasites.

Contrary to the commonly held view, the ruminant does not necessarily compete directly with humans for food. Instead, because the herbivore can utilize cellulose material that people cannot use, the ruminant should be regarded as an animal without which considerable quantities of high-quality food would be denied to humans. The most obvious function for these herbivores is their use of natural grazing lands that are not suited to cultivation (Figure 1.16).

[7]"Feeding the Expanding World Population: International Action to Avert the Impending Protein Crisis," Report of the United Nations Economic and Social Council of the Advisory Committee on the Application of Science and Technology to Development, 1978.

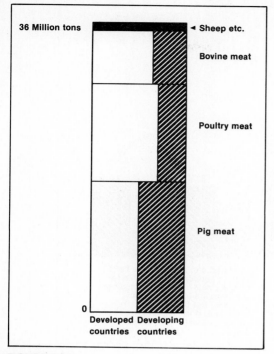

FIGURE 1.15
Increase in world meat production during the 10-year period 1970 to 1980. *(From* Commodity Review and Outlook 1980–81, *FAO Econ. and Social Develop. Series No. 30, FAO, Rome.)*

It is possible to realize good production from cattle and sheep fed solely forage rations. Moreover, it has been demonstrated that dairy cows and goats can maintain themselves and produce milk at creditable levels when fed rations consisting entirely of forages, pasture hay, and **silage.** An example of "grassland dairy farming" is the dairy production system of New Zealand.

Beef cattle and sheep can be raised from birth to market (or to an age suitable for finishing) with little feed other than the mother's milk and forages. It has been estimated that in a beef cow-calf operation, a cow and her offspring can be maintained, with the calves raised and fed to a slaughter weight of 454 kg (1000 lb), on rations consisting largely of forages supplemented with a molasses-urea mixture or, in the finishing period, with corn and cob meal and urea. The yield of meat protein from this operation approximately equals the yield when protein of oilseeds is fed to cattle. This feed regimen would free more plant protein for humans and illustrates how the production of meat need not compete with people for plant protein.

Most of the world's best lands are already farmed; future agricultural developments are more likely to result from intensification of the management

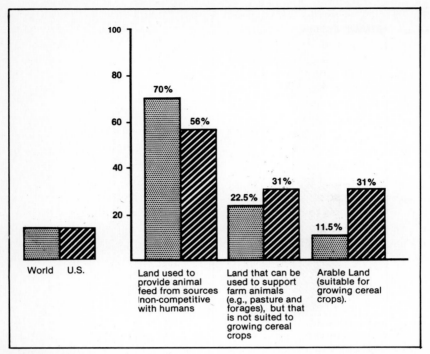

FIGURE 1.16
Land use in the United States and the world. Note that only 11.5 percent of the world's land is suited to the raising of cereal crops (bar graphs at right), whereas 22.5 percent of the world's land can be used for pasturing animals but is not suited for crops (center bar graphs). *(Data are from various FAO and USDA publications.)*

of marginal lands. Much such land is best suited to forage production, which suggests an expanded animal agriculture in the years ahead.

1.4.5 Energy and Efficiency in Animal Production

The price consumers pay for food is influenced heavily by the cost of energy. Approximately 5 billion gallons of fuel is used annually to raise, harvest, process, store and handle, transport, refrigerate, retail, and cook the food consumed in the United States. The recent threats to the United States energy supply have focused considerable attention on methods and levels of food production because petroleum is vital for motive power in modern agriculture; natural gas is needed to produce fertilizers and other chemicals; and energy is needed to irrigate and to produce farm machinery, vehicles, and other items used in modern food production. Future developments in animal agriculture will be greatly influenced by both the availability and the cost of fossil fuels. High prices of petroleum products will be reflected in higher costs of crop production, which in turn will result in higher-priced animal feeds, especially grain crops. Increasing worldwide demands for United States grains also can be expected to support high grain prices. Therefore, to produce meat and milk at marketable prices, it is probable that more and more ruminants will be fed proportionally greater quantities of forages in future years, a practice already common throughout most of the world.

Consumption of energy—a useful index of both resource consumption and impact on the environment—is growing worldwide at an annual rate of about 5 percent. This increase in energy consumption has caused some people to advocate a return to the sole use of animal power and organic fertilizers to produce humanity's food supply. It is true that animals currently provide an estimated three-fourths of the draft power for the world's agriculture. Additionally, many countries use minimal commercial fertilizer. However, an estimated one-fourth of humanity's total food supply is attributable to the use of chemical fertilizer. For example, more than 67 million metric tons of nitrogen fertilizer were used on 3 billion acres of cropland in 1981. United States corn yields have increased 188 percent since 1945 (32 versus 92 bushels per acre, bu/acre). Increased use of fertilizer accounted for an estimated one-half of this increase. It is possible, of course, to use green manures to reduce the high energy demand of chemical fertilizer. (For example, planting sweet clover in the fall and plowing it under 1 year later adds about 150 lb of nitrogen per acre to the soil.) This practice, however, removes land from the cropping system for a full year. Studies at the University of Illinois indicate that dried sewage sludge can be used to add nitrogen and other important plant nutrients to soil in support of crop production.

The displacement of approximately 25 million horses by tractors in the United States during the first half of this century released approximately 70 million acres once used to grow oats, corn, hay, and other feedstuffs needed to feed the nation's horses. This land now can be used to feed an approximately equal number of beef and dairy cattle that provide meat and milk for humans. Moreover, the increased labor efficiency that accompanied mechanization of farming (1 hour, h, of farm labor now produces more than 14 times as much food as it did in 1920) released millions of people for other forms of employment, thereby adding billions to the annual United States gross national product as well as providing goods and services important to our way of life.

If the United States returned to using animal power in agriculture, as is done in much of the world (Figure 1.17), it would need approximately 60 million draft animals, more than 20 times the number presently available. It would take about two decades to acquire this number, and it would take about 180 million acres (73 million hectares) of prime farmland to raise enough crops to feed them. This is about the amount of land now in field crops (other than forages) in the five corn belt states plus Wisconsin, Michigan, and California. To revert to labor-intensive agriculture would require an estimated 31 million farm workers, nearly one-third of the working population of the United States.

FIGURE 1.17
Animals provide a major source of power in the cultivation, harvesting, and marketing of food crops, as well as in providing transportation, in much of the world. The above photograph was taken by one of the authors in Latin America.

Of course, efficiency cannot be measured solely in terms of agricultural yields. Important, too, is the amount of energy required for the production of a given kind and quantity of food. Additional research is needed to determine the energy inputs required to produce various foods at varying levels of productivity.

1.4.6   Utilizing Animal Wastes

Current annual United States livestock manure production is estimated at 1.8 billion tons (10 to 12 times that of humans), of which more than 50 percent is produced in feedlots and confinement situations. Cattle and horses excrete approximately 15 lb of solids daily, pigs 1.7 lb, sheep 2.5 lb, chickens 0.12 lb, and turkeys 0.35 lb. Expressed another way, meat-producing animals excrete about 10 lb (including moisture) for each pound of weight gained, dairy cows excrete more than 1 lb of feces for each pound of milk produced, and laying hens excrete approximately 5 lb of feces per pound of eggs produced. Other animal wastes are associated with the processing of meat animals and with the manufacture of dairy products.

Animal wastes actually are resources out of place. Research and economies will help to put them into a place and/or form in which they will provide useful functions, some as fertilizers and others as feedstuffs.

The meat-slaughtering and meat-packing industry, for example, has converted animal wastes into useful products. Typical by-products and their uses include edible fats—tallow and grease; meat scraps and blood—tankage and other animal feeds; bone—bone meal; intestines—sausage casings and surgical thread;

glands—pharmaceutical products; and feathers—feather meal for animal feed.

The processing of milk and milk products requires large quantities of water, in which soluble milk solids accumulate in cleaning, for example. The milk solids can be reclaimed and used as animal feed, as food supplements, and as a growth medium for microorganisms used in producing certain pharmaceuticals.

The liquid waste whey is a by-product of cheese manufacture. Liquid whey from cheddar cheese production contains over half of the nutrients from milk, and the acid whey from cottage cheese manufacture contains approximately 70 percent of the nutrients of the nonfat milk used. Heretofore, over half of the whey has been discharged to waste-treatment plants and to streams or other bodies of water. This loss of milk nutrients increases the cost of cheese production and the national cost of water-pollution control. Complete utilization of whey would eliminate that source of pollution and concurrently increase humanity's food supply. Expanded uses of whey in the future include blending whey powder with basic food materials to produce new and/or less expensive foods such as processed cheese foods, fruit sherbets, custards, and bakery goods. Studies at the University of Missouri include recycling whey with various papers and/or nonprotein nitrogen through ruminants in the production of meat and milk.

Whey can be dehydrated by roller-drying and spray-drying. Whey powder contains approximately 11 to 13 percent protein, 70 to 75 percent lactose, and 7 to 8 percent minerals. Research under way is aimed at fractionating whey solids into pure protein and lactose.

Farm animals, particularly ruminants, will serve humanity in the future when their products that have previously been considered useless waste materials are recycled. Only 21.8 percent of the total animal waste nitrogen is identified as amino acids; the balance comes from nonprotein nitrogen compounds (31.6 percent from urea). Therefore, recycling of animal wastes can be expected to be most effective in ruminants, animals that can utilize nonprotein nitrogen.

Researchers at the University of Arizona and Texas A&M University are cooperating with scientists at the General Electric Company in using thermophilic bacteria to convert feedlot cattle manure into a high-protein livestock feed supplement. Gas containing 60 to 80 percent methane can be produced during the anaerobic digestion of animal wastes. Approximately 8 cubic feet (ft^3) of gas can be produced per pound of volatile solids added to the digester when cattle, swine, and poultry wastes are digested. The process holds great promise as a potential source of energy. Several studies have shown that dried animal wastes can be recycled. The utilization of dried chicken manure as part of the feed for chickens and ruminants has been demonstrated. Recent research shows considerable promise for recycling cattle manure ensiled with chopped corn and other forages.

Studies at the University of Arkansas showed that the amino acids of hydrolyzed feather meal are more than 95 percent utilized by the chick. Other studies indicate that amino acids of properly hydrolyzed hog hair are available to the chick.

The nutritional value of poultry litter in the diets of cattle and sheep has been studied extensively at the University of Arkansas, Virginia Polytechnic Institute and State University, and other agricultural research stations. On a dry matter basis, broiler litter averages about 31 percent crude protein (23 percent digestible protein) and is a rich source of calcium and phosphorus. Feeding poultry waste to finish cattle has not affected carcass grade or meat flavor, nor has it affected the flavor or composition of milk when fed to lactating cows. One possible problem caused by feeding animal waste is copper toxicity, observed in sheep fed broiler litter containing high levels of copper. (Three heavy metals— arsenic, copper, and selenium—are commonly added in very low amounts to livestock and poultry rations.) Additional research is needed in this area.

Notwithstanding recent developments, the most effective agricultural use of animal wastes continues to be disposal on land in a crop production cycle. However, dried animal wastes have considerable potential as soil conditioners and fertilizers for the home gardener, florist, or nursery operator. Animal manure is high in salts and can contain pesticides, drugs, and/or toxic metabolites. Consideration of these aspects must be made in future studies. Effective animal-waste management can be achieved only when food production is maintained with minimal threat to environmental quality at economical prices.

1.4.7 Food and Agriculture Organization (FAO)

The FAO of the United Nations was formally created in 1945. FAO collects, analyzes, interprets, and disseminates information relating to worldwide nutrition, food, and agriculture. It promotes and recommends national and international action with respect to scientific research, the improvement of education, administration, and the spread of public knowledge of nutritional and agricultural science; the conservation of natural resources and the adoption of improved methods of agricultural production; the improvement of processing, marketing, and distribution of food and agricultural products; the adoption of policies for the provision of adequate national and international agricultural credit; and the adoption of international policies with respect to agricultural commodity arrangements. Additionally, FAO furnishes technical assistance to governments when requested. It consists of approximately 120 member nations and is headquartered near the Colosseum in Rome. Information pertaining to its activities and publications may be obtained by writing FAO, via delle terme di Caracalla, 00100 Rome, Italy. FAO publications may be obtained in the United States from UNIPUB, Inc., 650 First Avenue, P.O. Box 433, Murray Hill Station, New York, New York 10016.

1.5 ANIMAL AGRICULTURE AND THE UNITED STATES ECONOMY

On the front page of the December 20, 1920, issue of *The New York Times* there appeared this prediction: "The United States will have a population of

197,000,000 people, the maximum which its continental territory can sustain in about the year 2100; Professor Raymond Pearl of the Johns Hopkins School of Hygiene and Public Health estimated in a Lowell Institute lecture last night. To support such a population, he said 260 trillion **calories** of food a year would be needed, and judging from the production of the last seven years, when the maximum population was reached, it would be necessary to import about half the calories necessary for sustenance."

The United States population passed the 197 million mark in 1966. As a result of American agriculture's advancing more in the past 50 years than in all previous years of its history, there is no problem in feeding that number. In 1800, 94 percent of the people in the United States lived and worked on farms, only 6 percent being urban dwellers, whereas in 1983 almost the reverse was true, the farm population representing only 3 percent and the urban about 97 percent. Japan is experiencing a similar, although somewhat delayed, trend. In 1880, 85 percent of Japan's people lived on farms and only 15 percent were urban dwellers. But by 1980, just one century later, the reverse was true: 85 percent of Japan's population were urban dwellers and 15 percent lived on farms. It is interesting that China feeds nearly one-fourth of the world's population with only about 8 percent of the world's arable land (acreage-wise, China is about the size of the United States, but it feeds, domestically, about 4.5 to 5.0 times as many people). Approximately 80 percent of China's billion-plus people live on farms and communes.

Meat, milk, eggs, wool, mohair, leather, and other fibers have long been staples of the American economy. The magnitude of providing Americans with these substances is made evident by awareness of the following facts:

Land Area   Animals are the largest users of the nation's land mass. Some 64 percent of the total land area of the United States is devoted to the production of animal feeds (36 percent for grazing and 28 percent for the production of **hay** and other forage crops and grain).

Comparative Cash Income   The sale of livestock and their products accounts for about 48 percent of the total cash income of United States farmers. (Of the 50 states, 25 derive half or more of their cash farm receipts from the sale of livestock and their products, which amounted to approximately $70.2 billion in 1982.)

Livestock Inventory   There are more than 200 million farm mammals and over 500 million turkeys and chickens, valued at more than $78 billion, on United States farms. The beef industry is the largest segment of animal agriculture. Sales of cattle and calves totaled $31.2 billion in 1980, or about 23 percent of cash receipts from all farm marketing. Nearly one-half of all farms in the United States have cattle on them. Every dollar of cattle sales generates an additional $5 to $6 of business activity in the farm supply and food business.

Creator of Employment   One out of every five jobs in private employment is related to agriculture; e.g., the meat and poultry industry (including meat-packing, prepared meats, and poultry-dressing plants) employs about 350,000 workers and has an annual payroll of over $5.2 billion. The dairy industry (including fluid milk, concentrated and dried milk, natural cheese, creamery butter, ice cream, and special dairy products) employs 175,000 people and has an annual payroll of more than $2.7 billion.

Feed Industry   This is the ninth-largest manufacturing industry in the United States and distributes about 527 million tons of grain and forage annually, valued at over $35 billion.

Farm Purchases   Altogether, United States farmers spent over $142 billion in the production of food and fiber in 1981. Farmers purchased approximately $10 billion of fertilizer and lime, pesticides, and pharmaceuticals. Their purchases were responsible for the employment of over 6000 chemical industry workers. Another $13 billion went for farm vehicles, machinery, and equipment; $16 billion was spent on fuels and lubricants; $23 billion went for feed and seed; approximately $8.8 billion went for taxes; and $23 billion went for interest. Think of the large number of jobs dependent directly on farm purchases. Thus, there is a close association between prosperity on the farm and prosperity in the city.

A century ago, the United States was an undeveloped country with great natural resources. Its agriculture was primitive and its population small. With approximately 90 percent of the population gaining its livelihood from farming and with minimal use of technology, much early agriculture exploited the land resources. Farmers moved from old to new soil as fertility decreased. As land frontiers were expanded, the realization developed that agriculture greatly affects economic development. Indeed, a great industrial system depends in large part on an efficient agriculture releasing humanpower to the labor pool and concurrently providing a dependable source of food and fiber.

At the beginning of the twentieth century, 7 of every 10 American workers were producing goods and 3 in 10 workers were providing services for others. Today, the figures have reversed: 7 of 10 American workers provide services, and 3 in 10 produce goods.

Food Production   Nearly one-half of the total food supply of humanity is contributed by mammalian, **avian,** and aquatic life. Farmers of the United States represent less than 0.1 percent of the world's population and yet annually produce 27 percent of the meat (about 70 billion lb), 30 percent of the fluid milk (about 133 billion lb), and 27 percent of the eggs (about 6 billion dozen) of the world.

Recreation   Although the horse no longer provides our power and transportation, its popularity for recreational activities is at an all-time high. There are more than 8 million horses in the United States, on which their owners annually spend $8.5 billion for feed and **tack.** Also, more people (77.5 million) attend horse races annually than see baseball games (both minor and major leagues; 56.3 million) or automobile races (51.0 million), the number two and number three spectator sports, respectively. Dog racing attracts more spectators than professional football, basketball, or hockey. Although dog racing is legalized in only 10 states, it is the seventh most popular sport and attracts more than 21 million spectators annually in the United States.

The productivity of agriculture is the main source of strength in the progressive American economy. In no other nation do consumers have so varied and nutritious a diet or buy their food for so small a fraction of their income.

1.5.1 Economy of Animal Products

United States consumers spend about 33 percent of their food budget for meat, poultry, and fish; 14 percent for milk, cheese, and ice cream; and 4 percent for eggs. Total expenditures for animal products represent about 51 percent. However, consumers receive substantially more than 51 percent of their food nutrients from animal products. Moreover, as noted in Figure 1.18, farmers receive a greater proportion of the consumer retail food dollar through animal

FIGURE 1.18
Farm share of retail food prices, based on the payment to farmers for the farm products equivalent to foods in the market basket and the retail price. Fruits and vegetables include both fresh and processed foods. *(USDA, 1981.)*

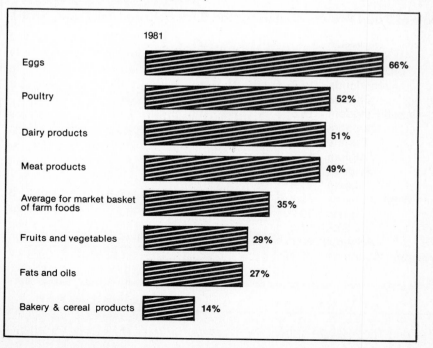

products than through other foods. North Americans are blessed with an average of 1 quart (qt) of milk and 1 egg per capita daily, in contrast to the undernourished people of Asia, who have only 1 qt of milk every 14 days and 1 egg every 10 days per capita (see Section 3.3).

One hour's work in a factory buys more food today than ever before or in any other country of the world; e.g., in 1982, it would buy 3.1 lb of round steak or 15.2 qt of milk compared with 2.0 lb or 5.7 qt in 1942 and 1.5 lb or 4.1 qt in 1932.

Although expenditures for food in the United States total nearly $300 billion, Americans earn a week's food supply by 4 P.M. on Monday. Thus, Americans still spend only about 16 percent of their disposable income for food, less than that spent for food in any other nation of the world. It should be appreciated that approximately 30 percent of the money spent in supermarkets goes for nonfood items. If expenditures for food purchased in restaurants and hotels and related institutional purchases are stripped out of the reckoning, the private consumption expenditures for food in the United States amounted to only 12.7 percent in 1980 (Figure 1.19).

FIGURE 1.19
Percent of total private consumption expenditures for food in 1980.

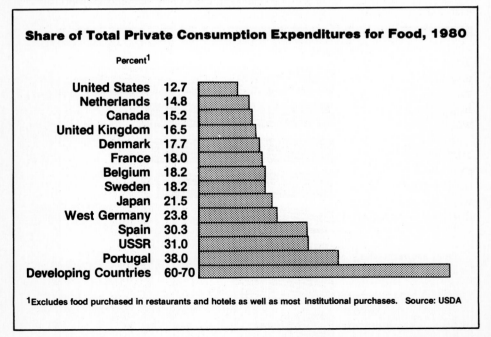

Share of Total Private Consumption Expenditures for Food, 1980

Percent[1]

	Percent
United States	12.7
Netherlands	14.8
Canada	15.2
United Kingdom	16.5
Denmark	17.7
France	18.0
Belgium	18.2
Sweden	18.2
Japan	21.5
West Germany	23.8
Spain	30.3
USSR	31.0
Portugal	38.0
Developing Countries	60-70

[1]Excludes food purchased in restaurants and hotels as well as most institutional purchases.   Source: USDA

1.5.2 Food Supplier to the World

The United States is the world's largest exporter of agricultural products. Through the Food for Peace (Public Law 480) programs, exports of high-quality animal proteins, such as those in dry milk, have been increased substantially. The United States contributed $1.44 billion in food aid to needy countries in 1980 and has contributed more than $34 billion in total food aid to developing countries since 1954. This represents approximately 46 percent of the world's total food aid contributed to needy countries during that period of time. United States agricultural exports go to more than 130 countries. Major regional foreign markets in 1981 were Asia ($16.2 billion), western Europe ($11.8 billion), Latin America ($6.9 billion), Africa ($2.9 billion), Canada ($2.1 billion), eastern Europe ($2.0 billion), and the U.S.S.R. ($1.7 billion). Farm exports in 1981 were valued at nearly $44 billion (approximately 40 percent of United States cropland production and about 19 percent of animal agriculture production is exported). United States farm exports are expected to increase in future years because of increased prices, tight world supply of grain, improving world economic conditions, and expansion of trade with the U.S.S.R., the People's Republic of China, the Republic of Korea, Mexico, West Germany, Taiwan, Spain, Egypt, and Japan. Japan imports about half of its total food requirements and currently is the largest market (over $6 billion annually) for United States agricultural exports.

The United States exports large numbers of breeding animals (especially beef and dairy cattle, goats, and swine), which are important to the genetic improvement of the livestock industries of other countries. Additionally, the United States supplies more than one-third of all protein meal consumed abroad. These exports support at least 1 million jobs, both on and off the farm. They would fill more than 1 million freight cars and were carried to 130 countries and territories. American agricultural abundance and technology are powerful forces for world peace. Our food and farm products are helping to relieve hunger and to promote economic growth in many developing countries of the world.

Future developments in animal agriculture of the United States will be affected greatly by foreign trade. Increased export demands for feed grains, for example, will tend to increase domestic prices, thus driving up and restricting domestic use. Unrestricted imports of certain animal products (e.g., dairy foods) would weaken domestic markets and could, over a period of time, decrease domestic production due to lower market prices.

However, expansion of United States agricultural exports benefits other sectors of the economy. For example, for every $100 increase in production of feed grains, wheat, rice, and oilseeds for export, an additional $110 output occurs in other sectors of the economy, such as transportation, storage, handling, and marketing.

In the more affluent countries, such as the United States and Canada, animal products can be expected to become more expensive to produce as world demands for cereals and protein supplements increase. Additional research is needed to discover ways and means of producing meat and milk more efficiently with proportionately greater amounts of forages and cereal by-products. Moreover, increased public concern with environmental pollution as well as animal welfare aspects of intensive animal agriculture can be expected. There will be increasing competition from modified or engineered foods, such as textured vegetable protein analogues, which simulate animal food products in appearance, taste, and nutritive value.

1.6 SUMMARY

Food ranks first among the needs of the human race, and it is humanity's most important renewable resource. Human anxieties about food are as old as history. So are people's interest in animals, which have contributed to human welfare since prehistoric times. Domestication of animals was a part of agricultural growth and development. Today, the primary importance of domestic animals for people is as a source of food and other products, but animals are also important for companionship and other purposes.

Of all the ills afflicting the human race, none seem more solvable—but concurrently more intractable—than hunger. The keenest competition for food is not between people of different regions or economic groups, but rather between people and the pests and diseases that attack food crops and plague animal agricultural production. Worldwide, insects and diseases claim an estimated one-fourth of all crops before they are harvested, and another estimated 15 to 20 percent are lost to insects and other pests following harvest (see Chapter 19). Similarly, diseases and internal and external parasites sharply reduce the potential productivity of animal agriculture (see Chapter 18).

Food production is the nation's, indeed the world's, largest business. Agriculture is an indispensable base for the United States and world economies. Production, processing, and distribution of food employs more people than the automobile, steel, transportation, and utilities industries combined. More than two-thirds of the people of the developing countries are directly involved in food production.

American agriculture has advanced more in the past 50 years than in all previous years of American history. However, evidence presented in this chapter illustrates clearly that agriculture is on a balance (Figure 1.20). From the beginning of agriculture, it has been a constant struggle to maintain this balance and thereby provide people with satisfaction of their wants and needs. Development of new breeds of livestock and varieties of plants, coupled with continuing research, has increased humanity's chances for achieving this goal, but there is an ever present threat that the precarious balance will tip.

The balance may be disturbed by a number of forces, which include the ever increasing population, reduction of the amount of acreage available per person, and the threat of economic distress, represented by low purchasing power. Disease, poor selection of breeds, inefficient management of resources, nutri-

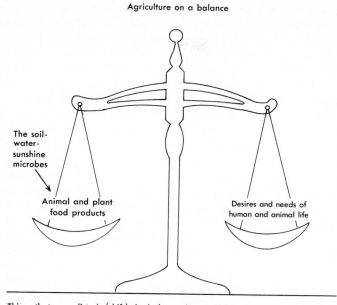

Agriculture on a balance

The soil-
water-
sunshine
microbes

Animal and plant
food products

Desires and needs of
human and animal life

Things that may disturb (shift) the balance that agriculture can achieve in nature

1. Disease
2. Wide birth-death ratio
3. Wide population-acreage ratio
4. Economic distress (low purchasing power)
5. Poor selection of animal and/or plant breeding stock
6. Inefficient management and use of resources
7. Nutritional imbalances
8. Predators
9. Weather
10. War

FIGURE 1.20
Diagram depicting the important role of agriculture in the delicate balance of nature.

tional imbalances, predators, weather, and the threat or actuality of war inflict their various individual and/or combined effects on the comfort and nutritional well-being of humanity.

The world is short on food *and* time. Today's hungry countries must compress the progress of decades into years if they are to adequately feed their rapidly increasing populations. There are three basic benchmarks to which the rate of increase in food production can be usefully related: (1) the rate of increase needed to keep pace with population growth, (2) the rate of increase needed to attain target rates of economic growth while maintaining stable prices, and (3) the rate of increase needed to eliminate the serious malnutrition common to many developing countries.

There is a deep and growing concern throughout the world over the outcome of the food-population race. During the next 9 years the world must prepare to feed an additional billion people. Never before have so many been added in such a short period of time. Even more significantly, fully four-fifths of the billion persons will be added in the already food-deficient developing countries. This growing imbalance between food and people threatens the economic and political stability of the developing countries.

The present generation is the first to possess the capability to essentially eliminate hunger. It will earn the thanks of future generations by perceiving and acting on this possibility. Efforts to meet world food deficits should not overlook the need for the development of a balanced agricultural economy in each country. A viable animal agriculture is of vital importance in this respect. Animal production is important to maximal effective use of available natural resources as human needs continue to be served through animal agriculture.

Conditions favoring animal production include (1) the requirements of a rapidly expanding human population; (2) the merits and/or special qualities of animal products; (3) the need for animals as a source of power; (4) the need for animals for mental health and personal satisfaction; (5) the role of animals in maintaining soil fertility and water conservation; (6) the flexibility of animals as transformers of feed into food and other useful products; and (7) the economic, social, and institutional forces that favor greater utilization of animal products and the practices of animal husbandry.

The key to sustained animal agriculture is the proper use of resources: air, water, land, and energy. Farm animals are kept for food production under a broad diversity of systems. The extremes vary from ruminants grazing rangeland to highly integrated confinement systems for poultry, ruminants, and swine.

In this chapter, present trends in human and animal populations were surveyed. An attempt was made to present an overview of animal agriculture. It was noted that since forages and roughages can be grown on land where tillage is impossible or impractical, ruminants are not necessarily competitive with humans for agronomic crops. Instead, by utilizing the grasses of such rough land, ruminants provide an added source of food for humans. It was noted that malnutrition is both a consequence and a cause of underdevelopment. Improved nutrition among children of developing countries is important to their growth and subsequent contributions to their respective national economy.

> Changes of diet are more important than changes of dynasty or even of religion.
> *George Orwell*

Inadequate nutrition is one of humanity's oldest problems. Today, the greatest problem facing people is not nuclear warfare, pollution, taxes, or inflation; instead, it is the problem of what to eat. Current political and military problems will fade as the importance of world food supplies comes into sharper focus.

Give a hungry child a cup of milk and he will be nourished for a day, but give his family a heifer and show them how to care for it and they will drink milk the rest of their lives.

M. E. Boyer

The flow of life is a continuum, delicate in its balances, intricate in its nuances, demanding in its observances. For animal agriculture, indeed for all agriculture, to survive, it must be in harmony with its natural surroundings. It is up to people to control their agriculture and thereby feed themselves and their kind, who will inherit the earth. People must somehow combat and defeat their foes and achieve an agricultural balance in which their wants and needs are met by an ample supply of nutritious animal and plant foods. Human needs are best served through a viable, dynamic animal agriculture.

STUDY QUESTIONS

1 Did you read the Preface to this book? If not, please do so.
2 Where were animals first domesticated? When?
3 Which farm animals were domesticated first?
4 Of what significance was the Land-Grant College Act of 1862?
5 What Act of Congress provided for the establishment of state agricultural experiment stations? When?
6 What was the Smith-Lever Act of 1914? The Smith-Hughes Act of 1917?
7 What is the threefold objective of agricultural education?
8 What proportion of the world population is engaged in agriculture? Is this true in the United States? Why? Of what significance is this to our way of life?
9 Is there an even distribution (and ratio) of people, land, agricultural production, and income throughout the world? Is the position of the United States favorable?
10 Which major areas of the world are increasing the fastest in population? Is their food supply adequate? Why should *we* be concerned about *their* problems? Explain.
11 Which animal products are produced in the greatest amounts throughout the world?
12 Can world livestock numbers be significantly increased? What are some possible constraints?
13 How does the United States rank in the world production of animal products?
14 Is the ocean fish supply presently being depleted by humans? What is aquaculture? Discuss its potential as a source of human food.
15 Can more people be fed if *(a)* they consume cereal grains (corn, oats, wheat) themselves or *(b)* the grain is first fed to animals and then people consume their products? Which is preferred? Which farm animal is most efficient in converting feed into food protein and energy?
16 What is a ruminant? (See also the Glossary.)
17 Do the people in the United States obtain most of their dietary protein from animal or plant sources? What about the people in other regions of the world?
18 Which countries are generally the most progressive, *(a)* those consuming largely animal protein or *(b)* those consuming largely plant protein? Which source of protein do you prefer?

19 Is the United States population increasing faster or more slowly than was predicted in 1920? What about food production?
20 What proportion of the total land area of the United States is used to produce animal feeds?
21 How much of the United States farmer dollar (income) is derived from the sale of livestock and their products?
22 Of each 1000 jobs in the United States, how many are related to agriculture?
23 United States farmers represent what proportion of the world's population? Are they producing their share of animal products? What proportion of United States crop production and animal production is exported annually?
24 What major spectator sport in the United States is closely related to agriculture?
25 What proportion of the United States consumer dollar is spent for animal products? Are animal products a good purchase as far as dietary nutrients are concerned?
26 What country in the world has the cheapest food in relation to wages earned? Why?
27 What country exports the largest amount of agricultural products?
28 Define agriculture. (See also the Glossary.) Discuss agriculture's impact on the United States economy.
29 Are animals necessarily in direct competition with humans for food? Discuss.
30 Cite examples of how animals may serve humans by recycling waste products.
31 What is FAO? What are some of its functions and activities?

BREEDS OF LIVESTOCK AND POULTRY

The sculptor works with lifeless clay that responds to his every touch, reflecting his degree of genius. The livestock breeder works with flesh and blood. His art is affected by the laws of heredity. The drag of Nature is his constant deterrent. Ideas, perceptions, facts, and counsel are his guides. His rewards for developing breeds and creating superior seedstock are financial returns, public recognition, and personal satisfaction.

F. S. Idtse

2.1 INTRODUCTION

A *breed* is a group of animals possessing certain characteristics that are common to the individuals within the group and that distinguish them from other groups of animals within the same species. These characteristics are the trademark of that breed and are transmitted from one *generation* to another. Hundreds of breeds of livestock and poultry have been developed throughout the world. It would be impossible to list and describe the characteristics of all the breeds here. Moreover, new breeds are being developed on a continuing basis. Therefore, only the major breeds of livestock and poultry of the United States and Canada will be given here to help the reader identify them.

2.2 DEVELOPMENT OF BREEDS

Older breeds of livestock originated many decades ago. They originated, as a general rule, within a certain region from individuals that often had few, if any, common characteristics (i.e., they varied in coat color, feather color, or other characteristics). For many years no particular attention was given to developing a breeding group with common characteristics. In time, however, breeders in certain areas or localities began to breed animals toward a specific objective, so that a herd had certain common characteristics, such as coat color. In some instances, two or more breeders combined their efforts to develop a given breed.

Careful selection by a producer eventually resulted in a group of animals that had enough common characteristics to be identified as a breed. Sometimes such animals were referred to as purebreds, meaning a pure breed. The term **purebred** is really a misnomer from the genetic standpoint. It is doubtful if there has ever been a single individual within a breed that was genetically pure. To be genetically pure, each of the thousands of pairs of genes within an individual must consist of like genes (or be *homozygous*). When a breed is referred to as a pure breed, it simply means that individuals within that breed possess certain characteristics, such as coat color, and that all offspring within the breed are from parents of that particular breed. For example, all Hereford cattle have a red body and a white face, and their offspring have the same coloring. The red coat color may, however, vary from a light yellow red to a dark red. The white face may also be associated with a large amount of white on the face, head, shoulders and neck, or with a small amount of white on those body parts. These colors are determined by genes, and since individuals within the same breed differ so greatly in the degree of white and the intensity of the red color, they are not pure genetically.

The black coat color of black Angus cattle is another illustration of the lack of genetic purity within a breed. It is true that most Angus cattle are black, but some are not. Of Angus calves born in the United States to black parents, about one out of 200 is red. To produce this proportion of red calves at birth, about 13 percent of the black parents must be carriers of the red gene, or *heterozygous* (*Bb*, and thus impure). The genetics of this phenomenon will be discussed in Chapter 4.

2.2.1 The Pedigree

A *pedigree* is a record of an individual's ancestry. Usually a pedigree includes only the names and registration numbers of the ancestors, thereby indicating that the ancestors were purebreds. However, in recent years pedigrees have included certain records of the performance of ancestors and also **carcass** information on the ancestors and/or their progeny and close relatives. A four-generation pedigree of a Hereford female is shown in Figure 2.1. The name and registration number of the sire (father) are placed at the top of the pedigree, whereas those of the dam (mother) are placed at the bottom. This pedigree shows the names and numbers of both the *maternal* and the *paternal* grandparents, great-grandparents, and great-great-grandparents. Since all of the ancestors were registered as purebreds, the offspring is a purebred.

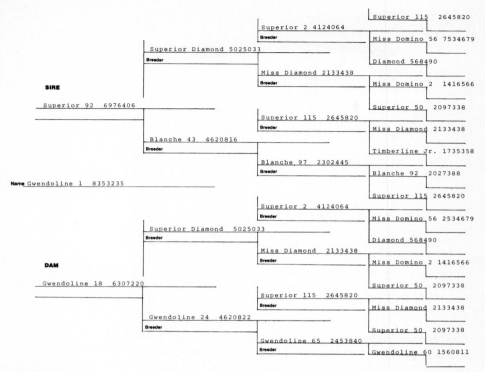

FIGURE 2.1
Pedigree of a purebred beef cow showing the format and the names and registration numbers of ancestors.

2.2.2 Pedigree-Record Associations

Each pure breed has a pedigree-record association. This is an organization of breeders who cooperate to improve the breed, preserve its purity, and protect and promote its interests. Each breed association has a secretary and a board of directors, elected from members, which is in charge of the association's business. This business includes recording pedigrees and issuing certificates of registry (Figure 2.2) and ownership for each registered animal. Registration certificates are provided for each purebred animal for a nominal fee. Funds from these fees support the work of the association. Breed associations adopt a standard of perfection for the breed, stating specifically the desirable and distinguishing breed characteristics. They also have certain rules that set forth requirements for admission of an animal to the registry. They publish a herd book that gives the name, registration number, breeder, owner, *sire, dam,* date of birth, and other information pertaining to each animal registered. Breed associations also establish rules for the advanced registry of animals recognized for special merit;

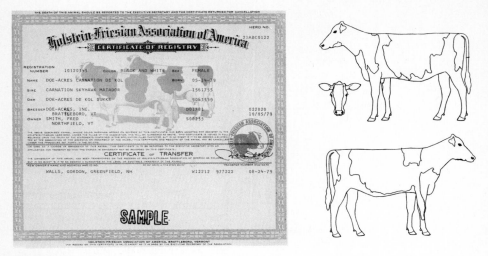

FIGURE 2.2
Sample registration certificate for a purebred animal (left). As a means of further identifying
registered animals, certain breeds require either a photograph or a sketch of the animal (right).
(Courtesy of Holstein-Friesian Association of America.)

offer prizes for the breed at certain fairs and shows; issue booklets and folders
containing detailed information about the breed and how to care for the animals;
and advertise in farm publications, calling attention to the merits of the breed.
Some breeds also are represented by a special magazine in which breeders can
advertise animals they have for sale, and that publishes articles dealing with the
sale, care, and promotion of animals within the breed. Some of these magazines
are owned and published by the breed associations; others are published by
private individuals.

2.2.3 New Breeds

New breeds of livestock and poultry have been developed over time in various
parts of the world. This is particularly true of swine, sheep, and beef cattle. Two
examples are Santa Gertrudis cattle and the Minnesota Number 1 breed of
swine.

New breeds are developed by first crossing two or more established breeds.
Each of the older breeds used to develop a new breed has certain characteristics
that the other lacks, and the genes for several traits not present in a single breed
are combined by crossbreeding. Selection for certain traits within the cross is
then concentrated after several generations. For example, the King Ranch
developed the Santa Gertrudis breed of cattle in Texas by crossing Shorthorns
and Brahmans. Shorthorns possess desirable beef qualities but lack heat and
insect resistance. Brahmans are not as desirable in beef qualities as Shorthorns
but possess a high degree of resistance to heat and insects. By crossing
Shorthorns and Brahmans, and by selecting within the crosses for desirable beef

qualities and heat and insect resistance, a new breed was developed that combined the desirable characteristics of both parent breeds. The same principle has been used to develop new breeds in other species.

Synthetic breeds are usually new breeds developed by crossing two or more older breeds, in much the same way as the Santa Gertrudis breed was developed. Large commercial companies have developed synthetic breeds of poultry and swine. At least one such company is currently developing several synthetic breeds of beef cattle. The object is to develop superior new breeds that perform well, and then to cross the new breeds in a systematic manner to obtain hybrid vigor, or **heterosis,** in the crossbred offspring. The different breeds to be crossed are usually unrelated genetically, and they are purposely developed in this way. The further apart the parents are in relationship, the more hybrid vigor can be expected, in those traits that show hybrid vigor, in the crossbred offspring.

Very little crossbreeding is practiced with dairy cattle. Moreover, there has been little effort expended to produce new breeds of dairy cattle in recent years. Instead, breeding efforts have been directed largely toward improving milk production in breeds such as the Holstein, for example, by intense selection. These efforts have been highly successful, especially when used in conjunction with artificial insemination and progeny testing (see Chapter 10).

The term exotic breed is commonplace among beef cattle groups today. Purebred cattle that have been shipped from their country of origin to another country in which they are not native are called *exotic breeds*. Simmental, Limousin, and Chianina cattle breeds are examples of exotic breeds of cattle shipped to North America in recent years.

2.2.4 Form and Function of Breeds

Most breeds of livestock were originally developed for the purpose of improving the livestock so that it assumed a particular form or performed a certain function. Breeding and selecting for specific characteristics dates back to the early history of the domestication of animals, but it was not until after the 1800s that most registry associations were formed and breed associations organized (Table 2.1). Information pertaining to selected breeds of poultry is given in Table 2.2.

Results from the selection and development of breeds of animals for a particular form or function are evident among many breeds of farm animals today. Cattle breeds have been developed for the production of meat or milk. The form and structure (body conformation) for these two functions are quite different (Figure 2.3).

Some breeds of cattle are known as *dual-purpose* breeds, meaning they are used for both milk and meat; others, known as *triple-purpose* breeds, are used for milk, meat, and draft. Types of swine have been developed for lard or bacon, and more recently miniature pigs, weighing about 40 lb at 5 months of age, have been developed especially for medical research purposes (Section 22.12). Breeds of horses have been developed for many purposes, including draft, racing,

TABLE 2.1
SELECTED BREEDS OF LIVESTOCK IN NORTH AMERICA

Breed	Color	Country or state of origin	Date of origin of herd book or association
		Swine	
American Landrace	W	Denmark	1950
Berkshire	Black (B) and white (W)	England	1884
Chester White	W	Pennsylvania	1894
Duroc	Red (R)	New York and New Jersey	1872
Hampshire	B and W (belt)	England or United States	1893
Hereford	R and W	United States	1934
Minnesota No. 1	R	Minnesota	1946
Minnesota No. 2	B and W	Minnesota	1948
Montana No. 1	B	Montana	1948
Palouse	W	Washington (state)	1956
Poland China	B and W	Ohio	1860
Spotted Poland China	B and W (spotted)	Indiana	1912
Tamworth	R	England	1897
Yorkshire	W	England	1893
		Beef cattle	
Aberdeen Angus	B	Scotland	1862
American Brahman	Steel gray	India	1924
Beefmaster	Many colors	United States	1949
Brangus	B	United States	1949
Charbray	Dun	United States	1949
Charolais	W	France	1864
Galloway	B	Scotland	1878
Hereford	R, white face	England	1846
Polled Hereford	R, white face	United States and Canada	1900
Polled Shorthorn	R, W, roan	United States	1889
Santa Gertrudis	R	Texas	1951
Shorthorn	R, W, roan	England	1835
Simmental	R, white face	Switzerland	1969
		Dairy cattle	
Ayrshire	R and W or mahogany and W	Scotland	1875
Brown Swiss	Gray brown	Switzerland	1880
Guernsey	Fawn and W	Guernsey Isle	1877
Holstein-Friesian	B and W	Netherlands	1885
Jersey	Light R or fawn	Jersey Isle	1868
Milking Shorthorn	R, W, roan	England	1912
Red Poll	R	England	1874

TABLE 2.1
SELECTED BREEDS OF LIVESTOCK IN NORTH AMERICA (CONT.)

Breed	Use	Country or state of origin	Date of origin of herd book or association
Sheep			
Cheviot	Mutton	England and Scotland	1891
Corriedale	Mutton and wool	New Zealand	1911
Dorset	Mutton	England	1891
Hampshire	Mutton	England	1889
Merino	Wool	Spain	1879
Oxford	Mutton	England	1888
Rambouillet	Wool	France	1889
Shropshire	Mutton	England	1883
Southdown	Mutton	England	1882
Suffolk	Mutton	England	1892
Horses			
American Quarter Horse	Saddle	Southwestern United States	1940
American Saddle Horse	Saddle	America	1891
American Trotter or Standardbred	Trotter	America	1871
Appaloosa	Saddle	United States	1938
Arabian	Saddle	Arabia	1908*
Belgian	Draft	Belguim	1887*
Clydesdale	Draft	Scotland	1877
Hackney	Carriage	England (Norfolk)	1891*
Morgan	Saddle	America (Vermont)	1909
Palomino	Saddle	United States	1941
Percheron	Draft	France	1876*
Pinto	Saddle	United States	1956
Shetland pony	Driving and saddle	Shetland Isle	1888*
Shire	Draft	England	1878
Suffolk	Draft	England	1880
Tennessee Walking Horse	Saddle	Tennessee	1935
Thoroughbred	Racing	England	1791

*When herd book was established in the United States.

riding, trotting, pacing, and working cattle (Chapter 21). Breeds of sheep have been developed for production of wool, mutton, or both. Breeds of chickens have been developed either for high egg production or for the production of meat (Chapter 12). Other breeds have been developed for fighting (game birds), and others for ornamental purposes.

Certain body conformations are in a general way best suited for certain functions. For example, a draft horse must be large, with heavy bone and muscle

TABLE 2.2
SELECTED BREEDS AND VARIETIES OF FOWL

American fowl (29 varieties)

Barred Plymouth Rock	Silver Laced Wyandotte
Buff Plymouth Rock	White Wyandotte
Partridge Plymouth Rock	Buff Wyandotte
White Plymouth Rock	Columbian Wyandotte
Rhode Island Red	Silver Penciled Wyandotte
Dominique	

Asiatic fowl (16 varieties)

Dark Brahma	Partridge Cochin
Light Brahma	White Cochin
Buff Cochin	Black Langshan
Black Cochin	White Langshan

**Belgian, Dutch, and German fowl
  (20 varieties)**

Silver Campine
Golden Penciled Hamburg
Silver Penciled Hamburg
Silver Spangled Hamburg

English fowl (18 varieties)

Silver Gray Dorking	Buff Orpington
White Dorking	White Orpington
Black Orpington	

Mediterranean fowl (16 varieties)

Ancona	Black Leghorn
Andalusian	Brown Leghorn
Black Minorca	White Leghorn
White Minorca	Buff Leghorn

to endure the rigors of the plow, whereas a racehorse must be thin and trim. A draft horse cannot race well, and a racehorse is not very useful for draft purposes. Beef cattle are *blocky* and muscular, whereas dairy cattle are less blocky and more angular (Figure 2.3). High-egg-producing breeds of chickens are smaller and thinner and produce little meat, whereas the heavy meat breeds are commonly poor egg producers.

2.2.5 Breed Differences

Breeds within the same species differ in many traits, including coat color, conformation, milk production, egg production, and rate and efficiency of gain. *Breed differences are genetic differences.* The statement is often made that there are greater differences among individuals within a breed than between breeds.

FIGURE 2.3
A Hereford female of excellent beef type (left) is compared with a Holstein cow of excellent dairy character (right). Note the difference in body conformation. *(Courtesy of American Hereford Association and Holstein-Friesian Association of America.)*

This is to be expected since there are greater differences among individuals within a breed than there are among breed averages. The average of all individuals within a breed (breed average) varies less, of course, than the total population of all individuals of all breeds. The reduced variation is a direct function of the number of individuals included in a given breed. For most traits, individuals of equal merit exist in all breeds. Therefore, in selecting for most traits, it is more important to select the best individuals within a breed than it is to choose the best breed.

The differences among breeds within a species are largely due to the kind of *genes* each breed possesses, and to the way these genes express themselves alone or in different combinations. Breeds may differ genetically for several reasons. Two breeds may differ because one breed may be *homozygous* for each of several pair of genes (*AABBccddEE*), and the other breed may be homozygous for the opposite gene of a pair or of several pairs (*aabbCCDDee*). Thus, one breed may have a gene the other lacks. Complete homozygosity within a breed, in which the genes of a pair are always the same (e.g., *aa*) probably seldom occurs. Instead, what is more likely is that for a pair of contrasting genes (*allelomorphs*) such as *A* and *a,* one breed possesses a high percentage of *A* (e.g., 85 percent) and a small percentage of *a* (e.g., 15 percent), whereas the second breed possesses a low percentage of *A* (e.g., 15 percent) and a high percentage of *a* (e.g., 85 percent). Hence, the main difference between the two breeds may be that they differ in the frequency with which a gene or genes are found in those two breeds. Principles of genetics are discussed in Chapter 4.

Breeds of animals within the same species are different genetically, as are individuals within a given breed. Breeds may differ from one another because they have started from a different group of ancestors, because humans have selected them for different purposes, or because they have drifted apart genetically as a result of being kept separate for many generations.

Efficiency of production within a breed may be increased by finding and mating the best individuals. For other traits, improvement comes more quickly when two or more families or strains are formed within a breed and then crossed to take advantage of hybrid vigor (heterosis) in the progeny. Because different breeds within a species are unlike in many of the genes they carry, crossing them results in more pairs of unlike genes (heterozygous) in their crossbred offspring, which is conducive to the expression of hybrid vigor.

2.3 BREEDS OF LIVESTOCK AND POULTRY

Numerous breeds of livestock and poultry are present in North America, and more are being developed or imported from other countries. Some breeds are present in larger numbers than others because they have been here longer and/or have become popular because of their particular breed characteristics and production traits.

2.3.1 Breeds of Swine

New breeds of swine developed in the United States include the Minnesota Number 1, the Minnesota Number 2, the Montana Number 1, and the Palouse. These new breeds were developed from crosses of certain older breeds with the Landrace. The Landrace breed was first introduced into the United States many years ago, but in order to introduce it, an agreement was reached that the breed would not be sold and reproduced as a pure one, because European swine breeders wanted to protect their purebred herds from competition with United States swine breeders. Some of the first imported Landrace lived in the USDA research facilities at Beltsville, Maryland. Another group was at Ames, Iowa, in the Iowa Experiment Station herd. Since it was not possible to breed and distribute purebred Landrace, they were crossed with some of the older, established breeds such as the Chester White, Black Poland, and Tamworth, and new breeds were developed from these crosses that carried a considerable percentage of Landrace blood. It was agreed that these new breeds could be reproduced and distributed within the United States free of restrictions.

Many commercial swine producers are currently using a three-breed cross of Hampshires, Yorkshires, and Durocs. Certain other older breeds are sometimes used for crossing, but not as often as the three breeds mentioned above. Selection for faster and more efficient gains and for more lean and less fat in the various breeds of swine during the past several years has been highly successful. This has been reflected in thinner back fat, as shown in Figure 2.4. Photographs of selected breeds of swine in the United States are shown in Figure 2.5.

2.3.2 Breeds of Beef Cattle

New breeds of beef cattle have been developed in the United States, mostly from Brahman crosses. These include the Brangus, Beefmaster, Charbray, and Santa

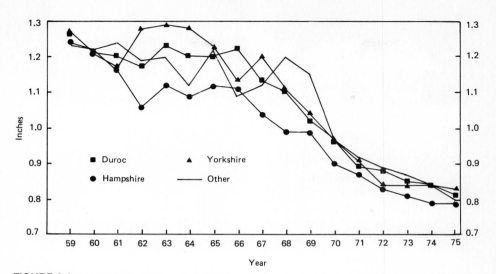

FIGURE 2.4
Trend in thickness of backfat, in inches at 220 lb body weight, among four major breeds of swine at the Missouri Boar Evaluation Station, 1959 to 1975. *(From* Mo. Agr. Exp. Sta. Res. Bull. *1021.)*

Gertrudis breeds. Other new breeds are currently being developed that do not contain Brahman blood. Breeds referred to as exotic breeds include the recently imported Blonde d'Aquitane, Gelbvieh, Maine-Anjou, Marchigiana, Normande, Pinzgauer, Romagnola, Simmental, and Tarantaise. Established beef breeds that have enjoyed considerable popularity in the United States over the past 40 to 50 years include the Angus, Hereford, and Shorthorn. Photographs of selected beef breeds of the United States are shown in Figure 2.6.

2.3.3 Breeds of Dairy Cattle

The dairy cattle breeds of North America have been established on this continent for many years. The Holstein-Friesian breed is present in the largest numbers. Other breeds of dairy cattle in the United States and Canada are the Ayrshire, Brown Swiss, Guernsey, Jersey, and Milking Shorthorn. Photographs of these six major dairy breeds are presented in Figure 2.7.

2.3.4 Breeds of Sheep

The sheep population in the United States has gradually decreased in recent years, so that the number of registered sheep now approximates 110,000 head. Breeds showing the greatest decrease in numbers include the Shropshire, Southdown, Hampshire, Corriedale, Merino, and Cheviot. The Suffolk breed has shown the greatest increase in recent years; it numbered approximately

(*a*) Berkshire

(*b*) Chester White

(*c*) Duroc

(*d*) Hampshire

(*e*) Landrace

(*f*) Poland China

(*g*) Spotted Poland China

(*h*) Tamworth

(*i*) Yorkshire

FIGURE 2.5
Selected breeds of swine in the United
States. *(Courtesy of the respective breed
associations.)*

(a) Aberdeen-Angus

(b) American Limousin

(c) American Polled Hereford

(d) Brahman

(e) Brangus

(f) Charolais

(g) Chianina

(h) Devon

(i) Galloway

(j) Hereford

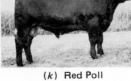
(k) Red Poll

(l) Santa Gertrudis

(m) Shorthorn

(n) Simmental

FIGURE 2.6
Selected breeds of beef cattle in the United States. *(Courtesy of the respective breed associations.)* *(Continued on page 56.)*

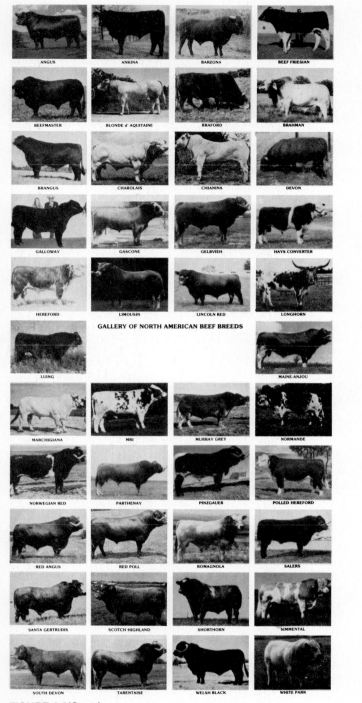

FIGURE 2.6(Cont.)
Gallery of selected breeds of beef cattle in North America. *(Courtesy of* Better Beef Business.)

(a) Ayrshire

(b) Brown Swiss

(c) Guernsey

(d) Holstein-Friesian

(e) Jersey

(f) Milking Shorthorn

FIGURE 2.7
Major breeds of dairy cattle of North America. *(Courtesy of the respective dairy breed associations.)*

50,000 in 1980, approximately twice the population of 1955. Photographs of selected breeds of sheep now in the United States are shown in Figure 2.8.

2.3.5 Breeds of Horses

Horses were first brought to America by Columbus on his second voyage in 1493. Spanish colonists and explorers later introduced additional animals.

(a) Cheviot

(b) Columbia

(c) Corriedale

(d) Horned Dorset

(e) Polled Dorset

(f) Hampshire

(g) Montadale

(h) Oxford

(i) Rambouillet

(j) Shropshire

(k) Southdown

(l) Suffolk

FIGURE 2.8
Selected breeds of sheep in the United States. *(Courtesy of* Sheep Breeder *and* Sheepman, *Columbia, Mo.)*

Horses were first used in the United States for either draft or riding purposes, and very few were from pure breeds. With the advent of tractors in the late 1930s and early 1940s, the horse population declined markedly. It has, however, increased in recent years. Light horses for recreational purposes have become very popular in the United States. Photographs of selected breeds of horses in the United States are shown in Figure 2.9. Chapter 21 is devoted entirely to horses and how they serve humanity.

2.3.6 Breeds of Chickens and Turkeys[1]

There are about 200 breeds and varieties of chickens and turkeys listed in the *American Standard of Perfection,* a publication of the American Poultry Association that lists and describes the recognized breeds and varieties of fowls. Of these, only the Single Comb White Leghorn, White Plymouth Rock, New Hampshire, and White Cornish (Figure 2.10 *a* to *d*) have been used in recent years to develop the strains of birds used by the poultry industry to produce table eggs and **broilers** (fryers). Similarly, the Bronze and White Holland turkeys have served as the parent stock for developing the strains used to produce turkey meat (Figure 2.11 *a* to *c*). Examples of typical egg-type lines used to produce white- and brown-shelled eggs are given in Figure 2.12 *a* and *b*. Typical female and male breeders used to produce commercial broiler chicks are shown in Figure 2.13 *a* and *b*.

Most commercial egg-type pullet chicks are produced by crossing inbred or otherwise highly selected lines of birds that originated from Single Comb White Leghorn stock. The desirable characteristics of the White Cornish, New Hampshire, and White Plymouth Rock breeds have been combined into special lines of birds that are crossed to produce commercial broiler chicks. The present trend is to produce broad-breasted, white-feathered market turkeys (Figure 2.11*a*) by using lines of birds that originated from the Bronze and White Holland breeds (Figure 2.11 *b* and *c*).

Due to the high reproductive rate of chickens and turkeys, relatively small numbers of birds are needed to produce hatching eggs. Therefore, breeders can be highly selective in deciding which individuals will be mated to maintain the breeds and lines, and to replenish breeding stocks. Also favoring rapid genetic progress in the breeding of poultry is the relatively short generation interval (see Chapter 5).

2.4 SUMMARY

Breeds of livestock and poultry have been developed, for the most part, to provide humanity with increased production of meat, milk, eggs, and wool.

[1]The authors are grateful to Dr. D. J. Bray, formerly professor of Animal Science, University of Illinois, and currently with the USDA in Washington, D.C., for his contributions to this section.

(a) American Quarter Horse

(b) American Saddle Horse

(c) American Albino Horse

(d) American Paint Horse

(e) American Paso Fino

(f) American Trotter or Standardbred

(g) Appaloosa

(h) Arabian

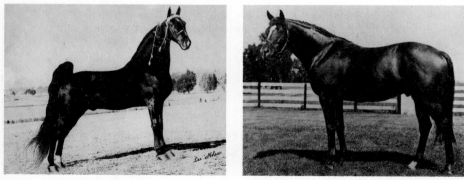

(i) Tennessee Walking Horse

(j) Thoroughbred stallion

FIGURE 2.9
Selected breeds of horses in the United States. *(Courtesy of the respective breed associations.)*

60

FIGURE 2.10
Most lines of chickens presently used for producing market eggs and poultry meat have been developed from breeds and varieties of (*a*) the White Leghorn, (*b*) the White Plymouth Rock, (*c*) the New Hampshire, and (*d*) the White Cornish. *(Courtesy of Watt Publishing Company.)*

FIGURE 2.11
(*a*) Modern white-feathered broad-breasted turkeys have been developed from highly selected lines originating from the (*b*) Bronze and (*c*) White Holland breeds. *(Part a, Courtesy of Nicholas Turkey Breeding Farms, Inc., b and c, Watt Publishing Company.)*

61

(a)   (b)

FIGURE 2.12
The Dekalb XL Link is typical of hybrid and strain-cross egg-type chickens used to produce white table eggs (a), and the DeKalb Sex-Sal-Link G is a typical egg-type line used to produce brown eggs (b). *(Courtesy of DeKalb AgResearch, Inc.)*

Exceptions include the development of certain breeds of chickens for game purposes and of horses for the track and pleasure riding.

In this chapter it was shown how breeds are developed, how they may differ genetically, and why there are different breeds. The major breeds of livestock and poultry of North America were noted. Collectively, they provide animal

FIGURE 2.13
Typical female (a) and male (b) breeders used to produce commercial broiler chicks. *(Courtesy of Hubbard Poultry Farms.)*

(a)   (b)

products for humanity—the foremost function of animal agriculture. The stage is now set to go on to learn more about the availability and nutritional contributions of animal products to the human race in Chapter 3.

STUDY QUESTIONS

1 What is a breed of livestock or poultry?
2 Why were breeds developed?
3 What is meant by the term purebred?
4 What is a pedigree? What information does it commonly include?
5 What is a pedigree-record association? Give examples of its activities and functions.
6 Name and give the color and country of origin of at least three breeds of swine, beef cattle, dairy cattle, and sheep. Name at least three breeds of horses and domestic fowl.
7 What is a dual-purpose breed? A triple-purpose breed?
8 How are new breeds of farm animals developed?
9 Is the practice of crossbreeding being followed in developing greater productivity among all farm animals? Cite examples.
10 Cite examples of exotic breeds of cattle imported in North America in recent years.
11 Give one or more examples of the relation of form (body conformation) and function (work and/or productivity).
12 How may differences among breeds be explained genetically?
13 Why was the imported Landrace breed crossed with established breeds of swine in the United States?
14 Name a breed of sheep that has enjoyed increased numbers in the United States in recent years.
15 When were horses first brought to America? By whom?
16 How useful and how important is crossbreeding to the poultry industry of the United States? Cite examples.
17 Give one or more factors that have favored rapid genetic progress, as measured in terms of productivity, in the breeding and development of commercial poultry in North America.

ANIMAL PRODUCTS AND HUMANITY[1]

Tell me what you eat, and I will tell you what you are. The fate of nations depends upon how they are fed.

J. A. Brillat-Savarin (1755–1826)

3.1 HISTORY OF AVAILABILITY

The long step from gathering food in the wild to growing crops and from hunting wild game to the **domestication** of animals went unrecorded. These important changes began in the early Neolithic period, long before people appreciated the importance of recorded events. In the new stone age, humans produced selected animal products: **poultry** and eggs; fish; fats from animals; and dairy foods, including whole milk, sour milk, butter, and cheese. People consumed most of their food near its origin. Then someone invented the wheel, and soon ox-drawn carts were carrying animal products to towns in exchange for tools, metals, and other goods. Later, oceangoing vessels sailed with cargoes of fish, meat, butter, and cheeses.

More recently, animal products have been available to a limited extent in developing nations and to a fuller extent in developed countries of the world. In general, the more highly developed nations, whose people are healthier and have a longer life expectancy than those in developing countries, are consumers of liberal amounts of animal products. The favorable effects of animal products on body development, muscle power, and general health have been noted in two African tribes that had considerable intermarriage but whose diets differed

[1]The authors acknowledge with appreciation the contributions to this chapter of Dr. R. T. Marshall, Department of Food Science and Nutrition, University of Missouri, Columbia.

greatly. The Masai tribe consumed a diet high in protein, with liberal amounts of meat[2] and milk, whereas the Kikuyu tribe consumed a vegetarian diet of cereals (chiefly maize), legumes, sweet potatoes, and green leaves. In the latter group, bone deformities, dental caries, **anemia,** pulmonary diseases, and tropical ulcer were prevalent, whereas the tribe consuming large quantities of meat and milk were 5 in taller and 50 lb heavier, had 50 percent greater muscle power, and were comparatively free from **disease.**

When the children of Japan began consuming more animal products (meat, milk, and eggs), they began growing taller and stronger and having better teeth. Of course, the importance of milk to sound teeth has been appreciated for thousands of years.

His eyes shall be red with wine, and his teeth white with milk.

Genesis 49:12

The average height of Japanese 14-year-olds increased more than 4 in during the three decades that followed World War II.[3] The average daily per capita Japanese intake of calcium in 1976 was approximatey 295 mg compared with only 20 mg in 1946. Until this change in diet, many Japanese people believed their ultimate height was merely a genetic factor. Now they realize more fully the importance of nutrition in genetic expression of adult size.

Other recent studies have shown that children adapted to a low calcium intake respond to increases in dietary calcium with increases in linear growth.[4] This result indicates that diets in many developing countries do not allow the full genetic growth potential of the peoples of those countries. Recent studies of **malnutrition** among preschool children in developing countries indicate that lack of nutrients during periods critical for growth and development of the central nervous system may cause irreversible cellular damage, thus inhibiting subsequent normal mental development. Malnutrition, especially protein deficiency, inhibits the development of protective antibodies and lowers resistance to disease. Furthermore, it is well established that malnourishment during early years of life results in delayed physical maturity, even if the deficiency is temporary and a normal diet is restored later. Even more ominous is evidence that protein deficiency in infancy and early childhood results in permanent impairment of the brain.[5]

[2]The association of broken animal bones with remains of humanity's forebears suggests that the meat-eating habit of people was acquired early.

[3]The typical Japanese fourth grader no longer fits in the chair his or her parents occupied a generation ago. Twice since the end of World War II, desks in Japanese schools have been moved out for larger ones, and architects are now designing doorways for homes and offices 6 in higher than in yesteryear.

[4]The human dietary requirement for calcium is considered to be a function of age; sex; hormonal status; absorption efficiency; intake of protein, fiber, and perhaps other nutrients; and physical exercise.

[5]In humans the dietary protein requirement is low during the first 4 months of infancy, when proteins account for only about 11 percent of the new tissue formed, but it increases sharply during the next 8 months, when protein represents about 21 percent of body weight gain.

The current availability and economy of animal products have already been discussed (Sections 1.4.3 and 1.5.1), but it should be indicated here that animal products have, over the past three decades, become a good food buy relative to wages earned. For example, recent USDA studies showed that the average factory worker could buy considerably more animal products with an hour's pay in 1982 than in 1950, as noted below.

Animal product	1950	1982
Frying chicken	2.5 lb	11.9 lb
Milk	8.0 qt	15.2 qt
Eggs	2.4 doz	9.8 doz
Pork	2.7 lb	4.7 lb

The above data suggest that the greatest recent strides in the efficiency of producing animal products have been made in poultry. One important reason for this is that the reproductive rate of poultry is much greater than that of beef or swine. For example, the average cow produces only about 0.7 progeny per year, which, in terms of slaughter weight, is about 70 percent of her body weight, whereas the average sow produces 14 progeny per year and those pigs represent a total market weight of 8 times the sow's body weight. The average meat-type hen produces 150 progeny per year, or about 100 times her body weight. These differences in reproduction rate have a profound effect on the relative costs of producing different meats. They also greatly affect per capita consumption trends, which are a function of retail prices. For example, in the early 1950s, the retail price of chicken was 80 percent that of beef, whereas in recent years, chicken prices have averaged only about 30 percent of those of beef.

The beef industry is the largest segment of United States agriculture, constituting about 23 percent of all farm marketings. But with per capita consumption of beef decreasing in recent years, the industry faces some tough challenges, including (1) increasing the rate of reproduction; (2) expanding the competitive advantage it holds in utilizing forage crop residues, by-products, and nonprotein nitrogen; (3) using genetic engineering to produce more useful microflora that, in turn, can provide increased potential for metabolizing feedstuffs into usable dietary nutrients; (4) improving its production and distribution efficiency; and (5) providing consumers with a wide range of conveniently packaged and competitively priced products.

3.2   COMPOSITION AND COMPARATIVE NUTRITIONAL CONTRIBUTIONS OF SELECTED ANIMAL PRODUCTS

Animal products provide abundant sources of all known food nutrients; however, some products are especially rich in certain dietary essentials and will be noted in discussions of major classes of **nutrients.** The composition of selected animal products is given in Table 3.1.

TABLE 3.1
COMPOSITION OF SELECTED ANIMAL PRODUCTS

	Water, %	Food energy*	Protein, %	Fat, %	CHO, %	Ash, %
Beef						
Round	66.6	197	20.2	12.3	0.0	0.9
Porterhouse	50.2	370	15.3	33.8	0.0	0.7
Rump roast	59.4	271	18.3	21.4	0.0	0.8
Hamburger	60.2	268	17.9	21.2	0.0	0.7
Lamb						
Rib chops	53.4	339	15.1	30.4	0.0	1.1
Shoulder roast	59.6	281	15.3	23.9	0.0	1.1
Pork						
Ham	56.5	308	15.9	26.6	0.0	0.7
Loin chops	57.2	298	17.1	24.9	0.0	0.9
Chicken						
Light meat	73.3	117	23.4	1.9	0.0	1.0
Dark meat	73.7	130	20.6	4.7	0.0	1.0
Turkey	64.2	218	20.1	14.7	0.0	1.0
Milk and milk products						
Whole	87.7	64	3.3	3.6	4.7	0.7
Skim	90.8	35	3.4	0.2	4.8	0.8
Nonfat, dried	3.2	362	36.2	0.8	52.0	7.9
Ice cream	60.8	202	3.6	10.8	23.9	1.0
Cottage cheese (creamed)	79.0	103	12.5	4.5	2.7	1.4
Cheddar cheese	36.8	403	24.9	33.1	1.3	3.9
Butter	15.9	717	0.9	81.1	0.1	2.1
Yogurt (plain, low-fat)	85.1	63	5.3	1.6	7.0	1.1
Eggs						
Whole	74.6	158	12.1	11.2	1.2	0.9
Whites	88.1	49	10.1	Trace	1.2	0.6
Yolks	48.8	369	16.4	32.9	0.2	1.7
Whole dried yolk	4.7	687	30.5	61.3	0.4	3.2
Fish, cod	81.2	78	17.6	0.3	0.0	1.2
Honey	17.2	304	0.3	0.0	82.3	0.2

*Calories per 100 grams (g) foodstuff.
Source: "Composition of Foods," USDA Handbook 8, 1963; and "Composition of Foods: Dairy and Egg Products," Agriculture Handbook No. 8–1; ARS, USDA. Revised November 1976.

3.2.1 Proteins

Upon what meat doth this our Caesar feed that he is grown so great?

William Shakespeare

Meat, milk, and eggs are excellent sources of **protein.** The quality of proteins is determined by their ability to support growth and maintenance. Proteins are composed of amino acids, which serve as building blocks from which the body fabricates its own special protein tissues. Proteins that contain the indispensable (essential) amino acids in about the proportions needed to build body tissues are known as *complete proteins,* as distinguished from *incomplete proteins,* which lack or are deficient in one or more of the essential amino acids. Animal proteins are superior to plant proteins for humans and other **monogastric** animals because they are better balanced in essential amino acids. Zein (a corn protein) is an example of an incomplete plant protein (Table 3.2); it is deficient in the amino acids lysine and tryptophan.[6] Animal proteins are excellent sources of lysine, and many, especially the proteins in milk and eggs, are rich in tryptophan. The **biological values** (also referred to as net protein utilization) of animal proteins are much higher than those of plant proteins; e.g., the biological values of whole eggs and milk are 94 and 85, in contrast to whole corn or navy beans (cooked) at 60 and 38, respectively. As a result, the nutritional value of proteins of corn, potatoes, white bread, or navy beans is substantially increased when they are consumed with animal products rich in the amino acids that are deficient in the vegetables. Recent research indicates that it may be possible to develop a high-lysine corn.

Thus animal proteins might be called "supplements" for vegetable proteins. The high nutritive value of milk protein makes it of special value in the treatment of **kwashiorkor,** a type of protein malnutrition found among young children in areas where people subsist on plant foods. Nonfat dry milk (36.2 percent protein) is used in many bakery and canned foods as a protein supplement as well as a functional ingredient. It is also used to fortify many foods now being exported to other countries as a part of the United States P.L. 480 Food for

---

[6]In diets containing few animal products, additional dietary protein is needed to compensate for the lower-quality vegetable proteins.

TABLE 3.2
PERCENTAGE AMINO ACID MAKEUP OF
SELECTED ANIMAL FOODS (AND THE PROTEIN OF CORN)

Amino acid	Beef	Pork	Chicken	Milk	Eggs	Zein
Arginine	6.4	6.7	6.7	4.3	6.4	1.8
Cystine	1.3	0.9	1.8	1.0	2.4	0.8
Histidine	3.3	2.6	2.0	2.6	2.1	1.2
Isoleucine*	5.2	3.8	4.1	8.5	8.0	4.3
Leucine*	7.8	6.8	6.6	11.3	9.2	23.7
Lysine*	8.6	8.0	7.5	7.5	7.2	0.0
Methionine*	2.7	1.7	1.8	3.4	4.1	2.3
Phenylalanine*	3.9	3.6	4.0	5.7	6.3	6.4
Threonine*	4.5	3.6	4.0	4.5	4.9	2.2
Tryptophan*	1.0	0.7	0.8	1.6	1.5	0.2
Tyrosine	3.0	2.5	2.5	5.3	4.5	5.9
Valine*	5.1	5.5	6.7	8.4	7.3	1.9
Total	52.8	46.4	48.5	64.1	63.9	50.7

*Essential to humans.
Source: Modified from M. L. Scott, *J. Amer. Diet. Assoc.,* **35**:248 (1959).

Peace program. The balance of essential amino acids in milk and its potential for providing daily dietary amino acids are shown in Figure 3.1. Milk—the nectar of mammals—is virtually a balanced meal in itself. Per unit of dry matter, milk is nutritionally equivalent to meat; yet when fed to animals, only about 10 percent of the milk solids are recovered in the form of meat solids. Therefore, it is nutritionally wasteful to feed milk to older animals. Instead, milk produced in excess of that needed to raise the young should be available for human consumption.

Because high-protein animal foods promote growth but do not fatten like other foods, they fit well into a reducing diet. Although excessive protein is somewhat wasteful and increases the kidney load, it serves to satisfy hunger cravings of the overweight without adding much to their fat stores. Proteins consumed in excess of normal needs are **deaminized,** and only a part of the original caloric value of the protein is retained. The balance is voided in the urine. By comparison, dietary **fats** and **carbohydrates** are utilized more fully, and their metabolic end products are stored in the body as fat. It has been estimated that whereas carbohydrates yield approximately 95 percent of their potential energy to the body, only about 70 percent of the potential energy of protein is useful in meeting energy needs of the body.

3.2.2   Fats

Although few animal products are purchased for their fat content, many contribute essential fatty acids, the fat-soluble **vitamins** A, D, E, and K, and

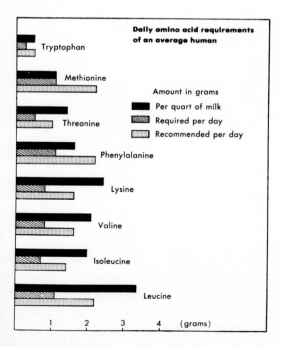

FIGURE 3.1
Daily amino acid requirements of and recommendations for an average human adult, with quantities furnished per quart of milk. *(R. G. Hansen and D. M. Carlson, J. Dairy Sci.,* **39***:665, 1956.)*

precursors of vitamins A and D. Since body fat can be produced from the carbohydrates of food, one might expect that dietary fat is not important. On the contrary, fats are essential dietary elements. Fatty foods contain fatty acids that are essential for good health and that cannot be synthesized in the body. Rats soon cease to grow when fed a low-fat diet containing all known vitamins but resume growth when small amounts of the essential fatty acids are added to their diet. **Saturated fatty acids** (palmitic and stearic) may be derived from carbohydrates and stored in the body, but they offer no protective action. The unsaturated (essential) fatty acids of body fat may be derived from eggs, milk, butter, and plant oils. These fatty acids include linoleic (two double bonds), linolenic (three double bonds), and arachidonic (four double bonds). Arachidonic acid is found only in animal tissues and milk. Any one of these fatty acids apparently can meet the body's need for essential fatty acids (Section 14.2.3).

The mean saturated and unsaturated fatty acid content for selected animal products is given in Table 3.3. Diet influences the composition of animal fats; e.g., lard from hogs fed a low-fat diet may have a linoleic content of only 1.2 percent in contrast with the normal of 7 percent. Animal fats have shorter-chain fatty acids than do plant fats and are more easily digested. (The **coefficient of digestibility** of milk fat is 97 for humans.) Fat aids in calcium absorption, and since milk contains abundant calcium, the complementary effect of fat in milk is especially important.

3.2.3 Carbohydrates

Meat and eggs are low in carbohydrates, whereas milk and honey are excellent sources (Table 3.1). When the Hebrews were promised "a land that floweth with

TABLE 3.3
FATTY ACIDS AND CHOLESTEROL IN SELECTED ANIMAL PRODUCTS

Food	Amount	Weight, g	Cholesterol, mg	Saturated fatty acids, g	Unsaturated fatty acids, g
Beef, round steak	3½ oz	100	125	4.70	5.75
Butter	1 pat	7	20	3.25	2.17
Cheese, cheddar	1 oz	28	32	5.38	3.27
Chicken, fryer	3½ oz	100	60–90	2.33	4.51
Eggs	1 medium	54	340	1.64	3.58
Lamb, chop	3½ oz	100	70	17.85	12.93
Liver, beef	3½ oz	100	320	0.87	2.14
Milk, whole	1 cup	244	33	5.24	3.79
Pork, loin or chop	3½ oz	100	70–105	9.50	14.25
Salmon	3½ oz	100	60	4.97	10.71
Turkey, breast meat	3½ oz	100	8–15	1.31	3.10
Brains, beef	3½ oz	100	2233	Not available	Not available

Source: Modified from O. B. Hayes and G. Rose, *J. Amer. Diet. Assoc.*, **33**:27–28 (1957).

milk and honey" (Joshua 5:6), they were being offered both a necessity of life, *milk,* and a luxury, *honey.* It would be remiss not to mention some of the unique characteristics of these two animal products.

Honey   A jar of honey on the table was once considered a mark of great wealth. The ancient Egyptians are believed to have used honey in embalming. As noted in Table 3.1, honey contains more carbohydrates (82.3 percent) than does any other animal product. A unique property of honey is its predominant sugar, fructose (levulose). This sugar is primarily responsible for the sweet taste. Honey contains a large amount of dextrose (glucose) but is low in sucrose (less than 8 percent by legal standards).

Most pure honeys granulate, or develop sugar crystals. An exception is tupelo honey, which seldom granulates because it contains a large amount of fructose in relation to dextrose. It is the dextrose that granulates, either partially or completely. Industrial users heat honey to prevent granulation.

Honey is heavier than water (the specific gravity of honey is 1.41), but the wax portion is lighter (sp gr 0.96). Most honeys will darken in color if stored at a temperature above 59°F.

Uses of Honey   Honey is an excellent energy food because it contains simple sugars that can be used quickly by the body. Honey contains mineral salts and other materials needed by the body. It is the only form of sugar food that does not need to be refined for human use. Bakers sometimes use it in place of sugar for their products. Many cough medicines and laxatives contain honey. The wax portion is used in making candles, lipsticks, polishes, waterproofing compounds, adhesives, chewing gum, ointments, and other materials. **Honey butter** is a popular spread and is made by beating honey and butter together.

Kinds of Honey   The flavor and color of honey are influenced by the kinds of flowers from which the nectar comes. Honey ranges in color from white to dark amber. Usually, the light-colored honeys have the mildest flavor. The commonest honey plants are alfalfa and clovers. Minnesota, where clovers are common, is the leading United States producer of honey. California ranks second in honey production; white sage and orange blossom honeys predominate there. In the east, buckwheat flowers are used, whereas in the south, tupelo, mesquite, sourwood, and gallberry supply nectar for honey. There are approximately 300 honeys available in the United States.

How Honey Is Made   Research indicates that bees recognize honey-yielding flowers first by color and second by scent. Honey is made by bees from flower nectar. Bees sip it from the blossoms and carry it to their hives. Each worker stores the nectar it collects in a special pouch, called a *honey stomach,* inside its body. In the stomach the sugar and nectar are broken down by a process called *inversion* into two simple sugars, *fructose* and *dextrose.* After honeybees deposit nectar in the hive, they allow most of the water to evaporate, and the liquid becomes thick. (The nectar may contain as much as 70 percent water as collected. The workers fan the nectar with their wings, reducing the water content to about 17 percent.) They also add **enzymes** that enhance the flavor. A special gland in the abdomen of the young worker produces beeswax. It is interesting that honeycomb has very thin walls (1/80 in) that can support 30 times their own weight.

The Honeybee Colony   Bees are the only insects that produce food commonly consumed by humans. There are some 20,000 **species** of bees, but only honeybees make honey and wax that humans can use. They form a colony that includes three classes of honeybees: the queen, which lays eggs; the workers, which gather food and care for the young; and the **drones,** which fertilize the queen. Honeybees have a highly developed society.

The *queen* honeybee lays eggs that hatch into thousands of workers. Laying eggs is the queen's only responsibility. If two queens hatch at the same time, they fight until one stings the other to death. When the young queen makes her maiden flight, she may mate with one or sometimes several drones. After mating, the young queen returns to the hive and begins laying eggs 2 days later. During her reproductive surge, she may lay 2000 eggs daily (more than her body weight) or more than 200,000 in a single season. With a life expectancy of about 5 years, the queen may lay 1 million or more eggs in her lifetime. Bee grubs are fed a mixture of honey and pollen known as **beebread.** The length of time they receive *royal jelly* (also called *bee milk)* determines whether the female grub will develop into a queen or a worker. Queen **larvae** are reared in special large cells and receive royal jelly throughout their entire grub phase. Royal jelly has recently enjoyed a fad as a food for humans. It is a creamy substance, rich in vitamins and proteins. It is formed by a **gland** in the head of the young worker bee.

The *workers* are all females and do all the work except the laying of eggs. In a queenless colony, the workers can lay eggs; however, since they are unfertilized, only drones are produced. The worker has a long tongue for gathering nectar. It uses its hind legs to carry pollen, which is used for food. After the worker gathers as much nectar as it can carry, it takes the shortest route (hence *beeline)* back to the hive. During the busy summer season, a worker usually lives about 6 weeks. A strong colony may have from 60,000 to 80,000 workers, each collecting only about 1/10 lb of honey during its entire lifetime. To make a pound of honey, a bee colony may have to extract nectar from a million or more blossoms, make 37,000 trips, and travel 50,000 miles.

All female bees (queens and workers) have a **diploid** complement of **chromosomes,** whereas the males (drones) are **haploid.** Workers have barbed stingers and can sting only once,[7] since the stinger pulls out of the bee's body. A worker bee dies a few hours after losing its stinger. The queens have smooth stingers but use them only to get rid of a rival queen. Bee **venom** has been used in arthritis therapy and other disorders of humans. The favorite medicine of Hippocrates was honey. He said, "The drink to be employed should there be any pain is vinegar and honey. If there be great thirst, give water and honey."

Drones are burly, clumsy creatures. They do not work and have no sting. Drones develop from unfertilized eggs **(parthenogenesis).** The only function of a

---

[7]Bees usually do not sting at temperatures below 55°F or on rainy days.

drone is to mate with a young queen. An unmated queen can lay only drone eggs; she must be fertilized in order to lay worker eggs. In autumn, when the honey flow is over, the workers allow the drones to starve to death. This is done because they are no longer useful to the colony and would eat too much of the stored honey.

Bees are especially beneficial in the pollination of fruits, vegetables, and seed crops. Worldwide, the profit derived from pollination by bees is estimated at 25 times higher than that obtained from bee products.

The bee helps the garden, the garden helps the bee, and people reap the harvest of both.

Anonymous

The bee is more honored than other animals, not because she labors, but because she labors for others.

St. John Chrysostom (fourth century)

In the past century, apiculture (beekeeping) has come to be considered a branch of animal science, and systematically controlled breeding, selection, and crossing of bees, using techniques commonly employed with other domestic animals, have been employed successfully. World production of honey is estimated at 1.8 billion lb and valued at about $0.9 billion. Leading honey producers include the U.S.S.R., Australia, Argentina, West Germany, the United States, Mexico, Canada, Norway, Cuba, and China. The annual per capita consumption of honey in the United States is about 1½ lb.

Pesticides are of special concern to beekeepers. Since bees roam a great deal in search of pollen and nectar, they may collect food from treated or drift-contaminated plants.

Lactose—Milk Sugar Is Unique Milk is the only substance in nature that contains the sugar known as *lactose,* and it is the principal carbohydrate or sugar present in milk. (There are only traces of other sugars.) It is amazing that the milks of all mammals, some 10,000 living species, from reptilian **monotremes** to moles, mice, monkeys, and humans, contain lactose. It must therefore be very important or it would not occur naturally in the milk of all these species.

Early Indians believed that the longer the child received breast milk, the longer would be his or her life, and it was not uncommon for children of American Indians to be breast-fed to ages 7 to 9. It has been reported that students who were breast-fed as infants make significantly higher scores on college examinations than their bottle-fed contemporaries.[8] Why or how could this be possible? Human milk is especially rich in lactose (7 percent), which correlates with the large brain size (as related to total body mass) of humans (Table 3.4). Lactose constitutes 56 percent of the dry matter of woman's milk, in

---

[8]J. R. Campbell, *In Touch with Students . . . A Philosophy for Teachers,* Educational Affairs Publishers, P. O. Box 248, Columbia, Mo. 65205, 1972.

TABLE 3.4
BRAIN SIZE AS RELATED TO
BODY WEIGHT AND LACTOSE IN MILK

Animal	Brain weight, g	Brain size as percentage of total body weight	Lactose in milk,%
Human	1400	2.5	7.0
Horse	600	0.25	5.9
Elephant	5000	0.20	3.4
Whale	2050	0.0025	1.8

Source: World Book Encyclopedia, vol. 2, Field Enterprises Educational Corp., Chicago, 1965, p. 460B.

contrast with 36 and 6 percent in cow's and rabbit's milk, respectively. Since lactose contains galactose, a constituent of the central nervous system (as galactosides and cerebrosides of nerve and brain tissue), milk lactose may be a "brain food," a special nutrient for growth and development of the central nervous system of mammalian young. (The human brain reaches about 70 percent of its future adult weight by 12 months of age.) Worldwide, infant nutrition could be improved by reversing the trend toward less breast-feeding because mother's milk is a valuable natural resource.

> The breasts were more skillful at compounding a feeding mixture than the hemispheres of the most learned professor's brain.
>
> *Oliver Wendell Holmes*

One of the most desirable qualities of lactose is its hygienic value. Unlike sucrose, lactose passes the **ileocecal valve,** to be slowly absorbed from the intestine. Its presence in the intestine stimulates the growth of microorganisms (especially *Lactobacillus acidophilus)* that produce organic acids and synthesize many B-complex vitamins. The high acid concentration suppresses protein **putrefaction** and the growth of many **pathogenic organisms** because these organisms are sensitive to high acidity. Consumption of *L. acidophilus* via certain dairy products, such as sweet acidophilus milk, has a favorable influence during **oral** administration of **antibiotics** and encourages reestablishment of normal intestinal flora after therapy ceases.

Assimilation of calcium, phosphorus, magnesium, and barium is enhanced by lactose in the intestine. This unique quality of lactose also makes milk an excellent antirachitic (rickets-preventing) food even when the milk is low in vitamin D. Moreover, woman's milk is more antirachitic than cow's milk, which may be explained by the higher level of lactose in woman's milk (7 versus 5 percent). These unique properties of lactose would not be predicted from its chemical composition.

Lactose is involved in the antipellagric (pellagra-preventing) properties of milk. The recommended daily consumption of niacin (nicotinic acid) for an average person is 14 to 18 mg. Since a quart of milk contains but 1 mg, it would be logical to conclude that a human would have to consume 14 to 18 qt of milk daily to obtain the needed 14 to 18 mg from milk. Instead, milk provides lactose, which enables microorganisms of the intestine to synthesize nicotinic acid. Thus milk is, in fact, an excellent antipellagric food. Additionally, milk contains tryptophan, and it is recognized that 1 mg of niacin is derived from each 60 mg of dietary tryptophan. A quart of milk contains approximately 540 mg of tryptophan, equivalent to about 9 mg of niacin.

Although lactose and sucrose have the same empirical formula, they differ in many respects. Lactose is about one-fifth as sweet as sucrose and is less quickly metabolized to acids than other sugars. This makes lactose less irritating to the stomach and intestinal mucosa than the more quickly metabolized sugars. Thus, milk is valuable in the diets used in the treatment of ulcers of the stomach and duodenum. (Approximately 15 million people in the United States have or have had ulcers of the stomach or duodenum.)

Lactose may have some value in reducing diets because of its less rapid absorption, compared with the more quickly metabolized carbohydrates, and its ability to reduce fat deposition in the body. In one study, **carcasses** of animals fed high-lactose diets were found to contain only 58 percent as much fat as those that received high-glucose diets.

Lactose is frequently used in the manufacture of **placebos,** which means "I shall please" in Latin. After reviewing the foregoing discussion of the unique properties of lactose, one might conclude that a patient receiving a placebo receives more than a psychological lift.

3.2.4 Minerals

Animal products provide abundant sources of essential dietary minerals. The total ash content of selected animal products is given in Table 3.1. Data on the content of six essential dietary minerals in animal products are presented in Table 3.5. Milk, milk products, and fish are excellent sources of dietary calcium and phosphorus. Dairy products, beef, chicken, lamb, liver, pork, and turkey are good sources of phosphorus. Liver, beef, and whole eggs are especially rich in iron, whereas milk is deficient. It is interesting that most newborn mammals are provided with liver stores of iron, copper, and manganese to last them beyond the normal nursing period. (The pig is not.) Animal products are good **supplements** to plant foods such as cereals, potatoes, and white bread, which are deficient in calcium. Phosphorus, sodium, potassium, and magnesium are other minerals important to human health. Diets rich in animal products that provide sufficient calcium are likely to supply enough phosphorus as well.

Milk is the best nutritional source of calcium, not only because of its richness in calcium but also because of the favorable Ca/P ratio. Of special interest is the parallel between milk composition and the maturing speed of newborn mammals

TABLE 3.5
MINERAL CONTENT OF SELECTED ANIMAL PRODUCTS, mg/100 g, EDIBLE PORTION

Animal product	Calcium, mg	Phosphorus, mg	Iron, mg	Sodium, mg	Potassium, mg	Magnesium, mg
Recommended daily allowance, human[*]	800	800	10.0 (18.0, women)		Est. 1875– 5625	350 (300, women)
Beef, good grade	10	152	2.5	65	355	18
Cheese, cheddar	750	478	1.0	700	82	45
Chicken, white meat	11	265	1.3	64	411	19
Eggs, whole	54	205	2.3	122	129	11
Honey	5	6	0.5	5	51	3
Ice cream	146	115	0.1	63	181	14
Lamb, choice grade	10	147	1.2	70	290	23
Liver, beef	11	476	8.8	184	380	18
Milk						
Whole	118	93	Trace	50	144	13
Skim	121	95	Trace	52	145	14
Human	33	14	0.1	16	51	4
Nonfat, dry	1308	1016	0.6	532	1745	143
Pork, ham	10	236	3.0	65	390	17
Salmon, pink	196	286	0.8	387	361	30
Turkey, cooked	8	251	1.8	130	367	28

[*]National Academy of Science–National Research Council, Food and Nutrition Board, Washington, D.C., 1980.
Source: "Composition of Foods," USDA Handbook 8, 1963; and "Nutritive Value of American Foods," USDA Handbook 456, 1975.

among species (Table 3.6). Nature provided milk having a relatively low protein and mineral content for humans, who mature slowly. Contrast this with the milk of the cow, which contains much more mineral and protein. **Bovine offspring** grow rapidly and thus need large quantities of calcium and phosphorus for bone development and protein for muscle development. **Sow's** milk contains even more mineral than that of the cow, which correlates with the very fast early growth of the newborn pig.

An interesting and important fact regarding the nutrition of the young is that nature designed the milk of a given species specifically for the growth and development of the offspring of that particular species. A German scientist, Bunge, found that dog's milk had an ash content of exactly the same composition as the ash of the newborn puppy. The mineral of the milk was therefore perfectly adapted for the construction of new puppy tissue. However, it was very different in composition from woman's, cow's, or other milk. Only in the case of iron is the quantity lower than corresponds to the composition of the offspring, but this is offset by the fact that the body is richer in iron at birth than it is at any other period of life.[9] The caseins (milk proteins) of different milks are different in chemical behavior. Moreover, the renninlike enzymes (Section 15.7.2) of the stomach are adapted for coagulation of the casein produced by the female of the same species.

Furthermore, the concentration of constituents in milk important to normal growth and development (especially minerals and protein) is dependent on the rapidity with which an animal grows. It is interesting that a higher percentage of milk ash is absorbed by infants receiving woman's milk than by those fed cow's milk. In one study, babies were found to absorb about 80 percent of the mineral from their mother's milk and only 60 percent from cow's milk. Adults were observed to absorb about 53 percent of the ash of cow's milk. Of interest, too,

---

[9]The iron reserve at birth for the various animal species is roughly proportional to the normal weaning age.

TABLE 3.6
MILK COMPOSITION AND GROWTH RATES OF SELECTED ANIMALS

Animal	Milk composition, %		Time, in days, for the newborn to double its weight
	Protein	Ash	
Woman	1.6	0.2	180
Mare	2.2	0.4	60
Cow	3.3	0.7	47
Goat	3.7	0.8	19
Sow	4.9	0.9	18
Dog	7.1	1.3	8
Rabbit	14.4	2.5	6

are the findings from a series of Japanese studies showing a significantly lower incidence of stomach cancer among persons who consumed two glasses of milk daily.

3.2.5 Vitamins

Animal products vary in their vitamin content (Table 3.7). In supplying dietary needs for both fat-soluble and water-soluble vitamins, liver is likely the best-balanced animal product source of vitamins. Aside from thiamine (B_1), ascorbic acid (C), and vitamin D, a serving of beef liver (100 g) would provide the approximate daily vitamin needs for the average human as recommended by the National Research Council (Table 3.7). Meat is an excellent dietary source of several B vitamins. The vitamin content of fresh pork reflects the amount of vitamins in the animal's ration. This ration effect is not observed in the meat of ruminants, presumably because of the synthesis of B vitamins by rumen microorganisms (Section 15.3.2). Only animal products provide reliable natural sources of vitamin B_{12}, which is essential in small amounts for the proper growth of humans, production of red blood cells, and functioning of the central nervous system.

Cheese, eggs, milk, and certain organ meats offer good sources of vitamin A. Unless fortified, all animal products except eggs are poor sources of vitamin D. Furthermore, liver is the only animal product with an appreciable amount of vitamin C. Ham, liver, eggs, and milk are good sources of thiamine (B_1), whereas liver, eggs, and milk products provide liberal amounts of riboflavin (B_2). Dietary niacin may be readily obtained from meat and milk (niacin equivalent, Section 3.2.3). For a reliable daily source of ascorbic acid, an orange or other citrus fruit is desirable.

3.3 PURCHASING FOOD NUTRIENTS VIA ANIMAL PRODUCTS

North Americans consume liberal amounts of animal products. These foods provide approximately two-thirds of our dietary protein, calcium, phosphorus, and riboflavin; about one-half of our fat and niacin; and about two-fifths of our iron, vitamin A, and thiamine (Figure 3.2). Animal products provide only 6.2 percent of our dietary carbohydrates, but of more interest to weight-conscious Americans is the fact that only one-third of our food energy is derived from animal products. Moreover, animal products are high in protein as related to total calories and therefore fit well into reducing diets. In addition, the nutritious balance of most food nutrients present in animal products makes them a valuable purchase for the consumer.

For the economy-minded consumer who wishes to purchase food nutrients on a least-cost basis, costs of various food nutrients can be computed in several ways; e.g., figuring a pound of cottage cheese with 61.7 g of protein at 95¢ one could provide all the needed dietary protein daily (56 g for an adult male) for

TABLE 3.7
VITAMIN CONTENT OF SELECTED ANIMAL PRODUCTS, mg/100 g OR IU/100 g, EDIBLE PORTION

Animal product	Fat-soluble vitamins		Water-soluble vitamins			
	Vitamin A, IU	Vitamin D, IU	Thiamine, mg	Riboflavin, mg	Niacin, mg	Ascorbic, mg
Recommended daily allowance, humans*	5000†	200– / 400‡	1.1– / 1.5	1.3– / 1.7	14– / 18§	60
Beef, good grade	60	...	0.07	0.15	4.0	...
Cheese, cheddar	1310	...	0.03	0.46	0.1	...
Chicken, white meat	60	...	0.04	0.10	11.6	...
Eggs, whole	1180	93	0.11	0.30	0.1	...
Honey	......	...	Trace	0.04	0.3	1
Ice cream	440	...	0.04	0.21	0.1	1
Lamb, choice grade	......	...	0.15	0.20	4.8	...
Liver, beef	43,900	...	0.25	3.26	13.6	31
Milk						
Whole	140	—‡§	0.03	0.17	0.1	1
Skim	Trace	—‡§	0.04	0.18	0.1	1
Human	240	...	0.01	0.04	0.2	5
Nonfat, dry	30	...	0.35	1.80	0.9	7
Pork, ham	....	...	0.51	0.23	4.6	...
Salmon, pink	70	...	0.03	0.18	8.0	...
Turkey, cooked	.....	...	0.05	0.18	7.7	...
One orange	200	...	0.10	0.04	0.4	50

*National Academy of Science–National Research Council, Food and Nutrition Board, *Recommended Dietary Allowances*, 9th rev. ed., Washington, D.C., 1980.

†The allowance for vitamin A is also expressed in microgram (μg) retinol equivalents (RE). One retinol equivalent = 1 μg retinol or 6 μg β-carotene. Children and adults are warned against regular, excessive daily intake of vitamin A [more than 7500 RE or 25,000 international units (IU)].

‡The recommended dietary allowance (RDA) for vitamin D is also expressed as μg of cholecalciferol. 10 μg cholecalciferol = 400 IU of vitamin D. Milks commonly have 400 IU of vitamin D added per quart.

§1 mg of niacin = 1 NE (niacin equivalent) or 60 mg of dietary tryptophan.

Source: "Composition on Foods," USDA Handbook 8, 1963.

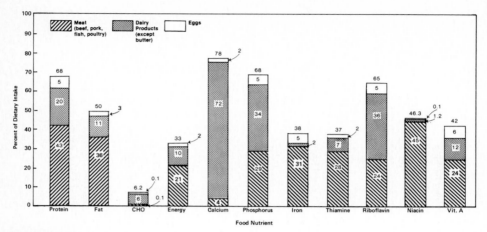

FIGURE 3.2
Contributions of animal products to the human dietary intake of selected food nutrients in the United States. *(Adapted from* National Food Review, *USDA, ERS, NFR-13, Winter 1981.)*

about 86¢, whereas if one chose porterhouse steak at $2.80/lb to provide that same amount of protein (at 60.8 g protein per lb), it would cost $2.58.

3.4 TRENDS IN PER CAPITA CONSUMPTION OF ANIMAL PRODUCTS

Before examining the trends in consumption of specific animal products in the United States, it is well to review the trends in total daily per capita nutrient consumption during the past 70 years. As is shown in Table 3.8, Americans

TABLE 3.8
DAILY NUTRIENT CONSUMPTION PER CAPITA IN THE UNITED STATES

Nutrient	1910	1920	1930	1940	1950	1960	1970	1980
Food energy, cal	3490	3290	3440	3350	3260	3140	3300	3420
Protein, g	102	93	93	93	94	95	100	101
Fat, g	124	123	134	143	145	143	157	162
Carbohydrate, g	495	457	474	429	402	375	380	396
Calcium, g	0.80	0.84	0.87	0.92	0.99	0.97	0.94	0.91
Phosphorus, g	1.55	1.47	1.48	1.50	1.53	1.51	1.53	1.52
Iron, mg	15.3	14.6	14.2	14.2	16.5	16.3	17.2	17.2
Vitamin A, IU	7600	7900	8000	8500	8400	8000	7800	8000
Thiamine, mg	1.63	1.52	1.54	1.55	1.90	1.85	1.84	2.19
Riboflavin, mg	1.82	1.82	1.84	1.90	2.29	2.27	2.28	2.39
Niacin, mg	19.3	17.5	17.3	17.8	20.2	20.9	22.5	26.1
Vitamin C, mg	107	104	103	115	105	108	109	124

Source: "U.S. Food Consumption," *USDA Bull.* 364, 1965; and *Agricultural Statistics,* USDA, 1981.

consume about the same amount of protein, phosphorus, and total calories; fewer carbohydrates; and more fat, calcium, iron, vitamins A and C, thiamine, riboflavin, and niacin than in 1910. The increased fat is in the form of vegetable fats and oils (Table 3.10). The protein/calorie ratio in 1910 and 1980 was 1:34. Thus, Americans are consuming approximately the same amount of protein per calorie today as in 1910. When the dietary nutrients furnished by major food groups are considered (Table 3.9), the amount of protein being obtained from animal products is found to be increasing and the amount of fat to be decreasing.

One of the hardest things to sell American shoppers is animal fat. They prefer to do their cooking in vegetable oil and want their steak lean. Through selective breeding and feeding, meats having more protein with less fat and fewer calories are now available (Figure 3.3).[10] Table 3.10 shows consumption of lard and butter to be about one-fourth that in 1910. Consumption of beef increased until 1970 in the United States, whereas consumption of **lamb, mutton,** and veal has decreased and that of pork has had a modest increase during the past 70 years. Total meat consumption per capita remains high (27.9 percent above that of 1910).

[10]Dramatic progress has been made during the past three decades in reducing external fat on pork and beef carcasses. Data on swine at the Ohio Agricultural Research Center reveal that the energy per 100 g of edible portion from the ham, loin, and shoulder decreased from 343 calories in 1955 to 248 calories in 1970 (a 27.6 percent reduction). Concurrently, energy per 100 g of edible portion of the entire carcass of beef decreased from 442 to 377 calories (a 14.8 percent reduction).

TABLE 3.9
PERCENTAGE OF DAILEY NUTRIENT INTAKE FURNISHED BY MAJOR FOOD GROUPS

Nutrient	Meats and eggs		Dairy products		Flour and cereal		Fruits and vegetables	
	1909–1913	1980	1909–1913	1980	1909–1913	1980	1909–1913	1980
Food energy	17.8	22.8	13.9	9.9	37.6	18.8	4.4	8.9
Protein	35.1	47.8	16.5	20.2	35.8	17.1	3.7	7.1
Fat	41.2	38.8	29.0	11.2	3.8	1.3	0.6	1.0
Carbohydrate	0.2	0.2	4.7	5.7	56.1	34.0	7.5	17.4
Calcium	6.8	6.6	68.3	71.6	7.4	3.3	9.1	9.5
Phosphorus	25.4	33.8	27.9	32.6	26.8	12.0	6.1	10.9
Iron	32.7	36.2	1.6	2.4	26.8	27.4	14.1	18.8
Vitamin A	28.1	29.4	19.6	12.2	2.0	0.4	28.3	46.9
Thiamine	34.1	29.8	9.1	7.2	25.6	41.0	10.7	16.6
Riboflavin	30.2	28.4	42.0	36.3	11.3	20.5	9.0	8.6
Niacin	43.5	45.5	1.5	1.2	22.6	26.9	8.5	14.8
Vitamin C	1.1	2.1	4.4	3.2	0.0	0.0	58.6	91.4

Source: "U.S. Food Consumption," *USDA Bull.* 364, 1965; and "National Food Review," USDA, Econ. Res. Service, NFR–13, Winter 1981.

Ham of Yesterday   Ham of Today

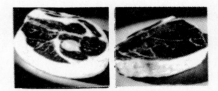

Roast of Yesterday   Roast of Today

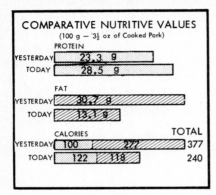

COMPARATIVE NUTRITIVE VALUES
(100 g — 3½ oz of Cooked Pork)

PROTEIN
YESTERDAY   23.3 g
TODAY   28.5 g

FAT
YESTERDAY   30.7 g
TODAY   13.1 g

CALORIES   TOTAL
YESTERDAY   100   277   377
TODAY   122   118   240

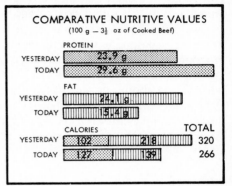

COMPARATIVE NUTRITIVE VALUES
(100 g — 3½ oz of Cooked Beef)

PROTEIN
YESTERDAY   23.9 g
TODAY   29.6 g

FAT
YESTERDAY   24.1 g
TODAY   15.4 g

CALORIES   TOTAL
YESTERDAY   102   218   320
TODAY   127   139   266

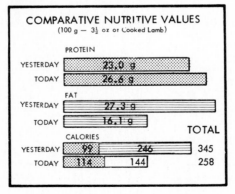

Chop of Yesterday   Chop of Today

COMPARATIVE NUTRITIVE VALUES
(100 g — 3½ oz of Cooked Lamb)

PROTEIN
YESTERDAY   23.0 g
TODAY   26.6 g

FAT
YESTERDAY   27.3 g
TODAY   16.1 g

CALORIES   TOTAL
YESTERDAY   99   246   345
TODAY   114   144   258

FIGURE 3.3
Comparative nutritive values of old- and new-type meats. *(National Livestock and Meat Board.)*

TABLE 3.10
UNITED STATES PER CAPITA CONSUMPTION TRENDS OF SELECTED ANIMAL PRODUCTS AND
VEGETABLE FATS, lb

	1910	1920	1930	1940	1950	1960	1970	1980
Meats (carcass wt),								
total	139.0	130.5	123.3	135.9	137.8	146.7	186.3	177.8
Beef	70.4	59.1	48.9	54.9	63.4	85.0	113.7	103.4
Veal	7.2	8.0	6.4	7.4	8.0	6.1	2.9	1.8
Lamb and mutton	6.5	5.4	6.7	6.6	4.0	4.8	3.3	1.5
Pork (excluding lard)	62.3	63.5	67.0	73.5	69.2	64.9	66.4	71.1
Fish (edible wt), total	11.2	11.8	10.2	11.0	11.8	10.3	11.8	12.8
Fresh and frozen	4.5	6.3	5.8	5.7	6.3	5.7	6.9	8.0
Canned	2.8	3.2	3.4	4.6	4.9	4.0	4.5	4.5
Poultry products (meat),								
total	18.0	16.0	17.9	17.5	25.1	34.7	49.6	60.6
Eggs (8 per lb)	37.1	37.4	41.4	39.9	50.0	43.7	39.9	34.1
Chicken	15.5	13.7	15.7	14.1	20.6	28.1	41.4	50.1
Turkey	1.0	1.3	1.5	2.9	4.1	6.2	8.2	10.5
Dairy products								
Total milk fat								
solids	29.7	28.9	32.1	32.5	29.4	24.5	20.6	19.9
Total nonfat milk								
solids	31.1	33.9	35.7	38.1	43.7	43.2	41.4	36.6
Cheese, cheddar	2.8	2.8	3.2	4.4	5.5	5.4	6.8	9.7
Cheese, cottage	0.6	0.6	1.2	1.9	3.1	4.7	5.1	4.5
Condensed and								
evaporated milk	6.5	10.2	13.6	19.3	20.1	13.7	7.1	3.8
Fluid milk and cream	262.1	289.1	280.8	275.6	308.8	295.1	271.6	250.0
Ice cream and frozen								
products	1.9	7.6	9.8	12.3	19.4	25.7	28.6	26.3
Nonfat dry milk	.....	0.2	1.3	2.2	3.7	6.2	5.3	3.0
Yogurt	.....	.....	.....	.....	0.1	0.3	0.9	2.6
Fats and oils, total	41.9	39.7	48.9	50.2	49.0	48.6	56.4	59.1
Butter (actual wt)	18.3	14.9	17.6	17.0	10.7	7.5	5.2	4.5
Lard	12.5	12.0	12.7	14.4	12.6	7.6	4.7	2.4
Total animal fat	30.8	26.9	30.3	31.4	23.3	15.1	9.9	6.9
Margarine (actual wt)	1.6	3.4	2.6	2.4	6.1	9.4	11.0	11.3
Shortening	8.0	7.6	9.8	9.0	11.0	12.6	17.3	18.2
Other edible fats and								
oils	1.5	1.8	6.2	7.4	8.6	11.5	18.2	22.7
Total vegetable fats								
and oils	11.1	12.8	18.6	18.8	25.7	33.5	46.5	52.2

Source: "U.S. Food Consumption," *USDA Bull.* 364, 1965; *Agricultural Statistics,* USDA, 1972 and 1981; and *Food Consumption, Prices, and Expenditures 1960–1981,* USDA, ERS, Stat. Bull. No. 694, November 1982.

Consumption of fish has held steady; however, there have been fluctuations in the amount of fresh, frozen, and canned fish consumed.

The per capita consumption of eggs in 1980 was slightly less than that of 1910; however, the consumption of poultry meats has increased tremendously.

Americans are currently eating 10 times more turkey and 3 times more chicken than in 1910. This fact may be attributed largely to improved efficiency of production of these two meats. Also, the modern turkey is plumper, meatier, and more compact; it has a larger proportion of breast meat than ever before. Chicken meat ranks high for use in canned, precooked-frozen, and dried foods. It is also used extensively in soups and for flavoring in other foods.

The consumption of dairy products on a **milk equivalent** basis has decreased. However, there have been substantial increases in the consumption of cheeses and frozen dairy desserts and, more recently, in the use of yogurt and nonfat dry milk (Table 3.10).

The big shift has been away from butter. Table 3.10 shows that consumption of milk-fat solids decreased from 29.7 to 19.9 lb per capita, whereas consumption of nonfat solids increased from 31.1 to 36.6 lb during the 1910–1980 period.

3.5 ATHEROSCLEROSIS

Heart disease is the nation's number one killer,[11] accounting for about 55 percent of all deaths in the United States today as contrasted with only 20 percent in 1900. **Atherosclerosis** is the underlying cause of coronary heart disease (CHD). **Cholesterol, phospholipids,** fats, iron, and proteins are deposited within the inner lining of the blood vessel walls. These compounds form a mass **(atheroma)** that protrudes into the opening of blood vessels, and later, calcium may be deposited in the atheroma, causing "hardening of the arteries," or the disease called *atherosclerosis.*

After three decades of nationwide publicity and public concern, cholesterol has become a dietary ingredient feared by many. Because of the cholesterol in meat, milk, and eggs, there have been those who associate these foods with atherosclerosis and advise against their consumption. However, hundreds of research studies have failed to show solid evidence of a cause-and-effect relation between CHD and consumption of animal products. Actually, egg yolks and brains are the only animal foods that contain more than 1 percent cholesterol (Table 3.3).

Synthesis from an important metabolic intermediate (acetyl coenzyme A) and degradation of cholesterol go on simultaneously in the animal body. Studies of cholesterol **metabolism** in humans have demonstrated that the liver synthesizes about 0.8 (0.5 to 2) g daily or approximately twice as much cholesterol as is consumed in the average daily diet. Cholesterol is an essential part of the structure of cell membranes in the body. It is also the starting material from which the body makes its own supply of sex and adrenal hormones. The body also can convert cholesterol into vitamin D, which is essential for proper calcification of bones and teeth. Approximately 80 percent of the cholesterol metabolized is transformed into various bile acids. In healthy individuals, the

---

[11]There are about 659,000 deaths annually in the United States from complications of atherosclerosis and approximately 2.2 million from all types of cardiovascular involvements.

body tends to maintain **plasma** cholesterol concentration by compensating for dietary intake through adjustment of synthesis and also through degradation and excretion of cholesterol and its products.

Of special interest is the fact that **herbivorous** animals (especially the cow) eat little or no cholesterol, yet their blood **serum** cholesterol level approximates that of humans. Elephants eat no animal fats. Yet in a recent study of the hearts and aortas of 415 elephants, it was found that 72 percent of the aortas and 27 percent of the coronary arteries had visible atherosclerotic lesions similar to those in human atherosclerosis. Atherosclerosis has also been observed in red deer on an island near Scotland, African buffalo, cattle, sheep, goats, and caribou (all ruminants).

Why do some people have elevated blood cholesterol levels? There are two main reasons. First, a relatively small number of people suffer from such diseases as hypothyroidism and nephrosis, and these individuals almost invariably have elevated blood cholesterol. But for the majority, the explanation is hereditary, just as it is for diabetes and high blood pressure.

Nevertheless, a popular hypothesis in the United States, commonly accepted as fact, is that diet is responsible for increased blood cholesterol and consequently for the prevalence of coronary heart disease. The reasoning seems logical. It is known that CHD results from the accumulation of fatty material on the walls of the coronary arteries. Furthermore, this material is largely cholesterol. It is generally assumed that the cholesterol of circulating blood settles onto the arterial walls much like silt from a stream of running water. It is contended further that the amount of blood cholesterol reflects the amount of saturated fat and cholesterol in the diet.

The above thinking is supported by the fact that CHD is rarely seen in some areas of the world, especially in the developing countries of Africa and Asia. Moreover, blood cholesterol concentration is low in the people of those areas, and their diets are very low in saturated fat. The main source of nourishment is whole-grain food with a high fiber content.

On the other hand, there are as many areas where people consume extremely large amounts of saturated fat and their blood cholesterol levels are nevertheless exceedingly low. Typical are the tribes of northern Kenya, where the main dietary substances are cow's milk and meat, with animal fat providing 60 percent of the total caloric intake. Yet blood cholesterol levels are about one-half of those in people in the United States, and CHD is virtually unheard of.

Farmers in the Swiss Alps live primarily on dairy products. Yet they, too, have low blood cholesterol and a low incidence of death due to heart disease.

3.5.1 Are Animal Fats at Fault?

Exogenous cholesterol intake from animal fats was at first thought to be responsible for the genesis of atherosclerosis, but this theory has been proved erroneous. Records on trends in per capita consumption of animal fats (Table 3.10) are pertinent and of special interest. During the 70-year period from 1910 to 1980, annual per capita consumption of butter and lard, the principal animal fats, decreased 78 percent (30.8 to 6.9 lb), whereas consumption of vegetable fats and oils increased 370 percent (11.1 to 52.2 lb) during the same period (Figure 3.4). Concurrently, livestock producers were producing leaner meat (Figure 3.3), and our percentage of dietary fat obtained from meat, milk, and eggs decreased (Table 3.9). [Since cholesterol is a component of the fat fraction of milk, skim (fat-free) milk contains no cholesterol.] Therefore, it would not seem logical to blame the consumption of animal fats for the increased incidence of atherosclerosis during recent years. Nor would it seem desirable to advise consumers to avoid animal products that contain fats. Low-cholesterol diets may be recommended for persons known to be genetically predisposed to high blood cholesterol or other possible high-risk factors associated with CHD.

Most research studies indicate that persons with very high blood cholesterol levels have more atherosclerosis and a greater risk of heart attack. However, so far, research does not indicate that decreasing the intake of dietary cholesterol will significantly lower blood cholesterol and reduce the risk of CHD. Thus, blood cholesterol level may be a symptom of the disease but not the cause itself.

Although much of the research related to dietary influences on blood cholesterol levels has been directed toward lipid components of the diet, recent studies indicate that the type of protein may influence blood cholesterol levels. For example, in rabbits used as test animals, soy protein diets were associated with low plasma cholesterol and extracted egg diets with high plasma cholesterol concentrations.

FIGURE 3.4
Per capita consumption of animal fats, vegetable fats, and oils in the United States, 1910 and 1980. *(Adapted from* Agricultural Statistics, *USDA, 1981.)*

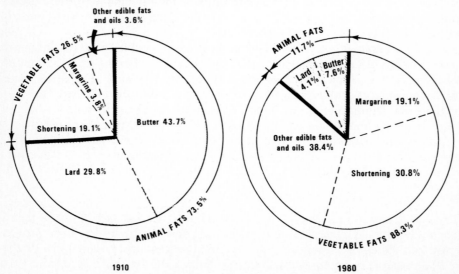

Eggs are high in cholesterol. Yet nature designed the egg to nourish a developing embryo. The blood cholesterol concentration of pregnant animals is about 50 percent above normal. Moreover, nature designed milk to nourish the newborn, and milk contains cholesterol. This leads one to conclude that nature had a purpose in providing cholesterol.

Although eggs contain high levels of cholesterol, recent research at the University of Missouri at Columbia and at the University of California School of Medicine and Public Health at Los Angeles showed that the consumption of one or two eggs daily caused no significant difference in average blood cholesterol levels between the test group (one or two eggs daily) and the control group (no eggs) when persons with normal blood cholesterol levels were used as research subjects.

Boiled shrimp has considerably more cholesterol than most meats from farm animals (e.g., the cholesterol content per ounce of boiled shrimp is about 43 mg compared with 25 mg for chicken, 26 mg for beef and pork, and 28 mg for lamb and veal).

3.5.2 Evidence That Cholesterol Alone Does Not Cause Heart Disease

The United States heads the list among countries in number of deaths from CHD but stands sixteenth in per capita consumption of milk fat. Moreover, in many countries having greater life expectancy than the United States (Netherlands, Sweden, Norway, Canada, Denmark, and New Zealand), the people consume liberal amounts of animal products. The Navajo Indians consume considerably more saturated fat than their fellow Americans, yet they have very little CHD. The Masai, an east African tribe, consume about 7 liters of milk per capita daily, and saturated fatty acids constitute about 60 percent of their diet. Yet they rarely develop atherosclerosis. It would seem, then, that the condemnation of saturated fat as the hypercholesteremic agent in animal fats is an excellent example of guilt by association.

The most extensive study of CHD ever undertaken extended over a period of more than 20 years and involved regular biennial examinations of more than 5000 people in Framingham, Massachusetts. The following quotations from "The Framingham Diet Study" report are germane:

No association between percent of calories from fat and serum (blood) cholesterol level was shown, nor between ratio of plant fat intake to animal fat intake and serum cholesterol level.

There is no indication of relationship between dietary cholesterol and serum cholesterol level. There is, in short, no suggestion of any relation between diet and the subsequent development of CHD in the study group, despite a distinct elevation in serum cholesterol in men developing CHD. Diet is not associated with concurrent differentials in serum cholesterol level.

The famous Framingham study, which was an important study in determining risk factors associated with CHD, concluded further that:

> With one exception, there was no discernible association between reported diet intake and serum cholesterol level in the Framingham Diet Study Group. The exception was a weak *negative* association between caloric intake and serum cholesterol concentration in men.[12]

In the Framingham study, the average daily caloric intake of the 437 men and 475 women was 3156 and 2142, respectively. Approximately 40 percent of these calories were from fat, and 70 percent of the fat was of animal origin.

Increasing numbers of medical and nutritional scientists are concluding that heart disease is caused by a complex of factors and that the indictment of animal fats was not only premature but also probably incorrect. Increased incidence of atherosclerosis appears to parallel prosperity. Heredity, obesity, physical inactivity, hypertension (multiple stresses such as the strain or pressure of modern living), too many cocktails, excessive smoking (according to the Framingham study, smoking one pack of cigarettes a day increases the risk of CHD by 80 percent), and other socioeconomic factors appear to be related to heart disease.

The above facts notwithstanding, research still does not permit unequivocal conclusions as to the possible relations between diet and atherosclerosis. Considerable evidence suggests that elevated levels of blood cholesterol and triglycerides are associated with accelerated rates of atherogenesis and CHD. Cholesterol and triglycerides are transported in the blood with specific aggregates of lipids and proteins called lipoproteins. Triglycerides are present mainly in a very low-density (90 percent fat) lipoprotein fraction (VLDL), whereas cholesterol is carried mainly in the low-density (75 percent fat) lipoprotein fraction (LDL), with some also present in the high-density (50 percent fat) lipoprotein fraction (HDL). Elevated blood concentration of LDL has been associated with an increased incidence of CHD, and intake of high levels of saturated fats tends to increase blood LDL and cholesterol levels. Physical exercise, dietary modifications, abstinence from smoking, and weight control may help change the LDL/HDL ratio to favor a lower CHD risk.[12a]

With so much at stake in terms of human health and longevity, as well as much of animal agriculture and associated businesses, further long-term, comprehensive, carefully controlled investigations of possible causative or contributory dietary factors of multifactorial CHD are urgently needed.

Carnitine: A Required Nutrient   Carnitine deficiency in humans was first described in 1973—66 years after its discovery. Deficiency symptoms include muscle weakness, cardiomyopathy, and hypoglycemia.

Carnitine is required for the transport of long-chain fatty acids across the

---

[12]W. B. Kannel and Tavia Gordon (eds.), *The Framingham Study . . . An Epidemiological Investigation of Cardiovascular Disease*, National Institutes of Health, U.S. Dept. of Health, Education and Welfare, Washington, D.C., December 1970.

[12a]Report of Nutrition Committee: "Rationale of the Diet-Heart Statement of the American Heart Association," American Heart Association, Dallas, TX, Circulation 65, No. 4, 1982.

mitochondrial membrane. When carnitine is lacking, cells cannot use fat for energy. A carnitine deficiency can occur when diets are low in this nutrient and if essential factors for carnitine biosynthesis (lysine, methionine, ascorbic acid, iron, vitamin B_6, or niacin) are also in short supply. Under normal circumstances, newborn infants, pregnant or lactating women, and adults who consume diets low in lysine and methionine would be the most likely to develop a carnitine deficiency.

Meat, milk, and eggs are rich sources of carnitine. Most cereals, fruits, and vegetables contain little or no carnitine. It is interesting that the carnitine in human milk is much more available to infants than that in cow's milk.

While the authors do not wish to imply that carnitine is related to CHD, it is possible that the lack of one or more essential dietary nutrients is associated with the disease.

3.6 THE ROLE OF DOMESTIC ANIMALS IN THE ACCUMULATION OF RADIONUCLIDES IN FOOD

Prior to the nuclear age, there was little or no interest in the strontium 90 (^{90}Sr) content of foods. As a result of **radioactivity** (fallout) from the testing of nuclear weapons, there now exists a minute amount of radioactive **isotopes** in many foods. It is clear that animal products serve as routes of **radionuclide** contamination in humans. However, the ability of the cow and hen to reduce the ^{90}Sr in milk, meat, and eggs relative to that in feed is an important barrier to human ingestion of radionuclides. The hen filters out more than 95 percent of the ^{90}Sr ingested into her body. Most of what reaches the egg is deposited in the shell, which is thrown away. Using accepted discrimination factors from diet to milk and to meat in the bovine, researchers have made estimates as to the contribution to humans of ^{90}Sr from milk, meat, and vegetables. About 82 percent of the ^{90}Sr humans consume comes from plant sources, 17 percent from milk, and a negligible amount from meat.

The Cow as a Radionuclide Filter Strontium and calcium are similar chemically, since they are in the same group in the periodic table. Plants take up radioactive strontium along with their necessary calcium from soil and show little discrimination between the absorption of those two minerals. However, animal and human bodies show a preferential absorption of calcium in bone formation and the preferential secretion of calcium into milk.

The cow secretes into milk less than 13 percent of the ^{90}Sr consumed in her feed. The balance is voided in **feces** and urine. (A negligible amount is stored in her body tissue.) Moreover, levels of ^{90}Sr and other radionuclides can be kept low by feeding **forages** that have been under storage prior to fallout. Also, it should be remembered that ^{90}Sr retention is lower when levels of calcium are optimal. Therefore, if humans are to deposit a minimum amount of ^{90}Sr in their bones, they should consume liberal amounts of high-calcium foods, of which milk and milk products are the best.

Excellent progress is being made on milk processing to remove ^{90}Sr. Present levels do not warrant such processing; however, should a nuclear event occur, the technique of removing as much as 98 percent ^{90}Sr from milk has been developed and could be employed.

3.7 FORTIFICATION OF ANIMAL PRODUCTS

The purposes of fortifying animal foods are to improve their nutritional qualities[13] and to provide a more uniform product throughout the year. The vitamin A potency of milk, for example, may vary as much as threefold, on the average, from its high in May through September, when carotene (the precursor of vitamin A) is abundant in grasses and forages, to its low in November through April, when forages are often low in carotene.

Fortification of foods with vitamins was first reported in 1924, when it was found that antirachitic properties (vitamin D) could be imparted to selected foods by irradiation with ultraviolet light. Liquid vitamin A concentrates (carotene) were introduced in about 1930, and pure, crystalline vitamin D became available in 1932. From 1932 to 1943, industry utilized advances in science to fortify milk so that it became the first and only food to provide humans with the minimum daily vitamin requirements, with the exception of vitamin C. Recently, multivitamin-mineral milk was introduced for test marketing. It includes a recommended level of ascorbic acid (vitamin C), as well as at least 4000 IU vitamin A, 400 IU vitamin D, 1 mg thiamine, 2 mg riboflavin, 10 mg niacin, 10 mg iron, and 0.1 mg iodine per qt.

3.7.1 Legal Aspects of Fortifying Animal Products

Certain states restrict the amount and kind of vitamin fortification. Milk is the only animal food approved and recommended for vitamin fortification jointly by the American Medical Association's Council on Foods and Nutrition and the National Research Council's Food and Nutrition Board. Present laws regulating nonfat dry milk preclude vitamin restoration or fortification for the United States domestic market. However, the fortification of nonfat dry milk with vitamins A and D is approved for overseas shipment, with the Agency for International Development (AID) paying the cost (about 3¢ per 100 qt of reconstituted nonfat milk).

Labeling   Milk being sold in the United States must have a statement on the container indicating the name and quantity of the product, its grade, and the name and address of its manufacturer. Federal Food and Drug Administration regulations require nutritional labeling.[14] The declaration of nutritional information on the label must contain the following information: (1) serving size; (2) servings per container; (3) caloric content per serving; (4) protein content per serving (nearest gram); (5) carbohydrate content per serving (nearest gram); (6) fat content per serving (nearest gram); and (7) percentage of U.S. Recommend-

---

[13]*Fortification* is *the addition of other nutrients to a given food,* in contrast with *enrichment,* which is *the restoration of lost nutrients, as in the milling of cereals.*
[14]Code of Federal Regulations, Title 21, Part 101—Food Labeling. Revised April 1983.

ed Dietary allowances (USRDA) for proteins, vitamins (A, C, thiamine, riboflavin, and niacin), and minerals (calcium and iron). In determining the percentage USRDA for protein, the **Protein Efficiency Ratio** (PER) is important. The USRDA is 45 g if the PER is at least equal to that of casein (milk protein), but 65 g if the PER of the total protein in the product is less than that of casein. Fatty acid composition and cholesterol content, sodium concentration, and contents of other nutrients may be declared.

A food is deemed to be an imitation and is subject to federal labeling controls if it is a substitute for and resembles another food but is nutritionally inferior to the food imitated. If it is not nutritionally inferior and bears a common or usual name that is not false or misleading, it need not be labeled as imitation.[15]

3.7.2 Fortifying with Vitamins

Because vitamins A and D are fat-soluble and therefore not found in nonfat milk, skim milk commonly is fortified with these two vitamins. So that uniform levels of vitamins A and D in whole milk are ensured, it is almost always fortified with vitamin D and often with vitamin A. Fortification with vitamins A and D adds about 0.05¢ to the cost of a quart of milk.

Vitamin A The daily vitamin A requirement for a human is 5000 IU. So that a high vitamin A potency in milk is ensured for babies and adults alike, milk is commonly fortified with a synthetic, crystalline vitamin A palmitate or acetate in the amount of 4000 to 5000 IU per qt.

Vitamin D The recommended daily dosage of vitamin D for humans is 200 to 400 IU. Since few foods contain appreciable amounts of vitamin D, it is the vitamin most likely to be deficient in an otherwise balanced diet.[16] It is closely related to the **absorption** and metabolism of calcium and phosphorus and is therefore commonly used to fortify milk, which has an abundant supply of these two important dietary minerals. The most commonly used method of application to milk is in the form of **irradiated ergosterol** (vitamin D_2, a plant **sterol**). Earlier, milk was exposed briefly to ultraviolet light, which converts 7-dehydrocholesterol (an animal sterol) into vitamin D_3. This explains why vitamin D is often referred to as the "sunshine vitamin," since it is the ultraviolet radiation from sunlight that converts the animal sterol (provitamin D) present in the skin into biologically active vitamin D (Section 16.3.2).

Dangers of Vitamin D Excessive doses of vitamin D mobilize calcium and phosphorus from tissues; this reverses the effects of normal doses. Vitamin D poisoning may cause nausea, **diuresis,** headache, **asthenia,** loss of appetite, and low retention of calcium and phosphorus. Dr. Cooke of Johns Hopkins University found that excessive vitamin D intake by pregnant women can lead to birth defects and mental abnormalities in infants.

---

[15]Code of Federal Regulations, Title 21, Part 101.3. Revised April 1, 1983.

[16]Only a minimal amount of vitamin D is stored in the body. Therefore it should be consumed daily.

3.7.3 Fortifying with Nonfat Milk Solids

In recent years, there has been a trend in the consumption habits of United States consumers to increase their use of nonfat and low-fat milk and milk products. To add nutritive value (especially protein, lactose, and minerals), as well as flavor and body, to these low-fat products, many milk-processing firms now add about 2 percent nonfat dry milk solids to fluid skim milk prior to pasteurization. Any milk labeled "with added milk solids not fat" must contain at least 10 percent milk-derived nonfat solids, and the ratio of protein to the total nonfat solids of the food and the **protein efficiency ratio** of all protein present must not be decreased as a result of adding such ingredients.

3.7.4 Milk Toning

In view of the limited world supply of milk and protein, many private and government-supported programs involve shipment of nonfat dry milk to countries and territories abroad. Much of this skim-milk powder (with an appropriate amount of pure filtered water) is added to the native milk, such as high-fat buffalo milk, to *tone* it to 3 percent milk fat. Further reduction of the fat level to 1.5 percent is called *double toning*. Toning is practiced on a large scale in India.

3.7.5 Natural Foods

The words "natural," "no preservatives," and "no artificial additives" are embraced by many consumers, whereas anything that smacks of chemical treatment is viewed with alarm. However, concerns about additives and preservatives are, in some cases, ill-founded and overdrawn. A natural product, just because it is "natural," is not necessarily safe for human consumption. Certain plants used as foods contain naturally occurring toxicants. They usually occur in minute amounts that the human body can tolerate safely. Yet, if subjected to the same Food and Drug Administration (FDA) testing as food additives, those "natural" plants would not be approved for human use. Few people would question the nutritional wisdom of consuming many, if not most, foods in their natural state. However, various state and federal agencies have as their main mission the monitoring of foods and food additives to assure the consuming public of a safe and wholesome food supply. This includes extensive, rigorous experimental tests of the safety of food additives.

3.8 PRESERVATION OF ANIMAL PRODUCTS

To preserve animal products is to enhance their keeping qualities and thereby increase their shelf life. There are several reasons for preserving animal products, which include (1) the constant demand for the perishable animal products, meat, milk, and eggs; (2) the seasonal supply of certain animal products; (3) the necessity of transporting from the producing area to areas of use; and (4) provision of an abundant and safe public supply.

Until early in the twentieth century, when transportation and refrigeration provided ways of moving and preserving humanity's food supply, many food items were available on only a seasonal basis. Milk, eggs, and poultry meats were available the year round, but cattle, swine, and sheep could be slaughtered only in cold weather so that the meat could be used, smoked, salted, or fried down before it spoiled. The principal meat for the farm family was pork; cattle served as a cash crop since they could be driven, sold, and slaughtered in nearby cities. About Thanksgiving time, each farmer would butcher a number of hogs in accordance with family size. Hams, shoulders, and side meat were packed in salt for about 6 weeks and then removed and cured with hickory smoke (or that of another hardwood). Each farm had a smokehouse. Sausage was made from trimmings of the shoulders, hams, hearts, livers, and other bits of meat. The women fried down the sausage, backbones, and spareribs. (They fried the meat, packed it in earthenware crocks, and covered it with melted lard rendered from the fatter parts.) This process preserved the meat for at least a year.

3.8.1   Methods of Preservation Applicable to Animal Products

The foremost concern in preserving animal products is the control of microorganisms that cause spoilage or illness. One or more of the following methods may be employed:

High Temperature   In 1864, Louis Pasteur invented the process of pasteurization, or killing bacteria in fluids with heat. The thermal treatment is the most widely used method of *killing* spoilage and potentially toxigenic microorganisms in animal products. It is employed in pasteurization and in most canning.

Low Temperature   Refrigeration is the most widely used method of *inhibiting the growth* of microorganisms in animal products without killing them. Low temperatures discourage the growth of bacteria. Most milk products are held at a temperature of 40°F or below.

Removal or Destruction of Microorganisms   The process for removing bacteria from liquid (e.g., milk) by means of centrifugal force is called *bactofugation*. The bactofuge utilizes differences in specific gravity between bacteria and other constituents of milk. The specific gravity of bacteria varies between 1.07 and 1.13, as compared with 1.032 and 1.036 for whole and skim milk, respectively.

Dehydration   Sun-drying is the oldest and least expensive type of **dehydration** and is still used for certain foods. Air-drying of meat has been practiced for centuries. Marco Polo reported early in the 1300s that every Tartar Mongol warrior carried 10 lb of dried milk as part of his food rations. Mechanical driers were first mentioned by Marco Polo in the fourteenth century, but these were not used commercially until the twentieth century. American troops in the Revolutionary War ate dried beef, sometimes called **jerky.** This consisted of thin strips of beef or game that were dehydrated by hanging in the sun to air-dry. An adaptation of jerked beef, known as *pemmican,* has been used by North American Indians and Arctic explorers. More recently, cooked meat has been dehydrated in hot-air driers. Milk and certain poultry products may be spray-dried or dried with drum driers having heated rollers. Dried eggs were imported from China for decades, but not until 1927 were eggs dried commercially in the United States. Foam- and puff-drying have been introduced and, along with vacuum freeze-drying, are discussed in Section 3.8.6.

Chemical Preservatives   Curing agents for meat include a mixture of sodium chloride, sodium nitrate, and sugar. Sorbitol is added to assist in meat preservation. Sorbic acid and propionates are often incorporated into packaging materials to inhibit the growth of molds inside the package. Sorbic acid migrates from the wrapper into the product. Other chemical preservatives include ascorbic acid (ascorbate) and certain phosphates (e.g., tripolyphosphate). Unfortunately, many chemicals in concentrations that adversely affect growth of microorganisms may be harmful to humans.

With rapid processing techniques and improved refrigeration, the practice of curing meat as a means of preservation per se has decreased. The principal reason for curing meat today is to impart various flavors.

Radiation   Meats can be preserved by means of **radiation** doses that *sterilize* (kill all organisms) or *pasteurize* (kill most organisms). Radiation can be used in combination with other methods of preservation, such as chemical additives or heat. The undesirable secondary reactions associated with radiation present problems, however. These reactions include changes in flavor, color, odor, and texture and the loss of some nutrients. To date, these undesirable effects have restricted the commercial use of **ionizing radiations** as a means of meat preservation. Radioactive particles strike and kill microorganisms in or on food without appreciably raising the temperature of the product. The process is therefore referred to as *cold sterilization.*

Poultry products were approved for irradiation in 1982 to extend shelf life, which will enhance exports. The reader is encouraged to see the glossary for pertinent information on the following terms: **rad, radicidation, radurization, radappertization,** and **gray.**

Environmental Preservation   Maintaining conditions unfavorable to the growth of microorganisms, e.g., using sealed or evacuated containers and substituting a gas such as nitrogen or carbon dioxide for oxygen in sealed and impermeable containers, may preserve animal products.

Low pH (High Acidity)   Low pH is an important method of preserving many cultured milk, cheese, and meat products. It is normally accomplished by fermentation. Following slaughter, muscle glycogen is fermented, with the resultant production and accumulation of lactic acid, which in turn retards bacterial growth. Biologically produced lactic acid enhances flavor and gives stability to fermented sausages, Lebanon bologna, and dry summer sausages Fermentation is accomplished through the addition of a fermentable sugar and a starter **(culture).** Similarly, a culture (controlled bacterial population) may be added to certain dairy products to develop lactic acid. Sour cream and cottage cheese are also frequently produced by direct acidification with a food-grade acid or with glucono-delta-lactone. This lactone slowly hydrolyzes, producing gluconic acid. Artificial flavors must be added to directly acidified milk products to make them taste similar to their cultured counterparts.

Many chemical preservatives are more effective at a low **pH**; e.g., the minimum inhibitory concentration of sodium nitrite is reduced over 40-fold when the pH is reduced from 6.9 to 5.

3.8.2 Providing Safe Honey

The chief sources of microorganisms in honey are the nectar of flowers and the honeybee. The high sugar content of honey, however, produces high osmotic pressure that restricts the growth of most microorganisms (especially bacteria). Since the pH of honey is low (3.2 to 4.2), the growth of osmophilic yeast presents the major microbiological problem. If the moisture content of honey is held to 21 percent or less, the problem of yeast growth is greatly reduced. Honey is pasteurized at 160 to 170°F for 5 min and promptly cooled to 90°F. This treatment kills the yeasts and prevents fermentation of the honey.

3.8.3 Providing Safe Meat

In the early days, pork was considered "unclean"; this idea may have resulted from the observation that pork meat could cause illness. In 1847, Joseph Leidy discovered the nematode **parasite** *Trichinella spiralis* in pork when cutting a cold ham sandwich. This discovery led to the conclusion that humans might become infected by eating raw pork.[17] All meat (slaughtered and processed) intended for interstate trade must be inspected for wholesomeness (freedom from disease) under the supervision of the USDA.

Trichinosis Trichinosis is caused by the small roundworm *T. spiralis,* which is killed when meat is (1) heated to an internal temperature of 137°F with no holding time, (2) refrigerated 20 or more continuous days at 0°F or below, or (3) smoked at 80°F for 40 h or more, followed by 10 days in a drying room at 45°F.

Trichinella larvae are rendered incapable of completing their life cycle in the new host when infected carcasses are exposed to gamma irradiation at a dosage of 30,000 rads. Such treated meat is said to retain its palatability and nutritional

---

[17]Recently, numerous cases of trichinosis have occurred among Eskimos who consumed undercooked bear meat.

contributions. However, the initial cost of facilities to provide effective irradiation (using cesium 137) is too high to allow economical use.

Curing of Meat Methods of curing meat include (1) dry cure, in which the curing agent is rubbed into the meat, as in bacon, ham, and beef trimmings intended for sausage making; (2) pickle cure, in which the meat is immersed in a solution of ingredients (e.g., 15% NaCl); (3) injection cure, in which the internal injection of curing agents (e.g., 24% solution of NaCl) results in a rapid and uniform distribution of the cure throughout the tissues (the curing agent is introduced by a single-needle injection into the vascular system of hams and by multiple-needle injections in smaller cuts, such as bacon, **jowl,** and shoulder); and (4) direct addition, in which the curing agent is mixed and ground in the meat, as in sausage making.

Smoking Smoking and heat processing of meat products are accomplished concurrently. When oak and/or hickory sawdust are used, smoking imparts a characteristic flavor. The combination of heat and smoke aids in preservation by reducing the bacterial population and by causing surface dehydration. Smoked poultry meat is frequently held 6 to 8 h at a smokehouse temperature of 170°F and then at 185°F until the internal temperature of the meat reaches 160°F. Few **red meat** foods are processed in which smoking constitutes an important role in preservation against bacterial spoilage. "Ready-to-eat" hams have been heated to an internal temperature of 137°F, whereas smoked hams are commonly processed at lower temperatures.

Canning Nicholas Appert, who was born in France in 1752, is credited with the invention of canning in response to Napoleon's needs. In 1795 the French Directorate offered a prize of 12,000 francs for a preservation process that would not seriously impair the natural flavor of fresh food. Appert, a confectioner, wine maker, and chef, set at the task. He heated the food in closed containers using Papin's invention of the pressure cooker. The goal was achieved in 1804, and he was given the 12,000 francs in 1809.

In 1810 he published *The Book for All Households on the Art of Preserving Animal and Vegetable Substances for Many Years*. The same practice of thermal processing after packing in containers is followed for preserving most canned meats today.

Preservation and Stability of Meat Nutrients Most vitamins of meat are relatively stable to processing. However, thiamine is partially destroyed in the course of curing, smoking, cooking, canning, heat dehydration, and treatment with ionizing radiations.

Meat Tenderizers Tenderness is perhaps the most important feature of high-quality meat. It depends largely on the condition of muscle fibers, which, in turn, consist of protein fibers interwoven and grouped together by a delicate sheath called the *sarcolemma*. Certain enzymes may be used to break the sarcolemma, resulting in greater tenderness.

The use of enzymes to help tenderize meat is not new. At least 800 years ago, Mexican Indians tenderized meat by wrapping it in papaya leaves. Early explorers found this practice among American Indians and South Sea islanders also.

Several proteases of animal, plant, and microbial origin are used as meat tenderizers: trypsin from the pancreas, bromelin (from pineapple), ficin (from figs), papain (from papaya), and *Aspergillus* fungal protease. The cooking temperature is of great importance since enzymatic action must occur before the enzyme is denatured (at 60 to 85°C). Enzymes used to tenderize meat are inactivated during later stages of cooking. Hams are also tenderized by proteases. Commonly, a heat-sensitive protease is added through the pickling solution in combination with phosphate salt. The enzyme is inactivated by heating the ham at 60 to 70°C.

Proteases are also used to tenderize animal casings for sausages and other processed foods and to prepare hydrolysates (products of hydrolysis) of meat, milk, fish, and plant proteins for use in manufacturing special diet foods, condiments, and animal feeds.

Certain proteolytic enzymes aid in skinning fish by selectively liquefying gelatin under the skin of fish. Old roosters and tom turkeys may be tenderized by injecting a protease solution intravenously about 5 min before the birds are killed. Another enzyme, glucose oxidase, combined with catalase, is used to ferment glucose of eggs prior to drying to prevent the browning of egg powder.

Recent research in Australia indicates that meat may be tenderized by high-pressure treatment at elevated temperatures. The process seems to act like an accelerated aging.

Poultry Meat All poultry meat must be inspected by USDA personnel for wholesomeness and processed under sanitary conditions. In 1970, USDA inspection programs of red meat and poultry meat were combined.

Broilers are usually marketed as fresh (but refrigerated) poultry. The **shelf life** of broilers in the display case and home refrigerator is directly proportional to the number of bacteria on the meat surface after processing. Therefore sanitary processing, rapid chilling, and subsequent refrigeration are important adjuncts to maximal shelf life. Any salmonellae that may be present on poultry meats are destroyed during cooking; however, there is some risk of cross-contamination of uncooked items exposed to the poultry before cooking.

Processing considerations applicable to broilers are also important for turkeys. However, most turkey meat is frozen, which minimizes bacterial spoilage. The problem of lipid oxidation and the resultant off-odors has been largely overcome by using packaging films that are virtually impermeable to oxygen.

More than one-third of all poultry meat marketed in the United States enters "further processed" items, such as pot pies, TV dinners, canned poultry rolls, and frankfurters. Bony parts (necks and backs) are deboned mechanically so that meat losses are minimized. Up to 15 percent of this meat emulsion can be combined with red meat in the manufacture of sausages and weiners. Recently, there has been good market acceptance of 100 percent turkey and poultry "franks."

Pasteurization of Deboned Poultry Meat   Rapid development of further processed poultry products and the strong demand for animal protein have accelerated developments in mechanically deboned poultry meat (MDPM)—a finely ground product that has been processed through an automatic deboner—obtained from turkey frames, poultry backs and necks, or the entire carcass of hens that have completed commercial egg laying. The deboner grinds the meat and bones and, through a series of screens, separates the edible portion from the nonedible portion (bone and gristle).

Because of the finely ground nature of MDPM, and because it is an excellent medium for bacterial growth, it deteriorates quickly if not handled and/or stored properly. Spoilage caused by microbial growth is the major problem in nonfrozen MDPM, whereas oxidative rancidity of the lipid portion is a major cause of deterioration of frozen MDPM.

Research at the Pennsylvania State University demonstrated that the storage life of MDPM can be extended by heat pasteurization, up to 6 min at temperatures ranging from 59°C (138°F) to 71°C (160°F), which reduces the bacterial load in the meat.

Since MDPM is frequently used in frankfurters.and other meat products that depend on meat protein to bind water, and since high pasteurization temperatures result in denaturization of the protein in MDPM, it is important that the times and temperatures used in pasteurizing MDPM be controlled closely.

3.8.4 Providing Safe Poultry Products[18]

Refrigeration, pasteurization, freezing, and drying are important processes in providing consumers with safe poultry products. Of course, healthy birds and sanitary practices are basic in this regard.

Shell Eggs   Most eggs are sterile when laid, but the shell surfaces immediately become contaminated with microorganisms. Less contamination occurs when eggs are produced in cages than in nests. Although most contaminating microorganisms have limited **public health** significance, *Salmonella,* a genus of bacteria often causing food poisoning in humans, may be present in cracked eggs. Eggs should be cooled to 55°F soon after being laid so that quality is preserved and bacterial growth is impeded. Shell eggs are washed and sanitized

[18]The authors are grateful to Dr. O. J. Cotterill, professor of Food Science and Nutrition, University of Missouri at Columbia, for his contributions to this section. For additional information, the reader is referred to W. J. Stadelman and O. J. Cotterill, *Egg Science and Technology,* 2nd ed., AVI Publishing Co., Westport, Conn., 1977.

in mechanical washers before reaching consumers. Cartoned shell eggs should be refrigerated in the store and in the home.

Pasteurization (thermostabilization) of shell eggs can be accomplished by heating at 130°F for 15 min in water and oil. Pasteurization kills bacteria, stabilizes egg whites, and pasteurizes the shell, thereby prolonging shelf life. The oil prevents moisture loss and retains CO_2, thereby maintaining the character of the albumen. Commercially, a fine mist of mineral oil is sprayed on the large end of shell eggs before packaging to minimize loss of moisture and CO_2.

Egg Products   Almost 1 billion pounds of liquid egg products are produced annually in the United States. Some are used immediately; others are frozen or dried. Since 1971, all egg products must have been processed under USDA inspection. Moreover, they are pasteurized to protect consumers against salmonellae. Because egg products vary in composition, several pasteurization conditions are employed. For example, liquid whole egg is pasteurized at 140°F for 3½ min. Salted (10 percent) egg yolks require higher temperatures to effect pasteurization. (As high as 154°F is recommended.) Eggwhite powder is pasteurized by holding at 130°F for 7 days. Quality-assurance personnel routinely check all lots of egg products for the presence of salmonellae prior to shipment.

Spray-drying is the commonest method of dehydrating eggs. Stabilized egg-white powder (the glucose is removed) has virtually unlimited shelf life. Dried products containing egg yolk are subject to lipid oxidation and often are refrigerated if stored for extended periods.

Both pasteurization and drying affect the foaming properties (reduce the cake volume) of egg white and whole egg. Freezing causes egg yolk to thicken (gelatinize). These changes are minimized by the addition of certain edible chemicals during pasteurization. The nutritive value of egg products is not affected by pasteurization or by freezing. However, minor nutrient loss may occur during dehydration and prolonged storage.

3.8.5 Providing Safe Milk Products

Milk and certain milk products are pasteurized to protect against pathogens. The temperatures and times appropriate to the *holding* and *high-temperature–short-time* (HTST) methods of milk pasteurization are 145°F for 30 min and 161°F for 15 s, respectively. Because of the protection afforded bacteria by higher fat and milk-solids contents, cream, ice cream, and certain other milk products are pasteurized at higher temperatures. These times and temperatures were established to provide a safe margin of time and/or temperature in the destruction of *Brucella abortus* (the bacterium that causes brucellosis), which may be shed into milk from an infected cow and transmitted to humans, causing undulant fever; *Mycobacterium bovis,* which may be transmitted through milk and cause tuberculosis in humans; and *Coxiella burnetii,* which also may be transmitted through milk and cause Q fever in humans (Sections 18.3.2 and 18.3.3).

Pasteurization might be termed the final protective measure for the consuming public. All cows are routinely tested for tuberculosis and brucellosis. **Reactor** animals must be disposed of (Section 18.3.3). Close control is maintained over milk production, handling, and processing methods to prevent contamination with microorganisms and to prevent multiplication of those that may get in. Bacteria normally found in well-protected milk are nonpathogenic, and most are readily killed by pasteurization. Pasteurization does not significantly affect the vitamin content of milk. (Actually, there is a slight reduction of vitamin C.)

A substantial amount of milk is preserved and marketed in sealed cans as condensed or evaporated milk. These products have about 60 percent of the water removed, and sweetened condensed milk has approximately 45 percent sugar added, which aids in preservation (changes the **osmotic pressure**), as well as adding sweetness and food solids.

Vitamin Loss in Storage   Milk or milk products sealed in airtight, dark containers and stored for extended periods of time show little or no loss in vitamin potency. However, milk bottled in clear glass and exposed to sunlight for only 1 h may lose 30 percent of its vitamin A and a considerable amount of ascorbic acid and may develop an oxidized flavor. Paper containers have essentially eliminated this problem in the United States; however, an oxidized flavor may develop in milk in translucent plastic containers exposed to a strong source of fluorescent light. The fluorescent-light-induced reaction destroys part of the riboflavin and vitamin C of milk. A yellow filter covering the fluorescent light or pigment in the plastic bottle will prevent passage of the offending wavelength, which is between 400 and 500 μm.

3.8.6   Recent Trends in the Preservation of Animal Products

Advances in equipment design and fabrication have made more economical and desirable several new methods of preserving animal products, particularly dehydration. Cooperation between equipment manufacturers and health officials leading to the adoption of the 3-A sanitary standards for the manufacture of dairy equipment has been a major contribution. E3-A standards apply to egg-processing equipment.

Foam-Mat- (Puff-) Drying   In this process, food is caused to foam, and while in this state, it is dried under vacuum at a moderate temperature. This puff-drying results in products that reconstitute well. Whole milk dried as a foam with low temperature and high vacuum has excellent flavor and solubility, superior to that produced by other methods. However, the shelf life is relatively short, because of **oxidation** and the chemical interactions (combination) of milk protein and lactose (Browning or Maillard reaction).

Agglomeration   By a process of slightly wetting and redrying previously dried products, an agglomerate can be produced that has excellent dispersibility and solubility properties. The result is "instant" powders. New products include combinations of milk and fruit and vegetable components.

Freeze-Drying (Lyophilization) Commercial application of freeze-drying (**lyophilization**) to animal products is of recent origin; however, the Incas of South America dried potatoes by this process many centuries ago. In the winter months, they spread their potatoes on the ground and let them freeze overnight. The next day's sun warmed the potatoes, but it did not thaw them since the temperature was cold at that altitude. Moisture in the ice crystals left the potatoes and sublimed into the air as water vapor. This evaporation was possible because of the high altitude and low air pressure (low vapor pressure). This type of drying is called *freeze-drying.* It simply means that the food is dried while it is frozen (just as a snowbank will sometimes disappear without melting). Today, a vacuum (pressure of less than 1 millimeter of mercury, mmHg) is substituted for the low air pressure of the high mountains. Freeze-drying is a **sublimation** process that removes moisture from frozen products without appreciably changing their shape, color, or taste. *Sublimation* means that *ice in the food goes directly from solid to vapor, bypassing the liquid phase.* This process is based on Le Châtelier's second principle, which states that a given mass of a substance occupies a larger volume as a gas than as a solid, and that if the outside pressure is below the vapor pressure, the equilibrium will shift from the solid to the gaseous state.

The first lyophilized foods were marketed in about 1960. Today steaks, roast beef, pork chops, ham, poultry meats, scrambled eggs, cream cheese, and even entire meals are available in this form. Many "space foods" have been lyophilized so that quality, convenience, and safety from spoilage are ensured. Dried animal foods are presently used by the armed services, campers (dry soups), food processors (as ingredients), restaurants, vending machines, hospitals, and consumers.

The foremost limiting factor in the widespread acceptance and use of lyophilized animal foods is cost. Poultry meat is expected to be the animal product lyophilized to the greatest extent during the next decade, followed by red meats, dairy products, and fish.

Lyophilization is especially desirable for the preservation of egg proteins because coagulation is reduced to a minimum by freeze-drying. The high temperatures required for killing bacteria will coagulate egg albumen.

3.9 THE FUTURE OF ANIMAL PRODUCTS

With the world population increasing at the annual rate of about 85 million, and the nutritional level of a large percentage of the populace presently below desired standards, the projected need for animal products presents a tremendous challenge to all involved in activities related to their production. Present trends toward increased consumption of animal products will continue as the economic condition of the world's people becomes more favorable. Efficiencies of converting feed nutrients into animal products will continue to improve. This will result in a reduction in cost to consumers (relative to wages earned and per capita real income) and an increase in the consumption of animal products. All these trends support an optimistic outlook for the producer, processor, distributor, and consumer of animal products. We expect increased nationwide, indeed worldwide, emphasis on human nutrition, and this has important implications for animal products. For the very dietary essentials so frequently deficient in human diets—protein, calcium, riboflavin, and iron—are all found in abundance in animal products.

In a recent USDA study of 14,500 men, women, and children of the United States, the nutrients most commonly found below recommended dietary allowances were calcium and iron. (They averaged more than 30 percent below recommended allowances set by the Food and Nutrition Board of the National Academy of Sciences–National Research Council.) The iron in the diets of infants and children under 3 years of age was about 50 percent below recommended amounts. Diets of adolescent girls and of women were below recommended levels of calcium, iron, and thiamine and, for some age groups, below recommended levels of vitamin A and riboflavin. Older men had diets low in calcium, vitamin A, riboflavin, and ascorbic acid.

The above study clearly indicates a great need for additional consumer education in human nutrition. In this statement it is assumed, of course, that more persons in the United States are malnourished because of nutritional ignorance and misinformation than because of poverty, although isolated instances of the latter also prevail. Nonetheless, increased intake of animal products could alleviate the deficiencies observed.

The future of animal products is interwoven with the costs of production, processing, and distribution. Recent trends have been toward greater processing, longer storage, and further shipment of foods. Yet these activities require additional energy. Increases in energy costs could impede the trend toward more convenience foods.

The prospects of increased food formulation and the so-called engineered foods may significantly affect future consumption of animal products. Several supermarket chains are selling hamburger mixed with vegetable proteins made from soybeans. These mixtures usually have a lower fat content than regular hamburger and cost less. The mixture tends to bind and hold more moisture than does pure ground beef. The competition between soybean products and animal fats, as well as between soybean products and certain meat products, in supermarkets is of great interest to livestock producers (and to soybean growers) and will be important to the future of animal products.

In the end, of course, decisions of consumers will govern the demand for and future of animal products.

> Unless our product has consumer demand, we could well be sitting on a bushel of gold and starve to death.
>
> *John A. Moser*

If livestock producers continue to increase production efficiency, breed leaner-type meat animals, and market milk having a higher protein/fat ratio, the future looks bright; however, failure to meet consumer demands will encourage the increased use of synthetic animal products. Changes in food habits within the United States are frequently brought about by advertising and promotion. The producers of animal products may enhance demand for their products through increased advertising. The automobile industry spends approximately 10 times more on the research and development of new models than the entire food industry invests in new-product development.

3.9.1 New Animal Products

Prepackaged, portion-controlled, fully prepared foods are perhaps the ultimate in *convenience,* which is the desire of American consumers. Many fresh meats will be converted into frozen form in the years ahead, and many will be packaged in a vacuum or controlled atmosphere. Sausage products and luncheon loaves (table-ready meats) will increase. There will be an increasing trend toward ready mixes, bake (heat) and serve, and instant types of animal products that can be prepared by microwave heating. New types of ripened low-fat cheeses will be developed and marketed. Instant milk and cheeses, dry butter, smoked cheeses, frozen whipped cream, and cottage whey sherbet are of recent origin, and more dairy foods will follow.

Restructured Meat Products   Sales of restructured meat products have increased in recent years. Meat products commonly referred to as restructured include flaked and formed, sliced and formed, and sectioned and formed. Such products, texture-wise, have characteristics somewhere between those of ground meats and those of intact muscle cuts. They have the advantages of (1) resembling the preferred whole muscle products such as cutlets, chops, and steaks, and having similar storage stabilities; (2) being less expensive than whole muscle products; (3) having controlled fat content; and (4) better utilizing secondary carcass parts. Restructured meat products have proved highly acceptable in consumer tests.

Processed Meats   The per capita consumption of processed meats increased from 42 lb in 1930 to 61 lb in 1980. In 1981, 60 to 70 percent of the pork and 15 percent of the beef and poultry in the United States were sold in processed forms. These trends are expected to continue.

Frozen Scrambled Egg Mixes[19]   The traditional scrambled egg is a mixture of yolk and white, with milk frequently added. This simple product is satisfactory for immediate consumption. However, many uses of such a product are more demanding, and special mixes have been developed for these purposes.

[19]O. J. Cotterill, "Unscrambling Frozen Egg Mixes," *Poultry Tribune,* **89**(10):10–11 (1983).

Institutions, such as dormitories and cafeterias, require a longer holding period between cooking and consumption. The two commonest problems with holding scrambled eggs on the steam table are (1) green discoloration, which is caused by a reaction of hydrogen sulfide formed with iron during cooking to form the green ferrous sulfide; and (2) weeping (properly termed syneresis), which is the separation of liquid from the coagulated egg protein after cooking.

The addition of other food ingredients improves scrambled eggs for institutional uses. Scrambled eggs made from these mixes have improved color and textural characteristics. Each ingredient added provides a useful function. For example, the commonest additions are (1) nonfat dry milk, which improves the nutritional quality by supplying additional calcium and phosphates to provide better color retention; (2) vegetable oil, which produces a softer scrambled egg coagulum; (3) gums, which reduce weeping by binding water; and (4) citric acid and/or phosphates, which reduce the pH and chelate (bind) iron to prevent greening.

Cooked-Frozen-Thawed-Reheated Egg Products   Omelets, quiches, scrambled eggs, and soufflés are examples of frozen egg products included in this category. They all contain cooked egg white. Unfortunately, when cooked egg white is frozen, the gel-like structure is destroyed by the formation of ice crystals and the liquid separates from the coagulum. To demonstrate this freeze damage, the reader is encouraged to freeze and thaw a hard-cooked egg. Note that the egg white is laid down in layers. Two methods are used to overcome this textural defect. First, freeze-stable starches can be added to the raw egg white to bind the watery liquid. This is the technique used in the "long egg" produced by the Ralston Purina Company. This "egg" is about 1 ft long, about 1½ in in diameter, and has a yolk center. A second method is to rapidly freeze the product at a low temperature, by using liquid nitrogen, carbon dioxide, or Freon. The technique of fast freezing retards ice crystal growth and reduces the freeze-thaw damage.

Hard-Cooked–Peeled Eggs   Hard-cooked eggs were probably the first processed egg product consumed by humans. For example, a forest fire could have cooked the eggs. Cooking an egg improves the flavor and aroma compared with those of a raw egg.

Since about 1970, hard-cooked–peeled eggs have been marketed commercially. Except by special order, some institutions use only these prepeeled, hard-cooked eggs. They are commonly distributed in 5-gallon, polyethylene buckets. The shelf life under refrigeration is about 1 to 2 months.

Normal cooking procedures are used to cook the eggs in the shell. If they are cooked too long at a high temperature, the surface of the yolk will turn green. This results from the reaction of hydrogen sulfide (produced in the white during cooking) with iron from the yolk to form iron sulfide. After the eggs are peeled, they are packed in a solution containing sodium benzoate and citric acid. To the users of this product, there is a savings in labor and energy and no loss due to incomplete or ineffective peeling.

3.9.2 Synthetic Animal Products

During the past four decades, the dairy industry lost most of its butter market to the more economical product margarine. More recently, **filled milk** and **imitation milk** have been marketed in several states. *Filled milk has nonfat milk solids as the nonfat base and vegetable oil as a source of fat. Imitation milks are produced from vegetable protein or sodium caseinate, vegetable fat, corn solids (or other sugars), stabilizers, and emulsifiers.* At present, most imitation milks are lower in protein, minerals, and unsaturated fatty acids and higher in carbohydrates and **saturates** than cow's milk. More synthetic foods that will compete with animal products are certain to follow. Yet it should be noted that the interrelationships existing among food nutrients are an important consideration. The nutritional merits of milk as a whole, for example, are greater than the total individual values of its components. The same is true of most animal products.

The use of meat substitutes is not new. It was reported[20] that the protein mainstay of the German army during World War II was not meat but rather "brattling," a soybean–skim-milk sausage, and, further, that peanut and cottonseed meals (after the oil had been extracted therefrom) were used in Germany during World War II as a meat substitute in the form of sausage. More recently, considerable interest has been generated in the production of synthetic meats. The technology of spinning fibers from protein powder makes it possible for simulated meats to be produced from vegetable proteins. Thus far, these products are made mainly in dry or frozen form and have a longer shelf life than chilled meats. Isolated soybean protein has been fabricated into meatlike cuts (Figure 3.5) and is currently available for consumption. Soy concentrates are used in certain manufactured products not requiring the texture of meat and where a deviation in flavor would not likely be noticed. It is estimated that more than 50 million people in the United States do not eat meat some or any of the time for reasons of religion, doctor's restrictions, or personal preference. Eating nonmeat foods resembling frankfurters, hamburgers, pork sausage, dried beef and roast beef, fried chicken, turkey loaf, and other items may appeal to this group. Current regulations allow the addition of soy protein concentrate up to 3.5 percent in properly labeled sausage products.

The retail cost per pound of protein has not generally favored soy-based **meat analogues** over natural animal products. Soy flour is higher in crude protein than is meat, but it is not utilized as well in the body and has a somewhat less nutritionally desirable amino acid profile than has meat protein. However, the pressing need for additional protein throughout the world has stimulated rsearch

---

[20]N. G. Horner, "The Soya Bean," *Brit. Med. Jour.*, **2**:269–270 (1941).

FIGURE 3.5
Synthetic turkey loaf of flavored spun soybean protein. *(Worthington Foods, Inc., Worthington, Ohio.)*

in the production of amino acids from **organic molecules,** in the hope that quality protein can be produced economically on a mass basis. Approximately 200 million pounds of soy foods for meat substitutes were manufactured in 1980.

3.9.3 Pesticides and Animal Products

Prior to the twentieth century, the consumer knew the producer personally. If anything was wrong (sour milk, cracked eggs, or other questionable standards of quality), the difference was quickly resolved or future sales were lost. Today, the public expects (and rightly so) agriculture to provide foods that are wholesome and safe. This means that food is free from harmful pesticide residues, disease agents, or other toxic substances.

Pesticide chemicals have played a significant part in increasing agricultural productivity both in the United States and worldwide. About 5 lb of pesticide is applied per person annually for pest control in the United States. Uptake of pesticides by animals, leading to residues in animal products, can result either from direct application of the pesticide to an animal or from the animal ingesting feed carrying pesticide residues.

It is known, for example, that DDT (dichlorodiphenyltrichloroethane) and certain other pesticides can be found in animal products. The current permissible level of DDT in manufactured dairy products is 1.25 parts per million (ppm) on the milk fat basis.[21] Thus, in milk containing 4 percent milk fat, the tolerance

---

[21]*Code of Federal Regulations*, Title 21, Part 193.120, 1981.

is 0.05 ppm. The use of DDT was banned in the United States when research revealed that daily doses as low as 0.3 mg per kg of body weight increased the incidence of liver tumors in mice. However, worldwide it has been agreed by joint meetings of **FAO/WHO** that there are particular circumstances where the benefits to humans arising from the proper use of DDT outweigh the possible risk from exposure (e.g., in control of malaria); therefore, a daily tolerance level of 0.005 mg per kg of body weight was established.

Concentrations of pesticides and other agricultural chemicals in animal products (and in other foods) are monitored closely by government agencies. Based on research findings to date, present constraints on tolerable amounts of these organic chemicals in foods are more than adequate to assure consumers of safe and wholesome animal products.

3.10 SUMMARY

The nutritive value of food is a positive factor in determining the health and quality of life from conception until death. Provisions for adequate nutrition in early years contribute to a more productive and enjoyable life in later ones. Most animal products are rich in high-quality protein and relatively low in calories. Animal products also make important contributions to humanity's needs for fats, carbohydrates, minerals, and vitamins. These dietary essentials are available at a reasonable cost to consumers.

For many centuries, domesticated farm animals have made a major contribution to human welfare as sources of food, clothing, transportation, power, soil fertility, fuel, and pleasure. The most important of these is food. Foods from animals are the most nearly complete foods known. In the United States, they provide more than half of the protein, calcium, and phosphorus and several vitamins in the diet. Animal proteins are of high quality because they contain the amino acids required for human nutrition in proportions similar to those in the human body. Meat and poultry are good sources of iron.

As "natural" foods, animal products effectively balance and supplement plant foods. Meat, milk, and egg proteins are high in biological value and provide essential amino acids often deficient in plant foods. Additionally, milk and milk products are rich in calcium and riboflavin; meats are rich in iron, niacin, and nicotinic acid. Milk and eggs also are important sources of vitamin A. Moreover, animal foods contribute to palatability and variety in the diet of humans.

Improved methods of food preservation increase the availability of animal products, and fortification ensures a more uniform and nutritious product composition. Mixtures of foods from animals and plants, especially fabricated foods, are likely to become more important in the diets of Americans and will provide opportunities for the development of new foods and new menus.

Modern production, processing, packaging, and distribution practices, coupled with close supervision from regulatory agencies, assure today's consumers of having animal products that are safe in terms of disease.

STUDY QUESTIONS

1 What is the relation of the consumption of animal products to the following characteristics of a nation, its people, or both: *(a)* health, *(b)* life expectancy, *(c)* progressiveness?
2 Cite an example of how diet influenced the stature of a nation's people.
3 Why are animal proteins superior to vegetable proteins for humans and other monogastric animals?
4 Corn is especially deficient in which two *essential* amino acids?
5 What are the *essential* fatty acids for humans? Are they found in animal products?
6 Are animal products good sources of carbohydrates?
7 Why does honey taste so sweet? Is it heavier or lighter than water?
8 What are several uses of honey?
9 What state leads in the production of honey?
10 What is the function of *(a)* the queen bee, *(b)* workers, and *(c)* drones?
11 Of what importance is royal jelly to the honeybee colony?
12 How much honey does the average worker bee gather in its lifetime?
13 How many times can the female bee sting a human?
14 What happens to the drones during the winter months?
15 Considering your answer to Question 14, how can you explain the presence of drones in the spring?
16 What is lactose? What plant products contain lactose?
17 Relate the percentage of lactose in the milk of a given species to the brain size of that species (as related to total body mass).
18 Discuss the unique properties of lactose.
19 What is the relation of lactose to calcium, phosphorus, magnesium, and barium assimilation? Does milk contain significant amounts of these minerals?
20 Does woman's milk or cow's milk contain the most lactose? Relate this to the prevention and therapy of rickets.
21 Is milk a good source of niacin? Explain the antipellagric effect of milk as a food for humans.
22 Compare lactose with sucrose in sweetness.
23 What is a placebo? (See Glossary.)
24 List several animal products that are good sources of calcium and/or phosphorus.
25 Is milk a good source of iron? Of copper?
26 Does cow's milk contain more or less mineral (especially calcium and phosphorus) than woman's milk? What correlation was made between the ash content of milk and the growth rate of a given species?
27 Animal products are somewhat deficient in which vitamins?
28 Do animal products have a favorable protein/calorie ratio for those concerned with excessive tissue?
29 Which animal products are economical sources of protein?
30 Do American consumers ingest more or fewer calories now than in 1910? What about the consumption of other major nutrients?
31 What is the trend in American consumption for *(a)* beef, *(b)* fish, *(c)* eggs, *(d)* poultry meats, *(e)* nonfat milks, *(f)* butter, *(g)* total animal fats, *(h)* vegetable fats?
32 Does the body synthesize cholesterol? Is it needed in the body?
33 Do people live longer in the United States than in all other countries of the world?
34 Strontium is similar to what element found in milk?

35 How much ^{90}Sr ingested by the cow is secreted into her milk?
36 Why do food processors fortify animal products?
37 Can milk be economically fortified with vitamins A and D?
38 Is nonfortified skim milk a good source of the fat-soluble vitamins? Why?
39 Is an excessive intake of vitamin D beneficial? Harmful?
40 What is milk toning?
41 Why are people concerned with preserving animal products?
42 What is the effect of high and low temperatures on bacterial growth?
43 What is bactofugation?
44 Is dehydration a recent innovation in the preservation of animal products?
45 What is a limitation of the use of chemical preservatives in animal products?
46 What is cold sterilization?
47 What natural component of honey restricts the growth of most microorganisms?
48 What are the times and temperatures used to effect the pasteurization of honey?
49 How may pork be rendered safe for human consumption (safe from trichinosis)?
50 Why is meat smoked?
51 Who is credited with the invention of canning?
52 Are most vitamins of meat destroyed in processing?
53 Why should eggs be heat-treated?
54 What are the times and temperatures used to effect the pasteurization of eggs?
55 What is thermostabilization of eggs?
56 What are the times and temperatures used to pasteurize milk?
57 Why is milk pasteurized? Does pasteurization destroy many vitamins of milk?
58 What is the effect of sunlight on the flavor and vitamin A potency of milk?
59 What is puff-drying?
60 How are instant animal product powders manufactured?
61 What is lyophilization? Sublimation? Are you using the Glossary?
62 For what are lyophilized animal products being used?
63 What is the primary limiting factor in the expanded use of lyophilized animal products? Which ones are likely to increase the most in use during the next decade?
64 What factors might influence the expanded use of animal products?
65 What are some trends in new animal products?
66 Do you think synthetic foods are a threat to animal products?
67 What is mechanically deboned poultry meat? Why is it pasteurized?
68 What problem may arise from the use of a strong source of fluorescent light in refrigeration cases in which milk is being marketed in translucent plastic containers?
69 Cite advantages of the marketing of restructured meat products. Cite examples of such products.
70 Are "natural" foods always safe for human consumption? Discuss.

FUNDAMENTAL PRINCIPLES OF GENETICS

If a man leave children behind him, it is as if he did not die.

Moroccan proverb

Recent discoveries in physiological genetics show rather clearly that **genes,** the determiners of **heredity,** play an important part in all the biochemical reactions in an animal's body. Since these reactions are necessary for proper form and function of the body, the importance of heredity in animal production and human welfare can readily be seen. The purpose of this chapter is to outline and discuss the basic principles of animal genetics.

4.1 THE CELL THEORY OF INHERITANCE

The **cell** theory states that all plants and animals are made of small building blocks called *cells.* Each plant cell is bounded by a tough outer cell wall and an inner membrane, and each animal cell is bounded by a membrane. Both plant and animal cells usually have a nucleus. Between the **nucleus** and the cell wall is the cellular material called **cytoplasm,** which contains certain components necessary for the function of that particular cell (Figure 4.1).

All cells originate from other cells through the process of cell division, and many of these new cells are capable of growing and dividing to produce other cells like themselves. Within the nucleus are **chromosomes,** which carry the hereditary material called *genes.* In body cells, these genes occur in pairs on chromosomes that also occur in pairs and are similar in their size, shape, and proportions. A pair of such chromosomes are known as **homologous chromo-**

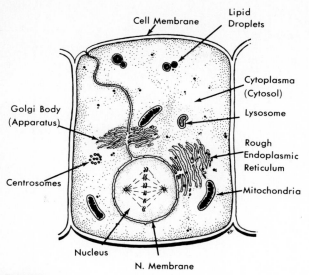

FIGURE 4.1
Present-day conception of a cell. *(Courtesy Dr. Robert Godke, Louisiana State University.)*

somes (*homo* meaning "alike" or "equal" and *logous* meaning "proportion"). The number of pairs of homologous chromosomes in body cells is constant in normal individuals within a **species.** Different species, however, often have a different number of chromosomes, as shown in Table 4.1. In animals, the body cells possess one pair of chromosomes known as **sex chromosomes,** since they are related to the sex of the individual. They are usually referred to as the X and Y chromosomes, the X chromosome being two to three times larger than the Y. In farm mammals and humans, the male is XY whereas the female is XX. In chickens, the male is XX and the female is Xw, which means that the female has a w chromosome much smaller than the X. This situation is the opposite of that found in mammals (i.e., the Xw chromosome arrangement in female chickens differs from the XX chromosomes in female farm mammals).

All chromosomes in body cells other than the sex chromosomes are known as **autosomes.** Thus, in humans, there are 22 pairs of autosomes and one pair of sex chromosomes, or a total of 23 pairs. For each pair of chromosomes an individual possesses in the body cells, he or she received one chromosome from his or her father and the other from his or her mother. This means that each parent normally contributes one-half of the chromosome number to each **offspring.** Each offspring, when it becomes sexually mature and reproduces, will transmit only one of each pair of chromosomes to any one of its offspring. Which one it transmits will be determined by the law of probability.

Pairs of chromosomes in body cells are referred to as the $2n$ number, or **diploid** (*di* meaning "two") number. In **gametes,** or sex cells, where only one of

TABLE 4.1
CHARACTERISTIC NUMBERS OF
CHROMOSOMES IN SELECTED ANIMALS

Animal	Chromosome number (2n)
Donkey	62
Horse	64
Mule	63
Swine	38
Sheep	54
Cattle	60
Human	46
Mink	30
Dog	78
Lion	38
Domestic cat	38
Bengal tiger	38
Chicken*	78

*It was once thought that the female was XO, having one fewer chromosome than the male. Recent studies, however, show that the female has a very small sex chromosome, called the w chromosome. Thus the female is Xw (sometimes called Zw).

each pair of chromosomes is found, these are referred to as the $1n$, or **haploid,** number of chromosomes.

4.2 CHROMOSOMAL ABNORMALITIES

Chromosomal abnormalities are usually referred to as *chromosomal aberrations.* They can cause drastic changes in the appearance **(phenotype)** of an individual, or they may even cause its death. Death can occur shortly after the fertilization of the **ovum** by the **spermatozoon,** during **intra**uterine life, or even some time after the individual is born. Several different types of chromosomal aberrations have been described in humans. A few have been described in animals other than humans.

Nondisjunction is one general form of chromosome aberration. This term means that the homologous chromosomes fail to separate when the sex cell is formed and that both members of a particular pair of homologous chromosomes go into the sex cell rather than the normal one-half of each pair. Two general forms of nondisjunction have been observed. The first, known as **aneuploidy,** results when an individual has the basic $2n$ number of chromosomes but also has one or more chromosomes duplicated or missing. Thus the individual possesses either one (or more than one) more chromosome than is normal, as in humans with Down's syndrome, or one (or more than one) fewer chromosome than the normal number. An extra chromosome often causes abnormal conditions in animals, whereas the lack of a chromosome is usually **lethal,** or causes that individual's death. A second form of nondisjunction is known as **polyploidy,** or **euploidy.** This term means that entire sets of chromosomes are duplicated, giving, instead of the normal $2n$ number in body cells, a $3n$, $4n$, etc., number. Polyploidy is quite common in plants but apparently is rare in higher animals, including humans.

Translocation is another form of chromosomal aberration. This term includes those instances in which a chromosome is broken and the smaller piece attaches itself to another chromosome not homologous to it. This condition has been reported in humans, cattle, and swine and likely occurs in other species of animals.

A chromosomal aberration is called a **duplication** when a portion of a chromosome attaches itself to the chromosome to which it is homologous. Duplication is probably due to the fact that the homologous chromosomes that *synapse* (come together) in **meiosis** do not completely separate at the reductional division, leaving a portion of one chromosome attached to its homologous mate.

Deletion refers to a chromosomal abnormality in which a portion of a chromosome is lacking. Deletions are usually lethal because the portion of a chromosome that normally carries genes that perform a function vital to the life of an individual is lacking.

Another form of chromosomal abnormality is known as an **inversion.** This is illustrated in Figure 4.2. An *inversion* means that *a portion of a chromosome has been rearranged so that genes occur on the chromosome in a new or inverse order from their original sequence.*

Changes in chromosome numbers, chromosome breakages, and the rearrangement of the sequence of genes on chromosomes are reflected in changes in the phenotypes of individuals.

4.3 DETERMINATION OF SEX

One pair of chromosomes in body cells are known as *sex chromosomes* because they determine the sex of the individual. In humans and larger farm animals, the female possesses two X chromosomes whereas the male possesses an X and a Y (Figure 4.3). In chickens, the male possesses two X chromosomes whereas the female possesses only one. The female chicken does not possess a Y chromosome but rather a small chromosome known as w, as mentioned previously.

In farm animals and humans, the X chromosome is several times larger than the Y and carries more genes, or hereditary material. This important point will be discussed later in the chapter.

4.4 CELL DIVISION

An animal's body consists of millions of building blocks called *cells.* Cells multiply or increase in numbers by undergoing division. Two general kinds of cell division are known, mitosis and meiosis. **Mitosis** refers to the kind of cell division in which each cell divides and forms two cells, both of which possess two

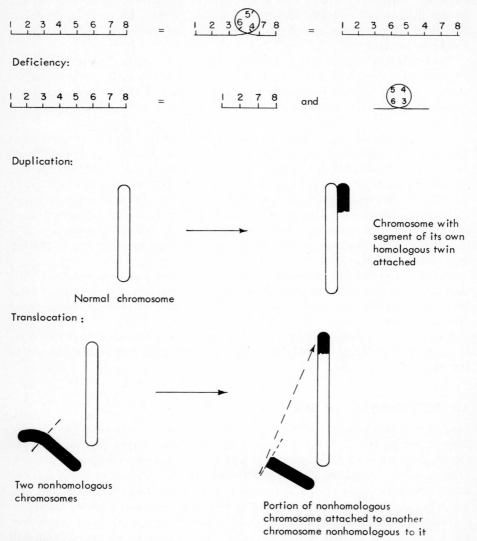

FIGURE 4.2
Illustration of various kinds of chromosome abnormalities in animals.

complete sets of chromosomes just like those found in the mother cell. **Meiosis** refers to a type of cell division in the sex cells, the sperm and the egg, in which the chromosome number is reduced by one-half from the diploid ($2n$) to the haploid ($1n$) number. The reduction is necessary so that the union of the sperm

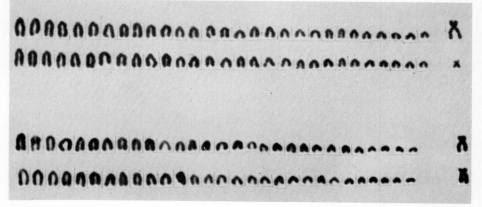

FIGURE 4.3
The chromosome **karyotype** of cattle. Males possess X and Y chromosomes (top); females possess two X chromosomes (bottom). These are actual photographs of chromosomes at the metaphase, where each chromosome appears doubled. *(Missouri Agr. Expt. Station.)*

and egg in fertilization will restore the diploid number of chromosomes normally found in body cells but will not exceed this number. It also ensures that one-half of the chromosomes of the **progeny** are contributed by each parent.

4.4.1 Mitosis

Mitosis has several different phases: the prophase, metaphase, anaphase, telophase, and interphase. Only two pairs of homologous chromosomes will be used to illustrate this type of cell division, although there are many more pairs than this in the body cells of animals.

The **prophase** is the beginning of mitosis. In early prophase, the chromosomes begin to shorten and thicken, and each chromosome appears as a double strand of identical **chromatids,** or sister chromosomes, connected together by means of a centromere (Figure 4.4). During midprophase the nuclear membrane disappears, and in late prophase the spindle fibers form and the centromeres become attached to a spindle fiber.

During **metaphase,** the chromosomes (chromatids) line up on the equator of the spindle fibers, the mitotic centers (**centrioles**) appearing at each pole.

The **anaphase** is characterized by each centromere separating into two centromeres, each with a chromatid. Each **centromere** then moves toward the mitotic center, dragging its chromatid with it.

In the **telophase** the nucleus is re-formed, and the nuclear membrane reappears around each mitotic center, resulting in two new nuclei. The chromosomes grow longer and disperse into a network of fine threads. This phase of mitosis is completed when the cytoplasm divides (**cytokinesis**) forming two new cells that possess the same chromosome complement as the original mother cell. The period of time between the telophase of one cell division and the prophase of the next is known as the **interphase.**

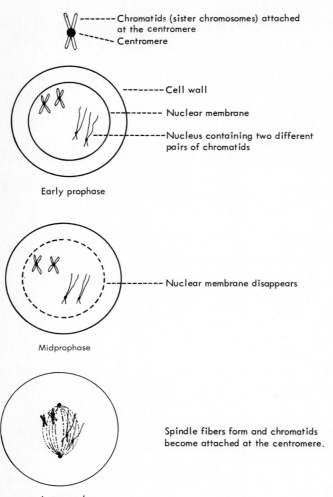

FIGURE 4.4
The different stages in mitosis. Chromosomes are not visible in
stained slides in all stages, but they are presented here as if they
were in order to illustrate the duplication of cells and chromosome
numbers.

4.4.2 Meiosis

Meiosis is similar to mitosis in some respects but is different in several others of
great importance. Meiosis is the type of cell division that forms the gametes (sex
cells) with the haploid ($1n$) number of chromosomes. It involves a **first** and a
second meiotic division.

In the prophase of the first meiotic division (Figure 4.5), the chromosomes
shorten and thicken, each chromosome appearing as two chromatids connected

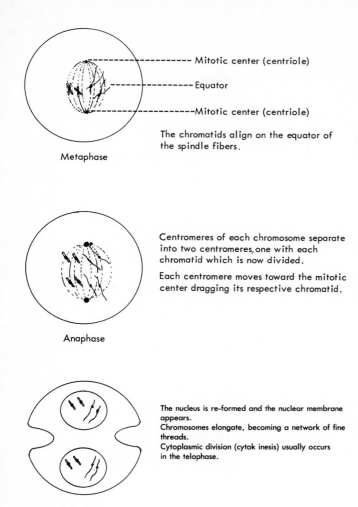

Mitotic center (centriole)

Equator

Mitotic center (centriole)

The chromatids align on the equator of the spindle fibers.

Metaphase

Centromeres of each chromosome separate into two centromeres, one with each chromatid which is now divided.

Each centromere moves toward the mitotic center dragging its respective chromatid.

Anaphase

The nucleus is re-formed and the nuclear membrane appears.
Chromosomes elongate, becoming a network of fine threads.
Cytoplasmic division (cytok inesis) usually occurs in the telophase.

Telophase

The last phase is known as the interphase. It is the period between the telophase of one division and the prophase of the next.

FIGURE 4.4
(cont.)

by a centromere, as in mitosis. A distinguishing feature of meiosis, however, is the pairing up (**synapsis**) of two homologous chromosomes, giving the appearance of four chromatids. These synapsed pairs are known as **tetrads.** Synapsis seldom, if ever, occurs in mitosis. During synapsis in meiosis, the homologous chromosomes may exchange parts, which is known as **crossing over.** The spindle fibers then appear, as in mitosis, and the synapsed chromosomes become

FIRST MEIOTIC DIVISION

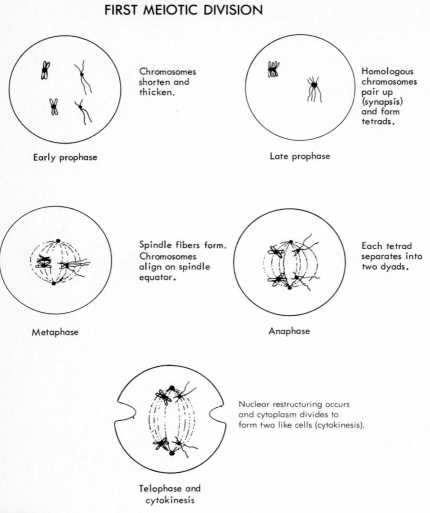

Chromosomes shorten and thicken.

Early prophase

Homologous chromosomes pair up (synapsis) and form tetrads.

Late prophase

Spindle fibers form. Chromosomes align on spindle equator.

Metaphase

Each tetrad separates into two dyads.

Anaphase

Nuclear restructuring occurs and cytoplasm divides to form two like cells (cytokinesis).

Telophase and cytokinesis

FIGURE 4.5
Different stages within the nucleus of the first meiotic division. (Only the nucleus is shown.)

attached to the spindle fibers at the centromeres. The chromosomes again align themselves at the equator of the spindle fibers during the metaphase.

In the anaphase of the first meiotic division, each tetrad separates into two **dyads** (a pair of chromatids connected by a centromere). In the telophase, the nucleus and cytoplasm re-form, resulting in the production of two new cells, each one of which contains only one chromosome of each original homologous chromosome pair. Thus it is a reductional division.

SECOND MEIOTIC DIVISION
(nucleus only)

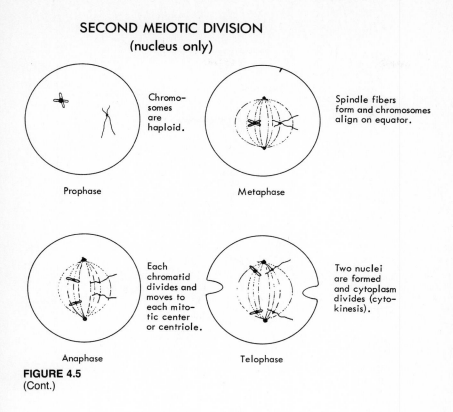

Prophase — Chromo-somes are haploid.

Metaphase — Spindle fibers form and chromosomes align on equator.

Anaphase — Each chromatid divides and moves to each mito-tic center or centriole.

Telophase — Two nuclei are formed and cytoplasm divides (cyto-kinesis).

FIGURE 4.5
(Cont.)

In the prophase of the second meiotic division, each chromosome in the nucleus is present in the haploid state. Each chromosome appears, however, as a pair of sister chromatids connected at the centromere. These sister chromatids align themselves at the equator of the spindle fibers in the metaphase and then separate in the anaphase, the two chromatids going to opposite poles of the spindle fibers. In the telophase, the nucleus re-forms and the cytoplasm divides, forming two cells that each have the haploid number of chromosomes.

The process of meiosis in the male **gonads** produces four spermatozoa from each spermatogonium, as is shown in Figure 4.6. Each oocyte, in meiosis, produces an ovum and three polar bodies, as shown in Figure 4.7.

4.5 THE GENE AND HOW IT FUNCTIONS

There have been many new discoveries about the gene and its functions in recent years. A better understanding of the processes involved in gene action indicates that there will be many new and exciting discoveries in the future that should result in a better life for humanity.

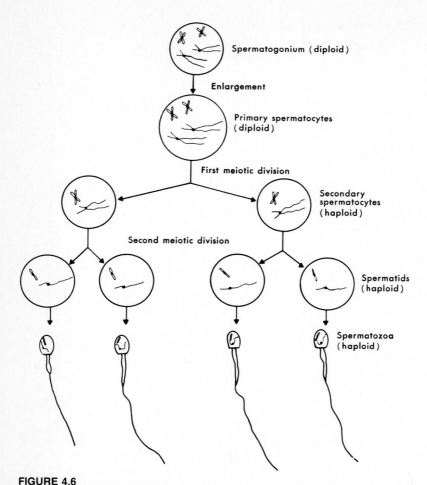

FIGURE 4.6
Outline of spermatogenesis (using only two pairs of homologous chromosomes and showing the nucleus only).

4.5.1 The Gene

Many years ago a gene was often defined as *the smallest unit of inheritance.* This definition probably was used because the actual chemical composition of the gene was not known. In recent years, research has shown that the gene is a portion of a **DNA** (deoxyribonucleic acid) **molecule.** Much information is now available on the chemistry and functions of genes.

The DNA molecule may be described as the backbone of the chromosome, analogous to the vertebral column, which is the backbone of the vertebrate animal body. The DNA molecule resembles a long, twisted ladder in which the

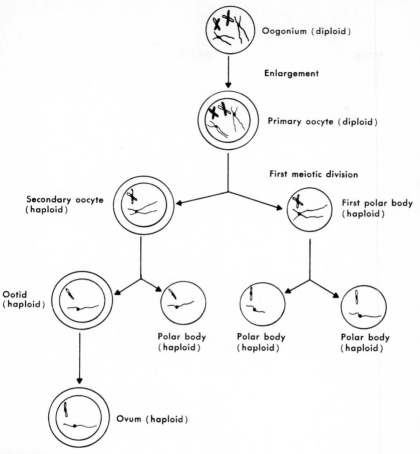

Oogonium (diploid)

Enlargement

Primary oocyte (diploid)

First meiotic division

Secondary oocyte (haploid)

First polar body (haploid)

Ootid (haploid)

Polar body (haploid)

Polar body (haploid)

Polar body (haploid)

Ovum (haploid)

FIGURE 4.7
Outline of oogenesis (using only two pairs of homologous chromosomes).

two strands (sides) are joined together by rungs (Figure 4.8). Each strand is called a *polymer* (*poly* meaning "many"; *mer* meaning "part") because it is composed of many repeated units, called *nucleotides*.

A **nucleotide** is composed of a nitrogenous base (either a purine or a pyrimidine) linked to a sugar, which is in turn linked to a phosphoric acid molecule (Figure 4.9). The sugar in **RNA** (ribonucleic acid) is ribose whereas in DNA it is deoxyribose. The various chemical constituents of RNA and DNA molecules are shown in Figures 4.10 and 4.11. A schematic arrangement of several nucleotides in a single strand of DNA is shown in Figure 4.12. The two single strands (nucleotides) of a double-strand molecule of DNA are joined together by hydrogen bonds between the bases. Adenine (A) always joins to

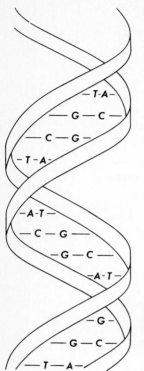

FIGURE 4.8
Diagram of a portion of the DNA molecule. The letters A, C, T, and G represent certain bases, or cross-links, connecting the two strands. The code sent by the gene to the cytoplasm depends on the kind and arrangement of these bases.

FIGURE 4.9
A nucleotide, adenylic acid. It consists of a combination of the base adenine, the sugar deoxyribose, and phosphoric acid.

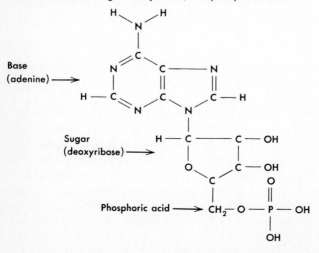

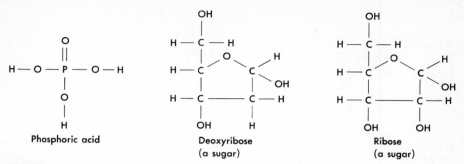

FIGURE 4.10
Chemical formula of phosphoric acid and sugar molecules found in DNA and RNA.

FIGURE 4.11
Purine and pyrimidine bases found in RNA and DNA.

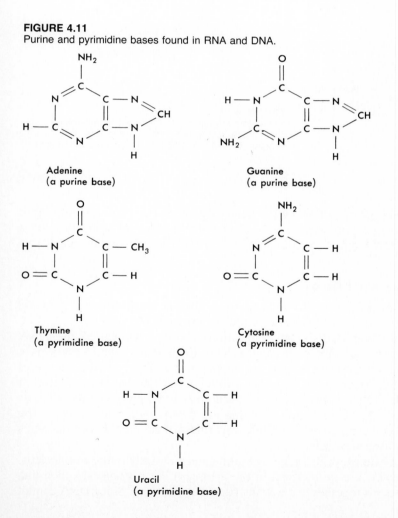

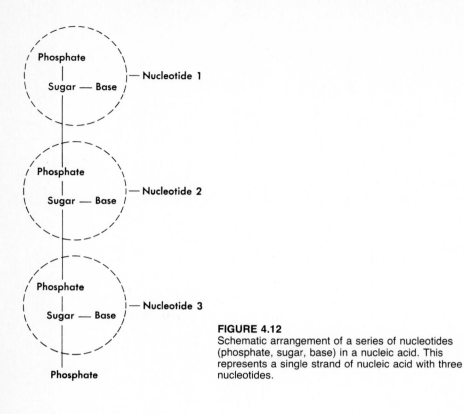

FIGURE 4.12
Schematic arrangement of a series of nucleotides (phosphate, sugar, base) in a nucleic acid. This represents a single strand of nucleic acid with three nucleotides.

thymine (T), and guanine (G) always joins to cytosine (C), as shown in Figure 4.13. The average gene (cistron) is thought to consist of a portion of a double-strand DNA molecule containing about 600 consecutive base pairs.

4.5.2   Functions of the Gene

The functions of the gene are to replicate itself when new cells are produced and to send the code to the cytoplasm to build certain proteins.

It is now thought that during replication, the two strands of the DNA molecule separate, as shown in Figure 4.14. Each strand then serves as a template, or mold, for synthesizing the missing part. Thus, if the sequence of bases in the single strand is T, G, C, G, it follows that the sequence of bases in the missing strand should be A, C, G, C, because A and T always go together, as do G and C. This base pairing allows the DNA molecule to copy the double-strand DNA molecule accurately, regardless of the number of times each cell divides to form new cells.

DNA, the substance of the gene, is found almost entirely in the cell nucleus. Proteins, however, are synthesized in the cytoplasm by ribosomes, which link various amino acids together in a way instructed by the gene. Since DNA cannot

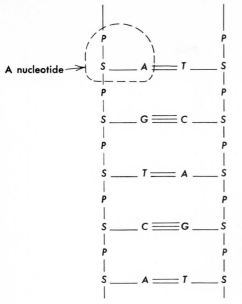

FIGURE 4.13
Schematic representation of a double-strand DNA molecule. P, phosphoric acid; S, sugar deoxyribose; A, base adenine; G, base guanine; T, base thymine; and C, base cytosine. Base A always pairs with base T and G with C. A nucleotide is the combination of P + S + a base, as shown by the dotted circle.

leave the chromosome, it synthesizes messenger RNA (mRNA), another nucleic acid, which carries instructions from DNA to the cytoplasm as to the kind of protein to build. Besides differing from DNA in the sugar molecules it contains, RNA also differs from DNA in containing the nitrogenous base uracil instead of thymine. Thus uracil is found in RNA but not in DNA, whereas thymine is found in DNA but not in RNA.

Proteins are organic compounds that are found in all body cells. They consist of the 20 naturally occurring amino acids listed in Table 4.2. These amino acids,

TABLE 4.2
AMINO ACIDS UTILIZED IN
PROTEIN SYNTHESIS IN THE
ANIMAL BODY

Alanine	Leucine
Arginine	Lysine
Asparagine	Methionine
Aspartic acid	Phenylalanine
Cysteine	Proline
Glutamic acid	Serine
Glutamine	Threonine
Glycine	Tryptophan
Histidine	Tyrosine
Isoleucine	Valine

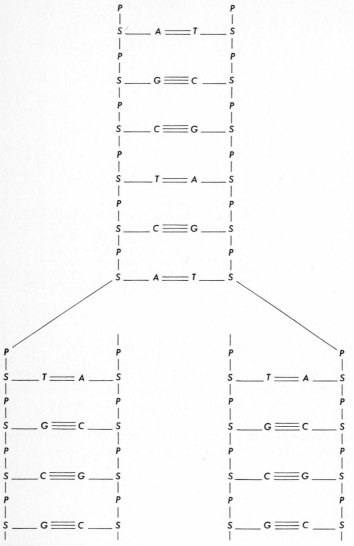

FIGURE 4.14
Illustration of how the double-strand molecule exactly duplicates itself.
P, phosphoric acid; S, sugar deoxyribose; A, base adenine; T, base
thymine; G, base guanine; and C, base cytosine.

joined together, form thousands of different proteins. The specific protein
produced depends on the kind, number, and arrangement of amino acids in the
protein molecule, just as a specific English word depends on different kinds,
numbers, and arrangements of the 26 letters in the English alphabet.

4.5.3 The Genetic Code

Genetic information is stored in the DNA molecule in the form of a triplet code, called a *codon*. A codon consists of *a sequence of three (triplet) nitrogenous bases that code only one specific amino acid.* Since the DNA molecule contains a total of four different bases, and any combination of three specifies only one certain amino acid, the total possible combinations of these four bases in groups of three would number 4^3, or 64. These 64 groups are more than enough possible combinations to specify the 20 amino acids. In fact, more than one combination of three nitrogenous bases can specify the same amino acid, as shown in Table 4.3.

The gene is thought to carry the code for the formation of a single polypeptide chain of a protein, as well as instructions to start and stop the sequence of amino acids in that polypeptide chain. Part of one of the two strands of the DNA molecule (gene) transcribes the code, or message, for a protein into a long single strand of mRNA. This message is then carried to the cytoplasm.

In the ribosomes, mRNA forms the template (or mold) for a sequence of particular amino acids. Completion of the message carried by mRNA is the work of another group of molecules, called *transfer RNA* (tRNA). Each tRNA molecule specifies only one of the 20 amino acids. Each group of tRNA molecules also recognizes a specific enzyme that joins it to a specified amino acid. Transfer RNA molecules also contain a series of three bases, called *anticodons,* and each anticodon seeks out and complements a codon on the mRNA molecule, carrying its particular amino acid with it, as shown in Figure

TABLE 4.3
MESSENGER CODONS AND THE AMINO ACIDS THEY ARE BELIEVED TO SPECIFY

UUU	Phenylalanine	CUU	Leucine	AUU	Isoleucine	GUU	Valine
UUC	Phenylalanine	CUC	Leucine	AUC	Isoleucine	GUC	Valine
UUA	Leucine	CUA	Leucine	AUA	Isoleucine	GUA	Valine
UUG	Leucine	CUG	Leucine	AUG	Methionine	GUG	Valine
UCU	Serine	CCU	Proline	ACU	Threonine	GCU	Alanine
UCC	Serine	CCC	Proline	ACC	Threonine	GCC	Alanine
UCA	Serine	CCA	Proline	ACA	Threonine	GCA	Alanine
UCG	Serine	CCG	Proline	ACG	Threonine	GCG	Alanine
UAU	Tyrosine	CAU	Histidine	AAU	Asparagine	GAU	Aspartic acid
UAC	Tyrosine	CAC	Histidine	AAC	Asparagine	GAC	Aspartic acid
UAA*	End	CAA	Glutamine	AAA	Lysine	GAA	Glutamic acid
UAG*	End	CAG	Glutamine	AAG	Lysine	GAG	Glutamic acid
UGU	Cysteine	CGU	Arginine	AGU	Serine	GGU	Glycine
UGC	Cysteine	CGC	Arginine	AGC	Serine	GGC	Glycine
UGA	?	CGA	Arginine	AGA	Arginine	GGA	Glycine
UGG	Tryptophan	CGG	Arginine	AGG	Arginine	GGG	Glycine

*UAA and UAG signal the end of polypeptide chains.

4.15. This ensures the correct order of amino acids as specified by the DNA. The specific amino acids are then combined into a protein molecule. Two bases in the codon pair with two bases in the anticodon, with adenine (A) pairing with uracil (U) and guanine (G) with cytosine (C). The third base in the codon, corresponding to the first base in the anticodon, has a certain amount of uncertainty in its pairing (the "wobble" theory). It can be seen from this discussion that protein synthesis is very complex and some details are not fully understood.

4.5.4   Control of Gene Function

Each body cell that has a nucleus contains a sample of all the genes the individual possesses. Not all these genes function at any given time. Thus there must be something within the cell that controls the function of some genes. A

FIGURE 4.15
Diagram showing how proteins are built in the cell. Messenger RNA is synthesized by DNA in the nucleus and then moves to the cytoplasm (ribosomes), where transfer RNA moves amino acids to fit the mold of the mRNA. Amino acids are then arranged together in a sequence, as instructed by the DNA. Ribosomal RNA is also present in the cytoplasm, but its function is not known.

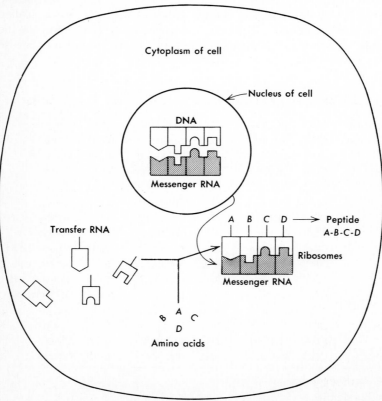

recent theory advanced by Jacob and Monod, derived from studies with microorganisms, has been proposed to explain the regulation of gene action. It is not known for certain if the same regulators are present in higher forms of animals.

The present theory of gene regulation proposes that two principal kinds of genes exist: the *structural* gene, which functions to synthesize specific proteins and enzymes, and *control* genes (regulator and operator genes), which are not directly concerned with protein synthesis but regulate the activity of the structural genes and the amount of protein synthesized. The *operator* gene is thought to be on the same chromosome as structural genes and adjacent to them. Each operator gene may control several structural genes in the way of an off-on switch. The *regulator* gene, on the other hand, may be on a different chromosome from the structural gene. The regulator gene is thought to produce substances that may combine with the proteins produced by the structural genes to form other substances called *repressors*. The repressor substance is thought to proceed to the operator gene and to keep it turned off, thereby inhibiting enzyme formation by the structural gene. If the repressor substance is blocked by a metabolite (inducer), the operator gene then turns on the structural gene that produces gene products. This complete gene complex is called the *operon*.

Genes may modify the phenotypic expression of other genes. This is especially true in epistatic gene action, of which many examples in higher organisms are known. Most of these examples involve qualitative traits such as coat color that can be readily observed. Other examples probably occur that are not so evident. Nongenetic factors also may modify the phenotypic expression of genes.

Various endocrine secretions are known to affect the phenotypic expression of some genes. The male hormone appears to be involved in a group of sex-influenced traits that are dominant in males and recessive in females. Certainly, genes for milk or egg production do not express themselves in males, although there is no doubt that males carry genes for these traits. Hereditary baldness in the human male is associated in some way with the male hormone testosterone, although the genes that produce hair on a man's head appear to function for a period of time, or until the man is between 20 and 30 years of age, when he may become bald. The relations between certain endocrine secretions and some traits in livestock are also of interest. Physiological age at maturity differs in breeds of livestock, showing that genes are involved in regulating the onset of maturity.

4.6 GENES AND EMBRYOLOGICAL DEVELOPMENT

The fertilized egg divides into two cells, then four, then eight, etc., for a considerable length of time. At first, the cells in the new embryo are the same, mother and daughter cells being identical. Later, however, the mother cells produce daughter cells that are different, thus forming the different body tissues and organs. Why they begin to produce cells unlike themselves in certain stages of embryonic development is not completely understood.

Some scientists believe that embryonic development depends on few, if any, genes until the midembryonic period. However, genes must act throughout the embryonic period to cause formation of various body organs and tissues. Evidence for this is suggested by the fact that different species of animals vary in their rate of embryological development in the uterus. These differences must be controlled by genes. Also, many **lethal genes** (killer genes) are known to have their effects during various stages of embryonic development. Some defects of the embryo are known to be due to environmental factors that resemble inherited defects. Viral infections and certain substances (e.g., thalidomide) consumed by the pregnant mother at a specific stage of pregnancy cause defects in the developing fetus similar to genetic defects, probably because they block cell division.[1] Such defects may be much more extensive in farm animals than is usually recognized.

4.7 VIRUSES

Viruses are of interest because of their similarities in some ways to genetic materials. All plant viruses probably contain only RNA, although exceptions are possible. Animal viruses may contain either RNA or DNA, but not both. In some viruses the nucleic acids are single-strand; in others, double-strand. The nucleic acids of viruses are enclosed in a protein coat, but they possess nothing resembling the cytoplasm of cells.

Viruses cannot reproduce without first invading a cell. Once they invade a cell, they use the substrates and metabolic system of that cell for their own reproduction. Infected cells do not divide but rather degenerate and die. Virus infections in unborn young during intrauterine life, when tissues and organs are being formed, often cause abnormal tissue development, and the young may be defective at birth. The frequency of occurrence of birth defects in animals due to intrauterine virus infections is not known.

4.8 GENETIC ENGINEERING[2]

The potential of genetic engineering is indeed enormous. This budding new technology has been called one of the four most significant scientific break-throughs of the twentieth century, on a par with unlocking the atom, escaping the earth's gravity, and the computer revolution.

Genetic engineering is coming to be a frequently used, yet loosely interpreted, term. Here, the term means the use of unconventional methods of modifying the animal genome, basically, transferring a gene from one individual to another. Some of the genetic engineering approaches that have been proposed for animal improvement include genetic transformation using isolated DNA, haploid animal production through cell cultures, transfer of organelles from one animal

---

[1]The present technique of amniocentesis (examining the fluid taken from the womb of a pregnant woman) can reveal any of some 70 possible genetic disorders in a woman's fetus.

[2]The authors acknowledge with appreciation the contributions to this section of Dr. R. D. Shanks, Department of Dairy Science, University of Illinois, Urbana.

to another, fusion of protoplasts of different species, cellular selection, and variant production by animal regeneration from protoplasts or the fusion of protoplasts. The possibility of selecting for many desired traits at the cellular level holds considerable promise as a useful tool in the genetic improvement of animals.

The technique of inserting genes into other cells (recombinant DNA or genetic transformation technology) has enabled scientists to code genes for desirable compounds and insert them into microorganisms, thus facilitating large-scale fermentation production of the compounds.

4.8.1 Microbe Engineering

The results of recombinant DNA technology, one form of genetic engineering, will make many important contributions to humans and domesticated animals in areas such as aging, cancer, immunology, and vaccines.

A piece of DNA that codes for a particular chemical can now be inserted into the genetic material of bacteria or other microbes such as yeast, indeed even into animal cells grown in tissue cultures. These new or altered forms of life are then multiplied into billions of cells that produce the chemical for which they have been engineered. Already, several biological products are being produced in this way, including human and cattle growth hormone, insulin, and interferon. Research is under way to develop organisms that will produce proteins that can be used as vaccines against viral hepatitis, influenza, and foot-and-mouth disease.

The most commonly used "factory" for manufacturing these products is the bacterium *Escherichia coli,* which has been one of the organisms most studied by microbiologists. It is precisely this background of knowledge about the organism that enables scientists to remove a circle of DNA called a plasmid from *E. coli,* cut the circle by using specific enzymes, and splice pieces of DNA from another organism into the gap in the circle (hence the term gene-splicing). When the reconstructed plasmid is reinserted into the bacterium, it produces the protein product for which it was coded.

It is possible that strains of bacteria will be developed that can produce single-cell proteins from waste petroleum or other products. Perhaps cellulose-degrading genes could be introduced from certain bacteria into *E. coli,* which normally inhabits the digestive tract of animals. This could be helpful to some animals in more effectively utilizing cellulose, which is not digested by the enzymes normally present in single-stomached animals (see Chapter 15).

Another approach to genetic engineering is to chemically synthesize pieces of DNA from the genetic code for the desired amino acids and insert the synthesized genes into a plasmid. This approach was used, in part, for producing pure insulin.

Genetic engineering offers many advantages over the current methods of producing vaccines. One of the most important is safety. To produce a conventional vaccine, the disease organism must be grown in large numbers, then inactivated. Both procedures can be dangerous and/or difficult. For example, in countries where the incidence of foot-and-mouth disease is controlled by vaccinating cattle, recent outbreaks have been caused by the escape of the virus from the vaccine production laboratories or by the inoculation of animals with improperly inactivated vaccines. Strict guidelines for genetic engineering studies have been developed. These are designed to help ensure the continued safety of such research. In addition to safety, genetically engineered vaccines will likely have the advantage of being considerably less expensive than vaccines produced from whole virus.

4.8.2 Gene Transfer

Recombinant DNA technology is receiving widespread attention and research funding. The introduction of mammalian genes for various proteins, e.g., insulin, into bacteria has resulted in the bacteria manufacturing the protein as a by-product. This procedure avoids the costly and time-consuming one of purifying a protein from an animal tissue source. In addition, human insulin so produced is not recognized as a foreign substance by humans, as is the insulin of cattle or swine, and thus patients do not tend to develop immunity against it.

A possible use for genetic engineering in cattle has emerged from work in the laboratory of Dr. D. E. Bauman of Cornell University. His research team observed that injections of growth hormone into lactating cows resulted in an increase of about 15 percent in milk production. Suppose an additional gene, or genes, for growth hormone could be introduced into an embryo. The adult cows would then secrete more growth hormone and would not need hormone injections.

Mouse embryos successfully received a gene for rabbit hemoglobin, and they were able to manufacture the hemoglobin. Perhaps the most significant genetic consideration for gene transfer is that once a gene is introduced successfully, it is transmitted to succeeding generations. This new potential of science opens a wide range of possibilities for introducing superior production and conformation traits and disease resistance.

4.8.3 Designer Genes

Green plants function as solar energy machines: they convert sunlight and plant nutrients into food, fiber, oil, and biomass (energy) for an ever-expanding world population that currently exceeds 4 billion people. Crop plants are the heart of a vast food and agricultural system that represents one of humanity's most precious natural resources. Recent discoveries in gene-splicing raise the exciting possibility that virtually any gene from any source can be engineered into plants.

Traditionally, plant breeders have combined the desirable genes or traits from two different parents into one descendant through selective matings. However, breeding has been possible only between closely related plants. Corn, for example, will not cross with wheat. The technologies being developed through genetic engineering may ultimately make it possible for the plant breeder to take genes from any plant species and insert (splice) them into any other plant.

For example, several genes that each confer resistance to a different disease-causing fungus could be isolated from several plant species, each possessing natural resistance to one of the fungi. These several genes could then be attached together in a test tube, multiplied in the *E. coli,* and put into a crop plant such as corn. The corn would then possess multiple resistance to those several fungi.

Another example of gene designing involves the ability of crop plants such as soybeans, alfalfa, and clovers to enter into a symbiotic association with certain bacteria. These bacteria take nitrogen from the abundant supply in the air and "fix" it into a chemical form that plants can use in making plant protein. The need to apply expensive nitrogen fertilizer to crops that have a symbiotic association with bacteria is thereby greatly reduced and, in some cases, virtually eliminated.

Genetic engineering techniques may enable scientists to isolate the genes from soybeans that make possible this association. These genes may then be transferred into such crops as corn and wheat—crops that now consume large quantities of commercial nitrogen fertilizer. The engineered ability of corn and wheat to associate with nitrogen-fixing bacteria would greatly reduce the need to apply expensive nitrogen fertilizer.

Genes currently being isolated from plants include traits for disease resistance, herbicide tolerance, drought tolerance, increased efficiency of photosynthesis, increased efficiency of fertilizer use, and symbiotic nitrogen fixation. Special pieces of DNA called vectors are used to carry genes from one plant to another. At least three different vectors for plants are currently being studied. Once the genes are "loaded" in their vectors, they must then be delivered to suitable target cells. One approach is to use isolated plant cells stripped of their thick outer wall; these cells will develop into mature plants that harbor the new trait. Difficulty with the delivery step is an important bottleneck currently being researched.

Protoplast fusion offers promise in genetic engineering because it facilitates the mixing together of mitochondria and plastids from two parents. This mixing could allow the recombination of the genetic information carried in these organelles. The end result could be new genetic combinations with useful qualities.

Selection for traits through the use of cells or protoplasts has the advantage that very large populations can be screened for expression of the trait. Thus, geneticists should be able to select for many traits at the cellular level. Examples of desirable traits for which it should be possible to easily construct screening systems in plants are resistance, or tolerance, to herbicides, salt, heat, cold, water stress, heavy metals, and disease toxins. Even plant yield might be selected for if some correlation could be found with a trait such as vigor or another cellular process.

Designing genes is an exciting field of science that holds great promise for humanity's pursuit of ways and means to feed a hungry world. It is a great challenge to scientists, and a large amount of venture capital is being committed to the field. Ultimate success is based on rapid advances in isolating and cloning genes, in developing insertion vectors, and in implementing new strategies for cell culture.

4.8.4 Cloning of Embryos

Cloning is not restricted to plants but is also possible with animals. In a technical sense, identical twins are clones. They are derived from a single cell, but the embryo split early in development to yield what is essentially two carbon copies. Embryonic cloning has involved the use of chemicals, but more commonly microsurgery, to divide the embryo. The greatest success to date has been obtained by splitting morulae to yield two blastocysts. Most current efforts involve splitting two-cell or four-cell embryos. Success is markedly reduced by the eight-cell stage, presumably because the embryo has begun to differentiate at the eight-cell stage of development.

Theoretically, it would be useful to be able to split a four-cell embryo into four identical individuals. Each of those individuals could, in turn, be cultured to the four-cell stage, split again, and so forth. However, each time an embryo is split, the cells of the next generation are one-half the size of that of the starting cell. Apparently there is a critical ratio of nucleus to cytoplasm in cells, which requires additional study before this theory can ever be put into practice on a commercial scale.

Researchers at the University of Wisconsin have successfully produced mice by separating two-cell embryos into individual cells and then culturing the clones (blastomeres) into blastocysts before transferring them into recipient mice.

The genetic potential of cloning is indeed great. For example, if 10 identical daughters produced one lactation record in 10 dairy herds, the balance of the stored, frozen identical clones could be marketed with approximately 80 percent repeatability for milk production potential. Producers and distributors would both benefit from such knowledge. In the artificial insemination industry, if twin, triplet, or quadruplet bulls could be produced by cloning, the carbon copies of the clones could be saved in frozen storage. For the outstanding bulls remaining following the computation of their progeny proofs, their identical genotype would still be in storage for use by the artificial insemination (AI) organization. An inventory of embryos would require considerably less space than identical bulls being fed and housed in another facility. It is very unlikely that scientists will soon be able to use mammalian cells from the skin, or some other organ, to use in generating identical individuals, although embryonic cloning of certain mammals does hold promise.

4.8.5 Nuclear Transfer

Nuclear transfer currently holds more promise than cloning for obtaining identical offspring. The method involves the use of microsurgery to collect cells from the inner-cell mass of the trophoblast. Approximately 200 cells can be obtained. Each nucleus is then transferred to the inside of an unfertilized egg that has had its own nucleus or chromosomes removed. All the cells that develop successfully become identical individuals.

4.8.6 Nuclear Fusion

Nuclear fusion is the union of nuclei from two gametes. Since gametes are either eggs or sperm, the following combinations are possible: (1) uniting nuclei from two eggs, resulting in all the offspring being daughters; (2) uniting nuclei from two sperm, resulting in one-half of the offspring being daughters, one-fourth being sons, and one-fourth bearing YY chromosomes, which are lethal, and so, of the surviving embryos, twice as many daughters as sons would result from fusing nuclei from two sperm; and (3) uniting nuclei from an egg and a sperm as occurs normally at fertilization. The potential possibilities of this procedure are significant. Two outstanding bulls (or cows) could be mated to each other by uniting gamete nuclei into an unfertilized egg and chemically activating the egg to begin division. This procedure has been performed successfully in laboratory species, but not in cattle.

4.9 SEGREGATION AND RECOMBINATION OF GENES

Genes occur in pairs in the body (**somatic**) cells. This statement is true except for **sex-linked** traits, the genes carried on the portion of the X chromosome that is not homologous to the Y chromosome. Another relatively rare exception might occur in cases where the body cells of an individual possess one or three chromosomes rather than the usual homologous pair. Of each pair of genes an individual possesses, one gene comes from the father and one from the mother. The individual will transmit one gene or the other, but not both, to any one offspring.

Mendel's law of **segregation and recombination of genes** states that when two genes are paired in body cells, they segregate independently of each other in the gametes. This may be illustrated by the following example:

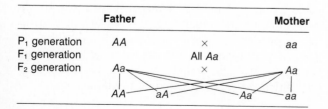

	Father		Mother
P₁ generation	AA	×	aa
F₁ generation		All Aa	
F₂ generation	Aa	×	Aa
	AA aA		Aa aa

This example shows that in the **F₁** (first filial) **generation** the genes *A* and *a* were paired. They are called **allelomorphs** because they are carried at the same locus (location) on the same pair of homologous chromosomes and affect the same trait but in a different way. When they produced sex cells, the F₁ individuals put gene A into their gametes about 50 percent of the time and gene a about 50 percent. Thus, even though they were paired in the F_1 individuals, they did not stay together in the gametes but rather segregated independently. They also recombined in pairs at random, giving a $1AA:2Aa:1aa$ ratio in the F_2 (second filial) **generation.**

Linkage of two or more genes prevents segregation of the different pairs of genes in gametes. *Linkage* means that *two or more pairs of genes are carried on the same homologous chromosomes,* as shown below:

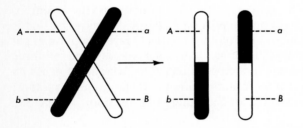

Linkage means that genes A and B would segregate together in the gametes, as would genes a and b. An exception to this, however, occurs when homologous chromosomes come together (synapse) in the process of meiosis. Sometimes during synapsis the homologous chromosomes exchange parts, so that A and b, as well as a and B, may get together, as shown in the following diagram. This is known as *crossing over* and occurs more frequently in genes carried farther apart on homologous chromosomes than in those carried closer together.

4.10 LAWS OF PROBABILITY AND ANIMAL BREEDING

The laws of probability apply to the segregation and recombination of genes. For example, an individual of **genotype** Aa will transmit gene A to approximately one-half of its offspring. The same is true for gene a. This leads to the assumption that the probability that an individual of genotype Aa will transmit either gene A or gene a to a particular offpsring is one-half, or 0.5. The probability that an individual of genotype AA will transmit gene A to any particular offspring is 1, whereas the probability that it will transmit gene a to its offspring is zero. These probabilities are valid provided no **mutations** of genes A and a occur.

The laws of probability also apply to the segregation of two or more pairs of genes in the gametes of an individual, provided each pair of genes is carried on a different pair of homologous chromosomes. The law involved states that the probability of two or more independent events occurring together is the product of each separate probability. To illustrate, what is the probability that an individual of genotype *AaBb* will transmit genes *AB* to a single offspring? If the two pairs of genes *A* or *a* and *B* or *b* are carried on different pairs of homologous chromosomes, they segregate independently of each other according to the law of probability. The segregation of each pair would be a single independent event. Thus the probability that individual *AaBb* will transmit gene *A* in a gamete is 0.5, and the probability that it will transmit gene *B* in a gamete is also 0.5. The probability that both genes *A* and *B* will be transmitted through the same gamete is 0.5 × 0.5, or 0.25. Very obviously, an individual of genotype *AABB* could not transmit combinations of genes such as *aB, Ab,* or *ab* through the same gamete unless a mutation occurred, and so the probability of this transmission is zero. The probability of the segregation of a certain combination of genes from an individual whose genotype possesses many different pairs of alleles can be calculated in the same way as long as each pair of allelic genes is on a different pair of homologous chromosomes. For example, the probability that an individual of genotype *AaBbCcDd* will transmit genes *ABCD* in the same gamete is 0.5 × 0.5 × 0.5 × 0.5, or 0.0625.

Laws of probability can also be applied to the recombination of genes in the **zygote.** For example, if two individuals that are of genotypes *Aa* are mated, the probability of offspring of the four possible genotypes would be

$$
\begin{aligned}
AA &= 0.5 \times 0.5, \text{ or } 0.25 \\
Aa &= 0.5 \times 0.5, \text{ or } 0.25 \\
aA &= 0.5 \times 0.5, \text{ or } 0.25 \\
aa &= 0.5 \times 0.5, \text{ or } 0.25
\end{aligned}
\left.\begin{aligned} & \\ & \end{aligned}\right\} \; 0.5
$$

This recombination gives the familiar 1:2:1 ratio between the cross of two individuals that are heterozygous for one pair of **alleles.**

The laws of probability can also be used to calculate the possibility of a recombination of two or more genes in the zygote. For example, the probability that the mating of two individuals of genotypes *AaBb* will produce an offspring of genotype *AABB* is 0.5 × 0.5, or 0.25, for genes *AA*; and 0.5 × 0.5, or 0.25, for genes *BB;* or 0.25 × 0.25, or 0.0625, for genes *AABB.* The probability of certain other combinations of genes in the offspring of such parents can be calculated in a similar way.

The laws of probability can be expanded to cover many other aspects of gene combinations in gametes and zygotes. For example, the probability that parents, both of whom are of genotype *Aa,* will produce a homozygous recessive *aa* offspring is 1 out of 4, or 0.25 (0.5 × 0.5). The probability of four offspring, all of which are of genotype *aa,* from such parents, would be 0.0039 (0.25 × 0.25 × 0.25 × 0.25).

4.11 MUTATIONS

Genes possess the remarkable property of exactly duplicating themselves generation after generation. Occasionally, however, mistakes are made in the gene-duplicating process and a new gene (allele) is born. Basically, most if not all mutations result in a change in the code sent to ribosomes by the gene by means of mRNA to form a particular protein. If the wrong code is sent, a different protein is formed. The missing protein or the new protein may cause a defect or a new genetic trait to appear. Differences that can be seen or measured between individuals are due to an accumulation of different mutations (old or new) within populations. These mutations are responsible for differences in coat color, size, shape, behavior, and other traits in various species. These differences among individuals are the raw material with which the animal breeder has to work.

4.12   PHENOTYPIC EXPRESSION OF GENES (NONADDITIVE)

The genotype of an individual refers to *its actual genetic makeup*. For example, individuals that are *AA, Aa,* or *aa* represent three different genotypes. *Phenotype* refers to *those differences in individuals that can be measured by means of the senses*, such as black or white, tall or short, or horns or no horns (polled).

Genes almost always segregate in gametes and recombine in zygotes in the same general way as long as they are carried on different chromosomes. However, even though these genes segregate and recombine in the same way, they can express themselves in the phenotype in many different ways. This is what is meant by different kinds of phenotypic expression of genes.

4.12.1   Dominance and Recessiveness

A gene is said to be **dominant** when it covers or hides the expression of its allele. For example, in Angus cattle the gene for black *B* is dominant to the gene for red *b,* its own allele, because an individual of genotype *Bb* is black. With two alleles such as *B* and *b,* there can be three genotypes and two phenotypes, as follows:

Three genotypes	Two phenotypes
BB	Black
Bb	Black
bb	Red

Three genotypes exist because the two genes have combined in pairs in three different ways. Two phenotypes (black and red) exist because both the *BB* and the *Bb* individuals are black and the *bb* individuals are red. The two black genotypes are quite different in their breeding ability, however. The homozygous black individuals *BB* will always produce black offspring, even if mated to red *bb* individuals. This is illustrated as follows:

Kinds of matings	Offspring	
	Genotype	Phenotype
$BB \times BB$	BB	Black
$BB \times Bb$	$1BB{:}1Bb$	Black
$BB \times bb$	Bb	Black

The heterozygous black Bb individuals do not always breed true, however. This is illustrated as follows:

Kinds of matings	Offspring	
	Genotype	Phenotype
$Bb \times BB$	$1BB{:}1Bb$	Black
$Bb \times Bb$	$1BB{:}2Bb{:}1bb$	3 black to 1 red
$Bb \times bb$	$1Bb{:}1bb$	1 black to 1 red

If red bb individuals are mated together, they should always produce red offspring. However, they do not breed true when mated to black individuals, as illustrated below.

Kinds of matings	Offspring	
	Genotype	Phenotype
$bb \times BB$	Bb	Black
$bb \times Bb$	$1bb{:}1Bb$	1 red to 1 black
$bb \times bb$	bb	Red

From the foregoing discussion, one can see that it is relatively easy to develop a purebreeding strain that is **recessive,** or red bb, in Angus. One merely has to mate recessives bb to recessives bb to get recessives bb. Developing a pure dominant or black strain is more difficult, however, because one cannot tell by observation whether individuals are homozygous dominant BB or heterozygous Bb. Special progeny tests must be conducted to distinguish between the two dominant genotypes. One test is to mate the black individual of the dominant phenotype (either BB or Bb; one would not know which) with homozygous recessive bb individuals. One red or homozygous recessive bb offspring from such a mating proves the black individual heterozygous Bb. If five matings are made to red bb individuals without a single red offspring being produced, the black individual is **homozygous** black BB, and this would be correct at the 95 percent level of probability.

Sometimes homozygous recessive *bb* individuals are not available for mating. In such a case, known **heterozygotes** *Bb* can be used for progeny tests. They are known heterozygotes *Bb* if they produce one homozygous recessive *bb* offspring or if one of their parents is homozygous recessive *bb*. One homozygous recessive offspring *bb* from such a mating proves the individual heterozygous *Bb*. Eleven such matings without a homozygous recessive offspring *bb* being produced proves the dominant, or black, individual homozygous *BB* at the 95 percent level of probability, and 17 such matings are proof at the 99 percent level.

The same principles of progeny testing used in these examples apply in all cases in which one pair of alleles, dominant and recessive, is concerned.

4.12.2   Lack of Dominance

Sometimes neither of two alleles is dominant to the other, and if both are present, each expresses itself in the phenotype. Coat color in Shorthorn cattle is a typical example of such inheritance. A Shorthorn is red if it has two genes for red (*RR*) and white if it has two genes for white (*WW*). A Shorthorn is roan, or a mixture of both red and white, if it has one gene for red and one for white (*RW*). In this case there are three genotypes and three phenotypes, and there is little or no difficulty in distinguishing between them.

4.12.3   Partial Dominance

Some genes are only partially, and not completely, dominant to their alleles. An example of this type of phenotypic expression of genes is the comprest gene in Hereford cattle. Two comprest genes (*CC*) cause the individual to be a dwarf. Two normal genes (*cc*) cause the individual to be of normal size or phenotype. One comprest gene and one normal gene (*Cc*) cause the individual to be a comprest, ranking midway between the normal and the dwarf in size or phenotype.

4.12.4   Overdominance

Overdominance refers to *a phenotypic expression of genes in which the heterozygote (a^1a^2) is superior in phenotype to either homozygote (a^1a^1 or a^2a^2).* Many examples are now known to illustrate overdominance. One of the better-known examples is sickle-cell anemia, found in many African blacks. In certain regions of Africa, the disease malaria is prevalent. This disease kills many individuals homozygous for normal adult **hemoglobin** (genotype *AA*). Those individuals homozygous for the sickle-cell gene *SS* die of **anemia.** No *AS* individuals die of anemia, and few of them die of malaria. The genotype *AS* resists, in some way, damage to the red blood cell by the malaria organism. Thus the heterozygote *AS* individual is superior in survival to either *AA* or *SS* individuals in this

environment. In the United States, where there is little or no malaria, the AA individuals would survive as well as the AS individuals, and there would be no heterozygote advantage.

Although heterozygotes are superior when overdominance is involved, they never breed true when mated together. They can always be produced by mating the two homozygotes together (a^1a^1 and a^2a^2), provided both of these genotypes are available for mating. Overdominance probably has its greatest effect on traits related to physical fitness, the heterozygote being superior to either homozygote.

4.12.5 Epistasis

Epistasis may be defined as a type of phenotypic expression of genes due to the interaction of two or more pairs of genes that are not alleles. Epistasis may be distinguished from overdominance in that the latter is due to an interaction between genes that are alleles. An example of epistasis is as follows:

a^t is the gene for tricolor in collie dogs. It is recessive to A^t, the sable and white gene.

M is the gene for merle in collies, giving a dilution effect; m is the gene for nonmerle.

A collie that has the genetic makeup a^ta^tmm is tricolored, or black, tan, and white. A collie with the genetic makeup a^ta^tMm is a blue merle, whereas one with the genetic makeup a^ta^tMM is a white merle. Thus the gene M influences the way the tricolor gene a^t expresses itself phenotypically even though the M and a^t genes are not alleles. Epistasis may affect the phenotype in different ways through various interactions between nonallelic genes.

Dominance and recessiveness, overdominance and epistasis are all types of nonadditive phenotypic expression of genes. A second general type of phenotypic expression of genes is known as *additive*.

4.13 PHENOTYPIC EXPRESSION OF GENES (ADDITIVE)

An additive type of gene action means that the effect of each gene that contributes something to the phenotype of an individual for a certain trait adds to the phenotypic effect of another gene that contributes something to the same phenotype. An example of additive gene action is skin-color inheritance in humans. It has been proposed that two different pairs of genes affect skin color in whites and blacks, although this may be an oversimplification because more than two pairs of genes may be involved. The proposed type of inheritance suggests that white persons are of genetic makeup *aabb,* and that when either an A or a B gene is present, each causes a darker skin color. Thus the possible genotypes and phenotypes would be:

Genotype	Phenotype
aabb	White
Aabb or *aaBb*	Light
AAbb, aaBB, or *AaBb*	Medium
AABb or *AaBB*	Dark
AABB	Black

Each contributing gene *A* or *B* makes the skin color darker, whereas neutral genes, *a* and *b,* do not affect skin color. Here it is assumed that genes *A* and *B* always contribute the same amount to skin color, but this may not be absolutely true. If they did differ slightly, it would not affect the validity of the example to any great extent.

Both additive and nonadditive types of gene action affect many economic traits in animals. These traits include rate of gain, **feed efficiency,** lactation, and egg laying as well as many other traits. Many pairs of genes may affect such traits. It is possible that some traits may be almost entirely affected by nonadditive genes whereas other traits may be affected almost entirely by an additive type of gene action. Still other traits may be affected by both additive and nonadditive gene action. How to determine what type or types of gene action affect various traits is discussed in Chapter 5.

4.14   SEX-LINKED INHERITANCE

Sex-linked inheritance refers to inheritance due to genes carried on the nonhomologous portion of X chromosomes. Sex-linked genes are usually recessive in their phenotypic expression.

It was pointed out earlier that each species of animal possesses a characteristic number of chromosomes and that the different species have different numbers of chromosomes (Table 4.1). For example, humans possess 23 pairs whereas cattle possess 30 pairs. In farm animals and humans, there is one set of sex chromosomes known as the X and Y chromosomes (except in poultry, which have a chromosome similar to the Y, called the w chromosome). In humans and farm animals, except poultry, the female possesses two X chromosomes whereas the male possesses one X and one Y. Poultry differ in this respect since the male is XX and the female is Xw.

The X chromosome in humans and farm animals is much larger and longer than the Y chromosome. This is illustrated in Figure 4.16. In the male, there is a portion of the X chromosome that does not pair with the Y, or, genetically speaking, that is not homologous (of like proportion) to the Y. Genes carried on this portion of the X chromosome are said to be *sex-linked.* Since only one X chromosome is present in the male, a gene on the nonhomologous portion of the X chromosome will express itself in the phenotype even if it is a recessive gene.

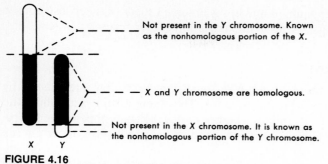

Not present in the Y chromosome. Known as the nonhomologous portion of the X.

X and Y chromosome are homologous.

Not present in the X chromosome. It is known as the nonhomologous portion of the Y chromosome.

X   Y

FIGURE 4.16
Diagram of the pairs of sex chromosomes in the male, which has one X and one Y chromosome. The female of such an animal has two X chromosomes.

In the female, where two X chromosomes are present, two recessive genes are required to cause a recessive trait to appear.

As is shown in Figure 4.16, a portion of the Y chromosome probably has no counterpart, or homologous portion, in the X. Genes carried there are transmitted only from fathers to sons. This type of inheritance is known as *holandric inheritance.*

Many sex-linked recessive traits, such as **hemophilia** and red-green color blindness, are known in humans, but very few are known in farm animals, possibly because studies have not been made in farm animals to determine the inheritance of such traits.

4.15 SEX-INFLUENCED INHERITANCE

Certain genes carried on the autosomes are present in body cells in pairs, but the sex of an individual determines whether the trait is dominant (in males) or recessive (in females). The gene for **scurs** (S_c) in European cattle appears to be inherited in this way. The inheritance of scurs in polled European breeds of cattle seems to be as follows:

	Phenotype in	
Genotype	Male	Female
S_cS_c	Scurred	Scurred
s_cS_n	Scurred	Not scurred
S_nS_n	Not scurred	Not scurred

Hereditary baldness in humans and horns in some breeds of sheep also appear to be sex-influenced traits, being dominant in males and recessive in females.

4.16   SEX-LIMITED TRAITS

Certain traits, such as lactation in cattle and egg laying in poultry, are affected by many genes but are expressed in only one sex. Thus there is something about the sex of the individual that determines whether or not the traits appear phenotypically. The appearance of these traits is related to the production of certain sex hormones.

Even though certain traits are limited to one sex, both sexes possess and transmit genes for such traits to their offspring. This means that both parents should possess superior genes for such traits if their offspring are to be superior. Since dairy bulls do not lactate and **roosters** do not lay eggs, it becomes a more complicated problem in males than in females to determine the kind of genes they possess for these traits. Estimates of their genetic merit must be based on the performance of female relatives such as their dam, sisters, or daughters (Chapter 5).

Single gene effects, such as cock-feathering or hen-feathering in chickens, are also limited in their phenotypic expression to only one sex. Such traits also appear to be related to interactions of genes and sex hormones.

4.17   SUMMARY

Cells are the building blocks of which the body is made, and each individual possesses trillions of them. The main parts of the cell include the nucleus and the cytoplasm. The nucleus is the heart and brain of the cell and contains the chromosomes. Each species of animal possesses the characteristic number of chromosomes for that species. Chromosomal abnormalities sometimes occur and usually result in an abnormal phenotype or death of the individual. Chromosomal abnormalities include variations from normal in structure and numbers.

Genes, the determiners of heredity, are carried on chromosomes. The gene is a portion of a DNA molecule and has the ability to replicate itself when new cells are formed. It also sends the code for protein structure and synthesis to ribosomes in the cytoplasm by means of messenger RNA. The genetic code and protein synthesis are complex, and functions of the gene in protein synthesis are a recent discovery.

Genes may express themselves in the phenotype in two general ways, known as additive and nonadditive phenotypic expressions. Genes also segregate into the gametes (sperm and ovum) and recombine in the zygote (the new individual resulting from the union of a sperm and an ovum) according to certain laws of probability.

A budding new field of science is genetic engineering. Already research findings have enabled scientists to code genes for desirable compounds and then to insert these genes into other cells (genetic transformation), such as microorganisms, thus facilitating large-scale fermentation production of the compounds. For example, plans are currently being implemented to produce interferon and insulin in microorganisms commercially.

Functional genes have been transferred successfully into mice through the one-cell zygote. These genes persist as part of the mature animal's genome and are transferred to future generations. The application of genetic transfer procedures to livestock species offers exciting possibilities for rapid improvements in farm animals.

STUDY QUESTIONS

1 Wild squirrels, which are usually brownish-gray in color, sometimes produce offspring that are albinos (white). When these white squirrels are mated, they always produce white offspring. What type of inheritance is involved in the production of albino offspring? Why do the albino squirrels breed true? Why are most wild squirrels of a brownish-gray color and not white? How did the albino characteristic originate?

2 The black gene B in Angus cattle is dominant to the red gene b. Breeders of black Angus cattle have never kept red animals for breeding. Why do red calves still appear in some matings of black Angus parents? How would you start a new breed of red Angus cattle?

3 How would you design a program to eliminate the red gene b from the black Angus breed?

4 The comprest gene C in Hereford cattle is only partially dominant to the normal gene c. Therefore normal individuals for this gene are cc, comprests are Cc, and comprest dwarfs are CC. How would you start a pure strain of comprest Herefords?

5 White coat color in some breeds of swine is dominant to black coat color, and the mule-footed condition is dominant to the normal split-hooved condition. The offspring of a certain white, mule-footed boar are always white and mule-footed, regardless of the phenotype of the sows to which he is mated. What is the probable genotype of the boar if white is W, black is w, the mule-footed condition is M, and the normal foot condition is m?

6 Another boar is white and mule-footed, but when mated to black sows, about one-half of the offspring are black, and when mated to normal-footed sows, about one-half of the offspring have normal feet. What is a possible genotype of this white, mule-footed boar?

7 Solid color in Angus cattle is dominant (S) to white spotting (s), found in Holsteins. What would be the expected color of a calf from an Angus bull mated to a Holstein cow?

8 Some breeds of chickens are white because of the presence of a dominant epistatic pigment inhibitor (I), whereas other breeds are white because of the presence of the recessive allele for no color (c) in the homozygous condition. The two pairs of genes are on separate pairs of homologous chromosomes. If a white male of genotype IICC is mated to white females that are iicc, what would be the phenotype of the F_1 offspring? If the F_1 individuals are mated among themselves, what would be the phenotypes of the F_2 offspring?

9 A certain kind of hemophilia in humans is a sex-linked recessive trait. A man who had hemophilia married a woman who did not have the disease but whose father did.

What kinds of sons, genetically, could they have for this trait? What kinds of daughters could be expected?

10 How many sons of the hemophiliac father mentioned in Question 9 would have hemophilia if their mother had not been a carrier of the recessive gene for hemophilia? How many of his daughters would have hemophilia in such a case? How many would be carriers of the gene for this disease?

11 Assume that you are a genetic consultant. A man comes to you with this problem: He works for an atomic energy plant, where he is exposed to radiation. Before he started working at the plant, he had three sons and two daughters who were normal in every way. Two years after he started working for this company, he had a fourth son who had hemophilia (sex-linked). He wants to sue the company for damages, insisting that it was a mutation from radiation that caused him to produce the son with hemophilia. What advice would you give him?

12 The A, B, and O blood groups in humans can be used at times to determine parentage cases in lawsuits. Both genes *A* and *B* are dominant to a gene for O, but they are not dominant to each other. Which of the following combinations would be possible?

Blood groups of		
Child	**Father**	**Mother**
(1) Group O	Group A	Group B
(2) Group O	Group AB	Group B
(3) Group A	Group B	Group A
(4) Group B	Group AB	Group AB
(5) Group AB	Group AB	Group O
(6) Group A	Group O	Group A

13 Assume, again, that you are a genetic consultant. A girl comes to you with a problem and wants some advice. She is Rh negative. She has three boyfriends of whom she is about equally fond. One is Rh negative; a second is Rh positive and both his parents were Rh positive; and a third is Rh positive, but his mother was Rh negative (Chapter 6). From a genetic standpoint, which boy would you recommend she choose for a husband? Why?

14 One hundred Indians in New Mexico were blood-tested, and it was found that 59 were of blood group M, 33 were MN, and 8 were NN. What was the frequency of the *M* and the *N* gene in this group of Indians? (See Section 5.3 before answering Questions 14, 15, and 16.)

15 Of 1200 American Indians that were blood-typed, assume 16 percent were of blood type N. What was the frequency of the *N* and the *M* gene in this population? How many should be blood group M and blood group MN? (M is not dominant to N, and vice versa.)

16 Suppose you bought 20 Angus cows that were homozygous black *BB* and a bull that was a carrier of the red gene *Bb*. If all cows produced a calf during the first calving, what would be the probable frequency of the red gene in the entire herd of 20 cows, one bull, and 20 calves?

17 A couple has just married. They plan to have four children and would like to have four boys. What is the probability their desire will be fulfilled if they have four children?

18 Joe Jones has a family of nine boys and no girls. *(a)* What is the probability that in families of nine all will be one sex? *(b)* What is the probability that the tenth child in this family will be a girl?

19 Henry Smith's father was not bald, but his mother was. What is the probability that he will be bald when he reaches 40 years of age?

20 A brown-eyed girl of blood type *MN* marries a boy who is brown-eyed and is of blood type *N*. The boy is not color-blind, but the father of the girl was. Their first child is a girl of blood type *N* with blue eyes and normal vision. What is the probability that their second child will be a boy of blood type *MN* who is color-blind and who has brown eyes? (The red-green color-blind gene is a sex-linked recessive.)

21 Andalusian chickens are of three types, black, splashed white, and blue. The black and splashed white breed true, but when mated together they always produce blue. When the blue Andalusians are mated together, they do not breed true but produce 1 black, 2 blue, and 1 splashed white. Explain why blue individuals do not breed true. How could blue Andalusians always be produced?

22 Several different kinds of combs in chickens are inherited. A single-combed individual is of genotype *rrpp,* a rose-combed is *R-pp,* a pea-combed is *rrP-,* and the combination of *R* and *P* in the genotype produces a walnut comb. What are the F_1 genotypes and phenotypes if a single-combed rooster is mated to rose-combed hens of genotype *RRpp*? What would be the phenotypic ratio in the F_2?

23 If rose-combed and pea-combed individuals, when mated, always produced walnut-combed chicks, what would be the genotype of the parents? What would be the genotypic and phenotypic ratios of the F_2 offspring, starting with the rose-combed and pea-combed parents in this question?

24 What kind of phenotypic expression of genes is necessary for the production of walnut combs in chickens? Explain. Would it be possible to develop a purebreeding strain of walnut-combed chickens? Explain.

25 In poultry, barring *B* is a sex-linked dominant trait and black *b* is its recessive allele. If barred *BB* males are crossed with black females, what would be the genotypic and phenotypic ratios of the offspring? If heterozygous barred *Bb* males are mated with black *bw* females, what would be the genotypes and phenotypes of the offspring? Remember that the female in chickens is Xw in chromosome composition.

26 In Question 25, if barred females were mated with black males, what would be the genotypes and phenotypes of the offspring? What would be the sex of the barred chicks? What would be the sex of the black chicks?

ANSWERS TO STUDY QUESTIONS

1 The albino gene *a* is a recessive gene. For an albino to be produced, it has to have genotype *aa*. Albino parents *aa* should always produce albino offspring *aa*. Most squirrels are brownish-gray because the gene for this color is dominant *A* to the albino gene *a*. Only a few albino genes *a* are present in the population. The albino gene probably arose by a mutation of gene *A* to *a*.

2 Because many black Angus still carry the red gene *b*, when black parents of genotype *Bb* are mated, one-fourth of their offspring should be red, or *bb*. A new breed of red Angus could be developed by mating red *bb* to red *bb*.

3 Discard all red Angus *bb* when they appear and do not use black parents of a red Angus calf for breeding because they are carriers of the red gene (genotype *Bb*).

Progeny-test bulls to determine if they are *BB* or *Bb*. This can be done by mating the bull in question to red *bb* cows. One red offspring proves the bull *Bb*. Five black and no red offspring out of red cows prove him *BB* at the 95 percent level of probability. Seven black and no red offspring from seven red cows prove him *BB* at the 99 percent level. Other progeny tests are to mate him to known carriers (*Bb*) of the red gene, or to at least 23 of his own unselected daughters. Actually to eliminate the red gene, both cows and bulls must be tested and proved *BB*. This would be difficult to do in cows.

4 It could not be done because comprests (*Cc*) do not breed true.

5 The genotype of the boar is probably *WWMM*.

6 The genotype of this boar is probably *WwMm*.

7 Solid black of genotype *Ss*.

8 The F_1 offspring would all be white. The F_2 offspring would be 13 white to 3 colored.

9 The woman is normal but a carrier of the gene for hemophilia. One-half of the sons would be normal and one-half would have hemophilia. Also, one-half of the daughters would be normal and one-half would have hemophilia. However, the normal daughters would be carriers of the recessive gene for hemophilia.

10 None. None of the daughters would have hemophilia, but all would be carriers of the gene for hemophilia.

11 The radiation was not at fault. The father transmits his X-chromosome only to his daughters. The sons always receive the X chromosome from their mother.

12 (1) is possible if the parents are *Aa* and *Ba*. (2) is not possible because the child is O, or genotype *aa*, and has to receive the *a* gene from both parents. The *AB* father does not possess the *a* gene. (3) is possible, (4) is possible, (5) is not possible, and (6) is possible.

13 If she chose the Rh negative/Rh negative boy, she would have no trouble with erythroblastosis fetalis in her children. However, all her daughters would be Rh negative/Rh negative, and she would just delay the problem for another generation. The Rh positive/Rh positive boy with Rh positive parents would be the greatest risk, genetically.

14 The frequency of the *M* gene is 0.755; that of the *N* gene is 0.245.

15 The frequency of the *M* gene is 0.6, and that of the *N* gene is 0.4. 432 should be of blood group M and 576 of blood group MN.

16 Frequency of the red gene *b* would be 0.134.

17 1 out of 16.

18 (*a*) 1 out of 512 families of 9. (*b*) 1 out of 2.

19 Probability is 1. His mother is of genotype *BaBa*, and he would receive a *Ba* gene from her. Since baldness is dominant in the male and recessive in the female, one *Ba* gene would be enough to cause him to be bald.

20 3 out of 32.

21 The blue individuals would be heterozygous. They could always be produced by mating black and splashed-white individuals.

22 The F_1 genotypes would all be *Rrpp*, and the phenotypes would all be rose-combed. The F_2 genotypes would be 1*RRpp*, 2*Rrpp*, 1*rrpp*. The phenotypes would be 3 rose-combed to 1 single-combed.

23 The rose-combed individuals would be *RRpp* and the pea-combed *rrPP*. The F_2 genotypes would be 1*RRPP*, 2*RRPp*, 1*RRpp*, 2*RrPP*, 4*RrPp*, 2*Rrpp*, 1*rrPP*, 2*rrPp*, 1*rrpp*. The phenotypic ratio in the F_2 would be 9 walnut-combed, 3 rose-combed, 3 pea-combed, and 1 single-combed.

24 Epistasis, because it is the interaction between nonallelic genes. Yes, it is possible to develop a purebreeding strain of walnut-combed chickens by finding those that are *RRPP* and mating them. The difficulty of developing such a strain would be very great because one would have to progeny-test to find those that were *RRPP*.

25 Both males and females would be barred, with the males *Bb* and the females *Bw*. Heterozygous barred *Bb* males mated to black *bw* females would give 1 *Bb* male that was barred, 1 *bb* male that was black, 1 *Bw* female that was barred, and 1 *bw* female that was black.

26 Genotypes = 1 *Bb* to 1 *bw;* phenotypes = 1 black to 1 barred. The barred chicks would be males, and the black chicks females.

PRINCIPLES OF SELECTING AND MATING FARM ANIMALS

Though reason is progressive, instinct is stationary. Five thousand years have added no improvement to the hive of the bee or the house of the beaver.

Colton

Many pairs of **genes** affect the expression of economic traits in farm animals. Economic traits include **fertility,** rate and efficiency of gains, milk and egg production, and **carcass** quality. These traits determine profit or loss in commercial livestock production. Since many genes affect these economic traits, there is usually no sharp distinction between **phenotypes** but rather a continuous variation from one extreme to another. This type of inheritance is referred to as **quantitative inheritance.** Conversely, **qualitative inheritance** refers to such traits as coat color or horns (or **polledness**), in which only one or a relatively few pairs of genes are involved and there is a sharp and distinct difference between phenotypes. Different selecting and mating schemes are required for these two general types of inheritance.

5.1 PHENOTYPIC VARIATIONS IN QUANTITATIVE TRAITS

In a large population of animals, a quantitative trait, such as rate of gain in the feedlot, varies from individuals that make slow gains to others that gain rapidly. The rate of gain for a majority of individuals, however, will tend to cluster around an average figure somewhere between the two extremes. If data for individuals making different rates of gain are graphically plotted, there will be a tendency for the rates of the entire group to fall into a curve resembling the shape of a bell, as shown in Figure 5.1. This phenotypic variation and the shape of the resulting curve are expected within a large group of individuals. Variation is the raw material with which animal breeders must work to improve their **livestock.** Without this variation there would be no hope of improvement through the application of breeding methods.

Qualitative inheritance involves genetic principles that deal specifically with the individual. This is because there are very distinct differences between phenotypes and because few genes are involved. *Quantitative* inheritance, conversely, involves the individual less and the total population more. Thus principles of genetics of populations are important for improving quantitative traits such as those of economic importance in livestock.

5.2 STATISTICAL EVALUATION OF QUANTITATIVE TRAITS

Statistical methods for describing the phenotypic variation of quantitative traits have been devised. The base, or starting point, is the average (mean) of the

FIGURE 5.1
Frequency distribution curve of the average daily gain of 176 beef bulls **full-fed** for 140 days. Rate of gain is a quantitative trait in beef cattle in which there is a continuous variation in the phenotype from one extreme to another, with no sharp distinction between phenotypes.

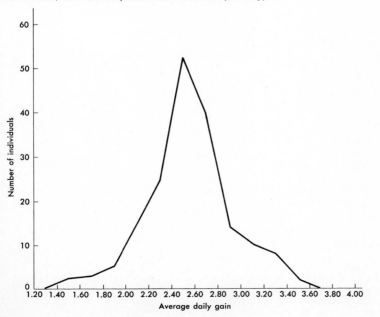

entire population being studied. This is determined by adding the phenotypic value of a trait for all the individuals in a population and dividing this sum by the total number of values in the population. If the letter X represents the phenotypic value for each individual, the Greek symbol Σ the sum of all values when added, and n the number of values in the population, the mean of that population may be described by the formula

$$\text{Mean} = \frac{\Sigma X}{n}$$

Within a population, quantitative traits vary widely between two extremes, a low value and a high value. This is called the *range*. A more useful statistical description of variation is the *variance*, which is an average of the squared deviations from the mean. The variance may be determined, as shown in Table 5.1, by calculating the phenotypic mean of the population and then subtracting each phenotypic value from the mean. Each deviation from the mean is then squared to remove the negative signs and to give more weight to the extreme values. The sum of the total squared deviations is then divided by $n - 1$, the value n representing the total number of phenotypic values in the population. The answer thus derived is the variance. The value $n - 1$ (degrees of freedom) is used because in small populations the use of n alone would tend to make the variance too large from a statistical standpoint.

TABLE 5.1
CALCULATION OF THE VARIANCE AND THE STANDARD DEVIATION FOR AVERAGE DAILY GAINS OF 10 BULLS ON A PERFORMANCE TEST

No. of bull	Av daily gain (ADG)	Mean for all bulls	Deviation from mean	Square deviation from mean
1	3.20	2.61	0.59	0.348
2	2.70	2.61	0.09	0.008
3	1.80	2.61	−0.81	0.656
4	2.40	2.61	−0.21	0.044
5	2.60	2.61	−0.01	0.000
6	2.90	2.61	0.29	0.084
7	3.00	2.61	0.39	0.152
8	2.20	2.61	−0.41	0.168
9	2.50	2.61	−0.11	0.012
10	2.80	2.61	0.19	0.036
Total (sum)	26.10		0.00	1.508

Mean for all bulls $= \dfrac{26.10}{10} = 261$

Variance $= \dfrac{1.508}{n-1} = \dfrac{1.508}{9} = 0.1675$

Standard deviation $= \sqrt{\text{variance}} = \sqrt{0.1675} = 0.4092$

The variance may be determined by a short-cut method, using a calculator and the formula

$$\text{Variance} = \frac{\Sigma X^2 - (\Sigma X)^2/n}{n - 1}$$

where X is each phenotypic value in the population, Σ the Greek symbol that means to sum all X or X^2 items, and n the total number of items. By using special and appropriate statistical techniques, the variance can be separated into several portions, giving an estimate of the proportion of various factors causing phenotypic variation in a trait for that particular population.

Sometimes the phenotype of a population for a quantitative trait is described as the mean plus or minus ($\pm$) the *standard deviation* (abbreviated SD). The standard deviation is the square root of the variance. The mean $\pm$ 1 standard deviation should include approximately 68 percent of the phenotypic values for the trait in question in the described population. The mean $\pm$ 2 or 3 standard deviations should include 95 or 99 percent, respectively, of all phenotypic values in any respective population.

5.3 FREQUENCY OF GENES IN A POPULATION

Genetic differences among populations, such as families, races, breeds, and even species, are largely due to differences in gene frequencies. The *frequency* of a gene refers to *how rare or abundant a particular gene is in a population as compared with its own* **allele** *(or alleles)*. In most examples used to illustrate genetic crosses, the frequency of the two alleles in each generation is kept equal. For example, the cross of a red *RR* Shorthorn **bull** with white *WW* Shorthorn **cows** should give the following ratios:

P_1	*RR* red	$\times$	*WW* white
F_1		All *RW* roan	
		1*RR*	
F_2		2*RW*	
		1*WW*	

If there were 20 roan individuals *RW* in the F_1 **generation,** there would be a total of 40 red *R* and white *W* genes, since genes occur in pairs in body cells. Of these 40 genes, 20 would be red *R* and 20 would be white *W*. Thus the frequency of the red gene *R* would be 20/40, or 0.5, and the frequency of the white gene *W* would also be 20/40, or 0.5. Gene frequencies are expressed as fractions of 1.

In the F_2 **generation,** where there is a 1:2:1 ratio, a total of eight red *R* and white *W* genes can be assumed. Of the eight genes, 4/8 would be red *R* with a frequency of 0.5 and 4/8 would be white *W* with the same frequency. Thus the frequencies of the red and the white genes are equal in the F_2 generation.

In populations of animals, the mating of particular **genotypes** is not controlled, since individuals of any one genotype could mate, by chance, with individuals of any one of the other genotypes. As a result, the frequencies of the two alleles in the offspring of a single mating may not be equal.

To illustrate the calculation of gene frequencies, it will be assumed that in a herd of 100 Shorthorns, 49 are red *RR*, 42 are **roan** *RW,* and 9 are white *WW.* This is not a 1:2:1 ratio as in the previous example (F_2). The reason the genotypic ratio in this example herd is not a 1:2:1 ratio is that there are many more red *R* than white *W* genes in this population. The frequency of the red and white genes in this **herd** of 100 Shorthorns can be calculated as follows:

Number of animals of different colors	Genotype	Total no. of genes in herd	
		Red	White
49 red	*RR*	98	0
42 roan	*RW*	42	42
9 white	*WW*	0	18
Total number of genes		140	60

Thus the frequency of the red gene *R* would be 140/200, or 0.7; the frequency of the white gene *W* would be 60/200, or 0.3.

In populations of animals the frequencies of two alleles are seldom equal, and in some populations one gene may be present at a much higher frequency than its allele. Obviously, if all individuals in a certain population are **homozygous** for a particular gene, the frequency of that gene is 1 and the frequency of its allele is 0. Genetic variation at a particular **locus** is present when the frequencies of the alleles are less than 1 and more than 0.

The examples previously used to explain the concept of gene frequencies included traits, such as coat color in Shorthorns, in which **dominance** was not complete and genotypes could be determined by observing phenotypes. The frequency of genes in a population can also be estimated for traits in which dominance is complete and phenotypes of the homozygous dominant and the **heterozygous** genotypes are the same. However, there are two prerequisites for calculating the frequency of alleles when dominance is complete: (1) matings must be at **random** (no selection) for at least one generation, and (2) the accurate proportion of the homozygous recessive individuals born within the population concerned must be determined. With this information available, the frequency of the **recessive gene** can be calculated by merely extracting the square root of the proportion of homozygous recessive individuals in a particular generation of **progeny.** The frequency of the dominant allele may be calculated by subtracting the frequency of the **recessive** allele from 1.

To illustrate, assume that in a large **herd** of black Angus cattle, 1 out of each 100 calves born was red. The black gene *B* is dominant in Angus, and the red gene *b* is recessive. Thus the frequency of the red gene *b* can be calculated by taking the square root of the frequency of the red calves born in the herd. This would be the square root of 1/100 (1/10, or 0.1). The frequency of the black gene *B* would be $1 - 0.1$, or 0.9. Knowing the frequency of both the black *B* and the red *b* genes, the proportions of black individuals that are homozygous and heterozygous in the calves can be calculated. The proportion of homozygous black individuals would be the frequency of the black gene (0.9) times the frequency of the black gene (0.9), or 0.81. The proportion of heterozygous individuals in this population would be twice the frequency of the red gene (0.1) times the frequency of the black gene (0.9), or 0.18 ($2 \times 0.1 \times 0.9 = 0.18$).

The above calculations can be made by applying the Hardy-Weinberg law. This law states that in large populations, where the frequency of one of two alleles is equal to *a* and the frequency of the other is equal to *b*, the sum of the frequencies $a + b$ equals 1. If matings are at random, the offspring of the three genotypes will occur in a definite ratio, or will be in equilibrium in the next generation at the frequencies of a^2, $2ab$, and b^2. Under these conditions, we can calculate the frequencies of many different pairs of allelic genes in large populations.

As stated previously, two alleles may not occur in equal frequencies in populations. Four factors are known to be responsible for these differences: (1) mixture of populations (migration), (2) **mutations,** (3) **selection,** and (4) genetic drift.

5.3.1 Mixture of Populations (Migration)

To illustrate the meaning of this factor, assume there is a herd of 20 white Shorthorn cows. The frequency of the white gene *W* in this group of cows would be 1, and the frequency of the red gene *R* would be 0. If a red bull *RR* were mated to these white cows and produced 20 calves, all these calves would be roan *RW,* and the frequency of the white gene *W* in the calves would be 0.5 as compared with 1 in their dams.

The same change in frequency of a particular gene would occur if one population in which the frequency of some particular gene was high was mixed and mated with another population in which the frequency of that same gene was low. The frequency of the gene in the offspring of the mixed population would be somewhere between the frequencies of the gene in the two parent populations, but it is not likely to be the same as that of either parent population.

5.3.2 Mutations

A **mutation,** as defined in Chapter 4, is a change in a gene that causes a sudden change in the phenotypic expression of that gene. Genes and their alleles mutate at different rates. Some seldom mutate, while others may mutate often. The difference between alleles in their mutation rate may cause a change in gene frequencies in a population over a long period of time. For example, if in a large population the frequency of gene *A* were 1, the frequency of its allele *a* would be 0. However, gene *A* may mutate to *a* twice as rapidly as gene *a* mutates to *A*.

Therefore, in this population, there would be a tendency over an extended period of time for the frequency of gene A to decrease and that of gene a to increase, until there were twice as many a genes as there were A genes in that population. When this occurred, the total number of new mutations of A and a would be approximately equal. The frequencies of these two alleles would remain in equilibrium in the population and would not change unless there were forces other than mutation rate involved.

5.3.3 Selection

Selection can be a potent force in changing the frequency of a gene, or genes, in a population. This is illustrated by the example

F_1	Aa	$\times$	Aa
F_2		$1AA$	
		$2Aa$	
		$1aa$	

The frequency of the recessive allele a in the F_2 is 0.5. If the homozygous recessive individuals aa were culled (selected), the frequency of the recessive allele a would be 0.33 instead of 0.5. Conversely, the frequency of the dominant allele A would have been increased to 0.67 by culling the aa individuals. Thus the main genetic effect of selection, if selection is effective, is to change gene frequencies.

5.3.4 Genetic Drift

Genetic drift is attributed to the sampling nature of inheritance and means that the frequency of a particular gene in a small population may be quite different than in the larger population from which the small population was derived. Suppose, for example, the frequency of genes A and a is 0.5 in a large population. If a small group of individuals leave this larger population (migrate) and settle in another area where they become isolated and interbreed among themselves, the frequencies of alleles A and a in this small group could be 0.7 and 0.3, or any other frequencies, depending on chance. Therefore the frequencies of genes A and a in the new isolated population could be quite different than in the original population from which they came.

5.4 CAUSES OF PHENOTYPIC VARIATION

Basically, there are three general causes of phenotypic variation in quantitative traits: (1) **heredity,** (2) **environment,** and (3) the joint action of heredity and environment.

Hereditary differences among individuals are those due to genotype, or genetic makeup, of individuals. The genotype of an individual is fixed at conception, when the **spermatozoon** and **egg** unite in the process of **fertilization.** An individual dies with the same genotype with which it was born (except for possible mutations). The phenotype of the individual, however, is subject to change and does change quite often throughout life. The genotype of an individual determines what it can and will transmit to its offspring. Genotype also affects phenotype for a particular trait, not only because of the kinds of genes the individual possesses, but also because of the way in which certain genes act in different combinations within the individual.

Environmental differences are those caused by anything other than heredity. These include differences due to disease, *nutrition,* management, accidents, and other factors. Superiority or inferiority due to environment will not be transmitted from parents to their **offspring.** Therefore it becomes important to the animal breeder to determine whether heredity or environment or both cause an animal to be superior.

For some traits, the interaction between heredity and environment may be an important source of phenotypic variation. What this interaction really means is that two **breeds** (genotypes) may perform quite similarly in one environment but quite differently in another. For example, in the semiarid regions of the southwestern United States, **steers** from the British breeds and steers from Brahman × British crosses will gain at about the same rate when fed at the same place and on the same ration in winter months when it is cool. In hot summer months, however, the Brahman **crossbred** steers will usually gain faster than the **purebred** British steers because they have the genetic ability to withstand hot dry weather.

5.4.1 Heritability Estimates

Heritability estimates indicate the proportion of total phenotypic variation that is attributable to heredity. These may be recorded in percentages. A heritability estimate of 40 percent means that 40 percent of the total phenotypic variation is due to heredity. If the heritability estimate is subtracted from 100 percent, the difference is an estimate of the total phenotypic variation that is attributable to environment. Heritability estimates for various classes of livestock are shown in Table 5.2.

The lower the heritability estimate for a trait, the slower the progress one can expect in improving the trait by finding and mating the best to the best. Also, the lower the heritability estimate, the larger the proportion of the total phenotypic variation that is due to environment. Extremely low heritability estimates for a trait suggest that this trait could most likely be improved in the offspring by crossing different families, inbred lines, or breeds.

5.5 SELECTION

Selection may be defined as *causing or permitting certain individuals within a population to produce the next generation.* The selection that the animal breeder

TABLE 5.2
HERITABILITY ESTIMATES FOR SELECTED TRAITS IN CATTLE,* SHEEP, SWINE, AND POULTRY

Trait	Percent heritability			
	Cattle	**Sheep**	**Swine**	**Poultry**
Fertility	0–15	0–15	0–15	0–15
Number of young weaned	10–15	10–15	10–15	....
Weight of young at weaning	15–25	15–20	15–20	....
Postweaning rate of gain	50–55	50–60	25–30	....
Postweaning efficiency of gain	40–50	20–30	30–35	....
Fat thickness over loin	40–50	....	45–50	....
Loin-eye area	50–60	....	45–50	....
Percent lean cuts	40–50	....	30–40	....
Milk production, lb	25–30			....
Milk fat, lb	25–30			....
Milk solids, nonfat, lb	30–35			....
Total milk solids, lb	30–35			....
Body weight	....	....	....	35–45
Feed efficiency	....	....	....	20–25
Total egg production	....	....	....	20–30
Egg shape	....	....	....	35–40
Egg weight	....	....	....	35–45
Eggshell color	....	....	....	40–45
Age at sexual maturity	....	....	....	30–40
Viability	....	....	....	5–10

*See also Table 11.3.

practices is usually referred to as *artificial* selection; that which nature does through the survival of the fittest is known as *natural* selection. Artificial selection by animal breeders is an attempt to increase the proportion of desirable genes in a population for a particular purpose. These genes may be dominant in their phenotypic effect, but they are often additive (plus genes). In a few instances, artificial selection is practiced for recessive genes. An example of this is selection for red Angus in the black Angus breed.

5.6 SELECTION FOR DIFFERENT KINDS OF GENE ACTION

Quantitative characters involve many different genes, which may express themselves in many ways in the phenotype. The phenotypic expression of genes may be divided into two general types, known as *nonadditive* and *additive*. Nonadditive include dominance and recessiveness, **overdominance,** and **epistasis,** as well as any other type of expression in which the phenotypic effects of genes are not of the additive type. The purpose here is to discuss methods of selection for the different kinds of additive and nonadditive gene action.

5.6.1   Selection for Dominance and Recessiveness

Selection against a dominant gene requires that all animals showing the dominant phenotype be culled. This assumes **penetrance** of the gene is complete. If penetrance of the gene is not complete (not 100 percent), it will not be possible to discard all individuals that possess the dominant gene, because its presence will not always be observed in the phenotype. Penetrance probably approaches 100 percent for most dominant genes in farm animals. The expression of certain dominant genes is quite variable, however, so that it is possible that a carrier of a dominant gene might not always be recognized.

Selection for a dominant gene means that individuals showing the presence of the gene by their phenotype are kept for breeding purposes. The homozygous recessive individuals are discarded. Effective selection for a dominant gene on the basis of phenotype is difficult, because it is not possible to distinguish between homozygous dominant *DD* and heterozygous dominant *Dd*. Thus some dominant phenotypes are selected that carry the recessive gene. **Progeny tests** must be conducted to find which individuals of the dominant genotype are heterozygous *Dd,* and these must be discarded.

Selection for a recessive gene is relatively simple if penetrance of the gene is complete. One merely keeps individuals showing the recessive trait, and then these are mated. Recessives mated to recessives should produce all homozygous recessive offspring, although mutations and rare interactions with other genes may occasionally affect the phenotype of recessive individuals.

Selection against a recessive gene requires that all individuals showing the recessive trait be discarded. Heterozygous individuals *Dd* that are of the dominant phenotype but are carriers of the recessive gene must be detected by means of progeny tests, and they must also be discarded. Thus selection for a dominant gene and selection against a recessive gene present the same problem; i.e., the distinction between homozygous dominant *DD* and heterozygous dominant *Dd* phenotypes must be considered in both.

Several kinds of matings can be made to determine if an individual is of the dominant phenotype but a carrier of the recessive gene. The kind and number of matings to prove an individual homozygous dominant *DD* at the 95 and 99 percent levels of probability are given in Table 5.3. Only one recessive offspring from any of these matings proves the individual being tested to be heterozygous dominant. Conversely, one cannot be positive that an individual tested is homozygous dominant, although the degree of confidence increases as the number of offspring from matings with the tester animals produces offspring only of the dominant phenotype.

5.6.2   Selection for Overdominance

Overdominance is a type of gene action in which heterozygous individuals are superior in merit to homozygous ones. Usually, the superiority is in vigor or performance. Many pairs of genes may be involved for an economic trait, but selection for this type of phenotypic expression of genes will be illustrated here

TABLE 5.3
NUMBER AND KINDS OF MATINGS REQUIRED TO PROVE AN INDIVIDUAL
HOMOZYGOUS DOMINANT *DD* AT THE 95 AND 99 PERCENT LEVELS OF
PROBABILITY

Animal of dominant phenotype but genotype unknown can be mated with	Number of matings producing no recessive offspring	
	95% level of probability*	99% level of probability*
Homozygous recessive individuals	5	7
Known heterozygotes	11	16
Unselected different daughters of a known heterozygous sire	23	35
A sire mated to his own unselected daughters (all different daughters)†	23	35

*Probability level at which the individual is proved homozygous dominant *DD*.
†Tests for any recessive gene the sire might be carrying and not only for a specific recessive gene.

with only two pairs. Assume that there are individuals of the following genotypes and that heterozygous individuals are superior:

Individual number	Genotype
1	$A^1A^1B^2B^2$
2	$A^1A^1B^1B^2$
3	$A^1A^2B^1B^2$
4	$A^2A^2B^2B^2$
5	$A^2A^2B^1B^2$
6	$A^2A^2B^1B^1$

Since heterozygous individuals are superior when overdominance is involved, individual 3 should be selected because it is heterozygous for two pairs of genes. When individuals of this genotype are mated, some offspring will be of quality equal to that of the parent, but many will be inferior. The average performance of offspring from several such matings would be considerably below that of heterozygous parents. Thus selection of superior phenotypic individuals would result in selecting those most heterozygous, and they would not breed true. The average of their offspring would tend to regress to a point below the average of the heterozygous parents. From this example, it can be seen that selection on the basis of an individual's merit would be disappointing.

Although heterozygous individuals will not breed true when mated together, the mating of the right genotypes should always give offspring that are heterozygous. For example:

Parent 1	×	Parent 2
$A^1A^1B^1B^1$		$A^2A^2B^2B^2$
	All offspring would be	
	$A^1A^2B^1B^2$	

Double-heterozygous offspring would also be produced if each of the parents were homozygous for the opposite alleles ($A^1A^1B^2B^2 \times A^2A^2B^1B^1$). Once heterozygous offspring are produced, however, one must return to the mating of homozygous parents (homozygous for different alleles) to produce all $A^1A^2B^1B^2$ offspring. Selection for overdominance, then, is selection among different lines, families, or breeds to find those that combine the best to give the most heterozygous offspring. This is done in the production of hybrid corn. Inbred lines of corn are formed by inbreeding to make all individuals within the lines homozygous for all genes. The many lines formed by inbreeding are then crossed in various combinations to find those that combine best to produce the most vigorous (heterozygous) offspring. Those that combine the best are probably homozygous in opposite ways. For example, line 1 might be $A^1A^1B^1B^1C^1C^1$, whereas line 2 might be $A^2A^2B^2B^2C^2C^2$. Crossing these two lines would produce offspring heterozygous for the three pairs of genes. In actual practice the **linecross** offspring may be made heterozygous for many pairs of genes, but there is no way of knowing for which pairs or how many.

5.6.3 Selection for Epistasis

Epistasis is a type of gene action in which one pair of genes may affect the phenotypic expression of another pair of genes that are not their alleles. These genes may vary widely in the way they interact with each other. In spite of this variation, the method of selection would be similar for all types of epistasis. To illustrate selection for epistasis, let it be assumed that two pairs of alleles (Aa and Bb) are involved and that a combination of genes A and B in the genotype gives the desirable phenotypic effect. The following gene combinations of these two pairs of alleles would be possible:

Individual number	Genotype
1	AABB
2	AABb
3	AAbb
4	aaBB
5	aaBb
6	aabb
7	AaBB
8	AaBb
9	Aabb

The desired combination of *A* with *B* would be found in genotypes 1, 2, 7, and 8, and selection for superior individuals would result in retention of individuals of these genotypes for breeding purposes. Obviously, the goal would be to save only those individuals that are *AABB*, because, when mated together, they would produce only *AABB* offspring, or would breed true. Selection on the basis of the individual's phenotype, however, would probably cause us to keep genotypes 1, 2, 7, and 8 in equal numbers, but all except genotype 1 would fail to breed true when mated together. Thus it is possible to establish a pure line that would breed true for a certain epistatic effect, but this is not highly probable in practice. It must be kept in mind that many pairs of alleles (more than two) may affect various economic traits in farm animals. This leads to the conclusion that selection for superior merit of the individual (selection on the basis of individuality) is not very effective when epistasis is involved.

Probably the most effective way to select for an epistatic effect would be to form **families** (inbred lines or breeds) and test them to find those that combine the best in crosses to give superior offspring. Those combining the most satisfactorily would be retained and crossed to produce superior progeny. The original lines (families or breeds) would be kept pure and crossed again and again to produce superior crossbred or linecross offspring.

Selection among families or lines is therefore probably the best method of selection when overdominance and/or epistasis is important.

5.6.4 Selection for Additive Gene Action

Additive gene action is a type of phenotypic expression of genes in which the phenotypic effect of one gene adds to the phenotypic effect of another.

Selection for additive gene action should be largely on the basis of individuality, although attention to the merit of the individual's close relatives should help make selection more accurate. To illustrate why additive gene action is selected for on the basis of individuality, the following examples will be used. Assume that genes *A* and *B* add two points to the superiority for a trait when the residual genotype *aabb* is given a value of 20 points. The value of each genotype would be

Genotype	Value of genotype, points
aabb	20
aabB	22
aaBB	24
aABB	26
AABB	28

If these individuals were selected in a comparable environment so that phenotypic variations due to environment were minimized, the individuals most likely to be selected would be those of genotype *AABB*, with a phenotypic score of 28 points. If such individuals were selected and mated, they should produce superior offspring. It is true that to a certain extent superiority or inferiority due to environment would be confused with that due to heredity, but selecting and mating the best to the best for several generations (mass selection) should improve the overall merit of individuals in the population.

Superiority for a trait when additive gene action is important, therefore, is dependent on the kind and number of desirable genes (plus or contributing genes) an individual possesses. It does not depend on a certain combination of alleles or nonallelic genes to give a **nicking** effect when only additive gene action is involved.

5.6.5   How to Determine If Additive or Nonadditive Gene Action Affects a Quantitative Trait

Qualitative traits are affected by relatively few genes, and there is a sharp distinction between phenotypes. Thus it is not too difficult to determine the type of phenotypic expression of genes involved when appropriate progeny-testing procedures are applied. Quantitative traits, however, may be expressed through the action of many pairs of genes, and there is no sharp distinction between phenotypes. Moreover, environment may also significantly influence the phenotypic expression of quantitative traits. This raises the important question of how to determine whether additive or nonadditive gene action has the greater influence on a particular economic trait. Information summarized in Table 5.4 shows, in a general way, how this can be accomplished.

Heritability estimates for various traits are summarized in Table 5.2; **hybrid vigor** effects (crossing) for various traits are summarized in Table 5.8. **Inbreeding** effects are usually the opposite of **crossing** effects. Thus, if crossing improves a trait, inbreeding should have a detrimental effect.

Selection in farm animals is usually concerned with several traits. Some of these traits may be largely affected by additive gene action, whereas others may be largely affected by nonadditive. Some traits may be affected by both, but in varying degrees. How can a selection program be designed if this is true? The

TABLE 5.4
HOW TO DETERMINE IF AN ECONOMIC TRAIT IS
AFFECTED LARGELY BY ADDITIVE OR NONADDITIVE GENE
ACTION OR BOTH

	Type of gene action indicated		
	Additive	Nonadditive	Both
Heritability estimate	High	Low	Medium
Crossbreeding effect	None	Large	Low-medium
Inbreeding effect	None	Large	Low-medium

recommendation is to select and improve the highly heritable traits (those affected largely by additive gene action) in purebreds and then cross the improved breeds to obtain hybrid vigor (nonadditive gene action) in the crossbred progeny.

5.7   SELECTION OF SUPERIOR BREEDING STOCK

Superior breeding animals are those that possess a large proportion of superior genes for a desirable trait, or traits, in a homozygous state. These superior genes may express themselves in different ways phenotypically, but usually the dominant and additive plus genes are the ones in which the animal breeder is interested when selection is on the basis of individuality.

The identification of superior breeding stock may be based on the phenotype of the individual and/or on that of its close relatives, especially if accurate records are available.

5.7.1   Individuality

Attention should be given to the type and performance of the individual when possible. An individual's own merit is most important when the trait or traits sought are medium to highly heritable. Such traits are determined largely by additive gene action, and individuals that are superior in a group tested are more likely to be superior because of their genetic makeup. The selection procedure in such a case becomes one of finding the best and mating the best to the best.

Traits of low heritability are those largely affected by environment and nonadditive gene action. Superior individuals for such traits are often disappointing in their performance as parents. As mentioned previously, traits of low heritability require the formation of families, inbred lines, or breeds. These must be tested in crosses to find those that combine best when crossed to produce superior linecross or crossbred progeny.

The main error encountered when selection is based on individuality is that effects due to heredity and environment are often confused. Confusion of environmental and genetic effects can be avoided by comparing the individuals being considered for breeding purposes at the same location, during the same time, and when they are fed and managed alike. These precautions will not eliminate all variation due to environment, but they will reduce it.

When many individuals are compared under standardized environmental conditions, those that are superior in performance are more likely to be superior because of their genetic makeup. Therefore they should prove to be superior parents. The principle involved is that when the proportion of the total phenotypic variation attributable to environment is reduced, more of the remaining variation is due to heredity. This may be illustrated by the example

$$\text{Percent of variation due to heredity} = \frac{\text{variation due to heredity}}{\text{variation due to environment} + \text{variation due to heredity}} \times 100$$

If the variation due to heredity is 6 and that due to environment is 6, obviously the proportion of variation due to heredity is

$$\frac{6}{6 + 6} \times 100, \text{ or } 50 \text{ percent}$$

If the amount of variation due to environment is reduced to 4 by appropriate methods, the proportion of total variation due to heredity is

$$\frac{6}{4 + 6} \times 100, \text{ or } 60 \text{ percent}$$

Rate and efficiency of gains in most livestock are moderately to highly heritable and therefore are affected mostly by additive genes. Performance testing of a group of individuals includes determining rate and efficiency of gains from a period of time shortly after weaning until they near market weight (140 days in cattle). To obtain efficiency of gain records on an individual, it is necessary to feed each animal and to record the amounts of feed consumed and its gain. From these records, the amount of feed required to make a pound of gain can be calculated. Because of the time and expense required to feed each animal individually, feed-efficiency records are seldom determined in a performance test. One can easily determine rate of gain for each animal within a group by merely obtaining the weights of each animal at the time they go on full-feed and then again when the full-feeding period is completed. Pounds of gain made during the feeding period divided by the number of days fed gives average daily gain **(ADG)**.

Performance testing as described above is done mostly with beef cattle and swine and to a lesser extent with sheep.

5.7.2 Records of Relatives

The performance of close relatives of an individual also helps to determine if an individual is genetically superior. The more closely an individual is related to its relatives, the more attention should be given to the merits of those relatives. The degree of relationship among various individuals is shown in Tables 5.5 and 5.6.

An individual has three kinds of relatives: (1) the ancestors in its **pedigree;** (2) its descendants, or progeny; and (3) its collateral relatives, which are those not related to it as ancestors or descendants. **Collateral relatives** include brothers, sisters, cousins, uncles, and aunts.

Selection for traits that can be measured only after the death of the individual, or in only one sex, often requires that criteria be based on the performance of relatives rather than the individual. The carcass quality of some farm animals cannot be measured accurately in the live animal. Use of the backfat probe and sonoray in swine, however, has helped increase the accuracy of selection for carcass quality in the living animal (Figure 5.2). Selection for carcass quality in swine can also be based on the carcass quality of **littermates** after slaughter. Littermates that are full-siblings (brothers and sisters) are 50 percent related. When single births predominate, as with beef cattle, usually the

TABLE 5.5
PROBABLE PERCENTAGE OF GENES (RELATIONSHIP)
THAT AN INDIVIDUAL HAS IN COMMON WITH HIS OR HER
COLLATERAL RELATIVES* ABOVE THE AVERAGE OF THE
POPULATION

Kind of relative	Percent relationship
Identical twins	100.00
Fraternal twins†	50.00
Full brothers or sisters *(full-siblings)*	50.00
Half brothers or sisters *(half-siblings)*	25.00
First cousins	12.50
Double first cousins	25.00
Second cousins	3.13
Third cousins	0.78
Aunts or uncles	25.00

*Collateral relatives are those not related as ancestors or descendants. In these examples it is assumed that none of the relatives are inbred.

†It is possible for fraternal twins to be from two different sires. They would be related by 25.00 percent.

closest relatives that can be slaughtered are half-siblings. Their relationship is 25 percent (Table 5.5).

Selection among dairy cattle for improvements in milk production today dictates that greater emphasis be given selection of bulls than selection of cows.

TABLE 5.6
PROBABLE PERCENTAGE OF GENES (RELATIONSHIP) THAT AN
INDIVIDUAL HAS IN COMMON WITH HIS OR HER ANCESTORS
ABOVE THE AVERAGE OF THE POPULATION

	Percent relationship	
Kind of ancestor	Not corrected for inbreeding	Corrected for inbreeding
No inbreeding involved		
Parent	50.00	
Grandparent	25.00	
Great-grandparent	12.50	
Great-great-grandparent	6.25	
Inbreeding involved		
Both a parent and a grandparent*	75.00	67.08
Double grandparent	50.00	47.14
Double great-grandparent†	25.00	24.62
Triple great-grandparent	37.50	36.38
Quadruple great-grandparent	50.00	47.14

*Individual resulting from the mating of a sire to his daughter or a dam to her son.

†The same individual is a grandparent of both the sire and the dam. If a double great-grandparent is in only the sire's or dam's pedigree, the relationship need not be corrected for inbreeding and would be 25.00 percent.

FIGURE 5.2
Measuring fat-cover and loin-eye area in the living hog by means of
the sonoray, an ultrasonic instrument. The instrument sends out
sound waves that go through tissue without harming the animal.
The time needed for the sound waves to pass through and bounce
back from different tissues varies with tissue thickness. The
measurement is quite accurate. *(Missouri Agr. Expt. Station.)*

Much more selection pressure can be applied to bulls because, through the use
of artificial insemination, one bull can be mated with thousands of cows
annually. Conversely, a cow commonly produces only five to 10 calves during
her lifetime (and many cows produce fewer than five).[1] Moreover, most females
must be raised and relatively few culled if herd size is to be maintained. This
situation results in far less intensity of selection among dairy females than among
males. However, selection pressure for milk production should be applied
through rigid culling of low producers when possible.

[1]The technology associated with superovulation and embryo transfer will enable the animal
breeder to raise many more offspring from cows in the future (see Chapter 9).

Selection of sires for improved milk production requires special methods since bulls do not secrete milk. They do, however, possess and transmit genes for milk production. A bull's genetic worth must be determined from the production records of closely related females, especially his daughters.[2]

Several *sire indexes* have been developed to estimate the genetic merit of dairy bulls.[3] Most of these have been based on production records of a sire's daughters compared with those of other cows in the same herd. Indexes used in the past include daughter average, daughter-dam difference, equal-parent index, daughter-contemporary herd difference, daughter-contemporary herd index, and herdmate comparisons. Presently, most sire evaluations are based on *modified contemporary comparisons* (MCC), or **predicted difference** (PD).

In modified contemporary comparisons, records of a sire's daughters are compared with those of cows (contemporaries) that calve and produce milk in the same herd during the same season and year and have primarily the same lactation as the daughters of the sire being evaluated. A sire's daughters and their contemporaries are theoretically provided with equal opportunity to produce milk. If the first lactation milk yield of cow A were 15,000 lb, and that of her first lactation contemporaries were 14,000 lb, the modified contemporary deviation would be +1000 lb. Modified contemporary deviations are especially valuable because they minimize environmental effects, e.g., season, lactation, and management.

For comparative purposes, production records of dairy cows are converted to reflect a 305-day (305D) lactation, two times (2X) daily milking, and mature equivalency (ME). When production records of unselected daughters of a sire exceed those of contemporaries, the bull is indicated to be transmitting superior milking ability.

Modified contemporary comparisons are an improvement over *herdmate comparisons* because MCC (1) account more accurately for the genetic level of **herdmates,** (2) account for genetic trend, (3) adjust more accurately for environmental effects, (4) use pedigree data on bulls, (5) consider production records in progress, and (6) account for within- and between-herd variation more accurately through improved repeatability. The MCC formula for estimating predicted difference (PD) for milk and fat yields is as follows:

$$PD = R(D - MCA + SMC) + (1 - R)\, GA$$

where R represents repeatability, $D - MCA$ represents how much a daughter's record deviated from the modified contemporary average, SMC is the average predicted difference for sires of modified contemporaries, and GA is a group average.

[2]Even when traits are highly heritable, as are rate and efficiency of gain in beef cattle, selection on the basis of individuality may be strengthened by selecting individuals with good-performing relatives. Therefore, when records on relatives are available, they are useful in all methods of selection.

[3]The authors acknowledge with appreciation the contributions to this section of Dr. R. D. Shanks, Department of Dairy Science, University of Illinois, Urbana.

The precision with which predicted differences (PD) for milk and fat yields will estimate a bull's transmitting ability for milk production, relative to other bulls of that breed, is called *repeatability;* it depends on (1) number of daughters in the **Sire Summary,** (2) number of herds in which the daughters are located, (3) distribution of the daughters among these herds, (4) number of lactations per daughter, and (5) number of contemporaries. The higher the repeatability, the more confidence can be placed in the PD. Repeatability increases as the number of daughters and number of lactations per daughter increases. Additionally, daughters in more herds reflects greater reliability.

Environmental effects are adjusted by measuring the deviation of each daughter's record from the modified contemporary average. Two groups of contemporaries are defined and each has a different effect, depending on the lactation of the daughter. The groups are first lactation herdmates and second or later lactation herdmates. If a daughter's lactation record is her first, then the group of second or later lactation herdmates count as just one contemporary, whereas each herdmate in the first lactation group is counted as a separate contemporary. The reverse is true if the daughter's record is from a second or later lactation; then the group of first lactation herdmates are counted as only one contemporary, and each herdmate in the group of second or later lactation herdmates is counted as a separate contemporary. These calculations are done first for each herd, and then combined for each bull, such that an additional daughter in a new herd has a larger contribution than an additional daughter in a herd with several daughters.

Adjustment for the genetic level of herdmates is the purpose of adding the predicted difference for sires of modified contemporaries to the deviation of a daughter's record from that of the modified contemporary average. Iteration is essential to complete this adjustment successfully. Predicted difference is estimated for all sires of modified contemporaries. The SMC for each herd is calculated and new PDs are estimated for all bulls. If changes between the previous PDs and the new PDs are large, the process is repeated. This technique is called *iteration.*

The last expression in the PD formula represents the contribution from pedigree information. First, a pedigree index (PI) for milk is calculated for each bull: PI = (sire's PD for milk)/2 + (maternal grandsire's PD for milk)/4. The group average for milk and fat is then a function of the pedigree index and birth date. The group average is regressed by 1 minus repeatability. This expression is most important when repeatability is low. As repeatability increases, the contribution to the PD from pedigree information becomes smaller and will eventually approach zero.

The most accurate and meaningful estimate of a dairy sire's genetic worth for milk production is his PD for milk and fat yields. It represents an estimate of the bull's transmitting ability based on the productivity of his daughters in comparison with a genetic base. The difference in PD between two bulls provides a reliable estimate of the expected production deviation between daughters of the bulls. If bull A has a PD milk of +2000 and bull B has a PD milk of +1000, then daughters of bull A are expected to outproduce daughters of bull B by 1000 lb of milk, assuming the daughters of bulls A and B are exposed to the same environmental effects.

The genetic base of PD 74 was a zero average PD of all dairy bulls summarized in the fall of 1974. This corresponded to the average cow calving in mid-1967 to early 1968. There are plans to change the genetic base effective January 1984. The average PD values for bulls of all dairy breeds summarized in July 1982 were +344 lb for milk yield and +9 lb for fat yield. The genetic trend is for milk and fat yield to increase. Active bulls of all dairy breeds in artificial insemination (AI) in July 1982 averaged +1191 lb PD milk and +36 lb PD fat. Thus, through the application of modern science and technology, tremendous genetic progress was achieved between 1974 and 1982 in the average predicted difference for milk (PDM) for all active AI bulls of all dairy breeds, as is shown in Figure 5.3.

The Animal Science Institute of the USDA publishes biannually a USDA-DHIA Sire Summary list that contains PDs for milk, fat, protein, solids-not-fat yield; for fat, protein, and solids-not-fat percentage; and for income on bulls of six dairy breeds. These genetic evaluations are based on information on the bulls' daughters in herds participating in official production-testing programs (*DHIA* and *DHIR*). A free copy may be obtained by writing the Animal Improvement Programs Laboratory, Building 263, Agricultural Research Center, USDA, Beltsville, Maryland 20705.

FIGURE 5.3
The genetic effect of predicted difference for milk (PDM), in pounds, of all active AI bulls of all dairy breeds (*a*) compared with the genetic effect of PDM of all bulls of all dairy breeds (*b*) both AI and non-AI.

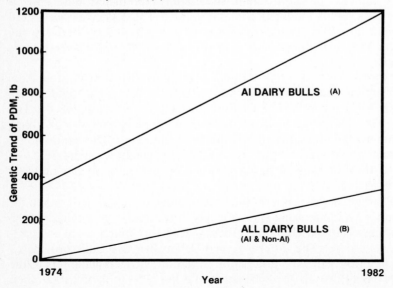

Various ways and means of estimating the genetic worth of dairy cows have been developed through the use of production records. For example, geneticists at Iowa State University have researched the use of *estimated average transmitting ability* (EATA), which estimates the breeding value a cow will transmit to her offspring.[4] Calculations are based on five sources of milk production records: (1) the cow, (2) her dam, (3) her daughters, (4) her paternal half-siblings (other daughters of her sire), and (5) her maternal half-siblings (other daughters of her dam). Information from each pedigree source is weighted in accordance with how much the information contributes to the EATA. Thus, emphasis given these sources of genetic worth is influenced by several factors, for example, the degree of relationship to the cow being evaluated. A cow is twice as closely related to her dam or daughters as she is to her maternal or paternal sisters. Therefore, in calculating EATA, greater weight is given to each production record of a cow's daughters than to each record of her half-sibling(s).

This measure of the cow's genetic merits for milk production is similar to the predicted difference (PD) applied to dairy sires. Combining the cow's EATA with the sire's PD provides a reliable estimate of the dairy merit of their progeny. For example, the expected EATA of females and the expected PD of males resulting from mating a cow having a +1800 EATA and a bull with a +1200 PD for milk would be +3000 divided by 2, or +1500 lb. The reason for the division by 2 is that each parent can be expected to transmit only one-half of its superiority (or inferiority) in dairy merit to its progeny.

The USDA annually compiles and publishes estimates of the genetic transmitting ability of cows identified (through the Dairy Herd Improvement Testing Program) as having demonstrated superior genetic merit for milk production. These cow index values are listed in the USDA-DHIA Cow Performance Index and are based on production records of the cow (modified contemporary deviation), her paternal half-siblings (sire's PD), and her dam's cow index. Free copies can be obtained from the Animal Improvement Programs Laboratory, Building 263, Agricultural Research Center, USDA, Beltsville, Maryland 20705.

5.7.3 State of the Art

An alternative method of sire evaluation was developed by Dr. C. R. Henderson and coworkers at Cornell University. Although the method is not used by the USDA in the calculation of predicted differences of milk and fat production, it is used for calving-ease evaluation of dairy sires, for estimated progeny difference of sires for the American Angus Association, and for predicted difference type for several dairy breeds. The method is often identified by the acronym BLUP, representing best linear unbiased prediction.

---

[4]For discussion of other means of appraising the genetic worth of dairy cattle the reader is referred to Chapter 6 ("Breeding Dairy Cattle") of the textbook *The Science of Providing Milk for Man*, by J. R. Campbell and R. T. Marshall, McGraw-Hill Book Company, New York, 1975.

A statistical explanation of the method is beyond the scope of this book. However, for references and applications see *Variance Components and Animal Breeding,* the proceedings of a conference in honor of C. R. Henderson, edited by L. Dale Van Vleck and Shayle R. Searle, Cornell University, Ithaca, New York, 1979.

The advantages of BLUP compared with herdmate comparisons and older methods are (1) it is easier to use because the values are stabilized from one sire summary to the next; (2) it is more accurate because BLUP accounts for both genetic competition and genetic trend; and (3) it offers greater flexibility, since it permits the use of all available information on relatives and progeny. Disadvantages of BLUP are the complexity, computer expense, and difficulty to explain.

5.8 PREDICTING THE AMOUNT OF PROGRESS POSSIBLE THROUGH SELECTION

The preceding discussion dealt with how to select for quantitative and qualitative traits. In this section factors useful in predicting selection progress will be discussed.

5.8.1 For One Generation

The amount of progress made through selection for a particular trait in one generation is equal to the heritability of the trait times the selection differential. The formula for this calculation is

Genetic progress = heritability estimate × selection differential

Heritability estimates for various traits are usually obtained from experimental reports (Table 5.2). Selection differentials, however, must be calculated from data from animals where selection is practiced.

The *selection differential* may be defined as *the average superiority of those selected as parents over the average of the population from which they were selected.* The selection differential is sometimes referred to as the *reach,* or the intensity, of selection. To illustrate the calculation of expected genetic progress through selection, let it be assumed that in a herd of cattle the average weaning weight of all calves adjusted to a "bull-calf basis" was 400 lb. From these calves, bulls that weighed 500 lb were selected for breeding, and the selected heifers weighed 450 lb. The selection differential for the bull calves would be 500 − 400, or 100 lb. For **heifers,** the selection differential would be 450 − 400, or 50 lb. The selection differential for both parents combined would be 100 lb (bulls) + 50 lb (heifers) ÷ 2, or 75. If the heritability of weaning weight is 25 percent, the expected genetic improvement (genetic progress) in the offspring of the selected parents would be 25 percent of 75 lb, or 18.75.

The magnitude of the selection differential is dependent on the number of individuals that can be culled. Table 5.7 presents selection differentials for boars

TABLE 5.7
SELECTION DIFFERENTIALS AMONG BOARS AND GILTS IN A SELECTION
EXPERIMENT FOR BACKFAT THINNESS

Generation of selection	No. of boars selected per generation	Selection differential for boars, mm	No. of gilts selected per generation	Selection differential for gilts, mm
P_1	7	2.9	30	2.8
F_1	7	3.3	20	2.0
F_2	6	2.7	26	0.8
F_3	7	2.8	28	1.4
F_4	6	2.2	25	2.6
Total and av	33	2.8	129	1.9

Source: Missouri Agr. Expt. Station.

and gilts in a selection experiment for backfat thinness in swine at the Missouri
Agricultural Experiment Station. The selection differential for **boars** was 2.8
mm, compared with 1.9 mm for gilts. Thus the selection differential for boars
was larger because fewer of them were kept for breeding purposes.

The more traits selected for, the smaller the selection differential. This is
because it is more difficult to find an individual that is excellent for two or more
traits than it is to find an individual that is excellent for only one. Selection
among farm animals should be limited to those few traits that are of the greatest
economic importance.

5.8.2 For Several Years

Genetic progress in selection over a period of years depends on (1) the degree of
heritability of the trait selected, (2) the size of the selection differential, and (3)
the length of the generation interval. This may be expressed in an equation as
follows:

$$\text{Annual genetic progress} = \frac{\text{selection differential} \times \text{heritability for trait}}{\text{length of generation interval, years}}$$

The **generation interval** may be defined as *the average age of parents when their
offspring are born.* It averages about 2 to 3 years in swine, 4 to 6 years in cattle,
and 30 to 35 years in humans. For some insects and small laboratory animals, the
generation interval may be only a few days or a few weeks in length. Since **sows**
and boars are often sold after the first litter is produced, the generation interval
for swine could be about 1 year. Progeny testing and the use of its results as a
basis for selection may double, or even more than double, the length of the
generation interval. This decreases the amount of progress made by selection in
a period of several years.

5.9 GENETIC CORRELATIONS

Genetic correlations may be determined mathematically, or they may be determined by selection for only one trait. If success is achieved in improving the single trait selected for, possible changes can also be observed in unselected traits. A genetic correlation means that two or more traits are affected by many of the same genes. Genetic correlations between two traits may influence the amount of progress made in selection for one of them.

Since most economic traits in farm animals are influenced by many genes, it is reasonable to expect that some of the same genes may affect more than one trait. This is probably because the same genes affect a certain physiological process that determines to a certain extent the expression of two or more traits. Many instances are known in which a single major gene may cause the appearance of two different traits in animals. This is known as **pleiotropy.**

The same genes may affect two different traits in a positive, or desirable, way. This is known as *a positive genetic correlation.* Experimental results suggest that selection only for faster gains in beef cattle usually causes a genetic improvement in the efficiency of gains. Thus these two traits are said to be influenced by many of the same genes. The same genes may also affect two traits in a negative manner, which means that as one trait is selected and improved genetically, another trait declines in genetic merit. An example of a negative genetic correlation is that high milk production appears to be correlated with a lower percentage of milk fat. Still another possibility exists in which none of the same genes affect two different traits. This is often referred to as *the lack of a genetic correlation.* If no genetic correlation exists between two traits, both traits would have to be selected for at the same time in order to obtain genetic improvement in both.

5.10 MATING SYSTEMS FOR LIVESTOCK IMPROVEMENT

The preceding discussion was directed toward selection principles and their use in animal improvement. Once superior animals are identified and selected, it is necessary to devise mating systems that will give the most genetic improvement. The mating systems that may be used are inbreeding, linebreeding, outbreeding, and crossbreeding.

5.10.1 Inbreeding

Inbreeding may be defined as *the production of progeny by parents that are more closely related than the average of the population from which they came.* A base (average of the population) to determine where inbreeding begins is necessary because, even though they are not referred to as relatives, all individuals within the same species have many genes in common. Humans recognize relatives as those individuals that have the same ancestor (or ancestors) in the pedigree of both father and mother. Individuals are considered to be relatives if they have a common ancestor within the last three or four generations.

The major genetic effect of inbreeding is to increase the number of pairs of genes that are homozygous in an inbred population. These genes are made more nearly homozygous by inbreeding regardless of whether they express themselves phenotypically as additive or nonadditive. Probably the most practical effect of inbreeding is that it pairs recessive genes, which are usually detrimental in their effect, and because of this pairing, they express themselves in the phenotype.

Inbreeding does *not* create recessive genes. If recessive genes are covered up by dominant genes in the original noninbred population, they will be paired by inbreeding. The more intense the inbreeding, the more pairs of genes will be made homozygous. Likewise, if few or no recessive genes are present in the original population, they will not be paired by inbreeding, and the results are not so detrimental.

Inbreeding has been used in laboratory animals to study numerous parameters among strains (e.g., disease resistance; immune responses; endocrine functions; and response to drugs, cancer, and variations in dietary requirements), as well as to study coat color patterns and body conformations (Figure 5.4).

Increased homozygosity due to inbreeding increases breeding purity. In the following example, individual 1 could transmit only one combination of genes (*ABCD*) to its offspring through the gametes. Individual 2 could also transmit only one combination of genes to its offspring (*abcd*). Individual 1, which is homozygous dominant for four pairs of genes, would cause all offspring to resemble it regardless of the genotypes of the other individuals to whom it is mated. Individual 1 is said to be **prepotent** because it is homozygous dominant for these four pairs of genes. Individual 2 is not prepotent because it transmits only recessive alleles. Thus, if individual 2 is mated to individuals possessing the dominant alleles, the offspring will resemble the other parent rather than individual 2. Individual 3 is heterozygous for four pairs of genes, and it could transmit 16 different combinations of genes to its offspring through its gametes. Therefore it would not breed true.

	Genotype	Genes in gametes
Individual 1	*AABBCCDD*	*ABCD*
Individual 2	*aabbccdd*	*abcd*
Individual 3	*AaBbCcDd*	16 possible combinations of genes

Increased homozygosity due to inbreeding also tends to fix recessive genes in a small inbred population. These recessive genes may be present at high frequencies before their presence is detected and their genetic nature is understood. For example, an autosomal recessive gene for hemophilia (failure of blood to clot) was discovered in a small inbred line of Poland China swine at the Missouri Agricultural Experimental Station. Many of the homozygous recessive boars bled to death when **castrated,** and some sows bled to death at **parturition** before the cause of death was determined. So many individuals within this small line were either homozygous recessive or normal carriers of the gene that when

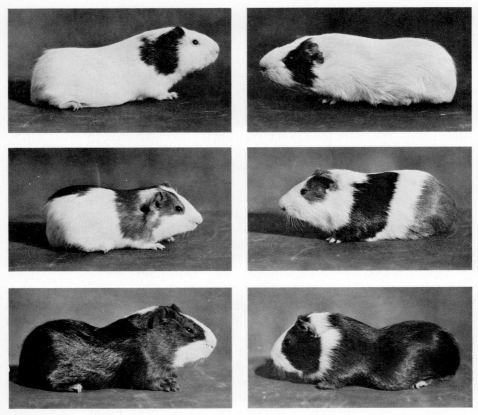

FIGURE 5.4
Illustration of varying coat color patterns and conformation in closely inbred strains of guinea pigs, both males (left) and females (right), in generations 18, 12, and 13 (top to bottom, respectively) of brother-sister matings. Note the relative degree of uniformity of the top, middle, and bottom individual strain pairs. *(Courtesy Dr. Sewall Wright, Emeritus Professor of Genetics, University of Wisconsin, Madison.)*

the genetic cause of the defect was discovered, the line had to be abandoned for breeding purposes. The recessive gene has been retained at the Missouri Station, however, mainly for studies of blood-clotting mechanisms.

Increased homozygosity due to inbreeding, when combined with selection, can be used to eliminate detrimental recessive genes from an inbred population. When inbreeding is practiced and recessive defects are paired and uncovered, the incidence of the recessive gene may be greatly reduced by discarding the individuals and their close relatives. However, a large inbred line is necessary before such rigid culling can be practiced, and the breeder must know what to look for as inbreeding progresses.

Increased inbreeding, accompanied by selection for certain performance traits such as **type,** conformation, and rate of gain, is a good method of establishing families within a breed that are quite different in their phenotype. The major phenotypic effects of inbreeding are to uncover recessive defects and to cause a decline in traits related to physical fitness. These traits include vigor at all ages, but especially from shortly after birth to weaning. They also include fertility.

The main reason for inbreeding is to produce inbred lines that may be used for crossing purposes to take advantage of hybrid vigor. A classic commercial example of this is found in the production of hybrid seed corn. About 99 percent of all corn produced in the United States today comes from planting hybrid seed that was produced by means of a systematic combination of certain inbred lines. The system used in hybrid-seed-corn production is also being used in the production of other crops, such as hybrid tomatoes and hybrid sorghum. A similar system is also used in producing certain kinds of commercial poultry.

5.10.2 Linebreeding

Linebreeding is *a form of inbreeding in which an attempt is made to concentrate the inheritance of one or more outstanding ancestors in the pedigree.* An example of linebreeding is shown in Figure 5.5. Other kinds of matings can be used to illustrate linebreeding, but all have as their objective the concentration of the inheritance of some desirable ancestor in the pedigree. Homozygosity (inbreeding) usually does not increase as rapidly when linebreeding is practiced, as is true of some forms of inbreeding, because usually only half-brother × half-sister matings are made. However, some systems of linebreeding can result in very intense inbreeding. As a consequence, intensely linebred individuals may lack vigor and show genetic defects, as is true in almost all cases of intense inbreeding.

FIGURE 5.5
A pedigree and its arrow diagram illustrating linebreeding. Note that the only common ancestor of the sire *S* and dam *D* is individual 5 and that there are four pathways connecting individuals *X* and 5 through the sire and dam. Hence it is called *linebreeding*.

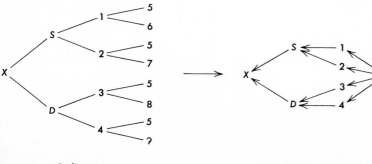

Pedigree   Arrow diagram of the pedigree

5.10.3   Outbreeding and Crossbreeding

Outbreeding refers to *mating unrelated families within the same breed.* **Cross-breeding** refers to *mating individuals from different breeds.* Crossbreeding is more intense in its genotypic and phenotypic effects than outbreeding, but these effects are similar in both types of breeding.

The main genetic effect of outbreeding and crossbreeding is opposite to that of inbreeding. They result in increased heterozygosity, or decreased homozygosity, within the population in which they are practiced. These systems of mating, because they increase heterozygosity, tend to cover up recessive genes, decrease breeding purity, and eliminate families in just one generation. Several cycles of crossbreeding may result in greater variability in the crossbred population than is found in purebreds for traits such as color and, possibly, body size and conformation. The first cross (the F_1) between two breeds (P_1, P_2), however, may be more uniform than purebreds for certain performance traits.

The main phenotypic effect of outbreeding and crossbreeding is to cause an improvement in traits related to physical fitness. This is the opposite of the effect of inbreeding. *The increased vigor of crossbreds as compared with the average of the purebred parents that make the cross is known as* **hybrid vigor (heterosis).** This is of great value in the commercial production of plants and animals.

Hybrid vigor may be defined as the superiority of crossbred offspring over the average of the purebreds used to make the cross. This may be expressed as follows:

$$\text{Purebred average} = P = \frac{P_1 + P_2}{2}$$

$$\text{Percent hybrid vigor} = \frac{\text{crossbred av} - \text{purebred av}}{\text{purebred av}} \times 100$$

$$= \frac{F_1 - P}{P} \times 100 = \frac{F_1 - (P_1 + P_2)/2}{(P_1 + P_2)/2} \times 100$$

Estimates of the amount of hybrid vigor expressed by certain traits in farm animals are given in Table 5.8.

Hybrid vigor is due to nonadditive gene action, such as dominance, over-dominance, or epistasis, but it is not due to additive gene action. In general, the farther apart the parent breeds are in relationship, the more hybrid vigor is expressed for those traits showing hybrid vigor. Theoretically, the reason for this is that one breed may be mostly homozygous for one allele (such as *DD*), whereas the other breed may be mostly homozygous for the other allele (*dd*). This could be true for several pairs of alleles; therefore the offspring from such crosses would be heterozygous for more gene pairs than would crosses of more closely related breeds or individuals.

Crossbreeding is the preferred system of mating for commercial production of many classes of livestock. This is particularly true of swine, because more pigs

TABLE 5.8
ESTIMATES OF AMOUNT OF HETEROSIS EXPRESSED FOR CERTAIN
TRAITS IN FARM ANIMALS WHEN A LINECROSS OR CROSSBRED
SYSTEM OF MATING IS FOLLOWED

Trait	Amount of heterosis			
	Cattle	**Swine**	**Sheep**	**Poultry**
Conception rate	Some*	Some*	Some	Some
Birth rate	Some	Medium	Some	....
Survival—birth to weaning	Large	Large	Large	....
Survival—weaning to market	Some	Some	Some	....
Weaning weight	Some	Some	Some	....
Postweaning growth rate	Little	Little	Little	....
Postweaning feed efficiency	Little	Little	Little	....
Milk production	Some	Some	Some	....
Carcass†	Little	Little	Little	Little
Egg production	....	....	....	Medium
Fertility and hatchability	....	....	....	Medium

*In a few crosses there is evidence that crossbreeding gives lower conception rate than does purebreeding. Perhaps, if true, it is due to incompatibility of sperm and egg or of embryo and mother.

†Crossbreds may be slightly fatter than the purebred parents.

survive from birth to weaning and because they may be heavier at weaning in the **three-breed cross.** Crossbred sows may wean 40 to 45 percent heavier *litters* (reflecting hybrid vigor) than purebreds. Similar figures are not available for sheep, but hybrid vigor probably approximates 20 to 25 percent. The increased pounds of calf weaned per breeding cow in the herd are probably near 10 to 20 percent for beef cattle. The percent hybrid vigor is probably less for this trait in beef cattle than in swine, because all cows, regardless of whether they are purebred or crossbred, usually produce just a single calf. Conversely, crossbred sows **farrow** and wean more pigs than purebreds. Crossbred cows, on the average, have heavier calves at weaning than purebreds, and more crossbred than purebred calves survive to weaning.

Several systems of crossbreeding may be used for the commercial production of livestock. These have been used more widely on a practical scale in swine production than with sheep and cattle.

The *single cross* refers to *crossing any two breeds.* The parents are purebreds, whereas the offspring are crossbreds. Hybrid vigor from this system of mating is limited to crossbred progeny. Such a system gives a considerable amount of hybrid vigor, but for best results it is necessary to use females from a breed that is known for its high prolificacy, good milking qualities, and mothering ability. Over 90 percent of the commercial **broilers** produced today come from the single cross of white Plymouth Rock females, rated fair in meatiness and good in egg production, with White Cornish males. Individuals of the Cornish breed are large-breasted and meaty individuals, but they are poor in egg production. The crossbred offspring combine the desirable characteristics of both parent breeds in their growth and body shape, plus some heterosis for physical fitness. A similar scheme is now being followed in some regions of the United States in commercial swine production.

The **backcross** (crisscross) is another crossbreeding system that may be used for commercial production of livestock. As an example, suppose Angus × Hereford females were crossed. The offspring of such a mating would be 1/2 Angus and 1/2 Hereford. Females of this cross could be kept for breeding and mated to an unrelated Angus bull. The calves produced would be 3/4 Angus and 1/4 Hereford. Heifers from this cross could then be mated to an unrelated Hereford bull, giving calves that were 5/8 Hereford and 3/8 Angus. This crisscross system of mating, using crossbred females mated in alternate generations to unrelated Angus and Hereford bulls, could be followed for many generations. Such a crossbreeding system has the advantage of using the more productive crossbred mothers, but some hybrid vigor may be lost in later generations, as compared with the first backcross between these two breeds.

The three-breed cross probably gives the optimum in heterosis of all proposed schemes used. This is because crossbred mothers are used, and the third breed may add additional hybrid vigor. An example of the three-breed cross would be the use of a Duroc boar on crossbred Hampshire × Yorkshire females from the original single cross mentioned previously. A rotation of purebred boars from the three breeds on subsequent generations of selected crossbred gilts could be continued indefinitely. Or a fourth breed could be used if desired. After the three-breed cross, one is merely attempting to retain as much of the hybrid vigor obtained as possible.

Regardless of the crossbreeding system used, only the best crossbred females and the best purebred males should be used. The reason for this is that heterosis, for traits that reflect it, adds to the average merit of the parents. For example, assume that hybrid vigor for litter size at weaning is 20 percent, and the litter size at weaning in each of three breeds is as follows:

Breed A	7.0
Breed B	8.0
Breed C	9.0

If the hybrid vigor is nearly the same for any combination of breeds (it may not be in actual practice), one would want to cross breeds B and C because the litter size of 20 percent from heterosis would add to the average of these two breeds (8.5), giving a total of 10.2 pigs as compared with 9.0 for the cross of breeds A and B.

5.11 SUMMARY

Quantitative inheritance, which involves many pairs of genes, affects most economic traits in farm animals. Phenotypic variation is the raw material with which the animal breeder must work to improve livestock and poultry. Methods of estimating the proportion of total phenotypic variation due to heredity have been developed, and from this knowledge, principles of improvement of livestock through breeding have evolved.

Tools the animal breeder has available to use in molding the raw material are selection, inbreeding, linebreeding, outbreeding, and crossbreeding. Proper use of these tools has greatly increased the efficiency of livestock production. More systematic use of these tools in the future should add to the income of the livestock producer.

STUDY QUESTIONS

1 In a herd of cattle the average weaning weight of all calves is 400 lb. Heifers weighing 450 lb and bulls weighing 500 lb are selected for breeding purposes. What is the selection differential for those animals selected for breeding?

2 For the situation in question 1, if only average heifers were selected for breeding, what would the selection differential be?

3 In one study, the correlation between husbands and wives for IQ was higher than the correlation between parents and children. What is a possible explanation for this?

4 In an experiment with chickens, selection was practiced only for longer shanks. Selection was effective, and shank length increased as the number of generations of selection increased. Since selection was based on individuality, or for individuals with longer shanks, what type of gene action was probably involved?

5 What are some possible explanations for different races of people with different skin colors?

6 What is responsible for the snowshoe hare in Alaska turning white in winter and brown in summer? Why do rabbits in the United States not change colors in winter and summer, as do snowshoe hares?

7 Assume that, in beef cattle, you have 10 heifers with the following 210-day weaning weights: 560, 550, 530, 510, 500, 450, 420, 410, 400, and 390 lb. If you saved five of the heaviest heifers for your herd, what would be their selection differential if the average heifers in this herd weighed 400 lb? If you saved three of the top heifers, what would their selection differential be? Since the three selected heifers had a larger selection differential than the five heifers, what important principle does this illustrate?

8 Calculate the generation interval in your family.

9 Compare the generation interval of Question 8 with those of four of your friends. From these data, what would you conclude about the length of the generation interval in humans?

10 In a herd of cattle in which the average daily gain in the feedlot for all calves is 2 lb per animal, heifers that gain 2.2 and bulls that gain 3 lb are kept for breeding. If the offspring gain 2.3 lb per animal daily, what is the heritability of rate of gain from these data?

11 Backfat thickness at 200 lb in an entire pig crop is 1.5 in. Boars measuring 1 in in backfat and gilts measuring 1.3 in are retained for breeding. If backfat thickness in swine is 45 percent heritable, what would be the expected backfat thickness in the progeny of these parents?

12 In Question 11, why was the backfat thinner in the selected boars than in the selected gilts?

13 Is the following statement true or false? "All humans are inbred." Explain your answer.
14 Mating a sire to 23 of his own unselected daughters is one system of progeny testing for the presence of a recessive gene in his genotype. What are the advantages and disadvantages of this kind of a progeny test?
15 Calculate the mean, variance, and standard deviation for the heifers in Question 7.
16 Is weaning weight in beef cattle a quantitative or a qualitative trait? Why?
17 Although black Angus breeders have never kept red Angus for breeding, about one out of 200 calves dropped among black Angus parents is red. Why does the red gene persist in the black Angus breed?
18 Outline a program that would help eliminate the red gene from the black Angus breed.
19 How would you start a pure breed of red Angus?
20 Assume you have a herd of red Angus and one of your cows produced a black Angus calf. How would you explain this?

ANSWERS TO STUDY QUESTIONS

1 The selection differential is 75 lb.
2 The selection differential, considering both heifers and bulls, is 50 lb.
3 It appears that there is a good correlation between the IQ of the husband and wife because a person with a high IQ tends to marry another with a high IQ. The IQ of parents is similar because of both hereditary and environmental effects. The correlation between IQ of parents and their offspring is mostly due to heredity.
4 Probably an additive type of gene action.
5 Mutations have been responsible for the genes for different skin color. Natural selection in a particular part of a country might have been effective in developing a certain skin type. In other words, one skin color might have caused better survival than another skin color in a given area.
6 Snowshoe hares in Alaska possess genes for changing coat color in different seasons. The ability to change coat colors probably arose from new mutations and increased in frequency, because the ability to do this made them better fit for survival in Alaska. Rabbits farther south probably do not change coat colors with seasons because they would be at a disadvantage when snow was not on the ground, since a white coat would make them more vulnerable to predators. In fact there probably has been selection against the ability to change coat color in the midwest and farther south in the United States.
7 The selection differential of the five heaviest heifers would be 130 lb. The selection differential for the three heaviest would be about 147 lb. This illustrates the general principle that the fewer animals kept for replacement, the larger the selection differential, and vice versa.
8 Answer depends on the individual.
9 Answer depends on the individuals.
10 About 50 percent.
11 Approximately 1.34 in in the progeny.
12 Because fewer boars are kept for breeding and the selection intensity would be greater than in gilts.
13 False. Inbreeding refers to an increase in the number of homozygous gene pairs when

relatives are mated as compared with the average of the population (or base population).

14 The main advantage is that such a progeny test is for any recessive gene the sire might be carrying, rather than a specific one. One disadvantage is that the offspring would be at least 25 percent inbred when from a father-daughter mating. This would result in a decline in vigor in the inbred offspring. Another disadvantage would be that the sire might be so old by the time the progeny test was completed that he might be out of **service** or dead.

15 The mean is 472 lb; the variance, 4262.22 lb; the standard deviation, 65.29 lb.

16 It is a quantitative trait because it is affected by many genes, and there is no sharp distinction between phenotypes with measurements for the trait in the population falling into a normal frequency distribution curve.

17 Perhaps there has not been enough emphasis on identifying and culling black carriers of the red gene *Bb*. Another remote possibility might be that selection favors heterozygous red *Bb* individuals.

18 Discard all red individuals and their parents. Progeny-test both cows and bulls to determine those that are homozygous black *BB* and keep only these for breeding. Homozygous black *BB* individuals would not produce red offspring unless a new mutation from black to red occurred.

19 Mate red individuals with red individuals.

20 The red gene might have mutated from red to black. Only one black gene is required to make an individual black. Also, one should check the records and make certain both parents were actually red.

CHAPTER **6**

ANATOMY AND PHYSIOLOGY OF FARM ANIMALS[1]

We are so accustomed to investigating the functions of different organs or parts of the body that we do not see the individual as a living whole.

Stewart Paton

6.1 INTRODUCTION

Although there is considerable variation among **species** within the animal kingdom, domestic farm animals show many similarities in form (anatomy) and function **(physiology).** In many instances, humans have altered their form and function for their own needs. In this chapter some anatomical and physiological characteristics of farm animals are discussed. A knowledge of these is important because it enables one to describe animals more clearly in judging and selecting for breeding purposes and to properly apply husbandry and veterinary practices.

Because of the complexity of the bodies of humans and animals, scientists divide anatomy into several branches. *Gross anatomy* is the study of structures that can be seen by the unaided eye. *Microscopic anatomy,* or *histology,* is the study of tissues by means of microscopes. *Comparative anatomy* compares the body structure of different species. **Embryology** is the study of the body as it develops **in utero** in farm mammals and in the fertilized eggs of **poultry.** A knowledge of body structure is essential in understanding the function of **morphology** in maintaining health and resisting disease. *Physiology* is the branch

[1]The authors acknowledge with appreciation the contributions to this chapter of Dr. P. C. Harrison, Department of Animal Science, University of Illinois, Urbana.

of biology related to studies of body-organ function, individually and collectively, in systems.

The anatomy and physiology of the reproductive system are presented in Chapter 9. The **endocrine** system is discussed in Chapter 7. The physiology of growth, **lactation,** egg laying, and digestion are discussed in Chapters 8, 11, 12, and 15, respectively. Certain aspects of environmental physiology are presented in Chapter 13.

6.2 EXTERNAL BODY PARTS

Various external body parts of selected farm animals are shown in Figures 6.1 to 6.6. Other terms, not given in the figures, are also used to describe and identify the location of body parts of an animal. These include dorsal, ventral, caudal (posterior), and cranial (anterior). **Dorsal** refers to the top or back side of an

FIGURE 6.1
External parts of a dairy cow.

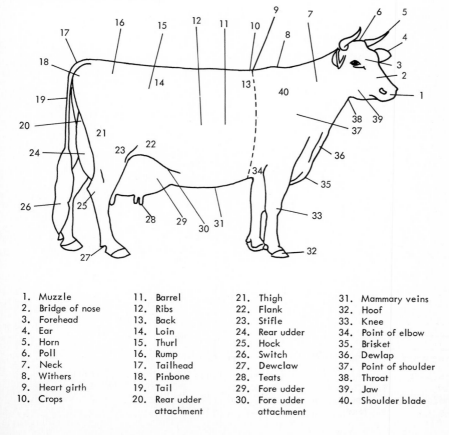

1. Muzzle	11. Barrel	21. Thigh	31. Mammary veins
2. Bridge of nose	12. Ribs	22. Flank	32. Hoof
3. Forehead	13. Back	23. Stifle	33. Knee
4. Ear	14. Loin	24. Rear udder	34. Point of elbow
5. Horn	15. Thurl	25. Hock	35. Brisket
6. Poll	16. Rump	26. Switch	36. Dewlap
7. Neck	17. Tailhead	27. Dewclaw	37. Point of shoulder
8. Withers	18. Pinbone	28. Teats	38. Throat
9. Heart girth	19. Tail	29. Fore udder	39. Jaw
10. Crops	20. Rear udder attachment	30. Fore udder attachment	40. Shoulder blade

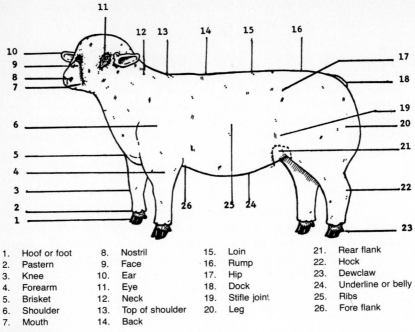

1.	Hoof or foot	8.	Nostril	15.	Loin	21.	Rear flank
2.	Pastern	9.	Face	16.	Rump	22.	Hock
3.	Knee	10.	Ear	17.	Hip	23.	Dewclaw
4.	Forearm	11.	Eye	18.	Dock	24.	Underline or belly
5.	Brisket	12.	Neck	19.	Stifle joint	25.	Ribs
6.	Shoulder	13.	Top of shoulder	20.	Leg	26.	Fore flank
7.	Mouth	14.	Back				

FIGURE 6.2
External parts of a sheep. *(Courtesy Dr. Robert Godke, Louisiana State University.)*

FIGURE 6.3
External parts of a market hog. *(Courtesy Dr. Robert Godke, Louisiana State University.)*

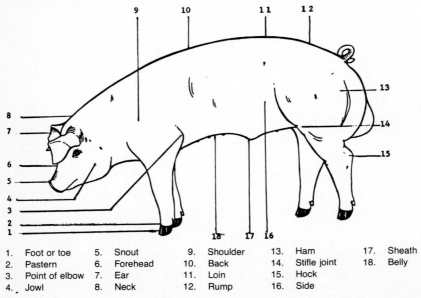

1.	Foot or toe	5.	Snout	9.	Shoulder	13.	Ham	17.	Sheath
2.	Pastern	6.	Forehead	10.	Back	14.	Stifle joint	18.	Belly
3.	Point of elbow	7.	Ear	11.	Loin	15.	Hock		
4.	Jowl	8.	Neck	12.	Rump	16.	Side		

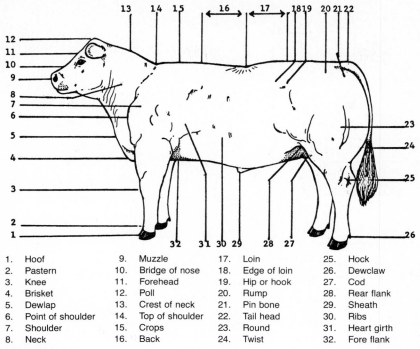

1. Hoof	9. Muzzle	17. Loin	25. Hock
2. Pastern	10. Bridge of nose	18. Edge of loin	26. Dewclaw
3. Knee	11. Forehead	19. Hip or hook	27. Cod
4. Brisket	12. Poll	20. Rump	28. Rear flank
5. Dewlap	13. Crest of neck	21. Pin bone	29. Sheath
6. Point of shoulder	14. Top of shoulder	22. Tail head	30. Ribs
7. Shoulder	15. Crops	23. Round	31. Heart girth
8. Neck	16. Back	24. Twist	32. Fore flank

FIGURE 6.4
External parts of a steer. *(Courtesy Dr. Robert Godke, Louisiana State University.)*

individual; **ventral** refers to the belly or underside. **Cranial** (anterior) means toward the front, whereas **caudal** (posterior) is the opposite, meaning toward the rear.

Skin is the exterior covering of the body and is continuous with the exterior membranes of the respiratory, **urogenital,** and digestive tracts. Hair, wool, horns, feathers, and hooves are considered to be modified appendages of skin. The skin functions in many ways, such as protecting against infections, regulating temperature, and helping the individual respond to its **environment** by means of **sensory** nerves. The skin also contains **glands** of secretion and excretion (Figure 6.7).

The skin consists of two layers, the **epidermis** and the **dermis.** The epidermis is the outer layer of epithelial cells, which contains no blood cells. The inner layer, known as the *dermis,* or corium, is composed of a connective tissue network and includes blood vessels, **lymph** vessels, nerves, glands, hair follicles, and muscle fibers. The principal glands of the dermis are sweat glands (sudoriferous) and sebaceous glands. The latter secrete oily substances that lubricate the hair and skin. The dermis also contains special types of glands, such as the modified sweat glands found in the snout of swine. Even mammary glands are considered to be highly modified glands of skin. Skin thickness varies with

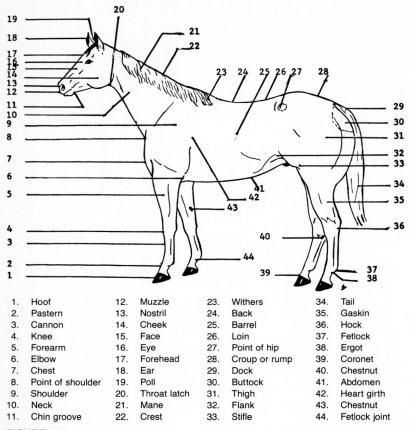

FIGURE 6.5
External parts of a horse. *(Courtesy Dr. Jack Kreider, Louisiana State University.)*

1.	Hoof	12.	Muzzle	23.	Withers	34.	Tail
2.	Pastern	13.	Nostril	24.	Back	35.	Gaskin
3.	Cannon	14.	Cheek	25.	Barrel	36.	Hock
4.	Knee	15.	Face	26.	Loin	37.	Fetlock
5.	Forearm	16.	Eye	27.	Point of hip	38.	Ergot
6.	Elbow	17.	Forehead	28.	Croup or rump	39.	Coronet
7.	Chest	18.	Ear	29.	Dock	40.	Chestnut
8.	Point of shoulder	19.	Poll	30.	Buttock	41.	Abdomen
9.	Shoulder	20.	Throat latch	31.	Thigh	42.	Heart girth
10.	Neck	21.	Mane	32.	Flank	43.	Chestnut
11.	Chin groove	22.	Crest	33.	Stifle	44.	Fetlock joint

species, **breed,** sex, and body location (e.g., eyelids are thin, whereas the sole of the foot is thick).

   Hair is the coat covering of cattle, goats, horses, and swine, whereas *wool* is the covering of sheep. Wool differs from hair by being finer-textured, soft, and curly; it is characterized by very small overlapping scales. In temperate or cold climates, the hair coat of animals grows long for the winter and is shed in the spring. The wool of sheep is usually shorn (clipped off) by the caretaker in the spring, before the onset of hot weather. This allows the sheep to be more comfortable in the summer months and also yields a product with economic value. The wool usually grows out before the following winter, thus protecting the sheep from cold temperatures. Also, wool growth increases during the spring and summer; it is an adaptive mechanism for insulation against radiation heat gain in hot, arid climates. *Feathers* are the body covering of poultry. Hens shed

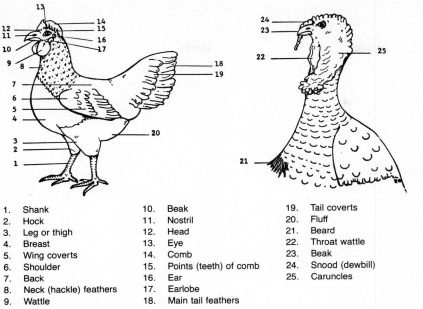

1.	Shank	10.	Beak	19.	Tail coverts
2.	Hock	11.	Nostril	20.	Fluff
3.	Leg or thigh	12.	Head	21.	Beard
4.	Breast	13.	Eye	22.	Throat wattle
5.	Wing coverts	14.	Comb	23.	Beak
6.	Shoulder	15.	Points (teeth) of comb	24.	Snood (dewbill)
7.	Back	16.	Ear	25.	Caruncles
8.	Neck (hackle) feathers	17.	Earlobe		
9.	Wattle	18.	Main tail feathers		

FIGURE 6.6
Parts of a chicken and turkey. *(Courtesy Dr. John A. Sims, Iowa State University; redrawn by Dr. Robert Godke, Louisiana State University.)*

FIGURE 6.7
Diagram of hair follicle of the skin.

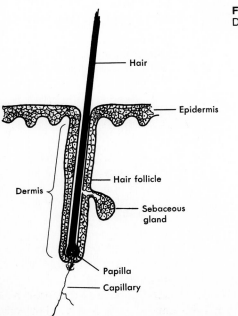

their feathers (**molt**) in late summer and grow new ones in time to provide a body covering in the winter. It is interesting that hair, wool, and feathers are proteinaceous in composition and are produced through the action of **genes** (Chapter 4).

Cattle, goats, sheep, and swine have cloven (split or divided) hooves; however, horses do not. A breed of swine known as the *mule-foot* (describes hoof appearance) does not have cloven hooves. Mule-foot in swine is genetically **dominant** over the normal cloven hoof. A similar heritable condition of cattle in which the hoof is not cloven has been described.

6.3   THE SKELETAL SYSTEM

The bony tissues that form the body's framework constitute the skeletal system. This system includes the long bones of legs, ribs, vertebrae, skull, and other bony processes. In farm animals and many other species, the skeletal systems are internal (endoskeleton) and basically alike. A giraffe's neck has the same number of bones as a mouse's neck; it is merely the length and size of the bones that vary. In many species, such as insects, the skeleton is outside the body (exoskeleton).

The outer layers of bony tissues are filled with mineral deposits, mainly in the form of calcium and phosphorus (Figure 6.8). The inner core is a soft tissue known as *bone marrow*. Some of this tissue is yellow and consists mostly of fat; it is called *yellow marrow*. The red tissue of bone marrow is known as *red marrow* and functions in blood cell and platelet formation.

The skeletal tissue increases in size and length as an animal grows. Growth of bones occurs at the ends in the region of the **cartilage** between the **epiphysis** (the

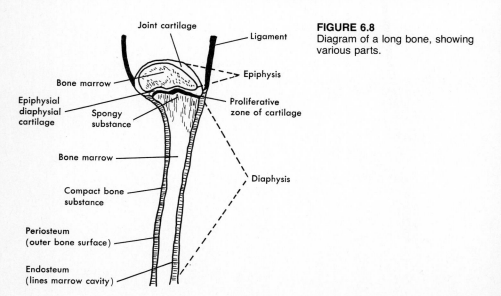

FIGURE 6.8
Diagram of a long bone, showing various parts.

Joint cartilage

Ligament

Epiphysis

Bone marrow

Epiphysial diaphysial cartilage

Spongy substance

Proliferative zone of cartilage

Bone marrow

Diaphysis

Compact bone substance

Periosteum (outer bone surface)

Endosteum (lines marrow cavity)

end) and **diaphysis** (the shaft). This cartilage, known as *epiphysial-diaphysial cartilage,* gradually becomes calcified and is replaced by bony material. Once the cartilage is completely replaced by bone, further bone growth ceases. **Calcification** of the epiphysial-diaphysial cartilage at the ends of long bones occurs at physical maturity. Further growth in body proportions (except fattening) then ceases (Chapter 8). It is interesting that bone is continually being reabsorbed and replaced in mature animals, although no net bone growth occurs. The diameter of long bones also increases as they grow and lengthen. This increase in size results from production of new bony tissue by the **periosteum** that surrounds the cortex of bone. As new bone is deposited, portions of the deeper, inner bone are removed, in turn increasing the size of the marrow cavity.

Hormones, vitamins, and other **nutrients** can affect proper bone growth. The mineral tissues of bone can be depleted in cases of nutritional deficiencies (Chapters 16 and 17), and they may become fragile and/or distorted. The skeletal systems of the cow and the chicken are shown in Figures 6.9 and 6.10. The skeletal systems of all farm **mammals** are similar, but only that of the cow is shown. Cattle normally have 13 pairs of ribs; however, some occasionally have 14 pairs.

Bony tissues are fairly rigid and become more brittle with age. For this reason, they are sometimes fractured or broken, but the body has the power to

FIGURE 6.9
Skeleton of a cow. *(Courtesy of Dr. P. D. Garrett and Dr. D. E. Rodabaugh, University of Missouri.)*

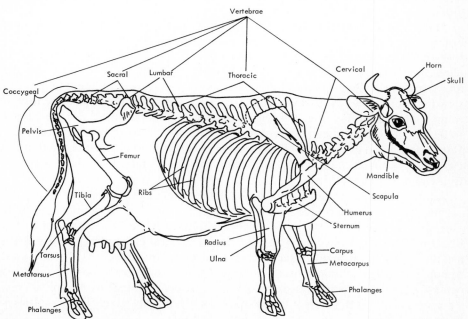

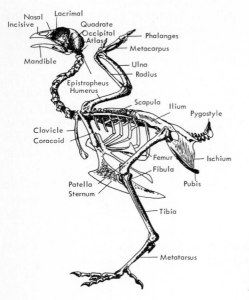

Nasal
Incisive
Lacrimal
Quadrate
Occipital
Atlas
Mandible
Epistropheus
Humerus
Clavicle
Coracoid
Patella
Sternum
Phalanges
Metacarpus
Ulna
Radius
Scapula
Ilium
Pygostyle
Femur
Fibula
Pubis
Ischium
Tibia
Metatarsus

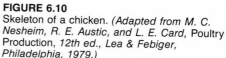

FIGURE 6.10
Skeleton of a chicken. *(Adapted from M. C. Nesheim, R. E. Austic, and L. E. Card,* Poultry Production, *12th ed., Lea & Febiger, Philadelphia, 1979.)*

mend or repair the break. Normal mending of bones occurs when both ends of the fracture are brought together and held securely in place by a splint, bone pin, or other device. The break is first filled by a fibrin clot, which is then invaded by capillaries to furnish a blood supply. The break is later filled with true bony tissue.

The skeletal system functions in body support. This is especially true of long bones. The skeletal system also provides leverage for muscular movement. Movement is made possible by several kinds of joints. *Ball-and-socket joints* allow movement in all directions. An example is the shoulder joint, where the arm is attached. *Hinge joints* allow certain body parts (e.g., the toes) to move in two directions. *Pivot joints* are found in the neck and allow the head to be turned in more than one direction. *Gliding joints* as found in vertebrae allow the body to be flexible so that it can be bent forward, backward, or in several directions. Bones are joined at the joints by **ligaments** outside the joint capsule. Within the capsule is **synovial** fluid, which lubricates the joints and allows them to move freely and without friction.

Another important function of the skeletal system is protection of vital body organs. The rib cage protects the heart, lungs, and certain abdominal organs. It also aids in breathing. The skull protects the brain. The importance of this is illustrated when a brain concussion (jarring of the brain by a blow) occurs. Such a concussion can cause unconsciousness and even permanent damage to this vital organ. The spinal column protects the spinal cord from injury. Such protection is essential for the life and well-being of an individual. Damage to the spinal cord by a serious injury may result in paralysis of certain body parts. This paralysis is often permanent because nerves cannot regenerate.

6.4   THE MUSCULAR SYSTEM

The external muscular system of the cow is shown in Figure 6.11. Muscle consists largely of **protein** and is the lean portion of the **carcass** in meat animals. Muscle cells usually occur in bundles or sheets, although they may occasionally occur as individual cells scattered throughout the tissue.

Muscles are commonly classified as voluntary or involuntary, i.e., under or not under the control of the individual's will. Voluntary muscles function when the animal calls on them. Muscles are also classified as *smooth* (unstriated) or *striated* (striped). When viewed microscopically, striated muscles show cross striations (stripes or streaks), whereas smooth muscles do not. These stripes result from the proteins *myosin* and *actin*. Voluntary muscles are striated; involuntary muscles may be either smooth or striated.

Skeletal muscles are voluntary. They are connected to bones by **tendons,** which consist of dense connective tissue. Muscles are usually attached to two bones and provide power for movement of various body parts. Skeletal muscles are also classified to indicate the type of movement they produce. *Extensor* muscles cause body parts to straighten; *flexor* muscles cause them to bend. *Abductor* muscles cause body parts to move away from a plane through the body, and *adductor* muscles draw them toward the body plane. Each voluntary muscle fiber is controlled by a branch of a voluntary nerve, or motor neuron. One motor neuron and its many branches that supply muscle fibers are collectively called a *motor unit.*

Involuntary (smooth) muscles are found in the walls of organs of many body systems, such as the digestive, urogenital, and circulatory systems. Moreover,

FIGURE 6.11
Selected muscles of the cow. *(Courtesy Dr. P. D. Garrett and Dr. D. E. Rodabaugh.)*

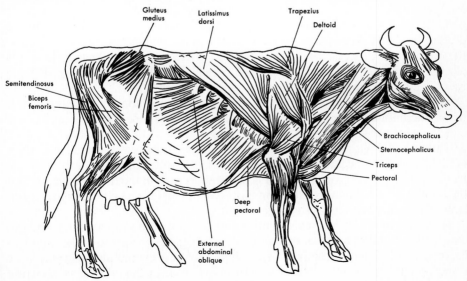

involuntary muscles are located in many secretory organs and function to expel or force out their secretions. (An exception is the endocrine glands.) Within the digestive tract, involuntary muscles provide impetus for movement of **ingesta.** In the female reproductive tract, involuntary muscles are responsible for uterine motility, which moves sperm up the tract to meet the descending **ovum.** They also aid in concert with voluntary muscles to expel the **fetus** at **parturition.** Involuntary muscles cause contractions of the heart and increase or decrease the diameter of blood vessels, thus regulating or controlling the amount of blood flowing to a particular body area. Contractions of involuntary smooth muscles are regulated by the autonomic nervous system. The hormones *epinephrine* (adrenalin) and *norepinephrine* (noradrenalin) stimulate autonomic nerves to cause relaxation and contraction, respectively, of the involuntary smooth muscles.

Cardiac muscle (an involuntary striated muscle) constitutes the larger mass of the heart. Although the heart beats spontaneously, its rate and force of contraction can be altered by the autonomic nervous system.

The physiology of muscle action has received considerable attention from research workers, yet it is not completely understood. Several years ago, muscle contraction was thought to result from energy released on the breakdown of proteins. If this were true, it would mean that following work, there should be an increased excretion of the end products of protein **metabolism** from the body. Research studies failed to show this. Rather, the energy utilized for muscle action came largely from nonprotein sources. However, if food intake during work was insufficient and the animal lost weight, both proteins and body fats were broken down and used as energy sources.

The details of the chemical reactions involved in muscle contraction are quite complex. Therefore only a simple summary will be given. The chief source of energy used for muscle contraction is now thought to be adenosine triphosphate (ATP). When a muscle is stimulated, ATP breaks down into adenosine diphosphate (ADP) and a phosphate radical, releasing energy, which is used in the contraction process. The breakdown of muscle glycogen to lactic acid serves to generate more ATP, thus providing a cycle for production of energy that can be utilized during muscle contraction.

Another step in the chemistry of muscle action includes the breakdown of phosphocreatine to creatine and phosphoric acid. This reaction also supplies the energy and phosphoric acid needed to generate ATP. Phosphocreatine is resynthesized from phosphoric acid and creatine, using energy available when ATP is in excess, and thus serves as a storage battery. About one-fifth of the lactic acid produced from glycogen is oxidized to carbon dioxide and water, and the energy released by this **oxidation** is used by the liver to resynthesize glycogen from the remaining four-fifths of the lactic acid.

Restoration of the original chemical condition of the muscle to that before contraction results in a loss of part of the glycogen by means of oxidation to lactic acid by an aerobic activity. Thus glycogen must eventually be replaced by some means for normal contraction to occur. Energy for muscle contraction comes from an anaerobic reaction, using the energy released by lactic acid formation. (Muscle glycogen is converted anaerobically into lactic acid.) However, this reaction cannot continue indefinitely, because there is an accumulation of lactic acid and a lack of oxygen to oxidize it (to combine it with oxygen). The deficit of oxygen needed to replace that used aerobically is called the *oxygen debt*. This debt must be repaid before there can be a normal resumption of muscular activity. Through the temporary anaerobic recharging of the muscle, nature has provided animals with a means of expending several times more muscular activity in a short time than would be possible if oxygen had to be supplied concurrently with muscle action. Studies at the Missouri Agricultural Experiment Station showed that horses and human males, during intense muscular work, may expend energy at 100 times the resting state. Such activity cannot be maintained for long because of the onset of fatigue. It would seem that individuals who can expend a greater amount of energy before their oxygen debt halts muscular activity would be those most likely to survive in environments that required rapid and prolonged muscular activity. An expenditure of energy 6 to 8 times greater than that of rest may be endured during prolonged hard work. Under these conditions oxidation keeps pace with energy expenditure.

Certain minerals are also involved in muscle activity. Calcium **ions** function as an **enzyme-**activating factor necessary for muscle contraction (Chapters 16 and 17). One theory states that calcium ions initiate the enzyme activation of myosin filaments in muscle. Relaxation of muscle is believed to result from the binding of calcium, which makes it unavailable in the myosin area. Magnesium ions can activate the calcium-binding enzyme. Thus magnesium ions are also involved in muscle relaxation.

6.5 THE CIRCULATORY SYSTEM

All history shows the power of blood over circumstances, as agriculture shows the power of the seeds over the soil.

Edwin Perry Whipple (1819–1886)

The circulatory system of animals consists of the heart, veins, capillaries, arteries, **lymph vessels,** and lymph glands.

6.5.1 Anatomy and Physiology

The *heart* is located in the thoracic cavity between the lobes of the lungs (Figure 6.12). The mammalian heart is a marvelous and complicated organ. It is the pump that circulates blood to all body parts. Failure to perform this function (heart failure) terminates life. It has been estimated that the heart of a person 70 years old has beaten at least 2.5 trillion times and has pumped more than 435,000 tons of blood. A typical human heart pumps blood through an estimated 60,000 miles of blood vessels daily (enough blood to fill a 4000-gal tank car).

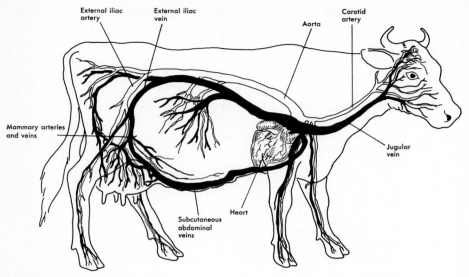

FIGURE 6.12
Systemic blood circulatory system of the cow (veins are dark and arteries are light). *(Courtesy Dr. P. D. Garrett and Dr. D. E. Rodabaugh.)*

Arteries are blood vessels that carry oxygen-rich blood from the heart to various body tissues. Arterial walls are thick and contain heavy muscle layers that can withstand the blood pressure resulting from the heart's beating. *Veins* are blood vessels that return blood from throughout the body to the heart and that convey it from the lungs, where carbon dioxide and oxygen are exchanged, to the heart. Veins have comparatively thin walls (compared with the walls of arteries), which are collapsible. In places, veins contain valves that aid the flow of blood to the heart.

Capillaries are minute blood vessels that lie between the terminal arteries and arterioles and the beginnings of the venules and veins. They are very tiny and thin-walled (one cell thick) and are widely distributed in all body tissues. Transfer of nutrients from blood to tissues, and waste products from tissues to blood, occurs in capillaries. It is interesting that every pound of excess fat contains an estimated 200 miles of capillaries. This partly explains how being overweight can overwork the heart and contribute to eventual heart failure.

Lymph vessels are accessories to the body's circulatory system. They originate in tissue spaces and converge to form larger ducts as they pass through lymph glands. Many such lymph ducts converge into a single duct, or vessel, which empties into the large blood veins of the circulatory system. Lymph glands filter foreign substances from lymph, preventing their passage into the bloodstream. They also produce **lymphocytes,** one type of white blood cells.

Contraction of the heart circulates blood through the circulatory system. The heart controls its own beat through the action of the sinoatrial node, called the

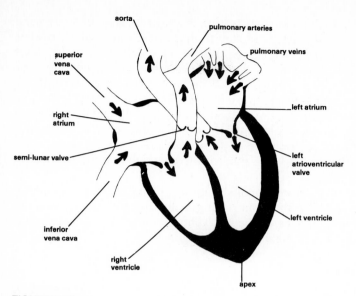

FIGURE 6.13
Diagram depicting the four chambers of a mammalian heart. The arrows indicate the direction of blood flow. Deoxygenated blood enters the right atrium through the vena cava. Blood is sent from the right ventricle to the lungs via the pulmonary arteries. Oxygenated blood is returned to the left atrium via the pulmonary veins and is pumped through the aorta to the general circulation by the left ventricle. *(Drawn by F. E. Staten, University of Illinois, 1983.)*

pacemaker, which is located in the right atrium. The heartbeat, however, is also influenced by accelerator and inhibitor nerves. In humans, faulty pacemakers have been replaced by an electronic timer that controls the heartbeat. Such an operation-and-replacement therapy has been employed successfully many times.

Contraction of the heart begins in the right atrium and quickly spreads to the left atrium. Blood is then forced into the ventricles, which contract, closing the atrioventricular (AV) valves. This forces the blood into the major arteries (Figure 6.13). The heart itself requires a very rich blood supply. It represents only about 0.5 percent of the body weight of humans but requires approximately 5 percent of the total blood supply.

The *systemic circulatory system* (somatic circulation) refers to the heart and the vessels that move oxygenated arterial blood to all body regions and return venous (unoxygenated) blood to the heart. Systemic circulation is quite complex, and it is dependent on particular tissue requirements (e.g., digestion versus exercise). Arterial blood in the systemic circulation is of a bright red color, but after passing to the veins, it becomes very dark or brownish red. The *portal system,* a part of the systemic circulatory system, conveys venous blood from the stomach, pancreas, small intestine, and spleen to the liver, where it passes through a second capillary bed before returning to the heart.

The vascular system that circulates blood through the lungs, in which it is oxygenated, is known as the *pulmonary system.* This system is an exception and differs from the systemic circulatory system in that veins, rather than arteries, carry oxygenated blood. The artery involved is the *pulmonary artery.* This artery divides after leaving the heart, sending one branch to the right lobe and another to the left lobe of the lungs. Each branch of the pulmonary artery subdivides many times, first forming *arterioles,* and finally capillaries of the lungs, where oxygen is taken into the blood and carbon dioxide is exhausted from the blood. Once blood is forced through these capillaries, it passes into *small veins (venules),* which converge to form *pulmonary veins.* These veins return oxygenated blood to the left atrium of the heart, where the pulmonary circulation ends.

The pressure required to pump equal volumes of blood through the pulmonary circulation is much less than that required in the systemic circulation. Consequently, the heart muscles of the left ventricle develop a much greater muscle mass than those of the right ventricle (Figure 6.13).

6.5.2 Blood Composition

Approximately 50 to 65 percent of the total blood volume consists of **plasma.** It is the golden-straw-colored liquid that remains after cells have been removed. To obtain plasma, a blood sample is treated to prevent clotting and let stand until the cells settle to the bottom of the container. (Plasma may also be obtained by centrifuging whole blood.) The corpuscles are suspended in blood plasma. Plasma usually contains approximately 90 percent water and 10 percent solids. The solids are composed of inorganic salts and organic substances such as **antibodies,** hormones, vitamins, enzymes, proteins, and glucose (blood sugar). The cellular, or nonplasma, portion of blood contains erythrocytes (red blood cells), leukocytes (white blood cells), and platelets. Erythrocyte and leukocyte values for selected species are presented in Table 6.1. **Serum** is the fluid remaining after blood has clotted. Serum, then, is plasma from which the fibrinogen (a clotting factor) has been separated in the clotting process.

TABLE 6.1
AVERAGE VALUES FOR CERTAIN BLOOD COMPONENTS IN SELECTED ANIMALS*

Species	Milliliters of blood per pound of body weight	Millions of erythrocytes per cubic millimeter of blood	Thousands of leukocytes per cubic millimeter of blood
Cattle	26.1	6–10	5–12
Chickens	25.0	2– 3	20–35
Dogs	42.7	5– 8	8–15
Goats	32.0	13–18	6–14
Horses	41.0	6–10	5–11
Sheep	30.0	9–11	5–10
Swine	30.0	6– 7	15–25

*Many factors, such as age and stress, affect values of blood components.

Red blood cells contain **hemoglobin,** which has the important function of transporting oxygen from the lungs to various body tissues. Red blood cells also have **antigens** (on their surface), which are responsible for the various blood types of different species.

6.5.3 Hemoglobins

Hemoglobin is responsible for giving red blood cells their characteristic red color. Chemically, hemoglobin is an organic compound consisting of heme (an iron-porphyrin complex) and globin (a protein). Hemoglobin readily absorbs oxygen from air in the lungs, forming oxyhemoglobin. Oxygen is held rather loosely by hemoglobin and is readily given to tissues within the body. Red blood cells in an adult mammal are formed in the red bone marrow. By the time they reach the bloodstream, their **nucleus** has been lost. In birds, however, the nucleus remains throughout the cell's life. The life of the red blood cells in the circulatory system usually varies from 90 to 120 days, after which the cells are removed by the spleen. **Anemia** is a condition in which the number of red blood cells, or the amount of hemoglobin, is reduced to a subnormal level. It may be caused by loss of blood from a wound or by **infestations** of bloodsucking **parasites.** Anemia may also result from a subnormal rate of red blood cell production because of faulty nutrition (Chapters 16 and 17) or from a shorter than normal life of red blood cells (about 60 days in sickle-cell anemia in humans).

Many types of hemoglobins in humans have been described. These are caused by gene **mutations.** When a new gene mutation occurs, a different protein (globin) is produced because of a change in the code sent from the genetic material (**DNA**) to the ribosomes in the **cytoplasm** of cells for building a particular protein (Chapter 4). At least two different types of hemoglobin are normally found in the red blood cells of animals. *Fetal* hemoglobin is present in the red blood cells of young at birth, but it is soon replaced by *adult* hemoglobin. Recent studies have shown that humans and animals sometimes possess abnormal types of hemoglobin. An abnormal type, known as *sickle-cell hemoglobin,* is found mostly in the Negroid race and has been of great value in studies designed to determine how genes function at the cellular level.

6.5.4 Blood Types

The red blood cells of humans and animals possess certain gene-determined antigens on their surfaces. **Antibodies** against certain antigens may also occur in blood serum (Section 18.2.5). Antibodies for a particular antigen are not found in the blood of the same individual because the reaction between the two would **agglutinate** the red blood cells and cause death. Such a reaction may be responsible for some early embryonic death losses in humans and animals.

The first blood group discovered was the A, B, and O series in humans. At least three genes (*A, B,* and *a*) are involved. Genes *A* and *B* are dominant to *a,* which is the gene for blood group O. However, *A* and *B* are not dominant to each other. Gene *A* produces antigen A; gene *B,* antigen B; and group O individuals possess neither antigen A nor B on their red blood cells. Individuals belonging to blood group A can be of **genotypes** *Aa* or *AA;* those of blood group B, of genotypes *Ba* or *BB;* those of blood group AB, of genotype *AB;* and those of blood group O, of genotype *aa.* Although the antibody for a particular antigen does not occur together with it in the same individual, the opposite antibody does occur. For example, individuals of group A possess antibody B in their serum, those of group B possess antibody A, those of group AB have neither antibody, and those of group O possess both antibodies A and B.

A knowledge of the A, B, and O blood-group series has been of great value in matching types of blood for successful whole-blood transfusions (Figure 6.14), for use in certain medical and legal problems, in parentage disputes, and in certain genetic studies.

The discovery of the Rh blood group and its importance in humans has been of widespread interest. It received its name because the antigen was discovered first in the rhesus monkey (hence Rh). About 85 percent of the white population in the United States possess this antigen. They belong to the Rh positive blood group. Research has shown, however, that the Rh positive group consists of several different alleles (genes). The other 15 percent belong to the Rh negative blood group. It has been found that the Rh negative gene is recessive to all Rh positive genes. Thus an Rh positive person can be of genotypes Rh positive Rh positive or Rh positive Rh negative, whereas an Rh negative individual can have only the Rh negative Rh negative genotype.

The antibody against the Rh positive antigen does not occur naturally, but an Rh negative person receiving a whole-blood transfusion of Rh positive blood can build antibodies against the antigen. The first transfusion of this kind would cause no complications; however, subsequent transfusions would cause agglutination of the recipient's red blood cells, possibly causing death.

The Rh positive antigen-antibody reaction was later found to be responsible for the condition known as *erythroblastosis fetalis* in human infants at birth. This condition results in jaundice, anemia, and often death. It occurs only when an Rh negative woman marries an Rh positive man and has Rh positive children.

FIGURE 6.14
Schematic diagram showing compatibility and incompatibility among blood types for whole-blood transfusions. The key to transfusions possibly depends on whether or not incoming cells are agglutinated. The direction of the arrows indicates which type can receive whole-blood transfusions from the other types. Blood type O is a *universal donor* and blood type AB is a *universal recipient.*

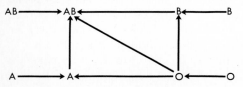

The Rh positive antigenic substance in some way crosses the placenta from the blood of the baby to that of the mother. The Rh negative mother builds antibodies against the Rh positive antigen from her child. These antibodies then cross the placenta from the blood of the mother to that of the child. When the mother produces enough antibody, many of the baby's cells are destroyed. Two-way blood transfusions at birth often save such a child. Usually, the Rh negative mother can have several Rh positive children before she builds enough antibody against the Rh positive antigen to cause serious difficulty. However, an Rh negative mother who has received a whole-blood transfusion with Rh positive blood may have enough antibody to affect her first Rh positive child. Rh negative children are not affected. When a father is Rh positive Rh negative and the mother is Rh negative Rh negative, approximately one-half of their children are Rh negative Rh negative. However, if the father is Rh positive Rh positive, all the children are Rh positive Rh negative.

A method for preventing erythroblastosis fetalis in human infants has been developed. This method is based on the discovery that an Rh negative mother of blood type O (genotype *aa)* always possesses antibodies A and B in her blood serum. If her baby carries blood antigens A or B on its red blood cells and some of these get into the mother's bloodstream at birth, they will be destroyed by means of the antibody-antigen reaction. This reaction also destroys Rh-positive-bearing blood cells at the same time and before they have an opportunity to stimulate Rh positive antibody production by the Rh negative mother. As an experiment, Rh negative mothers with newborn Rh positive babies were injected with serum that contained antibodies against the Rh positive cells of their children. This serum did destroy a high proportion of the Rh-positive-bearing cells, but it was found that the treatment later stimulated Rh positive antibody production. Additional experimentation led to the successful use of anti-Rh gamma globulin, which coats the antigen of the Rh positive cell so that it does not come in contact with antibody-forming cells. This treatment inhibits production of antibodies against the Rh positive antigens if given within 72 h after the birth of the first Rh positive baby. This protects the second child from erythroblastosis fetalis. Injections must be repeated with each successive Rh positive baby the Rh negative mother has thereafter. The treatment is ineffective if given to Rh negative mothers who have already given birth to Rh positive babies and therefore possess Rh positive antibodies. It will, however, greatly benefit Rh negative women who start families in the future.

Many blood types and groups have been discovered in farm animals. Some appear to cause diseases similar to the erythroblastosis fetalis in humans, but the antibody in animals is usually transferred from mother to young through **colostrum.** The placental membrane of nonprimate mammals is less permeable than that of **primates** and does not allow antibodies to pass through to the young within the uterus.

Blood-typing of farm animals has several uses on a practical basis. It may be used to identify individuals and their parents. (Blood-typing is required of bulls used in commercial artificial insemination.) Blood types appear to be associated with performance in some instances, and the mating of individuals having certain blood types may result in superior performance. Considerable blood-typing has been done in poultry in relation to the hatchability of eggs and egg production. In swine, blood-typing has been used to identify individuals that are genetically susceptible to porcine stress syndrome (PSS), a condition that can affect survival under stress conditions and can also cause pale, soft, exudative (PSE) carcasses to be produced.

All proteins in the blood of animals are genetically determined, and certain tests for their presence can be used for identification purposes. Those commonly used are albumins, haptoglobins, prealbumins, and transferrins.

6.5.5 White Blood Cells (Leukocytes)

Leukocytes include basophils, eosinophils, lymphocytes, monocytes, and neutrophils. They differ from erythrocytes in possessing a nucleus during their residence in the circulatory system. Neutrophils and lymphocytes commonly represent 85 to 90 percent of the leukocytes in farm mammals. The total number of these two types of cells in farm mammals is approximately equal, although the proportion varies somewhat among species. The biological significance of these different proportions, if any, is not fully understood. Temporary stress in farm mammals results in a very significant increase of neutrophils in proportion to lymphocytes. The proportion returns to the original level a few hours following relaxation or the removal of stress. The change in proportions of these leukocytes during periods of stress results from the interaction of a hormone from the anterior pituitary gland (ACTH) and adrenal cortical hormones.

Neutrophils are produced in bone marrow. They fight disease by migrating to the point of infection, engulfing (**phagocytizing**) bacteria, and destroying them. Leukocytes generally increase in number during bacterial infections, but viral infections cause their number to decrease (leukopenia). Thus a differential leukocyte count may be used to diagnose disease. Eosinophils are so named because they contain granules that turn red when stained with eosin (a red dye). Eosinophils normally constitute less than 5 percent of the total leukocyte count in farm animals. An increase from normal in eosinophils is indicative of an allergic reaction within the body or some type of parasitism.

Lymphocytes are produced in the lymph glands, spleen, thymus, and other lymphoid tissue. It is believed that they fight disease by producing and releasing antibodies when infections occur. Lymphocytes may also be involved in the production of antibodies necessary for the development of immunity to certain diseases. Lymphocytes undergo division readily when placed in a suitable culture outside the body. Dividing lymphocytes clearly show individual **chromosomes** during the **metaphase** stage of division. This principle is now widely applied in studies of chromosome numbers in many animal species.

Blood platelets (thrombocytes) are believed to be formed by large cells known as *megakaryocytes* in bone marrow. A cubic millimeter of blood contains 200,000 to 600,000 platelets. Platelets form a plug at the site of injury to a blood vessel and are associated with the blood-clotting mechanism.

6.6 THE DIGESTIVE SYSTEM

The digestive system consists of a tube that extends the full length of the body. It includes the mouth, pharynx, esophagus, stomach (or stomach compartments in **ruminants**), small intestine, cecum, large intestine, and anus (Figure 6.15).

Various modifications of the digestive system are found in different species. Ruminants (cows, goats, and sheep) have four stomach compartments. These include the **rumen (paunch), reticulum** (honeycomb), **omasum** (manyplies), and **abomasum** (true stomach). Poultry possess a crop, proventriculus (true stomach), and ventriculus (gizzard), as shown in Figure 6.16 (see also Chapter 15).

Accessory organs of the digestive system include the liver, pancreas, and salivary glands. The salivary glands, located in the throat region, secrete into the mouth and pharynx, which moistens and lubricates food to facilitate swallowing. Saliva also aids in initiating the digestive process. The liver filters blood, stores certain nutrients absorbed from the intestine, and secretes bile, which aids in fat digestion. The pancreas is located adjacent to the liver and secretes certain enzymes necessary for the digestion of foods (Table 15.2). A more complete discussion of the anatomy and physiology of the digestive system is presented in Chapter 15.

6.7 THE RESPIRATORY SYSTEM

The respiratory system of animals includes the lungs and the passageways through which air is brought into and exhaled from the lungs. These

FIGURE 6.15
Selected internal parts of the cow. *(Courtesy Dr. P. D. Garrett and Dr. D. E. Rodabaugh.)*

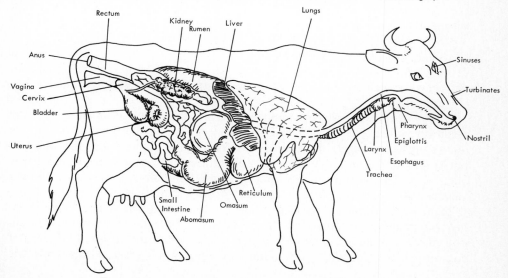

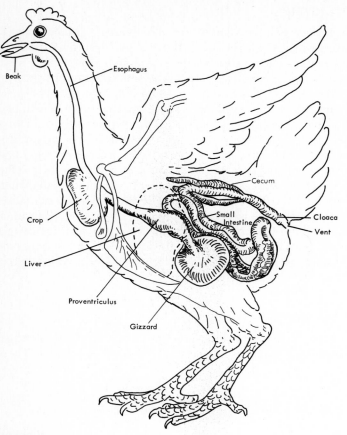

FIGURE 6.16
Digestive system of the chicken. *(Courtesy Dr. P. D. Garrett and Dr. D. E. Rodabaugh.)*

passageways include the nostrils, nasal cavity, **pharynx, larynx,** and **trachea** (Figure 6.15).

The *nostrils* are two in number and are the external openings of the respiratory tract. Each nostril leads to a nasal cavity, which is separated from the mouth by the hard and soft palates. Both food and air pass through the *pharynx,* but its structure is such that air cannot be inspired at the same time food is being swallowed. The passage of food and air through the pharynx is controlled by a valvelike structure, called the *epiglottis.*

In mammals, the *larynx* is known as the *voice box,* but it also controls the inspiration and expiration of air. Moreover, it prevents the inhalation of foreign objects into the lungs. The *trachea* is a continuation of the larynx. It consists of adjacent rings of cartilage, resembling the vacuum hose seen on many electric floor sweepers. These rings are rigid and prevent collapse of the trachea, so that it always remains open. The trachea continues as a single tube to the base of the heart, where it divides into two tubes, called *primary bronchi.* In birds, this is the area of the voice box syrinx. One bronchus passes into each lobe of the lungs. The primary bronchi branch into still smaller bronchi and finally into very small tubes, known as *bronchioles,* in the lungs. The ends, or terminal branches, of bronchioles are the *respiratory bronchioles,* which in turn open into alveolar ducts that lead to the final and smallest portions of the respiratory passageway, called *alveoli* (Figure 6.17). Birds differ from mammals in having relatively nonexpanding lungs and accessory air cavities in the body cavity and major long bones. The accessory air cavities decrease body density and aid in making birds air-mobile. Chickens have nine air sacs (four pairs and a single) and foramen (small cavities or perforations in bones) through which some respiratory action occurs. These air sacs also increase their buoyancy (ability to float on water or in air).

Each lobe of the mammalian lung consists of elastic, spongy material that is greatly expanded when filled with air. When expanded to full capacity, the lungs completely fill the space available in the thoracic cavity.

FIGURE 6.17
Diagram of the mammalian lung. Enlarged area depicts the terminal respiratory structures. *(Drawn by F. E. Staten, University of Illinois, 1983.)*

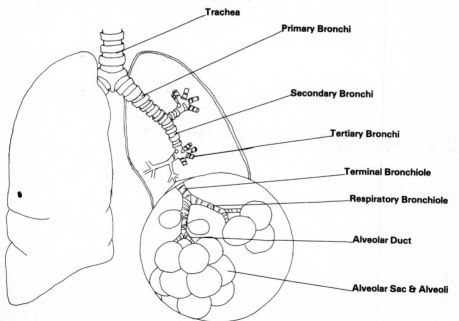

The primary function of the respiratory system is an exchange of gases with the atmosphere. Oxygen is the gas absorbed by the lungs from inhaled air, and carbon dioxide is the gas exhaled from the lungs. If carbon monoxide is inhaled, it unites with the iron of hemoglobin (the oxygen-carrying component of blood) and does not dissociate, forming stable compounds called *carboxyhemoglobins* that cannot carry oxygen. As a result, the animal dies from lack of oxygen or, in a sense, of suffocation. Other chemicals, such as nitrates, chlorates, cyanide, and prussic acid, are also poisonous to animals because they interfere with internal respiration or with the normal utilization of oxygen by tissues. (See Section 18.6.)

Air is brought into mammalian lungs (inspired) by contraction of the diaphragm and enlarging of the thoracic cavity, thereby creating a greater vacuum. This action causes the lungs to enlarge, drawing air into them. A partial vacuum exists within the thoracic cavity under normal conditions. If this vacuum is destroyed by a puncture, the lung collapses. Air is forced out of the lungs (expired) by a decrease in the vacuum in the thoracic cavity. This is caused by relaxation of the diaphragm, contraction of the muscles that decrease thoracic cavity volume, and retraction of the elastic fibers within the alveoli.

The rate of breathing is controlled by a group of nerve cells in the medulla of the brain. This region, known as the *respiratory center,* regulates inspiration and expiration of air by the lungs. It is influenced by the carbon dioxide content of blood, body temperature, and other brain centers.

Oxygen passes from the alveoli in the lungs to the red blood cells of the circulatory system by means of simple diffusion. A reverse diffusion occurs, resulting in the discharge of carbon dioxide. Oxygen and carbon dioxide in body tissues are exchanged in a similar way.

6.8   THE NERVOUS SYSTEM

Normal animals have the ability to adjust or to react to many stimuli in their internal and external environments. Their ability to do this depends basically on their nervous system. Thus the nervous system coordinates all the physical activities of the body and provides the basic pathways for the action of all the senses (hearing, sight, smell, taste, and touch).

The nervous system may be grossly divided into two major parts: the central nervous system (brain and spinal cord) and the peripheral nervous system (somatic and autonomic nerves).

Nerve cells are called *neurons.* Neurons have a single long fiber, called an **axon,** and several branched threads, called *dendrites.* The dendrites receive stimuli from other nerves or from a receptor organ, such as a sense organ. The impulse passes from the dendrite through the cell body to the axon. The axon conducts impulses to the dendrite of another neuron or to an *effector organ,* such as a muscle cell. The place where axons and dendrites come together is called a *synapse.* For an impulse to pass from a receptor organ to the brain or from the brain to an effector organ, it must travel over several neurons, crossing several synapses. Some impulses, however, do not pass through the brain.

A nerve may be either a single neuron (axon) or a bundle of several neurons (axons). Several nerve fibers bound together are called a *nerve trunk*. Some axons, or nerve fibers, are covered by a fat-containing *myelin*, or *medullary*, *sheath*. A small bundle of cell bodies gathered together outside the brain and spinal cord is called a *ganglion*. Nerve cells that *receive* stimuli are called *sensory*, or *afferent*, *neurons*. These neurons carry impulses from the sense organs to the central nervous system. Nerve cells that carry impulses from the brain or other nerve centers to muscles or glands are called *motor*, or *efferent*, *neurons*.

6.8.1 Gross Anatomy of the Brain

The *brain* consists of the cerebrum, cerebellum, pons, and medulla oblongata (Figure 6.18).

The *cerebrum* is the largest component of the brain. Its surface area consists of numerous folds and is wrinkled in appearance. It serves as the decision-making center of the brain and controls such mental activities as voluntary muscle control and interpretations of various sensations, e.g., hearing, seeing, and tasting. It is also involved in reasoning. The cerebrum is usually large relative to body size in animals of higher intelligence. Destruction of the cerebrum does not cause a complete cessation of body functions.

The *cerebellum* functions as a coordinator of the brain's other centers and is a mediator between them and the body. It functions as a coordinator of muscular activity in eating, talking, running, and walking. Damage to the cerebellum results in incoordination, which interferes with voluntary muscular action, but it does not cause paralysis. Farm animals have a large cerebellum, although elephants and whales probably have the largest relative to body size.

FIGURE 6.18
Gross anatomy of the mammalian brain.

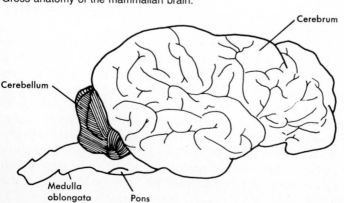

The *pons and medulla oblongata* control reflex actions such as breathing, swallowing, vomiting, and blinking of the eyelids. They usually act independently of the cerebrum and cerebellum.

6.8.2 The Spinal Cord

The spinal cord is located in the center of the vertebral column and is the main line through which messages are transmitted to and from the brain to various body parts. It is a continuation of the *medulla oblongata*. The spinal cord is segmented (divided), and each segment gives rise to a pair of spinal nerves. The spinal cord receives sensory, or afferent, nerve fibers, which transmit impulses from different parts of the body via the **dorsal** roots of the spinal nerves. They also yield efferent, or motor, nerve fibers, which transmit impulses from the brain and spinal cord to various body parts through the ventral roots of the spinal nerves.

6.8.3 The Peripheral Nervous System

Somatic Nerves   This system includes all nervous structures outside the brain and spinal cord. Peripheral nerves consist of bundles of nerve fibers that convey impulses from the brain and spinal cord to muscles (motor nerves) or carry stimuli from the external body to the spinal cord or brain (sensory nerves).

Autonomic Nerves   This is a relatively independent system of nerves positioned between the spinal cord and brain on one side and the various glands, blood vessels, viscera, heart, and smooth muscles of other body regions on the other. This system provides for semiautomatic regulation of involuntary functions subject to central influence. Whereas the peripheral nervous system is associated with what are usually known as somatic (body) structures, the autonomic nervous system is associated with **visceral** organs.

6.9 THE URINARY SYSTEM

The urinary system consists of the kidneys, ureters, bladder, and urethra. Major parts of the urinary system are shown in Figure 6.19.

The *kidneys* are paired and located ventral to the lumbar vertebrae in the abdominal cavity. Each kidney is closely attached to the abdominal wall by a **fascia,** vessels, and the **peritoneum.** The kidneys of the cow and hen are lobulated. The right kidney of the horse is heart-shaped and the left is bean-shaped. The kidneys of swine and sheep are bean-shaped. The gross structure of the kidney consists of an outer tissue layer, called the renal *cortex,* and an inner portion, called the renal *medulla.* The medulla connects with the expanded beginning of the ureter. Microscopically, the cortex of the kidney consists of thousands of tiny connections of vascular and urinary tubules. Blood is filtered from the capillary bed, called the *glomerulus,* into the blind sac of the

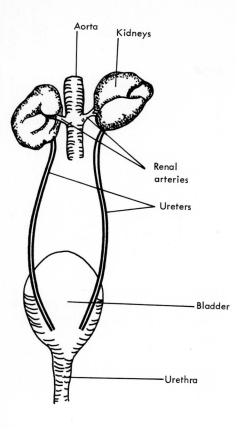

Aorta

Kidneys

Renal arteries

Ureters

Bladder

Urethra

FIGURE 6.19
Major components of the mammalian urinary system.

urinary tubule, called *Bowman's capsule* (Figure 6.20). Together these unions of vascular and urinary tubules are called *Malpighian bodies*. These tubules join together in the medulla, forming larger and larger tubes, which eventually empty into the central cavity (pelvis) of the kidney, continuous with the ureter. Blood is conveyed to the kidney through the renal artery and distributed to the glomeruli.

The *ureter* is a single tube leading from each kidney to the bladder. The *bladder* is a very elastic sac that stores urine until it is voided. Chickens do not have a urinary bladder, but rather the ureters end in the *cloaca*, and feces and urine are voided together. The *urethra* is a highly elastic tube that leads from the bladder to and through the penis in the male (Figure 9.2) and to the vagina in the female (Figure 9.7).

One function of kidneys is to filter waste products from blood. These are voided from the kidneys in urine that flows steadily into the ureters and then into the bladder. Urine normally contains various mineral salts; urea from protein metabolism; uric acid from metabolism of a modified protein, called nucleoprotein; creatinine from metabolism of protein and muscle; and many other materials.

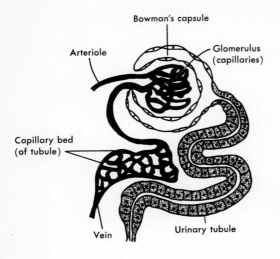

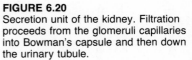

FIGURE 6.20
Secretion unit of the kidney. Filtration proceeds from the glomeruli capillaries into Bowman's capsule and then down the urinary tubule.

The end product of protein and nucleoprotein metabolism in poultry is uric acid. Most animals have enzymes that convert (oxidize) uric acid into allantoic acid, which is excreted in the urine. The exceptions are primates, Dalmation dogs, birds, and reptiles (Figure 6.21). Uric acid precipitates at very low concentrations and can form urate deposits in soft tissues and bone joints, thereby causing the condition called *gout*. The kidneys also function to regulate

FIGURE 6.21
Urinary end products of nitrogen and nucleic acid metabolism in birds and mammals. *(Courtesy Dr. P. C. Harrison, University of Illinois, Urbana.)*

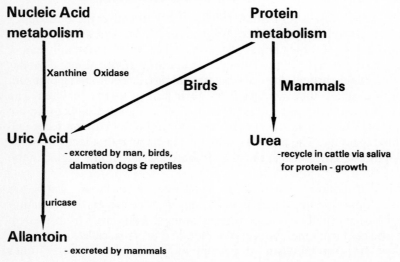

blood composition and to maintain the normal internal conditions necessary for maintenance of life. If both kidneys cease functioning, life ceases.

6.10 SUMMARY

In anatomy and physiology, farm animals are similar in many respects. All have skeletal, muscular, circulatory, digestive, respiratory, reproductive, and nervous systems. Each of these systems performs the same general function within each individual in the many different species. Abnormalities of anatomy and physiology may be of an inherited or an environmental origin. These can greatly impair proper functioning and morphology of the body and may even cause death.

The blueprint for proper development and function of each individual is carried within the genetic material (DNA) of the fertilized egg. These blueprints are followed closely in the anatomical and physiological development of an individual. However, they can be modified to a certain extent by factors within the environment, such as nutrition and disease. The blueprint in the DNA calls for a pig to be a pig and a calf to be a calf. This part of the blueprint is followed to completion without exception.

STUDY QUESTIONS

1 Define *(a)* gross anatomy, *(b)* histology, *(c)* comparative anatomy, *(d)* embryology, and *(e)* physiology.
2 What is meant by *(a)* dorsal, *(b)* ventral, *(c)* cranial, and *(d)* caudal?
3 Name the two layers of skin and the major glands of skin.
4 What is the coat covering for sheep? For horses? For poultry?
5 What farm animals do not possess cleft (split) hooves?
6 Distinguish between an endoskeleton and an exoskeleton.
7 What substances in a ration may affect bone growth?
8 Name the different kinds of joints found in an animal's body.
9 Distinguish between voluntary and involuntary muscles.
10 What are the major functions of the skeletal system?
11 What classifications are there for muscle other than voluntary and involuntary?
12 Describe a motor unit.
13 Describe the chemical reactions involved in muscle action.
14 Define the term *oxygen debt.*
15 Name the parts of the circulatory system, and give their functions.
16 In what way does the pulmonary system differ from the systemic circulatory system in its function?
17 Distinguish between blood serum and blood plasma.
18 Describe anemia in animals.
19 In humans, what blood type is a universal donor? A universal recipient?
20 Explain how the disease erythroblastosis fetalis may occur in some human infants.
21 What kinds of leukocytes are found in blood, and where are they produced?
22 Name the parts of the digestive system.
23 Name the four stomach compartments in the ruminant.

TABLE 7.1
HORMONES SECRETED BY THE ENDOCRINE GLANDS AND SOME OF THEIR MAJOR FUNCTIONS

Endocrine gland	Hormone secreted	Major physiological function
Hypothalamus	Gonadotropin-releasing hormone (GnRH)	Stimulates release of LH and FSH
	Corticotropin-releasing hormone (CRH)	Stimulates release of ACTH
	Thyrotropin-releasing hormone (TRH)	Stimulates release of TSH
	Growth—hormone—releasing hormone (GHRH)	Stimulates release of growth hormone
	Growth—hormone—inhibiting hormone (somatostatin)	Inhibits release of growth hormone
	Prolactin-releasing hormone (PRH)	Stimulates release of prolactin
	Prolactin-inhibiting hormone (PIH)	Inhibits release of prolactin
	Oxytocin	Causes ejection of milk, expulsion of eggs in hens, and uterine contractions
	Vasopression (anti-diuretic)	Causes constriction of the peripheral blood vessels and water resorption in the kidney tubules
Anterior pituitary	Growth hormone (GH or somatotropin)	Promotes growth of tissues and bone matrix of the body
	Andrenocorticotropin (ACTH)	Stimulates secretion of steroids (especially glucocorticoids) from the adrenal cortex
	Thyrotropin or thyroid-stimulating hormone (TSH)	Stimulates thyroid gland to secrete thyroxine
	Prolactin (Prl)	Initiates lactation and induces maternal behavior
	Gonadotropic hormones	
	Follicle-stimulating hormone (FSH)	Stimulates follicle development in the female and sperm production in the male
	Luteinizing hormone (LH)	Causes maturation of follicles, ovulation, and maintenance of the corpus luteum in the female. Causes testosterone production by the interstitial cells of the testes in the male

TABLE 7.1
HORMONES SECRETED BY THE ENDOCRINE GLANDS AND SOME OF THEIR MAJOR
FUNCTIONS (Cont.)

Endocrine gland	Hormone secreted	Major physiological function
Thyroid	Thyroxine, triiodothyronine	Increase metabolic rate
	Calcitonin	Lowers the concentration of calcium in the blood and promotes incorporation of calcium into bone
Parathyroid	Parathyroid hormone	Maintains or increases the level of blood calcium and phosphorus
Adrenal glands Cortex (shell)	Glucocorticoids	Mobilize energy, increase blood glucose level, have an anti-stress action
	Mineralocorticoids	Maintain salt and water balance in the body
Medulla (core)	Epinephrine (adrenalin)	Stimulates the heart muscles and the rate and strength of their contraction
	Norepinephrine	Stimulates smooth muscles and glands and maintains blood pressure
Ovaries Follicles	Estrogens	Cause growth of reproductive tract and mammary duct system
Corpus luteum	Progesterone	Prepares reproductive tract for pregnancy, maintains pregnancy, and causes development of mammary lobule-alveolar system
	Relaxin*	Causes relaxation of ligaments and cartilage in the pelvis, which assists in parturition
Testes	Androgens (testosterone)	Cause maturation of sperm; promote development of male accessory sex glands and secondary sex characteristics
Pancreas (islets of Langerhans)	Insulin Glucagon	Lowers blood glucose Raises blood glucose
Placenta (in some species)	Gonadotropins (pregnant mare serum gonadotropin, human chorionic gonadotropin), estrogens, and progesterone	Promote the maintenance of pregnancy

*Relaxin is produced by the corpus luteum in the pig but is produced by the placenta in the mare.

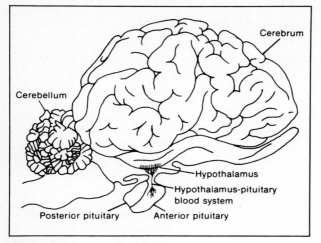

FIGURE 7.3
Major structures of the cow brain. *(From T. R. Troxel and D. J. Kesler, "The Bovine Estrous Cycle Dynamics and Control," Univ. of Illinois Coop. Ext. Cir. 1205, 1982.)*

gland by small blood vessels known as the portal blood vessels. Oxytocin and vasopressin, however, are produced by nerve cells in the hypothalamus and travel by way of the nerve cells to the posterior pituitary gland, where they are stored.

The anatomy and physiology of some other glands of internal secretion and the function of their various hormones that deal with reproduction, growth, and milk and egg production will be discussed in detail in subsequent chapters.

7.3 THE CHEMICAL NATURE OF HORMONES

Chemically, hormones fall into three general classes: (1) **proteins** and peptides, (2) phenolic derivatives, and (3) steroids. Proteins and peptides are composed of amino acids. The general chemical structure of amino acids is as follows:

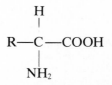

$$\overset{\displaystyle H}{\underset{\displaystyle NH_2}{R-C-COOH}}$$

The letter R is called the *side chain* of amino acids and consists of various combinations of carbon and hydrogen and, in some cases, nitrogen, oxygen, or sulfur. In the amino acid glycine, the side chain R is replaced by hydrogen (H), whereas in alanine it is replaced by CH_3, a methyl group. The dividing line

FIGURE 7.4
Chemical structures of epinephrine and thyroxine, which are examples of hormones that are phenolic compounds.

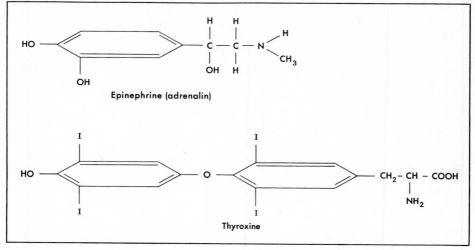

between proteins and peptides is arbitrary, with proteins having a molecular weight greater than 10,000.

Epinephrine, secreted by the adrenal medulla, and thyroxine, secreted by the thyroid, are phenolic compounds (Figure 7.4).

Steroid hormones (Figure 7.5) are secreted by the ovary, testes, adrenal cortex, and certain fetal membranes. Several kinds of estrogens and androgens are found in body fluids; they are all steroid in nature and closely allied in chemical structure. Acetate and cholesterol are important precursors in the biosynthesis of steroid hormones. Synthetic compounds such as diethylstilbestrol have been synthesized in laboratories and used in animal production (Sections 7.8.3 and 14.6).

7.4 FUNCTIONS OF HORMONES

Very small amounts of hormones are required to perform a particular function. Hormones affect growth, the shape of the body, and the way in which the body uses food, and perhaps most important, they allow the body to make proper adjustments to changes in the outside world. When overdoses are administered to animals, the effects may be detrimental. For this reason, proper dosage level for treatment is critical. In addition, two individuals may respond differently to the same hormone dosage. This is possible because two or more individuals differ in the amount of a hormone they secrete. Moreover, a given hormone may act on different tissues in different species. For example, prolactin stimulates the mammary gland to secrete milk and causes the pigeon's crop gland to **proliferate**

physiological potency when they came in contact with certain tissues of the body. The active substances caused muscle tissue to contract **in vitro** and, when injected into animals, caused a sharp lowering of blood pressure. These previously unknown substances were given the name *prostaglandins* because it was thought that they were secreted by the prostate gland. Later it was discovered that these substances actually came from the seminal vesicles and many other body tissues. They since have been found to constitute a family of substances with a wide range of biological effects.

Although PGs resemble hormones in their actions, they are quite different chemically. Prostaglandins represent a large group of chemically related 20-carbon hydroxy fatty acids. At least 37 have been identified. The two main series appear to be PGE and PGF (Figure 7.6).

Injection of PGs results in a variety of transient effects on blood pressure, lipolysis, gastric secretion, uterine contraction, and functions of circulating platelets. The specific physiological effect produced depends on the precise structure of the prostaglandins administered. Physiological effects may be species-specific and may depend on the dosage given. Effects also may vary with differences in the physiological environment.

The area in which prostaglandins have received the most attention in farm mammal research is in the synchronization of estrus. Prostaglandin $F_2\alpha$ will synchronize estrus in cattle, horses, and sheep. It has been approved by the FDA for use in cattle and horses. Prostaglandin $F_2\alpha$ will also induce parturition in pigs.

It is interesting to note that aspirin is an inhibitor of prostaglandin biosynthesis.

7.8   PRACTICAL USES OF NATURAL AND SYNTHETIC HORMONES

The discovery of hormones and the elucidation of their major functions in the body have led to their use in the control or regulation of certain body functions and in practical livestock production.

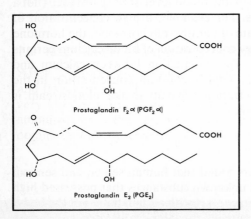

FIGURE 7.6
Chemical structures of selected prostaglandins. Note their configurational similarities.

7.8.1 Reproduction in Farm Animals and Humans

It was estimated in 1982 that over 50 million of the world's women of reproductive age were taking compounds orally in the form of pills as a means of birth control. A wide variety of steroid substances have been shown to be extremely effective in suppressing the production of gonadotropic hormones and thus in blocking ovulation. The most widely used product contains a combination of a synthetic progesterone and a synthetic estrogen.

Progestins have been used experimentally to inhibit and synchronize estrus and ovulation in farm mammals. When administered to a group of females for a period of several days, they inhibit ovarian follicle development and ovulation. When these compounds are removed, follicular growth is resumed and most females in the group will come into estrus and ovulate within a period of 3 to 5 days. Hence estrus and ovulation are caused to occur in a group of females at about the same time (synchronized). As already mentioned, $PGF_2\alpha$ will also effectively synchronize estrus and ovulation in cattle, horses, and sheep. $PGF_2\alpha$ acts by destroying the secretion of progesterone by the corpus luteum.

The ovulation rate in farm mammals may be increased (superovulation) by injecting gonadotropic-hormone preparations during the follicular phase of the estrous cycle. Reproductive performance of **gilts** receiving gonadotropins has been quite variable. Consistent improvements have been observed in reproductive efficiency in sows receiving gonadotropins at the time their pigs are weaned. Twinning has been induced in cattle by treatment with gonadotropins, and the treatment with gonadotropins of certain women of low fertility has resulted in multiple births. Gonadotropins are also used to superovulate females for embryo transfer (see Chapter 9).

The administration of various hormones to male animals has had little or no value for increasing sperm production. Certain chemicals do prevent testes development and sperm production. Estrogenic compounds have been used to caponize (chemically castrate) male chickens, but this method is not presently used on a practical basis.

Gonadotropin-releasing hormone (GnRH) is approved by the FDA for the treatment of cystic ovarian disease. Of cows with ovarian cysts treated with GnRH, 80 percent will reestablish ovarian cycles and become pregnant. GnRH has also been used experimentally to induce ovulation in anestrous cows. If GnRH is administered to dairy cows 2 weeks postpartum, they have a lower probability of developing ovarian cysts. GnRH has been used in combination with progestins and $PGF_2\alpha$ to synchronize estrus and ovulation. Experimentally, GnRH has also been used to increase conception rates in repeat breeder cows. For cows that have had difficulty in becoming pregnant, an injection of GnRH at estrus improves conception rates significantly. Therefore, GnRH has numerous therapeutic uses in farm mammals. More uses will probably be developed in the future.

Testosterone can be administered to female farm mammals to induce male sexual behavior. These testosterone-treated females act like males and can be used as aids in estrus detection.

7.8.2 Mammary Growth and Lactation

Many hormones are known to be involved in the growth of mammary glands and their secretion of milk. Administration of estrogen and progesterone for 7 days will initiate lactation in about 70 percent of cows treated. Use of these compounds for inducing lactation has not been approved by the FDA.

7.8.3 Growth and Fattening

Estrogens and synthetic estrogen compounds have been shown to increase the rate and efficiency of gains in **ruminants** (cattle and sheep). The mechanisms by which these compounds improve growth efficiency in ruminants are not fully understood.

The administration of hormones to swine during growth and fattening has not been very effective and therefore has not been employed commercially.

7.8.4 Hormones and Public Health

Synthetic hormones are known to affect reproductive and other processes in the body. If residues of these hormones are present in meat and other foods consumed by humans and animals, they may pose a health hazard. Sensitive methods have been developed for detecting minute residues in tissues. In general, residues of these compounds in animals do not appear to be a **public health** hazard. Great care is taken to prevent residues of these hormones from being present in human food. They are under constant surveillance by the FDA.

The relation of hormones to the occurrence of cancer has been given considerable attention by research workers. Estrogens have been investigated the most thoroughly. It is possible that estrogens may cause carcinoma in humans when given in large doses for prolonged periods of time. It appears very unlikely that this hazard would be encountered in humans under normal conditions.

7.9 SUMMARY

A remarkable **homeostatic** interrelation exists between hormones and related factors (including appetite, diet, health, **nutrition, stress,** and **environmental** responses). Caution should be taken in hormone application to avoid disturbing hormonal balance. As noted by Brody,[2] the following quotation is an excellent summary of this balance concept.

> The most important concept in endocrinology which emerges from the feverish activity of the past decade is the principle of endocrine balance. Discovered and rediscovered by several investigators, the paradoxical truth has dawned finally that man or beast may suffer less from the loss of several glands than from losing a single

---

[2]S. Brody, *Bioenergetics and Growth*, Reinhold Publishing Corp., New York, 1945.

one. For each of the precious juices the other secretions supply a partial antidote, so that health and personality may be preserved, delicately poised. Henceforth, the practicing consultant and laboratory investigator, or both, must think in terms of integrated hormonal effects.[3]

STUDY QUESTIONS

1 How do endocrine and exocrine glands differ?
2 What is the master endocrine gland of the body? Where is it located?
3 What is the function of the hypothalamus?
4 List the hormones secreted by the anterior pituitary gland and give their main functions.
5 What hormones are produced by the adrenal glands? What hormones are produced by the placenta in some species? Which species?
6 List the three general chemical classes of hormones.
7 What is meant by the feedback mechanism in hormone control?
8 What is the role of the hypothalamus in the release of certain hormones from the anterior pituitary gland?
9 What are prostaglandins? Give some of their functions.
10 List some uses of hormones in humans and farm animals.
11 Which hormones may be involved in the growth of mammary glands? In secretion of milk?
12 Which synthetic hormones may be used to increase the rate and efficiency of gain in beef animals?
13 In what ways could hormones be a hazard to public health? What precautions are taken to prevent them from being a health hazard?
14 If hormones were not well balanced within the animal body, what could be the consequences?

[3]W.T. Slater, *The Endocrine Function of Iodine,* Harvard University Press, Cambridge, Mass., 1940.

THE PHYSIOLOGY
OF GROWTH
AND SENESCENCE[1]

If we call the weight of the human egg one, then the weight of the individual produced from the egg is about 16 billion times that of the egg itself.

G. H. Parker

8.1   INTRODUCTION

All humans begin life as a single cell (the fertilized egg). This single cell is less than 1/700 in in diameter and is so small that its weight cannot be determined accurately. Through the process of *cell division* and **differentiation,** this cell develops into a mature individual weighing between 150 and 200 lb (mature human males) and possessing as many as 250 trillion cells. Few cells in the mature body resemble the original fertilized egg because, through the processes of differentiation and **morphogenesis,** changes occur that form all the organs and tissues characteristic of humans. Growth results from physiological processes that cause one cell to develop into a many-celled individual.

A knowledge of the factors responsible for body growth is of fundamental importance to students of animal science. A certain amount of growth must occur before animals can convert their foods into animal products useful to humanity. Rapidly growing animals are generally the most efficient in converting food into weight gains. However, proper production practices are essential to

[1]The authors acknowledge with appreciation the contributions to this chapter of Dr. P. J. Bechtel, Department of Animal Science, University of Illinois, Urbana.

maximize the efficiency of growth and the quantity and quality of products yielded by various farm animals. Few aspects of animal science are more intriguing than the development of a completely new animal originating from one cell. This development proceeds in such a precisely controlled way that all the intricate organization of cells, tissues, organs, and systems characterizing the functioning adult develops with rarely a flaw.

8.2 THE PHENOMENON OF GROWTH

If nature were our banker, she would not add the interest to the principal every year; rather would the interest be added to the capital continuously from moment to moment.

J. W. Mellor

All living things (plant and animal) grow. A giant redwood in California grows from a seed measuring 1/16 in in diameter to become a tree often 300 or more ft tall. A microscope is required to see a whale egg, yet a mature whale may be more than 100 ft long and weigh about 160,000 lb. In the lower animal forms, striking changes occur between the time the single cell starts growing and the time an adult individual emerges. The single cell of certain **insects** grows first into a wormlike **larva,** which in turn enters a **pupal,** or quiet, stage before the adult insect finally develops.

In the higher vertebrates (animals with backbones), the tiny cell steadily develops into a form that at birth resembles the adult. Although there are significant differences in the proportion of various body parts, the **neonate** looks like the adult. Growth is the normal process of increase in size produced by the accumulation of tissues that are similar in composition to that of the original tissue or organ. True growth involves an increase in the size and weight of structural tissues such as the muscles, bones, heart, and brain and of all other body tissues (except **adipose** tissue) and organs. From a chemical standpoint, true growth is an increase in the amount of **protein** and mineral matter accumulated in the body. Increases in weight due to fat deposition or to the accumulation of water are not true growth.

All the parts of an animal's body grow in an orderly manner. The legs and arms in humans grow in proportion to the height and length of the body. Following birth the head, however, grows proportionately much more slowly than the limbs (Figure 8.1). Nevertheless, there is wide variation in size and body proportions within a **species.** For example, a Clydesdale (**draft** horse) is many times larger than a Shetland **pony.** Even within a **breed** there is considerable variation in size and weight among individuals.

Although the various body parts grow in an orderly manner, the body does not grow as a unit, because various tissues grow at different rates from birth to maturity. For example, muscle, heart, and total body tissues may show a 45- to 50-, 12- to 15-, and 20- to 22-fold increase, respectively, in size and weight

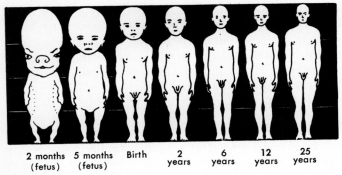

2 months   5 months   Birth   2   6   12   25
(fetus)   (fetus)   years   years   years   years

FIGURE 8.1
Changes in the relative size of body components during the fetal and
postnatal growth in humans. *(S. Brody,* Bioenergetics and Growth,
Reinhold Publishing Corp., New York, 1945.)

compared with the brain, which may show only a three- to fourfold increase
(Figure 8.2). The head and brain grow fast early in the lives of vertebrates.
Furthermore, their legs may increase in length faster than the balance of the
body. For example, **foals** have long legs relative to other body components.
Certain parts of an animal's body continue growing after others have stopped.
The incisor teeth of rats continue to grow throughout their life. Some animals
have shells that restrict their growth beyond a certain size. Others (e.g., crabs)
shed their shells and construct larger ones.

In animal production, growth is usually measured as an increase in body
weight. Although this is generally accepted as an adequate measure, technically
it is not accurate. It is possible to feed a young animal in such a way that its total
body weight does not increase while the skeleton and certain organs may show
considerable growth.

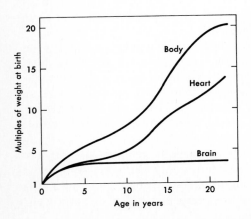

FIGURE 8.2
Graph depicting differences in relative growth
rates of body, heart, and brain tissues in
humans. *(Modified from D'A. W. Thompson,*
On Growth and Form, *Cambridge University
Press, New York, 1942; and W. T. Keeton,*
Biological Science, *3d ed., W. W. Norton &
Co., New York, 1980.)*

Regeneration is a special feature of growth. When a part of an animal's body is injured, the lost tissue may be (1) replaced by the same kind of tissue or (2) replaced by another kind, called *scar tissue.* Sponges are capable of almost complete regeneration. Starfish can grow whole new bodies from a single ray if a small amount of the center portion remains. A salamander that loses a leg will grow a new one. Worms can grow new heads. Several kinds of long-tailed lizards often break off their tails to escape the grip of an enemy, and they soon grow new tails. In higher animals, however, the power of regeneration is not as remarkable. When a dog's tail is cut off in the lawn mower or a bear's paw is lost in a trap, new parts do not grow. Similarly, a cow, chicken, or horse cannot grow a new leg to replace a severed one. Most higher animals can regenerate hair (feathers), nails (hooves), skin (hide), blood cells, liver, and certain other tissues.

8.2.1 Growth and Development of Humans

During the first stages of growth, the slowly developing human is called an *embryo.* At 2 months it is only about 1½ in long but has the form of a human being. Although all its parts are well formed, the head is quite large compared with the trunk and limbs (Figure 8.1). Beginning with the third month, the fully developed embryo is called a **fetus.** At 7 months it weighs about 2 lb and is approximately 15 in long. Two months later, or just prior to **parturition,** the fetus commonly weighs 6 to 8 lb and is from 19 to 21 in long. Thus the body is growing very rapidly at birth. The weight at birth may be influenced by (1) the mother's **nutrition** during pregnancy, (2) the number of previous pregnancies, (3) sex, (4) race, and (5) **environment.** Each successive pregnancy leads to an average increase of about 75 g (2.65 oz) in the weight of the child at delivery.

The rapid growth rate continues for about 2 years and then gradually declines until adolescence. For approximately 2 years during the adolescent period, the growth rate of boys and girls is accelerated. This is commonly referred to as the *adolescent spurt* in height growth. Boys usually exhibit this spurt of growth between the ages of 13 and 15 and may grow 4 to 12 in in height. Girls normally begin the adolescent growth spurt about 2 years earlier, and their maximum growth rate is somewhat less than that of boys. Men are taller than woman chiefly because of this difference in the adolescent growth spurt. Until that period is reached, the average heights of boys and girls are about equal.

When People Stop Growing Most humans stop growing between the ages of 18 and 30. (A person is usually tallest at age 26.) After attaining maximum height, a person begins an exceedingly slow decrease, which is usually not noticeable until the person reaches old age. It is caused by a thinning of the pads of **cartilage** that grow between the bones of the vertebral column (backbone). The tendency to be tall or short is governed largely by **heredity.** Thus tall parents usually have tall children, and short parents have short children.

8.3 THE CELL IS THE UNIT OF GROWTH

The living cell is to biology what the electron and proton are to physics. The two theories are independent exemplifications of the same idea of "atomism."

A. N. Whitehead (1861–1947)

Growth results from an increase in both the size and the number of body cells. An increase in the number of cells is called **hyperplasia;** an increase in cell size is called **hypertrophy.** *Accretionary growth* is an increase in size resulting from the accumulation of noncellular material, such as albumin in blood and calcium in bone. When cells hypertrophy, it is usually the protein in the **cytoplasm,** and not the **nucleus,** that increases in size. The process by which a single cell grows from within is also called growth by *intussusception.* It is interesting that the cell size of the mouse is about the same as that of the elephant, although the elephant, of course, has many more cells than the mouse.

The most significant aspect of hyperplasia is the increase in **DNA.** Since most cells usually have only one nucleus, and each nucleus contains the same content of DNA, the amount of DNA present in a tissue can be used as an index of the number of cells in that tissue. During hypertrophy, cell size increases, but the DNA content remains constant. In most cells, the increase in cell size is accompanied by an increase in cellular protein. Therefore, the protein/DNA ratio can be a useful indicator of growth. There is a maximum protein/DNA ratio which varies with tissue, species, age, and other factors. Under optimal environmental conditions and nutritional parameters, a maximum protein/DNA ratio is obtained; hence, if the animal is to become larger in size, it must synthesize additional DNA. During a mild nutritional setback, such as a brief period of starvation, the amount of tissue protein will decrease, whereas the content of DNA remains constant. Then, when proper feeding of the animal is resumed, the content of cellular protein quickly increases to the previous protein/DNA ratio. This rapid rate of growth is called *compensatory growth.*

Hyperplasia and hypertrophy of all cells in the body occur during embryonic life. In the adult animal, however, three different types of cells are found. *Permanent cells* cease dividing early in prenatal life. Their numbers remain constant thereafter. Both muscle and nerve cells are examples of this type. *Stable cells* continue to divide and increase in number during the major portion of the growth period, but division ceases and the number becomes fixed in the adult. Most body organs, such as the lungs and kidneys, contain cells that fit this category. A third group of cells is known as **labile** (apt to change) cells. Epithelial and epidermal tissues fit this category. They continue to divide and differentiate throughout the animal's life in response to the life-sustaining need to replace worn-out tissue and cells.

Some body cells are capable of undergoing hypertrophy during the different stages of life. For example, muscle tissues greatly enlarge when extra demands are placed on them by bodily exercise. Other cells are capable of hyperplasia but do not undergo this process unless there is a special demand for it. For example, in the healing of wounds the capillaries present in the surrounding tissue bud from endothelial cells, with new growth occurring in the direction of the wound. As growth progresses, the capillaries become hollow, forming channels through which blood circulates. This growth continues until the raw surface of the wound is covered. Along with the new capillaries comes a procession of **fibroblasts,** which produce connective tissue in the wounded tissue. They form fibers that heal the wound. Nerve cells cannot divide after they are formed in the developing fetus. Therefore, when a nerve cell is severely damaged, it will not be replaced by a new cell. However, the axis cylinder (fiber) can grow from the nerve cell end and find its way into the tissue formerly supplied by the severed nerve. This explains how some aspects of normal nerve tissue function may be restored.

Studies of cell division using radioactive **isotopes** show that only about 3 percent of the cells within the adult human body are capable of dividing and repairing tissues. A normal dividing cell produces two daughter cells, one of which is capable of dividing while the other is not. The latter perishes, leaving no descendants. Through this method of division, the total cell number is approximately constant in normal tissue, although there may be a reduction in cell numbers as the individual ages. Cancerous cells divide no more rapidly than normal cells, but they increase much more rapidly in numbers because there are fewer restraints and/or controls on cell division. What controls and eventually stops the division of normal cells is not clearly understood.

8.3.1   Growth and Development of Muscle, Fat, and Bone

The development of three different types of tissue will be examined briefly in this section. Skeletal muscle is an unusual tissue in that individual muscle cells are long and thin and contain thousands of nuclei. Muscle cells are formed by a distinct sequence of events (Figure 8.3).

FIGURE 8.3
The sequence of events accompanying the formation of muscle cells. *(Courtesy Dr. P. J. Bechtel, University of Illinois, Urbana.)*

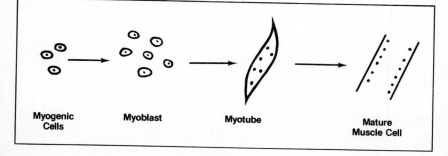

Myogenic Cells   Myoblast   Myotube   Mature Muscle Cell

During fetal development, single nucleated cells that are predestined to become muscle cells divide many times until another type of single nucleated cell, called a *myoblast,* is formed. Myoblasts do not divide but instead fuse together, forming a *myotube* (Figure 8.3), which is an immature muscle cell. Once the nuclei are inside the myotube, they are unable to divide further. Other single nucleated cells, called satellite cells, are adjacent to the myotube and continue to fuse with the maturing muscle cell, resulting in an increased number of nuclei in the muscle cell. The process of muscle development is interesting from another standpoint in that the number of muscle cells in a mature animal is fixed just prior to birth. Thus, the tremendous growth of muscle that occurs following birth is not an increase in the number of cells, but rather an enlargement in size of the existing cells.

Fat consists of fat cells and supporting connective tissue. The mature fat cell is called an *adipocyte.* These large cells contain a droplet of fat, which is surrounded by other cellular components. The mature fat cell is derived from an immature fat cell called an *adipoblast.* The adipoblast contains a number of tiny lipid droplets. The mature fat cell is commonly 50 to 100 times larger than the adipoblast. Fat tissue is very dynamic, and it increases or decreases in size in accordance with the animal's plane of nutrition, as well as with other factors. Once a fat cell has been formed, it can be reduced to a very small size, as in response to starvation. However, the cell will fill rapidly with lipid materials under the appropriate conditions. There are two types of adipose (fat) tissue, which are referred to as *white fat* and *brown fat.* Most body fat is white, and it functions as a depot of stored energy. Brown fat is very active metabolically and can be used to maintain body temperature, which is especially important in neonatal animals.

Bones grow during both the prenatal and postnatal periods. Bone of a mature animal is composed of approximately 50 percent mineral [$Ca_{10}(PO_4)_6(OH)_2$] and 50 percent organic material and water, on a weight basis. Volume-wise, mineral constitutes only about 25 percent of bone. Most of the organic material in bone is collagen, with a small amount of cellular material. Bone is formed by the interaction of three different cell types: chondrocytes, osteoblasts, and osteoclasts. *Chondrocytes* are cells that produce cartilage, *osteoblasts* produce bone collagen and other bone components, and *osteoclasts* break down bone during the process of resorption. Bones grow in length by ossification of the epiphysial cartilage (see Chapter 6). Once the epiphysial cartilage has been completely ossified, the bone will stop growing in length. Bones can grow in width, and they can also undergo some degree of remodeling through the process of readsorption of existing bone by osteoclasts and the formation of new bone by osteoblasts.

8.4 PERIODS OF GROWTH

Growth in farm **mammals** may be divided into two main periods: (1) prenatal and (2) postnatal.

8.4.1 Prenatal Growth

The prenatal period of growth occurs between the time the **ovum** is fertilized and the young are born. Only 21 days are required for the fertilized hen egg (a mass of yolk and albumen) to develop into a fully functioning baby **chick.** Prenatal development takes much longer in farm mammals.

The new individual arises from the union of a **sperm** and an ovum, forming one cell[2] with the **diploid** number of **chromosomes.** One-half of each chromosome pair comes from the father and one-half from the mother. The single fertilized cell divides to form two, then four, then eight, etc., cells, following a pattern in which each mother cell produces daughter cells, both of which are capable of dividing to form two new cells (Chapter 4). When they are separate, the ova and sperm do not have the power to grow and reproduce themselves, and they eventually die. After combining during the process of **fertilization,** they have the ability to divide and thus increase cell number.

Somatic cells formed by cell division have a diminishing power to grow by means of hyperplasia. The power to reproduce is lost in many highly differentiated cell types. For example, the mature red blood cell of mammals loses its nucleus before it enters the bloodstream, but its cytoplasm becomes filled with **hemoglobin,** which transports oxygen from the lungs to the various body parts. The red blood cell dies after about 90 to 120 days in the bloodstream, but bone marrow continues to produce replacements. Skin cells arise from the germinal layer of skin. **Keratin** accumulates in their cytoplasm; they lose their nucleus and power to grow. The end result of this differentiation is cellular death. Thus, in the body, when certain cells perform a function in which their ability to divide is lost, other cells must divide and increase in numbers to replace those that are destroyed. This is the process used to create new liver and skin cells. A problem occurs when a differentiated cell, such as a nerve or brain cell, is destroyed, because there is no mechanism to create new cells of these types. The result is an irreversible diminishing in numbers of such cells, and the organism then loses some of its ability to regulate itself.

The rate of cell division in the newly fertilized ovum varies among species. The number of cells in a fertilized rabbit ovum is approximately 32 within 40 to 50 h after mating. The fertilized ovum of swine is in the two- to four-cell stage at that same postmating time.

The developing fertilized ovum soon goes through a process of differentiation (growth and organization of cells into specific structures) in which the mother cell produces different kinds of daughter cells, such as those of the brain and heart. This process is irreversible in that brain cells cannot be transformed back into egg cells. More and more organs are formed as the ovum development proceeds, until all are complete. Then further differentiation ceases. The organization of various dividing cells into special organs with a particular

---

[2]It is interesting to note that the eggs of the Texas armadillo regularly divide into four parts, giving rise to four young. In some parasitic insects, one egg often forms as many as 2000 separate embryos.

makeup is known as *morphogenesis,* or **organogenesis.** During organogenesis differentiated cells synthesize new structural proteins and **enzymes** that are specific for the tissue being formed. How cells are instructed to differentiate into organ cells is not known, but the process has been repeated billions and billions of times. Body organs are formed in an orderly sequence. The head portion of the embryo develops before the tail, and the cephalochord (the head of the spinal column) develops before other organs.

Prenatal growth in farm mammals proceeds at varying rates among different species. The pig, for example, spends only 110 to 115 days within the mother's uterus, compared with 335 to 345 days for the **foal** (baby horse). All farm mammals are born at a fairly comparable degree of maturity, in spite of the great variation among species in the length of time spent in the uterus. However, the degree of maturity at birth varies greatly among animal species other than farm mammals.

The opossum is born 12 days after **conception.** The young are very immature at birth, and since there are no fetal membranes, many baby opossums find their way from the birth canal to their mother's pouch, where they attach their mouth to a teat and continue their development. The number of baby opossums born greatly exceeds the number of teats the mother possesses. Those not finding a teat perish. This is a form of natural selection for the more vigorous individuals.

The baby rat is born 21 to 22 days after conception. At birth it is quite helpless; it has no hair; its eyes are closed, and it is unable to walk. The guinea pig, on the other hand, is born about 67 to 68 days after its parents mate. The young are quite mature at birth; they possess a coat of hair, their eyes are open, and they can walk and run almost immediately (Figure 8.4).

The human infant is carried **in utero** about 9 months and is helpless at birth, even though its eyes are readily opened and the senses soon begin to function. Babies cannot walk, however, until they are nearly 1 year of age. In early prenatal life, the human female exceeds the male in biological growth rate. Thus a full-term male is somewhat more immature than a full-term female at birth.

FIGURE 8.4
Note the marked difference in the degree of maturity at birth in the rat (left) and the guinea pig (right). *(Missouri Agr. Expt. Station.)*

Moreover, males are more susceptible to death losses. Thus sex **hormones** appear to influence the degree of maturity of the human infant at birth.

Genes transmitted by the **sire** limit the birth size of **progeny** of large mothers, but in small mothers, the uterine environment of the mother is more important in limiting the size of the progeny at birth than are the genes transmitted by the sire. For example, a small Shetland **mare** mated to a large Shire **stallion** produces a very small foal at birth. Conversely, when the mother is a large Shire mare and the father a small Shetland stallion, the foal is much larger. Maternal effects on size tend to diminish with advancing age of the offspring, although differences often persist for many months. The persistency of maternal effects in later life depends on the time in life that the influences have had their effect. The earlier in life they occur, the longer they persist. This is probably also true of the effects of other environmental factors that limit growth.

The number of muscle cells an individual possesses is fixed at the beginning of the fetal stage. Thereafter, increased muscle size depends on an increase in size (hypertrophy) of individual muscle cells. Skeletal size may be modified to a certain extent by environmental factors, but it cannot exceed the limit set by the individual's genetic constitution.

8.4.2 Postnatal Growth (Growth after Birth)

Growth usually begins slowly, then goes through a more rapid period, and then slows down again or may even stop. This pattern yields the characteristic sigmoid (S-shaped) growth curve. The general course of growth after birth in all species of farm mammals is somewhat similar. If the body weight of an animal is plotted on the Y axis (ordinate) of a graph and its age to maturity on the X axis (abscissa), the curve resembles the letter S (Figure 8.5). However, the bend in the growth curve is about one-third of the distance from the baseline to the top of the curve. There are, apparently, few exceptions to the sigmoid form of growth curve. (Some organs, such as the **gonads** and mammary glands, are cyclic in growth.) It applies to the growth of populations of either unicellular or multicellular plants and animals. These curves are similar because the multicellular individual represents a population of cells that is similar to a colony of single-celled individuals.

The growth curve of farm mammals after birth may be divided into two principal segments. The first is represented by an increasingly steep slope that extends from the beginning of growth until one-third to one-half of the mature weight is reached. The length of this segment, however, depends on the species and the influence of environmental factors that may affect growth. The second segment extends from the end of the first to maturity, or to the end of the growth period. The general shape of the curve is molded by two opposing forces, a *growth-accelerating force* and a *growth-retarding force* (Figure 8.6). The growth rate at any given time depends on the magnitude of each of these opposing forces.

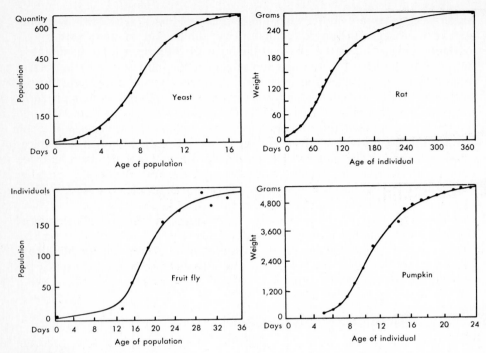

FIGURE 8.5
Typical S-shaped growth curves of unicellular and multicellular plants and animals. Note their similarity. *(Redrawn from* Missouri Agr. Expt. Sta. Res. Bull. *97, 1927.)*

The growth-accelerating force is present in body cells or in the individual cells constituting a population of single-celled individuals. The growth-retarding force is found in the environment surrounding the cells or the individuals in a population. This force involves (1) exhaustion of the supply of essential nutrients and/or (2) the accumulation of waste products, which results in a toxic

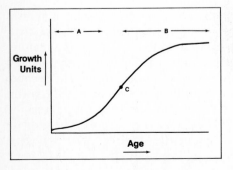

FIGURE 8.6
Standard postnatal growth curve: (*a*) growth-accelerating phase, (*b*) growth-deaccelerating phase, and (*c*) point of inflection.
(Courtesy Dr. P. J. Bechtel, University of Illinois, Urbana.)

environment. Results from studies of cell cultures appear to verify the importance of these two factors. When a piece of tissue was cut from a living chicken's heart and placed in a culture medium where nutrients were supplied and waste products continuously removed, the tissue lived for many years, doubling its mass every 24 h. In the live intact chick, heart cells soon reached the point where cell division (growth) ceased entirely. This experiment suggests that cells are potentially capable of infinite division under ideal conditions.

The point in the growth curve where the growth rate ceases to increase and from which it begins to decrease is referred to as the point of **inflection** (Figure 8.6). This point, according to Brody,[3] generally indicates (1) the period of maximum velocity of growth, (2) the age of **puberty,** (3) the beginning of the period of increasing specific **mortality,** and (4) the period of reference for the determination of age equivalents in different animals and populations. The point of inflection occurs at about 14 years in humans, 5 months in cattle, and 2 months in sheep.

Factors limiting the maximum size of a single multicellular individual include (1) its genetic constitution, (2) its **nutrient** supply, (3) the effects of certain **diseases,** and (4) the hormone supply, which may also be related to an individual's genetic constitution. A fifth factor, body-surface/volume ratio, may limit extremely large size. As size increases, animal volume increases much more rapidly than surface area. Within a species an extremely large size eventually makes it difficult to obtain enough food to sustain the animal and to eliminate body waste products efficiently.

The growth rate from birth to weaning may be greatly influenced by the amount of milk secreted by the mother and by the individual's health. Total growth during this period also depends on the individual's inherent growth impulse. Growth of body parts after birth proceeds at different rates in most farm mammals. Shortly after birth, skeletal parts generally grow more rapidly than other tissues in proportion to total body weight. This rapid rate of skeletal growth decreases by weaning time. Maximum muscular development occurs somewhat later than maximum skeletal growth, and it is occurring at an increasing rate by weaning time. Body-weight increases that occur after both skeletal and muscular tissues have passed their maximum rates of increase are due primarily to increased fat deposition. These body-weight increases occur much later than the usual weaning time, commonly near the usual market age and weight.

Inadequate nutrition delays peak muscle growth and slows the rate of fat deposition, whereas proper nutrition hastens the occurrence of peak rates of both. Early-maturing animals fatten at a younger age than late-maturing ones. Breeds within a species vary in their **maturation** rate, as do individuals within a breed. Typical changes in body weight with advancing age for selected breeds of dairy cattle and swine are shown in Figure 8.7.

[3]*Missouri Agr. Expt. Sta. Res. Bull. 97,* 1927.

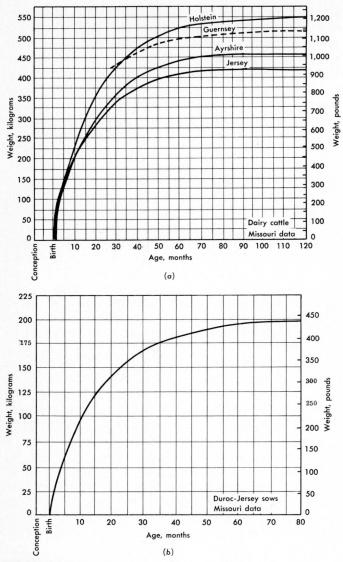

FIGURE 8.7
Average changes in body weight observed with advancing age of
(a) four breeds of dairy cattle; (b) Duroc sows. *(Redrawn from
Missouri Agr. Res. Bull. 96, 1926.)*

8.5 HORMONAL CONTROL OF GROWTH

Secretions of several **endocrine** glands are known to have a vital effect on growth
in most animal species. Research in which the endocrine glands were surgically
removed or where injections of extracted hormones were administered to young,

growing animals has clearly demonstrated hormonal effects. Some naturally occurring abnormalities of endocrine secretions have also provided information pertaining to these subjects.

8.5.1 Growth Hormone

The anterior pituitary gland, located in the skull at the base of the brain, secretes a hormone closely related to an individual's growth (Table 8.1). It is present in the pituitaries of all vertebrates and is known as *growth hormone* (GH) or *somatotropin* (STH). GH is a protein and is secreted by acidophilic cells (cells having an affinity for acid dyes). Highly purified preparations of active GH have been obtained from the pituitaries of many species, including the pure crystalline form from cattle and humans. The chemical structure of GH from several species of animals has been determined, and commercial preparation, utilizing the principles of genetic engineering, has become a timely reality.

Human growth hormone (HGH) consists of a long chain of 188 amino acid residues bridged at two points by sulfhydryl bonds. GH is species-specific, so that preparations usually are not effective unless prepared and injected within the same species. Thus the human pituitary dwarf must be treated with HGH to be **therapeutically** affected.

Removal of the anterior pituitary gland in young, developing animals causes growth to cease and body weight to decrease. The daily injection of anterior pituitary extracts will promote resumption of growth in such animals. Long-term injections of GH preparations in young, normal animals will produce extremely large individuals (giants). This has been demonstrated in rats and dogs. Secretion of excess HGH in young humans has been credited with the production of giants 7 to 10 ft tall. If excessive GH is administered to mature individuals, a condition called **acromegaly** results. It is a condition in which parts of the body, such as the head, hands, and feet, become enlarged. GH does not appear to have a major effect on growth during fetal life. When rat, rabbit, and

TABLE 8.1
GROWTH HORMONE CONTENT OF THE
PITUITARY GLANDS OF HOLSTEIN
HEIFERS OF DIFFERENT AGES

Age, weeks	Milligrams of GH per gram of anterior pituitary
16	11.67
32	10.90
48	8.18
64	6.98
80	7.73

Source: D. T. Armstrong and W. Hansel, *J. Animal Sci.*, **14:** 1242 (1955).

swine fetuses are **decapitated,** they continue to grow at a normal rate. However, GH is present in the anterior pituitary glands of fetal mice, chickens, and pigs, and it appears to be capable of stimulating growth when administered to postnatal animals. Removal of the maternal pituitary gland has little or no effect on fetal growth, perhaps because in many species GH from the mother cannot pass from her blood (through the placenta) to that of the young. Different breeds of cattle grow at varying rates before birth. In a Missouri study, **purebred** Angus, Hereford, and Charolais calves weighed 63, 71, and 82 lb, respectively, at birth. Part of this variaton in birth weight was due to differences in maternal size, but some of the variation was postulated to be due to differences in GH secretion or the response of the tissues to GH.

Even though growth finally ceases in mature animals, GH is present in their pituitaries (Table 8.1), although probably in reduced amounts. Perhaps skeletal growth ceases because the tissues are no longer responsive to GH. In mature animals, the **epiphysial** grooves of the long bones are closed, preventing further growth. A selection experiment at the Illinois Agricultural Experiment Station was effective in developing two genetically different lines of the Hampshire breed of swine. At 180 days of age one had a heavy and the other a light weight. Pigs from the *heavier* line had significantly greater amounts of GH per unit of anterior pituitary tissue at all ages than those from the *light* one. Similar results were reported by the Cornell Agricultural Experiment Station for fast- and slow-growing Holstein **heifers.**

Hypophysectomy causes skeletal growth to cease, resulting in dwarfism. The administration of GH causes resumption of skeletal growth. This is proof that GH is involved in skeletal size, especially in long-bone growth.

GH affects amino acid **metabolism.** This effect is apparent in the growth of young, in which muscle-mass increases require the synthesis of protein and thus an increased bodily retention of dietary nitrogen. GH apparently stimulates the synthesis of proteins from amino acids. How it does this is not fully understood. GH is thought to act directly on certain tissues; other effects are believed to be indirect, such as the stimulation by GH of the secretion of *somatomedins,* which then act directly on a target tissue.

GH reduces the amount of fat stored in the body. Following GH injections, there is usually an increase of free fatty acids in the blood, and it is thought that this rise is due to the release of fatty acids from adipose tissue. Some evidence suggests that GH enhances the conversion of carbon from fatty acids into carbon that may be incorporated into proteins. GH also has varied effects on **carbohydrate** metabolism, some of which are not fully understood.

Probably one of the best indicators of the functions of GH is illustrated when GH preparations are administered to normal, growing animals. The treatment results in an increased proportion of **carcass** protein and water and a reduction in the proportion of fat. The carcass content of a treated animal is similar to that of a very young animal. Increases in body size are due not to fat deposition but to an increase in tissue size.

8.5.2 Thyroid Hormone

The **thyroid** hormone of blood is almost entirely L-thyroxine, although there are traces of triiodothyronine. Production and release of thyroid hormones is controlled by thyroid-stimulating hormone (TSH) from the anterior pituitary gland (Chapter 7).

Thyroidectomy (removal of the thyroid glands) and subsequent thyroid therapy may be followed by disturbances in metabolism, development, and growth. The main physiological effect of thyroxine is to increase energy production and oxygen consumption of most body tissues. A lack or deficiency of thyroxine in the young results in a condition called *cretinism,* which usually includes dwarfism and idiocy. Thus a certain level of thyroxine secretion is essential for metabolic functions involved in normal growth.

Low thyroid activity (**hypo**thyroidism) is usually accompanied by (1) reduced food intake, (2) low blood-sugar concentration, (3) less liver glycogen storage, and (4) lowered nitrogen retention. Fat deposition may be increased. An overactive thyroid (**hyper**thyroidism) increases metabolic rate and, if severe, may cause a breakdown of certain tissues, accompanied by weight loss. Several inexpensive thyroid-active substances have been developed recently, and many of them have been tested to determine if they have an economic value (1) as growth stimulants in farm animals or (2) for increasing **lactation** and egg laying. Other substances have been developed that suppress thyroid secretion. In general, neither thyroid-activating nor thyroid-suppressing compounds appear to have a useful application in livestock production (Chapter 14). They should be effective in treating individuals suffering from hypo- or hyperthyroidism.

8.5.3 Androgens

These male hormones are secreted by the interstitial cells of the testes and to a lesser extent by the adrenal glands. Androgens stimulate growth in young farm animals, as shown by size differences in males and females, the males being significantly heavier at almost any age. Males also have better rates of body-weight gain and **feed efficiency** than either females or castrated males. **Castration** of males slows growth to some extent and accelerates the fattening process in cattle, sheep, and swine. The faster growth of noncastrated males appears to be due to a greater development of muscular and skeletal tissue. It has been shown in many animal species that androgens favor tissue nitrogen retention, which means that there is a greater amount of protein synthesized and utilized and/or a lesser amount of protein **catabolized** in the presence of these hormones.

Research with laboratory animals suggests that small amounts of testosterone are secreted by the testes and adrenals of prepubertal males. These small amounts appear to stimulate growth by functioning with GH to give a maximum growth response. At puberty, however, or when large doses of testosterone are administered, the male hormone appears to block the growth hormone effect on the epiphysial cartilage. When the epiphyses fuse to the long bones, linear growth ceases. It is known that as an animal matures, this fusion occurs and long-bone growth ceases. Perhaps the larger amounts of androgens secreted after puberty directly or indirectly enhance this epiphysial closure, resulting in the final mature height of the individual.

8.5.4 Estrogens

These female hormones are produced by the ovary and to a lesser extent by the adrenals. In pregnant females, estrogens are produced by the placental tissues. Estrogens are known to aid in regression and closure of the epiphysial cartilage plate of long bones, thus slowing growth. Increased estrogen secretion occurs with the onset of puberty. This fact partially explains why girls stop growing appreciably in height soon after the onset of menstruation (coinciding with the secretion of estrogen).

8.5.5 Insulin

Insulin is a protein hormone secreted by the beta cells of the islets of Langerhans in the pancreas (Chapter 7). This hormone increases the uptake of amino acids and glucose into muscle. Insulin also stimulates growth by increasing the synthesis of RNA and protein in a number of tissues, including muscle and bone.

8.5.6 Glucocorticoids

The glucocorticoids are inhibitors of growth. One of these hormones, cortisol, decreases the synthesis of DNA and protein in a number of tissues.

8.6 NUTRITION AND GROWTH

Although higher animals, including farm mammals, have very complex physiological systems, such as the endocrine glands and a central nervous system, they are unable to manufacture certain substances needed for their everyday life. These substances must be obtained from a source outside the body for proper growth to occur (Chapter 14).

The mother carefully protects the nutritional needs of the young as it develops in her uterus. She often draws on body reserves to supply the needs of the developing fetus. If nutrients supplied to the mother are severely deficient during pregnancy, birth weights of the young will be subnormal and their vigor may be reduced. The lack of **vitamins** and minerals in the mother's diet during pregnancy may have a marked effect on the vigor of the young without expressing a great effect on birth weights. A lack of vigor is usually followed by heavy death losses of the young at birth or shortly thereafter. Frequently poor prenatal growth rate (light birth weight) has little effect on mature size in animals if postnatal (after birth) nutrient supplies are adequate.

The effect of undernutrition after birth on mature size depends on (1) the age at which underfeeding occurs, (2) the length of the underfeeding period, and (3) the kind of underfeeding (energy, vitamin, or other deficiency) that is practiced. A diet may provide adequate energy but be inadequate in vitamins and minerals, resulting in undernutrition. Early Missouri studies (1919) showed the effect of underfeeding in cattle. **Yearling steers** were fed a diet to maintain, but not increase, their weight. Skeletal growth continued in the steers, and they increased in height and length. Their fat reserves were depleted, however, and they became very thin.

Research reports differ as to whether underfeeding can permanently stunt an animal, thus preventing it from attaining its potential mature size. Severe underfeeding beginning at birth and extending for several weeks may prevent the individual from attaining its normal mature size. If an underfed animal is placed on a full **ration** again, it often grows more rapidly than is normal for that particular species. (This type of growth is called *compensatory gains.*) Underfeeding when an animal is young results in a longer time required to attain normal mature size when it is again placed on a **full-feed.** This result suggests that limited feeding delays maturity in some way. When the growth of rats was restricted by feeding a low-**calorie** diet beyond the average life span of that species and the rats were then full-fed, they were able to resume growth, although they did not attain the normal body size of controls. The life span of rats may be extended by growth retardation (by caloric but not protein, vitamin, or mineral undernutrition).

The plane of nutrition can be used to adjust the growth rate in different stages of the growing-fattening period, especially in swine. It also can be used to limit the amount of carcass fat present at slaughter (Table 8.2). Since fattening in

TABLE 8.2
THE EFFECT OF LIMITED FEEDING ON GAINS, AMOUNT OF FAT, AND SKELETAL GROWTH IN THE CARCASS OF SWINE AT SLAUGHTER

	Full-fed	Limited-fed*
Initial weight, lb	38.7	39.1
Final weight, lb	205.1	216.7
Average daily gain, lb	1.5	1.3
Body length, mm	717	722
Leg length, mm	520	533
Backfat thickness, mm	46.2	42.9
Depth of fat on ham, mm	40.0	35.5
Weight of fat cuts, %	29.7	27.5
Age at slaughter, days	167	193

*Included 12 pigs per lot. Those limited-fed were given 85 percent of a full-feed.
Source: Missouri Agr. Expt. Station.

swine usually occurs quite rapidly between 100 and 200 lb, limiting the ration energy content of the diet at this time will also limit the amount of carcass fat.

To attain optimum inherent growth, then, a balanced diet is essential. We often think of certain races and nationalities as tall or short. However, children of Japanese parents born in California tend to be taller than their parents. Improved nutrition is credited for the increase in the height of these Japanese children over their parents (Chapter 3).

The energy consumed per kg in doubling birth weight is approximately the same for most mammals (except humans), namely, about 4000 calories (cal) (Table 8.3).

8.7   HEREDITARY MECHANISMS IN GROWTH

Variations in growth within a species reflect both hereditary and environmental influences, since both have large effects on growth in farm animals and humans. Environmental factors affecting growth include nutrition, exposure to disease, **parasites,** and injuries. Hereditary factors include those in which a single gene (or group of genes) may have a large effect on the size and growth rate of an individual. A good example is dwarfism, which occurs in several animal species. Hereditary factors also include those in which each of many genes has a small individual effect, while the overall effect of these may be quite large. An example of this type of heredity is the *large* and *small* strains of swine, sheep, cattle, horses, and **poultry.**

8.7.1   Prenatal Growth

Prenatal growth may be inhibited somewhat even though the developing fetus has a genetic potential for rapid growth and a larger size. Prenatal growth in

TABLE 8.3
ENERGY CONSUMPTION PER
KILOGRAM REQUIRED TO
DOUBLE BIRTH WEIGHT

Species	Energy required, cal
Swine	3754
Sheep	3926
Cow	4243
Dog	4304
Horse	4512
Cat	4554
Rabbit	5066
Human	28,864

Source: M. Rubner, "Das Problem der Lebensdauer und seine Beziehungen zum Wachstum and Ernährung," Berlin, 1908.

chickens is limited by egg size or, more specifically, by the amount of food available to the developing chick. In litter-bearing animals, extremely large litters may result in some developing fetuses being undersized at birth because of a lack of uterine space and/or available nutrients. The transplantation of fertilized rabbit ova from a small to a large race increased the birth weight of the developing fetuses about 40 percent.[4] However, young from the transplanted small race did not weigh as much at birth as young coming directly from the large race (large mothers) because their size potential was limited by heredity. Similar results have been obtained when fertilized eggs have been exchanged between small and large strains of female mice.

8.7.2 Growth from Birth to Weaning

Growth during this period may be affected by the maternal environment to which the young are exposed before and after birth. In chickens, effects of the nutrients supplied by the egg may be continued beyond **hatching,** but these effects appear to be largely **transitory** and disappear later in life if the individual has the inherent capability for fast growth. In lactating mammals, the inherited growth impulse possessed by the young may be temporarily overshadowed by the amount of milk supplied by the mother. Many studies with swine indicate that zero to 20 percent of the variation in weaning weight is due to the animal's own genes and 35 to 50 percent to the maternal environment to which the litter is exposed. This environment includes such factors as available milk, litter size, and other factors peculiar to each litter. The growth of cattle and sheep between birth and weaning appears to be more closely related to the genetic makeup of the individual calf and lamb (20 to 30 percent) than is the growth of litter-bearing animals. Research data suggest that birth weights in cattle are more highly heritable than weaning weights and that many of the genes responsible for fast growth before birth are also responsible for fast growth between birth and weaning.

8.7.3 Postweaning Growth

After weaning, maternal influences, some of which can be hereditary, become relatively less important than the individual's own genetic makeup, and they may disappear completely at maturity. Although the maximum genetic size of the individual is fixed at conception, many environmental factors—such as inadequate nutrition, disease, and accidents—may prevent the individual from achieving its genetic potential.

The genetic influence on the mature size of an individual within a species has been demonstrated many times by selection experiments. Large and small strains of chickens, mice, rats, rabbits, dogs, swine, sheep, and cattle have been developed by selection for size differences over a period of years. The mature size of farm mammals is related to the rate and efficiency of gain from birth to

[4]O. Venge, *Acta Zool. (Stockholm)*, **31**:1–148 (1950).

market weight, although other factors are also involved. In general, cattle that reach a large market weight are still growing at the usual market weight of 1000 to 1100 lb and do not possess as high a proportion of fat in their bodies as do animals of smaller, earlier-maturing strains and breeds at the same weight. It is interesting that some animals, such as beavers, never stop growing. Their life span in the wild is about 12 years, and captives have lived 19 years.

8.7.4 Genetic Control of Growth Mechanisms

Although it is known that genes affect growth in almost all animals, including humans, only fragmentary evidence is available to illustrate the physiological pathways of gene action. Experiments have shown that breeds of chickens, rats, and inbred strains of swine vary in their needs for certain nutrients, especially the B-complex vitamins. Some strains show deficiency symptoms for certain nutrients while others fed the same diet show no adverse effects.

Hereditary dwarfism in humans and mice has been shown to result from a lack of growth hormone production, indicating the relation between genes and the synthesis of protein hormones. Chickens have shown wide genetic differences by the response of the **comb,** thyroid, and gonads to administration of anterior pituitary hormones. This suggests that there are genetic differences in the sensitivity of certain target organs to hormones. In one strain of yellow mice, there is a greatly increased rate of fat deposition, which has been attributed to a hormonal alteration of carbohydrate metabolism. Genes are responsible for this hormonal abnormality. Some evidence suggests that increased rate and efficiency of gain in swine due to **hybrid vigor** result from a more efficient metabolic system. Hybrid vigor is due to a nonadditive type of gene action (Chapter 5).

8.7.5 Association between Growth and Other Traits

In some species, growth appears to be genetically associated with other vital body processes. An increase in degree of inbreeding is accompanied by a decline in growth and an increase in mortality rate. **Crossbreeding** usually increases the growth rate and reduces mortality. Both inbreeding and crossbreeding provide examples of a genetic effect. Many of the genes responsible for rapid growth are also responsible for efficient gains. Faster growth in several species appears to be related to a higher **ovulation** rate, although in some experiments in which selection was for either heavier or lighter mature weights, lowered fertility was observed in both. Among beef cattle on a New Mexico range, large-type Hereford cows produced significantly more and heavier calves in their lifetime than smaller, **blockier-type** Herefords.

It is interesting that the metabolic rate of animals does not change proportionately with body weight. Brody,[5] in his monumental text, concluded that a doubling of body weight was associated with a metabolic increase of approxi-

[5]S. Brody, *Bioenergetics and Growth*, Reinhold Publishing Corp., New York, 1945.

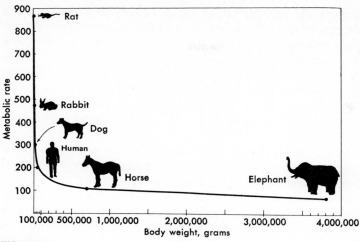

FIGURE 8.8
Graph depicting the inverse relationship of relative metabolic rate and body
size in selected mammals. Metabolic rate is given in cubic millimeters of
oxygen per gram of body weight per hour. *(W. T. Keeton,* Biological
Science, *3d ed., W. W. Norton & Co., New York, 1980.)*

mately 73 percent. This means that basal metabolism varies to the 0.73 power of
body weight $(W^{0.73})$. The relation of relative metabolic rate and body size is
shown in Figure 8.8.

8.8 SENESCENCE (AGING)

> And so from hour to hour we ripe and ripe.
>
> *William Shakespeare (1564–1616)*

> Life is most delightful when on its downward slope.
>
> *Lucius A. Seneca (4 B.C.–A.D. 65)*

Senescence is the process of growing old. People refer to *growing old,* suggesting
that it is recognized as a gradual and continuous process rather than one that
occurs suddenly. The term *aging* encompasses the complex of changes that
eventually lead to deterioration of the animal and ultimately to its death.

Problems of aging are especially important in human medicine, and much
research is currently being conducted in this field. Such problems are of less
importance in farm animals because breeding animals are often **culled** well
before they reach old age. This is particularly true of swine and poultry. Cows,
however, are kept to an older age if they are not removed from the herd for
reasons unrelated to aging. Many beef cows are kept until they are 9 to 12 or

more years of age. Nevertheless, there is an age of each farm animal at which productive performance peaks. This is shown in Table 8.4 for selected farm animals. These data show that egg-laying rate is highest during the hen's first year and declines gradually thereafter (Chapter 12). Maximum litter size of **sows** occurs between 3 and 4 years of age, although few sows are kept in production that long. This is because sows become extremely large and are more expensive to feed and manage. Moreover, large sows present more problems at **farrowing** than smaller and lighter ones, because they are more likely to overlay and smother their pigs after birth.

It has been said that an animal begins to die as soon as it is born. This is true in the literal sense, because each passing day brings it one day closer to the actual day of its death. It is true also in the physiological sense, because shortly after fertilization of the ovum, cells of certain tissue cease their division. Cell division ceases progressively among other tissues as the young develop, until division is finally limited to cells in tissues such as the skin surface (epidermis) and blood, in which periodic cell replacement is absolutely essential to protect the health and well-being of the individual.

Why does the mayfly live but a few hours while the pike (fish) may live two centuries? The rate of decline in the velocity of growth with increasing age is generally inversely proportional to the length of life (Figure 8.9). In a species that reaches maturity in a few years (cattle), the life span is shorter than in a species such as the human, in which many years are required to attain maturity. Certainly, species vary in their life span. The rat becomes old in 3 years, the cow in 20 to 25, the horse in 30 to 35, humans in 80 to 90 years, and the Galapagos tortoise in 500 years. As stated in the Bible, ". . . it is appointed unto men once

TABLE 8.4
AGE AND PERFORMANCE OF SELECTED FEMALE FARM ANIMALS

Age of females, years	Weaning wt of calves, lb (Hereford)	Av litter size farrowed (Duroc and Poland sows)	Egg production per year (Leghorn hens)*
1	360	7.8	169
2	389	9.4	146
3	411	**9.8**	124
4	427	9.5	109
5	438	9.2	95
6	**443**	9.2	86
7	**442**	7.3	66
8	434	...	67
9	421	...	51
10	402	...	41

Note: Boldface numbers represent peak productive performance.
*Present annual egg production is higher; however, the same trend prevails.
Source: Missouri Expt. Station bulletins.

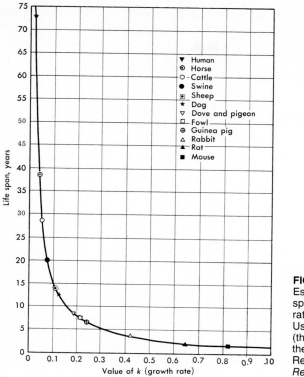

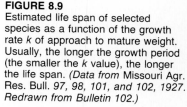

FIGURE 8.9
Estimated life span of selected species as a function of the growth rate k of approach to mature weight. Usually, the longer the growth period (the smaller the k value), the longer the life span. *(Data from* Missouri Agr. Res. Bull. *97, 98, 101, and 102, 1927. Redrawn from Bulletin 102.)*

to die . . ." (Hebrews 9:27), and "the days of our years are threescore years and ten; and if by reason of strength they be fourscore years, yet is their strength labour and sorrow; for it is soon cut off, and we fly away" (Psalms 90:10). Humans have always searched for the fountain of youth. In general, humans are not content with living just threescore and ten years, but rather would like to live forever. If science could extend the human life to 1000 years, think of the problems this would present. Even if this were scientifically possible, many persons would still think their life span too short.

The slowest beasts are always strongest and manage, too, to live the longest.
Benjamin Franklin (1706–1790)

Of academic interest is the belief of some scientists that the life span of individuals is more or less fixed at conception by the metabolic potential the individuals receive from their parents. Furthermore, the rate at which individuals utilize their metabolic potential determines to a certain extent their length of life. This suggests that the "candle should not be burned at both ends," or a shorter life will result. As noted by Brody,[6] Rubner[7] calculated the quantity of energy metabolized per kg of body weight from maturity to death and found that it is nearly the same in **homeothermic** animals (except humans). This is illustrated by the following estimates:

Species	Body wt, kg	Maximum life span, years	Calories expended during lifetime per kg adult body wt
Horse	450	38	170,000
Cow	450	28	141,000
Dog	22	13	164,000
Cat	3	8	224,000
Guinea pig	0.6	6	266,000

Unlike the above species, which expend about 200,000 cal/kg of body weight during the life cycle, humans were estimated to expend about 800,000 cal/kg. However, if humans extend their days and shorten their periods of rest, they must **metabolize** more energy per day. Some scientists theorize that this results in a decreased life of one or more body parts and/or systems.

Many physiological functions deteriorate with advancing age of humans and animals (Figure 8.10). The gonads secrete smaller amount of hormones, muscular strength and speed of motion decline, reaction time is increased, and there appears to be a longer time required for recovery when body substances become unbalanced. For example, when **intravenous** injections of glucose are administered to older persons, a longer period of time is required to restore the normal blood glucose concentration. Collagen becomes less contractile with increasing age, as shown by less elasticity of the skin and blood vessels compared with those of younger individuals. Some people believe that with advancing age there is a gradual loss of functional units in the organs or a gradual impairment of those remaining; both processes could be involved. A gradual breakdown of neural and endocrinal control in the aging animal may be responsible in large part for changes associated with senescence.

Many factors are known to affect the life span of individuals. Longevity is related to heredity. Benjamin Franklin, a great American statesman, lived to the age of 84, closely approximating his mother's and father's ages at death, 85 and 89 years, respectively. Females in most species live longer than males. Diet can lengthen or shorten life span. In Cornell studies, rats fed high-protein diets had decreased life spans. (In the studies, 14 percent dietary protein appeared to be the most favorable to longevity.) In **poikilothermal** animals (animals in which the body temperature varies with environmental temperature), life span is short-

---

[6]Ibid.
[7]M. Rubner, "Das Problem der Lebensdauer und seine Beziehungen zum Wachstum und Ernährung," Berlin, 1908.

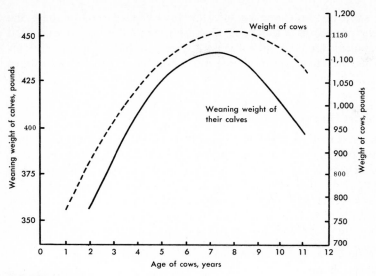

FIGURE 8.10
Graph showing the peak of physical maturity (maximum body weight of
cows) and of physiological maturity (maximum weaning weight of calves)
of Hereford cattle. Note the decline in both after 7 to 8 years of age.
(Missouri Agr. Expt. Station.)

ened by increasing the temperature and lengthened by lowering it. In humans
the vascular and nervous systems are usually the first to suffer from senescence.

8.9 SOME THEORIES OF AGING

Consequently to preserve life is to use meats and drinks according to the age of the
person. For the dyet of youth is not convenient for old age nor contrariwise.
Thomas Cogan, "The Haven of Health" (1596)

Several theories of why aging occurs have been proposed, although no single one
appears to explain it adequately. Aging is a complex process, with many factors
and physiological functions being involved.

8.9.1 Genetic Theory

The genetic theory states that in aging there is an accumulation of mutations
within body cells that interfere with their normal function. Some people believe
that rapidly dividing tissue cells are able to discard new mutations, whereas
slowly dividing cells have an accumulation of mutations that causes organ
degeneration. Evidence for this theory is that nonlethal doses of radiation

shorten the life span in some animals, presumably because they increase the number of somatic mutations within cells. Opposing this theory is evidence that aging changes occur in fixed postmitotic cells, whereas radiation primarily affects dividing cells. Also, radiation does not appear to affect the physical properties of collagen.

8.9.2   Immunological Theory

According to the **immunological** theory, organisms have the ability to produce **antibodies** against foreign cells and substances to which they are exposed if exposure occurs before a certain critical period in life. After this time, the individual gradually loses its ability to resist invasion of foreign substances. For example, with advancing age humans become more susceptible to certain infectious diseases to which they are immune at a younger age. This suggests that with advancing age they lose their ability to cope with an adverse environment.

8.9.3   Developmental Theory

The developmental theory states that aging results from what is often called *overdifferentiation* (extreme cellular chemical specialization). For example, nerve cells of an adult animal cannot divide, and there are no other cells that can divide and then differentiate into nerve cells. Thus, when nerve cells die or become nonfunctional, some aspect of control over the body is lost. This loss of ability to coordinate the regulation of the body may, in due time, result in its destruction.

8.9.4   Biochemical Theory

The biochemical theory of aging proposes that rare and irreparable nongenetic metabolic accidents occur. The products of such accidental chemical reactions gradually accumulate in cells and may be insoluble and have no function, although they may interfere with normal cellular metabolism. It is known that pigments accumulate in some aging cells, a fact that appears to support the theory.

8.10   SUMMARY

In livestock production, *growth* is often defined as *the normal increase in weight and body size over a period of time.* This is not an accurate definition, however. True growth from a chemical standpoint is *an increase in the amount of body protein and mineral matter.* Deposition of fat and accumulation of body water are not true means of growth.

The cell is the basic unit of growth. In growth, cells increase in numbers (hyperplasia) and size (hypertrophy). Early embryonic growth is due to an increase in cell numbers, whereas most growth near birth or shortly thereafter can be due to increases in cell size and number. Cell division ceases in some organs much before it does in others.

The postnatal growth curve of all animals is sigmoid (S-shaped). The growth rate of animals decreases shortly after puberty. Many factors are known to affect growth. These include (1) the genetic constitution of the individual, (2) nutrients and environmental conditions encountered during its lifetime, and (3) hormones secreted by the individual's endocrine glands. The life span in multicellular animals is generally related to their age at maturity. A species that reaches maturity more quickly has a shorter life span than one that matures at an older age. As the animal grows older, it loses some of its ability to cope with the environment. Although there is not complete agreement on the causes of aging or senescence, most authorities agree that cells either die or lose their ability to function properly. Both processes may be important. In any event, death occurs when one or more vital organs cease to perform their required functions.

Human beings are by nature predestined, after growth and maturation, to decline and finally die. Their structure and functions are more complex than those of any other living creature. Life is like the flame of a candle whose form is constant while each particle changes every moment. Death takes small bites during the aging process, and eventually destroys enough cells that the body tissues fail to function as an integrated entity, resulting in the death of the entire organism.

STUDY QUESTIONS

1 Define growth, differentiation, morphogenesis, regeneration, hypertrophy, hyperplasia, intussusception, permanent cells, stable cells, and labile cells.
2 Growth is not just an increase in body weight. Explain.
3 What factors may influence the weight of the young at birth?
4 How do muscle cells increase in size?
5 How is the protein/DNA ratio related to the growth of tissue?
6 Explain how bones grow in length and width.
7 How do cancer cells differ from normal cells in their division process?
8 What are some body cells that divide in a way similar to that of cancer cells?
9 What are the two main periods of growth in mammals?
10 Describe the postnatal growth curve in populations and in individual animals.
11 What forces are responsible for the typical kind of growth curve?
12 What factors are responsible for cessation of growth in populations of single-celled individuals?
13 List factors responsible for the maximum size of an individual.
14 What hormones are related to growth?
15 Describe the source, function, and chemical composition of growth hormone.
16 Is GH secreted in the normal mature individual?
17 What is the relation of nutrition to growth and life span of animals?
18 What evidence is there that hereditary mechanisms affect the rate and amount of growth in farm animals?

19 Is growth related to other traits in mammals? Explain.
20 What is senescence?
21 Describe some changes within the body that occur with advancing age.
22 List some factors that are known to affect the life span of animals.
23 Discuss theories of aging. Does any one theory adequately explain all the processes that occur in aging?

CHAPTER

ANATOMY AND PHYSIOLOGY OF REPRODUCTION IN FARM MAMMALS[1]

Let the earth bring forth the living creature after his kind, cattle, and creeping thing, and beast of the earth after his kind, and it was so.

Genesis 1:24

9.1 INTRODUCTION

Reproduction is a complicated process in all **species** of animals. It is complicated because it is so dependent on the proper functioning of biochemical processes of many organs. Also, the anatomy of the male and female reproductive systems must be compatible within each species to make efficient reproduction possible.

Physiological compatibility on the part of the male and female is also absolutely necessary for reproduction. The female must be willing to accept the male in the act of mating (**copulation**) when the egg is released from the **ovary** (**ovulation**). Similarly, the male must be willing and able to deliver **spermatozoa** to the proper site in the female reproductive tract at the proper time for **conception** to occur. Synchronization of **ovum** release from the ovary and introduction of spermatozoa into the female reproductive tract is necessary because the lives of the released ovum and the ejaculated spermatozoa are limited to a few hours within the reproductive tract of the female (Table 10.3). Maturation of the ovum, ovulation, **estrus,** the estrous cycle, pregnancy, **parturition,** and **lactation** are all dependent on proper function of various

[1]The authors acknowledge with appreciation the contributions to this chapter of Dr. D. J. Kesler, Department of Animal Science, University of Illinois, Urbana.

hormones and organs. Any abnormality in the anatomy or physiology of the reproductive system results in lowered **fertility,** infertility, or **sterility.** It is amazing, indeed, that the fertility rate in farm animals is as high as it is in view of the many complicated biological processes involved that are necessary for conception and birth of the new individual.

In this chapter the anatomy and physiology of reproduction in male and female farm **mammals** are discussed. The physiology of lactation and egg laying is discussed in Chapters 11 and 12, respectively. The anatomy and reproductive physiology of the male and female chicken are discussed in Chapter 12.

9.2   ANATOMY OF THE MAMMALIAN MALE REPRODUCTIVE TRACT

The role of the male farm mammal in reproduction is less complex than that of the female. His role is to produce large numbers of viable male sex cells (spermatozoa) and to deliver them to the proper place in the reproductive tract of the female at the proper time. Inability of the male to perform either of these functions results in infertility.

The male farm mammal contributes approximately one-half of the inheritance to each of his **offspring.** The possible exception to this is that he supplies something less than one-half of the inheritance to his male offspring, because he transmits to them only the **Y chromosome.** The Y chromosome is considerably smaller than the **X** chromosome and therefore probably carries fewer **genes** (Figure 4.3). After the male mammal delivers viable spermatozoa to the reproductive tract of the female and the ovum is fertilized, his role in reproduction is completed. He has little or nothing to do with the welfare of the developing young after that time.

The male reproductive organs in various species of farm mammals are similar in the parts they contain, but there are species differences in their size and degree of development (Figures 9.1 and 9.2).

The testicles (testes) in the male are the *primary sex organs*. They produce spermatozoa and the male hormone, *testosterone*. Each male normally possesses two testicles carried in the scrotum. The scrotum in the **ram, bull,** and **stallion** is located in the **ventral** portion of the body **anterior** to the hind legs. In the boar, the scrotum is located ventral to the anus (Figure 9.2). The scrotum functions as a heat-regulating mechanism in all male farm mammals, especially in the ram and bull. The testicles are drawn close to the body when the outside temperature is cold, and they are relaxed when the outside temperature is hot. The scrotum keeps the testicles 4 to 5°C below normal body temperature. This lowered temperature is essential for spermatogenesis.

In some male farm mammals, one testicle may fail to descend into the scrotum, being retained in the body cavity at normal body temperature. Such a male is a unilaterally **cryptorchid** animal *(crypto* meaning "one," *orchid* meaning "testes")*; however, since he has one testicle in the scrotum, he is usually fertile. The testicle in the body cavity does not produce spermatozoa, but

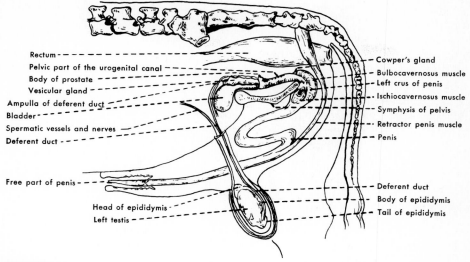

FIGURE 9.1
Diagram of the reproductive tract of the bull, showing major parts. *(University of Missouri Extension Service.)*

it does produce testosterone. Some males have both testicles in the body cavity and are known as bilateral cryptorchids. Bilaterally cryptorchid males do not produce spermatozoa and are sterile, but they do possess normal sexual activity and the outward appearances of the male because they produce testosterone.

Castration in the male consists of removing both testicles, usually within the first few days of life. Such males do not develop male characteristics and lack sex

FIGURE 9.2
Diagram of the reproductive tract of the boar, showing major parts.
(University of Missouri Extension Service.)

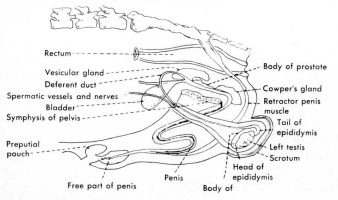

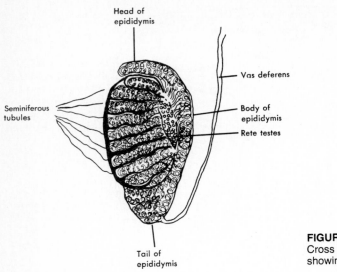

Head of
epididymis

Vas deferens

Seminiferous
tubules

Body of
epididymis

Rete testes

Tail of
epididymis

FIGURE 9.3
Cross section of a testis,
showing different parts.

drive **(libido)**. A castrated male lamb is called a **wether**; a castrated male calf, a **steer**; a castrated male pig, a **barrow**; and a castrated male horse, a **gelding**. A male that is castrated after sexual maturity, however, retains some of its secondary male characteristics and libido and is often known as a **stag.** When unilaterally cryptorchid males are castrated, usually the testicle in the scrotum is removed but not the one in the body cavity. When such a castrated male reaches the age of sexual maturity, the one testicle in the body cavity produces the male

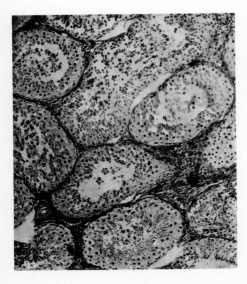

FIGURE 9.4
Cross section of seminiferous tubules from the testes of a mature ram. Within the seminiferous tubules are male gametes in different stages of development. Between the seminiferous tubules are the cells of Leydig, which produce testosterone, the male hormone. (Missouri Agr. Expt. Sta. Res. Bull. *265, 1937.*)

hormone, and the individual possesses normal sexual activity and sex character-istics even though he is infertile.

A diagram of a cross section of a testicle is shown in Figure 9.3. Each testicle contains long tubules, known as *seminiferous tubules.* The seminiferous tubules join in the *rete testis,* which in turn connects with the head of the epididymis. Between the seminiferous tubules are found connective tissue, blood capillaries, and interstitial cells (cells of Leydig).

A portion of the testes of a ram showing cross sections of seminiferous tubules is seen in Figure 9.4. The seminiferous tubules in the sexually mature male possess spermatozoa in all stages of development (Figure 4.6) and *Sertoli,* or nurse, cells. The **lumen** of seminiferous tubules possesses spermatozoa, which have the general shape of those in semen. Spermatozoa in the testes do not possess motility and are not capable of fertilization. The **morphology** of mature spermatozoa of different species of farm mammals is similar, and mature spermatozoa of different farm mammals possess the same parts as ram sperm, shown in Figure 9.5. Semen characteristics for the various farm mammals, together with other pertinent information, are given in Table 10.3

The *epididymis* is a long, continuous, tortuous tube measuring 400 to 500 ft long in some instances. It consists of a head, body, and tail. The head portion is connected with the testis (Figure 9.6). The tail appears as an enlargement on the

FIGURE 9.5
Microphotograph of spermatozoa of the ram. Spermatozoa from the semen of other farm mammals are similar in structure and appearance. *(Missouri Agr. Expt. Station.)*

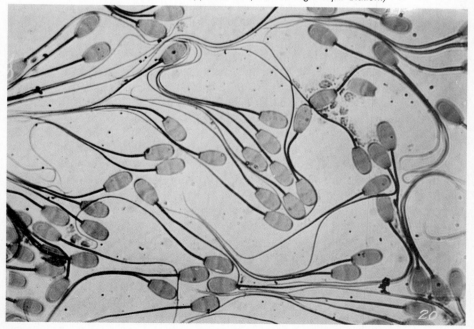

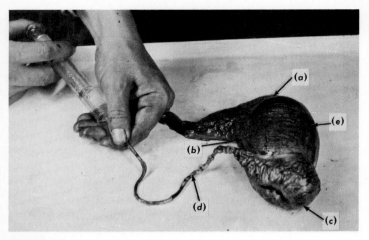

FIGURE 9.6
The testicle of a boar. (*a*) Head of the epididymis; (*b*) body of the
epididymis; (*c*) tail of the epididymis; (*d*) vas deferens; (*e*) testicle.
(Missouri Agr. Expt. Station.)

ventral portion of the testes in the scrotum of the ram and bull, but it is inverted,
or on the **dorsal** portion of the testes, in the scrotum of the boar (Figure 9.2).
The tail of the epididymis is a storage depot for spermatozoa, and as many as 200
billion sperm have been collected from the epididymides of a sexually mature
boar at castration. Spermatozoa mature physiologically as they progress through
the epididymis.

The tail of the epididymis widens into a larger tube leading to the *urethra*
(Figure 9.6). This long tube is known as the *vas deferens*. A male may be
rendered infertile if a portion of each vas deferens is surgically removed
(vasectomy). This prevents passage of spermatozoa into the urethra, and
therefore they do not get into the seminal fluids at **ejaculation.** If a vasectomy is
performed correctly, the male is infertile but retains all his sex drive and male
characteristics. Vasectomies are often performed in males for the purpose of
using them as estrous detector animals. A vasectomy may also be performed in
the human male as a means of birth control, where a pregnancy in his spouse
might be dangerous and/or unwanted.

The vas deferens widens and becomes enlarged in the ram and bull at the
point where it joins the urethra. This enlargement, known as the *ampulla,* may
serve as a temporary storage depot for spermatozoa in species, such as the bull
and ram, in which ejaculation of semen is rapid.

The *urethra* begins at the opening of the bladder and is continuous with the
penis. It acts as a passageway for both urine and spermatozoa. In the sexually
mature male, especially the bull, the **posterior** portion of the urethra is S-shaped
and is known as the *sigmoid flexure.* At the time the male mates with the female,
the penis becomes erect and engorged with blood. The sigmoid flexure straight-

ens, extending the penis so that semen can be deposited into the female reproductive tract. The retractor muscle relaxes at this time, allowing the sigmoid flexure to straighten at the time of copulation. After copulation this muscle contracts (retracts), drawing the penis back into the sheath, where it is protected from injury. It has been reported that some bulls cannot perform natural service because the retractor muscle will not relax and allow the extension of the penis. Even though such bulls may possess **viable** spermatozoa in the testes and epididymides, they are infertile because they cannot deliver spermatozoa to the proper site in the female reproductive tract.

The penis of male farm mammals varies in shape among species so as to be compatible with the reproductive tract of the female, in order to aid efficient reproduction. The penis of the boar possesses a corkscrew shape at its end so that it fits into the interlocking cervix of the sow during copulation. This causes much semen to be forced into the sow's uterus.

The male reproductive tract consists of several glands, known as the *accessory sex glands*. These include the prostate, vesicular glands, and Cowper's glands (bulbourethral glands).

The **prostate gland** is located near and around the opening of the bladder. Its secretions nourish and stimulate the activity of sperm. In the human male and in males of some other species, the prostate gland often becomes enlarged and inflamed, causing pain and difficult urination. The prostate gland is surgically removed when so affected.

The two vesicular glands enter the urethra near the vas deferens and near the neck of the bladder. The vesicular glands were called seminal vesicles for many years because it was thought that they were a storage depot for mature spermatozoa. This was later proved wrong. The vesicular glands are very large in the mature boar, measuring 2 to 4 in in diameter. They are much smaller in males of other farm mammals. Functions of the vesicular glands are to neutralize urine residues, add volume to semen, and possibly to stimulate the activity of spermatozoa.

The two *Cowper's glands* are located posterior to the bladder, prostate, and vesicular glands. They resemble small walnuts in shape in the bull but are about 4 to 5 in long in the sexually mature boar. Also known as the bulbourethral glands, Cowper's glands secrete a fluid that becomes a gel in the semen of the boar when it contacts secretions of other accessory sex glands. This gelatinous material is the last to be ejaculated and forms a plug in the cervix of the sow, which probably prevents the backflow of semen from the vagina at copulation and aids in more efficient reproduction. Secretions of the Cowper's glands also add volume to the ejaculate and neutralize the detrimental effects of urine in the urethra.

9.3 ANATOMY OF THE MAMMALIAN FEMALE REPRODUCTIVE TRACT

The female farm mammal plays a very important role in reproduction. Through the ovum she supplies one-half of the inheritance of each of her offspring, she

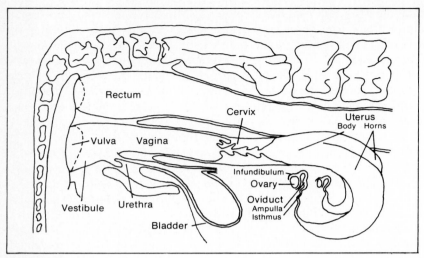

FIGURE 9.7
Drawing of the major reproductive organs of the cow. *(From T. R. Troxel and D. J. Kesler, "The Bovine Estrous Cycle: Dynamics and Control," University of Illinois Coop. Ext. Cir. 1205, 1982.)*

nourishes the young within her body (in the uterus) until birth, and she nourishes (through her milk) and cares for the young after birth until they are weaned.

The anatomy of the reproductive tracts of female farm mammals is mainly similar, but there are some differences in size and shape among species. The major differences are in the size and shape of the uterine horns.

The female reproductive tract (Figures 9.7 and 9.8) consists of the vulva, vestibule, vagina, cervix, uterus, uterine horns, oviducts, and ovaries (Figure 9.9 *a* and *b*). The *vulva* is the exterior portion of the female reproductive tract. It serves as the entrance to the internal organs, and it also accommodates the passage of urine during urination.

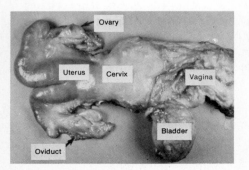

FIGURE 9.8
Photograph of the major reproductive organs of the cow. *(From T. R. Troxel and D. J. Kesler, "The Bovine Estrous Cycle: Dynamics and Control," Univ. of Illinois Coop. Ext. Cir. 1205, 1982.)*

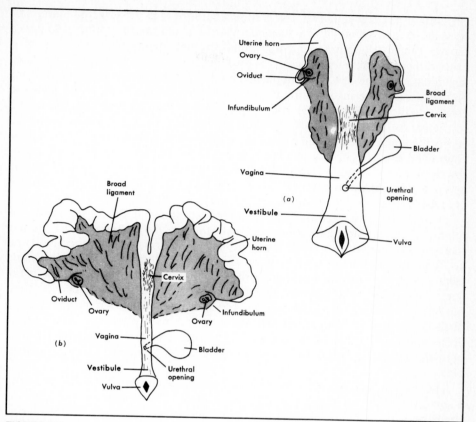

FIGURE 9.9
(a) Drawing of the reproductive organs of the cow, showing major parts; (b) drawing of the reproductive organs of the sow, showing major parts.

The *vestibule* is the general passageway to the urinary and reproductive tracts. It extends inward from the vulva for about 4 in (10 centimeters, cm) to where the urethra opens into its ventral surface from the bladder. The clitoris, which has the same embryonic origin as the penis, lies on the base of the vestibule.

The *vagina* is that portion of the reproductive tract between the vestibule and the cervix. It is about 12 in (30 cm) long in most mature female farm mammals. In the cow and ewe, semen is deposited by the male during copulation in the anterior portion of the vagina. In the mare and sow, at least a portion of the semen is deposited in the uterus.

The *cervix* is the opening into the uterus through which sperm must pass to meet the egg in the process of fertilization. It is also the opening through which the young must pass from the uterus at birth. The cervix is about 2 to 3 in (4 to 7 cm) long in the cow and ewe, but is somewhat longer in the sow and mare. The cervices of the cow and ewe possess circular folds that make it difficult to pass a pipette (an instrument for passing fluids) through them into the uterus. It is practically impossible to do this in the ewe, but it is possible in the cow. Artificial insemination in the cow is accomplished by the placement of semen in the cervix or the uterus. The inseminator places one hand into the rectum and feels the cervix via the rectal wall. The insemination tube is then inserted into the vagina to the cervix. A gentle pressure on the pipette and a rolling movement of the cervix through the rectal wall cause the pipette to pass through the cervix in most cows. The sow has an interlocking type of cervix that also makes it difficult to pass a pipette through it into the uterus. In the mare, the opening through the cervix is not difficult to pass, and even capsules an inch or more in diameter can be deposited in the uterus during estrus without difficulty.

The reproductive tracts of the cow and sow are shown in Figure 9.9 *a* and *b*. A female mammal normally possesses two *uterine horns* (except the human female), although one or both may be absent in abnormally developed individuals. The uterine horns of the sow are much longer than those of other female farm mammals. The sow's uterine horns may measure 5 to 7 ft (2 meters, m) in length when the broad ligament is removed and the horns are stretched to their full length. The sow possesses longer uterine horns because she normally gives birth to 10 or more young, whereas the cow and mare possess shorter uterine horns and usually give birth to only one young. The ewe often gives birth to one young, but it is not unusual for her to have twins, triplets, or even quadruplets.

The anterior end of each uterine horn becomes much smaller and blends into a small tube, known as the oviduct. The *oviduct* is lined with microscopic cilia (hairlike projections) that assist in the movement of the **ovum** toward the uterine horns. The oviducts are 4 to 5 in (9 to 11 cm) long and end in a funnel-shaped membrane known as the *infundibulum*. The infundibulum receives the ovum when it is released from the ovary at ovulation and guides it into the oviduct.

The normal female farm mammal possesses two ovaries, one ovary located in proximity to each oviduct. The *ovary* possesses thousands of female sex cells in different stages of development. Female farm mammals, as well as the human female, possess all the potential ova they will ever have at the time they are born. Only a relatively few of these ova are ovulated, however.

A diagram of a maturing follicle and ovum is shown in Figure 9.10, and an actual photograph of a sow's ovary is shown in Figure 9.11. Ovaries of mature female farm mammals possess *follicles* in different stages of development. During estrus the ovary possesses mature follicles within which the ovum matures. When fully mature, the follicle ruptures and releases the ovum (ovulation). The ruptured follicle then becomes filled with blood and is known as a *corpus hemorrhagicum* (blood-filled body). Cells then gradually replace the blood clot in the ovary at the site of the ruptured follicle, and the follicle is replaced by a *mature corpus luteum* (Figures 9.11 and 9.12). If pregnancy does not occur, the corpus luteum regresses, becoming small and white, and is known as a *corpus albicans* (white body). The corpus albicans gradually disappears and

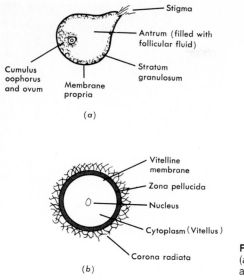

Stigma

Antrum (filled with follicular fluid)

Stratum granulosum

Cumulus oophorus and ovum

Membrane propria

(a)

Vitelline membrane

Zona pellucida

Nucleus

Cytoplasm (Vitellus)

Corona radiata

(b)

FIGURE 9.10
(a) Diagram of a maturing follicle; (b) diagram of a mature ovum.

can no longer be seen on the ovary. If pregnancy occurs, a functional corpus luteum remains throughout the pregnancy.

9.4 PHYSIOLOGY OF REPRODUCTION IN FARM MAMMALS

The physiology of reproduction in farm mammals involves the different parts of the reproductive tract mentioned previously, as well as the **endocrine glands** and their secretions, discussed in Chapter 7.

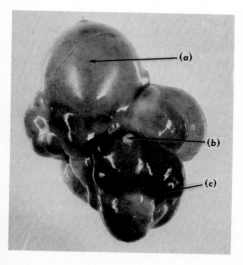

(a)

(b)

(c)

FIGURE 9.11
A sow ovary. (a) Large follicle; (b) corpus albicans; (c) corpus luteum. (*Missouri Agr. Expt. Station.*)

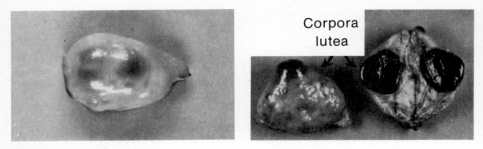

Corpora
lutea

FIGURE 9.12
Ovary of the cow with developing follicles (left) and corpora lutea (right). The ovary on the right
was cut to indicate how deep the corpus luteum grows. *(From T. R. Troxel and D. J. Kesler,
"The Bovine Estrous Cycle: Dynamics and Control," Univ. of Illinois Coop. Ext. Cir. 1205, 1982.)*

9.4.1 Reproduction in Males

Gonadotropic hormones secreted by the anterior pituitary gland are involved in
reproduction in the male. They are the same hormones from the anterior
pituitary gland that affect reproduction in the female. The follicle-stimulating
hormone (FSH) from the anterior pituitary gland stimulates sperm production
by the seminiferous tubules. The luteinizing hormone (LH) stimulates the
interstitial cells (cells of Leydig) to produce the male hormone, testosterone.

Testosterone causes the growth, development, and secretory activity of the
accessory sex glands, which include the prostate, vesicular glands, and Cowper's
glands. Testosterone also stimulates secretory activity of various other cells in
the male reproductive tract, including those in the epididymis. The hormone
also causes growth and development of the rest of the reproductive tract,
including the penis, and it is necessary for maturation and survival of spermatozoa stored in the epididymis. The secondary sex characteristics, such as the male
voice, the crest in the bull and boar, thick horns in the bull, whiskers in the
human male, as well as many other characteristics, are dependent on secretion
of testosterone. This is very evident when males are castrated at a young age,
because the secondary sex characteristics never fully develop in such individuals.
Sex drive, or libido, in the male is also dependent to a great extent on
testosterone production, although certain psychological factors may be involved.
The role of psychological factors is demonstrated in males castrated after sexual
maturity, because they often retain normal, or nearly normal, libido for several
weeks following castration. Testosterone secretion may also play a significant
role in social dominance (peck order) among animals (Chapter 20).

The amount of semen produced and the number of spermatozoa contained in
each semen sample vary among species. Semen consists of spermatozoa mixed
with secretions of various accessory sex glands and those of the epididymides.
The average volume of the normal ejaculate (semen) of the ram is about 1 cc; of
the bull, 5 cc; of the stallion, 75 to 100 cc; and of the boar, 200 to 300 cc. The
average number of spermatozoa per ejaculate varies from about 2.5 billion in the

ram to 20 billion in the boar (Table 10.4). Semen contains some spermatozoa that are dead, some that are alive but nonmotile, others that are weakly motile, and some that show strong, forward progressive motility. Proportions of different kinds of motile and nonmotile spermatozoa from many samples of normal bull semen before and after storage are shown in Figure 9.13. Spermatozoa of abnormal shapes often appear in the ejaculate; a high percentage of these is indicative of an upset in spermatogenesis in the testes and is often related to low fertility (Figure 10.3).

Spermatozoa may be stored and retain their fertilizing capacity for several days in the epididymis within the body of males. Their fertile life is limited, however, in some species when stored outside the body at temperatures slightly above freezing. Bull spermatozoa have been collected and stored outside the body with considerable success for use in artificial insemination (Chapter 10). Bull semen can also be frozen and stored indefinitely at very low temperatures. The fertile life of boar, stallion, and ram spermatozoa stored at temperatures slightly above freezing is limited to 48 h or less. Spermatozoa from these species have not been frozen and stored as successfully as those from bulls. The life of the spermatozoa of farm mammals in the female reproductive tract is limited to 24 to 96 h (Table 10.4).

Many experiments have shown definite adverse effects of thermal stress on reproduction in male farm animals. Adverse effects appear to be commoner in rams than in males of other species. Some rams are low in fertility or temporarily infertile during the summer months. Poor quality of semen during periods of

FIGURE 9.13
The percentage of dead and live spermatozoa in fresh and stored semen (0°C) of the bull as determined by means of the live-dead stain and actual counts of the percentage of motile sperm made with a hemocytometer. *(Missouri Agr. Expt. Station.)*

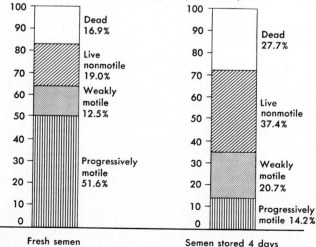

thermal stress is reflected in production of smaller numbers and in a larger percentage of abnormal spermatozoa. This indicates an adverse effect on spermatogenesis. Bulls also may show a decline in sperm production during the hot summer months, but infertility seldom results. Increasing the temperature of the testes by almost any means (cryptorchidism, insulation, etc.) causes decreased sperm production and sometimes infertility in males of several species.

9.4.2 Reproduction in Females

The ovary of the sexually immature female farm mammal is quiescent for a period of several months, and then suddenly follicles mature, the female comes into estrus (heat), the egg is ovulated, and the female has reached the age of sexual maturity (**puberty**). The approximate age of puberty for females of different species of farm mammals is given in Table 9.1. The cause of sudden onset of puberty in the female is not clear. Ovulation can be induced in sexually immature female farm mammals by injections of gonadotropins. This shows that the ovaries of these immature females are capable of responding to gonadotropins, but the hormones probably are not present in the anterior pituitary gland or are not released in sufficient amounts to cause the onset of puberty. Most females within a given species reach puberty near a certain general age and weight, but breed, family, and individual differences exist. The breed and family differences suggest a genetic effect on this trait.

Immature **gilts** near the age of puberty have been brought into estrus when stressed by moving them from one farm to another, by mixing them with other hogs that are strangers to them, or by removing them from a **full-feed,** dry-lot situation to a **limited-feed,** pasture situation. This evidence suggests that stress may be responsible for release of gonadotropins from the anterior pituitary gland if these hormones are already being secreted and stored.

Endocrine Control of the Ovary   FSH from the anterior pituitary gland, when released into the bloodstream, initiates development of ovarian follicles.

TABLE 9.1
AGE OF PUBERTY IN FARM
MAMMALS

Species	Approximate age of puberty, months	
	Males	Females
Cattle	6–9	5–14
Sheep	6–8	7–10
Horses	18–24	15–24
Swine	5–8	5–8

LH acts in synergism with FSH to stimulate maturation of the follicles. As a follicle develops and matures, the egg also matures, and the cells lining the follicle multiply and secrete fluid until the follicle becomes a fluid-filled cavity resembling a blister (Figure 9.12). At this point, the cells lining the follicle wall begin to produce estrogens.

Estrogens have several important functions in the body. These include (1) the induction of estrus (heat) in the female; (2) an increase of motility of the uterus, oviducts, and infundibulum during estrus, which aids in bringing the sperm and egg together in the process of fertilization; (3) dilation of the cervix; (4) synthesis and secretion of cervical mucus; (5) initiation of duct growth in the mammary gland; and (6) development of secondary sex characteristics.

In addition, estrogen acts in a unique manner in the hypothalamus and the anterior pituitary gland to stimulate a massive release of LH called the LH surge. The LH surge stimulates follicle(s) to rupture and to release the egg (ovulation). The infundibulum then picks up the egg and rapidly transports it down to the ampullary-isthmic junction for potential fertilization.

After ovulation, estrogen levels drop rapidly, and the cells in the ruptured follicle begin to grow and divide to form a new structure called the corpus luteum (Figure 9.12). The corpus luteum begins to produce progesterone. If conception occurs, the corpus luteum will be maintained and will continue to produce progesterone, since progesterone appears to be required to maintain pregnancy. If conception does not occur, progesterone levels will drop and the corpus luteum will regress. This regression may be due to an increase in a subtance, such as $PGF_2\alpha$, that destroys the corpus luteum, to the lack of a substance to maintain the corpus luteum, or to a combination of both. In any event, the decline in progesterone is followed by an increase in FSH which, together with LH, stimulates further development and maturation of follicles. Another ovulation then occurs.

Progesterone has several functions: (1) it maintains pregnancy by blocking uterine contractions, (2) it prevents ovulation, and (3) it combines with estrogens to stimulate growth of the lobule-alveolar system of the mammary glands.

Estrus and the Estrous Cycle   Estrus is the time the female will receive the male in the act of mating. The *estrous cycle* begins at puberty and is the interval between two estrous periods when the female is not pregnant. The length of estrus and the estrous cycle for selected farm mammals are presented in Table 9.2. Estrus and ovulation are synchronized in most species so that the introduction of spermatozoa into the reproductive tract of the female and the release of the ovum from the ovary occur at about the same time. This increases the probability of successful fertilization, because the life of the ovum and sperm in the female reproductive tract is limited to a few hours.

Some female farm mammals show very obvious behavioral changes associated with estrus. The cow will often have a clear mucus on the vulva, and she may also exhibit certain behaviors: she may become unusually nervous; have reduced appetite and milk production; and butt, bellow, and walk about more than

TABLE 9.2
LENGTH OF ESTRUS, TIME OF OVULATION, AND LENGTH OF
ESTROUS CYCLE IN SELECTED FEMALE FARM MAMMALS

Mammal	Length of estrus	Length of estrous cycle	Approximate time of ovulation
Cow	12–24 h	19–21 days	10 to 14 h after end of estrus
Ewe	24–36 h	15–17 days	24 to 30 h from beginning of estrus
Mare	4–9 days	19–21 days	24 to 48 h before end of estrus
Sow	24–72 h	19–21 days	35 to 45 h from beginning of estrus

usual. She may also attempt to mount other cows; however, since this behavior is also seen in cows that are not in estrus, it is not a reliable indicator of estrus. The most reliable sign of estrus is that the cow will stand to be mounted by other cows or by a bull. Mares commonly urinate frequently and exhibit rhythmic vulvar contractions called winking. Sows will show an immobility response: standing still if pressure is applied to their backs. Mares and sows demonstrate these behaviors especially in the presence of stallions or boars. Estrus is less obvious in ewes, but experienced producers can detect ewes in estrus.

The estrous cycle in farm mammals has four different phases: proestrus, estrus, metestrus, and diestrus. *Proestrus* is that phase of the cycle which occurs just before estrus. The reproductive system at that time is beginning preparations for release of the ovum from the ovary. There is a thickening of the vaginal wall, increased vascularity of the uterine mucosa, and maximum follicular growth. *Estrus* is the period of the estrous cycle when the female will accept the male in the act of mating and when the follicle matures and releases the ovum. *Metestrus* is the phase of the estrous cycle immediately following estrus, when the uterus makes preparations for pregnancy and the corpus luteum forms and begins to secrete progesterone. *Diestrus* is usually the longest phase of the estrous cycle, occurring between metestrus and proestrus. During the first part of diestrus, the corpus luteum becomes fully developed. In the latter part there is regression of the corpus luteum and uterine mucosa.

Nonpregnant female farm mammals normally express both inward and outward signs of the estrous cycle throughout the year, or during the breeding season if they are seasonal breeders. In some species of animals, such as most primates (including humans), inward signs of the estrous cycle occur but no specific outward signs are evident. Growth of follicles on the ovary and **proliferation** of the endometrium of the uterus to prepare for pregnancy occur periodically even though no estrus occurs. Menstruation (a period of bleeding) is characteristic of primates and is due to a cyclic sloughing off of the portion of the endometrium of the uterus that was built up to receive and nourish the fertilized ovum in anticipation of pregnancy. Such cycles are called menstrual cycles. When the human female reaches the age of 45 to 50 years, the menstrual cycle gradually ceases. This is known as *menopause* (a pause in the occurrence of menstruation), or the *change of life,* after which the normal events of the menstrual cycle no longer occur and the female is infertile. Menopause does not occur in female farm mammals.

Continuous estrus occurs in the nonpregnant, sexually mature female rabbit, at least during the breeding season. Copulation in the rabbit, as mentioned in Chapter 7, induces ovulation by releasing LH from the anterior pituitary gland. When ovulation occurs, corpora lutea are formed on the ovary and the reproductive system is brought under the influence of progesterone as in other species.

Monoestrous and *polyestrous* are terms sometimes used to describe the yearly sexual activity of females in some species. Some wild animals, such as the fox, have only one estrous period each year and thus are referred to as *monoestrous* (*mono* meaning "one") animals. Other animals have many estrous periods annually (if they do not become pregnant) and are called *polyestrous* (*poly* meaning "many") animals. Other species, such as sheep, are definite seasonal breeders, because they show sexual activity during only a particular season of the year. Many breeds of sheep do not have estrous cycles during the late spring and summer months, but sexual activity begins abruptly in late summer and early fall, during a period when daylight is decreasing. Some breeds of sheep mate during most seasons of the year. Other species of animals, such as birds, show sexual activity in early spring, when daylight is increasing, but they are sexually inactive at other times of the year (Chapter 12). Seasonal breeding activity appears to be related to the effect of day length on secretion or release of gonadotropins by the anterior pituitary gland. This is not the sole reason for seasonal breeding, however, because when day length is held constant in some experiments, a seasonal pattern of reproduction still persists in most females. It appears, then, that an inherent internal rhythm of reproduction exists, separate and apart from daylight.

Regardless of the season of breeding, wild animals mate at a time when their young will be born in the most favorable environmental circumstances for survival; that is, in the spring, when there is an ample supply of food and it is warm. The **domestication** of animals has tended to reduce seasonal breeding in some species, such as swine and cattle, so that young can be produced at any time of the year. Even in these species, however, there appears to be a seasonal pattern of greater sexual activity and more efficient reproduction at the period of the year corresponding to the natural mating season in wild animals of that particular species.

Estrus usually does not occur during pregnancy, although exceptions have been observed in the cow. After pregnancy is terminated by birth of the young, estrous periods occur once more, and the normal estrous cycle is initiated in most female farm mammals.

Estrus occurs in sows 1 to 5 days after they **farrow.** If bred at this time, they seldom **settle** (become pregnant) because they usually do not ovulate. Estrus accompanied by ovulation occurs again in some sows about 6 to 8 weeks following farrowing if they are nursing pigs. When pigs are weaned, however, sows usually come into estrus within 3 to 5 days, provided that an estrous cycle has not been initiated before the pigs are weaned.

Mares usually come into estrus 5 to 10 days after foaling (giving birth to their young). This estrous period is called *foal heat* and is accompanied by ovulation. Foal heat lasts from 1 to 10 days, and mares often are bred on the ninth day after foaling.

Cows nursing calves will usually show their first estrus 40 to 80 days after parturition. When feed is limited, the period between calving and the first estrus may be prolonged to 80 to 100 days, or even longer. Milked cows will generally show their first estrous period 20 to 30 days after parturition.

Most breeds of sheep are seasonal breeders and cannot be bred immediately following lambing (giving birth to young). The first estrous period following lambing in many breeds of sheep usually does not occur until the next breeding season begins, in late summer or early fall.

Abnormalities of estrus and the estrous cycle are sometimes observed. Some females show almost continuous estrus for many days rather than the normal period of rest between estrous periods. Such females are called *nymphomaniacs,* a term that indicates abnormal sexual desire. The condition is caused by cystic follicles that fail to ovulate.

Ovarian cysts are structures that develop from follicles that fail to ovulate. Cystic follicles appear to result from the lack of certain hormones or from an imbalance of two or more of them. Actual photographs of ovarian cysts in cattle and pigs are shown in Figure 9.14. They occur in about 5 to 15 percent of dairy cows, but seldom in heifers and beef cows. Single or multiple ovarian cysts have also been detected in sows and gilts. In cattle, the cysts are usually 2.5 to 4.0 cm in diameter, but they may be much larger. Although some cows with ovarian cysts are nymphomaniacs, most are anestrus. Estrous cycles can be reestablished for about 80 percent of cows in this condition by treating them with 100 μg of GnRH. It is important to keep in mind, however, that the tendency to develop ovarian cysts is probably heritable.

In some females (especially in cows) there is an interval of only 8 to 10 days between estrous periods, rather than the usual 21 days. This is due to the abnormally early degeneration of the corpus luteum, with an insufficient amount of progesterone being produced to inhibit production of LH by the anterior

FIGURE 9.14
Ovarian cysts on an ovary from a dairy cow (left) and multiple ovarian cysts on ovaries from a gilt (right). *(From D. J. Kesler and H. A. Garverick, "Ovarian Cysts in Dairy Cattle: A Review,"* J. Animal Sci., **55:**1147, 1982. Photographs courtesy Univ. of Missouri, left, and Univ. of Illinois, right.)

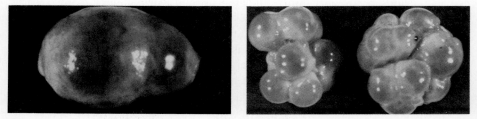

pituitary gland. An early degeneration of the corpus luteum allows growth of a new group of follicles sooner than normal, and estrus occurs within a shorter period of time. Other females may not show estrus at the first expected estrous period following breeding, but they will show estrus at the second. Thus one estrous period appears to be skipped. One explanation for this situation is that pregnancy actually might have been initiated, but the developing embryo died and was resorbed after a few days. Another possible explanation is that the female came into estrus and it was not observed because estrus was limited to a very short period of time or was mildly expressed.

Exposure to high **ambient temperatures** often lowers reproductive efficiency of the females of several species. This is especially true if their body temperature is increased above normal. Adverse effects from such exposure are observed in tropical areas, in regions where summer temperatures normally exceed 100°F, and in other regions in which abnormally high temperatures, often associated with periods of prolonged drought, may occasionally occur during the summer months.

Adverse effects of high ambient temperatures on reproduction have been noted in several species, but especially in the ewe. Extreme thermal stress may cause the ewe to fail to exhibit estrus. Thermal stress at the time of mating or shortly thereafter can cause a failure in normal development of the fertilized ovum. Later in gestation, thermal stress may cause some fetal deaths and resorptions, abortions, and birth of abnormally formed young. Under laboratory conditions, extremely high temperatures have caused abortions in cows. Abortions were noted in some sows at the Missouri Agricultural Experiment Station in the summer of 1954, when ambient temperatures during a heat wave reached 112 to 114°F. Smaller litters were farrowed that fall, more mummified fetuses were observed at birth, and variation in litter size at birth was much greater than when the same sows had farrowed the previous spring. Some experiment stations have reported that litter size at farrowing was larger among sows that had been cooled by sprinkling during the hot summer months than among controls.

Ovulation   *Ovulation* is *the process by which the ovum (egg) is released from the ovary.* The usual time of ovulation in various species of farm mammals is given in Table 9.2. Among farm mammals, the cow and mare usually ovulate just one ovum per estrous period, although multiple births sometimes do occur because of the ovulation of two or more ova. It is also possible for two or more offspring to develop from a single ovum. Such individuals are genetically alike and thus are identical.

Multiple births are common among sheep and are the usual type of birth in swine. Ewes may ovulate from one to five ova during a single estrous period. In swine, gilts may ovulate 10 to 12 and sows 15 to 20, or more. The number of ova developing into young at birth depends to a certain extent on the amount of space and nutrients available to the ova in the uterus and on other factors, some of which are unknown. Production of more than the usual number of ova may be induced in females by administration of gonadotropins containing FSH during

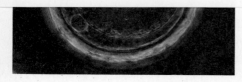

proestrus, at the time follicles normally develop. Production of more than the usual number of ova is known as *superovulation.*

Ovulation among different species is often referred to as *spontaneous or*

FIGURE 9.16
Inbred Yorkshire pigs from purebred Hampshire sows. They resulted from the transplantation of fertilized ova from Yorkshire to Hampshire sows. *(Missouri Agr. Expt. Station.)*

Implantation is a gradual process in farm mammals. Present evidence suggests that implantation occurs 10 to 18 days after fertilization in the ewe, 12 to 24 days after in the sow, 20 to 32 days after in the cow, and 35 to 60 days after in the mare. In some wild species, such as the weasel, mating occurs in the summer, but implantation does not occur until late winter. This situation is known as *delayed implantation* and is another timing mechanism in the reproductive process of animals to ensure the birth of young when the environment is favorable for survival.

Several types of abnormal implantations or pregnancies have been observed in humans and may occur in other animals. These types of abnormal implantations include ovarian, tubal, and abdominal pregnancies. The young are seldom, if ever, carried to full term in such pregnancies.

Early in pregnancy, the young develop certain membranes to provide for their protection and nourishment. These are known as *fetal (extra-embryonic) membranes;* those of the pig are shown in Figure 9.17. The **amnion** is the innermost membrane that surrounds the fetus. It is filled with a fluid known as *amniotic fluid,* within which the developing fetus is suspended. The outer layer of fetal membranes is called the **chorion,** and it makes contact with the maternal uterine tissues. In **ruminants,** which include the cow and ewe, the chorion attachment to the uterus is of a **cotyledonary type** (Figure 9.18), in which contact is made only at certain points on the uterus rather than over its entire surface area. The placental attachment in the mare and sow is of a diffuse type (Figure 9.19), in which contact with the uterus is made over most of the surface area of

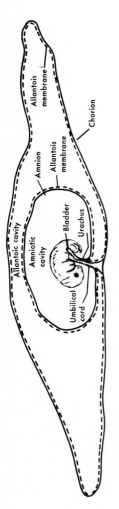

FIGURE 9.17

Drawing of the fetal membranes of the pig. Placental membranes are of the diffuse type, making contact with the uterus over the entire surface of the chorion.

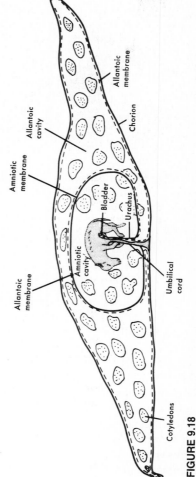

FIGURE 9.18

Drawing of the fetal membranes of the calf. Placental membranes of ruminants (cattle, sheep, and goats) are of a cotyledonary type, making contact with the uterus only at certain points (the cotyledons).

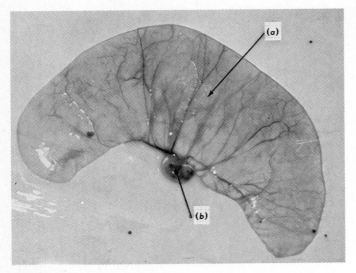

FIGURE 9.19
Fetal membranes of the pig at 25 to 30 days of pregnancy. (a) The allantoic sac filled with fluid; (b) the fetus encased in the amnion. Note the blood supply to different parts of the chorion, leading to the umbilical cord of the fetus. *(Missouri Agr. Expt. Station.)*

the chorion. The outer portion of the allantois is fused with the chorion, and the inner layer with the amnion, forming a sac, or space, filled with allantoic fluid.

The placenta is formed by the fusion of the chorion and the uterine mucosa. It has several important functions: (1) transmission of **nutrients** from mother to young, (2) transmission of wastes from young to mother, (3) protection of young from shock and adhesions by means of amniotic fluid, (4) prevention of the transmission of bacteria and other large molecular substances from mother to young, and (5) secretion of certain hormones. The placental barrier in domestic mammals prevents large **molecules** such as **antibodies** and some of the fat-soluble **vitamins** (vitamin A) from passing in large amounts from mother to young. Some **viruses** are small enough to penetrate the placental barrier and may cause defects in the young if they reach it in the stage of pregnancy when certain body parts are being developed. The developing young are particularly susceptible to viral infections because they have not produced antibodies of their own and have received few, if any, from the mother. Certain other chemical substances in the mother's ration penetrate the placental barrier in some farm mammals and may cause defects during fetal development.

Each developing embryo usually has its own set of membranes, although those of two individuals may fuse, resulting in their having a common blood supply. Twins in **cattle,** and probably in other species, have common membranes and a common blood supply. Having common membranes and a common blood supply is reported to cause the **freemartin** condition in a heifer born twin with a bull calf. A heifer born twin with a bull is a freemartin about 90 percent of the time. The reproductive system of the freemartin female does not develop normally, and she is sterile. A scientific explanation of this phenomenon has not been elicited. This condition either does not occur or is very rare in other species of farm mammals.

In the sow, where many young are nourished within the uterus during the same pregnancy, the young are spaced at definite intervals within the uterine horns (Figure 9.20). If the young are too close together or if the uterus is crowded, some of the developing young fail to survive at birth. Another phenomenon also occurs in the sow to help ensure an approximately equal number of young in both uterine horns: if one ovary produces only one ovum and the other ovary 12, some fertilized ova from the side where many ova were produced will move across to the uterine horn on the opposite side where only one ovum was produced. This mechanism tends to distribute the young evenly within the two uterine horns and is known as *intrauterine migration*.

Even though the gestation period varies greatly in length among species, all young in female farm mammals are born at about the same degree of development or maturity. This is not true of all species of animals, however. For example, baby opossums are born about 12 days after the ova are fertilized. They have no fetal membranes at birth. They find their way, in some way, into

FIGURE 9.20
Radiograph (x-ray) of the uterine horn of the sow at midpregnancy, showing spacing of the fetuses. Only bony parts of the body of the fetuses show in this radiograph. *(Missouri Agr. Expt. Station.)*

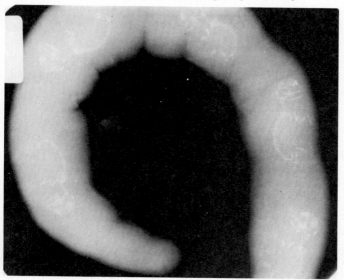

the pouch of the mother after birth, fasten their mouths to one of the nipples, and remain attached to that nipple until they are fully developed. The female opossum normally ovulates 45 to 50 ova and gives birth to 25 or more young, but there are only 12 to 13 nipples available in the pouch, so that no more than this number of young survive. The baby mouse, rat, and rabbit are more mature than the opossum at birth; nevertheless, they are totally helpless. Their eyes are closed, they have no hair, and they cannot walk or run. They grow and develop quickly, however, if properly nourished by the mother. They may be weaned at about 21 days of age. The young of the guinea pig are much more mature at birth and can care for themselves, if necessary, when they are only a few hours old (Chapter 8).

Parturition   *Parturition* is *the act of giving birth to young by the mother.* It is preceded by certain activities in females of some species. Females of most species lose their appetite and usually try to find a secluded area where they will have privacy during the birth process. Sows will often build a nest before giving birth to young (Chapter 20). The **ligaments** and muscles around the pelvic region of most females (especially the cow and mare) relax, and usually milk is present 24 to 48 h before the young are born.

Considerable new knowledge has been published recently regarding the endocrinology of parturition. It is now known that parturition in farm mammals is initiated by the fetal anterior pituitary gland. The role of the fetal anterior pituitary gland was first demonstrated to be involved in parturition in cattle with prolonged gestation. Pregnancy in these cattle was generally prolonged from 3 to 4 weeks, resulting in birth weights of calves varying between 110 and 168 lb (the usual birth weight is 65 to 95 lb). In such instances the calves were dismembered or delivered by cesarean operation. It was demonstrated that this prolonged gestation was caused by the absence of the fetal anterior pituitary gland. It was later shown that experimental destruction of the pituitary gland also caused prolonged gestation. Animals without a pituitary gland, however, would deliver their young if they were administered ACTH. Therefore, ACTH produced by the fetal anterior pituitary gland appears to be the hormone that initiates the process of parturition. The same phenomenon has also been demonstrated in sheep.

Other hormones involved in parturition are estrogen, $PGF_2\alpha$, oxytocin, relaxin, progesterone, and glucocorticoids. $PGF_2\alpha$ is released to destroy the corpus luteum. This is necessary since high levels of progesterone inhibit uterine contractions. After progesterone levels are reduced, oxytocin, and $PGF_2\alpha$ stimulate uterine contractions. Estrogen sensitizes the uterus to oxytocin, and relaxin causes relaxation of the birth canal.

Females should be observed very carefully at parturition to see if the birth process is progressing properly. This is especially important in young females producing their first offspring. The normal presentation of the young in cattle, horses, and sheep is front feet first, with the outstretched head resting on the feet. In swine, both head- and hind-feet-first presentations are normal. Various abnormal presentations are encountered among animals. These include one or both legs folded back, the neck bent backward, or a posterior presentation (except in swine, where the young are normally born hind feet first). Posterior presentations in cattle are of particular importance because when the connection of the umbilical cord with the uterus is severed, fetal blood flow from the mother ceases and the young has no oxygen supply. Unless delivered promptly, it may perish from lack of oxygen.

Abnormal presentations are indicated when labor lasts several hours without progress being made in delivering the young. One may assist the female when it is needed by pulling gently, but firmly, on the young at the time *labor pains* (uterine contractions) occur in the mother. If this assistance is not effective, an experienced person should be called. It is sometimes necessary to dismember the young or deliver them by a cesarean operation (named after Caesar, who supposedly was delivered in this way). Cesarean operations have been successful in recent years, since the advent of **antibiotics** that control postoperative infections.

After parturition, it is important that the newborn receive colostrum that is especially rich in antibodies and vitamin A (Chapter 11). The young particularly need antibodies from their mother's milk to combat infections, because they possess few, if any, antibodies at birth and it is several days before they can build their own. It is also important to observe mothers (especially cattle and sheep) closely following parturition to make certain that the afterbirth is completely expelled from the uterus. The **afterbirth** is *the placenta and the extra-embryonic membranes with which the fetus is connected.* In some instances, retained afterbirths can cause infections and subsequent reproductive problems.

9.5 APPLICATION OF RECENT RESEARCH FINDINGS IN THE PHYSIOLOGY OF REPRODUCTION

As was discussed in Chapter 7, hormones and other biological compounds have been used to control various aspects of reproduction in farm mammals. Certain hormones and chemicals used experimentally have not been approved by the FDA for commercial use on the farm or ranch. In Sections 9.5.1 through 9.5.9, several applications of recent research discoveries and technologies in the field of reproductive physiology will be discussed.

9.5.1 Synchronization of Estrus and Ovulation

Estrous synchronization involves manipulating the reproductive processes so that females can be bred during a short, predefined interval with normal fertility. This application of technology facilitates breeding in two important ways: (1) it reduces, and in some cases eliminates, the labor of estrous detection; and (2) it enables the producer to schedule the time of breeding. For example, if all the animals in a herd can be induced to exhibit estrus at about the same time, the producer can arrange for a few days of intensive artificial insemination.

Although the total amount of labor involved with insemination may not be reduced significantly, it is concentrated into a much shorter period of time. Other advantages of estrous synchronization include: (1) it creates a more uniform group of offspring, (2) it enables the producer to breed more females to a select male, (3) it reduces the time span of breeding and parturition seasons, and (4) with certain compounds, heifers can be bred without estrous detecting.

One compound that has been used to synchronize estrus and ovulation in cattle, horses, and sheep with consistent success is $PGF_2\alpha$, which is effective in synchronizing estrus by lysing the corpus luteum. If $PGF_2\alpha$ is injected 6 days after ovulation, it lyses the corpus luteum and causes estrus to occur 2 to 3 days after treatment.

Progestins will also synchronize estrus in farm mammals. These compounds prevent maturation of follicles and the occurrence of estrus and ovulation. After the hormone is withdrawn, females will come into estrus and ovulate in a few days. Several progestin compounds and procedures have been developed, but gaining FDA approval for commercial use of these compounds has been slow and difficult.

One progestin procedure approved for use in cattle by the FDA in 1983 was Syncro-Mate B. The Syncro-Mate B procedure involves placing norgestomet (a progestin) implant in the ear for 9 days and injecting 3 mg of norgestomet and 5 mg of estradiol valerate at the time of the implant insertion. Cows or heifers can be bred at estrus for 2 to 3 days after the implant is removed, or by a predetermined breeding 48 to 52 h after implant removal.

The Syncro-Mate B treatment suppresses estrus and ovulation after the corpus luteum has regressed. If Syncro-Mate B is administered during the luteal phase of the cycle, the corpus luteum should regress on days 16 or 17 as is normal, but the release of norgestomet from the ear implant will suppress estrus and ovulation until the ear implant is removed. If Syncro-Mate B is administered shortly after ovulation (when the corpus luteum is developing), the injection of norgestomet and estradiol valerate appears to cause corpus luteum regression; once again, the ear implant will suppress estrus and ovulation until it is removed.

9.5.2 Superovulation

The ovaries of female farm mammals possess many thousands of potential ova at birth and produce no new ones thereafter. Since female livestock produce and utilize a limited number of ova to produce young during their lifetime, most of the original ova in the ovary do not produce young and are in effect wasted.

Superovulation consists of injecting the female with fertility drugs causing the larger follicles—each of which contain one oocyte, or egg, in the ovaries—to mature and ovulate (rupture and release the egg). Injections are administered in the middle part of the cycle (9 to 14 days after estrus in cows). Follicle-stimulating hormone is injected twice daily (4 to 5 mg per dose for 4 to 5 days). Additionally, on day 3 or 4 of FSH injection, $PGF_2\alpha$ is injected, which results in destruction of the corpus luteum on the ovary, thereby triggering the next estrus, commonly 2 days after administration of $PGF_2\alpha$.

When the bovine female is treated with a fertility drug, she will usually produce and ovulate more eggs than normal. The eggs can be fertilized inside the female, or removed and fertilized in vitro where "matings" can be planned for each individual egg.

If the eggs are fertilized inside the female, the resulting embryos can be left to develop into twins or triplets, or they can be collected and transferred to other females called *surrogate mothers*. When surrogate mothers are not readily available, the embryos can be frozen to await transfer at a later time.

Even a single embryo can be made to produce additional offspring. A single-cell fertilized egg divides into two cells, which divide into four cells, eight cells, and so on. At this very early stage of development, all cells in the embryo are exactly alike genetically. This means that splitting a four-cell embryo into four separate cells by microsurgery and then transferring these cells to surrogate mothers could potentially result in identical quadruplets.

The potential of this technology is indeed tremendous. For example, a bovine female on fertility drugs can produce eight or more embryos. If these embryos are separated at the four-cell stage, each cell is grown in culture to the four-cell stage, and the process repeated 6 times, theoretically more than 4000 potential offspring could be produced, all genetically identical. But the potential for manipulating embryos extends even further. For example, cells from a one- or two-cell embryo could have their genetic component removed or destroyed. A nucleus from a body cell of an adult animal could then be inserted into the embryonic cell, resulting in an identical copy (clone) of the adult animal.

Variations of cloning could also involve the possibility of replacing the original nucleus with the nuclei from just sperm, or from just eggs. The offspring would have two genetic fathers, or two genetic mothers. It could be the equivalent of crossing a bull with himself, or a cow with herself. And the latter would ensure that all the offspring were female.

9.5.3 In Vitro Fertilization (Test-Tube Babies)

The first reported so-called test-tube human baby was born in July 1978 in Great Britain The procedure involves, to sum: (1) recovering a mature egg from the ovary; (2) placing the egg in a petri dish enriched with a culture medium similar to that of the oviducts; (3) collecting and placing viable spermatozoa in the culture medium and incubating at body temperature for 18 h; (4) transferring the fertilized egg to a culture medium simulating the uterine environment and incubating for another 18 h, during which cell division occurs, resulting in the formation of the morula; and (5) placing the morula intravaginally into a recipient uterus, following proper hormonal preparation of the recipient female to accept the new embryo. This procedure is complicated and difficult for many reasons. For example, the matter of the timing of egg maturity and collection with proper preparation of the recipient uterus is critically important.

An estimated 300,000 married women in the United States have blockage of the oviducts, and an estimated 1.5 million men have a condition known medically as oligospermia (insufficient sperm count to effect conception). For such infertile couples, the above procedure of embryo culture and transfer offers promise as a means of bearing children that are genetically their own.

Scientists at the University of Wisconsin and at other experiment stations have developed procedures for successful in vitro fertilization in cattle. Eggs are aspirated from immature follicles on the ovary and placed into culture with hormones to prepare them for penetration by sperm.

The future for this method of mating in cattle is quite promising. If eggs are obtained from immature follicles with the use of a laparoscope, the donor female can continue to produce eggs and lead a normal reproductive life. In fact, eggs could even be collected from the ovaries of pregnant cows. In vitro fertilization permits an individual sire to be used to fertilize each egg. Adoption of this practice on a wide scale would enable scientists to progeny-test cows, as is now done with bulls. Additionally, far fewer sperm would be required compared with the number required with the current practices employed in artificial insemination. For example, a bull collected now that yielded 600 straws from a single ejaculate could be used to fertilize 36,000 eggs in vitro with that same ejaculate. This procedure could extend greatly the reproductive potential of outstanding bulls.

9.5.4 Embryo Transfer (ET)

Embryo transfer, a recent application of science, is the placing of an embryo into the lumen of the oviduct or uterus. More broadly, ET includes the sequence of steps for transferring embryos from one female to another, including superovulation, embryo recovery, and storage of embryos in vitro. The donor is the genetic mother, from which embryos are recovered, and the host or surrogate mother is called a recipient (Figure 9.21).

The first step in embryo transfer involves estrous synchronization and superovulation using the procedures discussed previously. After estrus and insemination, the developing embryos are ready to be transferred. The embryos of pigs and sheep are generally transferred by surgical methods, whereas those of cattle are usually transferred utilizing nonsurgical techniques.

Although the first ET was performed successfully in rabbits nearly a century ago at Cambridge University, the first successful bovine ET was reported in 1951 at the University of Wisconsin. More recently, ET has been successful in humans, sheep, goats, and horses.

Normally, cows produce only one calf per year; however, with embryo transfer it is common for a cow to be the genetic mother of 12 to 15 or more calves born within a few days of one another (Figure 9.21). Some cows have

FIGURE 9.21
The Holstein cow (upper right) is the genetic mother of the 10 calves shown above. She was superovulated and the embryos were recovered from her uterus 7 days after conception. After 3 to 10 h of culture in vitro, the embryos were transferred to the uteri of the 10 recipient cows (left) for gestation to term. The surrogate mothers may be of any bovine breed; however, before the embryos can be transferred successfully, the estrous cycles of the surrogate mothers must be in synchrony with that of the genetic (donor) mother. *(Photograph was kindly provided by Dr. G. E. Seidel, Jr., Embryo Transfer Laboratory, Colorado State University, Fort Collins.)*

provided eggs that were fertilized and resulted in the birth of more than 50 calves in a 12-month period. Thus, embryo transfer enables the animal breeder to proliferate offspring from cows identified as genetically superior, and to selectively mate each cow with several superior sires in a given year. Another use of embryo transfer is to obtain calves from otherwise infertile donors. Such infertility often results from a senescent uterus, which is circumvented by transferring the embryos to a younger, more active and environmentally sound uterus. Further application of this technology is the export of embryos, which is less costly and involves less risk of transmitting disease than exporting live animals.

Somewhat analogous to the placing of whooping crane eggs into nests of the sandhill crane, embryo transfer could be used to spare rare species or breeds from extinction. An example is the current effort to save the Angora breed of sheep from extinction in Australia. The breed is being reestablished from only a few animals by transferring Angora embryos into other breeds of sheep. The same technique might be used with endangered species such as the black-footed ferret and the mountain bison.

Other merits of ET include (1) the use of valuable or "elite" genotypes to produce a larger proportion of desirable phenotypes, such as cows producing milk of a more favorable protein-to-fat ratio, and twinning in some species of livestock could be accomplished more rapidly. Similarly, the impact of sex-linked genes could be increased through ET. (2) ET techniques can improve the intensity of selection of cows to be the dams of replacement heifers and of bulls used in artificial insemination. (3) ET techniques can improve the accuracy of selection through progeny testing of females. (4) The application of ET could extend the reproductive life of females via frozen embryo processes. Such a procedure would facilitate increased selection intensity, since elite dams would be available for producing sons beyond their lifetime. (5) ET procedures may also offer the opportunity to decrease generation interval, if such a change would improve genetic change per unit of time. (6) One additional benefit of ET application to the production of young sires is increased concomitant production of females with good pedigrees (see Chapters 4 and 5).

Potential benefits from ET using present technology are most important when applied to the production of young sires for sampling. The application of ET procedures for producing female replacements is somewhat questionable in view of current economics, if additional milk yield is the only source of added income. Progeny-test results for cow evaluation improve the accuracy of selection.

Twinning with ET   Since the total calf crop is the main factor affecting the productivity of a cow herd, improving reproductive efficiency by increasing the number of cows that produce twins would have a major economic impact on beef production. Hormonally induced twinning frequently results in two embryos in the same uterine horn, which leads to more abortions than when one fetus implants in each uterine horn.

Embryo transfer is clearly an effective way to induce twins. Pregnancy rates range from 65 to over 90 percent. Methods involve placing one embryo into each uterine horn, or placing one embryo into the open uterine horn of a previously bred female. California scientists obtained 72 percent twins (of recipients pregnant) when two embryos were transferred into previously inseminated females, and 70 percent twins when one embryo was transferred.

Testing for Mendelian Recessive Alleles   If a homozygous recessive individual is mated to a suspected carrier of a Mendelian recessive allele, and eight normal offspring result, the probability that the individual is not a carrier is 99.6 percent. Without superovulation and embryo transfer, testing for a particular undesirable trait would require essentially the full reproductive life of a cow. This compressing of an 8- or more-year test period into 1 year is especially important when an outstanding cow is a daughter of a carrier.

Using ET in Genetic Testing   Embryo transfer has been used to diagnose and study the inheritance of both syndactyly and polydactyly in cattle. Syndactyly, or mule foot, is the commonest genetic defect of cattle in the United States. It has been observed in Angus, Chianina, Hereford, Holstein, Simmental, and cross-bred cattle. Since it is a single-gene recessive, congenital defect, it can be carried and transmitted by adults normal in appearance. Thus, all suspect carrier AI sires should undergo test matings before being used extensively. Using embryo transfer from syndactylous donors bred to suspect sires provides a relatively economic and fast way of progeny-testing bulls. Suspect females can be tested by inseminating them with semen from a mule-foot bull before embryo recovery. Embryos recovered from syndactylous donors or suspect cows are transferred into recipients for fetal development. Two embryos can be transferred into each recipient to reduce the cost of the test. Cesarian removal of the fetuses from the recipients at 60 days following conception reduces the time required to complete the test.

Bovine polydactyly (three or more toes per limb) is much less common than syndactyly. It has, however, been observed in Holstein and Simmental cattle. The mode of inheritance of polydactyly is not well understood, but it appears to involve at least three genes, one dominant and two recessive.

Collecting Ova and Embryos from Donor Cows   Nonsurgical methods are utilized to recover ova on days 6, 7, or 8 after estrus. The donor cow is restrained and given an epidural block in the area of the tail head. The local anesthetic injected into the spinal column prevents the animal from straining when the arm is in the rectum. A flexible rubber tube (Foley catheter) with three passageways is then passed through the cervix and into the uterus (Figures 9.22 and 9.23). It is temporarily stiffened with a metal or plastic rod to enable it to pass through the cervix and into the body or horn of the uterus. A rubber balloon, which is built into the anterior end of the tube, is then inflated, via passageway B, with approximately 20 to 25 milliliters (ml) of air (to about one-half the size of a golf ball), so that it extends to fill the uterine lumen and prevents fluid from escaping around the edges. There are two holes in the tube anterior to the balloon, which lead into separate passageways: one (A) for fluid entering the uterus, and the other (B) for fluid draining from the uterus. Embryos (about 1/200 in in diameter) are collected in a balanced salt solution to which antibiotics and heat-treated serum are added. Dulbecco's phosphate-buffered saline is the most commonly used balanced salt solution. The solution is held in a container about 3 ft above the cow, where it is connected to the Foley catheter by an inflow tube (Figure 9.23). A second tube is connected to passageway C of the catheter to collect the medium that has flushed (washed) the ova and embryos out of the uterus. This solution is then collected in cylinders holding approximately 2 pints (908 ml) of fluid. Unfertilized ova and embryos settle to the bottom within 30 min, and the bulk of the medium is siphoned off. The last cupful of fluid is searched under the microscope for the treasured embryos. Since bovine embryos form no intimate attachment to the uterus before day 18, they can be recovered nonsurgically for up to about 14 days with no apparent damage, although research indicates that a larger number of normal embryos can be obtained 6 to 8 days after estrus than at other times.

The embryos are identified, classified, washed through sterile medium, and then stored until transfer. They commonly live in this solution at 37°C for about

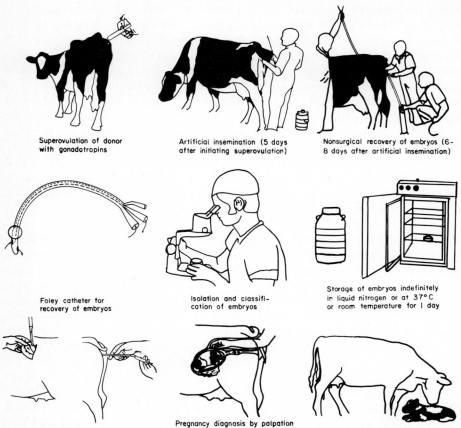

Superovulation of donor with gonadotropins

Artificial insemination (5 days after initiating superovulation)

Nonsurgical recovery of embryos (6-8 days after artificial insemination)

Foley catheter for recovery of embryos

Isolation and classification of embryos

Storage of embryos indefinitely in liquid nitrogen or at 37°C or room temperature for 1 day

Transfer of embryos to recipients surgically or nonsurgically

Pregnancy diagnosis by palpation through the rectal wall 1 - 3 months after embryo transfer

Birth (9 months after embryo transfer)

FIGURE 9.22
Schematic presentation of bovine embryo transfer procedures. *(From G. E. Seidel, Jr., Science,* **211:***351–358, January 23, 1981. Copyright 1981 by the American Association for the Advancement of Science. Used by permission.)*

24 h without loss in viability, or they can be frozen to −196°C in liquid nitrogen. The current state of the art has not perfected the freezing and thawing process, so that up to half of the embryos may be killed. Those retaining viability can be stored indefinitely.

For high pregnancy rates, the recipient, or surrogate mother, must be at the same stage of the reproductive cycle as the donor. Embryos are usually transferred nonsurgically using artificial insemination equipment. Pregnancy rates are increased slightly if the embryos are transferred surgically through a small incision in the flank of the recipient under local anesthesia. With the

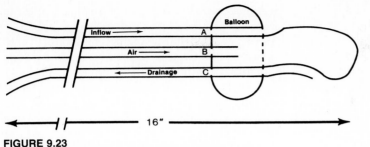

FIGURE 9.23
Diagrammatic depiction of a three-way Foley catheter.

surgical method, pregnancy rates of 60 to 70 percent can be achieved, approximating those obtained through normal breeding.

Embryo transfer currently costs from $500 to $1500 per pregnancy. Firms providing such services have grown rapidly in recent years. An estimated 50,000 calves were born as a result of this technology in North America in 1982.

Embryo transfer techniques have been used in many species, including humans. The most important commercial species after cattle are horses and swine, although a limited amount of commercial work is also being done with sheep, goats, and rabbits.

Many future technologies will depend on embryo transfer, including cloning. Procedures already developed in the laboratory include sexing the embryos; fertilization in a test tube; twinning in cattle by transferring two embryos; and even dividing embryos into two, three, or four parts to produce identical twins or multiplets. It is probable that some or all of these procedures will be used commercially in the decade ahead.

Summary of ET   In less than three decades, embryo transfer has evolved from a research tool into an important component of the livestock industry. Although it is not expected to have the same genetic impact as AI on the improvement of cattle, ET offers the animal breeder an opportunity to accelerate the breeding improvement program. Improvements in freezing and sexing of embryos will further increase the usefulness of ET as a tool available to animal breeders of tomorrow.

9.5.5 Freezing of Embryos

The freezing of bovine semen has been perfected. However, the freezing and subsequent thawing of viable embryos has been more difficult, since embryos are composed of masses of cells, rather than being a single cell as is the case with spermatozoa. When embryos are frozen and thawed, external cells are exposed to the environment before internal ones. This can result in cell-to-cell junctions becoming weak and lead to disaggregation of the cell mass, which can result in embryonic death.

Currently embryos 6 to 7 days of age (0.2 mm in diameter) are recovered from superovulated donor cows using nonsurgical collection methods. The embryos are pooled in a supportive growth medium such as Dulbecco's Phosphate-Buffered Saline (PBS), or Ham's F-10 supplemented with blood serum or bovine serum albumin. Dimethyl sulfoxide (DMSO), glycerol, and 1,2-propanediol are protective agents used to minimize the damage of freezing to embryos.

With the embryo-freezing technology outlined above, embryos can be recovered, .stored, and transported almost anywhere in the world. As the methodology of freezing embryos is perfected even further, prolonged storage and long-distance transport of frozen embryos will become commonplace.

Sophisticated thermal devices have been engineered that hold promise for better controlling the temperature required for embryo survival. The advent of microwave technology presents a new approach in thawing the embryo from the inside out. Research efforts in the preservation of whole organs, such as kidneys and skin, have provided scientists with useful information resulting in significant improvements in the freezing and storage of embryos. With present technology, however, when embryos are frozen, thawed, and transferred to recipients, the success rate is only about 50 percent of that with fresh, unfrozen embryos.

9.5.6 Microsurgery with Embryos

An exciting area of research involving large animals is the micromanipulation of embryos and gametes. The process requires microsurgery and micromanipulation of day-5 or day-6 morulae-stage embryos. The embryos are separated into halves, resulting in identical twins. Although the process is tedious and time consuming, it holds great promise for research in which the use of identical twins greatly facilitates comparative research studies.

Micromanipulation and microsurgery used experimentally with several species have given rise to some exciting possibilities regarding male and female gamete combinations. It may be possible to get dam-dam or sire-sire matings from the same or different cows or bulls. Such matings may be accomplished by taking an egg (oocyte), removing its nucleus, and replacing it with two nuclei from eggs of two different cows, or the same cow, thus producing a "mating" from two different cows or a "mating" from just one cow. A sire "mating" could be accomplished by removing the nucleus from an egg and injecting two sperm cells from the same bull or two different bulls. Such a union produces a "mating" from the same bull or two different bulls.

It may also be possible to perpetuate replicas of proved mature animals by cloning. Cloning requires removing nuclei from cells of the proved animal and transferring them to prepared eggs. On transfer to recipients, offspring identical to the desired animal could be produced. Although these possibilities are a bit futuristic, they represent areas of science currently under investigation.

9.5.7 Sex Control

For centuries, humans have been interested in controlling the sex of the young in mammals, especially in livestock and human beings. Many methods have been proposed but none yet has shown promise of being entirely successful. If the sex of the offspring could be predetermined, it could be used to advantage in certain circumstances. For example, detrimental and lethal sex-linked recessive traits could be controlled by producing only female offspring, although many of the females produced would be carriers of the recessive sex-linked gene. Also, it would enable families to have a boy or a girl offspring in accordance with their wishes. Or when a herd of livestock is being expanded, producing more females would result in a more rapid increase in herd numbers. This would be particularly advantageous in a dairy herd. Even when herd size is being maintained, producing mostly female offspring would make it possible to apply greater selection pressure among the females used for replacement purposes. On the other hand, producing mostly males would be an advantage to the commercial livestock producer, because males are usually heavier and may bring more per pound at market time. Producing mostly males in a particular calf, lamb, or pig crop would permit more intensive selection among the male offspring used for breeding purposes.

Sex control is theoretically possible in mammals because two different kinds of sperm are produced by the male. One sperm carries the X chromosome, and one the Y. Thus methods used to control sex attempt to separate the X- and Y-bearing spermatozoa.

Recently, a technique has been developed for identifying the X and Y sperm in semen. The Y spermatozoa possess a component called the F body that can be stained by special techniques, whereas the X spermatozoa do not. This makes it possible to study the success of the methods of separating the X- and Y-bearing sperm without awaiting the results of mating tests. To date, however, this technique is well-established only for the human, gorilla, and vole. Because reports for the use of the technique in the bull are unconfirmed, its extension to use in the bull and other species requires additional research (cf. Section 10.14).

9.5.8 Sexing Embryos

Although it is not yet possible to separate X- and Y-bearing bovine spermatozoa reliably, the sex of embryos can be determined **karyotypically.** Cells can be obtained by biopsy of morulae or by cutting off the tip of an elongating blastocyst. Although most embryos continue normal growth following biopsy, pregnancy rates are lower than those of unsexed embryos. The bovine sex chromosomes are easily distinguished, so that sex determination is highly accurate when suitable metaphase chromosomes are available for examination. A secondary benefit of karyotyping is the detection of chromosomal abnormalities. An alternative to karyotyping could be detection of the gene product of the Y chromosome, H-Y antigen. Procedures developed to date for sexing embryos are time consuming and impractical on a commercial basis.

9.5.9 Induced Parturition

Glucocorticoids, a family of hormones secreted by the adrenal glands, will induce parturition in the ewe, mare, sow, and cow. However, very high dosages are required in the mare and sow.

Dexamethasone and flumethasone are two synthetic glucocorticoids now available commercially. They induce calving in the cow at the rate of 85 percent or greater. Induced parturition should not be attempted until the last 2 to 3 weeks of pregnancy, since the calves are more likely to live at that stage of gestation. Cows may be expected to calve 34 to 60 h after treatment.

Induced parturition in cows is used to control the time of calving. This makes it possible to observe the cow or heifer closely, so that assistance can be rendered if needed. It is especially useful in heifers that are calving for the first time and in those that were bred to large bulls of another breed. Pregnancy may also be terminated when abnormalities have been detected. Induced parturition can be used to lower birth weights and shorten gestation periods when these traits appear to be greater than normal, thus helping to reduce death losses at calving. Properly induced parturition has no significant detrimental effects on milk production, survival of calves, or their subsequent growth rate.

One disadvantage of induced parturition is that retained placentas are commoner than with normal gestation periods and parturitions. This is one reason the procedure has not been widely accepted by cattle producers.

Prostaglandins have also been used to induce parturition in female livestock. Intravenous infusions or muscular injections of $PGF_2\alpha$ have been used to induce parturition in sows. Treatment is followed by birth of the first piglet in 30 to 35 hours. The gestation period is also shortened, but the survival rate of the piglets appears to be normal. Injections of $PGF_2\alpha$ in heifers also commonly shortens the gestation period and induces parturition. Prostaglandins seem to increase the incidence of retained placentas in cows and heifers.

9.6 SUMMARY

Efficient reproduction is one of the most important aspects of animal production. A profit from livestock production requires that a large number of young be marketed per breeding female. Therefore efficient livestock production requires a knowledge of the production and management of farm mammals, as well as of their anatomy, physiology, genetics, and nutrition and of disease and parasite control among them. After all, young farm mammals must be born before they can live to a desirable age to supply food for humans.

STUDY QUESTIONS

1 Define the following terms: FSH, LH, seminiferous tubule, cells of Leydig, estrus, estrous cycle, ovulation, fertilization, freemartin, gestation, parturition, follicle, progesterone, estrogens, cryptorchid, and vasectomy.
2 Draw a diagram of the male reproductive system and label the parts.

3 What are the primary sex organs in the male? In the female?

4 What are the secondary sex organs in the male? The accessory sex organs?

5 What are the main functions of the testes? The scrotum?

6 What does production of a large proportion of abnormally formed spermatozoa indicate?

7 Can a male have all the normal sex characteristics and still be sterile? Explain.

8 What hormone stimulates sperm production by the testes?

9 What hormone stimulates testosterone production in the male?

10 Diagram the reproductive tract of the female and name the parts.

11 Describe differences between the reproductive tracts of the cow and the sow.

12 Outline the endocrine control of the ovary.

13 What hormone does the follicle secrete?

14 What hormone is secreted by the corpus luteum?

15 Outline the functions of LH and FSH in the female.

16 List functions of estrogens, progesterone, and testosterone.

17 What is meant by sperm capacitation?

18 What are some irregularities of estrus and the estrous cycle in farm mammals, and what causes each?

19 If a sexually mature female remains in almost constant estrus, what are a probable cause and a suggested treatment?

20 If a sexually mature female does not exhibit estrus, what are the probable cause and suggested treatment?

21 What is the normal length of the estrous cycle in the cow, sow, ewe, mare, and human female?

22 Describe changes that occur within the female reproductive tract between estrous periods.

23 Which species may be classed as a monoestrous animal? As a polyestrous animal?

24 What is the menopause? Does it occur in female farm mammals?

25 Name a spontaneously and a nonspontaneously ovulating species.

26 What is meant by superovulation? By superfetation? (cf. glossary also).

27 How do the sperm and egg meet in the process of fertilization?

28 Where is the usual site of fertilization in farm mammals?

29 Name a species of animal that has follicles on the ovary but no corpus luteum.

30 Describe some abnormal pregnancies that may occur in animals.

31 What is polyspermy? Why may it occur in farm animals? What is the usual result when it does occur?

32 What is parthenogenesis?

33 What is meant by intrauterine migration?

34 What are some functions of the placenta?

35 What are the fetal membranes in the young of farm animals?

36 Describe differences between the placental membranes of the calf and the pig.

37 Is it possible to determine the number of ova produced in female farm animals at a single heat period? Explain.

38 What is a cesarean operation? Why is it so named?

39 Why is it necessary for newborn farm mammals to get colostrum? Is this necessary in humans? Why?

40 Why is a posterior presentation abnormal in cattle?

41 What is meant by the synchronization of estrus and ovulation? Discuss the fundamental physiology involved.

42 What is superovulation? Discuss its merits and limitations.
43 What are the basic procedures used in the technology of in vitro fertilization, or so-called test-tube babies?
44 Discuss the procedures and applications of embryo transfer.
45 What is the current state of the art of freezing embryos?
46 Of what potential significance is microsurgery with embryos?
47 What is the current state of the art of "sexed semen"? Of sexing embryos?
48 What is meant by induced parturition? Discuss its merits and limitations.

CHAPTER

ARTIFICIAL INSEMINATION[1]

The environment fosters and selects; the seed must contain the potentiality and direction of the life to be selected.

George Santayana (1863–1952)

10.1 NOMENCLATURE AND DEFINITION

Artificial insemination is the introduction of male reproductive cells into the female reproductive tract by an artificial means. It is commonly abbreviated AI when associated with domestic animals. In humans, artificial insemination is abbreviated AIH when the husband's semen is used and AID when the semen is that of a donor.

The foremost value of artificial insemination in farm animals lies in its use as a tool for the genetic improvement of **livestock** (especially traits having economic importance) on a mass basis.

10.2 HISTORY AND DEVELOPMENT OF ARTIFICIAL INSEMINATION

Humans may not have originated the practice of artificial insemination. Males of certain **species** of spiders possess no **copulatory** organs. They deposit semen on a small mat spun by themselves, dip their feelers into the semen, and plunge the **sperm**-laden feelers into the abdomen of the female.

[1]The authors acknowledge with appreciation the contributions to this chapter of Dr. J. D. Sikes, Department of Dairy Science, University of Missouri—Columbia.

An old Arabian document dated 700 of the Hegira, which corresponds to the year 1322 of our era, records that an Arab chief of Darfur, who owned a prized **mare,** introduced a wad of wool into the animal's **genitals** and left it there for 24 hours. He knew that a neighboring hostile tribe (specifically an enemy chieftain) had an excellent **stallion** in its possession. At night he crept into their camp and held the odorous wad under the stallion's nostrils, whereupon the horse became sexually excited and subsequently ejaculated on a piece of cotton held in readiness by the owner of the mare. He then hurried home and introduced the cotton into the mare's vagina. The mare then came in **foal.**

A second incident recorded in the fourteenth century further substantiates the early use of AI. While at war with a neighboring tribe, an Arabian sheik, knowing that his enemies had exceptionally good mares that gave them a military advantage, sent some of his men into the enemy's camp by night with orders to fertilize the mares with semen of an old, lame, stabled stallion of inferior genetic merit.

It was Anthony van Leeuwenhoek, a Dutch lens maker in Delft, Holland, who first discovered human spermatozoa through a microscopic lens in 1677. He described what he saw as "man swimming in his own pool." In reporting his findings related to sperm cells and semen research to the Honorable Viscount Brouncker, president of the Royal Society, he said, "If your Lordship should consider that these observations may disgust or scandalize the learned, I earnestly beg your Lordship to regard them as private and publish or destroy them, as your Lordship thinks fit."

In 1777, Lazzaro Spallanzani, a priest of Modena and a **physiology** professor at the University of Pavia, began a series of successful experiments using AI on reptiles. In 1780, he used AI successfully to inseminate a Spanish **bitch** that had produced one previous **litter.** He later reported:

> When she showed signs of coming in **heat,** I shut her up, fed her myself and kept the key on me. When she had been thus isolated for 13 days, I saw clear signs of heat, the external sex organs becoming moist and excreting engorged fluid. On the 23rd day she seemed to me to be ready for an artificial **fecundation. By spontaneous ejaculation,** a young dog of the same **breed** provided me with 19 grains (about 1.2 cc) of seminal fluid which I immediately injected into the uterus by means of a syringe. As fecundation depends on the natural warmth of the semen, I took care to have the syringe at the same temperature as that of the animal, namely 30°C. After two days, I saw that she was no longer in heat and after 20 days that her belly was swollen; so on the 26th day I set her free. It was 62 days after my injection that she **whelped** three living pups, two males and one bitch, resembling the parents in color and shape. I found that six grains of the seminal fluid had remained behind in the syringe and concluded from this that a very small amount of semen is required in nature. This discovery has persuaded me that we shall be able to do the same thing with larger animals.

He further concluded:

> The day will come when this discovery will acquire immense importance for the human society.

In subsequent studies, Spallanzani found that the fertilizing power of semen resided in sperm carried by the spermatic fluid. When semen was filtered, he observed that the liquid that passed through was incapable of causing fertilization, but the residue on the filter was high in fertilizing capacity. This discovery gave rise to intensive investigations of sex cells.

An American dog breeder, Everett Millais, artificially inseminated 19 bitches and successfully impregnated 15 during the period from 1884 to 1896. Concurrently, Professor Hoffman of Stuttgart recommended AI following natural matings in horses. He recovered semen deposited by the stallion into the vagina of the mare by using the speculum-and-spoon technique. The semen was then diluted with cow's milk and injected into the uterus, using a syringe. He concluded, however, that this was impractical and discontinued his studies.

In 1899, Elias I. Ivanoff, a Russian researcher, began a series of studies using AI. He was successful in pioneering the artificial insemination of birds, horses, cattle, and sheep and apparently was the first to artificially inseminate females of the latter two species successfully. Mass breeding of cows through artificial insemination was first accomplished in Russia, where 19,800 cows (an average of 100 cows per bull) were bred in 1931. The Russian scientist Dr. V. K. Milovanov stated that the year 1931 marked the end of the experimental stage in artificial insemination of cows and its beginning as a "powerful zootechnical tool." Thenceforth, he concluded, the objective of AI was utilization of spermatozoa from the best sires on the largest possible number of cows. Its other uses, **disease prophylaxis** and the therapy of sterility, became of secondary importance.

In Denmark in 1936, the first cooperative artificial-breeding association in the world was organized.

In 1935, Dr. C. L. Cole and Professor L. M. Winters artificially inseminated sheep at the University of Minnesota. In May 1936, Dr. Cole discussed the use of artificial insemination with Dr. L. O. Gilmore (Ohio Experiment Station), who had a breeding problem—his albino **bull** would not serve an albino **heifer.** Cole offered to collect semen from the bull and inseminate the heifer, and Gilmore accepted. On February 25, 1937, the first calf to result from artificial insemination in the United States was born.

Professor E. J. Perry visited one of the cooperatives in Denmark and saw the possibilities of artificial insemination. In 1938 he established the first cooperative artificial-breeding association in the United States. It was organized by the Extension Service of the New Jersey State College of Agriculture with the aid of the New Jersey Holstein breeders. However, artificial insemination of cows was already being practiced in a few herds in the United States at that time.

In 1939, a Jersey **cow** on exhibition at the World's Fair in New York was artificially inseminated and conceived to a bull that was on exhibition at the San Francisco Fair. This event generated much interest and, as shown in Table 10.1, the use of artificial insemination in cattle in the United States has since been widely accepted. In 1981, 16,544,000 units of dairy and 2,800,000 units of beef semen were sold and custom frozen in the United States (Table 10.2).

TABLE 10.1
SIRE USE THROUGH ARTIFICIAL INSEMINATION IN THE UNITED STATES

Year	No. of sires	Cows per sire	Sires per stud
1938*			
1939	33	227	4.7
1943	574	318	9.7
1948	1745	982	19.2
1953	2598	1865	27.1
1958	2676	2483	37.7
1963	2559 (incl. 401 beef)	3250	50.2
1968	2380 (incl. 352 beef)	3303	72.1
1971†	2514 (incl. 347 beef)	3402	96.8

*Initiated in May.
†The above data have not been compiled since 1971.

Artificial insemination is currently used in dairy and beef cattle, goats, sheep, swine, horses, turkeys, bees, dogs, red fox, fish, mink, humans, and many other species. A summary of its worldwide use in cattle is given in Table 10.3.

Research leading to the development of artificial insemination on a commercial basis in the United States (especially in dairy cattle) is almost exclusively a contribution of the colleges of agriculture of the land-grant universities. Pioneering research was done at the Illinois, New Jersey, Missouri, Michigan, Minnesota, New York, and Wisconsin stations.

TABLE 10.2
UNITS OF BEEF AND DAIRY SEMEN SOLD AND CUSTOM FROZEN IN THE UNITED STATES (THOUSANDS)*

Year		Domestic use	Export sales	Total units	Units custom frozen	Grand total
1971	Beef	2077	159	2236	1055	3291
	Dairy	10,877	468	11,345	336	11,681
	Total	12,954	627	13,581	1391	14,972
1976	Beef	1369	179	1548	1274	2822
	Dairy	10,753	1390	12,143	570	12,713
	Total	12,122	1569	13,691	1844	15,535
1981	Beef	973	176	1149	1651	2800
	Dairy	13,332	2233	15,565	979	16,544
	Total	14,305	2409	16,714	2630	19,344

*Source: Dr. G. A. Doak, Technical Director, National Association of Animal Breeders, Columbia, Mo., personal communications.

TABLE 10.3
WORLD USE OF ARTIFICIAL INSEMINATION (1973–1974)*

Region	Number of cows
Europe	63,952,150
U.S.S.R. (Europe and Asia)	25,679,000
North America	10,876,000
Central and South America	4,533,617
Asia	4,758,396
Africa	365,551
Oceania (Australia and New Zealand)	2,034,000
World totals	112,198,714

Selected country	Percent cattle bred with AI
Denmark	99
Finland	99
Israel	99
Hungary	98
Japan	98
Czechoslovakia	94
Bulgaria	92
Norway	90
Poland	87
Sweden	80
France	75
U.S.S.R.	75
West Germany	70
Netherlands	69
East Germany	65
Great Britain	65
United States (dairy)	60–65
(beef)	3
Canada (dairy)	55
(beef)	11

1973–1974 World use of AI in selected species

Cattle	112,198,714
Sheep	67,421,811
Goats	68,322
Swine	1,714,414
Mares	926,117
Total	182,329,378

*Sources: H. A. Herman, "Improving Cattle by the Millions—NAAB and the Development and Worldwide Application of Artificial Insemination," University of Missouri Press, Columbia and London (1981); various publications of Foreign Agr. Service, USDA; and Dr. G. A. Doak, Technical Director, National Association of Animal Breeders, Columbia, Mo., personal communications.

10.3 IMPORTANCE AND IMPLICATIONS OF ARTIFICIAL INSEMINATION

It is now well known that artificial insemination provides an important tool in livestock breeding. It was initially organized in the United States as a means of making available the service of superior **purebred** dairy **sires** to all dairy cattle breeders, especially those with **grade** cattle. However, it is currently commonplace in purebred **herds.** Also, more recently, its use in breeding beef cattle, swine, goats, and horses throughout the United States has increased. Since 1956, thousands of ampules of frozen semen have been exported[2] annually to other countries. The need for improvement of cattle, incuding water buffalo, in many lands offers a great challenge.

10.3.1 Advantages and Benefits

Artificial insemination increases the usefulness of superior sires. The average bull in commercial AI service is currently being mated to an estimated 5000 cows annually. Many are mated to 50,000 or more, in contrast with only 30 to 50 cows normally bred to bulls under natural mating conditions annually. One Hereford bull produced nearly 300,000 ampules of semen during his 12 years of AI service. Artificial insemination provides for a fast increase in production potential. A recent USDA study indicates that when superior sires are used, 3 to 4 times more rapid genetic improvement can be made through AI than through natural service (Figure 5.3).

It is estimated that the genetic superiority of dairy cows sired by bulls in AI service, compared with the same number of cows sired by natural service, results in approximtately 8 billion pounds of additional milk being produced annually in the United States. Furthermore, the genetic superiority of slaughter steers currently sired by AI results in an estimated production of over 40 million additional pounds of edible meat annually. This extra milk and meat increases the efficiency of production and results in lower per serving cost to consumers.

The time required to establish a reliable sire proof on a young bull is greatly reduced through AI (see Chapter 5). Therefore the transmitting ability is determined more quickly, and as a result of bulls being mated to cows in many herds the proof is more meaningful.

Artificial insemination prevents the exposure of a healthy sire to diseases such as vibriosis, trichomoniasis, vaginitis, leptospirosis, brucellosis, and tuberculosis. It makes possible the use of sires unable to serve a cow naturally and the mating of large sires to small females without injury or mating problems. The danger to the caretaker in handling bulls is greatly reduced and often eliminated.

Artificial insemination permits small purebred breeders and all grade and/or commercial breeders to have the services of expensive sires. Some bulls have been brought into artificial insemination service at a cost well in excess of

---

[2]U.S. exports of bull semen totaled approximately $23 million in 1982. Leading markets were West Germany, the Netherlands, Mexico, Italy, Canada, Brazil, and Argentina. Hungary, Chile, Colombia, Venezuela, South Africa, and Egypt are viewed as major emerging markets.

$100,000. Yet their semen is commonly available for only $5 to $25 per mating. The average artificial insemination cost per mating (semen plus inseminator fee) in the United States is about $20 to $25. Furthermore, AI permits mating of outstanding individuals though great distances apart. Also, the mating to an outstanding sire after his death is made possible through the freezing of semen and use of artificial insemination. One example of this is cited in which an AI **stud**[3] achieved a successful mating more than 25 years after the sire's death.

Frozen semen of production-tested bulls can be used as a control in studies of genetic progress. Progeny resulting from periodic matings to the bull initially used are compared with progeny of production-tested sires selected later. For example, assume a selection experiment was initiated in 1965 using semen from production-tested bull A. Progeny resulting from matings in 1975 and 1985 using semen from bull A can be compared with progeny of bulls in service in 1975 and 1985 to determine the extent of progress in selection.

Because semen is examined regularly, infertile bulls are likely to be detected earlier than with natural mating. This procedure encourages the keeping of better breeding, calving, and fertility records. Artificial insemination increases fertility in some species, e.g., turkeys. It serves as a research tool for studies in reproductive physiology, fertility, sterility, and cellular physiology.

Artificial insemination offers a wider choice of sire selection and individual matings. Many problems and decisions associated with buying a bull are eliminated. Furthermore, it provides a means of attaining more uniformity of market cattle and permits better control of seasonal reproduction. In beef, AI has provided the only means of access to certain of the recently introduced beef breeds.

10.3.2 Limitations

Artificial insemination can quickly disseminate undesirable genetic traits. Recent USDA studies show that only about one-third of the **proven sires** improved herds in which they were used. As a result of mating more cows per sire, the frequency of a specific detrimental **recessive gene** is more likely to be increased. Moreover, mass mating with a given bull tends to narrow the genetic base and thereby restrict variation, which is a tool important to animal breeders. Although these are potential limitations, the authors wish to point out that most males used in AI programs are carefully selected (often through inbred progeny testing) to minimize the spread of lethal and undesirable traits.

Special training in semen collection, processing, storage, and use is essential with artificial insemination. In addition, artificial insemination requires special equipment and facilities (especially in working with beef cattle). Artificial insemination requires experienced and well-trained inseminators to achieve the same **conception** rate as with natural service. Extreme caution is essential in

---

[3]A bull stud, or semen-producing, business is any individual or business entity owning or leasing one or more bulls from which the individual or business entity collects, processes, and distributes semen for use in the insemination of animals owned by others.

processing semen to prevent errors in labeling, handling, shipping, and using frozen semen.

Although New Year's Day, Easter, the Fourth of July, Labor Day, Thanksgiving, and Christmas are the only "no service" days, the demands during periods of heavy breeding may cause undesirable delays in obtaining service. Furthermore, it is often difficult to detect cows in heat, especially among beef cattle, whereas the bull maintains a 24-h vigil for animals in heat and is naturally much better adapted to recognize symptoms of **estrus** than are humans. It should be noted, however, that two bulls may chase the same cow while other cows in estrus are nearby.

Artificial insemination requires more labor than does natural service. It requires a high cow population per square mile to be profitable for the inseminator and economical for the livestock breeder. It reduces the demand for bulls for breeding purposes. However, many bulls are not of sufficient genetic merit to be used for breeding purposes.

10.4 SEMEN COLLECTION[4]

The primary objective in semen collection is to obtain maximum output of high-quality spermatozoa per ejaculate. As the frequency of ejaculation increases, the volume, sperm concentration, and total number of spermatozoa per ejaculate generally decrease; however, total weekly sperm harvest increases. Proper stimulation and preparation of the bull before semen is collected increases the number of spermatozoa obtained per ejaculate.

Several studies indicate that oxytocin and certain other hormones influence sperm output. Sperm output in **rams,** rabbits, and certain other species increases when males are injected with oxytocin. Research at Michigan State University showed that anterior pituitary hormones are released into the blood at ejaculation; luteinizing hormone increased slightly and growth hormone concentration doubled within 5 min after ejaculation. Curiously, growth hormone was higher in mature than in young bulls and highest when ejaculation followed sexual preparation. Prolactin increases sixfold and testosterone twofold within 5 min after ejaculation.

Cornell researchers found that weekly sperm output of Holstein bulls may average between 30 and 35 billion. Sperm output was positively correlated with maximum testes-scrotal circumference.

10.4.1 Semen Output and Frequency of Ejaculation

Factors known to influence semen output are age of the bull (maximum output occurs between 4 and 8 years of age), season of the year (December is the best

---

[4]For additional information, the reader is referred to Peter Crowe-Swords, *Bovine Semen Collection and Processing Techniques,* edited by Dr. John Taylor, 5012 Norquay Dr., NW, Calgary, Alberta, Canada, June 1979.

month for total sperm production), and frequency of ejaculation (increased frequency of ejaculation results in more total sperm being obtained, but with fewer sperm per collection). Bulls collected 6 times per week ejaculated 3.3 times more sperm per week than those collected only 1 time weekly. In Cornell University studies, first ejaculates produced 78 percent more sperm than second ejaculates.

Researchers at The Pennsylvania State University found that when semen was frequently collected from bulls, they could be ejaculated up to 77 times in 5 hours with no apparent ill effects. In recent studies at The Pennsylvania State University, a continuous high frequency of ejaculation (6 times weekly) from bulls 1 to 7 years of age greatly increased the harvest of semen from the bulls without impairing their growth, reproductive capacity, or fertility.

Since approximately 95 percent of an ejaculate of average concentration (1000×10^6 sperm per milliliter, ml) is accessory gland secretions, it is safe to assume that the initial ejaculate temperature approximates that of the bull's body temperature.

10.4.2 Testicular Development

Testicular weight is an important trait that provides an accurate estimate of the amount of sperm-producing parenchyma in the testis. Since testicular weight cannot be measured directly in breeding bulls, a scrotal circumference (SC) measurement is used as an accurate and repeatable estimate. The relation between SC and paired testes weight is $r = 0.95$. Canadian scientists found that the heritability of testicular size in yearling beef bulls completing 140-day growth-performance tests is 0.69.

Factors affecting the rate and extent of testicular development include age, breed, body weight, and nutritional regime. The use of beef bulls with above-average testicular development can be expected to result in female progeny reaching puberty at a young age. Such heifers should result in a somewhat more fertile cow herd with higher lifetime productivity.

Recent research at North Carolina State University suggests that selection for increase in scrotal size at 365 days should increase testes size and weight and sperm number in bulls. Because sperm production by a male is, in large part, a function of testis size, the number of units of semen that can be obtained is closely related to testicular size. Research is under way at Louisiana State University to develop ways of increasing testicular size in bulls using techniques involving immunization and hormones.

Recent research at Colorado State University also showed that scrotal circumference or testicular size is correlated positively with fertility ($r = 0.58$). Since the correlation of SC with actual testicular weight is $r = 0.95$, this means that SC is a reliable predictor of the amount of sperm-producing tissue within the testes. Therefore, the larger the testes, the greater the sperm production potential. This is of great economic significance, whether a bull is used on a conventional or on an artificial insemination basis.

Another interesting finding of the Colorado studies was a negative genetic correlation of -0.71 between age at first estrus in heifers and SC of half-sibling males. This suggests that the female progeny from bulls with above-average testicular size for their age would begin to cycle and exhibit estrus earlier than average, which would be beneficial in breeding programs where heifers are bred to calve as 2 year olds. Additionally, it suggests that selection for early age at puberty in heifers should result in a more fertile cow herd with higher lifetime productivity.

10.4.3   Preejaculation Sexual Preparation

Sexual preparation of bulls may be defined as *prolonging the period of stimulation beyond that adequate for mounting and ejaculation.* Studies at Michigan State University showed that motile sperm output per ejaculation in bulls can be increased as much as twofold when two or three false **mounts** (and active restraint) are given as rapidly as possible before ejaculation.

Stimuli other than false mounting may also be used to increase sperm output in bulls. These stimuli include moving the stimulus animal, exchanging stimulus animals, changing locations of preparation, changing the personnel handling the bulls, and combinations of these. Additionally, undefined stimuli from a teaser animal near a bull cause the bull to ejaculate more sperm. The undefined stimuli are not visual since blinded bulls also respond to sexual preparation with increased sperm output. Part of the stimuli must involve olfactory (sense of smell) mechanisms since bulls mount **estrous** cows twice as fast as nonestrous cows and yield larger volumes of semen and greater sperm numbers.

Apparently, sexual preparation is less effective, as measured by time to first mount, in beef than in dairy bulls. Pennsylvania studies showed that sexual preparation significantly increases both sperm concentration and semen volume for dairy and beef bulls. However, it took nearly 10 times longer to stimulate beef (10.9 min) than dairy (1.1 min) bulls, based on time to first mount. Semen harvest and related data are presented in Table 10.4.

TABLE 10.4
SEMEN PRODUCTION IN SELECTED MATURE MALES

Class of animal	Semen vol/ ejaculate, ml	Sperm concen- tration, $\times 10^6$/ml	No. of sperm/ ejaculate, $\times 10^9$	Life of sperm in female tract, h	Survival time with fertilizing capacity (fresh)	No. of females per ejaculate
Boar	200–300	25–1000	20	24–40	1–2 days	15–25
Bull	5–6	800–1200	4–6	28–30	4–6 days	300–500
Cock	0.2–2.0	0.5–60	0.03	168–504	1–2 days	8–12
Human	2–6	50–150	3.5	24–96	1–3 days	.....
Ram	0.7–2.0	800–4000	2–3	34–40	5–7 days	40–100
Stallion	50–150	30–800	6	96–144	1–2 days	8–12

10.4.4 The Artificial Vagina (AV)

In 1930, Russian scientists developed the artificial vagina (Figure 10.1). The AV has proved to be the most practical and satisfactory means of collecting semen in most domestic animals. Its use aids in obtaining a normal ejaculate in which the semen is clean and free of extraneous secretions. Additionally, a quick measure of quantity is obtained in the graduated semen-collection vial, and the spermatozoa are of high **viability.**

The artificial vagina should have an internal temperature of about 43 to 50°C and should be the appropriate length for each bull. It is important that the bull ejaculate near the end of the AV into a directacone to reduce sperm loss on the surfaces of the AV and to prevent semen from being in contact with the warm temperature of the AV.

During collection, the artificial vagina should be positioned parallel to the cow in a slanting position near the anticipated path of the bull's penis. The operator should guide the penis into the AV by grasping the sheath well behind the orifice. To avoid possible retraction, premature ejaculation, and spread of disease, the operator should not physically touch the protruded penis. As the bull thrusts forward for ejaculation, the operator should allow the AV to move forward and tip it slightly to allow the semen to flow into the collecting vial. Special care must be taken to avoid bending the penis. This may injure or cause discomfort to the bull. Semen collection from the ram and stallion via the artificial vagina is similar to that in the bull; however, in the **boar** the emission of semen continues for 5 to 20 min and thus prolongs the collection period.

Freshly harvested semen must be protected against heat or cold shock to ensure motility and viability, which are essential to fertility. Sanitation is important with respect to equipment, personnel, handling, and processing of semen to protect against bacterial contamination and reduced semen quality.

10.4.5 Mechanical Manipulation

This method of collecting semen has had only limited sucess in cattle but is of more importance in fowls. In bulls, gentle pressure is applied through the rectum on the ampullae and vas deferens. The primary disadvantages are that semen is often contaminated with bacteria and extraneous matter, small ejaculates are common, and many animals do not respond well. However, this

FIGURE 10.1
An artificial vagina used to collect bovine semen. *(Missouri Agr. Expt. Station.)*

method may be used to obtain semen from bulls unable to mount and serve a cow or artificial vagina. Samples obtained by this method have only about one-half the number of sperm as semen collected using the artificial vagina.

10.4.6   Electrical Stimulation

Through the use of a voltmeter, transformer, milliammeter, and one or more electrodes, an alternating current of 3 to 30 volts may be produced that will cause an electrical excitation of the ejaculatory nerve centers, resulting in partial to complete ejaculation (Figure 10.2). Repeated rhythmical stimulation periods are alternated with short rest periods. This method of semen collection has been used routinely in rams and to a lesser extent in other farm animals. It results in the collection of clean semen and may be used on males that have been injured or are lacking in **libido.** It may also be used on animals who were "psychologically conditioned" by natural matings and refuse to serve the artificial vagina.

In the bull, it is possible to obtain accessory sex **gland** secretions at lower voltage than is required for ejaculation. As a result, large semen samples are often obtained from a bull with an electroejaculator, but these samples have a lower sperm concentration. The ram requires a low-peak voltage (2 to 8 volts) whereas the boar requires a higher (25 to 30 volts) voltage than the bull for ejaculation by this method.

Cornell studies showed that semen obtained by electroejaculation was lower

FIGURE 10.2
An electroejaculator used to collect semen from the bull. *(H. A. Herman and F. W. Madden,* The Artificial Insemination of Dairy and Beef Cattle, *6th ed., Lucas Brothers Publishers, Columbia, Mo., 1980.)*

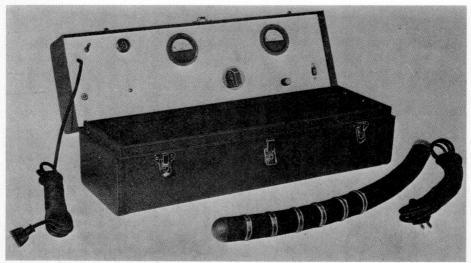

in solids than that obtained with an artificial vagina. (The total weight of solids of **bovine** semen, as measured by refractometry, averages about 10 g of solids per 100 g of seminal plasma.)

Tranquilization appears to aid in the collection of higher-quality semen in the bovine when electroejaculation is employed. Moreover, less electrical stimulus is needed to obtain an ejaculate in the tranquilized bull.

10.4.7 Vaginal Collection

Perhaps the oldest, yet least desirable, method of harvesting semen is to recover it from the anterior end of the vagina following copulation. This may be accomplished by using a syringe, sponge, or vaginal spoon. It is not recommended, however, because the semen is contaminated with mucus and urine, there is increased danger of spreading genital diseases, and it is impossible to recover an appreciable amount of the ejaculate.

10.5 EVALUATION OF SEMEN

The best indication of fertility in a bull is a live calf born to his mating. This evidence is never available in young bulls, and breeding records are often incomplete in older ones. Therefore certain physical and chemical tests have been developed to assist in determining semen quality.

Appearance This can be noted at the time of collection. The color of semen is usually milky white, and the fluid is quite viscous. Usually, the greater the viscosity, the higher the sperm concentration.

Enumeration of Spermatozoa The concentration, or number of spermatozoa per unit volume of semen, is an important consideration in determining the optimum dilution ratio of semen to be processed. It is primarily the number of spermatozoa, the percent of mobile spermatozoa, and the degree of progressive motility that are used to appraise the estimated semen fertility.

There are a number of methods for estimating spermatozoa number. In one, photoelectric colorimetry, the amount of light passing through a standard dilution of semen is used to estimate sperm number. This method is rapid and highly repeatable and is used extensively. Hemocytometry, which is also used to count red and white blood cells on a calibrated slide, is a second method of counting spermatozoa. A third method involves visually making an opacity rating, which is a rapid estimate of sperm concentration. In the process, semen is diluted and compared with previously prepared opacity standards of known concentration. The packed-cell method, which employs the hematocrit technique (also used extensively to determine the percent of red cells in blood), is a fourth method of estimating numbers of spermatozoa. Although the latter method provides an accurate count of spermatozoa, it requires more semen than the photoelectric colorimeter method.

Photometric techniques are quite variable in the determination of sperm concentration in the boar and stallion because their ejaculates vary substantially in the nonsperm cellular semen component and therefore decrease the accuracy of the readings. Researchers at Colorado State University found great monthly variation of seminal characteristics among stallions.

Motility This is rate of movement of spermatozoa and can be observed microscopically. A drop of semen is placed on a clean, clear glass slide that has been warmed to about 38°C. Using a low-power objective of the microscope, motility can be observed and a motility value may be assigned based on ratings of 0 to 5.

Because it is simple, quick, and inexpensive, percent motile spermatozoa is probably the most widely used assay for evaluating semen. However, evaluations of motility are subjective and are influenced by the thickness and the temperature of the sample on the slide. Research indicates that the accuracy of estimating motility of spermatozoa is increased by (1) using a semen sample of 8 to 10 μm in thickness, (2) viewing it near an air bubble and/or away from the edge of the slide, and (3) maintaining clean slides and coverslips. Most research indicates that motility estimates a lone are not reliable predictors of semen fertility.

Commonly used motility ratings with their respective descriptions are the following:

5 = *Excellent motility.* 80 percent or more of the spermatozoa are in highly vigorous motion. Swirls and eddies are rapid and changing.

4 = *Very good motility.* Approximately 70 to 80 percent of the spermatozoa are in vigorous rapid motion. Waves and eddies form and drop rapidly.

3 = *Good motility.* 50 to 70 percent of the spermatozoa are in motion. Motion is vigorous, but waves and eddies formed move slowly across the microscopic field.

2 = *Fair motility.* 30 to 50 percent of the spermatozoa are in motion. No waves and eddies can be observed.

1 = *Poor motility.* Less than 30 percent of the spermatozoa are in motion, resulting in sluggish motility.

0 = *No progressive motility.*

It is interesting that the motility of boar semen decreases rapidly when the semen is exposed to light but increases in the presence of oviductal fluids. Researchers at the University of Georgia found that the motility of boar semen increased from 70 percent at entry to 90 percent after being held 1 h in the oviduct.

Morphology A microscopic determination can be made of the percent and type of abnormal spermatozoa present in semen. High-quality semen contains a minimal number (5 to 15 percent) of abnormal spermatozoa, whereas low-quality semen frequently contains a larger number (up to 30 percent) of

morphologically abnormal spermatozoa. Deformed spermatozoa may result from cold or heat shock, x-ray, and nutritional or endocrine imbalances that disrupt normal **spermatogenesis.** First ejaculates of bulls following sexual rest are characterized by low motility and many abnormal sperm cells. Normal and abnormal spermatozoa are shown in Figure 10.3.

FIGURE 10.3
Normal and abnormal types of bovine spermatozoa. *(H. A. Herman and F. W. Madden,* The Artificial Insemination of Dairy and Beef Cattle, *6th ed., Lucas Brothers Publishers, Columbia, Mo., 1980.)*

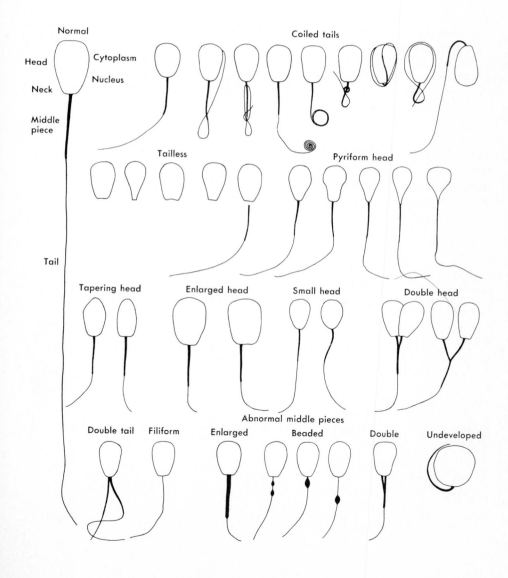

Dead-Alive Stain   This test was initially developed at the Missouri Agricultural Experiment Station using ram semen and is based on differences in permeability to a dye among living and dead spermatozoa. Live spermatozoa do not absorb dye and remain light in color, whereas dead sperm cells accept dye and appear dark under the microscope (Figure 10.4). The dead-alive stain is most meaningful when used with the microscopic motility examination. A fairly high percentage of dead sperm may not be readily apparent in the motility check because many inactive spermatozoa are swept about by movements of the live and active sperm cells.

Dead sperm dilute the concentration of fertile sperm of semen; they are also toxic to live sperm. However, addition of the enzyme catalase apparently eliminates the toxic effect of killed sperm on **livability** of bovine spermatozoa.

In general, morphological analyses are somewhat subjective and are influenced greatly by techniques and stains employed.

Scientists at Virginia Polytechnic Institute and State University showed a high correlation ($r = 0.60$) between fertility and average intact acrosomes, the part of the spermatozoan that releases egg-penetrating enzymes. The correlation between fertility and abnormal sperm morphology ranged from $r = 0.27$ to $r = 0.37$.

Cold Shock   This is a simple and quick check of spermatozoan vigor. The percentage of resistant spermatozoa may be determined by comparison of the number of sperm that are alive before and after cold shock. This test aids in estimating storage life and fertilizing capacity of semen. Cold shock of semen occurs mostly by a rapid decrease in temperature above freezing.

FIGURE 10.4
Stain showing dead (dark color) and alive (light color) sperm cells. *(H. A. Herman and F. W. Madden,* The Artificial Insemination of Dairy and Beef Cattle, *6th ed., Lucas Brothers Publishers, Columbia, Mo., 1980.)*

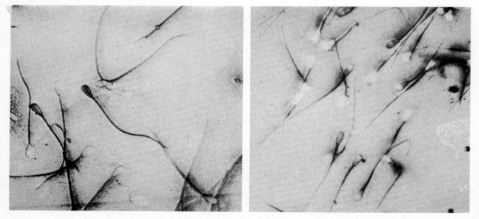

Reaction Rates   It is possible to measure glycolytic and respiratory rates of spermatozoa. **Metabolically** active spermatozoa participate in **glycolysis,** which is the **reduction** of glucose and other simple sugars to acids (lactic and others) by the action of cellular **enzymes.** The respiration rate (oxygen uptake and conversion into carbon dioxide) is closely related to the rate of glycolysis.

Methylene Blue Reduction   This test is based on the principle that active cells use oxygen. Thus semen containing a high concentration of active spermatozoa uses oxygen more rapidly than does low-quality semen, which has fewer active sperm cells. Use of oxygen results in an excess of hydrogen, which is free to combine with methylene blue (chloride) to form leukomethylene blue. The time required for the blue color to "bleach out" serves as an indication of the number and activity of spermatozoa in semen.

Metabolic processes tests (e.g., the methylene blue reduction test), enzyme analyses (e.g., for dehydrogenase or fumarase), and other chemical tests (e.g., assaying for citric acid and/or selected mineral salts) provide an objective approach to quality control of semen; however, repeatability of results and reliability for predicting fertility are not high.

Incubation Test   This test employs incubating semen at a high temperature for a relatively short period. By this method, the potential storage time of semen can be estimated. High temperatures increase reaction rates and all related biological processes of spermatozoa. The effect of high temperature on reaction rate is based on the van't Hoff–Arrhenius equation, which predicts a 10 percent increase in chemical reactions for each 1°C increase in temperature.

Hydrogen-Ion Concentration   Bovine semen has an average **pH** of 6.7. Semen obtained by the massage method often has above-normal pH (8), whereas semen collected in the artificial vagina after intense precollection teasing or long intervals between collections is lower than normal in pH (6.3). Changes in the pH of semen after collection may be regarded as a rough measure of the metabolic activity of spermatozoa, producing lactic acid from glycolysis. The resulting lowered pH causes a decrease in semen motility.

Sephadex Filtration   Research conducted at the University of Minnesota has shown that filtration of dead, immobile, or damaged sperm cells utilizing a Sephadex G-15-120 column and an electronic cell counter is an objective and repeatable procedure for analyzing semen in Plus-X and certain milk-yolk extenders (diluters). The assay is unsuitable for evaluating sperm quality in homogenized whole milk extenders because only about one-half as many sperm pass through the filters with homogenized whole milk compared with Tris-stepwise and Tris-complete extenders. One-fourth as many sperm pass through the filters with skim milk compared with egg-yolk citrate extender.

10.6   EXTENSION OF SEMEN

The main objective in extending semen is to increase the volume of the ejaculate so that a large number of females may be mated to a given male. In natural mating, one ejaculate, and often more, is used to inseminate one female, whereas through artificial insemination and the extension of semen, one ejaculate may be used for several hundred females. In the bovine, each ejaculate averages 5 ml of semen, containing between 0.8 and 1.2 billion spermatozoa per milliliter. Research has established that 12 million or less normal live spermatozoa in a volume of 1 ml will result in satisfactory conception rates among females inseminated artificially. It is apparent, therefore, that the average ejaculate from a normal bull can be used to inseminate 300 to 500 cows if the volume is increased by extension.

Semen of the ram is highly concentrated (Table 10.4), and even though ejaculate size is small, a large number of **ewes** (40 to 100) can be inseminated from one ejaculate. The volume of the boar ejaculate exceeds that of all other farm animals (Table 10.4); however, because sperm concentration is lower and spermatozoa must travel a long distance through the uterine horns to fertilize the ovum, one boar ejaculate can be used for only about 20 **sows** through artificial insemination. Semen from the stallion and **cock** also appear to have limited extension qualities, each ejaculate being useful in the breeding of about 10 females through artificial insemination.

10.6.1   Characteristics of a Good Extender

The ideal medium for semen extension will not only increase ejaculate volume but will also be favorable for both survival and longevity of spermatozoa and maintenance of their viability and fertility over extended periods of time. A number of media have been developed that provide adequate nutritional components, pH (6.5 to 6.7), **buffering** capacity, and protection against bacterial contaminants and temperature shock. In addition, the effects of electrolytes, nonelectrolytes, viscosity, and toxicity should be considered. An extender must be free of substances, bacterial products or infectious organisms, harmful to the spermatozoa, the female reproductive tract, the fertilization process, and the implantation and development of the fertilized ovum. An ideal extender should be simple to prepare, readily reproducible, reasonably inexpensive, and readily available.

The semen and extender are at the same temperature when mixed. Semen is extended to provide a desired number of services, each containing an adequate number of live sperm cells to ensure conception. For example, plastic straws are usually processed to provide 0.25 to 0.5 ml, and glass vials (ampules) 0.7 to 1.0 ml, of extended semen per service. Both contain approximately the same total number of spermatozoa.

Egg-Yolk Extenders   In 1934, Milovanov observed the beneficial effects of adding egg yolk to extending fluids. Today, egg-yolk citrate is one of the commonly used semen extenders. Only the yolk portion is used because egg white contains a substance (lysozyme) toxic to spermatozoa. Variations from 5 to 50 percent in the amount of egg yolk used appear to give good results; however, lower concentrations (20 percent or less is the most common mixture) result in more watery extenders that tend to drip from the delivery pipette during insemination.

Gelatin-Containing Extenders   These were developed by Danish workers. They employ a method of inseminating with semen gelled in cellophane straws; however, commercial artificial insemination in the United States does not employ the use of diluents containing gelatin. Gelatin capsules are often used as semen containers in artificial insemination of mares.

Milk Extenders   In recent years either boiled whole-milk or skim-milk diluents have been employed in commercial artificial insemination. These diluents are economical, are easily prepared, provide good protection for spermatozoa, and give satisfactory fertilizing capacity to semen. The foremost objection to whole milk as an extender is that the fat globules make microscopic examination difficult. This problem is avoided when skim milk is used and can be partially solved by the use of homogenized milk. Some processors use the milk of goats, which is "naturally" homogenized. Cow's milk has been successfully used as an extender for bull, ram, boar, and stallion semen. Glucose is often added to supplement the low sugar content of stallion semen. Studies at Texas A & M University indicate that evaporated milk can be used with glycerol as an extender in freezing stallion semen.

Fruit and Vegetable Juices   Juices have been used in some countries for extending bovine semen, with apparently acceptable results. Such juices as tomato broth and carrot juice, which are abundant and economical, have been used. Coconut milk is being used to extend fresh semen in the tropics, where refrigeration is at a premium and coconuts are plentiful. Bovine semen may be held at room temperature for 1 week in coconut milk.

Extenders for Frozen Semen   The use of glycerol and selected sugars is important in the extension of semen to be frozen (Section 10.7.2).

Others   Blood **plasma** and **serum,** Tyrode's solution, starch solution, seminal fluid, various solutions of alcohol glycerols, and sugars have been used. However, most of these are presently used only in research.

10.6.2   Coloring of Semen

To assist the artificial-insemination technician in identifying semen by breeds and thereby safeguard against error, many artificial-insemination organizations use certified food and vegetable dyes. The addition of one drop of coloring per 50 ml of extender does not affect livability and fertility of the sperm. The National Association of Animal Breeders (Section 10.10) has adopted the color designations at left for dairy bulls; colors used by some AI studs to help identify selected beef breeds are given at right.

Ayrshire—purple	Angus—orange
Brown Swiss—brown	Beef Shorthorn—magenta (purplish red)
Guernsey—uncolored egg yolk	Charolais—pink
Holstein—green	Hereford—tan
Jersey—red	Santa Gertrudis—blue
Milking Shorthorn—lime	Simmental—yellow

Uniform colors for semen of beef breeds will likely follow. Some artificial-insemination studs use colors to label semen vials (ampules) and straws.

10.6.3 Antibiotics

Antibiotics commonly used in frozen semen are penicillin, streptomycin, neomycin, and polymyxin B. Most extenders include **antibiotics.** Penicillin and streptomycin have been used successfully to enhance the keeping qualities of bull and stallion semen and to increase conception rate (5 to 12 percent) in the bovine.

Streptomycin and penicillin are good as prophylactic measures in controlling the spread of bovine vibriosis through semen of infected bulls. Although satisfactory for use in fresh semen, sulfanilamide is not recommended for use in frozen semen since it appears to be somewhat toxic to spermatozoa during the freezing process.

It is interesting that the microfloral content of stallion semen is greater than that of bull semen. This is because the prepuce of stallions is more highly contaminated with bacterial flora. Recent studies at Cornell University showed that the antibiotic amikacin inhibited bacterial growth in stallion and bull semen and had no significant effect on the motility of stallion and bull spermatozoa, even at relatively high concentrations.

Trichomonas fetus can survive semen freezing and processing (with antibiotics) procedures presently used in the AI industry. In 1979, 7.8 percent of 280 Oklahoma and 7.3 percent of 109 Florida range bulls were determined to be positive for *T. fetus,* and in 1980, 5.8 percent of 328 bulls tested in California were found to be positive for the organism.

10.7 SEMEN STORAGE

The cardinal concern in extending the service of superior males to large numbers of females is to preserve fertile spermatozoa successfully.

10.7.1 Fresh Liquid Semen

Semen of several species may be extended and successfully stored at 5°C for 1 to 4 days. During the infancy of artificial insemination, semen was shipped in thermos bottles filled with cracked ice. Often it was flown by light aircraft and dropped by parachute in the vicinity of the local inseminator. Currently, all artificial-insemination studs of the United States are freezing semen, and only a few continue to use fresh liquid semen in the insemination of cattle. Conversely, fresh semen is extensively used in the artificial insemination of sheep, swine, and turkeys. Since fresh stallion semen does not store well beyond 24 h and boar semen not beyond 48 h, shipment of these is somewhat impractical at present.

10.7.2 Frozen Semen

In 1776, Spallanzani observed that freezing stallion semen in either snow or the winter cold did not kill the "spermatic vermiculi" but rather held them in a motionless state until exposed to heat, after which they were motile for several hours. Davenport reported in 1897 that human spermatozoa survived freezing at $-17°C$.

In 1949, British scientists reported the discovery that extenders containing glycerol could be used to freeze fowl semen. This discovery resulted when a laboratory technician mistook a bottle containing glycerol for a diluting-medium bottle. He subsequently observed that the addition of glycerol enhanced the resistance of spermatozoa to freezing. This has proved to be the most revolutionary procedure relative to artificial insemination in cattle that has been developed to date. It has reduced the barriers of time, distance, selective matings, and semen losses in artificial insemination.

Frozen semen is currently being used successfully in cattle, goats, horses, dogs, fish, and humans; however, further research is needed before it can have practical significance in poultry, sheep, and swine. (A litter farrowed in Minnesota in 1970 was apparently the first successful artificial insemination of a sow using boar semen preserved by freezing.)

To date, pregnancy rates and live embryos per female in swine have been higher for females inseminated with fresh spermatozoa than for those inseminated with frozen-thawed spermatozoa. Artificial insemination of sheep with frozen semen does not result in the high fertility achieved with fresh semen or with natural service (52 percent as opposed to 86 percent in recent Canadian studies).

The principal method of storing frozen semen is in liquid nitrogen (N_2) at $-196°C$ ($-320°F$). Liquid nitrogen is supercold. It boils at $-196°C$ and turns into gas when released from pressure. It is the fourth-coldest known substance. Cornell studies indicate that there is little or no decline in fertility of frozen bovine semen stored in liquid nitrogen for 2 years. Apparently, after 2 to 3 years, there is a slight loss of fertility with time.

One concern in freezing spermatozoa is preventing crystallization of cellular water. Therefore an extender containing glycerol[5] is added to dehydrate the sperm cells prior to freezing and alter the formation of ice crystals, which would

---

[5]Minnesota studies showed that more of the enzymes important to the sperm's gaining entrance into the sow's ova for conception were retained when glycerol was not used in freezing boar semen.

kill the cells. Bulls vary greatly in the ability of their sperm to withstand freezing and storage. A substantial amount of research effort is being expended to elucidate the reason(s).

Current methods of freezing and thawing bovine semen recommend rapid rewarming which results in better spermatozoa survival than does slow rewarming. Thawing frozen semen in ice water for 8 to 10 min before use is no longer recommended, especially for semen frozen in straws.

The quality of frozen-thawed semen, particularly viability, is closely related to the biological performance of that semen as measured by heterospermic insemination. *Heterospermic insemination* refers to the use of semen from more than one male in the insemination dose. Here semen is commonly mixed so that sperm numbers from each male are estimated to be equal. *Homospermic insemination* refers to the conventional use of only one male in each insemination dosage. Performance of the male is expressed as conception rate. Studies at the Virginia Experiment Station and at several other stations show a clear relation between the heterospermic performance of males (i.e., the ability to sire offspring competitively) and their homospermic performance (i.e., the ability to sire offspring when used singly). Thus, males that are superior homospermically are also superior heterospermically. In cattle, the reported correlation between

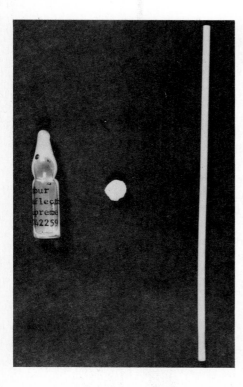

FIGURE 10.5
Methods of storing frozen semen (left to right): ampule, pellet, and straw. *(Missouri Agr. Expt. Station.)*

the logarithm of the heterospermic index and the homospermic index of stud bulls is positive and significant, $r = 0.69$.

Recent studies at the University of Minnesota have involved the collecting and freezing of spermatozoa from fish. Dr. E. F. Graham and his research associates have achieved an 80 to 90 percent fertility rate with frozen and thawed fish spermatozoa. Subsequent research has involved the freezing of fertilized fish eggs. Perfection of this technique would free fish culturists from the necessity of scheduling their seasonal work around the natural spawning seasons. It could also facilitate the use of larger fish to stock lakes, thereby giving a size advantage over natural predators and so increasing the rate of fish survival.

The world trend is toward freezing semen in the concentrated form. Several concentrations and techniques are being studied (Figure 10.5).

Ampule   This is a container with enough semen for only one service. Each ampule commonly contains 0.5 to 1 ml of diluted semen. Glass ampules have been used extensively in artificial insemination. However, the recent trend is heavily toward the use of straws.

Pellet   The pelleting of semen was first reported by Japanese researchers in 1962. It is a concentrated method of storing semen, but it presents problems in identification, automation, and sanitation. Pellets are thawed in physiological saline (0.9% NaCl).

Straws   Maximum use of spermatozoa can be obtained through inseminating from the straw in which the semen was frozen. Plastic straws are being used extensively in the United States (more than 95 percent of all units of frozen semen are marketed in straws today). They are about 2 1/2 to 5 in in length and commonly hold from 0.25 to 0.5 ml of semen. In comparison with glass ampules, plastic straws require less storage space, yield a higher recovery rate of motile sperm when thawed, and fewer sperm are lost during the insemination process (e.g., fewer sperm adhere to the container).

Rapid rates of thawing are beneficial to post-thaw motility, acrosomal retention, and fertility of bovine spermatozoa packaged in plastic straws. Straws are commonly thawed in clean water at 90 to 95°F for 40 s.

Shell-Freezing   This method was developed in 1966 by Graham at the University of Minnesota. It results in a higher percentage of live sperm than was possible using previous processes. Graham also found a means for removing dead and abnormal sperm from semen. This is accomplished by passing semen through a glass-fiber filter. Eggs fertilized by abnormal and dead cells usually abort prematurely and so it is important that those sperm be removed. This technique increases the viability of spermatozoa and ensures a higher conception rate.

Lyophilization of Semen   This technique has become a challenge to scientists. Using a high vacuum, semen moisture is changed from a solid to a vapor, bypassing the liquid form **(sublimation)**. Semen so treated for 24 h may lose 90 to 95 percent of its moisture. However, only 5 to 15 percent motile sperm are recovered when the semen is placed in an extender for use. To date, only one successful pregnancy has been reported from the use of **lyophilized** semen in the United States, and other scientists have been unable to duplicate successfully the procedure reported.

10.7.3 Custom Freezing

For a fee, many established bull studs collect, freeze, process, store, and dispense semen from bulls owned by private breeders. Such semen is generally used to service cattle belonging to the owner of the sire. This practice will likely continue to increase in the future.

10.8 REGULATIONS GOVERNING ARTIFICIAL INSEMINATION IN CATTLE

The artificial insemination of purebred dairy cattle must be done in compliance with certain regulations developed by the Purebred Dairy Cattle Association (a federation of the principal dairy-breed registry organizations in the United States) and the National Association of Animal Breeders. All bulls from which semen is frozen, along with their living untyped parents, must be blood-typed before offspring from the use of such semen are eligible for registration. Blood-typing is a means of verifying identity.

Semen must be labeled to include the complete registration name and number of the sire from which it was collected. Upon the death of a sire from which semen has been frozen, or termination of its lease, the owner or lessee must report to the breed registry organization involved the number of containers of frozen semen from the sire in his/her possession or ownership. Current inventories by sire and breed are required at all times. Other records relating to female ownership and identity are required. Health standards for bulls producing semen for AI have been set forth by the American Veterinary Medical Association and the National Association of Animal Breeders.

Registration requirements of selected beef-breed associations relating to artificial insemination are summarized in Table 10.5.

10.9 INSEMINATING THE COW

A cow usually remains in heat for 12 to 24 h and should be inseminated within 6 to 10 h following the first visible signs of standing estrus. Insemination equipment commonly includes disposable plastic insemination tubes, polyethylene bulbs or syringes, and plastic gloves. Semen may be deposited in the vagina, cervix, uterine body, or uterine horns. Semen deposition deep in the cervix is preferred because spermatozoa live longer in the cervix than in the uterus. Moreover, cervical semen deposition is preferred to uterine deposition, since conception rates are equal but there is less uterine injury, infection, and chance

TABLE 10.5
SUMMARY OF AI REGULATIONS FOR REGISTRY IN SELECTED BEEF-BREED ASSOCIATIONS

Association	Blood-typing of bull required	Must be owner of bull	Nonowner certificate required	Use of semen after death of bull
American Angus Assoc.	Yes	No	Yes	Yes
American Brahman Breeders Assoc.	Yes	No	No	Yes
American Chianina Assoc.	Yes	No	No	Yes
American Hereford Assoc.	Yes	No	Yes	Yes*
American-International Charolais Assoc.	Yes	No	No†	Yes
American Polled Hereford Assoc.	Yes	No	Yes	Yes
American Red Poll Cattle Assoc.	Yes	No	Yes	Yes
American Shorthorn Assoc.	Yes	No	Yes	Yes
American Simmental Assoc.	Yes	No	No	Yes
Devon Cattle Assoc.	Yes	No	Yes	Yes
International Brangus Breeders Assoc.	Yes	No	Yes	Yes
North American Limousin Foundation	Yes	No	No	Yes
Red Angus Assoc. of America	Yes	No	Yes	Yes
Santa Gertrudis Breeders International	Yes	No	Yes	Yes

*Frozen semen can be used by owner indefinitely. No changes in ownership allowed, but nonowner certificates may be sold.
†Must file semen transfer or owner's signature on calf application.
Source: "A Summary of AI Regulations of Beef Breed Associations," Certified Semen Services, Inc., a subsidiary of Nat. Assn. of Animal Breeders, Columbia, Mo., June 1983.

of spreading disease. Furthermore, pregnancy is less likely to be interrupted in cows showing signs of estrus during pregnancy when semen is deposited in the cervix rather than in the uterus. Semen deposition in the uterine horn is undesirable, because the nonovulating uterine horn might be selected. (Pregnancies occur more frequently in the right uterine horn of cows.) Deposition of semen in the vagina results in dilution, contamination, and lowered conception rate. Of course, when a bull inseminates a cow naturally, approximately 5 to 10 billion spermatozoa are deposited in the vagina. However, when semen is deposited artificially into the cervix, considerably fewer sperm are required to achieve conception.

10.10   THE NATIONAL ASSOCIATION OF ANIMAL BREEDERS, INC.

This nonprofit organization was founded in 1947 and has its national offices in Columbia, Missouri. The basic objective is to effect mass livestock improvement through artificial insemination. It publishes the *Advanced Animal Breeder* monthly and works closely with breed associations, AI studs, and the American Veterinary Medical Association to establish and publish codes for animal identification,[6] health, and other rules governing the use of artificial insemination. Nearly all organizations engaged in the commercial operation of AI businesses in the United States are members of NAAB.

10.10.1   Certified Semen Services, Inc. (CSS)

This organization was formed in 1976 as a subsidiary of the NAAB. Initially, CSS provided a semen identification auditing program involving an annual on-site visit by a CSS representative. During this visit, procedures and records relating to the identification of semen—from the time a bull is bought or custom collected through processing and distribution—are reviewed.

Subsequently, the CSS program was expanded to include sire health. The CSS Minimum Health Requirements outline specific testing procedures for bulls and mount animals before they enter isolation, during isolation, and for bulls housed in a central location following isolation. Specific diseases tested for in the CSS Sire Health Program are brucellosis, leptospirosis (five serotypes), paratuberculosis (Johne's disease), trichomoniasis, tuberculosis, and campylobacteriosis. Health tests related to the above diseases are conducted in accordance with procedures described in the "Recommended Uniform Diagnostic Procedures for Qualifying Bulls for the Production of Semen," published by the American Association of Veterinary Laboratory Diagnosticians. In the CSS program, semen packages containing semen from bulls meeting the CSS Minimum Health Requirements are designated as CSS Health Certified Semen.

10.11   ARTIFICIAL INSEMINATION IN POULTRY

The primary objective of artificial insemination in poultry is to achieve better fertility, in contrast with the objective of disseminating better genetic characteristics in other species. It is especially beneficial to breeders of broad-breasted turkeys, which have low fertility in natural matings.[7] More than 90 percent of the turkey breeders in the United States use artificial insemination in conjunction

---

[6]As part of positive identification, blood-typing of a bull to be used in AI and of his living parents that have not been blood-typed is required by breed associations.

[7]Research at the University of Illinois showed that roosters fed caffeine became sterile in about 3 weeks. Apparently, libido was unaffected by ingesting caffeine, which is believed to inhibit spermatogenesis. The Illinois studies showed that when roosters were fed caffeine for 63 days (three cycles) and then had the caffeine withdrawn from their diets, sperm production was resumed in about 3 weeks.

with natural mating to increase fertility substantially. Semen is "milked" from trained **toms** into thermos bottles. About 0.3 cc of semen can be obtained twice weekly. The semen is pooled and commonly used within 4 h following collection, since results with stored turkey semen have been disappointing to date. The use of artificial insemination in the chicken hen, using semen from the cock, is increasing.

10.12 ARTIFICIAL INSEMINATION IN BEES

Bee breeders are using artificial insemination to develop and maintain superior strains of bees. Double-**hybrid** bees resulting from artificial insemination may produce up to 50 percent more honey than comparable commercial lines. Bees possessing greater longevity and activity and less sting have been developed through artificial insemination. Semen is taken from the everted penis after the **drone** (12 or more days of age) has been placed in chloroform fumes so that partial eversion is induced. The semen is then deposited in the oviducts via a special syringe. A good technician can inseminate 100 queen bees hourly.

Anesthesia of queens with CO_2 hastens the onset of egg laying and keeps queens quiet during instrumental insemination. If CO_2 were administered to an unmated queen, she would begin laying eggs. Since such eggs would be unfertilized, the resulting offspring would be drones. Then, by collecting semen from these drones and inseminating the virgin queen artificially, an inbred generation of bees could be obtained. It is possible for four sperm cells to penetrate an egg laid by the queen bee. Therefore bee geneticists have been able to transmit eye characteristics of four different males in new bees.

When antibiotics are added to bee semen, it remains capable of fertilization for 16 to 18 weeks and has been shipped to Europe and Latin America.

10.13 ARTIFICIAL INSEMINATION IN HUMANS

It is every woman's heritage to bear children.

Frances Seymour

In the Bible, infertility is looked upon as a divine punishment (Genesis 30:1–2). The barren marriage is a problem nearly as old as humanity. In some countries, a barren woman met the fate of losing her life; in others, she was returned to her parents. Through the years, the onus of **barrenness** usually fell on the female. Within the past **generation,** however, people have learned to appreciate that childless marriages can be attributed to the male, female, or both. About 10 to 15 percent of all marriages are childless for reasons of infertility.

According to a 1977 publication of the American College of Obstetricians and Gynecologists entitled "Infertility Causes and Treatments," both the man and the woman have problems contributing to infertility in 15 to 20 percent of all infertile couples, whereas 35 to 40 percent of the cases of infertility are due to problems in the man and another 35 to 40 percent result from problems in the woman. In the balance of cases, no known cause can be determined.

Pregnancy by Proxy The first recorded case of pregnancy by proxy was biblical. Abraham's barren wife, Sarah, sent Abraham to mate with Hagar, her handmaiden, who bore him Ishmael. And notwithstanding all the emotional stresses and strains such arrangements may cause, childless couples have been quietly making them ever since. In recent years, however, AID has removed at least some of the sting from planned surrogate motherhood.

In 1785, J. Hunter recommended that the technique AIH be employed in a case where the husband (an English merchant) had a physical deformity preventing normal **coitus.** Results were successful.

Medical World published the first discussion of AID in April, 1909.

An estimated 50,000 children are born annually in the United States through artificial insemination (AIH and AID). However, there are legal, social, psychological, ethical, moral, theological, and cultural problems to be considered before AID is practiced.

10.13.1 Situations Encouraging AIH (AID)

Several factors might prompt a couple to consider artificial insemination: (1) The husband is infertile, and the couple prefers AID (or pooled semen) to an adoption. The woman is fertile by all standards of investigation. (2) The doctor may recommend AIH as a means of increasing conception. Several ejaculates of the husband's semen are centrifuged to concentrate the spermatozoa. The couple should have attempted spontaneous conception for at least 18 months prior to AIH. Multiple ejaculates from oligospermic (low sperm count) husbands can be collected, frozen, stored over a period of several months, and then thawed, pooled, concentrated by centrifugation, and inseminated (AIH) at spermatozoa concentrations many times greater than that of the original single ejaculates. This technique has proved successful when a reduced number of mobile spermatozoa was the primary cause of male infertility. Spermatozoa normally constitute about 5 percent of the volume of the human ejaculate. The normal concentration of spermatozoa in human semen varies between 50 and 150 million per ml (Table 10.4). (3) There may be an Rh incompatibility, the husband being **homozygous** positive and a series of pregnancies having been lost. (4) A somewhat rare situation of inborn errors of metabolism (transferred by recessive genes), which results in a condition such as phenylketonuria, or faulty metabolism of phenylalanine, may encourage the use of AID. (5) Social pressure may be involved, and the couple may feel inadequate in relation to their neighbors, family, and friends unless they bear children. (6) Stored semen collected prior to male sterilization (**vasectomy**) may be used through AIH to allow family planning to be continued.

The recent disclosure that a California businessperson has set up an exclusive semen bank named Repository for Germinal Choice, which offers the sperm of Nobel prize winners to carefully selected women, has rekindled the scientific and moral controversy sparked by advocates of human genetic engineering movements even before World War II. One sperm contributor, Dr. William B. Shockley, a 1956 physics laureate, said, "I welcome this opportunity—and endorse the concept of increasing the people at·the top of the population."

10.13.2 Donor Characteristics

It is essential that the donor remain anonymous. He must be fertile, free of disease, of aptitude equal to or above that of the couple in consideration, of similar physical proportions to the husband, and compatible with the Rh genotype of the wife. It is a common practice to mix donor semen with that of the husband.

10.13.3 Legal Agreements

For the protection of the doctor, husband, and wife, certain statements must be signed and witnessed. These statements commonly include an expressed desire for AIH (AID) by both husband and wife, an acceptance of the resulting child by the husband as a type of semiadoption, and the exclusion of adultery, rape, or any other involvement consideration that might be considered harmful to marital happiness. The husband is asked to be present when his wife receives an artificial insemination.

10.13.4 Freezing and Thawing of Human Semen

Both foreign and domestic researchers have successfully frozen human semen, with subsequent **impregnations.** Glycerol is commonly used to protect the spermatozoa during freezing, after which they are stored in liquid nitrogen. Agents used to increase sperm recovery after thawing include human follicular fluid, human milk, human blood serum, and egg yolk.

There were about 1500 births reported worldwide from the use of frozen human semen in 1981. The longest reported period that frozen human sperm have been preserved is 10.5 years.

10.13.5 Timing of AIH (AID)

It is impossible to perceive the time of **ovulation** in women, whereas signs accompanying estrus and ovulation are visible in most domestic animals. Therefore certain measures are often used to assist the detection of ovulation, which in turn increase the likelihood of conception. The **Papanicolaou** stain, basal-body-temperature charts, and formation of a **spinnbarkeit,** when interpreted by a competent **cytologist,** provide the most practical and accurate indications of ovulation. Most ovulations occur from day 9 to day 21 of the cycle, with about one-third occurring on day 14.

It was observed in 1867 by W. Squire in Great Britain that basal body temperatures (BBT) could be correlated with changes in the menstrual cycle of women. The concept that the biphasic response was related to ovulation was introduced by T. H. Van de Velde of The Netherlands in 1904. He later suggested that the temperature elevation observed in the second half of the cycle was caused by progesterone. In 1974, Zuspan and Rao correlated urinary catecholamine excretion with BBT patterns, suggesting that hypothalamic thermoregulatory control is mediated by norepinephrine. Several workers have proposed that BBT can be used either as an indicator of ovulation or to provide a means of fertility control. The consensus is that the procedure is useful clinically, but that the current state of the art does not render it a reliable method of fertility regulation.

At the time of ovulation in women, the cervical mucus becomes clearer, more watery, and less tenacious. Its "stringiness" and acidity can be measured.

Blood plasma progesterone—the hormone produced by the ovary (corpus luteum) following ovulation—can also be measured to determine if ovulation has occurred in women.

10.14 THE FUTURE OF ARTIFICIAL INSEMINATION

Few advances in science have had the impact that artificial insemination has had on animal breeding. During the past five decades, artificial insemination has advanced from an unfamiliar term to a well-accepted practice in the mating of domestic animals throughout the world. Yet the potential that artificial insemination presents as a means of improving animals on a mass basis remains virtually untapped in many species. Lack of communications, roads, transportation, and other factors impede the increased use of AI in developing countries.

It now appears that estrous synchronization among cattle, sheep, and swine will soon be practical on a large scale. With the synchronization of estrous cycles in many females, it will be feasible to artificially inseminate large numbers of females with fresh, as well as stored, semen from outstanding males in a relatively short period of time. This would be impossible through natural mating.

Estrous Synchronization and AI Artificial insemination has received wide acceptance and use in the dairy industry but only limited use in the beef industry, largely because of management constraints. A major portion of this problem has been resolved with the introduction of practical estrous synchronization programs that make possible more efficient use of labor. The availability of prostaglandin F2α and other materials for regulating ovulation in cattle will result in an increase in the number of beef cows inseminated artificially (see Section 9.5.1).

Unlike dairy producers, beef producers derive the major portion of their income from the calf crop, thereby making fertility the most important consideration. Results of a recent study placed the economic importance of (1) fertility, (2) growth rate, and (3) carcass quality in the ratio of 10:2:1. This indicates that to beef producers fertility is 5 times more important economically than growth and 10 times more important than carcass quality.

Progeny testing is well established in dairy-cattle breeding programs. Its use is expanding rapidly in the breeding of beef cattle and swine. Swine testing stations, for example, have identified boars capable of siring lean, meaty, well-muscled pigs that gain rapidly and use feed efficiently. Performance-tested boars can be mated with 2000 or more sows annually through AI, whereas by natural service, a given boar may sire only about 80 litters per year.[8]

Progeny testing will become even more important as a means of reinforcing the merits of artificial insemination in **domestic** animals. Breeders using progeny-performance data to select and breed improved livestock find their males in demand by artificial-insemination organizations. Computer matings have become a reality. Frozen semen will make germ plasm from outstanding males available to purebred and commercial breeders throughout the world.

Through fundamental research, many significant advances in the physiology of reproduction will be forthcoming. Among these advances will be measures and techniques giving an increased impetus to artificial insemination.

Genetic Engineering and AI With genetic engineering techniques for transferring specific genes within and among species, it may be possible to transfer the gene responsible for twinning in sheep to cattle and thereby greatly increase the reproductive performance of cattle. Even more exciting is the possibility of transferring genes in the armadillo that control cleavage of the ovum to produce identical quadruplets. It may also be possible to use genetic engineering techniques to transfer the genes that regulate the synthesis of the hormones that control growth and metabolic processes. Once these genes are transferred and become permanently fixed in the recipient animal, they can be spread quickly and widely throughout the population through the use of AI.

Sperm-Typing for Sex Preselection Recent studies at the University of California School of Medicine at San Diego indicate that it is possible to separate human X and Y sperm cells using a laminal flow method of sperm fractionation.

The X sperm, which carry the larger of the two sex chromosomes, swim in a more inclined path away from the flow axis and fractionate at the bottom of the flow cell, whereas the lighter Y sperm are enriched at the top fraction.

The potential merits of sex preselection techniques in livestock and poultry are indeed immense and could have a tremendously favorable impact on animal

---

[8]Recent studies at Ohio State University indicate that litter size may be increased in gilts (10.4 as opposed to 8.4 pigs per litter in the Ohio studies) by intrauterine treatment with boar semen 3 weeks before insemination.

food production and efficiency in both developed and developing countries. It is entirely possible that in the near future scientists will use genetic engineering techniques to isolate or synthesize genes controlling desirable production characteristics of farm animals in the laboratory. These genes could then be transplanted into Y sperm and transmitted via AI, along with the usual set of sperm chromosomes, to vast numbers of females. Once the desirable genes became established in the germ line of a breeding animal, they would have tremendous potential for propagation.

Milk Progesterone and Pregnancy of Cattle   It is important to determine pregnancy early. One means of doing this is assay for milk progesterone. Concentrations of estrogen and progesterone are inversely related in the peripheral circulation of cattle. Estrogen levels are highest during the follicular phase, whereas progesterone is elevated during the luteal phase. Measurement of either hormone will provide a reflection of ovarian status. Morphological changes in the corpus luteum (CL) are accompanied by fluctuations in progestin concentrations of peripheral blood. If pregnancy occurs, the CL persists, thus preventing cyclicity. By day 19 after breeding, plasma progesterone concentrations are significantly higher in pregnant animals. Cyclic changes in progesterone that occur in blood plasma also occur in milk; thus the testing of milk for progesterone 19 to 25 days after insemination is 80 to 90 percent accurate in diagnosing pregnancy of lactating cows, and 95 to 100 percent accurate in detecting animals not pregnant. Milk progesterone as a pregnancy test was offered commercially in Great Britain in 1975, and it is now available through several laboratories in the United States. Radioimmunoassay is the procedure commonly employed to measure milk progesterone.

Detecting Estrus in Cattle   The detection of estrus is extremely important in conjunction with the use of AI. The following two approaches recently reported in cattle are of interest.

In USDA studies utilizing dogs previously trained to detect a characteristic odor in the cow's reproductive tract at estrus, it was observed that the estrous odor emerges slowly during the 3 days before estrus, reaches a definite peak in intensity on the day of estrus, and disappears within 1 day thereafter. The dogs were more than 80 percent accurate in detecting estrus in cows.

Canadian scientists at the University of Guelph have developed a technique for detecting estrus in cows with 95 percent accuracy. It is based on the use of a mercury gauge to record the frequency of movements of the cow's neck.

Reproductive Physiology Research and AI   Additional research is needed to improve present and develop new techniques of heat (estrous) detection, ovulation detection, breeding or insemination per se (e.g., where is the best place and when is the best time to deposit semen in the female reproductive tract?), early pregnancy detection, evaluating and predicting the fertility of males, perfecting fertility matching in animals (e.g., are certain bulls more effective in achieving conception in certain cows?), contact capsule insemination (e.g., could a female be inseminated at any time during the estrous cycle with a fertile time-released dose of semen?), fertility prediction (e.g., is it possible to identify, and thereby cull, animals with important reproductive problems, both males and females, at an early age?), sexing semen, freezing of embryos, and embryo transfer.[9]

10.15 SUMMARY

Artificial insemination was used in mating horses and dogs at least two centuries ago. However, its commercial application for mating farm animals was not made until improved communications, refrigeration, transportation, and especially advances in research in reproductive physiology paved the way.

AI has helped facilitate tremendous improvements in the genetic merit of farm animals during the past five decades. The use of AI enables the animal breeder to allow the best to multiply the fastest.

Largely because of the sizable number of females that can be mated to outstanding males, many livestock breeders use AI as a tool in improving animals. They thereby increase profits and provide consumers with more acceptable animal products and at reduced costs. In poultry the use of AI increases producer profits through improved fertility; in horses it offers a means of mating females to outstanding stallions that are often great distances away.

In the effort to provide more meat and milk for humans, the use of AI to spread superior genetic germ plasm holds great opportunities, especially in light of the fact that only about 4 percent of the cattle of developing countries (in comparison with more than 30 percent of the cattle of developed nations) are artificially inseminated.

STUDY QUESTIONS

1 Distinguish between AID and AIH.
2 Who first observed human spermatozoa? When?
3 What contributions to the progress of artificial insemination did Spallanzani make?
4 What percentage of dairy cows in the United States are inseminated artificially?
5 Name 10 species in which artificial insemination is now used.
6 Which state agricultural experiment stations pioneered research relating to artificial insemination?
7 What are some merits of artificial insemination of farm animals?
8 What are some limitations of artificial insemination of farm animals?
9 Name four methods of collecting semen, and briefly list the advantages and disadvantages of each.
10 What is the best indication of fertility in a male?
11 Why is the enumeration of spermatozoa important?

[9]Because of the potential use of embryo transfer, a prized 9-year-old Holstein cow sold in 1982 for $1,025,000.

12 Of what use is a morphological study of semen?
13 What is the basis of the dead-alive stain of spermatozoa?
14 How many fold can the average ejaculate of bovine semen be extended? Boar semen? Stallion and cock semens?
15 Describe a good semen extender (diluter).
16 Name several semen extenders. Which ones are preferred?
17 Why is semen color-coded for use in artificial insemination?
18 Can fresh stallion semen be stored for longer or shorter periods than bull semen?
19 What is the principal method of storing frozen semen?
20 Name three ways of storing frozen semen in the concentrated form.
21 What is lyophilized semen? Is the technique practical?
22 Do all beef breeds allow semen to be used indefinitely after a bull is dead?
23 Discuss the purposes of the organization Certified Semen Services, Inc.
24 What is the primary objective of artificial insemination in poultry? Is it the same in cattle?
25 Describe several situations that might cause a couple to consider AID.
26 Why is a legal agreement important in AID?
27 Can human semen be frozen successfully?
28 What is the relation between estrous synchronization and artificial insemination?
29 Discuss the relation between artificial insemination and genetic engineering.
30 What are the implications of sperm-typing for sex preselection?
31 Why is the detection of estrus in farm mammals important to the use and success of artificial insemination? Discuss two or more recently reported techniques of detecting estrus in cows.
32 What are some areas of reproductive physiology in which additional research is needed to more fully realize the potential of artificial insemination in food-producing animals?
33 What, in the opinion of the authors, does the future hold for artificial insemination of farm animals?

CHAPTER 11

PHYSIOLOGY OF LACTATION[1]

Thy breasts shall be as clusters of the vine.

The Song of Solomon 7:8

11.1 INTRODUCTION

During the past century, science has exposed many of the secrets related to the **physiology** of **lactation.** There are more than 10,000 species of animals with true mammary **glands** that perform activities vital to the preservation of **mammalian** life. In humans and the elephant, there are only two mammary glands, which are located in the **pectoral** region. There are four mammary glands in the **inguinal** (groin) area of the mare; however, they are served by only two teats, each containing two **ducts** (one per gland). Sheep and goats possess two inguinal mammary glands and as many teats. The cow has four glands, which terminate in four teats. **Multiparous animals** (the **sow, bitch,** and others) often have 10 to 18 mammary glands along the abdominal wall from the pectoral to the inguinal region. Many of the lower mammals of the **monotreme** group have mammary glands consisting of a large number of small cutaneous glands, without a nipple, which resemble enlarged sebaceous glands. The **neonate** obtains nourishment by licking the region of the maternal abdomen in which these glands are situated. Mammary glands of women and bitches have several ducts opening on the body surface; those of cows, goats, and sheep have a single opening. Teats of the sow and mare are traversed by two streak canals. A woman's breast consists of some

---

[1]The authors acknowledge with appreciation the contributions to this chapter of Dr. W. R. Gomes, Head, Department of Dairy Science, University of Illinois, Urbana.

20 distinct lobes. Each lobe is a distinct gland, with a **galactophorous,** or milk, duct opening by a separate **orifice** onto the summit of the **mammilla.** There are no cisterns in the woman, sow, or bitch; therefore milk passages communicate directly with the surface of the nipple. The capacity of their mammary glands is increased by a dilation of the milk passages (milk sinuses).

The highly developed mammary gland of the cow is the sole basis for the existence of the dairy industry and contributes immeasurably to the health and **nutrition** of humans. It is no wonder the **bovine udder** has appropriately been called the world's greatest food factory. Since lactating cows produce more than 90 percent of the world's milk supply, discussion in this chapter will be primarily related to the physiology of lactation in the bovine species.

> The cow is the foster mother of the human race. From the day of the ancient Hindoo to this time have the thoughts of men turned to this kindly and beneficient creature as one of the chief sustaining forces of human life.
>
> *William Dempster Hoard (1836–1918)*

11.2   MAMMARY GLAND DEFINED

Mammary glands are the major features that distinguish mammals from other kinds of animals. Mammals derived their name from the Latin word *mamma,* which means "breast." Since the mammary gland secretes outward, it is an *exocrine gland.* The mammary gland is also generally called a *mesocrine gland,* even though parts of the cell may be secreted with the milk.

11.3   ANATOMY AND ARCHITECTURE OF MAMMARY GLANDS

The four mammary glands of the cow[2] are collectively called the *udder* and, exclusive of milk, commonly weigh about 30 to 40 lb (varying from 10 to 120 lb). A high-producing cow may accumulate 60 to 90 lb of milk in her udder prior to milking. (Only about 80 percent is normally removed at milking.) A desirable udder should be capacious, possess a relatively level floor, and be strongly attached. A high-quality udder should contain a maximum amount of secretory (glandular), a minimum amount of connective, and essentially no fatty tissue. The connective tissue might be likened to the walls, floor, and ceiling of a large house. The secreting cells fill the rooms. After milking, the normal high-quality udder feels soft and pliable, with no apparent lumps or knots, which would be indicative of connective tissue resulting from injury or **disease.** However, it is impossible to predict accurately how productive an udder is by visual inspection or by its size.

The principal supporting structures of the udder are the **median** and **lateral** suspensory **ligaments** (Figure 11.2). The median suspensory ligaments contain elastic fibers that are stretched laterally and vertically as the udder fills with

---

[2]As with many biological phenomena in nature, there are exceptions in the number of mammary glands of cows (Figure 11.1).

FIGURE 11.1
Although almost all cows have only four functional mammary glands, the above photographs depict a cow with six functional glands.

milk; this stretching allows the squarely placed, perpendicularly hanging teats to protrude obliquely outward and downward. Probably less than 50 percent of the milk secreted can be stored in the natural storage areas of the udder, and the balance secreted is accommodated only through stretching of the udder. High milk production may cause these ligaments to become permanently lengthened and may result in a "breaking away" of the udder from the body, causing a pendulous udder. This condition may be partially alleviated by affixing udder supports (Figure 11.3). The udder is not directly connected with the abdominal cavity, except through the inguinal canal.

11.3.1 Internal Structure

The udder is divided into halves (right and left) by a heavy membrane. Front and rear quarters, which normally represent about 40 and 60 percent, respectively, of the secreting capacity, are separated. The milk from each quarter can be removed only from the teat of that quarter. An udder injected with different-colored stains and sectioned will show a clear division between fore and rear quarters.

At the end of each teat is the opening (streak canal, or meatus) through which milk is removed (Figure 11.2). It is largely the strength (tone) of the **sphincter** muscles surrounding the streak canal that prevents milk from flowing out and determines the ease (difficulty) with which milk may be withdrawn. Weak and incompetent sphincters that fail to exercise control of the closing of the streak canal contribute to a condition called **patency** and increase susceptibility to **mastitis.**

The teat widens into its cistern and in turn into the gland cistern (Figure 11.2). There are 10 to 20 large ducts **(galactophores)** branching into all parts of the

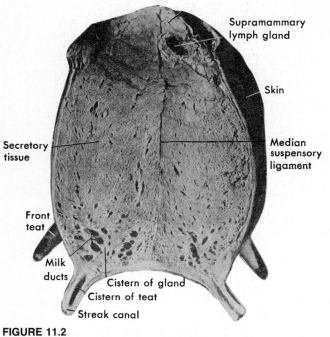

FIGURE 11.2
Cross section of the rear quarters of a cow's udder. (Missouri Agr.
Expt. Sta. Bull, *344, 1935.)*

gland to collect and convey milk downward. Each gland is divided into lobes separated by connective tissue membranes. The lobes are each drained by a single duct and further branch into lobules (sacs), which are drained by ducts (tubes). The lobules are composed of a large number of hollow spherical structures resembling bunches of grapes, called alveoli. Milk constituents are formed in the epithelial cells lining each alveolus and secreted into its **lumen** (Figures 11.4 and 11.5).

There are millions of "grapes," or alveoli, in high-quality, efficient mammary glands. An alveolus is very small (probably about 60,000 per cubic centimeter) and varies from 0.1 to 0.4 millimeter (mm) in diameter. Its epithelium varies from 0.007 to 0.010 mm in height, depending on its milk-secreting status. Completely surrounding an alveolus are groups of fibers (myoepithelial cells). These muscle fibers stretch out as milk is secreted, and when a cow is properly stimulated to let down her milk, they contract, forcing out the milk.

11.3.2 The Circulatory System

The blood supply to the udder during lactation is profuse. There are 300 to 500 volumes of blood passing through the udder for each volume of milk secreted.

FIGURE 11.3
Dairy cows on a farm near Copenhagen. Note the harness-like device
(cow brassiere) attached to minimize the stress given to the udder
attachments of the high-producing cow in the foreground.

Therefore the udder must possess an extensive **vascular** bed. Arterial blood
enters the base of each half of the udder through an external **pudic** artery (via
the inguinal canal). These branch into the **cranial** (supplies the forequarters) and
the **caudal** (supplies the rear quarters) arteries. Capillary blood collects in two
large veins, one at the base of each half of the udder. These are joined together
by a smaller vein in the rear. There are three primary routes for venous blood to
return to the heart for oxygenation. The first route is through the **subcutaneous**
abdominal vein (milk vein). The second route is the external pudic vein
(parallels the external pudic artery). The third possible route is the perineal vein
through the pelvic arch. Many have attempted to use the size and prominence of
the milk veins (superficial veining) as an index of milking ability; however, the
milk veins have been **ligated** experimentally without appreciably affecting milk
production.

Surrounding each alveolus is a network of tiny capillaries carrying blood to
the base of the epithelial cells that line the alveolus. The materials in blood from
which milk is made pass through the capillaries and are taken up by the
milk-making cells.

11.3.3 The Lymphatic System

The lymphatic system is composed of lymphatic vessels that carry **lymph** from
mammary tissues to the heart via the venous blood system. In the cow the flow is
upward and toward the rear of each half of the udder, where it passes through
the **supra**mammary lymph glands. The glands act as filters that remove or
destroy foreign substances (bacteria) that may have gained entrance into the
tissues. It is through the effective action of lymph glands pouring leukocytes
(white blood cells) into the blood vascular system that infections are kept

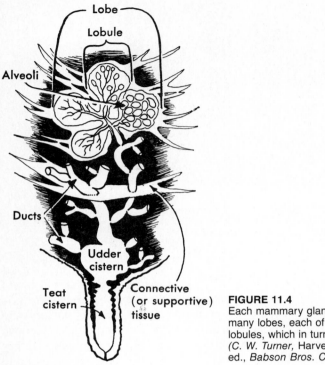

FIGURE 11.4
Each mammary gland in the cow consists of many lobes, each of which contains smaller lobules, which in turn contain many alveoli. *(C. W. Turner,* Harvesting Your Milk Crop, 3d ed., *Babson Bros. Co., Chicago, 1973).*

localized and invading microorganisms are destroyed. These lymph nodes (glands) and vessels are especially important in controlling **inflammation** at **parturition** and in removing sloughed or injured tissue. Congestion (swelling) in the udder at the time of parturition results from the accumulation of large quantities of lymph and is called **edema** (Figure 11.6 *a* and *b).*

The lymph capillaries surround the alveoli and absorb part of the **plasma** and products returning from the **intercellular** spaces into which they have flowed from the secreting cells.

Edema in the cow's udder tends to inhibit milk ejection. This is likely a built-in safeguard of nature that minimizes the intensity of milk secretion until the parathyroid gland can become fully functional in its regulation of blood calcium and phosphorus. This may help to prevent **parturient paresis** (Section 11.5.4) in high-producing animals.

11.4 GROWTH AND DEVELOPMENT OF MAMMARY GLANDS

Mammary glands of the bovine female begin developing in early **fetal** life. The teats are apparent at birth. The teat and gland cisterns are also present. As the

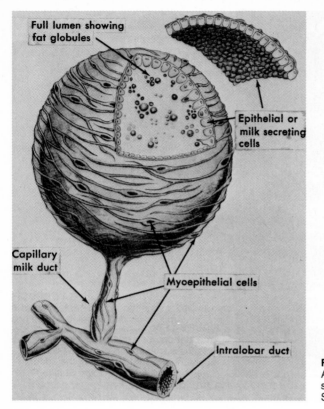

Full lumen showing fat globules

Epithelial or milk secreting cells

Capillary milk duct

Myoepithelial cells

Intralobar duct

FIGURE 11.5
An alveolus where milk is secreted. (Missouri Agr. Expt. Sta. Bull, *793, 1965.)*

young immature female grows, the udder increases in proportion to the increase in body size. Prior to **puberty,** there is little duct development or growth of glandular (secretory) tissue. When the **heifer** reaches sexual maturity, estrogen (produced by the follicles on the **ovaries**) stimulates development of the large duct system. With each recurring estrous cycle, the glandular tissue is stimulated to grow rapidly. After heifers have passed through a number of estrous cycles, the ducts show much branching into the udder (Figure 11.7). Heretofore it has been accepted that there is little or no lobule-alveolar growth prior to pregnancy. However, recent research indicates that there is some development of the lobule-alveolar system in nonpregnant females, since virgin heifers have been stimulated to secrete milk by the use of estradiol benzoate (0.3 mg per 100 lb body weight per day for 14 days). In retrospect, this observation is not so unexpected. When **ovulation** occurs, the follicle quickly develops into a corpus luteum and secretes progesterone, which causes development of the lobule-aveolar system. Endocrine control of the growth of the mammary gland is depicted in Figure 11.8.

believed to stimulate the secretion of prolactin, which in turn is discharged quickly into the blood following the stimulus of milking. It then passes via the blood to the udder to stimulate the epithelial cells to secrete milk during the interval between milkings.

11.5.2 Thyroxine

Lactation is influenced greatly by the hormone thyroxine, which is secreted by the **thyroid** gland. Thyroxine regulates the **metabolic** processes of the body and influences the general well-being of animals (Figure 11.10). It increases the appetite, heart rate, flow of blood to the udder, and rate of milk secretion. If the thyroxine secretion is low, a cow cannot secrete a large quantity of milk, regardless of the size of udder she possesses or what the other hormone secretion rates may be. Thyroidectomy of a cow causes a very marked decrease (up to 75 percent) in milk secretion, which can be restored by administering **desiccated** thyroid or thyroxine. Research workers have used sensitive radioimmunoassays to measure circulating levels of thyroid hormones. This technique gives the animal breeder another tool for selecting, mating, and **culling** animals. Research

FIGURE 11.10
Comparison of a Jersey heifer (left) thyroidectomized at 54 days of age with a control (right) when both were 40 months of age. Note retardation of growth and development. Thyroidectomy resulted in dry, scaly skin, short hair, little horn growth, and absence of estrus. *(Courtesy Missouri Agr. Expt. Station.)*

has revealed that thyroxine secretion rates show seasonal trends, with increases in the fall and winter and declines in the spring and summer. In Missouri, dairy cows secrete only one-third as much of this iodine-containing hormone in the summer as in the winter. This partially explains why milk secretion slows in hot weather.

11.5.3 Growth Hormone (Somatotropin)

The anterior pituitary gland secretes growth hormone. It largely regulates the growth rate in young animals and also influences milk secretion. This hormone appears to influence the substances from which milk is made by increasing the availability of blood amino acids, fats, and sugars for use by the mammary gland cells in milk synthesis.

Scientists at Cornell University reported recently that daily injections of bovine GH can increase milk production of dairy cows by as much as 15 percent over short periods of time. If this increase reflects improved efficiency of milk formation from blood-borne nutrients and not merely increased mobilization of body stores, future procedures for selecting cows, or changing cows through genetic engineering, may be based in part on GH levels.

11.5.4 Parathyroid Hormone

The hormone that regulates blood levels of calcium and phosphorus, which are abundant in milk, is parathyroid hormone. It is also related to parturient paresis (milk fever). On initiation of lactation, there is a rapid withdrawal of calcium and phosphorus from blood by the mammary glands. Animals with adequate parathyroid hormone secretion and good nutrition do not reveal symptoms of parturient paresis. Recent studies indicate that vitamin D (D_2 or D_3) will maintain normal lactation in parathyroidectomized animals. This result further documents the observation that feeding high levels of vitamin D (20 million IU 3 to 7 days prepartum) will significantly reduce the incidence of parturient paresis in cows. Stimulation of parathyroid hormone secretion by feeding a calcium-deficient diet during the 2 weeks prior to calving is also successful in preventing parturient paresis in cows.

11.5.5 Adrenals

Whereas small amounts of adrenal hormones are essential to milk production, the secretion or administration of high quantities may depress lactation. The administration of cortisone and deoxycorticosterone in adrenalectomized animals will maintain lactation. These hormones are secreted more profusely when animals are subjected to stress of various kinds. Above-normal adrenal hormone secretion rate is beneficial in overcoming **stress** conditions, at the expense of milk secretion.

11.5.6   Oxytocin

This hormone is secreted in the hypothalamic nuclei and stored in the **posterior** lobe of the pituitary gland. It is essential for the ejection of milk (Section 11.8). It also stimulates the uterine **musculature** to contraction, induces expulsion of the egg in the hen, and is used to induce active labor in women or to cause contraction of the uterus after delivery of the placenta.

11.5.7   Placental Lactogen

Studies in several parts of the world have led to the suggestion that developing fetuses—depending on the genes provided by their sires—can influence the amount of milk produced after calving by their dams. Physiologically this "sire of the fetus effect" on subsequent lactation is attributed to the production of a placental hormone by the developing young. In cattle, a bovine placental lactogen (bPL) has been identified. It has properties of both growth hormone and prolactin but is identical with neither. Since the source of bPL is lost when the placenta is expelled, its effects must occur prior to peak milk secretion in the subsequent lactation.

11.5.8   Insulin

Insulin, secreted by the pancreas, is involved in the movement of glucose into the cells of the body. Although an absolute requirement of insulin for lactation has not been demonstrated in vivo, the hormone is required to maintain mammary cells in tissue culture. These cells are not sensitive to insulin before conception or when the mammary gland is involuted, perhaps because of a lack of insulin receptors. During gestation and lactation, however, insulin promotes mammary cell growth and cell division. Although insulin induces mammary cell division, alone it is incapable of inducing synthesis of milk proteins by the daughter cell. Synthesis of milk proteins also requires hydrocortisone and prolactin.

11.6   HOW MILK IS MADE

One of the oldest theories advanced relating to the formation of milk was that the blood of menstruations suspended during pregnancy was transported to the mammary gland for the formation of milk. One of the best-known drawings of Leonardo da Vinci shows ducts connecting the mammary gland directly to the uterus of a woman.

Later, milk secretion was found to result from the combination of (1) filtration or transport of certain constituents from the bloodstream, (2) synthesis of other constituents by cellular metabolism, and (3) cell degeneration (leading to the concept that milk was liquid meat). Present concepts of lactation primarily involve a combination of the first two processes.

11.6.1 Techniques for Determining Milk Precursors

Arteriovenous Differences By collecting arterial and venous blood samples, it is possible to **assay** constituents of arterial blood supplying milk **precursors** to the mammary glands and constituents of venous blood as it leaves the udder. Then, by comparison of the difference, the quantity of blood constituents utilized by the milk-secreting cells can be determined.

If it is assumed that 400 volumes of blood pass through the mammary gland for each volume of milk secreted, an arteriovenous (AV) difference of 1 mg percent of a milk precursor passing in an unaltered form could account for 0.4 percent of this component in milk. Therefore small AV differences are important.

Perfusion of Excised Glands It is possible to remove **(excise)** the udder surgically and suspend it in a normal position in the laboratory. Blood can then be circulated through the glands **(perfusion)** by an artificial heart and purified by an artificial lung. This technique enables researchers to introduce substances believed to be essential to lactation into the system and to follow their uptake into milk. Bovine mammary glands can be kept functional for 6 to 8 h or longer using this technique.

Mammary Tissue Slices **In vitro** studies employing thin slices of fresh mammary tissue have demonstrated the utilization of selected nutrients for milk synthesis. The slices are incubated at body temperature in a medium rich in oxygen and potential milk precursors. Materials utilized for milk synthesis can be determined by analyzing the **substrate** before and after incubation. Recently the use of broken-cell preparations (homogenates) to study the biosynthesis of milk components has been employed.

Radioactive Isotopes A substance believed to be a milk precursor is labeled with a **radioactive isotope** and injected into an experimental animal. If the compound appears in the milk, it may be assumed to be a milk precursor (see Chapter 22).

Electron Microscopy The observation of milk precursors and milk constituents during their formation in the cell and of their secretion into the lumen of the alveolus is possible when cell images are magnified many thousandfold. Some constituents have been traced by labeling them with enzymes or antibodies.

11.6.2 Precursors of Milk

Protein Milk proteins appear to result partially from synthesis and partially from filtration. Since milk **casein,** lactalbumin, and lactoglobulin are not present in blood, they must be synthesized from blood precursors (amino acids). These proteins constitute about 94 percent of the protein nitrogen in cow's milk.

However, the immunoglobulins and **serum** albumin appear to be identical in blood and milk and therefore apparently diffuse or are transported into milk unchanged from the blood. It is of interest that normal milk contains only 0.05 to 0.11 percent immunoglobulins, whereas colostrum may contain 15 or more percent.

Using electron microscopic techniques, dairy scientists at the University of Illinois have demonstrated that immunoglobulins are selectively concentrated (at receptors?) between the mammary cells and transported into the alveoli, where colostrum is formed.

Lactose   The principal carbohydrate of milk is lactose, which consists of one molecule of glucose and one of galactose. Glucose is a normal blood component, whereas lactose is not. Determinations of arteriovenous differences have shown glucose to be taken up by the mammary tissues. (Arterial blood loses about 25 percent of its glucose content while passing through the lactating gland.) Incubated lactating mammary tissue slices have been shown to synthesize lactose from glucose. Moreover, glucose has been shown to be the principal precursor of lactose with experiments using **labeled** carbon **atoms.** It appears, then, that lactose is synthesized from glucose in the glandular cells of the mammary gland. Recent studies indicate that up to 20 percent of milk lactose is derived from short-chain fatty acids of the mammary gland.

Fat   Approximately 75 percent of milk fat is synthesized in the mammary gland. In the **ruminant,** acetate is the principal precursor of milk fatty acids of chain lengths up to and including the 16-carbon acid, palmitic. This explains why cows on high grain and low **forage** rations often secrete milk having a low fat content (this diet results in the reduced production of acetate in the rumen). Oleic, stearic, and higher fatty acids originate primarily from blood glycerides. Glucose is important in the synthesis of the glycerol portion of the fat molecule.

Human milk fat usually contains a higher concentration of unsaturated fatty acids than does cow's milk. Moreover, the fatty acid composition of human or cow's milk can be significantly altered without affecting either milk yield or total fat content. Milk fat closely resembles dietary fat during energy equilibrium, whereas on an energy-deficient diet, milk fat is nearer the composition of human or cow depot fat.

Minerals   The minerals of milk are derived from the blood and reach milk through filtration. Colostrum milk has a much higher mineral content than normal milk (Table 11.4, Section 11.11).

Vitamins   **Vitamins** enter milk unchanged from their form in blood. Those vitamins (B complex) synthesized by the **rumen flora** through fermentation processes are in rather constant amounts in milk, whereas the concentrations of fat-soluble vitamins, especially vitamins A and D, in milk are dependent on quantities in the ration and the body stores of the animal. Any variation in the concentration of B-complex vitamins in milk is determined largely by factors such as **breed** and stage of lactation rather than by diet.

Water The most prevalent milk component is water (87 percent by weight). It is filtered from blood. Milk and blood are **isotonic.** Both have 6.6 atmospheres (atm) of **osmotic pressure.** This means that there is no osmotic pressure developed on either side of the semipermeable membranes of the milk-secreting cells. Yet milk contains about 90, 13, 10, 9, and 5 times as much sugar, calcium, phosphorus, lipids, and potassium, but only one-half and one-seventh as much protein and sodium, respectively, as does blood plasma. This apparent paradox is possible because when many of these constituents are synthesized into milk components they are in a form that does not influence the osmotic pressure to any appreciable extent. For example, calcium is tied with casein to form calcium caseinate in milk. Nearly 75 percent of the osmotic pressure on the milk side results from the lactose (which is in true solution), whereas the minerals are in a nonionic state. On the blood side, the chlorides account for approximately 75 percent of the osmotic pressure.

It should be noted that when mammary tissue is injured or diseased, as with mastitis, the cell permeability is altered and certain blood constituents filter through into the milk in greater amounts, e.g., the chlorides and blood serum proteins. In addition, synthesis of certain constituents within the gland, e.g., lactose, is slowed.

11.7 HOW MILK IS DISCHARGED (SECRETED)

The concept of secretory activity within the epithelial cells is diagrammatically illustrated in Figure 11.12. As the cycle of secretion begins, the glandular cells are shallow and **cuboidal** in form. The secreting cells lengthen as they synthesize milk constituents, and fat begins to collect in the half of the cell lying next to the alveolar lumen. As the alveolar lumena fill with milk, they cause the mammary glands to become engorged, like a sponge. Essentially all the milk that can be obtained at the time of milking is present in the udder when milk evacuation begins. The secretion of milk is the most rapid immediately following milking. As milk secretion progresses and is discharged, there is a gradual rise in the **intramammary** pressure (Figure 11.11). This intramammary pressure slows the cycles of milk secretion and discharge. Milk secretion, each hour following milking, is in the magnitude of 90 to 95 percent of that of the preceding hour.

When intra-alveolar pressure reaches 30 to 40 mmHg (about that of capillary and about one-fourth that of systemic blood pressure), milk secretion is appreciably reduced, if not stopped entirely, and reabsorption begins. This is possible because the increasing milk pressure on the blood capillaries reduces blood flow through the udder and may also directly affect the synthesis of milk constituents in the mammary cell. Even in the absence of increased mammary

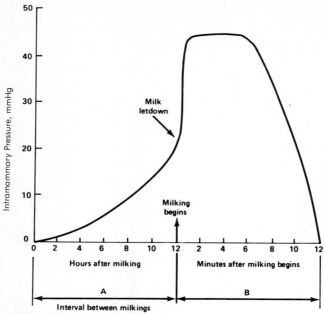

FIGURE 11.11
Changes in intramammary pressure in relation to milking times.

pressure, however, milk constituents may directly inhibit the secretory activity of the epithelial cells. For example, addition of small quantities of milk to in vitro cultures of rat mammary tissue significantly reduced synthesis of milk fat. When milk was injected daily into the body cavities of lactating mice, the level of milk production was reduced. Other data from both in vitro and in vivo studies support the hypothesis that the secretion of milk constituents, particularly milk fat, is inhibited by increasing levels of those constituents.

Research to date indicates that secretory vesicles (elaborated by the Golgi apparatus) are the major means by which milk serum constituents (milk minus the fat globules) are discharged from the cell. Specific areas of the cell synthesize casein and lactose and form vesicles containing these constituents surrounded by a membrane. After the secretory vesicles are formed and filled, they migrate to the apical region (closest to the lumen) of the cell, their membranes fuse with the plasma membrane, and their contents are discharged into the alveolar lumen (Figure 11.12).

Lipid droplets (fat globules) form near the endoplasmic reticulum. As they mature and migrate to the apical region of the secreting cell, they increase in size while they are progressively enveloped, or surrounded by the plasma membrane. Eventually the membrane-bound globule is "pinched off" into the alveolar lumen (storage area of the lumen where milk constituents are brought together). Formation of a fat globule requires approximately 5 h.

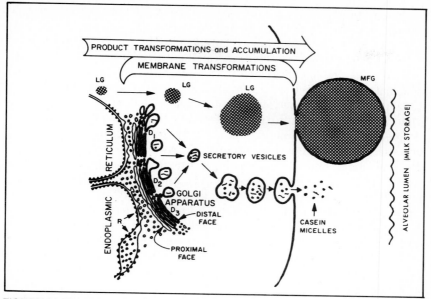

FIGURE 11.12
Diagrammatic summary of the structural and functional relations among the endoplasmic
reticulum (ER), Golgi apparatus, secretory vesicles, and plasma membrane in membrane
transformations and the secretion of lipid (fat) and protein (largely casein) of milk in the
lactating secretory (epithelial) cells of the alveolus. In the representation of Golgi
apparatus function diagrammed above, new membrane material is received from the
ER at the forming, or proximal, face and is transformed and utilized in the elaboration of
secretory vesicles at the distal, or maturing, face. Secretory proteins are contained in
these vesicles, and the caseins appear to undergo micelle formation or aggregate within
the secretory vesicle. The vesicles are also believed to contain lactose. Fusion
of secretory vesicles with the plasma membrane provides new plasma membrane to
replenish that lost from the envelopment of fat droplets during milk fat globule formation.
Lipid globules (LG) emerge covered by a layer of apical plasma membrane to form the
milk fat globules (MFG). Export milk proteins synthesized in the ER, shown above with
bound ribosomes (R), move to the proximal face of the Golgi apparatus. Secretory
vesicles bleb off the distal face of the Golgi cisternae and move to the apical surface,
where the vesicular membrane fuses with the apical plasma membrane. Contents of the
vesicles include casein micelles, other proteins, and most of the constituents of milk,
which are discharged into the alveolar lumen. D_1, D_2, and D_3 represent dictyosomes, or
Golgi bodies. (*Modified from T. W. Keenan et al., in B. L. Larson and V. R. Smith (eds.),
Lactation: A Comprehensive Treatise, vol. II, Academic, New York and London, 1974.*)

   Secretory vesicle membranes replenish that portion of the plasma membrane
given up to the fat globule, which indicates that the secretion of fat and that of
protein and lactose are closely related. Vesicles are abundant in mammary
secretory (epithelial) cells, and the amount of membrane they can contribute to
the cell surface appears to be more than adequate to replenish plasma
membrane lost to milk fat during the droplet secretion.
   Secretory vesicles are formed by the Golgi apparatus (Figure 11.12). The
vesicles arise along the distal (maturing or exit) face of the dictyosomes (Golgi

bodies) and are formed at the expense of cisternal membranes. (Cisternae are the flattened vesicles or tubules that make up the endoplasmic reticulum, ER.) They begin to fill with product while still attached to the cisternae. As cisternal membranes are lost through vesicle formation, new membrane material is added to dictyosomes as the steady state of secretion is maintained. This new membrane material appears to rise from the ER, where proteins and lipids of the membrane are synthesized. Small membrane vesicles bleb off (like blisters) from the ER and fuse to form the proximal or forming cisternae of the dictyosomes. Newly formed cisternae are gradually displaced toward the distal face through loss of cisternae resulting from vesicle formation. As the cisternae migrate across the stack, their membranes are gradually transformed from being ER-like to being vesicle- and plasma-membrane-like. The Golgi apparatus serves as the intermediate link between the ER and the plasma membrane.

Secretion of the three major components of milk (protein, fat, and lactose), are interrelated through membrane flow and differentiation. Thus, there is concomitant synthesis of both secretory and membrane protein in mammary glands, and milk fat secretion cannot proceed without concomitant synthesis and secretion of milk protein. Finally, lactose synthesis cannot occur without the synthesis of milk proteins.

Secretion of milk involves not only discharge of lactose, protein, and fat globules from the epithelial cell but also discharge of large quantities of cellular membranes. Although conceptions of milk formation and discharge are in a constant state of change, investigations utilizing high-pressure liquid chromatography, electron microscope autoradiography, and other sophisticated research tools and techniques of modern science have greatly expanded understanding of the correlation between the biochemical and cytological events associated with milk formation and discharge.

11.8   THE PHENOMENON OF MILK LETDOWN

The evacuation of milk from the udder involves more than pressure (as in hand milking) or vacuum (as in machine milking) to open the teat canal. Vacuum per se will cause the galactophores and larger ducts to collapse and prevent emptying of the fine ducts and alveoli.

Unless proper stimulation is provided, only a limited amount of milk may be removed at the onset of milking. This is because milk is held back in the gland by the small size of the capillary duct leading from each alveolus, the constrictions of each branch of the duct system, and the sag in the ducts at their points of suspension by connective tissue (due largely to the weight of milk). When a cow is properly stimulated, there is a sudden expulsion of milk from the alveoli, which causes a filling and distension of the large ducts and udder cisterns. This is commonly referred to as *letdown,* or milk expulsion. What phenomenon is involved?

Erection Theory   The theory was advanced early that erection, comparable to that occurring in the penis, was involved. Massaging of the teats, which are richly **innervated** and thus easily excited (much like the erectile tissue of the penis), was believed to be the stimulus that resulted in contraction of the smooth muscle cells. The veins would thereby be occluded, resulting in the udder becoming engorged with blood. The pressure created would force the milk from the alveoli and smaller ducts into the larger ducts and udder cisterns. However, subsequent studies have rendered this theory untenable.

The nipple of the breast in a woman is one of the most richly innervated structures in the body. Nerve fibers, both somatic and sympathetic, are found in profusion in and beneath the skin. The nipple contains smooth muscle that is arranged both circularly and vertically. The stiffening and erection of the nipple that results from mild stimulation is caused by the contraction of these strands of muscle. Nature likely provided nipple erection to facilitate grasping of the nipple by the baby during **suckling.** However, the shape of a woman's breast seems to be more for sexual excitement than for the convenience of feeding the infant. The nipple even when erected is too short for suckling, and the baby must get too close to the breast, which often makes breathing difficult. (Hence the long nipples used in artificial rearing.)

Neurohormonal Theory   Ott and Scott[6] injected extracts from the posterior pituitary gland into the ear veins of lactating goats and observed an increased milk flow through **cannulated** teats. Gaines[7] associated the nervous system with the expulsion phenomenon when he **anesthetized** a bitch and found that milk expulsion was inhibited. He then injected extracts from the posterior pituitary **intravenously** (IV) and observed milk letdown. These extracts were later purified, and the hormone involved was identified as oxytocin (pitocin).

The natural stimulus for milk letdown by a cow is the nursing, or suckling, act by the calf. However, massaging the udder and teats will cause the same reflex. Also, **auditory, olfactory,** and visual stimuli are often effective. Massaging the uterus and ovaries of the cow often causes the release of oxytocin. By definition, *oxytocin* means *rapid birth.* It was so named because it provokes vigorous uterine contractions and thus aids in expelling the fetus and placenta from the uterus. Obstetricians frequently use oxytocin to induce and assist labor. Uterine contractions often occur in women when their babies nurse. This effect has prompted some doctors to recommend that mothers nurse their babies, since it may speed the return of the uterus to normal size.

Milk may frequently be observed to flow freely from the teats of a cow or mare or the nipples of a lactating woman during **coitus.** This flow of milk results from the stimulus given the uterus that provokes the release of oxytocin. Nature made provisions for coitus during pregnancy in women by including an enzyme called oxytocinase (pitocinase) in their blood plasma that inactivates oxytocin. It

[6]I. Ott and J. C. Scott, *Proc. Soc. Exp. Biol. Med.,* **8**:48 (1910).
[7]W. L. Gaines, *Amer. J. Physiol.,* **38**:285 (1915).

first appears in blood about 3 weeks after fertilization of the ovum. From the fourth to the thirty-eighth week of pregnancy, there is more than a 1000-fold increase in plasma oxytocinase. Thus its concentration parallels placental growth, and it has been used to estimate the stage of pregnancy. Its activity disappears entirely very shortly after parturition. Evidence of the presence of this enzyme in other species has not been clearly demonstrated.

Contractile Tissue as the Basis for Milk Letdown On proper stimulus, oxytocin is discharged into the bloodstream and reaches the udder 30 to 40 s later. There it causes the myoepithelial cells surrounding the alveoli and ducts to contract (Figure 11.13). These smooth muscle cells in turn squeeze the alveoli, much like squeezing the bulb of a syringe, and expel the milk from the lumina down the ducts and into the cisterns of the glands and teats. Cistern milk pressure is almost doubled. (The pressure becomes the equivalent of 50 to 60 mmHg.)

FIGURE 11.13
The blood vessels and myoepithelial cells surrounding an alveolus. *(C. W. Turner,* The Mammary Gland, *Lucas Brothers Publishers, Columbia, Mo., 1952.)*

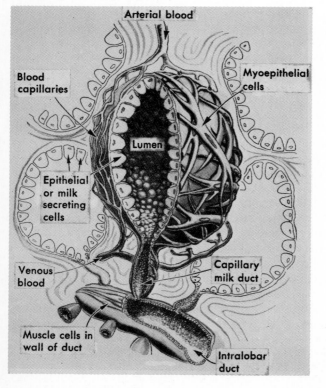

Oxytocin is usually effective in maintaining myoepithelial contraction for 8 to 12 min. The effect is thus **transitory.**[8] After this time, there is a decrease in the amount of milk obtained, a decrease in the rate of milk flow, and an increase in the percentage of fat in the residual milk. The *residual,* or complementary, *milk* is *that milk remaining in the mammary glands following the completion of milking,* and normally averages about 15 to 20 percent of the total yield and about one-fourth of the milk fat. It is possible to obtain residual milk by injecting 10 IU of oxytocin intravenously or by injecting 20 IU of oxytocin **subcutaneously.** The latter method of injecting the hormone is much less disturbing to the cow than the former method. The practice of removing residual milk is often followed by veterinarians prior to antibiotic therapy for mastitis. This practice permits maximum effectiveness of antibiotics and minimal dilution of them with milk. Residual milk increases in quantity with higher production and advancing age but decreases with advancing lactation.

In 1889, Babcock[9] observed that the last quarter milked yielded the least milk. This led to the recommendation of fast milking to obtain the maximum amount of milk. Does the transitory effect of oxytocin result from muscle fatigue of the myoepithelial cells or from **dissipation** or inactivation of oxytocin? Since administering additional oxytocin will result in the release of most residual milk, it does not follow that muscle fatigue can account for the transitory effectiveness of oxytocin. A portion of the residual milk may be accounted for by a gradual collapse in the duct system as milk is removed from the udder. A calf reduces the amount of residual milk in the collapsed duct system through a vigorous massaging during the suckling act.

Cows vary considerably in their rate of milk ejection. Peak flow (occurring between the first and second minutes of milking) may reach 12 to 15 lb/min in fast-milking cows and be as low as 2 to 6 lb/min in slow-milking ones. Research studies at Illinois, Missouri, and other experiment stations have shown this trait to be highly **heritable** (between 0.54 and 0.8). British scientists found that milking rate is largely controlled by the anatomy of the teat. Furthermore, they reported that the milking rate for individual cows remains rather constant throughout life.

Oxytocin Effects Neutralized A cow does not voluntarily withhold her milk from humans. In 1660, G. Markham offered the following advice: "The milkmaid whilst she is milking, shall do nothing rashly or suddenly about the cow which may afright or amaze her; but as she came gently, so with all gentleness she shall depart."[10] This statement acknowledged that fright, anger, noise, pain,

---

[8]There is sufficient oxytocin stored in the posterior pituitary to give 10 or more letdowns in succession. However, most cows will respond with only one letdown at a given milking. It is said that many cows in India will let down only a portion of their milk for humans at milking and thereby save milk for a second letdown for the suckling calf.

[9]S. M. Babcock, *Sixth Ann. Rept. Wisconsin Agr. Expt. Sta.,* 1889, pp. 42–62.

[10]G. A. Markham, *A Way to Get Wealth, Book II, The English Housewife,* E. Brewster and G. Sawbridge, London, 1660, pp. 143–144.

irritation, or embarrassment (in women) may block milk letdown. This inhibition results from the release of epinephrine (adrenalin) from the adrenal medulla. This hormone increases blood pressure, heart rate, and cardiac output. In the mammary gland, it causes the tiny arteries and capillaries to constrict and thus prevents oxytocin from reaching the myoepithelial cells in concentrations sufficient to cause contraction. Apparently, it is more potent on a per-unit basis than oxytocin and thereby quickly and effectively overrides the effects of the latter. Epinephrine may remain in the blood for a longer period of time than oxytocin. Hence a frightened cow should be allowed to stand for 20 to 30 min before one attempts to gain milk letdown.

The sow usually yields the most milk in the fore teats, and the amount usually progressively lessens from the front to the rear teats; hence a possible basis for the commonly held view that the **runt** of the **litter** nurses the rear teat. Could it be that less oxytocin reaches the rear teats? Is the quantity of oxytocin insufficient to attain maximum letdown in the total mass of mammary tissue? However, recent research at Cornell University indicates that there may be as many low-yielding mammary glands near the front as near the rear in sows (Figure 11.14).

It is possible for a cow to eject milk normally from three quarters and not from the fourth. One may observe this by placing a teat cannula in an injured

FIGURE 11.14
Milking machine for sows. This 12-teat-cup milking machine (left and right) was designed by Dr. D. A. Hartman during research at Cornell University on milk yields of sows, nursing frequencies and growth rates of piglets, and other pertinent parameters. Utilizing a movie camera–time clock apparatus, Hartman observed that piglets nursed on average at intervals of 43.5 min the first week, and that the interval gradually lengthened until the sixth week, when they nursed every 58.2 min. Piglets select a certain teat within a few days after birth and nurse it throughout the lactation. The larger, more vigorous pigs did a more complete job of extracting milk and thereby stimulated the mammary glands to higher milk production. Smaller, weaker piglets did not empty the glands as completely and consequently those glands did not produce as much milk, resulting in small pigs at weaning. Hartman noted that the runts (smallest pigs) do not always nurse the hind teats, a commonly held view. There is considerable variation in milk production among mammary glands, but low-producing glands are as likely to be near the front as the rear of the sow. Milk production per sow ranged from 398 to 723 lb of milk during the 6-week experimental periods. This variation explains in large part why certain litters are healthier and grow much more than others. *(Courtesy Dr. D. A. Hartman and Cornell University).*

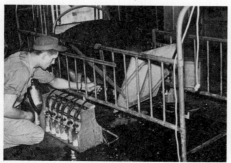

teat. If one then stimulates milk letdown, the cow can be milked from the normal quarters, and concurrently milk will flow out through the cannula. Then, when the injured teat is touched, an abrupt cessation of milk flow results for about 30 s, while the other three teats continue to give milk normally. Milk flow was interrupted by a nervous reflex contraction and not by the release of epinephrine.

11.9 REGRESSION (INVOLUTION) OF THE MAMMARY GLAND

Failure to evacuate milk from the mammary gland causes **involution,** even when adjacent glands are evacuated or suckled. An example is the **atrophy** of nonnursed glands of sows having small litters of pigs. Moreover, following peak production after parturition, there is a gradual reduction in mammary size and in the quantity of milk secreted, until the animal goes dry (nonlactating). Peak production in cows normally occurs about 6 to 12 weeks into the lactation and may reach 120 to 180 lb daily in very high-producing cows, whereas in women peak flow is from 1000 to 3000 cc daily at about the twenty-fifth week of lactation. Some women lactate for 15 or more years. Sows yield from 6 to 18 lb of milk daily, peak production occurring about 3 to 4 weeks postpartum. Illinois studies showed that the average nursing interval in sows was about 67 min; the average yield was 347 g/h. Piglets normally nurse about 30 ml from their mothers hourly. Research at Cornell University showed that the frequency of nursing decreased as the piglets became older (Figure 11.14).

The gradual reduction in milk secretion may result in part from a slow decline in the secretion rates of various hormones or a reduction in hormone receptors, but there is also a gradual loss of secreting cells into the milk. It is possible that one or more hormones causes a reduction in the rate of cellular division following maximum milk yield, resulting in slow multiplication of epithelial cells. Thus, after several months of lactation, the total number of secreting cells would be appreciably reduced. The slow loss of cells could continue until the end of the lactation period. Following the cessation of milking, there is a rapid shrinking of the mammary glands due to disappearance of the alveoli (desquamation). This process is called *involution.* In women degeneration, or involution, of mammary glands occurs rapidly after menopause. In the nonlactating gland, a few alveoli are to be found, since the gland then consists largely of ducts and connective and fatty tissues. It is interesting that women have nearly full maintenance of the lobule-alveolar system during the nonlactating state, whereas the cow has intermediate maintenance and the rat minimal maintenance of the lobule-alveolar system during the dry period.

When milk withdrawal is incomplete, the drying-off process is hastened. When alveoli are not emptied, secretion is further reduced, thus reducing the total amount of milk secreted between milkings. Three common methods employed to convert a lactator into a nonlactator are intermittent milking (e.g., alternate days), incomplete milking, and abrupt cessation of milking.

Cows with small udders likely require less hormone to stimulate maximum milk-secretion intensity. Their persistency is usually higher, because adequate amounts of prolactin are secreted. Animals are usually more persistent during first lactation than at maturity, at which time the mammary glands have fully developed with recurring pregnancies and have a higher ratio of prolactin to milk-secreting tissue.

11.10 FACTORS AFFECTING LACTATION

Several factors influence the intensity of lactation. The role of hormones was discussed previously (Section 11.5).

Inheritance   The ability to secrete milk is dependent on the genetic constitution of the animal. Control of the concentration of enzymes directing milk synthesis in the secretory cells resides in the pituitary and other endocrine glands, which are directly or indirectly related to the genetic makeup of an animal.

Secreting Tissue   The most fundamental factor limiting lactation is the amount of glandular tissue. The milk-producing potential is proportional to the amount of milk-secreting tissue. Small mammary glands are likely to be a disadvantage in lactation because of their inability to secrete and to store enough milk. British researchers found a significant correlation between mammary-gland weight and milk yield at peak lactation in goats.

Stage and Persistency of Lactation   There is considerable variation in the persistency of milk secretion following peak production in early lactation. Some cows are very persistent, and their rate of milk-secretion decline is slow (2 to 4 percent of their previous month's production). The production of other cows declines very rapidly (6 to 8 percent of their previous month's production), so that they show poor persistency. Figure 11.15 illustrates a second method of expressing persistency. The average percentage decrease in milk production per month is used to indicate lactation persistency. Cows of high persistency produce more milk in a lactation than do cows of low persistency when their maximum production is equal. Cows that milk out rapidly are usually more persistent. One cow at the Missouri Agricultural Experiment Station milked for about 8 years without calving, and a New Zealand cow milked continuously and without a pregnancy for 12 years.

Frequency of Milking   As milk accumulates in the lumen of the alveoli and storage areas of the udder, it tends gradually to inhibit milk secretion. The more frequent removal of milk permits maximum intensity of the milk-manufacturing process. The completeness of milk removal at milking is also important. Milk left in the udder at milking time tends to check the secretion of milk during the interval between milkings. This explains the recommendation of incomplete

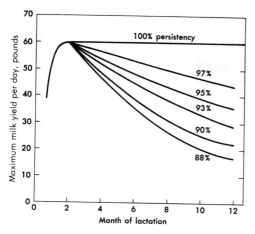

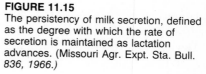

FIGURE 11.15
The persistency of milk secretion, defined as the degree with which the rate of secretion is maintained as lactation advances. (Missouri Agr. Expt. Sta. Bull. 836, 1966.)

milking for a few days postpartum; it lessens the requirement for calcium and allows the parathyroid gland to become active in producing the parathyroid hormone, which is believed to be related to prevention of parturient paresis (Section 11.5.4).

Pregnancy During the first 5 months of pregnancy, the decline in milk yield in pregnant cows is similar to the equivalent lactation period in nonpregnant cows. However, following the fifth month of pregnancy, cows begin to decline more rapidly in milk yield (decreased persistency). Missouri studies showed a 450-lb difference in milk yield allegedly due to the increasing divergence of nutrients from the mammary glands to the uterus for growth and maintenance of the fetus. Most scientists now believe it is progesterone that puts the brakes on milk secretion with advancing pregnancy.

Age It is believed that there is a slight additional growth of secreting cells during each pregnancy until cows reach about 7 years of age. This is manifested by the increase in yearly milk production to that age. After about 7 years, there usually is a gradual decline in total milk secreted during each subsequent lactation. Some cows remain productive to 20 or more years of age (the record is believed to be 27 years); however, most are retired from active production by the age of about 10 to 12. In women, milk output also decreases with increasing age.

Animal Size Larger cows normally secrete more milk, but they may not be any more efficient in converting food nutrients into milk than are smaller animals. Cows normally will not secrete more milk daily than the equivalent of 8 to 10 percent of their body weight, whereas goats may secrete enough milk daily to equal 20 or more percent of their body weight. Missouri studies have found milk production to vary with approximately the 0.7 power of body weight. Thus a 1 percent increase in body weight among animals tends to be associated with an increase of 0.7 percent in milk production.

Estrus   The activity of a cow when in **heat** generally reduces milk secretion; however, this is temporary. To minimize milk loss during **estrus,** cows should be confined.

Disease   Any one of many diseases may significantly reduce the amount of milk secreted. Disease may affect heart rate and, therefore, the rate of blood circulation through the mammary gland, which influences milk secretion.

Dry Period   Cows are normally bred 60 to 90 days after parturition. It is expected that they will lactate about 305 days and then be given a 60-day dry period before the next calving. The optimum length of dry period occurs when an extra day of dry period causes the loss in milk secreted in the previous lactation period to be balanced by a gain in milk secreted in the following lactation period. Of interest is one milk record in which 42,805 lb of milk was produced in 365 days without the benefit of a dry period. As breeders select dairy cows with greater persistency and learn to provide the most appropriate amount and balance of nutrients, the dry period will likely be shortened. There is no known physiological or endocrinological reason for believing that milk-secreting cells are benefited by a lengthy rest period.

Body Condition (Fatness) at Parturition   The storage of body fat during late lactation and the dry period and its subsequent use in the following lactation is an inefficient use of feed energy. It is perhaps only two-thirds as efficient as the direct use of dietary energy for the production of milk (energy). Additional body weight (body **condition**) at parturition generally results in additional milk secretion during the following lactation. Cornell workers found that 100 lb of body weight can yield about 400 lb of milk and 100 lb of body fat can yield about 880 lb of milk.

USDA workers have shown that it is possible for a high-producing cow to secrete more calories into her milk than she consumes in her feed. This is explained by a loss in body weight. They observed that while a Holstein cow was housed in an energy-metabolism laboratory, she produced 19,331 lb of **FCM** (4 percent fat-corrected milk) in 305 days. During the first 89 days of lactation, she secreted 9433 lb of milk, which required 15,000 more **Calories** (kilocalories) of estimated net energy per day than she received from her feed. Her weight loss during this period was 330 lb. This was calculated to account for nearly 3000 lb of milk that could not have been secreted from energy realized from her feed. During days 90 to 175, her body weight remained steady, and from day 176 to 305, she consumed substantially more feed than was needed for maintenance and milk production. Her **ADG** (average daily gain) was 4.1 lb during that period. Her ration was 80 percent **concentrates.**

Subsequent USDA studies in an energy metabolism laboratory demonstrated that high-producing dairy cows can utilize body fat for up to half of their energy needs during early lactation. However, the amount of protein reserves that can be utilized is not proportional to the amount of body fat. Therefore, the dietary energy/protein ratio should be considered when cows are withdrawing body reserves during intense lactation.

Environmental Temperatures Temperatures above 80°F for Holsteins and above 85°F for Jerseys reduce milk production. This may result partially from a decreased appetite and/or reduced thyroxine secretion. Optimum temperature is about 50 to 60°F, the **comfort zone** being 50 to 80°F for most lactating cows (see Chapter 13).

Feed The speed of synthesis and diffusion of various milk constituents is dependent on the concentration of milk precursors in blood, which reflects the quality and quantity of the food supply. Nature provides for maintenance, growth, and reproductive needs before energy is made available for lactation. Inadequate feed **nutrients** probably limit the secretion of milk more than any other single factor in the dairy cow.[11] Although good nutrition alone cannot guarantee high milk production, poor nutrition can prevent attainment of a cow's full potential just as surely as poor management, low genetic potential, or an unfavorable environment. The maintenance of lactation (galactopoiesis) is closely related to an adequate feed intake by the lactating animal. Inadequate nutrition limits milk secretion in cows to a greater degree than in women. For example, inadequately nourished, low-income mothers of India often yield as much milk daily as the average cow of the area (about 2 qt daily).

Preparation for Milking Of foremost significance in obtaining complete letdown of milk is proper preparation of the cow for milking. Improper preparation will result in incomplete milk evacuation, which increases residual milk and decreases the intensity and amount of milk secretion. Studies at Cornell University showed that premilking stimulation caused an earlier release of oxytocin from the neurohypophysis, which resulted in less milking time.

Stress Recently, more attention has been focused on the role of **stress** in the secretion of milk. As animals are selected to secrete higher levels of milk, stress will play an increasingly important role in lactation.

Suckling Stimulus Many cases have been reported in which women who have previously borne children have initiated lactation by allowing a baby to

---

[11]This is also true in sows. For example, recent research at the University of Illinois showed that milk production was significantly ($P < 0.05$) reduced in sows as dietary protein was reduced from 18 to 10 percent. There was, however, little change in the composition of milks among sows fed 10, 14, and 18 percent protein diets.

suckle their breasts. Such lactation occurs more commonly in young women but has been reported in postmenopausal women. The same phenomenon has been observed in virgin heifers suckled by other animals.

11.11 FACTORS AFFECTING THE COMPOSITION OF MILK

Milk may be defined as *the whole lacteal secretion obtained by the complete milking of one or more healthy lactating females.* According to G. Seiffert, "Mother's milk is a treasure impossible to replace." The term *milk,* unqualified, has come to be accepted as meaning cow's milk. The same constituents are present in the milks of all species of mammals; however, as shown in Table 11.1, there is considerable variation in the average milk composition of various mammals. Variations in the composition of milk may result from one or more of a number of causes.

Breed   As shown in Table 11.2, milk composition varies among breeds. The within-breed variation in milk composition is proportional to the percent of a given milk component for a given breed. Jerseys, therefore, because of their high fat test, have a greater variation in percent fat than Holsteins. As fat content changes by 1 percent, average changes in nonfat solids are Holsteins, 0.55 percent; Jerseys, 0.25 percent; and Guernseys, 0.36 percent.

Inherited Factors   Milk of some cow **families** contains greater quantities of certain milk components than others. Heritability estimates for milk yield and selected milk constituents are given in Table 11.3. Since the protein, fat, and nonfat solids (NFS) components are highly heritable (about 0.5), fast genetic progress in selective matings for these milk constituents might be expected. About one-half of the superiority of selected parents should be reflected in records of their **progeny.** A negative genetic correlation is given for the pounds

TABLE 11.1
AVERAGE COMPOSITION OF MILK FROM VARIOUS MAMMALS, PERCENT

Mammal	Milk fat	Protein	Lactose	Minerals	Total solids
Woman	3.7	1.6	7.0	0.2	12.5
Cow	4.0	3.3	5.0	0.7	13.0
Mare	1.3	2.2	5.9	0.4	9.8
Sow	5.3	4.9	5.3	0.9	16.4
Cat	3.3	9.1	4.9	0.6	17.9
Sheep	5.4	4.8	4.6	0.9	15.7
Goat	4.1	3.7	4.2	0.8	12.8
Elephant	15.2	4.9	3.4	0.8	24.3
Reindeer	18.7	11.1	2.7	1.2	33.7
Whale	22.2	12.0	1.8	1.7	38.1

TABLE 11.2
AVERAGE COMPOSITION OF MILK AMONG MAJOR DAIRY BREEDS, PERCENT

Breed	Fat	Protein	Lactose	Ash	NFS	Total solids
Ayrshire	4.0	3.5	4.7	0.68	8.9	12.9
Brown Swiss	4.0	3.6	5.0	0.73	9.3	13.3
Guernsey	5.0	3.9	4.9	0.74	9.5	14.5
Holstein	3.4	3.3	4.9	0.68	8.9	12.3
Jersey	5.4	3.9	4.9	0.71	9.5	14.9

Source: Missouri Agr. Expt. Sta. Bull. 365, 1936.

of protein, fat, and NFS, which means that as the total pounds of milk increases per lactation, the percentage of these components in milk decreases but the total pounds of these milk components increases. A high genetic correlation usually exists, therefore, between total milk yield and yield of any given component. It is thus evident that milk composition can be changed by selection.

Stage of Lactation   As shown in Table 11.4, colostrum milk is especially high in NFS because of the high protein content.[12] Following peak milk production, the protein and NFS content remains relatively stable until late lactation, when a gradual rise begins (see pregnancy effects, p. 364). Most cows' milk contains 0.5

[12]Researchers at Rutgers University found that the composition of colostrum of the sow can be affected by the prepartum protein intake. However, they concluded that sow's milk is not affected materially in composition by dietary factors, especially protein. They reported that weight gains of piglets nursed by sows fed a 5 percent protein diet were significantly less than those of piglets nursing sows fed 10 and 15 percent protein diets for the first 14 days. Weight gains among piglets nursing sows fed 10 or 15 percent protein rations were not significantly different.

TABLE 11.3
HERITABILITY ESTIMATES AND GENETIC CORRELATIONS OF MILK COMPONENTS

Milk yield and component	Heritability	Genetic correlations		
		Milk	Fat	NFS
Milk yield (ME)	0.32	. . . . . .	. . . . . .	. . . .
Fat, %	0.58	−0.20	. . . . . .	. . . .
NFS, %	0.59	−0.20	+0.50	. . . .
Protein, %	0.50	−0.20	+0.48	0.94
Lactose, %	0.53	. . . . . .	. . . . . .	. . . .

Source: Adapted from J. R. Campbell and R. T. Marshall, The Science of Providing Milk for Man, McGraw-Hill, New York, 1975, p. 127, and R. C. Laben, J. Dairy Sci., **46**:1293–1301 (1963).

TABLE 11.4
AVERAGE COMPOSITION OF COLOSTRUM
AND NORMAL MILK OF THE COW,
PERCENT

Component	Colostrum	Normal
Water	71.7	87.0
Milk fat	3.4	4.0
Casein	4.8	2.5
Globulin and albumin	15.8	0.8
Lactose	2.5	5.0
Minerals	1.8	0.7
Total solids	28.3	13.0

to 1.5 percent less fat during the first 2 months than during the last 2 months of lactation. The lactose content of milk is highest early in lactation and declines linearly during the balance of lactation.

Condition of the Cow at Parturition   Cows carrying considerable body fat at the time of parturition will secrete more fat and NFS into their milk than when in marginal body condition.

Season of the Year   The fat percentage of milk usually increases in the fall and decreases in the spring. Protein and NFS show less seasonal fluctuation than fat; however, both normally reach a high for the year in May and June and decrease in late summer. There appears to be an inverse relation between the protein and lactose contents of milk. Since there are no data available on milk composition for a full year when the feed supply is held constant, the apparent seasonal fluctuations may be reflecting changes in feed, environmental temperature, and/or other contributing factors. The unsaturated fatty acids are higher in fat produced in summer than in winter milk fat.

Completeness of Milking   The fat content of milk varies appreciably between the first and last milk evacuated, whereas the other constituents of milk are fairly constant at different stages of milking. The first streams of milk may contain only 1 to 1.5 percent fat, whereas the strippings may contain 6 to 8.5 percent. Where milking is incomplete or when cows are improperly stimulated prior to milking, the quantity of residual milk will increase and be richer in fat. The percentage of NFS will not be appreciably affected.

There is also a difference in fat content in the successive portions of milk evacuated from women and other species whose milk is capable of *creaming* (fat globules can be aggregated into more compact clusters). This property is not found in milk of the sow, goat, and water buffalo, in which the fat is quite evenly distributed throughout (much as in homogenized milk).

Interval between Milking Cows are normally milked at 12-h intervals. Uneven milking intervals affect the fat more than the protein or NFS percentages. With a longer interval, the milk will be lower in fat and lactose, it will be slightly higher in protein, and, as a result of the counterbalancing (fluctuations of protein, lactose, and minerals), the NFS content will remain stable.

Age of the Cow Protein, fat, and NFS decline with age. The NFS declines more than fat. In one study the decreases from the first to the ninth lactation were 0.08, 0.19, 0.25, and 0.34 percent for protein, fat, lactose, and NFS, respectively.

Feed Currently, there are no feeds, additives, or feeding practices that will profitably change the composition of milk. Most changes in milk composition resulting from feeding are temporary and limited in nature. Vegetable oils have been shown to cause a temporary increase in milk fat. Conversely, fish oils will depress the fat content of milk from 1 to 0.5 percent for as long as they are fed. Underfeeding reduces the protein and NFS content somewhat, but especially lowers the milk yield. Researchers at Michigan State University found that milk protein was higher among cows fed high-energy rations than among those fed low energy rations. High grain feeding coupled with very low **forage** feeding results in a significant reduction of milk fat in most cows.[13] The basis for this was explained previously (Section 11.6.2). Completely pelleted rations, or grinding of the forage, also reduces the fat content. NFS and protein are not appreciably affected.

One of the wonders of the cow is that her milk is nearly the same day after day whether her feed is good or poor. It is chiefly the *quantity* of her milk that varies with her feed, not the *quality*. A few exceptions have been noted, but most of these are minor. In addition, the vitamin A potency of milk reflects the body stores of a cow and the amount of carotene or vitamin A in the **ration.** While pasturing on fresh, green, lush grasses, the cow may secrete milk containing a lower fat percentage but as much as 2500 IU of vitamin A per quart, whereas when dry feeds predominate, the vitamin A may fall to 400 to 800 IU/qt. Vitamin D may vary in milk from 5 IU/qt in winter light to 50 IU in summer sunshine and 160 IU/qt when cows are fed 200 g of irradiated yeast daily. **Fluid milks** commonly have vitamins A and D added (see Chapter 3).

---

[13]Researchers at Michigan State University found that the fat-depressing effects of high-grain-restricted roughage rations can be alleviated by feeding sodium bicarbonate (baking soda), magnesium oxide, and/or calcium hydroxide, although the latter was less acceptable to cows. Feeding 1 lb of sodium bicarbonate or 0.5 lb of magnesium oxide per cow daily increased fat test from 2.7 to 3.1 percent in the Michigan studies. Research at South Dakota State University, Virginia, and other Agricultural Experiment Stations has shown that feeding concentrates containing 10 percent partially delactosed whey will maintain milk fat synthesis among cows fed fat-depressing rations. Studies in New York and Wisconsin showed that rations containing about 15 percent crude fiber (approximately 17 percent acid-detergent fiber) maintained normal fat percentages.

Excitement This may result in incomplete milk removal and thus a lower fat test. The NFS content is unchanged.

Disease Very little is known about the influence of certain diseases on protein and NFS. An increase in body temperature frequently will be accompanied by an increase in fat percentage and a decrease in milk yield and NFS content. Udder infections, such as mastitis, cause a decrease in milk fat, NFS, protein, and lactose and a notable increase in the mineral and chloride contents of milk. Mastitic milk frequently has a salty taste, which is attributed to a high concentration of chlorides. Scientists at Texas A&M University and others have found that mastitic milk contains more chloride, sodium, copper, iron, zinc, and magnesium, but less calcium, phosphorus, molybdenum, and potassium than normal milk.

Pregnancy The increase in NFS and protein late in lactation must be associated with pregnancy, since **open** (nonpregnant) cows do not show increases in these milk components with advancing lactation. Pregnancy has no apparent effect on the fat content of milk.

Environmental Temperature Temperatures above 70°F and below 30°F cause an increase in the fat content of milk, whereas protein and NFS quantities decline at higher temperatures and increase at lower temperatures.

Exercise Slight exercise apparently increases the fat content from 0.2 to 0.3 percent without reducing the quantity of milk secreted. Moderate to heavy exercise in high-producing cows, however, will result in reduced milk secretion and a corresponding increase in fat test. Exercise causes no apparent change in the NFS content of milk.

Drugs Most drugs have little or no effect on milk composition, although several will cause a decrease in the amount secreted. Many drugs and other compounds (e.g., **pesticides**), when consumed in the feed, may be secreted into the milk. Also, cows treated with **antibiotics** may shed them into milk. Generally, when antibiotic **therapy** is used in the treatment of mastitis, the milk should be discarded from cows so treated for at least 72 or 84 h following intramammary or intramuscular treatments, respectively.

11.12 IMMUNOLOGICAL ASPECTS OF COLOSTRUM

The manner in which developing animals achieve **immunologic** competence and the timing of this development have attracted much interest for theoretical, as well as practical, reasons. The transfer of immunity from mother to offspring differs among animal species. Young mammals depend on a **passive immunity** (acquired immunity produced by administration of preformed **antibodies**) for their resistance to infectious diseases. Mumps, measles, diphtheria, scarlet fever, and certain other infections of humans and animals rarely attack the very young. This suggests that the mother has an active **immunity** against the specific infective agent and that this immunity is transmitted to the offspring. **Neonatal** mammals are unable to produce antibodies within their own bodies for some time after birth, and they must acquire these antibodies from their mother either while in the uterus prepartum or through colostrum postpartum. The route of transfer varies among species (Table 11.5). Farm animals should receive colostrum during the first 12 to 24 h after birth if they are to acquire passive immunity. As shown in Table 11.4, colostrum is especially high in immunoglobulins. About 12 to 24 h postpartum, **gut** closure occurs, and the neonate digests these proteins, thus losing their immunization properties. This apparently results from the neonate being unable to absorb the large protein molecule (molecular weight about 900,000 compared with 6000 for the protein hormone insulin). Humans apparently do not benefit immunologically from colostrum milk, since they acquire passive immunity while **in utero.**

In women colostrum accumulates in the mammary glands in late pregnancy; true milk does not appear until 2 to 3 days after parturition. Milk secretion is then usually profuse and dramatic. Relief of tension by suckling is imperative, but if not allowed, lactation will cease in a few days. The act of suckling (or otherwise emptying the breasts of milk) is, of course, followed by further milk secretion. It has been demonstrated that the act of suckling is a specific stimulus to lactation, acting indirectly through hormones of the anterior pituitary gland.

Since newborn calves, foals, kids, lambs, and piglets have low innate resistance to disease, and since these animals do not acquire passive immunity while in utero, transfer of immunoglobulins via colostrum is of special significance. However, since intestinal permeability to immunoglobulins persists for only 12 to 24 h, it is imperative that newborn of these farm animals receive colostrum early if they are to acquire disease resistance. Should the mother die at parturition, the calf should receive colostrum from another cow or from a supply that has been frozen for such emergencies. Since the calf can absorb antibodies prepared in other animals, such as sheep or horses, colostrum from those sources may also be fed, and vice versa. However, certain diseases are species-specific, and colostrum from another species might not afford the desired protection. In research conducted at the Max Planck Research Institute,

TABLE 11.5
TIME OF TRANSFER OF PASSIVE IMMUNITY FROM
MOTHER TO OFFSPRING

Animal species	Prenatal	Postnatal
Horse, pig, ox, sheep, and goat	−	+ to 36 h
Dog and cat	+	+ to 10 days
Human	+	−
Rat and mouse	+	+
Rabbit and guinea pig	+	−

Mariensee, West Germany, and by Dr. J. G. Lecce at the North Carolina State University, it was discovered that piglets artificially raised and fed cow's milk or cow's colostrum outgained control animals that suckled sows naturally (Figure 11.16). This research finding could have practical significance in the future, since cows secrete milk with greater **feed efficiency** than do sows. Additionally, such a management practice could free sows to farrow three rather than the usual two litters annually.

Recent studies indicate that cows may be injected intramuscularly or infused intramammarily with an antigen (e.g., tetanus toxoid) to produce a specific antibody. The milk or blood from the treated cow can then be administered to laboratory animals, affording them protection against the disease for which the antibody was produced.

A substance foreign to the tissues of an animal stimulates the formation of a specific antibody. The foreign substance is called an *antigen* and reacts specifically in vivo or in vitro with the formed homologous antibody. This field of study presents a new challenge to researchers interested in preventive medicine (Sections 18.2.4 through 18.2.7).

11.13 SUMMARY

The presence of mammary glands and their ability to secrete milk distinguishes mammals from all other forms of animal life. It is fortunate, indeed, that nature

FIGURE 11.16
An automated device used to raise piglets from birth to 2 weeks of age at the Max Planck Research Institute, Mariensee, West Germany. Milk is warmed and fed mechanically at preset time intervals.

provided humans with many alternative means of obtaining the product of lactation. But especially important are the mobile, miniature milk-making factories that constitute the udders of cows and provide over 90 percent of the milk supply of humans.

Although percentages of milk constituents vary among species, the milk of all species contains the same components. Milk of each species was designed by nature to be especially beneficial in rearing the young of that species. Many factors influence the composition of milk and the intensity of its secretion. However, milk composition of a given animal is nearly the same day after day and throughout the world.

STUDY QUESTIONS

1 How many mammary glands does the mare have?
2 Give two anatomical characteristics in which mammary glands of cows and women differ.
3 Which species provide the most milk for humans?
4 Define a mammary gland.
5 What proportion of milk is commonly secreted in the fore and rear quarters of the cow?
6 Distinguish between the secretory and connective tissues of mammary glands.
7 Is there a possible relation between high milk production and a pendulous udder? Explain.
8 Can milk from the right forequarter be removed by milking the left forequarter? Why?
9 What determines how easy it is to remove milk from a cow?
10 What are galactophores?
11 Distinguish between epithelial and myoepithelial cells.
12 Where is milk secreted? What is an alveolus?
13 How much blood is circulated through the mammary gland for each volume of milk secreted?
14 Is the size of milk veins a good indicator of a cow's milk-secreting ability?
15 In what direction does lymph flow in the cow's udder?
16 What is edema? Is it totally undesirable at parturition? Why?
17 What are the roles of estrogen and progesterone in the growth and development of mammary glands?
18 How may one experimentally quantitate the degree of mammary gland growth?
19 What is "witch's milk"?
20 How is it possible to obtain milk without motherhood?
21 What is the role of prolactin in lactogenesis?
22 Do dairy cows secrete more or less thyroxine in winter than in summer? Relate this to the level of milk production in those two seasons.
23 How is growth hormone related to milk secretion?
24 Which vitamin is related to the parathyroid hormone?
25 Name three species in which oxytocin is important. What are its effects?
26 How is milk made? Discharged?
27 Give five techniques for determining milk precursors. Which appears the most practical? The least practical?

28 From where or from what are the following milk constituents derived: *(a)* protein, *(b)* lactose, *(c)* fat, *(d)* minerals, *(e)* vitamins, and *(f)* water?

29 How can milk and blood be isotonic when milk contains *more* sugar, calcium, phosphorus, lipids, and potassium and *less* protein and sodium than does blood plasma?

30 When is milk secretion the most rapid? Why?

31 Why are the strippings richer in milk fat than the first few streams of milk?

32 Can vacuum per se be used to fully evacuate milk from mammary glands? Why?

33 How did the erection theory explain milk letdown?

34 What contributions to the understanding of milk letdown were made by Ott and Scott?

35 What is oxytocinase? Is it present in all species?

36 What is meant by the transitory effect of oxytocin?

37 What is complementary (residual) milk?

38 To obtain the most milk, should one milk cows quickly or slowly? Why?

39 What is peak flow? Is this trait heritable?

40 What is the relation of epinephrine to oxytocin (consider milk letdown)?

41 What is involution? How may it be hastened?

42 Briefly discuss factors affecting lactation.

43 Is it possible for a cow to secrete more calories into her milk than she consumes in her ration? Explain.

44 What factors affect the composition of milk?

45 Is the percentage of milk protein, lactose, and fat of low or high heritability? Of what practical significance is this?

46 Why is it important for the calf to receive colostrum milk?

47 Does the neonatal human benefit immunologically from colostrum milk? Why?

48 Distinguish between in vivo and in vitro. (See Glossary.)

CHAPTER

12

PHYSIOLOGY OF
EGG LAYING[1]

Whereas the nourishment milk is produced for mammals in the breast, nature does this for birds in the egg.

Aristotle (384–322 B.C.)

12.1 INTRODUCTION

Most students think of an egg only as a kind of food produced by chickens. Actually, all birds lay eggs, but few other than chicken eggs are used for human consumption. Egg laying by birds results from a complex natural endowment whose prime aim is procreation. Many other kinds of living organisms, including insects, worms, fishes, reptiles, and **mammals** such as the cow and sow, produce eggs. In all levels of animal life (above the very lowest), it is impossible for young to be produced except from eggs.

> It has, I believe, been often remarked that a hen is only an egg's way of making another egg.
>
> *Samuel Butler (1612–1680)*

The foremost purpose of eggs in all **species** is to perpetuate life. Eggs, therefore, are an essential link in the reproductive cycle of animal life. An egg, like milk, is a secretory product of the reproductive system, and the **endocrine, metabolic,** and physiochemical mechanisms of egg laying are similar to those of

[1]The authors acknowledge with appreciation the contributions to this chapter of Professor Emeritus Q. B. Kinder, Dr. J. M. Vandepopuliere, and other members of the Poultry Science Department, University of Missouri-Columbia.

lactation (Chapter 11). The **bird's egg** is much larger than the mammal's egg because the bird's egg must contain food for the embryonic development of the young while it is growing outside its mother's body. Conversely, most young mammals develop embryonically within the mother's body and obtain food from their mother (after implantation) until they are ready to be born. Moreover, the **avian** embryo is more dependent on egg **nutriment** than is the mammalian infant on milk nutriment since the infant possesses many **nutrients** in the liver and other body tissues. The **cleidoic** (closed environment) arrangement of the egg demands that the egg nutriment be very concentrated. Since fat is more concentrated calorically (per unit bulk) than sugar, the egg is rich in fat rather than in sugar (as in milk).

12.2 EGG COLORS, SHAPES, AND KINDS

Eggs vary greatly in color, shape, and size. However, nature provided these several kinds of eggs with one common objective: to provide for the future form and existence of the species. (There are about 9000 species of birds.)

12.2.1 Egg Color

The wide diversity in the pigmentation of eggshells has interested ornithologists for a long time. Why do birds lay eggs that are white or colored, blotched, or speckled? Could eggshell pigmentation represent a natural **adaptation** for shielding the egg from sunlight? Are adaptive processes responsible for the laying of colored or spotted eggs that blend with their surroundings and thus are more nearly concealed from **predators**? Could colorful eggs attract the **brooding** bird to the nest or intensify the maternal instinct for the incubation period?

It is interesting that reptile eggs are almost always white and frequently thinshelled, with tough, flexible membranes, whereas those of the avian species are characterized by an almost infinite variety of tones and colors.

The eggs of wild birds have a wide range of colors, from the white of the flicker's egg to an almost solid-black egg laid by some ducks. There is a great variety of color patterns and markings of blues and browns among the eggs laid by other wild birds. Leghorn **hens** lay white-shelled eggs, although most **breeds** of chickens lay brown-shelled eggs. The Araucana hens of South America lay blue-shelled eggs. Contrary to the belief of some, there is no relation between eggshell color and the nutritional content contained therein.

12.2.2 Egg Shape

Fowl eggs differ substantially in shape. Most wild birds' eggs are similar in shape to those of **domesticated** hens. However, some vary in shape. The plover's egg is pear-shaped, the owl's egg is round, and the egg of the sand grouse is cylindrical.

Aristotle believed that the **rooster hatched** from the more pointed chicken eggs and the hen from the more nearly round ones. His theory was not accepted by those who quickly pointed out that owl eggs are almost spherical and yet both male and female young hatch from them.

Since albumen is added to the young yolk as it traverses the oviduct, why are eggs not all spherical? In general, the smaller the egg laid by a given hen, the more nearly round it is. This shape likely results from the ease with which a small egg passes through the oviduct. Large eggs are subjected to greater pressure exerted by oviductal muscles, so that they frequently have conical, oval, elliptical, or biconical shapes. Normal egg shape is determined in the magnum (Figure 12.3); however, the shape may be modified by abnormal or unusual conditions in the isthmus or uterus.

12.2.3 Bird Eggs

Usually, the larger the bird, the larger the egg. For example, the African ostrich, the male of which may be 8 ft tall and weigh 300 lb, lays an egg much larger than the adult female hummingbird, whose egg weighs about as much as a copper penny (Table 12.1). However, there are exceptions. The Indian Runner duck is smaller but lays a larger egg than the Plymouth Rock hen. The Coturnix quail, on a comparative body weight basis, produces an egg 2 times as large as the Leghorn hen used in commercial egg production.

The number of eggs that the different kinds of wild ducks lay varies greatly. The hornbill lays one egg a year. Pigeons usually lay two to four eggs a year, and gulls lay four. The graylag goose lays five to six eggs, the mallard duck 9 to 11, the ostrich 12 to 15, and the partridge 12 to 20 eggs annually. Domesticated hens and ducks may lay 350 or more eggs in 1 year.

12.2.4 Other Kinds of Eggs

There are many kinds of eggs. The earthworm lays eggs enclosed in capsules filled with a nourishing milky fluid. The female water flea lays summer eggs (carried in a pouch on her back) and winter eggs (enclosed in capsules). The eggs of moths, butterflies, and other insects are very small and of many forms and colors.

The oyster may produce 500 million eggs a year; however, most of them are eaten by many forms of water life. The enormous number of eggs laid annually by the oyster provoked the following statement by R. E. Coker: "The great object in the life of an oyster is to convert the whole world into oysters. The **biotic** potential of the oyster is limited more by the outside forces than by its own lack of biotic ambition." Fish also lay large numbers of eggs. The sturgeon lays some 7 million eggs annually. The common toad produces as many as 6000 eggs at a time in two long strings of jelly. Egg-laying snakes lay about 20 to 30 eggs, and lizards about 12 eggs annually.

Table 12.1
Proportional Parts of Selected Bird's Eggs

	Weight of egg, g	Incubation period, days	Proportional parts, %		
			Albumen	Yolk	Shell
Precocial birds*					
Ostrich	1400	42	53.4	32.5	14.1
Goose	200	28	52.5	35.1	12.4
Turkey	85	28	55.9	32.3	11.8
Duck	80	30	52.6	35.4	12.0
Chicken	58	21	55.8	31.9	12.3
Pheasant	32	24	53.1	36.3	10.6
Partridge	18	24	50.8	37.0	12.2
Average	268	.....	53.4	34.4	12.2
Altricial birds†					
Golden eagle	140.0	28–36	78.6	12.0	9.4
Dove	22.0	13–14	72.4	18.1	9.5
Pigeon	17.0	14–18	74.0	17.9	8.1
Starling	7.0	11–14	78.6	14.3	7.1
Robin	2.5	14	70.3	24.2	5.5
Wren	1.0	10	71.0	24.1	4.9
Hummingbird	0.5	16	69.7	25.3	5.0
Average	27.1	....	73.5	19.4	7.1

*Quite mature at birth. Note the high level of yolk in their eggs.
†Very undeveloped and immature at birth. Note the relatively low percentage of yolk in their eggs.

Sources: Complied from several sources, including J. R. Tarchanoff, *Arch. Ges. Physiol.,* **33:**303–378 (1884); F. Groebbels and F. Möbert, *J. Ornithol.,* **75:**376–384 (1927); and A. L. Romanoff and A. J. Romanoff, *The Avian Egg,* John Wiley & Sons, Inc., New York, 1949.

12.3 THE STRUCTURE OF AN EGG

When the contents of a fresh egg are exposed by careful removal of a sizable portion of the shell and shell membrane from the upper half, an **opaque,** circular white spot is commonly visible on the yolk's surface. In the unfertilized egg, this spot is called the *blastodisk,* and its homologous structure in the fertilized egg is called the *blastoderm.* The blastodisk is only 3 to 4 mm in diameter, and within it are the chromosomes. The most important part of an egg is the **nucleus** (germ). This part develops into the new animal. The other egg components provide food and protection for the young animal.

A bird's egg has five principal parts: (1) the shell, (2) the shell membranes, (3) the albumen (from the Latin *albus,* meaning *"white"*), called *egg white* because of its color after coagulation, (4) the yolk, and (5) the germinal disk (Figure 12.1).

The shell is composed of two main layers. These layers contain pores, so that water and gases can pass through the shell. It is also possible for bacteria to pass through the shell. As many as 8000 pores have been observed in an egg. The

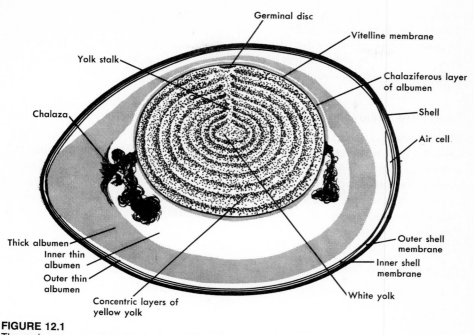

Germinal disc

Vitelline membrane

Yolk stalk

Chalaziferous layer
of albumen

Chalaza

Shell

Air cell

Thick albumen
Inner thin
albumen
Outer thin
albumen

Outer shell
membrane

Inner shell
membrane

Concentric layers of
yellow yolk

White yolk

FIGURE 12.1
The major component parts of a bird's egg. *(After F. B. Adamstone.)*

shell of a hen's egg is **translucent** as laid but becomes opaque as soon as it dries. A thin film called the *bloom* (cuticle) covers the outside of a fresh egg. The bloom soon dries and tends to seal the pores of the egg, thereby reducing the loss of water and gases and invasion by bacteria. During the shell-formation process, pores are formed in places where the eggshell is in contact with the uterine epithelium. The shell is more porous on the large end of the egg. During the 3-week period of hatching, about 15 percent of the egg's moisture evaporates. There is a direct linear relation between egg volume and shell thickness. The shell of a hummingbird egg is only 0.06 mm thick, whereas that of the hen egg is about 0.33 mm thick.[2] The eggshell must be thick enough to support its contents and yet fragile enough to crack easily when the young bird hatches.

Just inside the shell are two thin *shell membranes,* which surround the white (albumen) portion of an egg. The outer shell and inner shell membranes are

[2]Hens often lay proportionately more soft-shelled eggs during periods of elevated environmental temperatures. Breakage is a multimillion dollar problem of the poultry industry, and hot weather, particularly above 30°C (85°F), is a major cause.

Chickens cool themselves by panting, and this in turn changes the chemical composition of their blood. Carbonate is lost during panting, and as a result less calcium is available for deposition in eggshells.

Recent research at the University of Illinois showed that more calcium can be made available for eggshell production by providing carbonated water to laying hens in hot weather.

bonded together, except at the large end of the egg, where they separate to form the air cell. When laid, the egg contains no air cell until it cools and contracts, forming a space between the two membranes. The air cell increases in size with time (especially in a dry, warm place). Enlargement of the air cell results when water and gas escape from the egg.

12.3.1 The White (Albumen)

The albumen is an elastic, shock-absorbing, insulating, semisolid protein mass having a high water content. It has four principal parts: (1) a layer of outer thin white, (2) a layer of thick white, (3) a layer of inner thin white, and (4) a layer of thick white that surrounds the yolk. This innermost layer of white is twisted in a ropelike structure at each end of the yolk and forms the *chalaza*. The chalazas help anchor and stabilize the yolk near the geometric center of the egg but allow it to turn and twist (Figure 12.1).

12.3.2 The Yolk

The yolk is composed of a series of concentric, alternating dark and light layers and is contained in a thin yolk sac, called a *vitelline membrane*. The specific gravity of the yolk is less than that of the white, and therefore it stays slightly above the center in a freshly laid egg. The *germ cell* (germ spot) is a tiny area on the upper surface of the yolk. It is lighter in color than the rest of the yolk and is thus easily identified. The germ cell (blastoderm) in a **fertile** egg develops into the embryo under proper environmental conditions. A two-yolk (double-yolked) egg may be fertile but seldom hatches. Double-yolked eggs probably result from the almost simultaneous **ovulation** of two yolks. Infertile eggs are preferred for table consumption. Fertile ones should be held at low temperatures ($<10°C$) to prevent embryonic development.

Nature provided the egg to maintain continuity of life. Therefore its contents give a well-balanced diet to the developing embryo. (Development of the embryo takes a period of less than 2 weeks for the wren but nearly 8 weeks for the emu.) The shell is largely (93 to 98 percent) calcium carbonate ($CaCO_3$) and provides the embryo with calcium through diffusion for bone formation and other body-building purposes. The shell also contains a small amount of protein and other minerals.

When the eggs of birds are grouped according to the relative amounts of yolk and albumen, there are two broad classes. Those in which the yolk constitutes about 35 percent of the total weight belong to *precocial* birds; eggs in which the yolk constitutes a lower percentage (about 20 percent) of the total weight belong to the *altricial* group (Table 12.1). Since the yolk has the greatest food value, a relatively large yolk ensures a more advanced stage of development in the young at birth (hatching), which characterizes precocial birds. In the small-yolked eggs of altricial birds, the young are helpless nestlings. Moreover, most altricial birds lay eggs that have relatively thin shells. In proportion to total egg weight, the eggshells of precocial birds are considerably heavier.

12.4 REPRODUCTION AND EGG FORMATION

The formation of eggs and their subsequent utility is an essential link in the reproductive cycle. In many species, **fertilization** is not an essential preliminary to egg laying. The female pigeon, for example, may form and lay eggs without mating or may even lay eggs in the total absence of the male. Likewise, the domestic hen can lay eggs continuously without being mated or without being stimulated to lay by the presence of a male. This biological phenomenon has been utilized advantageously by humans in producing infertile eggs for human consumption. However, in nature, eggs must be fertilized to perpetuate life within each species. Hence the need for the rooster (male fowl) remains.

12.4.1 Anatomy and Architecture of the Avian Male Reproductive System

The avian male possesses two testes (**gonads**). They are normally yellow or gray and are situated high in the abdominal cavity near the **anterior** ends of the kidneys (Figure 12.2). Unlike those of the human male and other higher animals, the gonads of male birds do not descend into an external scrotum. The testis consists of a large number of slender convoluted **ducts,** from which the lining gives off the male reproductive cells (**spermatozoa**). These narrow ducts (seminiferous tubules) lead to the paired vasa deferentia which are tubes that convey the spermatozoa and seminal fluid outside the body.

Each vas deferens terminates near the small papilla, which together serve as an intromittent organ. This rudimentary **copulatory** organ is used to classify baby chicks according to sex on the basis of cloacal examination. Additionally, geneticists have introduced the fast-feathering sex-linked characteristic in many heavy meat strains. In chickens with this characteristic, the primary wing feathers grow faster than the covert feathers in the female. In the male, the covert feathers are the same length or longer than the primary feathers.

During the process of mating, the spermatozoa are introduced by the papillae into the oviduct opening in the cloacal wall of the female.

12.4.2 Development of the Bird's Egg

The reproductive organs of the hen normally consist of a *left* ovary and a *left* oviduct. The right ovary and oviduct are present early in life but later regress (**atrophy**).

Egg development begins with the formation of the germ and yolk in the female **ovary** (from the Latin *ovarium,* meaning *"egg holder").* The fowl ovary contains a large number (3000 or more) of spherical bodies (Graafian follicles) and resembles a cluster of grapes.

Each reproductive cell is called an *ovum* (from the Latin *ovum,* meaning *"egg").* It is contained in a thin envelope of the ovary called the *follicle.* Concurrent with egg development is the deposition of yellow and white yolk granules. When the yolk is fully formed, the follicle that houses it ruptures along a streak (a nonvascular suture line) called the *stigma.* The yolk is freed from the

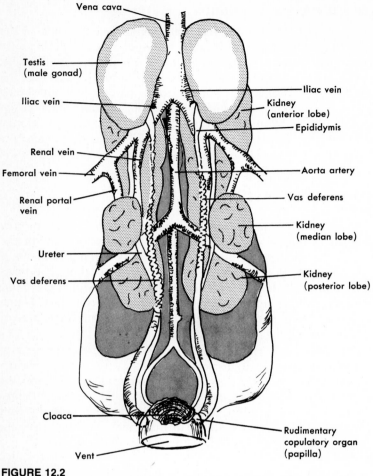

FIGURE 12.2
The avian male reproductive system. *(After L. V. Doom, University of Chicago.)*

ovary and, with its germinal disk, escapes into the infundibulum (funnel) of the oviduct.

Fertilization occurs in the infundibulum, and within half an hour of the egg's arrival it departs for a journey of about 25 h in assembly-line fashion. The yolk is grasped by the mouth of the oviduct (infundibulum) and travels slowly through the magnum (about 3 h), isthmus (about 1¼ h), uterus (about 20 h), and vagina (about ¼ to ½ h).

As the yolk traverses the magnum, most of the albumen is secreted and deposited around it. It is interesting that most of the proteinaceous albumen is deposited in only 3 h. The albumen is filtered from the bloodstream into the glands of the magnum. The magnum is incapable of distinguishing between an egg yolk and a foreign body. Thus it will deposit albumen around a marble, a cork, or even a round ball of paper.

The shell is formed in the lower region of the oviduct (uterus).[3] In the domestic hen, about 24 to 26 h elapse from the time the yolk enters the oviduct until the completed egg is laid (Figure 12.3). It should be noted that there is *no cervix* in the reproductive system of birds. The lower end of the oviduct opens into the cloaca.

The ancestors of domestic hens normally laid eggs during the spring months only. Through scientific selection and mating systems (Chapter 5) and modern, improved poultry management, domestic hens now lay eggs throughout the year. Under natural conditions egg laying is most intense during the spring months. Nearly all wild birds lay in the spring. Birds respond to the gradual increase in the amount of daylight during the spring months and thus exhibit **photoperiodism.**

The bird's eye is sensitive to light intensity. As the light increases, it causes an increase in the activity of the pituitary gland and its hormonal secretion. The higher **hormone** levels speed ova development and subsequent egg laying. The spectral quality of light is important. Ultraviolet light apparently has no effect on sexual maturity. Some species of birds are more responsive to red than to green light.

Sex Reversal   As noted previously, it is the left ovary and oviduct that are normally functional in the hen. However, after a **sinistral** ovariectomy (removal of the left ovary), the right gonad develops and grows into an organ resembling a testis. If an ovariectomy is performed on young chicks, a few will later produce sperm and male hormones, as evidenced by the development of secondary male sex characteristics (large **comb,** male plumage, and crowing).

This phenomenon suggests that both male and female reproductive organs were originally present in poultry. Both continue to be present in some lower forms of animal life.

12.4.3 How Avian Embryos Develop

Eggs are fertilized by the male within the female body in most insects, mammals, snakes, and turtles. Frogs and toads also mate, but the eggs are fertilized after they are deposited. Frog eggs will develop into tadpoles if the surface of each egg is punctured with a sharp needle. Unfertilized sea urchin eggs will develop when treated with certain chemicals. Few fish mate; rather, the male deposits his reproductive cells near the laid eggs.

In contrast to most other mammals, whose embryonic development occurs within the female, the duckbill platypus lays an egg with a large yolk and shell that develops embryonically outside the female body.

---

[3]Recent research at The Pennsylvania State University indicates that calcite crystal (calcium carbonate) seeding begins in the isthmus. Eggshell ash is 98 percent calcite.

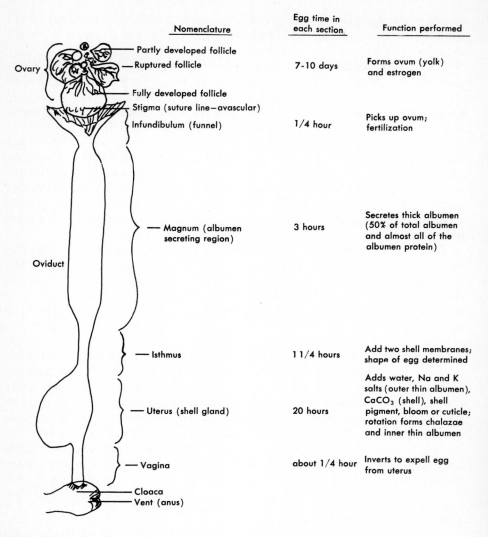

Nomenclature	Egg time in each section	Function performed
Ovary		
Partly developed follicle		
Ruptured follicle	7-10 days	Forms ovum (yolk) and estrogen
Fully developed follicle		
Stigma (suture line−avascular)		
Infundibulum (funnel)	1/4 hour	Picks up ovum; fertilization
Oviduct		
Magnum (albumen secreting region)	3 hours	Secretes thick albumen (50% of total albumen and almost all of the albumen protein)
Isthmus	1 1/4 hours	Add two shell membranes; shape of egg determined
Uterus (shell gland)	20 hours	Adds water, Na and K salts (outer thin albumen), $CaCO_3$ (shell), shell pigment, bloom or cuticle; rotation forms chalazae and inner thin albumen
Vagina	about 1/4 hour	Inverts to expel egg from uterus
Cloaca		
Vent (anus)		

The time from oviposition to ovulation = about 1/2 hour

The time from ovulation to oviposition = about 25 hours

FIGURE 12.3
Reproductive system and egg formation in the hen.

Mating habits among birds vary from monogamy to promiscuity. Pigeons are monogamous; the male and female share incubation, brooding, and feeding the young; both secrete crop milk. Wild geese pair up and form matings that may last for life, although the male is less faithful and generally accepts several mates. Other species of birds are more or less promiscuous in mating. Chickens and turkeys and **drakes** usually mate with 10 to 20 or more females. Chickens mate at any hour of the daylight period but most frequently immediately following laying (when the uterus is empty).

The fertilization of bird eggs occurs in the upper portion of the oviduct. Normally, only one male reproductive cell fertilizes each female **gamete** to form the basis for the development of the embryo. For good fertility in poultry, the ratio of males to females (natural matings) should be about 1:15 and 1:10 in the light and heavy breeds of chickens, respectively. Turkeys have been selected for broad breasts and large size. Natural mating is ineffective in today's commercial turkeys, which necessitates an artificial insemination program. Fertility is excellent for 7 to 10 days after AI and may continue for 50 or more days. It is interesting that fertility in the honeybee may extend for a year or more beyond mating. In lizards and snakes, fertility may extend 3 to 5 months after breeding, and in the diamondback terrapin it has been reported to last as long as 4 years.

Eggs of the domestic hen must be kept at the proper constant temperature (37 to 38°C) for 21 days before chicks will hatch. Turkey, guinea, and most duck eggs incubate for 28 to 30 days. (Muscovy duck eggs require 35 days.) Goose eggs incubate about 28 to 32 days, quail eggs 22 to 24 days, and pheasant eggs 23 to 28 days, depending on the variety. Eggs may be incubated under hens (natural incubation) or in incubators (artificial incubation).

12.5 HORMONAL REGULATION OF EGG LAYING

Both the physical appearance and the functioning of birds are profoundly affected by hormones. (The various endocrine glands and their secretions were discussed in Chapter 7.) Few endocrine effects result from the direct action of a single hormone. Instead, the physiological activity of the chicken, particularly the female, is dependent on a complex interrelation of glandular effects, as exemplified in the complex hormonal control of ovulation and egg formation.

The interrelations of these various hormones are shown diagrammatically in Figures 12.4 and 12.5. The entire process of egg laying is dependent on hormone synchronization and balance. If an individual endocrine gland releases a hormone without awaiting the proper signal, abnormalities such as yolkless eggs, soft-shell eggs, and eggs within eggs are likely to result.

12.5.1 FSH and LH

Follicle-stimulating hormone (FSH), from the anterior lobe of the pituitary gland, stimulates growth of ovarian follicles with their contained ova. When a follicle becomes mature, another anterior pituitary hormone (luteinizing hormone, or LH) is released and causes ovulation. Functional ovarian activity in the mature hen can be inhibited by starvation. This effect is attributed to a failure of the pituitary gland either to produce or to release gonadotropins, since the administration of these substances will cause follicular growth and ovulation.

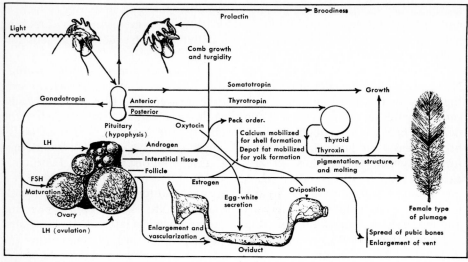

FIGURE 12.4
Diagram showing the principal effects of endocrine secretions and their interrelations in the female fowl. *(Adapted from L. E. Card and M. C. Nesheim,* Poultry Production, *11th ed., Lea & Febiger, Philadelphia, 1972.)*

The oviduct is also under hormonal control and is stimulated at the most appropriate time to receive the released ovum. Ovarian follicle secretions are responsible for (1) enlargement of the oviduct to functioning size, (2) spread of the **pubic** bones and enlargement of the vent, (3) mobilization of depot fat for yolk formation and of calcium for shell formation, and (4) female plumage. The secretion of albumen is apparently under the control of a hormone (androgen) secreted by the ovarian interstitial tissue.

12.5.2 Parathyroid Hormone

Formation of the eggshell is partially controlled by hormones secreted by the parathyroid glands. As discussed in Chapter 7, these hormones maintain normal calcium and phosphorus levels in the blood.

12.5.3 Thyroid Hormone

In addition to influencing body growth and feather color and formation, the **thyroid** gland is partially responsible for seasonal changes in egg laying, body weight, and egg weight. The latter effect presumably results from the varying amount of light to which hens are subjected with the changing seasons. The thyroid gland increases in size and hormone secretion during that portion of the year when **molting** normally occurs. The feeding of large doses of either fresh or

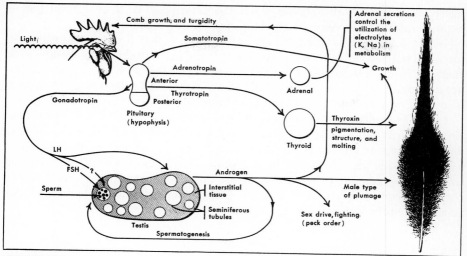

FIGURE 12.5
Diagram showing the principal effects of endocrine secretions and their interrelations in the male fowl. *(Adapted from L. E. Card and M. C. Nesheim,* Poultry Production, *11th ed., Lea & Febiger, Philadelphia, 1972.)*

desiccated thyroid results in rapid molting. Thyroxine also regulates body metabolism (Chapter 7).

12.5.4 Oxytocin

This hormone is secreted in the hypothalamus but is stored in and released from the posterior pituitary gland. It causes **oviposition,** or the contraction of muscles of the shell gland (uterus), and hence expulsion of the fully formed egg.

12.5.5 Prolactin

This hormone is responsible for the manifestation of the maternal instinct. Prolactin is believed to give a hen the urge to seek a nest and lay eggs in it. If prolactin is administered to a rooster in sufficiently high doses, he will become broody and set on eggs as a normal broody hen does. If discontinued, the effect soon disappears, and such roosters are likely to kill the **chicks** that they previously cared for and defended.

12.5.6 Sex Hormones

The sex hormones affect the social status of birds. The male sex hormone (androgen, or testosterone) is secreted by the testis and is responsible for the red, waxy comb and wattles. In the normal laying hen, the female hormone (estrogen) controls the secondary sex characteristics, such as normal female plumage, absence of spurs on the legs, and female sexual behavior.

Occasionally, a hen develops an ovarian tumor that destroys the estrogen-secreting portion of her ovary. In such cases, the hen will gradually assume male characteristics. Her comb and wattles will become large and coarse, after a molt her new plumage will closely resemble that of a male, and she may even crow. It is interesting that chickens are the only·birds that have combs.

Chickens (unlike farm mammals) *never exhibit estrus* and are *never pregnant.* Because birds produce no corpora lutea following ovulation, they can ovulate daily; in farm mammals, corpora lutea are present and secrete progesterone, which prepares the uterus for the reception and development of the fertilized ovum by a glandular **proliferation** of the endometrium. The secretion of the corpus luteum (progesterone) also prevents further ovulations when pregnancy occurs (Chapter 9).

12.5.7 Adrenals

Secretions of the adrenal glands are involved in the metabolism of **carbohydrates,** sodium, and potassium and in the regulation of blood pressure.

12.6 HOW AN EGG IS LAID (OVIPOSITION)

Eggs are usually formed small end **caudal,** i.e., with the small (sharp) end first, as they move in assembly-line fashion down the oviduct. This is true of wild birds and of domestic fowls. However, if the hen remains quiet during the act of laying, most eggs are presented large (blunt) end first. What explains this apparent paradox?

Just prior to being laid, the fully formed egg is turned horizontally (not end over end) 180°. For this to be accomplished, the egg must drop from its normal position high between the ischia (analogous to the hipbones) to a point opposite the tips of the pubic bones. This preliminary voyage is necessary because the normal egg is too long to turn in a horizontal plane within the pelvic arch. After the egg has been thus turned, it is in position ready to be laid. The hen accomplishes this egg-turning feat in only 1 to 2 min. Should the hen be physically disturbed as she raises her body slightly when the egg is about to turn, she is likely to expel it prematurely, and in that event it will be presented small end first. Occasionally, the egg is retained and goes back up the oviduct (reverse peristalsis), in which case the hen puts on a second shell. If it reverses to the magnum, the hen will add a second white and shell (an egg within an egg), as shown in Figure 12.6.

The act of laying an egg is made possible through the influence of the release of the hormone oxytocin from the **posterior** pituitary gland. Oxytocin causes the uterine **musculature** to contract vigorously and expel the egg. It seems reasonable to conclude that the uterine muscle pressure required for expelling an egg is

FIGURE 12.6
An egg within an egg. *(Courtesy* Columbia Missourian.)

more effectively applied to the small end; hence the basis for the reversal of the egg just prior to laying. The uterus expands to about 3 times its normal size during passage of an egg. The uterine wall everts **(prolapses)** through the vagina and cloaca, allowing the egg to drop from the cloacal **orifice.** The egg does not physically contact the wall of either the vagina or the cloaca. The vaginal musculature contracts voluntarily about the prolapsed uterus, thereby assisting in oviposition.

Both egg laying and ovulation are subject to the external influence of light and darkness. Normally, about 30 min after a fully developed egg is laid, a new egg is ovulated from the ovary. It is at this time that a hen cackles, not when she lays the *hard-shelled* egg. This delay from the time of oviposition to ovulation allows the bird time to get away from the nest before she calls for a mate.

By counting the number of ruptured follicles on a bird's ovary, one can estimate the number of eggs that the bird has laid over a period of several months. Reabsorption of the ruptured follicles, however, precludes an accurate determination of the total number of eggs laid throughout the bird's life.

12.7 FACTORS AFFECTING EGG LAYING

The most basic factor limiting the number of eggs a hen lays is her genetic makeup, which greatly influences her physiological efficiency and metabolic activity.

The average annual number of eggs laid per hen in the United States increased from about 140 in 1935 to 245 in 1984, which is an average increase of about 2.1 eggs per hen each year from 1935 to 1984.

12.7.1 Removal of Eggs from the Nest

The regular removal of eggs increases the rate of egg laying. Even wild birds such as the sparrow have the inherent ability to lay several times their usual seasonal egg output if the eggs are removed from the nest. In the wild, a hen laid about a baker's dozen (13 eggs) in a nest, and the pressure of the eggs against her breast caused her to realize that that was as many eggs as her body could cover and hatch effectively. This recognition caused her endocrine system to switch from secreting the gonad-stimulating hormones to secreting prolactin, which is the hormone responsible for the manifestation of the maternal instinct. When the chicks are hatched and the effect of the eggs' pressure on the breast is removed, the bird's pituitary resumes the secretion of FSH and LH, and egg laying continues. This phenomenon explains why the removal of eggs from the nests of laying hens aids in maintaining egg laying.

12.7.2 Age at Sexual Maturity

Sexual maturity is reached when a **pullet** lays her first egg (commonly at about 5 months of age). Early sexual maturity is usually associated with intense reproductive activity, and in general, the earlier a pullet begins laying, the more eggs she produces in a **biological year.** Chicks hatched earlier in the year tend to mature earlier than those hatched later. This difference in maturity is attributed to the increasing daylength during the growing period earlier in the year. Thus increasing daylength hastens sexual maturity, whereas decreasing daylength delays sexual maturity (Figure 12.7). Modern methods of light control in poultry rearing and laying can largely eliminate seasonal effects on age to sexual maturity and minimize variations in egg-laying patterns.

12.7.3 Persistency

Hens that lay throughout the year are more persistent and lay more eggs than seasonal layers. Lack of persistency is characterized by broodiness and an early molt. (The fall molt normally marks the end of a laying year.) **Culling** early molters removes inferior layers from a flock. Some birds enter a "winter pause" consisting of a 2- to 10-day (short pause) or a 10- to 60-day (long pause) nonproductive period. High-producing hens possessing high inherent persistency do not enter a winter pause.

12.7.4 Season and Light Patterns

Normally, domestic fowl lay eggs only during the day. Thus the laying cycle of a hen is related to light. Under normal conditions, ovulation occurs early in the morning and about 30 min after laying. However, if oviposition occurs late in the day (4 to 5 P.M.), the next **ovum** is not released for some 10 to 12 h (unless the normal lighting schedule has been reversed or the hen is kept under continuous 24-h light of constant intensity), and a nonproductive day follows. One would

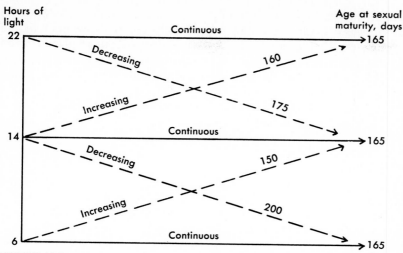

FIGURE 12.7
Diagram illustrating the control of sexual maturity in chickens by daylength (hours of light). Absolute light level is not as important as the change in the amount of light. (1) Continuous light at the same level results in about the same age at sexual maturity, (2) increasing light hastens sexual maturity, (3) decreasing light delays sexual maturity, and (4) increasing or decreasing light below 14 h has a greater effect than when light exposure is more than 14 h daily. (*Courtesy Professor Q. B. Kinder, University of Missouri.*)

expect hens to lay both day and night when lights are on in hours of darkness. However, this does not happen, because the extreme light sensitivity of the hen enables her to differentiate between the daylight and night hours on the basis of differences of light intensity. If, conversely, hens are housed in the total absence of natural light and are subjected to constant 24-h artificial light, they will lay approximately one-half of their eggs in day- and one-half in nighttime. Under such circumstances, there is no onset of darkness to delay ovulation and terminate a **clutch.** Light does not initiate ovulation in the bird, and darkness delays it.

If hens are exposed solely to artificial light from 6 A.M. to 6 P.M. and have no light the other 12 h, they will lay all their eggs during the day. However, if the light exposure is reversed (light from 6 P.M. to 6 A.M.) with no natural light during the solar day, they will soon (in about 3 days) shift their ovulation cycle and lay all their eggs at night.

Chickens were domesticated from jungle fowls that laid most of their eggs in the spring. In spite of domestication and selection, chickens tend to revert to the natural periodicity displayed by their jungle fowl ancestors and produce more eggs in the spring and summer and fewer in the fall and winter. Because of variations in the daily amount of light in different parts of the world, seasonal changes in egg laying are influenced by latitude. Spring is the breeding season for birds in the temperate zones of both hemispheres. Spring extends from March to May in the northern hemisphere and from September to November in the southern hemisphere. Therefore peak egg laying in New Zealand occurs at the time of lowest egg laying in the United States or in southern Europe. In southern India (about 10°N), where seasonal changes are almost imperceptible, egg laying is nearly uniform throughout the year.

The extreme heat of late summer decreases ovulation rate in hens, which is reflected by a decline in egg laying.

12.7.5 Broodiness

This trait is undesirable in egg producers since the hen neglects egg laying and favors **setting** on a nest when she goes broody. The Mediterranean breeds of fowls (e.g., Leghorns) are less broody than those of Asiatic origin (Brahmas).

Ancestors of the domestic hen laid only 25 to 30 eggs a year (two clutches). After laying the first clutch, the hen hatched and reared her chicks.[4] If time, weather, and circumstances permitted, she repeated this feat. However, through selective breeding of Leghorn hens, broodiness has been virtually eliminated.

Even though the natural tendency to incubate eggs (broodiness) has been decreased through selective matings, it still recurs cyclically in heavy meat birds and turkeys. If allowed to run its full course, each broody period lasts about 3 weeks, which is the length of time corresponding to the chicken's natural incubation period. Since broody hens seldom lay eggs, this nonproductive period is undesirable in commercial egg production. Broodiness may be induced by one or more of the following factors: (1) the presence of baby chicks, (2) high temperature (about 32°C), (3) constant darkness, and (4) the administration of prolactin and certain steroids (e.g., testosterone).

Wire cages or slatted nests delay broodiness. It is interesting that hens continue to lay in complete isolation, whereas with pigeons the presence of another pigeon (male or female) is required to provide the stimulus for ovulation. Apparently this is psychological, since the female pigeon will ovulate if she can see herself in a mirror. When two eggs are in a pigeon's nest, she becomes broody and ceases to ovulate.

12.7.6 Clutch Length and Intensity of Egg Laying

The interval between the ovulation of two successive eggs is usually 24 to 26 h. A hen normally lays one egg a day for several successive days before missing a day. This uninterrupted series of successive eggs laid is referred to as a *clutch*. This

---

[4]Eggs have been incubated by artificial means for thousands of years. Both the Chinese and the Egyptians are credited with having orginated artificial incubation procedures. The Chinese developed a method in which they burned charcoal to supply the heat. They also used the "hot-bed" method in which decomposing manure provided the heat. The Egyptians constructed large brick incubators that they heated with fires in the rooms where the eggs were being incubated. The Chinese are currently using incubators heated with methane produced from human, animal, and plant wastes. The incubators resemble large brick ovens. The eggs are covered with blankets and turned periodically by hand.

Table 12.2
Egg Laying by Selected Birds

Species	Average clutch size (no. of eggs)	Maximum no. laid annually when eggs are removed from nests
Chickens		
Egg-producing	10–14	300–365
Meat-producing	10–14	190–230
Game or fancy	10–14	60
Ducks		
Egg-producing	14–20	250–310
Meat-producing	14–20	120
Turkeys	15–20	220
Geese	12–15	100
Ostriches	12–15	100
Pheasants (in confinement)	10–12	104
Quails (in confinement)	12–20	130
Pigeons	2	50
Canaries	4–6	60

trait is heritable, and hens laying more eggs per clutch (six or more) produce more eggs than hens laying fewer eggs (one to three) because they have fewer nonproductive days. Hens having a short interval (less than 2 h) between oviposition and ovulation lay more eggs per clutch and have higher annual egg production.

Most wild birds lay only one clutch of eggs during the breeding season. However, should the clutch be destroyed, the bird may lay another clutch. Some wild birds, such as the robin and bluebird, lay, incubate, and brood two, and sometimes three, clutches in one season. Some wild birds may abandon the nest if any eggs are removed from it, whereas others may continue laying in an attempt to establish a clutch for hatching. In one study, when the eggs of a flicker were removed from the nest as soon as they were laid, the bird produced 71 eggs in 72 days. Data on normal clutch size and annual egg laying are presented in Table 12.2.

12.7.7   Individual Variation

There is little variation in clutch size among wild birds of the same species, whereas there is extreme variation in the number of eggs laid annually by domestic birds.

12.7.8   Age of Bird

Egg laying is most intensive in the pullet year of chickens but decreases almost linearly with each successive year. This decrease in egg laying in later life results from the progressively diminishing metabolic activity of the organs and tissues.

The fertility of eggs is also affected by **senescence.** Geese show little, if any, decline in reproductive rate for several years.

The life span of birds varies greatly. The kite, for example, may live 120 or more years, whereas songbirds (the **passerine** groups, such as warblers and buntings) and chickens may live only 12 years. The life span is usually inversely related to birthrate, as represented by the number of eggs in a clutch. Thus a bird that lays only one egg per clutch is likely to live considerably longer than a bird that lays 10 eggs per clutch. Egg laying seldom extends beyond the tenth year. In one study, a hen laid 1515 eggs during the first 8 years of her productive life.

Not only does the annual number of eggs laid decrease with advancing age, but the seasonal distribution of egg laying also becomes accentuated. Therefore, with advancing age, laying begins later in each successive year and terminates earlier; consequently, an old hen returns to the habits of her ancestors and lays eggs only during the spring and summer.

12.7.9   Molt

The shedding and replacing of feathers is a natural physiological phenomenon in birds, associated with a decline in the functional activity of the reproductive organs. Egg laying ceases as molting begins, and there is a marked regression in the secondary sex characteristics. Therefore molting corresponds to a period of physiological rest. Persistent layers molt last and require the shortest time to shed their old feathers and grow new ones. Moreover, birds that do not exhibit a complete molt may lay without lengthy interruptions for 2 or more years.

Molting can be induced by one or more of the following conditions: (1) withholding feed and/or water, (2) an abrupt decrease in the daily amount of artificial light, and (3) administration of derivatives of thyroxine (these also cause depigmentation of the feathers). No completely satisfactory means has been developed to prevent molting.

Recent studies at Texas A&M University indicate that feeding high levels of zinc (20,000 parts per million, ppm, of zinc as zinc oxide) increases egg laying after molting. In the Texas A&M Agricultural Experiment Station tests, groups of hens receiving high zinc levels for 8 days resumed egg production an average of 7 days after zinc was removed from the diet. Egg production of birds receiving the high zinc level for 10 days or less was significantly greater than that of birds molted by a conventional method.

12.7.10   Physical Constitution and Vigor

Irrespective of the inherent capacity for intense and persistent egg laying, a hen must possess superior vitality, **constitution,** and vigor to withstand the strenuous demands of sustained egg laying. This **stress** on her system and body may adversely affect her health and resistance to disease, speed senescence, and thus decrease her life span.

Body weight is a criterion of good health in the hen. An underweight hen cannot withstand long and persistent laying, and an excessively fat bird stops laying.

12.7.11 Nutrition

The egg laying of the domestic hen's ancestor, which laid only 25 to 30 eggs annually, was probably not particularly affected by **nutrition.** However, the hen of today is expected to lay 300 or more eggs annually and must therefore receive special nutritional considerations. To lay an egg a day, a hen needs approximately twice as much protein,[5] carbohydrate, and fat as is required for body maintenance, and she needs a substantial increase in dietary mineral intake. During the year, a 300-egg hen secretes approximately 4 times as much dry matter as is contained in her body. In the form of the 38 to 40 lb of eggs that she lays, it is composed of about 4.6 lb of protein, 4.1 lb of fat, 4.2 lb of minerals, and 25.1 lb of water. Thus adequate nutrition and a balanced diet are imperative for the laying hen. Of the 103 known chemical elements, 36 have been found in bird eggs.

The blood proteins, calcium, phosphorus, and **lipids** in the **laying** hen are about two-, two-, three-, and fourfold, respectively, higher than in the nonlaying hen. The increase in these blood constituents is explained by the need for great quantities of proteins for yolk and albumen formation, for an abundance of calcium and phosphorus in shell formation, and for lipids in the formation of egg yolk. When the hen ceases to lay, these blood constituents are soon reduced to the level prior to laying.

12.7.12 Drugs

Certain drugs (anthelmintics) used to control internal **parasites** in chickens may cause a decline in egg laying. Sulfanilamide inhibits egg laying if administered in concentrations of 0.25 percent or more of the total diet.

Although not considered a drug, cow **manure** contains certain androgenic substances and may cause a marked decrease in egg laying when ingested by the fowl.

12.7.13 Environmental Changes

Extreme heat or cold, rain, wind, or sudden changes in temperature and humidity may have a temporary adverse effect on the reproductive activity of the hen by (1) causing physical discomfort, (2) lowering vitality, (3) reducing food intake, or (4) restricting exercise and thereby resulting in reduced egg laying.

---

[5]Recent studies at the University of Missouri-Columbia demonstrated that the protein level in diets of laying hens could be reduced to 15 percent when proper amounts and balance of amino acids were provided.

12.8 FACTORS AFFECTING THE COMPOSITION AND CHARACTERISTICS OF EGGS

The average gross composition (by weight) of hen eggs is about 10 percent shell, 30 percent yolk, and 60 percent albumen (white). Egg composition was further defined in Table 3.1. Large eggs contain proportionally more albumen and less yolk than small eggs. The percentage of shell remains nearly constant, except that the first pullet eggs contain relatively more shell and less yolk than is characteristic of hen eggs in general. The last egg of a clutch (for clutch sizes of three to seven) is usually smaller and has a thicker shell than the egg that preceded it. The quantity of shell secreted per egg remains relatively constant; consequently, the larger the egg, the thinner the shell.

It is interesting that hen eggs are of fairly constant composition from day to day. Should the actively laying hen be deprived of important dietary essentials, nature allows her to withdraw from body reserves to produce an egg of full nutritional value. However, when calcium is not provided in the diet, body stores are soon depleted to such an extent that the hen stops laying.

A medium-sized hen producing 300 eggs annually may deposit about 3.75 lb of calcium carbonate (1.5 lb of calcium) in the eggshells. This represents about 30 times the amount of calcium in her entire body and points to the need for an abundance of calcium in her feed. Present evidence indicates that only about 60 percent of the calcium in a particular egg comes directly from the feed; the balance comes from deposits in the long bones. In wild birds, which lay only a few eggs, all the calcium may come from the bones. However, when laying hens are suddenly deprived of all feed calcium, their eggs promptly show thin shells, and the hens stop laying in 10 to 14 days. Of all egg components, the shell is the most variable. In one study, shell weight decreased about 25 percent when the hen's environmental temperature was increased from 19 to 39°C. The albumen also decreased significantly, whereas the yolk weight remained about constant.

12.8.1 Yolk Color

Yolk color is influenced mostly by the hen's diet. Highly pigmented feed ingredients, especially alfalfa, can produce dark orange yolks, whereas less pigmented feeds result in a relatively light yellow yolk, which is preferred by most egg consumers.

Gossypol (a toxic dye) is present in raw cottonseed meal and may cause olive-colored egg yolks, epecially in stored eggs. This color is attributed to the reaction between the ferric iron in the yolk and the yellow pigment gossypol. Apparently, the ferric **ions** in fresh eggs are insufficiently dissociated to enter into a visible reaction.

Artificial colors (red, blue, green, and others) may be incorporated into egg-yolk fats when various alcohol and oil-**soluble** dyes are fed to laying hens. One feeding of dye causes a colored layer in every yolk subsequently laid for a considerable period of time. This results from the dyed material being deposited simultaneously on a number of ovarian yolks having different degrees of maturity.

Blood spots (meat spots) probably occur in about 3 percent of all chicken eggs. They are more frequent in eggs of the heavier breeds (e.g., Rhode Island Reds) than in those of the lighter breeds (e.g., White Leghorns). Aristotle postulated that blood spots resulted from the premature expulsion of the yolk. Later, the theory was advanced that they result from ovarian bleeding at ovulation time. More recently, there is some evidence indicating that blood spots result from intrafollicular hemorrhage several days prior to ovulation.

12.8.2 Feed

Recent interest in the possible association of **cholesterol** and **atherosclerosis** (Chapter 3) has prompted those involved in egg production to seek ways to lower the cholesterol content and/or increase the degree of unsaturation of the fatty acids of egg (Section 3.2.2). The latter change has been accomplished by feeding diets containing higher levels of unsaturated **fats** to hens.

The cholesterol content of eggs is not so easily changed by alterations in the hen's diet. Feeding a high-cholesterol diet to a hen significantly increases the amount of cholesterol in the egg (by 20 to 30 percent); however, feeding a low-cholesterol diet will not significantly reduce the level of cholesterol in eggs. Working with normal human subjects, a scientist at the University of Missouri-Columbia demonstrated that the body reduces cholesterol synthesis to compensate for increased dietary intake, thereby maintaining a normal level of blood cholesterol (section 3.5.1).

Birds eat more in relation to their body size than do most humans. The smaller the bird, the more it eats in proportion to its weight. A 14-oz pigeon eats food equal to one-twentieth of its weight daily. To equal this, a 180-lb man who "eats like a bird" would have to eat 9 lb of food daily. (The average is 3 lb daily for a man.) The greater food intake per unit of body weight required by small animals results from an increased metabolic rate (see Chapter 8).

12.9 FACTORS AFFECTING EGG SIZE

The size of eggs is commonly expressed in terms of weight, which provides a more convenient basis of comparison than dimensions or volume. Egg weight varies immensely among species (Table 12.1). Hen eggs normally weigh from 40 to 80 g each.[6] The highest recorded chicken-egg weight was 320 g, and the smallest was 1.3 g.

[6]Minimum egg weights per dozen, determined by the USDA, are Jumbo, 30 oz; Extra Large, 27 oz; Large, 24 oz; Medium, 21 oz; Small, 18 oz; and Pee Wee, 15 oz.

12.9.1   Genetic Influences

The average egg weight per hen is highly heritable (about 40 percent), and the trait of laying large eggs has become rather permanently fixed in domestic birds. The average egg weight of the Dark Brahma fowl is 68 g compared with an egg weight of 40 g in its wild prototype, the Red Jungle fowl (a 70 percent increase).

12.9.2   Age of the Bird

Pullet eggs normally are about three-fourths the maximum size reached at maturity. Size of the first eggs is closely associated with body growth, which is related to time of hatching. Chicks hatched early (March and April) attain maximum egg size in less time than birds hatched later in the year (October and November). The first egg laid by a pullet is almost always her smallest, and the size of the first egg is a good indication of the relative size of eggs that a bird will lay in the future. If the first egg is large, the bird usually continues laying large eggs. Data related to hatching date, age at sexual maturity, and egg size are presented in Table 12.3.

12.9.3   Environmental Temperature

Eggs laid by mature hens are larger in the colder months (December to February) and smaller in hotter months (June to August).

12.9.4   Size of the Bird

Larger hens usually lay larger eggs. Of interest is the kiwi bird of New Zealand (weighs about 2000 g), which lays an egg weighing nearly one-fourth of its body weight. Since adult bird body weight is highly heritable (about 50 percent), the influence of bird size on egg size is closely related to the genetic makeup. Much

Table 12.3
Relation of Hatching Date to Days to Sexual Maturity and
Egg Size in Chickens

Month hatched	Light per day in month of hatch, h	Days to sexual maturity (natural lighting)	Large eggs per year, %
January	9.5	164	87
March	11.2	184	89
May	13.8	189	94
July	14.9	200	94
September	13.0	190	93
October	11.8	179	72

Source: W. C. Skoglund, New Hampshire Expt. Station.

of the early improvement in domestic-hen egg size resulted from selection for larger body size; however, today's commercial egg strains have been successfully selected for small body and large egg size concurrently. The greatest increase in egg size has been in fowls that have been under domestication the longest (some more than 5000 years); the least is observed in turkeys, which were domesticated more recently. An increase in egg size throughout the first year is associated with a gain in body weight.

12.9.5   Ovum Size

Egg size is largely determined by the size of the ovum passing through the oviduct. Relatively less albumen is deposited around a small yolk (as in pullets), and the finished egg is smaller than an egg laid by a mature hen, which normally ovulates a larger yolk.

The commonest abnormality in eggs is the phenomenon of two yolks. Aristotle stated that twin chicks hatched from double-yolked eggs, but this is highly unlikely. Instead, two ova are ovulated simultaneously; however, double-yolked eggs may also result from the premature ovulation of one of the two ova or from the retention of one yolk in the body cavity for a short period of time.

12.9.6   Intensity of Egg Laying

Although many contradictory reports regarding the relation between egg weight and the intensity (rate) of egg laying have been published, there is sufficient evidence to warrant the conclusion that if intensity only is selected for, egg size tends to decline, and if large egg size only is selected for, the intensity, or laying rate, declines.

12.9.7   Nutrition

A deficiency of feed nutrients tends to reduce egg weight and curtail the number of eggs laid. A deficiency in vitamin D tends to reduce egg weight. This may be related to the role of vitamin D in calcium metabolism, so important for eggshell formation. The feeding of desiccated thyroid results in hens laying smaller eggs.

12.10   IMMUNOLOGICAL AND MEDICAL ASPECTS OF EGGS

The egg possesses antigenic properties. Egg **antigen-antibody** reactions are manifested by such commonly observed phenomena as allergic symptoms, **anaphylactic shock** (caused by an exaggerated reaction of the organism to a foreign protein), **complement fixation,** and precipitation (the formation of a visible precipitate, caused by an antibody to soluble antigen that specifically aggregates the **macro**molecular antigen). Thus egg proteins are used to test whether a person is hypersensitive to certain substances.

Egg albumen may cause **allergy.** The term *allergy* refers to *clinical symptoms in humans that result from antigen-antibody reactions within the body.* Skin tests may be employed to diagnose possible adult and child allergy to egg albumen.

The hen transmits some antibodies to the chick through the egg. If diphtheria **antitoxin** is present in egg protein before incubation, it may be detected in the blood of newly hatched chicks. Tetanus antitoxin is similarly transmitted to the chick, as are antibodies against fowl pox and Newcastle disease. This **transitory immunity** disappears within the first month of life.

Egg albumen is often used as an antidote when a person orally consumes a poison. The albumen apparently prevents absorption of the poison by coating the mucous membrane of the stomach and inactivates the poison by combining with it chemically. Egg albumen is also used to increase the reliability of positive reactions in the Wassermann test (a test for syphilis based on the fixation of complement). Eggs and chick embryos are also used as **culture** media for the biological manufacture of over 30 **viral vaccines.**

When heated (above 57°C) with a liquid, egg albumen either settles to the bottom or forms a scum at the top. Egg albumen may be added to coffee to settle it. Egg albumen is also sometimes used in fruit and alcoholic beverages as a clarifying agent. A 2 percent solution of dried albumen is added to the beverage, and the mixture is then heated. The albumen coagulates and settles out, carrying with it the fine particles that otherwise would cause cloudiness of the beverage.

12.11 SUMMARY

There are many colors, shapes, and kinds of eggs. Nature designed the eggs of each species especially for the perpetuation of that species. Eggs of precocial birds contain more yolk (it contains the greatest food value) than those of altricial birds. This correlates with the greater maturity of precocial birds at birth.

Humans have found the egg both to possess fine flavor and to be a rich source of important nutrients. By means of domestication and continued genetic improvement through **selection,** humans have developed highly productive birds. Unlike their ancestors, they lay generous quantities of eggs throughout the year and thereby enhance the nutritional status of humanity. The egg-laying patterns of birds are closely related to changes in light. Through research, humans have discovered ways of applying artificial light and thereby minimizing seasonal fluctuations in egg production.

Although percentages of egg constituents vary among species, the eggs of all species contain the same components. Many factors influence the composition and size and the intensity with which eggs are laid. However, egg composition of a given bird species is nearly the same day after day and throughout the world.

STUDY QUESTIONS

1 Why are bird eggs larger than those of mammals?
2 Why do you think nature provided for variation in eggshell color?
3 Differentiate between precocial and altricial birds; also, the relative levels of albumen and yolk in their eggs. Review both Table 12.1 and Section 12.3.2.
4 What is the blastodisk? Blastoderm?
5 Name the five principal parts of a bird's egg.
6 What is the function of the chalazas?
7 Why does the yolk remain above center in a freshly laid egg?
8 Can double-yolked eggs be fertile? Will two chicks hatch from them?
9 Is mating essential to egg laying? To egg fertility?
10 How is it possible to classify baby chicks according to sex?
11 Are both the right and left ovaries and oviducts functional in most domestic birds?
12 Briefly relate how an egg is developed as it traverses the oviduct. How much time is required for this process?
13 Do domestic birds have a cervix?
14 What is photoperiodism? Is the spectral quality of light related to sexual maturity and egg laying in domestic birds?
15 What effect would ovariectomizing young chicks have on their secondary sex characteristics?
16 When do chickens most frequently mate?
17 What is the length of the incubation period in chickens? In turkeys?
18 How are the hormones FSH and LH related to egg laying?
19 Which hormone is closely related to eggshell formation?
20 Briefly discuss the functions of thyroxine in poultry.
21 Which hormone is involved in the expulsion of an egg?
22 What is prolactin? What is its relation to hatching of eggs naturally?
23 What may cause a hen to develop male sex characteristics? Do hens ever crow?
24 Are hens ever pregnant? Why?
25 How is it possible to have an egg within an egg?
26 When does a hen ovulate? When does she cackle?
27 Does the regular removal of eggs from a nest influence egg laying?
28 What is a biological year?
29 What is the effect of increasing light on sexual maturity in chickens? Of decreasing light?
30 Are persistency of egg laying and broodiness related in chickens?
31 Under normal conditions are eggs laid during the day or night? How can this laying pattern be changed?
32 What is broodiness? How may it be induced in hens?
33 What is a clutch? Is clutch size heritable?
34 Is there more variation in clutch size among wild or among domestic birds?
35 At what age is egg laying most intense in chickens? Is this true of geese?
36 What is the relation of clutch size to life span of birds?
37 Is molting natural? How can molting be induced? Can it be prevented?
38 Approximately how much of the dietary protein, carbohydrate, and fat intake of a laying hen is used for body maintenance? For egg laying?
39 What is the average gross composition of hen eggs? Does the composition vary much from day to day? Which egg component is the most variable?

40 Does a hen's feed affect egg-yolk color?

41 Does dietary fat affect the degree of unsaturation of fatty acids of eggs? Explain.

42 Do birds eat more or less in relation to their body size than humans? Why?

43 Briefly discuss factors afffecting egg size.

44 How can the formation of double-yolked eggs be explained?

45 Can the hen transmit antibodies to the chick through her egg?

ECOLOGY AND ENVIRONMENTAL PHYSIOLOGY[1]

The biotic power of an animal is limited only by the repressive forces in the environment.

Samuel Brody (1890–1956)

13.1 INTRODUCTION

More than 2000 years ago, Aristotle said that certain conditions of **climate** and weather produce the most energetic people and the finest **livestock;** that the rise or fall of mighty empires is due to changes in climate or to the knowledge, or lack of it, of controlling climate and weather; and that the nation that will lead the world is the one that will best control its weather. The pertinence of this ancient prediction is evidenced today by the urgency of such problems as a tendency toward higher summer temperatures and the phenomenal growth of populations throughout the world. In tropic and arctic climates, **adaptation** is difficult for the best and most productive livestock and poultry. In these areas the population/food ratio is unfavorable to the nutritional well-being of the people. Moreover, the increase in population is often greater than the rate of increase in food production. Aristotle's foresight can be further illustrated by the development of only a few mighty empires, because approximately three-fourths of the world's population lives outside the temperate zones in areas where the temperatures often rise above 100°F. This temperature is far above the maximum of the **thermoneutral,** or **comfort, zone** for humans and most

[1]The authors acknowledge with appreciation the contributions to this chapter of Dr. S. E. Curtis, Department of Animal Science, University of Illinois, Urbana.

animals. Today war and peace efforts are global in scope, embracing weather extremes that stress humans and machines to the limits of their endurance.

A **symbiotic** relation exists between humans and animals. Although each may live independently of the other, a reciprocal benefit exists. The plight of humanity is better because of animals, and vice versa. Animals tend to serve as a buffer between humans and their **environment.** Because of this relation, there is a need for additional knowledge of the symbiotic relations between human beings and animals and their environment.

Human ecology and animal ecology have developed in curious contrast to one another. Human ecology has been concerned almost entirely with the effects of man upon man, disregarding often enough the other animals amongst which we live.

Charles Elton

Progress during the twentieth century in acquiring knowledge and a better understanding of the relation between climate and humans and animals has been very significant. One of the first practical applications of controlling the environment was the use of artificial light in egg-laying houses to increase the hours of light and thereby stimulate fall and winter egg laying (Figure 13.1). Furthermore, artificial light may be used to hasten or delay sexual maturity in poultry (Chapter 12). Conversely, decreasing light stimulates sheep to mate in the fall, allowing the **lambs** to be born in the spring, when the plant **nutrient** supply and weather are more favorable for growth and development of the young. When sheep are transported across the equator from one hemisphere to another, they alter (by 6 months) the time of the calendar year at which they breed to conform with the seasons in their new environment. Thus light is known to influence reproduction in poultry and sheep and is believed to influence reproduction in goats, horses, cattle, and swine. Could the effects of increasing light be reflected in hormonal changes that accompany the unexplained increase in **lactation** when cows are put on green grass in the spring?

Samuel Brody, a keen-minded pioneer in the field of environmental physiology and animal energetics at the Missouri Agricultural Experiment Station from 1921 to 1956, gave both leadership and scholarship to the science of comparative animal physiology. The methods he and his research colleagues developed to measure physiological responses of horses and mules to various work loads under outdoor conditions are classic. Figure 13.2 illustrates the use of an ergometer to measure work rate and a respiration apparatus to measure energy expended for the work.

Data obtained demonstrated that there were significant individual differences among animals in the levels of pulse rate, work efficiency, respiration rate, and other cardiorespiratory measures, and further that these differences were correlated with the degree of oxygen debt observed following work. Such individual differences could be used in subsequent animal selection and breeding

FIGURE 13.1
This hen is believed to hold the world record for egg laying in 1 year. In recent research at the Missouri Agricultural Experiment Station, Dr. Harold Biellier selected birds that had the biological capability of responding, productionwise, to light-dark cycles of less than 24 h and, in addition, had the genetic capability of laying eggs at intervals of less than 24 h over an extended period of time. Dr. Biellier demonstrated experimentally that the length of environmental light to which birds are exposed has a profound effect on the level and persistency of egg production. When exposed to 22- and 23-h light-dark cycles, rather than the usual 24-h day length for part of the year, this hen laid 371 eggs in 365 days and went on to lay 448 eggs in 448 days. Such research has many important applications, as well as economic and production implications, in providing nutritious eggs at economical prices for the consuming public. *(Courtesy Dr. H. V. Biellier, University of Missouri.)*

programs. These experimental techniques and others that followed gave impetus to additional research studies related to the physiology of growth, reproduction, lactation, nutrition, and senescence, as well as to the endocrine aspects of farm animal production. Much of the material presented in this chapter had its genesis in the early research endeavors of Brody.

The need for more food has resulted in the migration of both humans and animals from one region to another, as will be discussed in Chapter 14. This migration, in many cases, exposes humans and animals to a new environment,

FIGURE 13.2
Field apparatus used in early research studies to test cardiorespiratory and metabolic functions in horses and mules during rest, work, and recovery from work. Data collected or generated included oxygen consumption, carbon dioxide production, pulmonary ventilation rate (volume of air exhaled and inhaled per unit of time), pulse rate, rectal temperature, energy expended, and work efficiency. *(Missouri Agr. Expt. Station.)*

which requires certain adaptations. There are many interesting and challenging aspects of **ecology,** which is the branch of biological science that deals with the relation of living things to their environment, and of *environmental physiology,* which deals with the surrounding conditions that affect structures and organ functions of humans and animals. An attempt has been made to include many of the interesting aspects of ecology and environmental physiology in this chapter. Behavioral adaptations of animals to their environments are explored in Chapter 20.

13.2   HEREDITY AND ENVIRONMENT

When the early American settlers first brought animals from Europe, many of the animals lacked resistance to the environmental extremes and died prematurely. Yet how many plants and animals would there be in America today if the pioneers had concluded that they could not exist and grow here because none were here at the time they came? The common farm animals are among the most adaptable birds and mammals on earth; that is why they could be domesticated in the first place.

Animals are products of **heredity** and environment. A given hereditary pattern is itself, however, the product of environment, because **mutation** rates and selective survival are conditioned by the environment. In this way the different climatic regions cause animals to adapt themselves to life in them. This adaptation may be illustrated by the fact that animals evolved in cool regions are adapted to cold weather, having an abundance of wool or hair and **subcutaneous** fat, whereas animals evolved in hot regions are adapted to hot weather by the sparsity of wool or hair and subcutaneous fat. The fat resources of animals in hot climates are stored in the humps (of cattle) and tails (of fat-tailed sheep). Such fat stores do not interfere with heat **dissipation** from the body.

As will be discussed in more detail later, dark materials tend to absorb more solar radiation than light-colored ones. If dark skin absorbs more radiant heat than light skin and if it is true that humans and animals best suited to the environment of a particular region have survived, how can the apparent paradox and anomalous adaptation be explained that the darker-skinned people commonly populate the hottest regions? Could this be due to a more uniform distribution of pigment throughout the **epidermis** or to a greater thickness of the corneum?

Skin pigments include **melanin** (black to yellow), which is produced in the body. Light-skinned persons develop temporary melanin pigmentation (freckling) when exposed to sunlight, whereas dark-skinned persons possess melanin or melanoid pigments as a genetic characteristic. Ultraviolet (UV) light is **bactericidal** and toxic to skin and nerve tissues. Thus the melanin protects nerves and skin from the injurious UV radiation. In this sense, pigmentation of tropical races has adaptive value.

The nature of an animal's life is shaped by environmental and hereditary forces. Environment includes the physical, chemical, and biological elements that surround animals. With the exception of feedstuffs, which in the case of food animals are commonly studied by themselves, all the components of the environment are included in the realm of animal ecology.

Various components of the environment may either promote or impair animal performance by facilitating or inhibiting productive and reproductive processes. Scientific investigations of environmental effects on animal production are of recent vintage, probably because most people believed that breeding, feeding, and management research had greater economic impact than improvements in environmental factors. However, as animal production becomes more intensive in space and time, environmental aspects of food-animal management become more and more important. The research of Dr. Stanley E. Curtis and his coworkers at the University of Illinois in this rapidly growing field of science is both timely and noteworthy, and his recent monumental textbook provides an excellent overview and summary of the present state of the art.[2]

13.2.1 Regional Adaptation of Cattle

Animals are now transported long distances and in large numbers from temperate to tropical zones and from semitropical zones to areas where cold

---

[2]S. E. Curtis, *Environmental Management in Animal Agriculture,* Iowa State University Press, Ames, Iowa, 1983.

conditions prevail. Moreover, frozen semen is now available for international shipment (Chapter 10), making it easier to transport new genetic materials from one area to another. However, transporting the animals or germ plasm (semen) does not change the inherent genetic characteristics that make it possible for the animal to adapt to its environment.

Most breeds of livestock were developed in temperate zones and were naturally selected to fit the environment. Several features distinguish Indian from European cattle: (1) Indian cattle have excellent radiators—enormous **dewlaps** (loose, pendulous skin under the throat extending back between the legs and along the belly) and sheaths (navel flap in females) and long ears.[3] Their bodies tend to be small, resulting in a large surface area per unit weight, and their skin has little hair. Conversely, European cattle have fat, hairy, tight hides, undeveloped dewlaps, and relatively small ears. (2) Indian cattle have a more extensive system for evaporative cooling, which allows them to adapt better to hotter climates. Missouri studies showed that in the zone of **thermoneutrality** (Section 13.9), Shorthorns had higher skin, respiratory, and total vaporization rates than Brahmans. However, when exposed to temperatures high enough to effect a rise in rectal temperature, the opposite was true. (3) The hair color of Indian cattle tends to be lighter; thus it reflects more sunlight than does that of European cattle. (4) Indian cattle are more resistant to **ticks,** flies, mosquitoes, and other pests than are European cattle.

13.2.2 Genetics and Heat Tolerance

Is it possible to combine the *high productivity* of the *heat-intolerant* European breeds with the *high* **heat tolerance** and *low productivity* of the Indian-evolved cattle to obtain a highly productive, heat-tolerant animal? Wide genetic differences are determined by a series of **genes** of the multiple type, some genes having an *additive* and others a *nonadditive* effect. When breeds are crossed, the genes recombine into new patterns. By combining genes from two or more breeds, backlogs of the evolutionary resources of both may be utilized in developing new breeds. An example is the Santa Gertrudis cattle developed on the Kleberg King Ranch in Texas, which possess the heat tolerance of their Indian ancestors (Brahmans) and the high-meat-production characteristics of their European ancestors (Shorthorns).

Genetic combinations having favorable hereditary characteristics may also be obtained within a breed, but progress is slower. As discussed in Chapter 5, the more homozygous the breed, the more limited the progress.

---

[3]McDowell (R. E. McDowell, *Improvement of Livestock Production in Warm Climates,* Freeman, San Francisco, 1972) reported that removal or reduction of the appendages did not bring about a significant change in the response of animals to thermal stress. The hump of Zebu cattle is high in fat and not well supplied with blood. Moreover, the Zebu has less subcutaneous fat than the European breeds. This suggests that one reason the Zebu is more heat tolerant than European cattle may be that it stores much of its depot fat in the hump, rather than as subcutaneous fat, which tends to impede the flow of warm blood to the surface for cooling.

13.3 ADAPTATION TO ENVIRONMENT

Complete adaptation to environment means death. The essential point in all response is the desire to control environment.

John Dewey (1859–1952)

Climate (from the Greek *klima,* meaning *"slope"* or *"tilt")* depends on topography, on distribution of vegetation (especially forests), and on large bodies of water. (Seventy-one percent of the earth's surface is covered with water.) The environment, therefore, is largely reflected in the climate of a particular area.

Animals themselves are simply entities through which materials and energy flow before eventually returning to the environment. Without the ability to adapt, animals are at the mercy of the environment. When animals are repeatedly or continuously exposed to major environmental changes, they may develop functional and structural changes that result in an increase in their ability to live without **stress** in the new environment. These changes are collectively referred to as the process of **acclimatization.**

Animals best adapted to a particular environment have survived and, unless "transplanted" by human beings, populate areas best suited for their survival. However, modern transportation facilities have increased the movement of animals and have placed the selection of breeding stock on a global basis, forcing consideration of the question, How do the changed conditions affect the productivity, health, and longevity of animals?

Environmental temperature affects the secretion of certain hormones. For example, Missouri research studies show that domestic fowl secrete over twice as much thyroxine in winter as in summer, and the same trend has been observed among lactating dairy cows. There is a parallelism between egg laying, lactation, growth, and, indeed, virtually all metabolic processes.

The depressing effects of hot weather on body activities of humans and animals may be termed "hot-weather laziness"; however, when viewed **homeostatically,** the decreased activity is a biological mechanism for preventing overheating of the body. Truly, this is an adaptive mechanism enabling humans to exist in hot weather. The responses of an animal's body to environmental stresses are more immediately crucial to life than the productive and reproductive processes exploited by humans in animal production.

13.3.1 Physical Environmental Factors in Adaptation

Physical environmental effects on humans and animals may be classified as *natural* and *artificial.* The *natural factors* include (1) temperature (heat and cold), (2) air **humidity,** (3) air movements (wind), (4) barometric pressure, (5) rainfall and water effects on the skin, (6) altitude, (7) dust, (8) radiation (visible, ultraviolet, and infrared), and (9) **cosmic radiations** and atmospheric electricity.

The *artificial factors* include (1) atmospheric pollution from industry, smog, and gases; (2) toxic compounds in water; (3) mechanical factors (noise, ultrasonics); (4) **ionizing radiation (isotopes, x-ray);** and (5) artificial **ionization** of air. Humans and animals have certain inherent mechanisms that enable them to adapt to changes in one or more of the natural and, to a somewhat lesser degree, the artificial physical environmental forces about them.

Altitude   Did you ever wonder why the trees and animal life thin out as you drive up the mountains in the western parts of the United States? Did you ever climb a high mountain? If so, you noted a decrease in temperature, vegetation, and animal life, but more noticeable was the increased respiration rate necessary to acquire sufficient oxygen to continue the climb. Certainly, the climatic effects of altitude changes are important and warrant further study. Atmospheric pressure is reduced by about one-half at 18,000 ft compared with that at sea level. The llama is well adapted to high altitudes (up to 17,500 ft).

13.3.2   Natural Adaptations of Animals to Their Environment

The razorback conformation of the pig was a natural adaptation, enabling it to penetrate dense vegetation in search of food. Desert animals commonly must traverse much land in search of food because of the effects of seasonality of rainfall on vegetative growth; thus they have relatively long legs for travel. The large, flat foot of the camel enables it to move readily over shifting sand. Similarly, the surefootedness of the mountain goat and sheep, which enables them to graze on steep and rocky hillsides, is not an accident of nature.

Animals having a relatively small surface area with respect to body mass are best adapted to cold climates. Such is the case of many European breeds of cattle. Conversely, **Zebu** cattle have a large amount of surface area in relation to body mass (especially considering the hump and the extra folds of loose and pendulous skin) and are better adapted to warmer climates.

Coat thickness and hair characteristics in mammals and plumage in birds are greatly influenced by climate. Some desert reptiles and birds have scaly coverings that are definitely reflectors of **radiant** energy and enable the animals to survive in a hot environment.

13.3.3   Natural Variations in the Heat and Cold Tolerance of Animals

The long-haired yak cattle of Tibet and western China and the woolly cattle of the Scottish Highlands are essentially as cold tolerant as the arctic-dwelling caribou and reindeer. Even the European-evolved cattle are cold tolerant. Conversely, the Indian-evolved cattle, called Zebu or Brahman, are heat tolerant but cold intolerant.

Missouri studies disclosed that the thermoneutral, or comfort, zone (temperature interval in which no demands are made on the temperature-regulating mechanisms) of European cattle is between 30 and 60°F, whereas for Indian cattle it is between 50 and 80°F. Therefore, beginning at about 60°F in European and about 80°F in Indian cattle, the thermoregulative mechanisms become active, as demonstrated by the increase in respiration and vaporization rates. Beginning at about 80°F in European and 95°F in Indian cattle, these mechanisms become incapable of meeting the demands for heat dissipation, resulting in (1) a rise in rectal temperature, (2) a decrease in feed intake, and (3) a decrease in milk secretion and other productive processes. Often, the animal loses weight as a consequence.

13.3.4 Comparison of the Climatic Adaptive Physiology of Humans and Cattle

The close symbiotic association between humans and cattle causes one to wonder what climatic adaptations they have in common. Because of modern science and technology, human beings are climatically the most adaptable species. In the nude, however, humans face problems at temperatures below approximately 60°F. (Critical temperature depends on humidity, air velocity, activity, and other factors.) Humans are much more heat tolerant than cattle. This is illustrated by their near-normal and the cow's 5 to 7°F above-normal rectal and skin temperatures when exposed to an environmental temperature of 105°F. Missouri studies showed that in a 105°F environment, the cow's skin and rectal temperatures were 105 and 107°F, respectively, whereas those of humans were 95 and 99°F, respectively.

Human sweating increases at the rate of 9 to 13 percent per 1°F temperature rise from 85 to 105°F. This accounts for their ability to withstand high temperatures. Conversely, the cow has very limited means of evaporative cooling in the same temperature range, which results in a rapid rise in rectal temperature. Swine are even less able to cool themselves evaporatively in a hot environment.

13.4 STRESS

A large number of environmental factors may cause an animal to respond homeostatically. Exposure to stressful conditions sometimes results in nonspecific increases in secretion by the adrenal glands and frequently in other disorders as well.

In farm animals, one of the most important stress factors from an economic standpoint is *heat stress.* When exposed to high temperatures, the animals exhibit **polypnea.** This condition of increased respiration rate results in the increased dissipation of heat in two ways: (1) by warming the inspired air and especially (2) by increasing **evaporation** from the respiratory passages and lungs.

At high temperatures livestock and poultry lower their food consumption and thereby reduce heat production. As a result of decreased food consumption, productivity (egg, meat, milk, and wool) is also reduced.

Measuring Thermal Stress   There are several means of detecting whether an animal is under thermal stress. The most obvious stress index is body-temperature response. An elevated body temperature in turn reflects certain other detectable indices, such as heart rate, respiration rate, and **heat production.**

13.5 HOMEOSTASIS AND HOMEOTHERMY

Homeothermic animals of large body size have no need of annual migration. **Homeostasis** enables them to adjust to the winter cold, whereas small homeo-therms (e.g., birds) find seasonal migration necessary. This is because small animals have relatively more surface area in comparison with body weight than do large homeothermic animals, and heat dissipation is proportional to body surface area.

The disappearance of the large reptiles in the Cretaceous and Eocene periods may have resulted from unusual temperature changes that homeotherms but not **poikilotherms** survived. (The reader is encouraged to visit the Dinosaur National Monument in Utah and Colorado.)

Homeotherms vary in their normal body temperature from 36°C (96°F) in elephants to about 43°C (109°F) in the smaller **avian** species. The rectal temperatures of selected animals may be grouped as follows:

96–101°F: Elephant (96.0), mouse (97.3), human (98.6), rat (99.0), horse (100.0), and monkey (101.1)

101–104°F: Cattle (101.0), cat (101.5), dog (102.0), sheep (102.3), swine (102.5), rabbit (103.1), and goat (103.8)

104–106°F: Turkey (104.9), goose (105.0), owl (105.3), and duck (106.0)

107–109°F: Chicken (107.0), English sparrow (107.0), hummingbird (108.0), and robin (109.4)

13.5.1 Homeotherms (Warm-Blooded Species)

Animals in this category tend to maintain constant *internal* temperatures as *external* temperatures vary. For example, the annual atmospheric temperature range in Montana may be more than 150°F (−40 to +110°F), yet the body temperatures of cattle, horses, and sheep wintering outdoors are constant to within 1°F. How is it possible for animals to adapt to the vicissitudes of the weather?

As *cold weather* approaches, the cold-weather homeothermic mechanisms are (1) growth of insulating hair and subcutaneous fat; (2) increase in thyroid activity; (3) consumption of great quantities of food, which, due to their **heat increment,** warm animals; (4) seeking a protective shelter and warming solar **radiations;** (5) grouping together (huddling); and (6) increase in activity,[4]

---

[4]It should be noted that animals commonly decrease their activity in the huddle; so much, in fact, that even though they need to increase feed intake, they actually decrease it—preferring to huddle rather than eat.

voluntary and involuntary (shivering). All these mechanisms increase heat production. A means of *heat conservation* is the contraction of the superficial blood vessels (**vasoconstriction**), reducing the blood flow at the skin, from which it loses heat via **conduction**, evaporation, radiation, and convection.

As *hot weather* approaches, the cooling mechanisms become important. These include (1) moisture **vaporization**, (2) avoidance of the heating solar radiation, (3) depression of thyroid activity, and (4) refraining from work (includes the agriculturally important productive processes such as egg laying, lactating, and meat production, since they increase heat production).

Moisture vaporization is the most important cooling mechanism and is the only one in humans during periods when environmental temperature equals or exceeds that of surface temperature. Profusely sweating species can therefore withstand very high environmental temperatures. Most nonsweating species attempt to compensate for their inability to sweat by panting, often protruding their tongues and blowing air rapidly over the moist surface, thereby accelerating the vaporization rate. Swine will die when exposed to an atmospheric temperature of 100°F in a dry, sunny lot, whereas they can withstand that temperature indefinitely when given access to a mud wallow; water evaporates from mud on a pig's skin at a very high rate, and this cools the pig greatly.

13.5.2 Partly Homeothermic, Hibernating Species

Animals in this category include the bat, woodchuck, dormouse, ground squirrel, hedgehog, groundhog (marmoset), opossum, and prairie dog. These animals maintain a fairly constant body temperature during the summer, but as winter approaches and their temperature-regulating mechanisms begin to fail, they retire below the frost line into burrows or migrate into caves and enter their winter sleep (**hibernation**) until the coming of spring. The body temperature falls during hibernation to a few degrees above freezing, and the metabolic rate and pulmonary ventilation (respiration) may decline to about 1 percent of the normal summer levels.

In times of famine, human beings have been known to show **dormancy** during most of the winter months. Unlike that of the marmoset, however, the body temperature of humans probably does not fall appreciably below normal during their "winter sleep." The dormancy (not true hibernation) of the bear is probably of the same category—not a loss in ability to maintain normal body temperature, but rather a way of saving energy by sleeping. Hibernation is probably an evolutionary adaptation to prevent starvation during periods of food scarcity.

Cold Habituation[5] When suddenly subjected to a relatively severe cold stress, animals that have been experiencing a milder cold stress for an extended period sometimes delay, for several hours, their thermogenic reaction to the

---

[5]Portions of this section were adapted from S. E. Curtis, *Environmental Management in Animal Agriculture*, Iowa State University Press, Ames, Iowa, 1983.

acute stress. In fact, their body temperature may actually decrease for a short period of time. This *cold habituation* is useful because it enables animals to withstand acute cold stresses of short duration, which frequently occur in nature (e.g., at night), without raising the metabolic rate unnecessarily. This enables animals to conserve feed energy. If the stress persists for longer than a few hours, heat production increases to effect heat balance at normal body temperature.

13.5.3 Poikilothermic, Estivating Species

These animals, which sleep during the hot, dry summer, are exemplified by frogs, crocodiles, and alligators. Estivation is probably an evolutionary adaptation to periods of water scarcity. The hibernation and **estivation** of **insects** and soil inhabitants are of considerable agricultural interest.

Poultry embryos might be classed as cold-blooded until they reach a certain age. (Even after hatching they must be kept under **brooding** heat units until they perfect their homeothermic mechanisms.) Similarly, newborn rats, puppies, kittens, and pigs are very susceptible to cold temperatures.

It is fortunate indeed that humans and the animals that provide their food have the ability to maintain a near-constant body temperature. At extremely low temperatures, reactions slow down or cease, and at too high a temperature, total destruction of organic complexes occurs and death will ensue.

13.5.4 Age and Homeothermy

As stated by Brody,[6] homeothermic mechanisms are not required by the developing mammal before birth. Depending on **ambient temperature,** the body temperature of children stabilizes (becomes fully homeothermic) between 12 and 24 months, that of rats in about 3 weeks, that of poultry in 3 to 4 weeks, that of calves and lambs within a few hours after birth, and that of **piglets** 2 to 3 days **postnatum.** Homeotherms are much better able to protect themselves against cold than against heat.

13.5.5 Experimental Hypothermia

Until recently, it was assumed that nonhibernating mammals could not tolerate significantly reduced internal temperatures. However, the survival of "super-cooled" bats, cats, and rats prompted surgeons to attempt this procedure **(hypothermia)** in brain and **cardiac** operations. An example is cited of a woman who was cooled to an internal, or "core," temperature of 48°F and recovered. Her heart was at a standstill for 45 min. Several investigators have observed **ventricular fibrillation** in humans and dogs during cooling. (It usually occurs at a body temperature of about 79 to 66°F.)

---

[6]S. Brody, *Bioenergetics and Growth,* Reinhold Publishing Corp., New York, 1945.

13.6 TEMPERATURE REGULATION

To accomplish homeothermy, the body is able to utilize physiological, behavioral, and anatomical mechanisms. For example, the **endocrine** *system* aids the animal in adjusting to seasonal and to short-term temperature changes by its secretion of thyroxine and epinephrine. Thyroxine is associated with slow, seasonal temperature changes; epinephrine is available to cope with rapid temperature changes.

The *nervous* temperature regulating center is in the hypothalamus, which is located at the head of the spinal cord, just below the cerebrum. It is an astonishingly accurate thermostat. The hypothalamic thermoregulatory center is itself sensitive to the temperature of the blood flowing through it, and it integrates this information with other inputs on the body's thermal status, both at the surface and in deeper organs and tissues, as it decides whether thermoregulatory responses are needed and, if so, which ones.

Humans can maintain their normal body temperature when exposed for short periods to an atmosphere of 300°F or higher (a temperature that will grill a beefsteak), provided the air is dry, thus permitting evaporative heat loss at a very high rate. Conversely, a damp (high relative humidity) atmosphere with a temperature of 120°F causes the body temperature to rise rapidly, and it cannot be endured for more than a few minutes. This is because the difference between the vapor pressure at skin or respiratory passages and the air is too narrow to permit a high evaporative rate.

The nervous system also controls the "ruffling" of feathers or "raising" of hair to decrease heat loss, or **thermolysis.** (Contraction of smooth muscles of the skin gives rise to "goose bumps," or "goose flesh.") Furthermore, shivering (tensing of muscles), chattering of teeth, and other muscular activities involved in heat production are under nervous control.

It is desirable to reduce heat loss in cold weather by such devices as (1) reduced vaporization (no sweating); (2) lowered respiration rate; (3) shunting of blood from the surface to the interior of the body; (4) huddling; (5) production of warm coats of feathers, fur, hair, or wool; (6) deposition of subcutaneous fat; and (7) sheltering from the wind and from the cold.

Chemical heat regulation adds to the heat-preservation process through (1) increased **thermogenesis** (by exercise), (2) increased shivering and muscle tension, (3) increased food intake (increased **heat increment of feeding),** and (4) increased adrenal and thyroid activity. (Epinephrine and thyroxine are the metabolic accelerators.)

The heat-regulating mechanisms are depressed by **anesthetics,** hypnosis, and general body fatigue.

13.6.1 Body-Temperature-Regulating Mechanisms

Breeds of cattle developed in different climatic circumstances usually possess adaptive characteristics that harmonize with their respective climates. Certain physical characteristics, such as hair length and thickness and size of peripheral features (ears, dewlap, navel flap, vulva) are easily recognized, whereas others, such as neuroendocrine peculiarities, are not so readily apparent.

Brody[7] summarized the rules governing climatic adaptations and named them after four pioneer investigators:

The Bergmann Rule (1847)   This rule relates body size to climate. It is a geometric fact that small, *light* animals have larger surfaces per unit weight and therefore lose heat more rapidly than larger, *heavy* animals, which have a small surface area per unit weight. It is obvious, however, that a tall, thin person may have exactly the same weight as a short, fat person yet have much more surface area.

The breeds of larger size are usually found in colder regions and the smaller breeds in warmer climates. This is illustrated by the fact that a larger breed of dairy cattle (Holstein) is better adapted to the cooler northern areas of the United States, and a smaller breed (Jersey) is better adapted to the warmer southern areas. Further examples of the natural adaptation of small animals to warm climates are the pygmy races of elephants, hippopotamuses, and buffalo (also pygmy humans) found in the hotter regions of the world. Dairy producers, however, have successfully adapted the large Holstein cow to the south by providing shade and various other means of decreasing the problems of heat gain and heat dissipation.

A seemingly contradictory example of the above rule can be found in the large Brahman cattle, which inhabit hot areas, and the smaller (with respect to overall body conformation) Shorthorn cattle, raised in a cooler climate. However, a careful examination reveals that the Brahman, while possessing a large periphery and thereby a large surface area, is not an especially heavy animal, but it possesses a very extensive heat-dissipation system per unit weight. Conversely, in the more compact Shorthorn, the surface area is smaller in relation to body weight, and heat loss is thereby decreased. Moreover, the ears, dewlap, navel flap, and vulva of the loosely built Brahman are much larger and more corrugated, and therefore the animal is heat tolerant and cold intolerant, whereas the compactly built European cattle are cold tolerant and heat intolerant (Figure 13.3).

The Bergmann rule is modeled after Newton's law of cooling,[8] which indicates that the larger the surface area of a given body, the greater the rate of heat transfer. The same holds for heat transfer by conduction, **convection,** radiation, and vaporization. Newton's law of cooling may be written as $q = kA (t_1 - t_2)$, where q is the rate of heat flow between the body surface and the environment; A is the surface area of the body; t_1 and t_2 are, respectively, the temperatures of the body surface and of the environment; and k is the coefficient of heat transfer, defined by the equation and determined experimentally.

Although Brody did not mention *Allen's rule* (1877), it is closely related to Bergmann's rule. Whereas Bergmann's rule deals with body bulk, Allen's rule

---

[7] *J. Dairy Sci.,* **39**:715–725 (1956).
[8] Sir Isaac Newton, 1642–1727.

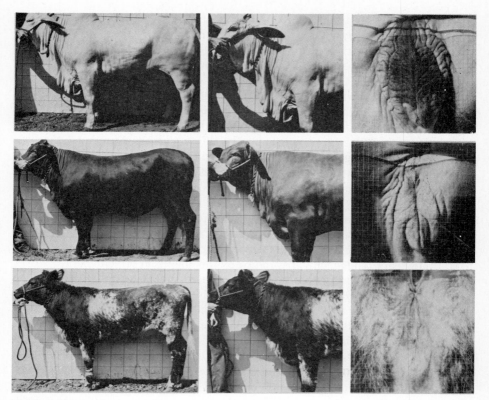

FIGURE 13.3
The differences in general conformation, and especially in relative heat-dissipating surface areas, in the Brahman (top), Santa Gertrudis (center), and Shorthorn (bottom) breeds of cattle at age 1 year. Note the larger and more corrugated and vascular vulva in the heat-tolerant Brahman. *(Missouri Agr. Expt. Station.)*

states that in a given species of warm-blooded animals, those living in cold areas tend to have shorter extremities than those in warm climates. This applies only to the living appendage and not to the fur or feathers covering the animal. This rule deals with the physics of heat loss from a warm body to a usually cooler surrounding. Allen's rule can also be stated: The nearer the body is to being a perfect sphere, the smaller is its surface area per unit volume. Each species has its own mechanisms for heat regulation, so that **inter**species comparisons should not be made.

The Wilson Rule (1854) This rule relates insulating cover to climate. Breeds that evolved in cold climates have a dense, heavy, thick external coat (300 g/m^2 in the Shorthorn), whereas those from warm climates have little hair (10 g/m^2 in the Zebu), which is short, glossy, stiff, and thin.

There is probably little difference in skin thickness between European and Indian cattle; however, there is a significant difference in the thickness of the subcutaneous fat (thick in European and thin in Indian cattle).

The Gloger Rule (1833) This rule relates color to climate. Skin pigments protect against ultraviolet radiations. Conversely, the hair of animals in the tropics tends to be light-colored and thereby reflects solar radiations and serves as a protection from overheating. Skin pigmentation of cattle gradually darkens with increasing temperature and humidity (Figure 13.11). Summer lightening and winter darkening of hair verifies this rule. Also, there is an increased secretion of **sebum** with increasing temperatures, which gives the hair a reflective and protective sheen against solar radiation.

The Claude Bernard Rule (1876) This rule relates climatic changes to internal body changes. Bernard observed that blood flow increased in rabbit ears during hot weather for cooling purposes and increased again during cold weather to warm the ears. The mechanism involved is analogous in principle to the valves in thermostatic regulators that control the steam supply to radiators. It is the vasomotor (controlling the size of blood vessels) dilating and contracting device and certain anatomical features that govern the rate of blood flow through the blood capillaries in the tissue layers near the skin in order to maintain skin temperature. By using these mechanisms, animals can tolerate very low temperatures at their extremities. Humans, however, do not have an effective vascular thermoregulator and would find it fatal to imitate the arctic gull, which parades its slim naked legs at −40°F, or the Eskimo dog or arctic bear, which walk on their naked footpads in regions where the temperature is −60°F. Livestock and poultry also possess some of these same mechanisms.

13.7 NUTRITIONAL CONSIDERATIONS OF ENVIRONMENTAL CONDITIONS

Brody frequently told his students that scientists often attempt to prove scientifically what farmers and other laypeople have known for many years.[9] This idea is illustrated by the practical observation that when exposed to cold conditions, farm animals tend to increase their heat production by increasing the amount of food consumed. Cattle will increase **forage** consumption (even if the forage is of inferior quality) during extremely cold weather. This allows them to take advantage of the **heat increment** (Chapter 14). They also tend to increase their tolerance to cold by increasing their deposition of an insulating subdermal fat layer. Furthermore, farm animals may increase heat production by shivering and conserve heat by grouping together, as will be discussed later in this chapter.

Changes are likely in the body's nutrient requirements at extremely low and high temperatures. In one study, the ingestion of pantothenic acid in amounts

[9]Personal experience of the authors.

far in excess of normal requirements improved the physiological response to the experimental stress caused by immersion in cold water. It is believed that stress disturbs **enzyme** activity.

13.7.1 Changing Diet Composition to Fit the Weather

Lowered feed consumption is one of the principal causes of decreased milk secretion in hot weather. In one Missouri study, dairy cows secreted 4 lb less milk and consumed 3 lb less **TDN** for each degree (°F) rise in rectal temperature.

Diets fed to lactating cows during periods of high ambient temperatures should be low in fiber and high in energy. Such diets have a lower heat increment, and because a smaller portion of the metabolizable energy is lost as heat, more **calories** are available for lactation. Possible dietary effects on milk production during the summer are depicted in Figure 13.4. The chief components of high-energy diets are cereal grains. However, grains are low in potassium (Chapter 17), and this, coupled with the increased urinary potassium loss (increased water consumption in hot weather), makes it necessary to provide additional potassium in the diet. Of course, economic considerations in diet composition are vital to profitable animal production.

The greatest heat increment of feeding is caused by **protein**-rich foods. When protein alone is fed to a fasting animal (in an amount possessing a heat value equivalent to the animal's estimated **standard metabolic rate, SMR),** the heat production is raised 30 percent or more above the standard level. **Carbohydrates** cause a rise of 6 percent and fats, 4 percent. These are average values, and the reader should appreciate that there is variation among diets and animals.

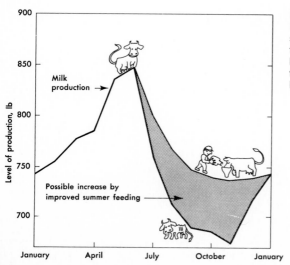

FIGURE 13.4
Milk yield declines during late summer, mostly as a result of high temperatures, but also in part because of the deterioration of the food supply, including the lignification of forages. *(USDA.)*

Actual heat production of 100 cal of net energy from protein, carbohydrate, or fat fed separately to a mature animal whose standard metabolism is 100 cal daily is 130, 106, and 104 calories, respectively. It is readily apparent, then, that a prizefighter should eat beefsteak after the fight and not before, because he will be sweating profusely during the increased activity within the ring and have no need for the added heat of protein digestion. High-protein foods are preferred in cold weather because of their high heat increment of feeding, whereas low-protein diets (but ones that meet daily protein needs) are more suitable in hot weather.

The current trend in animal production is to increase the energy/protein ratio in cold weather and to decrease it in warm weather, thereby sparing the burning of expensive protein as simple fuel. The actual crude protein concentration in this case is higher in summer and lower in winter.

13.7.2 Comparative Feed Efficiency of Homeotherms and Poikilotherms

Why do humans use **bird eggs** in preference to reptile eggs as food? Although it has a long **generation interval,** *Chelonia* (green turtle) produces eggs and meat that have been used by humans for many years. A green turtle can lay as many as 200 eggs in one day. Many tortoises that **graze** and do not compete with humans for cereals have been used for meat.

The **fowl** was **domesticated** largely for **cock**fighting and religious reasons (Chapter 2) and not for egg laying. The adaptation for efficient egg laying was the work of humans. Since poultry compete with humans for cereal grains, why not substitute a poikilotherm, which does not burn large quantities of food to maintain its body temperature? (The authors do not advocate this change in diet.)

Poikilotherms have low metabolic rates, but since they grow so slowly, their net **feed conversion** is not superior to that of the homeotherms (Table 13.1).

TABLE 13.1
TIME AND MAINTENANCE ENERGY EXPENDED FOR DOUBLING BODY WEIGHT
IN HOMEOTHERMS AND POIKILOTHERMS

	Body wt, g			Time required to double body wt, days			Maintenance energy expended (metabolism, cal)		
	Cat	Dog	Fish (pike)	Cat	Dog	Fish (pike)	Cat	Dog	Fish (pike)
Birth weight	87	225	70						
Doubling once	174	450	140	8	10	270	3200	3000	3300
Doubling twice	348	900	280	12	13	300	4000	3070	2350

Source: Modified from M. Rubner, *Biochem. Z.,* **148:**222, 268 (1924).

Table 13.1 shows that both growth rate and maintenance cost are higher in homeothermic animals (cat and dog) than in the poikilotherm (pike), but the overall efficiency is apparently about the same in both. This might be expected, since temperature probably affects the **anabolic** (productive) and **catabolic** (aging, destructive, and maintenance) processes to a similar degree. (In poikilotherms, the speed of life processes, such as aging, feeding, growth, and metabolism, increases and decreases with temperature.)

13.8 FEVER (PYREXIA)

When profound vasoconstriction occurs, one feels cold and may shiver. The body temperature is then elevated, causing a fever **(pyrexia),** which warms the skin vessels, causing their dilation (skin of face appears flushed). With recovery the temperature falls, and profuse sweating occurs to enable the body to dissipate heat rapidly. However, shivering and sweating should be considered emergency measures.

Fever has been recognized as a cardinal sign of disease since the time of Hippocrates. Most fevers reflect a disturbance of the normal equilibrium between thermogenesis and thermolysis. Malarial fever, however, illustrates another phenomenon through which heat is generated by shivering (heat production is increased about threefold) during a chill. As a chill commences, most body temperatures are normal, although skin temperature is subnormal.

Fever increases metabolic rate about 13 percent for each 1°C rise (7.2 percent per 1°F); hence the basis for the old adage, "Feed a fever." During severe infections, from 300 to 400 g of body protein may be destroyed daily (about 4 times the normal dietary intake; see Chapter 3). **Spontaneous** temperature variations of from 1 to 2°F are common during the first year of human life and should not be considered "fever" in the true sense.

Humans can tolerate a temperature limit for fever of about 107 to 110°F. (It may rise in **premortal** states to 112°F.) Most diseases have their characteristic temperature curves. Whereas the temperature rises steadily until the crisis is reached during pneumonia, the temperature in cases of malaria fluctuates considerably. Fever may be caused by (1) disease organisms (dead or alive), (2) chemical pyrogens (fever-producing substances) such as dinitrophenols, (3) thyroxine, (4) surgery, and (5) other foreign bodies.

An elevation in body temperature is often associated with the terms *heat exhaustion,* **dehydration** *exhaustion,* **heat stroke,** and *sunstroke.* These conditions may be precipitated by a hot environment, excessive heat production consequent to strenuous physical exercise, and/or exposure to sunshine (especially in the tropics) concurrent with high ambient temperatures. The collapse may result from the high temperature **per se,** but more likely from circulatory failure. The latter results largely from peripheral vasodilation, which necessitates accelerated circulation of the diminished available blood. (Too much blood is near the body surface for cooling and therefore is not available for oxygen transport.)

Heat cramps probably result from an excessive loss of water and salt (NaCl), although ascorbic acid (vitamin C) and other water-soluble vitamins and salts are also excreted in sweat. As a profusely sweating species, humans excrete considerable ascorbic acid in sweat; hence they have a greater need for vitamin C. (This serves as a scientific basis for athletes drinking additional orange juice—or other sources of vitamin C—during periods of vigorous exercise.) *Dehydration exhaustion* may occur after extended exposure to heat when the body fluid loss is not replaced.

In the nude, men have a higher skin temperature than women (partly due to less subcutaneous fat in men). When exposed to the same ambient temperature of 72°F, a relative humidity of 30 percent, and an air velocity of 20 ft/min, the mean skin temperature of women is 2°F below that of men, and that of the hands and feet is 5°F lower. This may be a partial basis for a commonly expressed saying about women, "cold hands but a warm heart." Room temperatures of 71.5°F for men and 76.0°F for women produce the same skin temperature (92°F) of both.

Clothing furnishes humans with a private climate. The degree of privacy depends on the amount worn and/or the amount of air that the clothing entraps. (Air held stationary enhances insulation.)

From the foregoing discussion related to skin temperature in man and woman and the potential protection from cold afforded each from their clothing, it would seem that women need extra protection from clothing. Actually, heat loss in women is about 10 percent less than in men, and because of their subcutaneous fat layer, women can withstand cold better than men can.

13.9 THE THERMONEUTRAL ZONE

The relative position of the thermoneutral (comfort) zone for different animals varies with many factors, including age and body size, nature of protective coverings, acclimatization, and capacity and opportunity for evaporative cooling, but depends largely on feed intake rate and the activity level of the body. For example, a resting, postabsorptive man in the nude feels most comfortable at about 85°F; after a meal of beefsteak, perhaps at 75°F; during a marathon race, possibly near freezing (32°F).

The thermoneutral zone, as defined by segment $B'B$ of Figure 13.5, may not be the best for the highest productivity. This is because productive processes involve a **heat increment** not easily dissipated in a warm environment. A cold environment (below temperature B' in Figure 13.5) is probably more stimulating to high activity and productivity than the warmer temperature B.

In Figure 13.5 heat production is plotted against ambient temperature to depict the relation between chemical and physical heat regulation. The physical temperature-regulation segment includes the zone of thermoneutrality *(BB')*, where the animal does not employ chemical thermoregulatory devices. The thermoneutral zone for humans is considered to be 72 to 85°F. Under standard

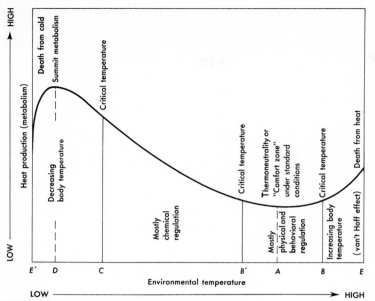

FIGURE 13.5
Generalized diagram of the environmental-temperature influence on heat
production in homeotherms (warm-blooded animals). Note the broad range
of accommodation to low temperatures in contrast with the restricted range
of accommodation to high temperatures. The increased heat production
with increasing cold (from *B'* to *D)* is a biological adaptation that provides
protection to the animal. The increasing heat production corresponding to
increasing heat (from *B* to *E)* results from a physiochemical necessity
expressed by the van't Hoff–Arrhenius generalization (Section 10.5) and is
destructive to the animals, ending in death *(E).* (Missouri Agr. Expt. Sta.
Bull. *423, 1948).*

metabolism (fasting and resting) conditions, the "critical" temperatures *(BB')* of
most farm animals are 60 and 90°F.

Thermogenesis starts increasing in direct response to cold as ambient
temperature decreases below *B'* (Figure 13.5) in an attempt to balance the
increasing thermolysis. When the environmental temperature reaches point *C,*
the temperature-regulating mechanism is no longer able to cope with the cold,
and body temperature starts decreasing, despite increased thermogenesis.

The French physiologist Giaja called environmental temperature *D* (Figure
13.5) the *summit metabolism* (maximum sustained heat production), after which
point a decrease in ambient temperature breaks down the homeothermic
mechanisms, resulting in a decline in both heat production and body temperature and eventually in death of the animal.

It is interesting that the thermoneutral zone of some poikilotherms is about
the same as that of some domestic homeotherms. For example, maximum

feeding by the grasshopper is interrupted when air and soil temperatures exceed 80 and 113°F, respectively. The grasshoppers then climb vegetation and remain relatively motionless, only occasionally nibbling the vegetation on which they are resting.

13.10   HEAT PRODUCTION

In cold conditions, humans and animals need to conserve and/or produce heat. They absorb heat from surrounding objects with temperatures higher than their own, from direct or reflected sunshine, and/or from stoves or open fires. Some heat is also gained by ingestion of hot foods or fluids. As shown in Figure 13.6, a delicate balance exists between the factors increasing heat production and those enhancing heat loss. Animals have a huge "maintenance cost" to supply energy for circulation, respiration, excretion, muscle tension, and many other processes. The maintenance-energy expense is eventually given off as heat.

As discussed in Chapter 14, water, vitamins, and minerals are essential nutrients for humans and animals; however, they provide no metabolizable energy and thus no potential body heat. *Basal heat production* is derived from *carbohydrates, fats,* and *proteins* (Figure 13.6). Factors increasing heat production over standard metabolic rate (SMR) are depicted in Figure 13.6 and include

FIGURE 13.6
Balance between the factors increasing heat production and those enhancing heat loss in humans.

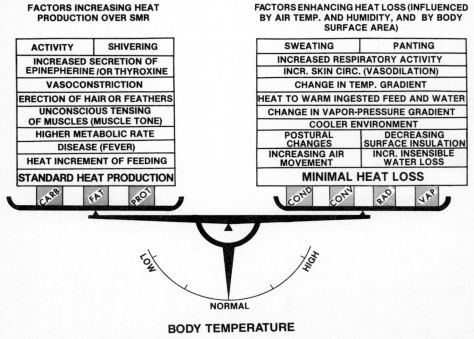

FACTORS INCREASING HEAT
PRODUCTION OVER SMR

ACTIVITY	SHIVERING
INCREASED SECRETION OF EPINEPHERINE /OR THYROXINE	
VASOCONSTRICTION	
ERECTION OF HAIR OR FEATHERS	
UNCONSCIOUS TENSING OF MUSCLES (MUSCLE TONE)	
HIGHER METABOLIC RATE	
DISEASE (FEVER)	
HEAT INCREMENT OF FEEDING	
STANDARD HEAT PRODUCTION	

FACTORS ENHANCING HEAT LOSS (INFLUENCED
BY AIR TEMP. AND HUMIDITY, AND BY BODY
SURFACE AREA)

SWEATING	PANTING
INCREASED RESPIRATORY ACTIVITY	
INCR. SKIN CIRC. (VASODILATION)	
CHANGE IN TEMP. GRADIENT	
HEAT TO WARM INGESTED FEED AND WATER	
CHANGE IN VAPOR-PRESSURE GRADIENT	
COOLER ENVIRONMENT	
POSTURAL CHANGES	DECREASING SURFACE INSULATION
INCREASING AIR MOVEMENT	INCR. INSENSIBLE WATER LOSS
MINIMAL HEAT LOSS	

LOW   HIGH
NORMAL

BODY TEMPERATURE

(1) physical exercise, (2) shivering, (3) erection of hairs (primarily conserves heat), (4) the unconscious tensing of muscles (muscle tone), (5) vasoconstriction, (6) fever, (7) disease, (8) specific dynamic activity of food (heat increment of feeding), (9) an increased secretion of epinephrine and/or thyroxine, and (10) an increased metabolic rate. Heat gains may also be derived from radiation and conduction (assisted by convection) and by consuming warm foods and drinks.

13.10.1 Shivering

Shivering is a short-term response to acute cold by an unadapted organism and may increase heat production in humans and animals by three- to fourfold. In one experiment men were exposed to cold for 9 days and nights. They shivered almost continuously during the experiment. However, the rate of shivering decreased at night, resulting in a drop in body temperature to as low as 93°F and to a toe temperature approximately that of the room, 56°F.

13.10.2 Variations of Heat Production among Animals

In one study the heat production per unit weight of Zebu **heifers** was observed to be about 20 percent lower than that of Brown Swiss heifers of about the same age and weight. This lower heat production of Indian cattle may partially explain their greater heat tolerance and lower cold tolerance.

Missouri studies demonstrated that Zebu or Brahman cattle are more heat tolerant than animals of European origin (Figure 13.7). This results from their lower heat production, lower productivity, and possibly lower standard metabolism.

Males normally have a higher metabolic rate than females or **castrated** males. Furthermore, a lactating dairy cow may produce twice as much heat as a nonlactating cow of the same size.

FIGURE 13.7
The heat tolerance of the Brahman (center). Note the panting and salivation in the Holstein (left) and to a lesser degree in the Jersey (right) at 105°F. *(Missouri Agr. Expt. Station.)*

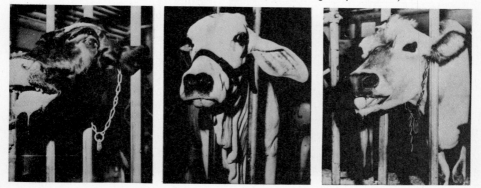

13.11 HEAT DISSIPATION

Heat produced in domestic animals and humans is primarily dissipated in four ways: (1) conduction, (2) convection, (3) evaporation, and (4) radiation (Figure 13.6). Also, a small amount of heat is lost by warming and humidifying inspired air and through the feces and urine. Since each of these four factors is important in heat regulation and is closely related to environmental temperature (Figure 13.8), they will be discussed individually.

13.11.1 Conduction

This means of heat dissipation is based on the principle that heat flows between warm and cold objects. Conduction, therefore, is dependent on (1) the physical contact of the animals with the surrounding surfaces or objects; (2) the temperatures of these surfaces (temperature gradient); and (3) the thermal conductivity and area of the contacting surfaces. High humidity in cold weather increases the feeling of cold because it increases the conductivity of clothing.

Cold water is an efficient and effective means of cooling by conduction. Metals *conduct* heat readily, whereas air, fat, feathers, hair, nylon, silk, wood, and wool conduct heat less readily. Hence people choose a metal frying pan (wood handle) and use a woolen or nylon carpet (full of air pockets) on the floor. Moreover, women protect their legs from the winter cold by wearing nylon or silk stockings. Heat loss via conduction, then, is minimized by the insulation of fur and clothing. Cattle and swine can increase their dissipation of heat via conduction by resting on cold concrete floors.

13.11.2 Convection

Factors influencing the effectiveness of convection in heat dissipation include (1) body surface area, (2) velocity of air movement (fanning), (3) temperature of

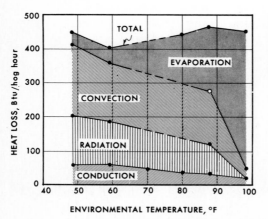

FIGURE 13.8
The effect of environmental temperature on heat loss in swine (75 to 125 lb) by the four major methods of heat dissipation. *(Modified from* Agr. Engr., **33***:150, 1952.)*

the animal's surface, and (4) ambient air temperature. The air immediately adjacent to an animal is usually warmer than the surrounding environmental air. Replacement of the layer of warm air surrounding the animal by the cooler surrounding environmental air removes heat from the animal by convection.

In still air, any object warmer than the air will cause the air near the object to be warmed. Because air that is warmer than the surrounding air is also of lower density, this warmer air will rise, carrying heat away from the warm object and simultaneously entraining cooler air around the object. Convection is enhanced by wind, which may either cool or warm the animal, depending on whether the wind is cooler or warmer than the surface temperature of the animal.

Convection, then, is the mechanism by which heat is transmitted from one molecule to another and then is carried away (heat is transferred to the air, which rises, taking the heat with it). Cooler air comes in to replace warmer air (and vice versa). Although not shown in Figure 13.8, a higher air velocity would have increased the rate of heat loss by convection, giving a lower hog surface temperature, thus decreasing the loss by radiation.

13.11.3 Evaporation

Evaporation is the most important means of heat dissipation during severe heat exposure because the temperature gradient between an animal's surface and the environment becomes increasingly narrow. Evaporation of water from the skin depends on (1) temperature and moisture of the skin; (2) skin covering (hair, wool, feathers); (3) humidity, velocity, and temperature of the surrounding air; (4) respiratory rate and volume; (5) water available for evaporation; and (6) surface area of the animal.

A thick hair coat may (1) reduce air movement about the skin and (2) entrap a layer of water vapor over the surface of the animal, thereby reducing the efficiency of evaporative heat dissipation.

Three types of perspiration are common to humans: (1) **insensible** perspiration (diffusion water), (2) thermal sweat (from sweat glands), and (3) nonthermal sweat (also called *emotional sweat*). *Insensible perspiration* is continuous (unless ambient humidity is 100 percent). This moisture diffuses through the skin and is exhaled from the lungs. *Thermal sweating* occurs at the sweat glands of the skin when humans are exposed to temperatures above their sweating threshold. Research has shown that humans, when unshaded, may perspire as much as 2 liters of moisture per hour at dry desert temperatures. There are over 2.5 million sweat glands in the average human residing in a temperate climate. Humans may increase sweating by 80-fold from the normal rate by immersing the body in water at 108°F. Vaporization of body water is the principal mechanism available for cooling the body when the environmental temperature is above that of the body. Actually, heat will be gained by conduction and radiation under these conditions. *Nonthermal sweating,* or "cold sweat," is of little importance in heat dissipation. It is commonest on the forehead, palms of the hands, and soles of the feet. Driving down some mountain passes, taking examinations, and other stress-producing situations, such as fear, fatigue, mental work, or nervousness, may provoke nonthermal sweating.

Humans produce skin moisture by the function of (1) **eccrine** sweat glands (above 80°F); (2) **apocrine** sweat glands, which open into hair follicles in the **axillae,** pubic, and anogenital regions; (3) sebaceous sweat glands (anatomically associated with the apocrine in that both enter the hair follicle), which produce sebum; and (4) passive cutaneous moisture diffusion. There is also vaporization from the respiratory tract (nose, mouth, **pharynx, larynx, trachea, bronchi,** and lungs). Humans vaporize about one-third of the total moisture from the respiratory tract and about two-thirds from the outer body surface at 95°F.

Persons having sweat glands that are less functional than normal must not expose themselves to high ambient temperatures or else they must find a means to keep their underwear moist. Otherwise, their body temperature rises rapidly on slight exertion at high environmental temperatures (85 to 105°F). It has been calculated that the body temperature of a human exposed to normal room temperature would rise about 1°C/h if no heat were lost evaporatively.

The proportionate amount of heat lost by vaporization can be altered in animals by **shearing.** For example, a normally feathered fowl loses about 50 percent of its heat by vaporization, whereas a frizzle-feathered fowl may lose only 15 to 20 percent of its heat by vaporization, because the frizzle-feathered fowl loses more heat by radiation.

Evaporative heat loss increases with an increase in environmental temperature (Figure 13.8), especially at temperatures higher than 85°F. In sweating species, the blood is shunted to the surface, where it is cooled by the vaporization of sweat. The internal blood deficiency resulting from this may be compensated for by an increased heart rate, and this is suspected as a major cause of fetal stunting in cows and ewes under heat stress during pregnancy.

Shade   Tree shade is very beneficial to humans and animals in hot weather. This is because moisture evaporates from the leaves, cooling the shaded area without interfering with air circulation. Exposure of humans to sunlight in an ambient temperature of 90°F approximately doubles their sweating rate over that in the shade.

13.11.4   Cooling in Nonsweating Species

There is little moisture loss from the skin of nonsweating animals. Only that which reaches the surface by passive diffusion or physical permeability, rather than by sweating (glandular activity), is evaporated. Moisture lost from the body surface of nonsweating species is called *insensible perspiration.*

Birds have no sweat glands and therefore accomplish evaporative cooling mostly by panting (begins at ambient temperatures between 80 and 90°F). Similarly, there are few functional sweat glands in the dog (except on the pads of the feet). Therefore the dog makes good use of evaporative cooling by panting, moving large volumes of air over its moist tongue and air passages. The blood is cooled as it flows through these areas.

Since overheating in nonsweating or lowly sweating species may result from the lack of skin moisture for evaporation, the external application of moisture would be beneficial. Under natural conditions of hot weather, swine resort to the moisture and coolness of mud, cattle stand in cool water, and sheep migrate to higher altitudes.

Fanning cools sweating animals because it accelerates the rate of vaporization. However, except when the ambient temperature is lower than the skin temperature, fanning is of little benefit to cattle, since they sweat very little. Arizona studies showed that evaporative coolers reduced peak temperatures 10 to 12°F below temperatures at corresponding locations under a conventional shade (Figure 13.9). Such a system moves the air, which is cooled to a temperature below that of the animal's surface, providing forced convection. Reproductive efficiency increased from 30 percent in controls to 60 percent in the "cooled" group during the summer months. Moreover, milk yield averaged about 4 lb more daily in the cows having access to the evaporative coolers.

Missouri studies showed that when Zebu cattle were exposed to temperatures of up to 105°F, moisture vaporized at a lower rate (per unit surface area and/or per unit weight) than from European cattle. At temperatures above 105°F the reverse was true. This demonstrates the reserve capacity of the Brahman to tolerate high temperatures.

The normal body temperature of the dairy cow is 101 to 102°F. She has a tremendous capability for producing heat but quite limited abilities to lose it. When exposed to an environmental temperature of 80 to 85°F, the lactating cow quickly responds with accelerated breathing and an elevated body temperature.

FIGURE 13.9
Cattle shades equipped with evaporative coolers. Conventional shades in the background. *(Courtesy Dr. G. H. Stott, University of Arizona.)*

When the ambient temperature exceeds body surface temperature, the animal is unable to dissipate heat to the surrounding air through its skin by sensible means (conduction, convection, and radiation). In this condition, vaporization becomes all-important. Respiratory vaporization is increased by panting; the animal's tongue protrudes, saliva flows, and the animal becomes heat-exhausted (Figure 13.7). An ambient temperature of 107 to 108°F for extended periods of time is fatal to cattle and sheep.

To date, the use of air conditioning by refrigeration in the commercial production of meat, milk, and eggs has been too expensive. In studies at the Missouri Experiment Station, cooling of only the heads of lactating dairy cows was found beneficial. Animals were exposed to an ·85°F ambient temperature, except for the heads of the test group, which were cooled to 65°F (Figure 13.10). Milk production of control cows decreased 25 percent from their level of milk production at 65°F, whereas that of the experimental group (cooled heads) decreased by only 10 percent. Zone cooling is used commonly at the sow's head in farrowing houses.

Evaporative cooling has also proved successful in environmentally controlled cage poultry-laying houses. When the ambient outside temperature is 100°F or above, the use of large exhaust fans to pull air through moistened surfaces (water

FIGURE 13.10
Cooling the head of a lactating dairy cow in an air-conditioned box. The balance of the cow is exposed to ambient temperature. *(Missouri Agr. Expt. Station.)*

is circulated through large areas of wood-wool or excelsior in the side walls) can reduce house temperature 10 to 15°F, depending on outside vapor pressure, which is reflected in an increase in egg laying.

13.11.5 Radiation

Radiation is a very important means of heat loss from animals to cooler objects and heat gain by animals from warmer ones. Like a fire, the sun radiates its energy in straight lines as **electromagnetic** waves of various lengths at the speed of light (186,000 mi/s). Radiant energy does not heat the air directly, but indirectly, by heating solid surfaces such as soil, water, buildings, trees, clouds, dust, and animals. In this way radiant energy is changed into thermal energy, which in turn heats the air by conduction and convection. Also, solid objects are heated by reflected radiation.

The loss (or gain) of heat transferred by infrared waves (wavelength above 700 nanometers) depends on temperature. When people sit near a large window (glass temperature of 0°F), they may feel cold. Having no appreciation for this principle, they frequently check the thermostat, only to find that it indicates a room temperature of 72°F. However, if they appreciate the radiation principle, they realize that heat rays are leaving their body and striking the cold glass window, resulting in a heat drain. Animals also receive infrared thermal radiation from their surroundings. All objects at a temperature above absolute zero (−273°C) radiate energy.

Factors influencing heat loss (or gain) by means of radiation include (1) surface area of the animal, (2) temperature of the animal's skin, (3) temperature of the air surrounding the animal, and (4) **emissivity** of the animal's skin (the ability to absorb and emit heat).

Most organisms have an emissivity between 0.95 and 1 and are therefore called *blackbodies.* The surface of the ground radiates very nearly as a blackbody. The atmosphere radiates infrared radiation toward the ground and also outward toward space. Water vapor, carbon dioxide, and ozone have strong absorption bands at infrared wavelengths and therefore provide an efficient means of radiating streams of energy (heat) groundward and spaceward. This explains in part why clear, dry nights are cold ones, whereas humid or overcast nights tend to be warmer because of additional infrared radiation from the sky. Additionally, clouds shield animals from excessive heat loss to outer space.

Highly polished metals, such as aluminum foil or copper, have a low emissivity of about 5 percent, as contrasted with 100 percent emissivity of solid-black material or 98 percent emissivity of skin (dark or light). This means that if the temperature of the object radiating in the far-infrared range is higher than that of the skin, the skin absorbs 98 percent and reflects 2 percent of the radiated heat. Conversely, if the skin temperature is above that of the environment, it approximates a heat-radiating blackbody and emits heat.

Aluminum paint absorbs only 40 percent of solar radiations compared with 70 percent absorption by asbestos shingles, unpainted wood, stone, brick, and red tile and 90 percent absorption by black surfaces such as slate or tar. A white surface may absorb only 20 percent of the visible radiation (only about one-half of the energy in the solar spectrum is in the visible portion) falling on it; hence the basis for changing the *color of the coat* with changing environmental temperatures (Figure 13.11). A white or yellow coat with a smooth, glossy texture is best to minimize the adverse effects of solar radiation.

When considering the possibility of animal life on other planets one must make reasonable estimates of the radiation spectrum and heat load relationships

FIGURE 13.11
The effect of environmental temperature on growth and hair color in the Brahman (top), Shorthorn (center), and Santa Gertrudis (bottom) breeds. Brahmans had darker hair at 50°F than at 80°F. The Shorthorns at 80°F were stunted. Other breeds did equally well at 50 and 80°F. *(Missouri Agr. Expt. Station.)*

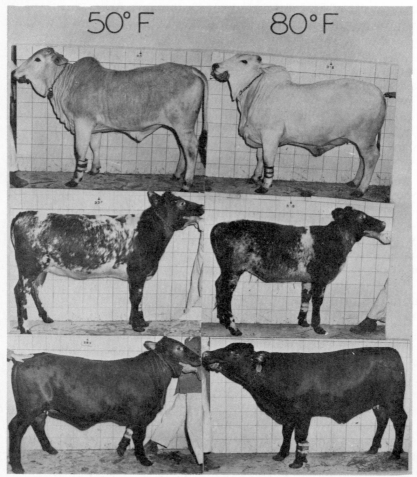

at the surface to be reasonably assured that *constructive* radiations are present and *destructive* radiations are absent.

Cosmic Radiation   In 1927, Herman J. Muller, a scientist, reported that the frequency of gene mutations was affected by **irradiation.** For this observation, he was awarded a Nobel prize. The earth's surface is constantly being bombarded with high-energy cosmic rays similar to those used by Muller. Could cosmic radiation have played a role in the natural selection of animals through its influence on gene mutation?

Animal life has been able to evolve over the earth's surface because of protection from the cosmic cold of outer space and from the ionizing radiations of the sun by an *envelope of atmosphere.* Furthermore, animals have been shielded from cosmic particles and solar storms by a magnetic field.

Sunburn   Although the ultraviolet portion (short wavelengths) of solar radiation does not greatly increase the heat load of humans, it is very important because after absorption by tissues, it causes damage to their cells (sunburn of the skin). From a practical viewpoint, it is well to remember that these short wavelengths may be significantly reflected from snow, light clouds, or still water.

Several pathological disturbances are associated with exposure of animals to intense sunlight. For example, when certain cattle are exposed to intense sunlight containing a relatively large ultraviolet component, they develop ocular squamous carcinoma (cancer eye). Curtis[10] reported that Hereford cattle—the breed most susceptible to cancer eye—vary in pigmentation of the eyelid and eyeball, and that there is a strong correlation between low pigmentation and high incidence of cancer eye. The **heritability** of eyeball pigmentation approximates 0.6, and that of eyelid pigmentation, 0.4.

Photosensitization can occur in any animal. It is a condition resulting from hypersensitivity of the skin to sunlight. Hypersensitivity is caused by the presence of a blood component that is derived either directly or indirectly from feed. When this blood compound is acted on by ultraviolet radiation, it is transformed into a toxic substance that harms and can even kill the animal. Heavily pigmented skin protects the blood from ultraviolet rays. One remedy for mild photosensitization is to shade the affected animal from direct sunlight.

A strain of Southdown sheep has, for example, been found to be congenitally photosensitive. These sheep have an impaired ability to clear phylloerythrin, a photosensitive end product of chlorophyll metabolism, from their blood. Phylloerythrin is formed from chlorophyll when certain substances are present in ingested feedstuffs. Newly sheared sheep are especially vulnerable to photosensitization and sunburn. Photosensitizing agents may be derived directly from feedstuffs. For example, the buckwheat plant is known to contain three such compounds.

---

[10]S. E. Curtis, *Environmental Management in Animal Agriculture,* Iowa State University Press, Ames, Iowa, 1983.

The seasonal variation of sunlight is enormous in the polar regions and slight in equatorial regions. The diurnal changes in sunlight are dramatic in the tropics, whereas they are slight in the polar regions.

13.12 EFFECTS OF CLIMATE ON PRODUCTION

It is now recognized, both commercially and experimentally, that high ambient temperatures depress growth, reduce production, and lower reproductive efficiency. During periods of hot weather, hens eat less and lay fewer and smaller eggs, with thinner shells (Chapter 12). A comfortable cow converts feed into milk more efficiently than one exposed to high ambient temperatures (Chapter 11). In one study, when the environmental temperature was increased from 60 to 95°F, the mean rectal temperature of cows increased from 101 to 104°F and the daily milk yield declined from 27 to 17 lb. Moreover, growth rate is greatly reduced in most European breeds of cattle during hot weather (Figure 13.11).

Missouri studies show that there is a change in the proportion of **VFA** (acetate, propionate, and butyrate) produced by cows when their **rumen** temperatures are elevated by circulating warm water through the rumen (Figure 13.12). This has practical significance because the VFA provide the **ruminant** with 60 to 80 percent of its energy needs (Chapter 15), and cows often drink warm water in the summer and cold water in the winter months.

High ambient temperatures have an adverse effect on **spermatogenesis** and are also detrimental to the **ova** and **sperm** in the female reproductive tract. The results are lower fertility and higher embryonic mortality. In one study, semen quality of dairy bulls was reduced more than that of the Brahman when the animals were exposed to high temperatures.

How single climatic factors such as temperature, relative humidity, air movement, and radiation (from fluorescent and incandescent lamps) affect humans and animals can be studied under controlled environmental conditions in **psychrometric** and other kinds of environmental *chambers* or *climatic laboratories* (Figure 13.13).

13.13 SUMMARY

Humans have long appreciated the value of additional clothing or skin (animal) for warmth in cold conditions and little attire when exposed to heat. They learned very early to modify their environment by using fire for heat and light, but only relatively recently have humans been able to cool themselves using mechanical refrigeration. Animals have many mechanisms that they can employ to conserve heat and increase heat production, but they possess limited means for increasing heat dissipation. Whether humans can genetically select animals to fit the areas where they want and need them or develop an economical means of cooling them remains unknown. However, the future success of the animal industry will largely depend on the effects of environment on producing animals.

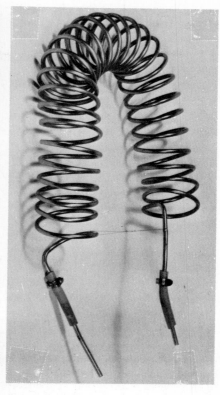

FIGURE 13.12
A technique for circulating water (warm or cold) through the rumen to study the effect of changing rumen temperature on VFA production and feed digestibility.
Water-storage tank with heater and temperature controls (upper left) is connected with a 24-ft (1/4-in) section of stainless-steel tubing (upper right) by high-pressure plastic tubing (bottom). *(Courtesy Dr. F. A. Martz and Missouri Agr. Expt. Station.)*

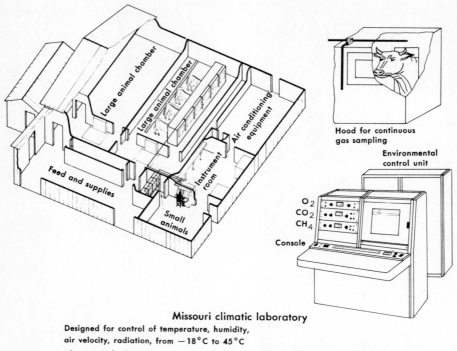

Hood for continuous gas sampling

Environmental control unit

O_2
CO_2
CH_4

Console

Missouri climatic laboratory

Designed for control of temperature, humidity,
air velocity, radiation, from −18°C to 45°C

(shown are facilities for continuous metabolism measurement of small and large animals)

FIGURE 13.13
Animal psychroenergetic laboratory for conducting studies under controlled environmental conditions of the effects of temperature (0 to 110°F), humidity (0 to 100 percent RH), air movement, light (fluorescent and incandescent), ventilation rate, and other environmental factors on the health and efficiency of growth, production, and reproduction of animals. (Missouri Agr. Expt. Station and USDA.)

Environment includes not only the climate complex but also nutrition, disease control, management, and other related factors. The genetic makeup of the animal determines its ability to produce, whereas the environment determines whether the animal will achieve the limits of its inherited capabilities. It is estimated that approximately 80 percent of **phenotypic** variability results from environmental factors.

The "climate complex" includes dozens of factors in addition to temperature, humidity, barometric pressure, wind velocity, and solar radiation, discussed here. *Ecology,* the interaction of animals with their environment, must be appreciated from the standpoint of energy flow, animal temperature, diffusion theory, chemical-rate processes, modern molecular biology, biological rhythms, air pollutants, social factors, pathogenic microbes, insects, and worms, to name a few.

The complexity of biological problems associated with growth, reproduction, productivity (meat, milk, eggs, wool), longevity, genetic mutations, genetic differences among individuals and species, physical activity and adaptation, and many others must be attacked with all the sophistication of modern science if humanity is going to continue to "be fruitful, and multiply, and replenish the earth, and subdue it: and have dominion over the fish of the sea, and over the fowl of the air, . . . and over the cattle, and over all the earth, . . . and over every living thing that moveth upon the earth," as instructed by its Creator (Genesis 1:26, 28).

STUDY QUESTIONS

1 What were Aristotle's thoughts pertaining to the effects of climate on people and their livestock?
2 Is there a symbiotic relation between humans and animals? Explain your answer.
3 What is the effect of light on reproduction in poultry? In sheep?
4 Define ecology. What does environmental physiology deal with?
5 Cite examples of animals that are better adapted to some climatic regions than others.
6 Why are Indian cattle better adapted to warm climates than European cattle?
7 What is acclimatization?
8 Distinguish between natural and artificial physical environmental factors affecting humans and the animals that serve them.
9 Cite several examples of the natural adaptation of animals to their environment.
10 What is polypnea? What causes it?
11 What is meant by homeothermy? Homeostasis?
12 Differentiate between estivation and hibernation.
13 What is meant by poikilothermy? Name some poikilotherms.
14 Why are homeothermic mechanisms not required by developing mammals before birth?
15 Where is (are) the heat-regulating center(s) of the body?
16 By what means can heat loss be reduced in cold weather?
17 Discuss briefly the following rules governing climatic adaptations: (1) the Bergmann rule, (2) the Wilson rule, (3) the Gloger rule, (4) the Claude Bernard rule.
18 What is meant by heat increment of feeding? By standard metabolic rate?
19 Would a high-protein diet be beneficial to humans in cold weather? Why?
20 Is there much difference between homeothermic and poikilothermic animals in their overall efficiency (maintenance energy expended)? Why?
21 What is thermogenesis?
22 What is meant by the thermoneutral zone? Cite some factors that influence it.
23 What are the principal factors that cause an increase in heat production over standard metabolic rate?
24 Discuss briefly the four primary means of heat dissipation in domestic animals.
25 What is insensible perspiration?
26 When exposed to a high ambient temperature, which animal would benefit most from an electric fan: (1) a chicken, (2) a sow, or (3) a cow? Why?
27 In general, is the production of farm animals in the United States depressed more by cold or hot weather?
28 For what purposes are psychrometric chambers used?

CHAPTER 14

PRINCIPLES OF NUTRITION— PLANT AND ANIMAL COMPOSITION[1]

So basically influenced are we by the matter of food and drink that revolutions, peace, war, patriotism, international understanding, our daily life and the whole fabric of human social life are profoundly influenced by it—and what is the use of saying "peace, peace" when there is no peace below the diaphragm. This applies to nations as well as individuals—men refuse to work, soldiers refuse to fight, prima donnas refuse to sing, senators refuse to debate, and even presidents refuse to rule the country when they are hungry.

Lin Yutang (1895–1976)

14.1 INTRODUCTION

Nutrition is the science that deals with food and the **nutrients** it contains. It includes the process of providing body cells with the proper amount and balance of nutrients to enable the cells to function in the many metabolic and endocrine processes involved in growth, body maintenance, work, production (meat, milk, eggs, wool), reproduction, and optimal formation of the body's defense mechanisms against disease. Nutrition, therefore, has a profound effect on the health of humans and animals.

Rapid strides have been made in recent years in establishing nutrient requirements of humans and animals. Basically, the science of nutrition tries to answer several questions about nutrients: What are they? What do they contribute? Why do humans and animals need them? How do they function

[1]The authors acknowledge with appreciation the contributions to this chapter of Dr. G. C. Fahey, Jr., Department of Animal Science, University of Illinois, Urbana.

within the body? How does the body utilize food substances for the various body processes?

A fundamental factor underlying the behavior of humans and animals has been their desire and need for food. Humans and animals prey on each other to satisfy this need. Availability of food has caused the migration of both humans and animals from one region to another, and it has been largely responsible for the settling and agricultural development of new lands. A scarcity of food has been the major cause of war among races of people. So important is the self-perpetuating instinct to meet food requirements of the body that many religious beliefs are based on concepts related to food.

Even though primitive humans knew that food was essential for survival, they knew nothing about the specific nutrients required by the body. They probably knew only that air, food, and water were the essentials of life. It has been said that the history of humanity perhaps could be written in terms of diet and the fulfilled promises of nutrition.

The modern era of the science of nutrition was pioneered by the French chemist A. L. Lavoisier, in the 1770s. Lavoisier was the first to recognize that animal heat is derived from the **oxidation** of body substance. He compared animal heat with that produced by a candle. The general form of the apparatus Lavoisier used in his experiments was illustrated in two drawings made by Madame Lavoisier. The methods of study he used, however, are unknown, because Lavoisier was executed on May 8, 1794, at the age of 51, by the Paris Commune. (He was found guilty of allowing the collection of taxes on water contained in tobacco.)

The animal body is a remarkable chemical laboratory that operates at relatively low temperatures and in which **genes** guide growth and development. Genes cause certain cells to produce **enzymes** that help digest food, enable muscles to contract, and assist **cells** in carrying out their varied and complex processes. Food is basic to life and is perpetually related to body chemistry and health. **Diseases** such as arthritis, **atherosclerosis,** and cancer have their genesis in the biochemistry of the body. Eventually, better understanding of how all these body processes work will be achieved. Only then will health and life itself be better understood.

The period of the industrial revolution has been paralleled by a biological revolution of comparable significance. During the past century, approximately 50 nutrients have been identified as essential for farm animals, and the integration of nutritional knowledge with genetic advances and modern production practices has produced remarkable achievements indeed. A notable example is the production of more than 25,000 kg (more than 55,000 lb) of milk in a single lactation by a Holstein cow.

Since the dietary needs of high-producing animals differ greatly from those of low-producing ones, future increases in the productivity of food-producing farm animals will depend in large part on the progress that can be made in the food and nutritional sciences. The challenge is great, but so is the opportunity to improve the nutritional status of animals that can in turn provide a reliable supply of high-quality foods beneficial nutritionally to humanity.

14.2 COMPOSITION OF PLANTS AND ANIMALS

All flesh is grass.

Isaiah 40:6

Plant and animal tissues are composed of water, **carbohydrates,** *proteins, lipids* (including **fat** and related substances), **vitamins,** and mineral matter. Plants contain the same substances as animals, but in different amounts. The main difference between plants and animals is that plants usually contain large amounts of carbohydrates whereas animals contain only traces. The cell walls of plants are composed largely of fibrous carbohydrates; cell membranes of animals consist mostly of proteins. Plants generally store their reserve food as starch; animals store theirs as fat. Animals depend on plants for energy; plants derive energy from the sun and manufacture certain nutrients that cannot be manufactured by animals.

14.2.1 Water

Chemically, water consists of two **atoms** of hydrogen and one of oxygen (H_2O). It is the most abundant and most important constituent of plant and animal tissues. Green plants contain 70 to 80 percent water, whereas animals contain from 40 to 75 percent. The amount of water an animal's body contains varies with its age and degree of fatness. For example, an embryonic **calf** is 90 percent or more water, a newborn calf is slightly more than 70 percent water, and a steer ready for market contains only 40 to 45 percent water.

Animals have three sources of water. The first is that which they drink. The second is that **ingested** as a component of food. The third is known as *metabolic water,* which is derived from the breakdown of carbohydrates, fats, and proteins. Metabolic water is the chief source of water for animals during **hibernation** (sleep or rest during winter).

Water has many functions in plants and animals. It transports nutrients from the soil to the plant leaves, where various organic substances are formed. It also transports nutrients from one part of plants and animals to another. Water distends the cells in plants and animals, thus helping them to hold their shape. It is used in many biochemical reactions in the body. It helps to regulate the temperature of plants and animals through **evaporation** and other processes. For example, when exposed to high environmental temperatures, humans sweat (or perspire) to maintain body temperature, whereas swine cannot sweat appreciably and so wallow in water to cool themselves. Water also is the principal constituent of the substances in the body that lubricate the joints. Moreover, an aqueous medium is necessary for most metabolic reactions of the body.

It is obvious that large amounts of water are needed by plants and animals in the performance of their various body functions. In fact, an animal will die more quickly from the lack of water than from the lack of any other dietary essential.

14.2.2 Carbohydrates

Carbohydrates are organic compounds that contain carbon (C), hydrogen (H), and oxygen (O). They are the main organic compounds found in plants and commonly represent 50 to 75 percent of the total dry matter in **livestock** feeds. Carbohydrates are abundant in most seeds, fruits, roots, and tubers of plants. They are formed by the process of photosynthesis, which involves the action of sunlight on chlorophyll. Chlorophyll is the active photosynthetic material in plants. It is a complex molecule with a structure similar to that of **hemoglobin,** which is found in the blood of animals. Chlorophyll contains magnesium; hemoglobin contains iron. More specifically, carbohydrates are formed of water (H_2O) from the soil, carbon dioxide (CO_2) from the air, and energy from the sun. A simple chemical reaction in which a carbohydrate (glucose) is synthesized by photosynthesis in the plant is as follows:

$$6CO_2 \quad + \quad 6H_2O \quad + \quad 673 \text{ cal} \quad \rightarrow \quad C_6H_{12}O_6 \quad + \quad 6O_2$$

| Carbon dioxide | Water | Energy from the sun | Glucose | Oxygen |

Photosynthesis is a very important process because through it comes all the food for humans and animals and most heat for houses and engines. Thus these sources of energy came originally from the sun and were made available by the process of photosynthesis.

Carbohydrates are classified as monosaccharides (simple sugars), disaccharides (two **molecules** of simple sugars), trisaccharides (three molecules of simple sugars), and polysaccharides (many molecules of simple sugars). Selected carbohydrates of food are grouped in Table 14.1.

Monosaccharides are simple sugars containing five or six carbon atoms in the molecule. They are water soluble. The monosaccharides that contain six carbon atoms have the molecular formula $C_6H_{12}O_6$. They include glucose (also known as dextrose), found in plants, ripe fruit, honey, sweet corn, etc. In animals, glucose is found mainly in the blood, where a certain concentration is vital for life as well as being an immediate source of energy. An ill person can be fed by having glucose infused directly into the bloodstream.

Disaccharides (*di* meaning "two") are carbohydrates that contain two molecules of simple sugars. They have the general formula $C_{12}H_{22}O_{11}$. Therefore they represent two molecules of simple sugar minus one molecule of water (two hydrogen atoms and one oxygen atom). The most important disaccharides are sucrose, maltose, and lactose. Sucrose is found in beet or cane sugar and contains one molecule of glucose (dextrose) and one of fructose (levulose). Sucrose is very sweet and is commonly used to sweeten foods. Maltose is found in germinating seeds and contains two molecules of glucose. It is only about one-third as sweet as sucrose. Lactose is milk sugar and is found only in milk (or milk products). It contains one molecule of glucose and one of galactose. It is even less sweet than maltose, which is one reason large quantities can be consumed.

TABLE 14.1
CLASSIFICATION OF SELECTED CARBOHYDRATES OF FOOD

Classification	Name	Monosaccharide unit	Proximate analysis portion
Monosaccharides	Arabinose Deoxyribose* Ribose Xylose	$C_5H_{10}O_5$	Nitrogen-free extract
	Fructose (levulose) Galactose Glucose (dextrose) Mannose	$C_6H_{12}O_6$	
Disaccharides	Lactose Maltose Sucrose	Galactose, glucose Glucose, glucose Glucose, fructose	
Trisaccharide	Raffinose	Galactose, glucose, fructose	
Polysaccharides (usually contain more than 10 monosaccharide units)	Dextrins Glycogen Gums Starch	$(C_6H_{10}O_5)_n$	
	Cellulose Hemicellulose	$(C_6H_{10}O_5)_n$	Crude fiber

*Deoxyribose has the formula $C_5H_{10}O_4$. The prefix deoxy- means minus one oxygen atom.

Polysaccharides have the general chemical formula $(C_6H_{10}O_5)_n$, which means that they contain many molecules of simple sugars. The two principal groups of polysaccharides are starch and cellulose, although there are other minor and less important ones.

Starch is the most prevalent polysaccharide in plant seeds, fruits, and tubers and is easily digested by animals. When starch is broken down, it yields many molecules of glucose. Glycogen, "animal starch," is found in small amounts in the liver, muscles, and other tissues of an animal's body. It also contains many molecules of glucose.

Cellulose is a polysaccharide that has the same general formula as starch $[(C_6H_{10}O_5)_n]$. It is found mostly in the cell walls and woody portions of plants. Cotton is almost pure cellulose. Cellulose and hemicellulose (fibrous carbohydrates) are not readily digested because animals lack the proper enzymes. Animals can utilize these carbohydrates only if they are first degraded by microorganisms that produce cellulose- and hemicellulose-degrading enzymes.

All animals have microorganisms in their intestinal tract. In most animals, there is only slight utilization of fibrous carbohydrates. However, **ruminants** (cattle, sheep, and goats) have evolved with a rumen (paunch) that harbors millions of microorganisms that degrade fibrous carbohydrates. These microbes digest the fibrous carbohydrates and make their bodies and their end products of digestion (largely volatile fatty acids) available for their host (Chapter 15).

The foremost functions of carbohydrates in animal diets are to supply energy and heat for body processes. (Certain carbohydrates, such as ribose, are also needed for certain substances in the animal's body, such as **RNA**.) Carbohydrates also aid in the utilization of body fat and exert a sparing effect on protein. Surplus energy is converted into fat, which is stored as a potential energy source for the animal. Crude fiber (including cellulose) is a source of heat and energy when digested. It also prevents **compaction** by providing a laxative effect and by maintaining the proper muscular tone in the alimentary tract. This helps prevent **bloat** in ruminants.

14.2.3   Fats and Oils (Lipids)

Fats and oils contain carbon, hydrogen, and oxygen, as do carbohydrates, but they contain more carbon and hydrogen in proportion to oxygen than do carbohydrates. For example, sucrose has the general formula $C_{12}H_{22}O_{11}$, whereas a fat containing stearic acid has the composition $C_{57}H_{110}O_6$ (see example on page 438).

A fat molecule is formed by the combination of three fatty acid molecules with one molecule of glycerol, as follows:

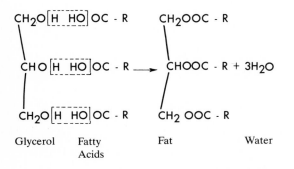

Glycerol   Fatty   Fat   Water
                 Acids

The letter R represents the fatty acid radical, which may be composed of several combinations of carbon, hydrogen, and oxygen. If R represents the hydrocarbon portion of stearic acid ($C_{18}H_{36}O_2$) minus the carboxyl group (COOH), the general chemical composition of the fat is $C_{57}H_{110}O_6$. Thus the fat tristearin is a triglyceride of the fatty acid, stearic acid, and glycerol.

Fats furnish about 2.25 times as much energy as carbohydrates when they are **metabolized** because they have a higher proportion of hydrogen to oxygen. For this reason they have a greater caloric food value per pound. Fats are insoluble in water but soluble in ether. When plant and animal tissues are extracted with ether, the substances that are soluble in this compound are called *lipids* (ether extract).

Nearly all fat found in food is digestible. However, it takes considerable time for digestive juices to break down fats. Therefore fried foods (coated with fats) will be digested more slowly than broiled foods. Fatty foods thereby keep a person from feeling hungry for a longer period of time than do nonfatty foods.

Fats have different melting points. Some are solids at room temperatures, but others, the oils, are liquids. Fats that are solids at ordinary room temperatures are composed largely of **saturated fatty acids.** Most oils are composed principally of *unsaturated fatty acids*. Unsaturated fats consist of fatty acids that have fewer hydrogen atoms attached to the carbon atoms. For this reason, they can readily accept hydrogen or other elements (such as iodine). The **iodine value** is a measure of the degree of unsaturation. The test is possible because an unsaturated fat can easily unite with iodine. (Two atoms of iodine are added to each double bond in the fat.) Examples of a saturated and an unsaturated fatty acid are as follows:

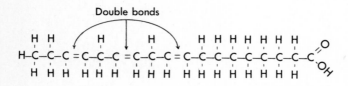

Stearic acid, saturated

Double bonds

Linolencic acid, unsaturated

Beef fat, or tallow, is composed primarily of saturated fatty acids and is a solid at ordinary room temperature. Linseed oil contains a high proportion of unsaturated fatty acids and is a liquid at room temperature.

It is generally true that unsaturated fatty acids (such as linoleic, linolenic, and arachidonic) are necessary for proper nutrition. They must be supplied in the diet because the animal cannot synthesize them. For this reason they are called *essential fatty acids*. The essential fatty acids are required for the construction of cell membranes and for the production of prostaglandins, which are hormone-like substances that have many effects on body cells and are involved in the immune response. A deficiency of linoleic acid leads to dermatitis, poor wound healing, impaired growth, and water loss from the skin.

Human infants on a fat-free diet have developed an **eczema** that was later cured by adding fat to the diet. A similar condition has been produced by feeding rats a fat-free diet. It is believed that certain essential fatty acids are necessary for the normal maintenance of the skin. Other experiments show that **calves, lambs, pigs,** and chickens need a minimum amount of fat for normal growth (Figure 14.1), and dairy cows secrete less milk when fed a diet too low in fat. Fat also appears to aid in **absorption** of fat-soluble vitamins. Adequate amounts of the essential fatty acids are supplied in most practical farm diets, however.

Nearly all fats in grains and seeds extractable with ether are true fats. In grass, hay, and other **forages,** ether-soluble substances that are not true fats are found in small amounts. These include **sterols,** carotenes, chlorophyll, **phospholipids,** waxes, and essential oils.

Sterols   These are complex alcohols found in both plants and animals. **Ergosterol** is found in plants and is especially important in livestock nutrition because, when activated by sunlight, it forms vitamin D_2. **Cholesterol** is widely distributed in animal tissues such as the brain, nerves, and blood. It seems to aid in the transport of fat from one part of the body to another. The action of sunlight on 7-dehydrocholesterol results in the formation of vitamin D_3 (Chapter 16).

Carotenes   These are yellow substances found in carrots, sweet potatoes, and milk fat. They are also present in green plants, although the yellow color is masked by the green color of chlorophyll. Carotenes serve as a precursor of vitamin A since animals have the ability to convert them into this essential vitamin.

FIGURE 14.1
The pig on the left was fed a diet adequate in fat (5 percent ether extract). The one on the right was fed a diet deficient in fat (0.06 percent ether extract). Note the loss of hair and scaly, dandrufflike dermatitis, especially on the feet and tail. (*Courtesy Dr. W. M. Beeson, Purdue University.*)

Chlorophyll   This is the green substance in plants. It has very little food value for animals because it is destroyed by digestive juices in the gastrointestinal tract. It has received considerable attention and publicity because of its deodorizing properties imparted to products designed for human use.

Phospholipids (Phosphatides)   These are present in small amounts in the bodies of animals. They contain various combinations of fatty acids, glycerol, phosphoric acid, and nitrogenous groups. Phospholipids are vital in living protoplasm. One of the most important of these compounds is lecithin, found in egg yolk, blood, and liver.

Waxes   These occur in small amounts in plants, forming tiny films on the surface of plant stems and fruits. Their main function appears to be in protecting the plant surface from the weather. They are relatively unimportant as a source of nutrients for livestock.

Essential Oils   These give plants their characteristic odor and taste. They probably function to protect the plant from insect and microbial predators. It is interesting that sagebrush is used to deodorize feces. (The compound responsible for the odor of sage is an essential oil.)

The essential oil of onion or garlic contains volatile flavor compounds high in sulfur. They include hydrogen sulfide, thiols, disulfides, trisulfides, and thiosulfinates. These compounds pass from the ingested plant via the animal's bloodstream into milk.

14.2.4   Proteins

Proteins are organic compounds that contain carbon, hydrogen, nitrogen, oxygen, sulfur, and phosphorus. They are the principal organic food compounds that contain nitrogen. Proteins are essential for life because they constitute the active protoplasm in all living cells. They are found in the reproductive and actively growing parts (such as the leaves) of plants. In animals, the muscles, internal organs, skin, hair, wool, horns, hooves, and bone marrow are composed of proteins, as are other body parts.

The functions of proteins in the body include (1) repair of tissue, (2) growth of new tissue, (3) metabolism (**deamination**) for energy, (4) metabolism into substances vital in body functions (these vital substances include blood **antibodies** that fight **infection**), (5) enzymes that are essential for normal body function, and (6) certain **hormones.**

Proteins consist of long chains of amino acids, which can consist of hundreds and even thousands of the 20 or more naturally occurring amino acids. Proteins are formed by the ribosomes in the cell **cytoplasm** according to the code transmitted from the gene by means of mRNA (Chapter 4). The chemical reaction in which amino acids are linked together in a chain (dipeptide) is as follows:

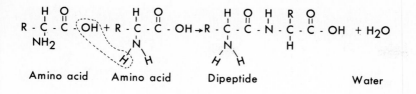

Amino acid   Amino acid   Dipeptide   Water

The $\begin{bmatrix} O & H \\ \| & | \\ - C - N - \end{bmatrix}$ linkage is known as the *peptide linkage.* The different proteins depend on the kind, number, and arrangement of the 20 or more amino acids linked together in the molecule, much as the meaning of different words in the English language depends on the kind, number, and arrangment of the 26 letters of the alphabet they contain.

Plants have the ability to form amino acids (and proteins) from nitrogen, sulfur, phosphorus, and water from the soil and carbon dioxide (CO_2) from the air through photosynthesis. Animals cannot synthesize amino acids. Therefore, they must obtain them directly from the feed they consume or from the digestion of bacteria that contain them and that are present within the animal's own digestive tract (in ruminants). Bacteria synthesize proteins from ammonia. The ammonia may come from the degradation of dietary protein or from nonprotein nitrogen compounds such as urea or ammonium salts.

Animals have a limited capacity to change surplus amino acids into those required by the body. They can convert only a few of the simpler amino acids into others needed by the body. Those that they cannot change and that are essential for growth must be supplied in the feed. Amino acids have been classified as *essential,*[2] *semiessential,* and *nonessential* according to their effect on growth in the rat. An **essential amino acid** is one that is not synthesized in the body at a rate necessary to meet the requirements for normal growth. Current nutrition literature indicates that the following 10 amino acids are essential (Figure 14.3): (1) lysine, (2) tryptophan, (3) phenylalanine, (4) leucine, (5) isoleucine, (6) threonine, (7) methionine,[3] (8) valine, (9) arginine, and (10) histidine. Whether an amino acid is essential depends on the **species,** the animal's age, and level of performance. For this reason, there may be differences within and among species (Figure 14.2 *a* and *b*).

Ruminants can obtain amino acids synthesized by microorganisms in the rumen from simpler nitrogen-containing compounds (Chapter 15). The microorganisms build these nitrogen compounds into proteins in their own bodies, and then they are digested by the animals. By this procedure, ruminants convert inferior proteins and even nonprotein nitrogenous compounds (such as urea) into the superior proteins present in milk and meat. Thus ruminants occupy a unique position in providing the best proteins (milk and meat) to humans.

---

[2]Note that *essential* means that the amino acid is *essential in the diet.* However, all amino acids are needed for protein synthesis. Thus it is important that the animal produce the nonessential amino acids also.

[3]Methionine is the least expensive and most commercially available amino acid.

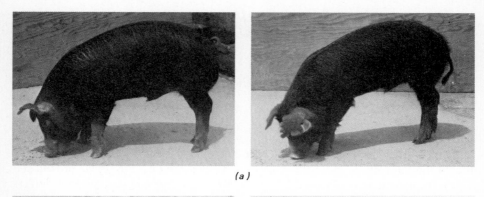

(b)

FIGURE 14.2
(a) The pig on the left was fed a diet containing adequate amounts of lysine (2 percent
DL-lysine). It gained 25 lb in 28 days. The pig on the right was fed a diet deficient in lysine. It
lost 2 lb in 28 days. *(b)* The pig on the left was fed a diet containing adequate amounts of
tryptophan (0.4 percent DL-tryptophan). It gained 25.5 lb in 21 days. The pig on the right was
fed a ration deficient in tryptophan. It lost 8 lb in 21 days. *(Courtesy Dr. W. M. Beeson, Purdue
University.)*

Horses do not possess a rumen but rather have an enlarged cecum, where some
protein synthesis is accomplished by microorganisms. The synthesis of proteins
from simple nitrogenous compounds by microorganisms is limited in nonruminant animals such as swine, poultry, and humans.

Ruminants are less likely to develop certain nutritional deficiencies than are
other farm animals because they obtain essential amino acids, the B-complex
vitamins, and vitamin K from microbial synthesis. They are usually followed in
this respect by the **herbivores** (e.g., horses and rabbits), then by **omnivores** (e.g.,
swine and poultry), and finally by such species as dogs and rats.

The quality of proteins in a feed is usually referred to as being high or low,
depending on whether the feed contains the essential amino acids in the proper
proportion or balance. Proteins from animal sources (meat, milk, eggs) are of
high quality, whereas those from plant sources, such as cottonseed meal, linseed

meal, and wheat shorts, are of low quality. An exception among plant proteins is soybean meal, which is of rather high quality (Chapter 3).

The nutritive value of a protein may be determined by evaluating responses obtained using experimental animals in the laboratory. Nutritive value is an expression of the percentage of protein in a feed digested by an experimental animal and the amount used for metabolic purposes. The nutritive value is called the **biological value** of a protein. A feed having a biological value of 100 percent would be one in which all the digested protein contained in it was utilized for metabolic purposes. Feeds seldom, if ever, have a biological value of 100 percent. Some have a value of 80 to 90 percent, and they are classed as a source of high-quality proteins. A low biological value (**BV**) indicates low protein quality in a feed or small quantities of protein usable for metabolic purposes (largely growth and maintenance).

Nonprotein nitrogen (**NPN**) substances are those found in plants and seeds that contain nitrogen but not in the form of protein. Urea, an end product of metabolism, is a compound containing nonprotein nitrogen. When fed to ruminants, it supplies part of the animal's protein needs because it is synthesized (an energy source such as corn or molasses is required) into protein by microorganisms in the rumen (Chapter 15). Urea is now produced on a commercial scale by processes that utilize nitrogen from the air.[4] Its structural formula is

$$\begin{array}{c} H \\ \diagdown \\ N \text{---} C \text{---} N \\ \diagup \qquad \diagdown \\ H \qquad H \end{array}$$
(with O double-bonded above C)
UREA

Protected Proteins Much of the dietary protein consumed by ruminants is destroyed through fermentative digestion in the rumen. On average, only about 30 to 40 percent of the dietary protein escapes destruction and is available for absorption from the gut. But the major source of protein for the ruminant animal consists of microbial cells produced in the rumen. The microbial cells contain about 50 percent protein and actually synthesize most of this protein from degraded dietary protein. The microbes eventually pass from the rumen to the intestine, where, like kamikaze pilots, they sacrifice their lives for the good of the cause. Approximately two-thirds of the protein of meat and milk resided at one time in microbial cells within the rumen.

This type of digestion facilitates the upgrading of low-quality dietary proteins to high-quality microbial protein. Additionally, bacteria can utilize nonprotein sources of nitrogen such as urea or ammonia to make microbial protein. It is possible, however, for rumen microbes to destroy more protein than they make,

[4]In 1828 a German chemist, Friederich Wöhler, discovered that he had made urea, a substance that a French chemist had found in urine 50 years earlier. Wöhler was trying to make another compound but instead synthesized urea.

resulting in a net loss of protein. It has been estimated, for example, that as much as 25 percent of the protein fed to high-producing dairy cows may be lost due to excessive destruction of dietary protein by rumen microbes.

One possible solution to this problem is to feed proteins less susceptible to breakdown by the rumen microbes. Certain feedstuffs, such as brewers' grains, distillers' grains, and corn gluten meal, have naturally resistant proteins. Approximately one-half of the protein in these feedstuffs will escape breakdown in fermentative digestion. Some of the most important and more commonly used protein sources, such as soybean meal, alfalfa, and cottonseed meal, are more extensively degraded (less than one-third of the proteins in these feedstuffs escapes destruction in the rumen).

Several approaches have been researched and used to decrease protein destruction in the rumen. Heat treatment of feeds will lower protein losses, and there is commercial interest in marketing heat-treated proteins designed specifically for ruminants. Chemicals, such as formaldehyde and tannins, have been used successfully to protect protein. It is easy, however, to overprotect protein, thus making it unavailable for absorption from the gut. This results in rich manure and poor producers.

Feeding protected proteins, or "escape" proteins, as they are called, can reduce the amount of dietary protein that must be fed to obtain the same animal performance. Another potential benefit derived from feeding relatively resistant proteins is that it affords greater opportunity for efficient use of NPN in the diet. When there is extensive degradation of protein in the rumen, the rumen microorganisms are content to use the resulting raw materials to make their body protein, and they do not make good use of the NPN that may have been added to the diet. If, however, less dietary protein is degraded, the rumen bacteria need, and do make use of, the nonprotein nitrogen added to their diet. The goal of an ideal protein-feeding strategy is to have most of the dietary protein escape destruction in the rumen and be used directly by the ruminant. Nonprotein nitrogen would be included in the diet for bacterial use and conversion into protein in the rumen.

14.2.5 Minerals

It has long been known that inorganic minerals play an important role in animal nutrition. Minerals constitute about 3 to 5 percent of the animal body. Animals cannot synthesize minerals and therefore must obtain them from food. Research has shown that these minerals must be provided in proper proportions as well as in adequate amounts (Chapter 17). An excess of certain minerals can be detrimental to the individual, resulting in poor performance in some instances. Fortunately, there is a wide latitude in the amounts of most minerals that can be supplied in the diet without harm to the individual. A mineral deficiency seldom causes death, but performance is often reduced to such a level that economic losses result. These losses are often **insidious,** which means that their effect is more serious than is readily apparent.

At least 15 mineral elements are recognized as performing certain essential functions in the body (Chapter 17). Proof of their necessity has been obtained through careful experimentation. New experimental evidence occasionally adds other minerals to this list. Some of the essential minerals are sodium, potassium, phosphorus, calcium, chlorine, magnesium, iron, sulfur, iodine, manganese, copper, cobalt, molybdenum, selenium, and zinc. Possible essential minerals include fluorine, chromium, barium, bromine, strontium, vanadium, and silicon (Figure 14.3). Many other mineral elements are present in the body because they are present in feed when it is ingested and are not completely excreted in the urine and **feces.** In general, good practical diets contain most minerals at the needed level. But it is often necessary to supply some minerals (mineral supplements) in amounts above those present in the feed. This is especially true of calcium, sodium, chlorine, and phosphorus. Additional mineral elements **(supplements)** must be included in diets in areas where they are known to be deficient in the soil and plants (Chapter 17).

Minerals perform many functions in the body. They (1) constitute a portion of the structural materials (skeleton), teeth, and hemoglobin; (2) function in maintaining the proper acid-base balance in body fluids and therefore are essential for life; (3) maintain the proper cellular **osmotic pressure** necessary for the transfer of nutrients across the cell membrane; (4) maintain the proper acidity of digestive juices so that digestive enzymes can perform their necessary functions; (5) maintain the proper contractility of muscles, especially those of the heart, and play an important role in the normal functioning **(irritability)** of nerves; (6) prevent convulsions; and (7) are related to the function of certain vitamins in bone building. Various minerals also have certain other specific and essential functions within the body, which will be discussed in Chapter 17.

Symptoms of mineral deficiencies include a general lack of thrift, slow and inefficient weight gains, impaired reproduction, and decreased production of milk, meat, eggs, wool, and work. These symptoms are usually not recognized until the deficiencies are extreme. Marginal deficiencies are often insidious and not recognized because no extreme symptoms are observed, yet they rob farmers and livestock producers of millions of dollars annually. Hence the old English proverb "An ounce of prevention is worth a pound of cure" is truly applicable. Mineral supplements are usually relatively cheap, so that deficiencies can be avoided by making the proper amounts available to animals. Mineral balances and interrelations are discussed in Chapter 17.

14.2.6 Vitamins

Vitamins are organic substances required in very small amounts, containing various combinations of carbon, hydrogen, oxygen, and nitrogen. Vitamin B_{12}, or cyanocobalamin, also contains the mineral cobalt. The substances called *vitamins* are not closely related chemically. They are divided into two groups: (1) fat-soluble vitamins (A, D, E, and K) and (2) water-soluble vitamins (the B

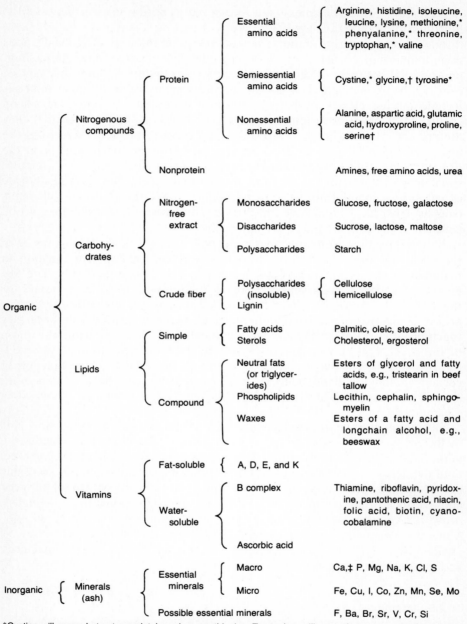

*Cystine will spare, but not completely replace, methionine. Tryptophan will spare, but not completely replace, nicotinic acid. Similarly, tyrosine will spare, but not completely replace, phenylalanine.
†Glycine is essential for young poultry (the chick can synthesize glycine, but not at a rate sufficient for maximum growth). Serine will spare glycine.
‡See Appendix E for elements represented by these symbols.

FIGURE 14.3 Classification of food nutrients.

complex and C). Vitamins differ in their physiological functions and distribution in feedstuffs.

Some vitamins are required by all species of animals, whereas others are required by only a few. It is not necessary to include certain vitamins in the food of some species because they are synthesized either by bacteria within the digestive tract or by the animal itself. Some foods contain **precursors** from which the animal derives the needed vitamin. For example, carotene in food is converted into vitamin A within the animal's body. Also, certain sterols are changed into vitamin D within the body through the action of sunlight. Synthesis of some B-complex vitamins, such as folic acid and inositol, is known to occur in the intestines of some animal species.

Vitamins function in many ways, but probably most important are their actions as organic **catalysts** or as the essential components of catalysts that play an important role in many biological **oxidations.** Vitamins function in normal digestion, absorption, and utilization of food. They function in growth and reproduction and are important for maintaining the general health of humans and animals. Certain vitamins are also essential for regeneration of visual purple in the eye and are necessary for the formation of prothrombin, which is essential for normal blood clotting. Some vitamins are coenzymes (work with enzymes), which are necessary for the utilization of energy and the synthesis of proteins from animo acids. More specific functions of each vitamin are summarized in Tables 16.1 and 16.2 and are discussed in Chapter 16. A classificiation of food nutrients is presented in Figure 14.3.

14.3 ANALYSIS OF FOODSTUFFS

Farm **mammals** depend on plants for the nutrients necessary for maintenance, growth, reproduction, lactation, and other body functions. Some foods from plant sources possess more nutrients than others. For this reason it is important to know their chemical composition. Such information can be used in formulating diets for livestock. The composition of feeds is determined and grouped according to the percentage of dry matter, crude protein, fat, crude fiber, **nitrogen-free extract** (NFE), and mineral matter (ash). Please refer to the Glossary for additional information pertaining to these compositional values.

14.3.1 Dry Matter

The percentage of dry matter is determined by finding the percentage of water in a food and subtracting that value from 100 percent. The percentage of water is determined by drying a finely ground sample of food in an oven until a constant weight is attained.

The calculations involved are as follows:

$$\% \text{ Water} = \frac{\left(\begin{array}{c}\text{wt of food sample} \\ \text{before drying}\end{array}\right) - \left(\begin{array}{c}\text{wt of food sample} \\ \text{after drying}\end{array}\right)}{\text{wt of food sample before drying}} \times 100$$

$$\% \text{ Dry matter} = 100 - \% \text{ water}$$

% Dry matter = 100

All foods contain some moisture, but the amount they contain depends on the length of time and the way they are stored and the amount of moisture in the air. Most dry forages and grains contain 8 to 12 percent moisture.

14.3.2 Crude Protein

The **crude protein** in a food includes all the nitrogenous compounds. To estimate the percentage of crude protein, the nitrogen content of a food is chemically determined and this figure is multiplied by the factor 6.25. This factor is used because nitrogen represents approximately 16 percent of a protein (100 ÷ 16 = 6.25). The approximate protein content is important, not only nutritionally, but because foods rich in protein are the most expensive.

A formula for calculating the percent of crude protein in a food is as follows:

$$\% \text{ Crude protein} = \frac{\text{wt of nitrogen} \times 6.25}{\text{wt of original sample}} \times 100$$

14.3.3 Crude Fat (Ether Extract)

The percentage of fat in a food is determined by extracting a finely ground sample continuously with ether for several hours in a suitable apparatus. The ether and the ether-soluble components are separated from the food. The ether is then evaporated, and the fatty residue is weighed and compared with the weight of the sample before extraction began. The fatty residue remaining after the ether is evaporated contains many substances other than true fat. This fatty residue is referred to as *crude fat.* Seeds of soybeans, flax, and cotton contain large amounts of crude fat (20 to 35 percent); those of corn and wheat contain only 4 to 5 percent.

14.3.4 Crude Fiber and Nitrogen-free Extract (NFE)

Crude fiber contains the more insoluble and nondigestible carbohydrates such as cellulose. The **crude fiber** in a food is determined by boiling a sample first in a weak acid, then in a weak alkali, and removing and washing out the dissolved material. The weight of the original sample and its dry weight after the dissolved material is removed give the basis for calculating the percentage crude fiber. **Concentrates,** such as corn and wheat, contain only 2 to 3 percent crude fiber; roughages may contain more than 20 percent.

Feeds having more *nitrogen-free extract* contain more digestible carbohydrates. The most important of these is starch. **NFE** is calculated by subtracting the percentages of water, ash, crude protein, fiber, and fat from 100 percent.

For example, the percent NFE of yellow corn would be calculated as follows:

	Percent
Water	15.0
Ash	1.2
Crude protein	8.6
Crude fiber	2.0
Fat	3.9
Total	30.7

The percent NFE of this corn would be $100 - 30.7$, or 69.3 percent.

14.3.5 Mineral Matter (Ash)

Mineral matter is determined by weighing a food sample and then burning (ashing) it until all the carbon has been removed. The weight of the residue compared with the original sample weight gives the mineral content. The mineral content of foods such as corn and wheat is low (between 1 and 2 percent); alfalfa hay contains about 8 percent mineral matter.

A chemical analysis is important because it relates something about the composition of a food. It is made even more valuable, however, when linked with a digestion trial (Section 14.4). A **proximate analysis** of food reveals nothing about its vitamin content, the *quality* of its protein, digestibility, palatability, the presence of injurious (toxic) substances, or whether the feed may have been spoiled before it was dried and cured.

14.4 DETERMINATION OF THE DIGESTIBILITY OF FEEDS

Although a chemical analysis of a food may relate something about its nutritive value for livestock, it does not actually indicate its degree of **digestibility.** This must be determined for a particular feed by conducting digestion trials (Figure 14.4 *a* and *b*). Many digestion trials have been conducted, and the percent digestibility of certain foodstuffs has been determined. These figures are average values, however, and may not apply exactly to an individual animal or to different species of animals. Additionally, it is important that the dietary intake be adequate when the nutritive value of feeds is being determined.

In a digestion experiment, the percentage of each class of nutrient present in a food is determined by chemical analysis. For a few days, an animal is fed weighed quantities of the food to be tested so that residues of previously fed foods are removed from the digestive tract. The experimental animal is then fed the same amount of weighed food each day. The feces are collected, weighed, and analyzed to determine their nutrient content. The difference between the nutrients in the food consumed and the nutrients in the excreted feces gives the amount retained in the body of the animal, or the amount of digested nutrients.

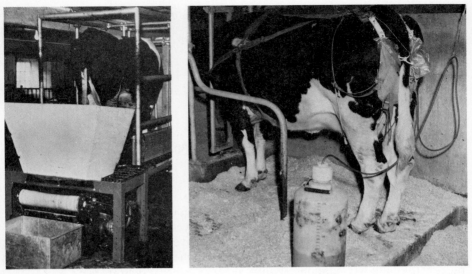

(a)   *(b)*

FIGURE 14.4
(a) Cow in digestion stall. Feed (amount offered and refused) is weighed to determine intake. Note the fecal collection pan at the rear of the belt. The urine flows into a collection pan at the front of the belt-motor apparatus (not visible). *(b)* An alternative method of urine and feces collection. *(Missouri Agr. Expt. Station.)*

The fraction of a nutrient that is digested is called the **digestion coefficient** and may be calculated for a feed (hay) as follows:

	Pounds of nutrients					
Amounts	**Dry matter**	**Protein**	**Fiber**	**Fat**	**NFE**	**Ash**
15 lb of hay	13.50	1.50	4.50	0.60	6.00	0.90
20 lb of feces	5.43	0.60	2.25	0.24	1.80	0.54
Pounds digested	8.07	0.90	2.25	0.36	4.20	0.36
Digestion coefficient	0.60	0.60	0.50	0.60	0.70	0.40

Thus

$$\text{Digestion coefficient for protein} = \frac{0.90 \text{ lb digested}}{1.50 \text{ lb consumed}} = 0.60$$

The digestion coefficient multiplied by 100 gives the percent digestibility, or 60 percent in the example above. Obviously, a determination of the percentage of each nutrient in a feed that is digested is a more accurate estimate of the value of a feed than is a chemical analysis. Many experiments have been conducted using different feeds, and the digestion coefficient has been determined and average values recorded.[5]

14.4.1 Nutritive Ratio (NR)

The **nutritive ratio** is the ratio of the sum of digestible carbohydrates and fat to **digestible protein.** It may be calculated as follows:

$$NR = \frac{\text{digestible NFE} + \text{digestible fiber} + (\text{digestible fat} \times 2.25)}{\text{digestible protein}}$$

A diet providing a large amount of protein in relation to other nutrients would possess a *narrow nutritive ratio,* whereas one containing a relatively small amount of protein and a large amount of fat and carbohydrates would be described as having a *wide nutritive ratio.* The correct nutritive ratio is necessary for maximum and efficient growth in animals.

14.5 THE ENERGY CONTENT OF FOODS

Energy is necessary for an animal to perform work and other productive processes. All forms of energy are converted into heat. Thus energy as related to body processes is expressed in heat units **(calories).** The *small* calorie is the amount of heat required to raise the temperature of one gram of water one degree centigrade (actually from 14.5 to 15.5°C). The **kilocalorie** (kcal) is the amount of heat required to raise the temperature of one kilogram of water one degree centigrade and is equal to 1000 small cal. A **megacalorie** (or **therm**) is 1000 kcal. One kilocalorie is approximately four British thermal units (Btu); 1 Btu is the amount of heat required to raise the temperature of one pound of water one degree Fahrenheit.

Several systems have been developed for expressing the energy content of foods and the energy requirements of animals. They will be discussed briefly.

14.5.1 Total Digestible Nutrients (TDN)

The **TDN** content of a food is expressed as a percentage and can be determined only by a digestion trial, as described previously. A formula for determining the percentage of TDN in a feed is as follows:

$$\% \text{ TDN} = \frac{\text{dig. protein} + \text{dig. crude fiber} + \text{dig. NFE} + (\text{dig. EE} \times 2.25)}{\text{unit of feed}} \times 100$$

---

[5]See F. B. Morrison, *Feeds and Feeding,* 22d ed., Morrison Publishing Co., Claremont, Ontario, Canada, 1959.

In this formula, lipids (ether extract) are multiplied by 2.25 because they possess about 2.25 times more energy per unit weight than do carbohydrates. This is because carbohydrates have sufficient **molecular** oxygen present to balance the hydrogen, so that heat is released only from the oxidation (combination with oxygen) of carbon. In lipids, there is much less oxygen present in the molecule, and oxygen outside the lipid molecule unites with both the carbon and hydrogen in the lipid, yielding more heat. The combustion of 1 g of hydrogen yields about 4 times more heat than does the combustion of a similar weight of carbon. Oxidations of carbohydrates, fats, and proteins yield about 4100, 9450, and 5650 cal/g, respectively.

A limitation of TDN as a measurement of food energy is that it does not account for losses such as those of combustible gases and the **heat increment** that occur when a feed is consumed by an animal. These losses are much larger for forages than for concentrates. This means that a pound of TDN from a forage has considerably less value for productive purposes in an animal than does a pound of TDN in a concentrate (grain). For example, 1 lb of TDN in corn yields about 1 megacalorie of net energy. In the better hays (forages), 1 lb of TDN yields about 0.75 megacalorie and in low-quality roughages about 0.5 megacalorie. The TDN content of feeds is commonly related inversely to the fiber content. Concentrates are low in fiber and high in TDN.

14.5.2 Gross Energy (GE)

The **gross energy** (heat of combustion) of a food can be determined by completely burning a sample of food until the oxidation products (carbon dioxide, water, and other gases) are formed. A *bomb calorimeter* is used to measure the heat liberated from such a combustion. A bomb calorimeter consists of a closed vessel (bomb) in which the food is burned. The bomb is enclosed in an insulated jacket containing water that absorbs the heat (calories) produced (Figure 14.5). The procedure is as follows: (1) the dried feed to be tested is weighed and placed in the bomb; (2) the cover is screwed down tight; (3) the bomb is charged with 25 to 30 atm of oxygen; (4) the bomb is then placed in the calorimeter jacket, where it is surrounded by a known volume of water at a known temperature; (5) the water is stirred until a constant temperature is attained; (6) the charge is then ignited by an electric current; and (7) readings are taken on the thermometer to determine the maximum temperature rise of the water. The gross-energy determination reveals the total caloric energy in the food being tested.

14.5.3 Digestible Energy (DE)

The digestible energy of a food is represented by that portion of the food consumed which is not excreted in the feces. The food is fed to an animal, and the energy present in the feces is determined by means of bomb calorimetry. The difference between the gross energy (Section 14.5.2) in the food and that in

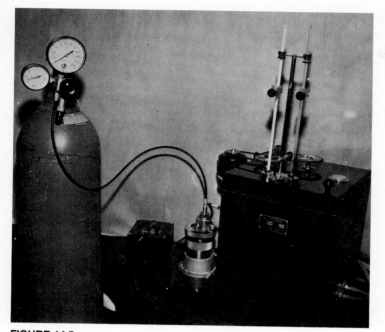

FIGURE 14.5
A bomb calorimeter. Note bomb being gassed (lower center). *(Missouri Agr. Ext. Station.)*

the feces is called *digestible energy*. Digestible energy can also be calculated from TDN values by using the factor of 2000 kcal/lb of TDN or from the digested nutrients by using the gross caloric factors for proteins, carbohydrates, and fats of 5.65, 4.10, and 9.45 kcal/g, respectively.

14.5.4   Metabolizable Energy (ME)

The **metabolizable energy** of a food is determined by subtracting the energy losses in the feces, urine, and combustible gases (primarily methane in ruminants) from the total energy in the food. A desirable characteristic of metabolizable energy is that more of the gross energy not available to the animal for productive purposes is accounted for than by the measurement of the TDN or the digestible-energy content.

14.5.5   Net Energy (NE)

Determination of the **net energy** content of a food goes one step beyond determining the metabolizable energy because the heat increment (heat resulting from the digestion and **assimilation** of feeds) is determined and subtracted from the total energy in the feed. Heat increment is measured by the amount of heat given off by an animal's body. Dr. Samuel Brody, world-renowned scientist in the area of bioenergetics and growth and formerly of the Dairy Science Department of the University of Missouri, called heat increment the "food utilization tax." Heat increment is approximately twice as high in lactating as in nonlactating cows, but this difference varies with the lactation level and the amount of food consumed by the **lactating** cow. During cold weather, a portion of the heat increment is used to keep the animal warm, and this is advantageous. Heat increment is a disadvantage to animals in hot weather (at least in daylight hours), when they have the problem of keeping cool (Chapter 13).

The net energy of a food can be determined in practical feeding experiments. A food, such as corn, in which the net energy is known, is compared with a food of which the net energy value is not known. NE can also be determined by the combined use of a bomb calorimeter and a respiration calorimeter. The former is used to determine the gross energy in a food and in the feces after the feed is fed; the latter is employed to determine the heat increment and losses of combustible gases.

Net energy, then, is that portion of the total (or gross) energy that is retained in the body for useful purposes, such as maintenance, growth, milk production, egg laying, and muscular work.

14.5.6 Estimated Net Energy (ENE)

The estimated net energy value of a particular food may be calculated by comparing the production from it with that of food whose net energy value is known (such as corn). Estimated net energy values are used in evaluating feeding programs for **DHIA** records processed by an electronic system. A factor favoring the estimated net energy system is that it is more accurate when the feeding values of concentrates and forages are compared in attempts to develop the most economical diet.

14.5.7 Uses of Energy by Animals

If you do not supply nourishment equal to the nourishment departed, life will fail in vigor; and if you take away this nourishment, life is utterly destroyed.

Leonardo da Vinci (1452–1519)

What happens to dietary food energy (gross energy) in the dairy cow is shown in Figure 14.6. This figure also summarizes the different kinds of energy discussed previously.

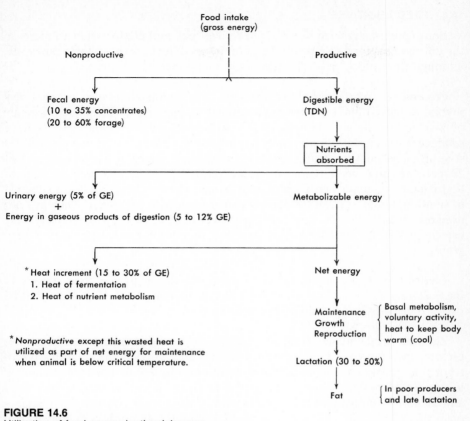

FIGURE 14.6
Utilization of food energy by the dairy cow.

Maintenance energy is that which is required to keep an animal in energy equilibrium and prevent the loss of body tissues or weight. The amount of energy consumed by an animal above that required for maintenance is used for *growth, reproduction, body fat, lactation,* and/or *egg laying* in the various farm animals. In relation to body weight, young, growing animals have much higher dietary requirements than mature ones for protein, energy, vitamins, and minerals. The amount of energy required for these various body processes has been determined, and average values have been summarized. Such information, along with data pertaining to the composition of foods, is quite helpful in formulating diets for different groups of animals.[6]

[6]The nutrient requirements of domestic animals are given in publications available through the National Academy of Science–**National Research Council,** 2101 Constitution Ave. N W, Washington, D. C. 20418.

14.6 FEED ADDITIVES

Various compounds have been added to livestock and poultry diets to increase the efficiency of food utilization. Most of these additives do not supply nutrients, although they affect food utilization in some species.

Protamone Also called thyroprotein, this **iodinated casein** product has thyroxine activity and has been fed to animals to increase the level and efficiency of production. Small amounts of this substance fed to young animals have produced, in some instances, more rapid growth. When fed to dairy cows at a stage of lactation when milk production normally declines, lactation increases of 20 percent have been reported. Some reports indicate that the feeding of thyroxine compounds may increase egg laying in hens. These drugs are effective in small amounts but can be detrimental if overdoses are given. They are not used on a practical scale in the feeding of livestock and poultry. Moreover, their use is prohibited in dairy cows enrolled in **official production**-testing programs.

Diethylstilbestrol This additive, a synthetic estrogen, was used earlier to increase rate and efficiency of weight gains in steers by 10 to 15 percent, but its use in meat-producing animals has been banned by the FDA.

Zeranol (trade name, Ralgro) This is an anabolic agent that promotes growth and improves feed efficiency in cattle. It is a chemical derivative of resorcylic acid lactone, a product of fermentation. Research indicates that one of its modes of action is to stimulate the pituitary gland to secrete increased amounts of somatotropin (growth hormone). Zeranol is implanted subcutaneously on the back side of the ear. Implanted animals commonly gain body weight about 10 percent faster and have a feed efficiency of about 8 to 10 percent greater than that of nonimplanted animals.

Monensin (trade name, Rumensin) Monensin is a polyether antibiotic useful as an anticoccidial agent for broilers and lambs. It is also used widely in the diets of beef cattle, in which it increases the efficiency of feed utilization by modifying rumen fermentation. The molar percentage of propionate is commonly increased about 25 to 40 percent, whereas that of acetate is usually decreased about 15 to 30 percent. Body-weight gain is usually not affected significantly, but the amount of feed required per unit of weight gain is commonly decreased by about 10 percent.

Another feed additive, *lasalocid* (trade name, Bovitec), apparently acts in a similar manner to Rumensin in improving feed efficiency; it is also used widely in feedlot cattle.

Antibiotics These compounds are widely used in swine and poultry diets. Antibiotics have been effective in increasing the rate and efficiency of gains in swine and poultry, especially when the incidence of disease is high. They have also been used in feeding baby calves before they have a fully developed rumen and have been fed in some instances to growing, fattening cattle. Their major use, however, has been at relatively low concentrations in diets of nonruminant animals, such as swine and poultry. Medicative levels in swine diets have been used effectively under some conditions.

Thiouracil This additive is a drug that has been experimentally used in efforts to increase the rate and efficiency of gain in animals. Favorable results have been reported, mostly for swine. Thiouracil depresses the production of thyroxine (the **thyroid** hormone) and lowers the metabolic rate of the animal. This conserves energy. Feeding thiouracil to swine for about 30 days after they reach a weight of 165 lb has significantly increased the rate and efficiency of gain. It has not been used on a practical scale on the farm.

Others Dozens of feed additives have been studied in various experiments. More will be tested in the future, but probably few will have an impact on livestock production equal to that of antibiotics and the synthetic estrogens.

14.7 SUMMARY

An understanding of the principles of nutrition is important because humans and animals are dependent on food nutrients for the processes of life. We have attempted to present the basic fundamentals of animal nutrition, giving special emphasis to the composition of plants and animals. A classification of food nutrients was presented in tabular form. Methods of analyzing foods, means of expressing their energy, and uses of food nutrients were discussed.

Much remains to be learned about the nutrient requirements of farm mammals and poultry. Many feeding standards were established early in the twentieth century, before recent selection techniques provided humans with a means of developing animals that possess greater inherent producing ability. Significant advances in animal and human nutrition will be forthcoming as humans learn how to better utilize the animals that serve them.

STUDY QUESTIONS

1 Is level of nutrition related to the behavior of humans and animals? Explain.
2 Who was Lavoisier? What did he contribute to the science of nutrition?
3 List the major substances found in the bodies of animals.
4 How do plants and animals differ in their chemical composition?
5 What are carbohydrates? What chemical elements do they contain?
6 List the main classifications of carbohydrates.
7 Which carbohydrates are easily digested by animals? Which ones are not?
8 What is the major nutritional contribution of carbohydrates?
9 How do fats and oils differ from carbohydrates in their chemical composition?

10 Why do fats supply more energy than carbohydrates when they are metabolized?

11 Describe the differences between saturated and unsaturated fatty acids. Which fatty acids are essential?

12 What is the major function of fatty acids in the body?

13 What are proteins? How do they differ chemically from carbohydrates and fats?

14 List the major functions of proteins in the body.

15 What determines if an amino acid is essential or nonessential?

16 Why do ruminants and nonruminants differ in their dietary protein requirements?

17 What is meant by the biological value of a protein? What is meant by the term "protected protein"?

18 List the major functions of minerals in the body. What are some symptoms of mineral deficiencies?

19 What are vitamins? What are their two major classifications?

20 What are some major functions of vitamins in the body?

21 Describe the major classifications of feedstuffs.

22 What major chemical substances are found in ether extract? In nitrogen-free extract?

23 A finely ground sample of food weighed 100 g. After being dried in an oven until it reached a constant weight, it weighed 88 g. What was the percent of water in the original sample? (*Answer:* 12 percent.) What was the percent dry matter (DM) in the feed? (*Answer:* 88 percent.)

24 The nitrogen content of a 200-g sample of food was determined and found to be 3.6 g. What is the crude protein content of this food? (*Answer:* 11.25 percent.)

25 A chemical analysis of a food sample gave the following results: ash, 2.6 percent; crude fiber, 10.6 percent; crude protein, 12.0 percent; fat, 4.1 percent; and water, 8.9 percent. What was the percent NFE in this sample? (*Answer:* 61.8 percent.)

26 A 100-g sample of food was burned until all the carbon was removed. The weight of the burned residue was 6.8 g. What was the percent of mineral matter (ash) in this sample? (*Answer:* 6.8 percent.)

27 A ration was found to contain 66.83 g of digestible NFE, 1.31 g of digestible fiber, 3.64 g of digestible fat, and 7.37 g of digestible protein. What is the nutritive ratio of this ration? (*Answer:* 1:10.35.) What does the nutritive ratio mean?

28 Another ration was found to contain 26.69 g of digestible NFE, 3.81 g of digestible fiber, 4.86 g of digestible fat, and 37.66 g of digestible protein. What is the nutritive ratio of this ration? (*Answer:* 1:1.1.)

29 A diet has the following composition:

	Percent
Digestible protein	7.6
Digestible crude fiber	1.7
Digestible NFE	66.0
Digestible fat (EE)	2.7

What is the percent TDN in this diet? (*Answer:* 81.4)

30 Define or describe the following terms: kilocalorie, megacalorie, gross energy, digestible energy, metabolizable energy, net energy, and digestion coefficient.

31 What is meant by the term feed additive?

32 What are some feed additives that have been used on a practical scale?

CHAPTER 15

THE PHYSIOLOGY OF DIGESTION IN NUTRITION[1]

Before hazarding a theory we propose to multiply our observations, to investigate the phenomena of digestion, and to analyze the blood both in health and disease.

A. L. Lavoisier (1743–1794)

15.1 INTRODUCTION

Some species of animals are vegetarians and depend entirely on plants for food. They are called **herbivores.** Other species feed almost entirely on flesh of other animals. They are called **carnivores.** Still other species consume both plants and flesh. They are called **omnivores.** Regardless of their eating habits, all animals depend on plants (directly or indirectly) for food. Moreover, it can be said that all animal life depends indirectly on the sun for food, because it is through the action of sunlight on the chlorophyll of plants and the process of photosynthesis that plants convert elements from the air and soil into the **nutrients** that supply food for animals. Thus, without energy from the sun, there would be no food for plants, animals, and humans.

Animals do not use all plant nutrients for the various body processes exactly as they come from the plant. Most complex nutrients must be broken down (digested) to simpler compounds before they can be absorbed and utilized. The different species of animals have digestive tracts adapted to the most efficient use of the type of food they consume. Thus herbivores differ from carnivores and omnivores in the anatomy and physiology of their digestive system.

[1]The authors acknowledge with appreciation the contributions to this chapter of Dr. G. C. Fahey, Jr., Department of Animal Science, University of Illinois, Urbana.

It is the purpose of this chapter to discuss the anatomy and physiology of the digestive systems of animals, especially of farm mammals and poultry.

15.2   TYPES AND CAPACITIES OF DIGESTIVE SYSTEMS

Anatomical and capacity differences in the digestive systems among species are more significant physically than nutritionally, because foods in the digestive tract are still, in a sense, outside the body. In the process of digestion, nutrients enter the body by **absorption** through the gut wall. The metabolic processes that then utilize the absorbed nutrients are essentially the same for all species.

15.2.1   Anatomy and Types of Digestive Systems

The alimentary tract extends from the lips to the anus. The principal parts include the mouth, pharynx, esophagus, stomach, and the small and large intestines.

The length and complexity of the tract vary greatly among species. The tract is relatively short and simple in carnivores but is much longer and more complex in herbivores. In some herbivores (horse and rabbit), the stomach is relatively simple and comparable with that of carnivores, whereas the large intestine, especially the cecum, is much more capacious and complex than that of carnivores. Conversely, in other herbivores (cow, goat, sheep), the stomach (polygastric system) is large and complex, whereas the large intestine is long but less functional.

The three general types of digestive systems and the foremost anatomical differences are depicted in Figure 15.1.

The anatomy of the **avian** digestive system is shown in Figure 15.2. The digestive system of birds differs from that of mammals in that birds have no teeth to physically break down their food. The glandular stomach in the bird is known as the *proventriculus*. Between the proventriculus and the mouth is an enlargement of the gullet, called the *crop*. The **ingested** food is stored temporarily in the crop, where it is softened before moving to the proventriculus. Food passes quickly through the proventriculus to the *ventriculus,* or *gizzard*. The main function of the gizzard is to crush and grind coarse food. This is aided by grit and gravel, which the bird accumulates from birth.

15.2.2   Capacity of Digestive Systems

The average capacities of the different parts of the **gastrointestinal** (GI) canal in selected animals and humans are given in Table 15.1.

The total capacity of the entire digestive system in the human, who is omnivorous, is only about 6 liters (6.34 qt). The adult human stomach holds about 1.2 liters, or about 5 cups. Even allowing for the prompt passage of many foods into the intestines, it is obvious that there is a physical limit to the quantity of food that humans can consume and process in a given time. The dog, a

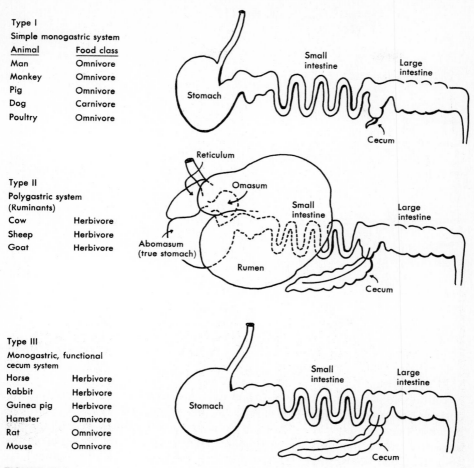

Type I
Simple monogastric system

Animal	Food class
Man	Omnivore
Monkey	Omnivore
Pig	Omnivore
Dog	Carnivore
Poultry	Omnivore

Type II
Polygastric system
(Ruminants)

Cow	Herbivore
Sheep	Herbivore
Goat	Herbivore

Type III
Monogastric, functional
cecum system

Horse	Herbivore
Rabbit	Herbivore
Guinea pig	Herbivore
Hamster	Omnivore
Rat	Omnivore
Mouse	Omnivore

FIGURE 15.1

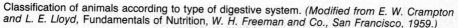

Classification of animals according to type of digestive system. *(Modified from E. W. Crampton and L. E. Lloyd,* Fundamentals of Nutrition, *W. H. Freeman and Co., San Francisco, 1959.)*

carnivore, also has a very small digestive system, although it is a little larger per unit of body weight than that of the human.

The pig, an omnivore, possesses a larger digestive capacity per unit of body weight than either the dog or human. Even so, the digestive tract in this species is limited in capacity. The digestive system of the pig is better adapted to the use of **concentrated** feeds such as grains, although limited amounts of forages can be included in the diet. The horse, a herbivore, possesses a much larger digestive system than the pig and is able to utilize large amounts of **roughages** in its diet because of a greatly enlarged cecum. The horse is considered a nonruminant animal, however.

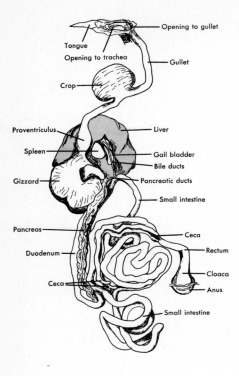

Tongue

Opening to gullet

Opening to trachea

Gullet

Crop

Proventriculus

Liver

Spleen

Gall bladder

Bile ducts

Gizzard

Pancreatic ducts

Small intestine

Pancreas

Ceca

Duodenum

Rectum

Ceca

Cloaca

Anus

Small intestine

FIGURE 15.2
Major components of the avian digestive
system. *(After F. B. Adamstone.)*

Cattle and sheep are herbivores and have greatly enlarged digestive tracts, designed to process bulky feeds. Both of these species are called **ruminants** because they possess three stomach compartments (**rumen, reticulum,** and **omasum**) not common to nonruminants. They are also called **polygastric** animals. Such a digestive system provides the space for processing large quantities of bulky **forages** necessary to provide energy and nutrients to sustain maintenance and high levels of production (milk, meat, and wool). For example, the cow has about 9 times more digestive tract capacity than humans in proportion to body weight. Not only is every digestive tract component that has a counterpart in humans 3 or 4 times more capacious, but the polygastric system of the cow has a total capacity per unit of body weight exceeding that of the simple human stomach by more than 30-fold. A cow weighing 1200 lb may have a stomach capacity of 300 lb.

The relative sizes of the four stomach compartments in the ruminant vary with age. The first three compartments are small in the newborn and very young ruminant, representing less than 30 percent of the total stomach capacity. As the young animal is switched from a milk diet to one containing cereal grains and forage, the rumen develops, and by the time it reaches maturity, it accounts for about 80 percent of the stomach capacity (Figure 15.3). In mature animals, food normally passes first into the rumen. Under normal conditions, the **calf** rumen

TABLE 15.1
CAPACITY, IN LITERS, OF THE DIGESTIVE SYSTEM OF SELECTED ANIMALS

	Animal					
Part	Ox (cattle)	Sheep and goat	Horse	Pig	Dog	Human
Gastric compartment						
Rumen	202	23				
Reticulum	8	2				
Omasum	19	1				
Abomasum	23	3	18	8	4	1
Subtotal	252	29	18	8	4	1
Small intestine	66	9	64	9	2	4
Cecum	10	1	33	1		
Large intestine	28	5	96	9	1	1
Total	356	44	211	27	7	6

Source: Adapted from H. H. Dukes, *The Physiology of Domestic Animals,* Comstock Publishing Associates, Ithaca, N.Y., 1970.

becomes functional in about 6 to 8 weeks. Evidence of a functional rumen is based on (1) rumen odor, which is indicative of fermentation; (2) a decline in blood glucose; and (3) the production of **volatile fatty acids.**

15.3   THE PROCESS OF DIGESTION

Digestion in higher animals includes all the activities of the alimentary tract and its glands in the conversion of food into materials available for absorption and **assimilation.** It also includes the rejection of unabsorbed food residues (**fecal materials**). Most foods, when they are consumed, are too complex to be absorbed into the blood and **lymph** without preliminary digestive changes. Glucose, soluble salts, water, and a few other nutrients are exceptions.

The factors of digestion are chemical, mechanical, microbiological, and secretory in nature. Chemical factors include **enzymes** and certain nonenzymic chemical substances (such as hydrochloric acid) produced by the digestive glands. The main mechanical factors are **mastication, deglutition, regurgitation,** gastric and intestinal motility, and **defecation.** Microbiological factors reflect the activities of bacteria and, in some species of animals, **protozoa.** The secretory contributions to digestion include the beneficial activities of the digestive glands.

15.3.1   Digestion in Monogastric (Nonruminant) Animals

Chemical and secretory factors of digestion, especially enzyme action, play an important role in the digestion of foods in humans, poultry, swine, and other nonruminant animals. Digestion of food (especially fiber) by means of microor-

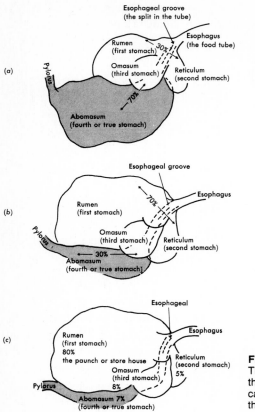

FIGURE 15.3
The development and relative capacities of the various stomach compartments of *(a)* the calf at birth, *(b)* the calf at 2 months, and *(c)* the mature cow.

ganisms is of lesser importance in these species. However, in the horse, digestion of fiber by microorganisms is of great importance, since most digestive activity occurs in the cecum.

15.3.2   Digestion in Polygastric Animals

Within the digestive compartments of the ruminant are large populations of microorganisms (approximately 1 billion bacteria and 1 million protozoa per milliliter) that break down **cellulose** to form the short-chain volatile fatty acids (acetate, propionate, and butyrate, commonly called **VFA**). These volatile fatty acids are absorbed largely through the rumen wall and provide the ruminant with 60 to 80 percent of its energy needs.

Although cellulose is digested in the rumen, we should hasten to add that starches, sugars, and proteins are also largely broken down here. In fact, an average of 60 to 90 percent of the total digestion within ruminants occurs in the rumen. It is, therefore, a remarkable fermentation vat.

In payment for their rumen-housing privileges, the microorganisms perform some special favors for their host. These include synthesis of all the B-complex vitamins (Chapter 16) and all the **essential amino acids** (Chapter 14). The latter can be synthesized from nonprotein nitrogen compounds (NPN), such as **urea** or diammonium phosphate, or from proteins that are deficient in one or more essential amino acids. Finally, the microorganisms give their life in payment for food and shelter, being digested by their host farther down the gastrointestinal tract.

15.3.3 Special Features of Ruminant Digestion

The nonglandular mucous membrane of the developed rumen is lined with papillae that aid in absorption from the rumen (Figure 15.4). Although Figure

FIGURE 15.4
The reticulorumen from a lamb fed a completely pelleted diet (left) and one fed shelled corn and chopped hay (right). Note that the papillae from the rumen epithelium of the lamb receiving the pelleted diet are longer and wider. *(Courtesy Dr. G. B. Thompson and Dr. W. H. Pfander, University of Missouri.)*

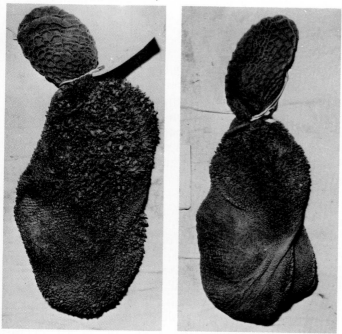

15.4 shows the increase in length and width of papillae in lambs fed completely pelleted diets, it does not clearly show that there is frequently a **keratinization** of the rumen epithelium on such diets. This latter effect may result in reduced absorption of nutrients from the rumen.

A substantial amount of gases (methane and CO_2) is produced in the rumen and, if allowed to accumulate, causes **bloat,** or an inflation of the rumen. However, these gases are normally expelled quite freely by **eructation** (belching) and to a lesser extent by absorption into the blood, from which they are eliminated through exhaled air from the lungs.

15.4 APPETITE

The most fundamental factor involved in the voluntary intake of food is appetite. A center in the hypothalamus is credited as being the chief factor in its control. Other factors probably involved include (1) the blood glucose level; (2) the amount of fill, or quantity of **ingesta,** in the stomach; and (3) the environmental temperature (decreased appetite in hot weather).

> When the appetite is given full control of what shall be eaten, it is surprising to note how pigs naturally select the specific feeds which swine herdsmen have long since approved of as the best, and, what is equally surprising, the pigs show a marked avoidance of those feeds usually considered as ill adapted to swine.
>
> *J. M. Evvard (stated in 1915)*

15.5 THE PREHENSION OF FOOD

Animals differ greatly in the manner by which they harvest and convey food to their mouths **(prehension).** Domestic animals use their lips, teeth, and tongue for the prehension of food. The upper lip is the chief prehensile structure in the horse, whereas the tongue performs that function in cattle. (Cattle have an especially strong tongue.) The sheep has a **cleft** upper lip (the goat does not) and uses its incisor teeth and tongue to procure food. A pointed lower lip, the teeth, and the tongue serve as prehensile structures in swine, whereas the beak is used for this purpose in poultry.

15.6 THE MASTICATION OF FOOD

Mechanical reduction of food size is accomplished in the mouth by chewing, or *mastication.* This act aids the digestive process by (1) reducing food particle size, resulting in a greater surface area for the digestive juices to act on; and (2) mixing the food with saliva to facilitate easier swallowing.

Remastication of food is important in ruminants, in which large quantities of food may be consumed in a relatively short period of time. The animal can then relax under a shady oak, regurgitate food (in the form of a **bolus**), and **ruminate.**

While most dentists agree that mastication is essential to digestion, Spallanzani proved nearly two centuries ago that certain foods, such as meat, could be digested without having been chewed. He observed meat to disappear from tiny linen bags that he swallowed and later recovered, following their tour of the digestive tract.

In birds the food enters the mouth and passes through the gullet to the crop, where it is temporarily stored and softened. It then passes rather quickly through the proventriculus (true stomach) to the ventriculus, or gizzard. Here the food is ground and crushed into smaller particles (since birds have no teeth).

15.7 ENZYMES OF THE DIGESTIVE TRACT

An enzyme is an organic compound produced by certain cells within the body that speeds biochemical reactions at ordinary body temperatures without commonly being used up in the process. Enzymatic activity is responsible for most of the chemical changes occurring in foods on their movement through the digestive tract. A summary of the enzymes contributing to the digestive process is presented in Table 15.2.

15.7.1 Digestion of Carbohydrates

In some species, the digestion of **carbohydrates** begins in the mouth, where the foods come in contact with ptyalin, an enzyme secreted by the salivary glands. The saliva of humans, swine, and dogs contains small amounts of amylase (or ptyalin) and that of the horse contains very small amounts, whereas that of ruminants (cow, sheep, and goat) contains none. Amylase digests starch to maltose and dextrins. Mucin in the saliva does not digest starch, but it lubricates the food so that it is easily swallowed.

As was mentioned previously, microorganisms in the rumen break down cellulose to form volatile fatty acids. They also digest starches, sugars, fats, proteins, and nonprotein nitrogen to synthesize microbial protein, B vitamins, and vitamin K. No enzymes from the ruminant gastric secretions are involved in microbial synthesis.

Amylopsin (amylase) from the pancreas is secreted into the first portion of the small intestine (duodenum), where it digests starches and dextrins to simpler dextrins and maltose. Farther along in the small intestine, other enzymes from the intestinal juices also digest carbohydrates. These enzymes include *sucrase* (invertase), which breaks down sucrose into glucose and fructose; *maltase,* which breaks down maltose into glucose; and *lactase,* which breaks down lactose into glucose and galactose.

Microorganisms in the cecum and colon also digest cellulose, forming volatile fatty acids. As is true in the rumen, enzymes secreted by the digestive tract of the animal are not involved in the digestion of cellulose by the microorganisms of the cecum and colon.

TABLE 15.2
DIGESTIVE PROCESSES IN FARM MAMMALS AND POULTRY

Region	Secretion	Enzyme	Substrate or function	End products
Mouth	Saliva	Amylase (ptyalin)	Starch and carbohydrates	Maltose and dextrins
			Urea (recycles)	
Crop (birds)	Mucin	...	Lubrication of food	
	Mucus	...	Lubricates food	
Rumen	...	Enzymes from micro-organisms	Cellulose	VFA
			Polysaccharides and starches	Microbial protein
				B vitamins, vitamin K
Stomach (abomasum in mammals and proventriculus in birds)	Gastric juices and acids	Pepsin	Sugars, fats, and proteins (urea)	Proteoses and peptones
			Proteins	
	Mucus	Lipase	Fats	Higher fatty acids and glycerol
		...	Coating of stomach lining and lubrication of food	
Gizzard (birds)	...	...		Ground foods
Nursing animals	Gastric juice	Rennin	Milk protein (casein)	Paracasein
Duodenum	Pancreatic juice	Trypsin	...	Peptones, peptides, and amino acids
		Chymotrypsin	Proteins, proteoses, peptones, and peptides	
		Amylopsin (amylase)	Starch, dextrins	Dextrins and maltose
		Steapsin (lipase)	Fats	Higher fatty acids and glycerol
		Carboxypeptidase	Peptides	Amino acids and peptides
	Bile (liver)	...	Fats	Emulsion of fats
Small intestine	Intestinal juice	Peptidase (erepsin)	Peptides	Amino acids and dipeptides
		Sucrase (invertase)	Sucrose	Glucose and fructose
		Maltase	Maltose	Glucose
		Lactase	Lactose	Glucose and galactase
		Polynucleotidase	Nucleic acid	Mononucleotides
Large intestine	...		...	
Cecum and colon	...	Cellulase from micro-organisms	Cellulose	VFA

468

15.7.2 Digestion of Proteins

The true stomach is the site of the beginning of protein digestion in several species. The abomasum of ruminants is comparable with the true stomach in other mammals and the proventriculus in birds. Hydrochloric acid is produced by cells in the true stomach, thus supplying an acid medium that activates pepsin and rennin, aiding in protein digestion. The initial step in protein digestion occurs when foods come in contact with the enzyme *pepsin,* from the gastric juice. Pepsin splits proteins into less complex compounds, known as proteoses and peptones. In young, nursing animals, the enzyme *rennin* causes milk to coagulate, forming *paracaseinate,* which stays in the stomach longer than if the milk had remained a liquid. Therefore more complete digestion results.

Pancreatic juice is secreted into the duodenum and contains the enzymes *trypsin, chymotrypsin,* and *carboxypeptidase.* These enzymes continue protein digestion initiated in the stomach by pepsin, breaking down the more complex substances into peptides and finally into amino acids.

15.7.3 Digestion of Fats

The enzymic digestion of fats begins in the stomach, where *lipase* from the gastric juice converts fats into higher fatty acids and glycerol. (This process is limited in many species.) In the duodenum, bile from the liver emulsifies fats, breaking them down into smaller globules and increasing the total surface area on which *steapsin* (lipase), the enzyme from the pancreatic juice, acts. Steapsin continues the digestion of fats, where it is not complete, breaking them down into higher fatty acids and glycerol.

15.7.4 Digestion of Other Nutrients

Minerals are dissolved from foods in the hydrochloric acid solutions of the stomach. They are also released from the organic compounds that are digested by the various enzymes. Water requires no digestion before being utilized by the animal. Little is known about the digestion of vitamins, but they probably can be used as such within the body without conversion into simpler compounds.

15.8 AVIAN DIGESTION

No digestive enzymes are secreted in the gizzard of fowls. The main function of this organ is to reduce the size of food particles. The digestibility of coarse foods is greatly reduced in gizzardectomized (gizzard-removed) birds, but in spite of this they remain healthy for many years.

From the gizzard, food passes through the intestinal loop, called the *duodenum,* which parallels the *pancreas* anatomically. The pancreas has an important role in avian digestion, as it has in other species. It secretes an abundant supply of pancreatic juice containing amylolytic, lipolytic, and proteolytic enzymes, which **hydrolyze** starches, fats, and proteoses and peptones, respectively. Liver bile also enters the duodenum and aids in the digestion of lipids.

Food materials move through the small intestine, whose walls secrete intestinal juices containing *erepsin* and some sugar-splitting enzymes. Erepsin completes protein digestion, yielding amino acids, and the sugar-splitting enzymes convert disaccharides into simple sugars (monosaccharides) that can be assimilated by the body. Absorption is accomplished through the villi of the small intestine.

No liquid urine is voided by birds. The urine is discharged into the cloaca and excreted with the feces. The white material in bird droppings is largely uric acid, whereas the urinary nitrogen of excreta of mammals is mostly urea. The relative shortness of the avian digestive tract is reflected in a rapid digestive process (about 4 hours).

15.9 ABSORPTION OF FOOD NUTRIENTS

The process whereby food nutrients, properly digested within the digestive tract by the digestive organs and enzymes, are transferred from the **lumen** of the gastrointestinal canal to the blood or **lymph** is called *absorption.*

Most absorption of nutrients in carnivores and omnivores, and a significant amount in herbivores, occurs in the small intestine. Probably no food nutrients are absorbed from the mouth and esophagus and very few from the stomach. (An exception is the absorption of volatile fatty acids across the rumen wall in ruminants.) Except for water absorption from the colon, very little absorption occurs from the large intestine of carnivores and humans. Conversely, the large intestine is the site of substantial absorption in many herbivores.

The mucous membrane of the small intestine is richly supplied with minute, finger-shaped projections, called *villi,* which absorb food nutrients. Within the villi are tiny capillary blood vessels and lymph **ducts,** which collect the absorbed nutrients. Fats are absorbed as fatty acids and glycerol, principally by the lymph, whereas water, inorganic salts, and the end products of carbohydrate digestion (monosaccharides and VFA) and of protein digestion (amino acids and peptides) are absorbed largely by the blood.

Absorption, then, completes the digestive process and makes nutrients available to support the process of life.

15.10 FACTORS AFFECTING THE DIGESTIBILITY OF FEEDS

A knowledge of factors affecting the **digestibility** of feeds is important, because this knowledge can be effective in increasing the efficiency of **feed conversion.**

15.10.1 Temperature

The environmental temperature may have a pronounced effect on the appetite of an animal and the amount of feed it consumes. This could have an indirect effect on the degree of digestibility of a food. Experiments with dairy cattle indicate that increases in temperatures up to 90°F are accompanied by corresponding increases in the apparent digestibility of feeds. This is largely because feed intake is reduced and food remains in the gastrointestinal tract for a longer period of time.

15.10.2 Rate of Passage through the Digestive Tract

If, for various reasons, the food consumed should pass through the digestive tract too quickly, there may not be enough time for complete digestion of the nutrients by the digestive enzymes. It is also possible, however, that if the **rate of passage** of the food were too slow, fermentation losses could be greater than desired. In general, however, experimental data indicate that a more rapid passage of food is related to a lower digestibility of the feed consumed.

15.10.3 Level of Feeding

The digestibility of feeds in ruminants is higher at a level of feeding near maintenance than it is when animals are on **full-feed.** Many experiments with swine suggest that the efficiency of food conversion is greater on a **limited-** than on a full-feed. In spite of these observations, experiments have failed to show any difference in the digestibility of feeds fed swine on either full or limited diets. Greater feed efficiency in animals fed limited amounts of feed could be the result of less wastage of feed or a more efficient utilization of the feed consumed rather than of improved digestibility.

15.10.4 Physical Form of a Feed

Whether grinding feed affects its digestibility depends on how well animals chew their grain before it passes through the digestive tract. Very young and very old animals, which do not have good teeth, do not masticate (chew) their food as well as mature animals with sound teeth.

Grinding Grain Grinding grain for animals exposes a larger surface to the digestive juices and could therefore increase its digestibility. Cattle do not chew their grain as thoroughly as sheep and swine, and much of the whole grain passes through the digestive tract into the feces without being digested. Therefore grinding grain for cattle usually increases its digestibility.

Grinding Hay Grinding hay increases its rate of passage through the digestive tract and thereby reduces its digestibility somewhat. Most forage-consuming animals chew hay well enough so that there is optimum exposure to the digestive process.

Pelleted Diets Pelleted diets have gained in popularity for many species of livestock in recent years. Pelleted or cubed diets have increased the rate and efficiency of gain in lambs and cattle (and possibly swine), but there is not complete agreement among research workers as to whether feed digestibility is improved.

Heating Heating per se does not greatly affect digestibility of feeds. However, heat may be used to destroy a growth inhibitor. For example, raw soybeans possess a trypsin-inhibiting factor that prevents the proper utilization of the protein by nonruminants such as pigs and poultry. Proper heating (but not overheating) of raw soybeans increases the availability of the proteins for these species by inactivating the inhibitor. Cooking most other feeds has little or no effect on their digestibility, and neither do the various processes of fermentation.

15.10.5 Composition of the Diet

In vitro experiments have indicated that the extent of digestion of cellulose in good-quality hay is higher than in poor-quality roughages, such as **corn stover,** wheat straw, corncobs, and mature timothy-bluegrass hay. The addition of complex minerals and **manure** extracts, along with a supply of available nitrogen, greatly increases cellulose digestion in poor-quality roughages but has little or no effect on the digestion of cellulose in good-quality clover, rye, and alfalfa hays.

Considerable attention has been given to the needs of ruminal microorganisms for maximal digestion of feeds. Urea has been added to ruminant diets in varying amounts for synthesis into proteins by rumen microorganisms. A certain amount of protein and urea in the diet appears to increase the digestibility of cellulose in roughages. Small amounts of minerals, such as phosphorus, iron, sodium, potassium, calcium, magnesium, sulfur, and chlorides, are essential for the growth of rumen microorganisms and, indirectly, are necessary for greater cellulose digestion. The addition of B vitamins (especially biotin) has increased the rate of cellulose digestion in the artificial rumen but has not shown a beneficial effect in **in vivo** studies.

Small amounts of readily available energy, such as are found in glucose and starch, appear to be required for the normal growth of rumen microorganisms, and thus aid in the digestion of cellulose. Large amounts of glucose and starch may decrease cellulose digestion, however.

The addition of corn oil or lard to diets of ruminants that contain large amounts of coarse roughages (such as cottonseed hulls) appears to decrease the digestibility of the roughage. A limited amount of fat can be included in diets of swine and poultry without affecting the digestibility of other nutrients.

Feeding **antibiotics** to swine and poultry has been shown in many experiments to increase rate and efficiency of gain. Such increases are due to a reduced incidence of **subclinical** infections rather than increased digestibility of feeds. Feeding antibiotics to young calves before their rumen has fully developed has also increased the rate and efficiency of gain. This is probably due to better utilization of feed nutrients rather than to increased digestibility of nutrients in the diet. Feeding various antibiotics to mature ruminants appears to be detrimental to the growth of rumen microorganisms, resulting in poor cellulose digestion. The level of antibiotics fed is, of course, important.

15.11 EFFICIENCY OF FOOD CONVERSION

Animals convert only a portion of the feed nutrients they consume into food for humans. For this reason, a larger population could be fed if humans ate plant products directly rather than consuming animals and their products (Chapter 1). Some plants, especially grass and other forages, cannot be utilized by humans. Therefore it is essential to use certain animals as "middlemen" for food production from these plants. Animal products supply large quantities of good-quality proteins and other important nutrients vital to the health and well-being of humans (Chapter 3). In addition, humans have had an appetite for animal products through the years. It is doubtful that they will lose this appetite in the years ahead, especially as standards of living improve.

The ability of animals to convert nutrients in the feed they eat into animal products is referred to as *efficiency of food conversion* and depends on (1) their ability to digest nutrients in feed; (2) their requirements for energy and protein for growth, maintenance, and other body functions; (3) the amount of these nutrients lost in metabolic end products and nonproductive work; and (4) the type of feed consumed.

The digestibility of air-dried feed varies with the species of animals that consume them. Nonruminant animals, such as swine and poultry, digest approximately 75 to 85 percent of the nutrients in concentrates but very little of the nutrients in forages such as hay and silage. Ruminants, however, digest about 50 to 55 percent of the total nutrients in alfalfa hay and 75 to 80 percent of the nutrients in concentrates such as ground corn. Furthermore, ruminants digest 75 to 80 percent of the dry matter in immature grasses.

Not all the digestible nutrients in a feed consumed by animals are used for productive purposes. Some of the digestible protein in a food is not incorporated into body tissues because there is not a proper balance of amino acids in the feed for building body proteins. Where there is a lack of balance, the extra amino acids are deaminated and used as a source of energy. The end products of protein metabolism, such as urea and uric acid, are eliminated in the urine and feces and are lost to the animal for productive purposes. In general, the higher the protein content of a feed, the larger the amount of energy lost through these channels.

As is true of proteins, not all the digestible energy in the other nutrients, such as carbohydrates and fats, is used for productive purposes. Some energy is lost in the feces, urine, combustible gases, and the heat of digestion (Figure 14.6). The proportion of digestible energy taken into the body that is used for productive purposes varies with the species, the individual, and the kind of feed consumed.

15.12 FACTORS AFFECTING THE EFFICIENCY OF FOOD CONVERSION

The efficiency of food conversion in meat-producing and **lactating** animals is usually expressed as the units of liveweight gain made or pounds of milk secreted by an animal per unit of feed consumed. A more accurate method would be to express the efficiency of food conversion in terms of the amount of feed required to produce a unit of milk or meat, or better yet, the amount of feed required to produce a unit gain of **carcass** or a pound of 4 percent **FCM.** Expressing efficiency on the basis of carcass weight, however, requires slaughtering the animal and recording accurate weights of carcass parts. This is time-consuming and expensive and has the disadvantage that breeding stock cannot be selected on such a basis. Recently, attempts have been made to develop measurements of muscle mass in the live animal by various means, including the low-level radiation counter (Chapter 22). If these attempts prove successful, it will be practical to express the efficiency of feed conversion on a carcass basis.

The efficiency of feed conversion into milk and eggs is rather easily measured. Perhaps this is one reason that the production efficiency of these two animal products has greatly increased recently. The greatest improvement in feed conversion in recent years among domestic animals has been in **broiler** and egg production. This improvement has been brought about by improved breeding, feeding, and management methods.

Many factors affect the efficiency of feed conversion. Some of them will be discussed in the following.

15.12.1 Inheritance

Data from many experiments show that feed conversion among farm mammals is from 30 to 40 percent heritable. This means that genes with an additive type of expression affect this trait and that selection of the best, and mating of the best to the best, should improve it. However, **genes** with a nonadditive effect have very little influence on the efficiency of feed conversion. This conclusion is based on the fact that feed efficiency declines very little when **inbreeding** is practiced and improves little when animals are **crossbred** (Chapter 5).

15.12.2 Age and Weight

Younger animals require less feed per unit of gain than do older and more mature ones. The main contributing factor to more efficient feed conversion in young animals is the composition of body mass, which constitutes the weight gain. As shown in Table 15.3, the percentage and pounds of fat increase as pigs grow older and become heavier. Since more energy is required to deposit a pound of fat than a pound of protein, less gain is made per unit of feed as a pig becomes older and heavier. Although not shown in Table 15.3, the protein mass

TABLE 15.3
EFFECT OF BODY WEIGHT ON EFFICIENCY OF FEED CONVERSION, DAILY
FEED CONSUMPTION, AND POUNDS OF FAT IN THE CARCASS OF SWINE

Weight of pig, lb	Feed per 100 lb gain	Pounds of feed consumed daily	Percent of fat in body	Pounds of fat in body
Birth (2.5 lb)	...	....	2.4	0.06
50	300	3.00	9.7	4.85
100	385	4.00	16.2	16.20
150	430	5.63	29.1	43.65
200	455	7.00	28.5	57.00
250	505	8.15	32.1	80.25
300	570	8.40	42.6	127.80

Source: Chemical-composition data from *Missouri Agr. Res. Bull.* 73, 1925. Percentage carcass composition may vary somewhat from these figures in modern meat-type swine.

in the pig also increases with advancing age and increasing body weight; however, it decreases as a percentage of total body composition.

Another limiting factor is that the maintenance requirement increases with the 0.75 power of body weight ($W^{0.75}$), which means that there is less feed, in addition to that required for maintenance, to use for growth as body weight increases. The amount of food consumed per unit of body weight also decreases as the animal becomes older and heavier.

15.12.3   Level of Feeding

Limited feeding, if not too drastic, often results in increased efficiency of gain. This is true in different species of animals and may result from an increased digestibility of the diet or from less feed wastage.

15.12.4   Average Daily Gain

Faster-gaining animals on full-feed usually make more efficient gains than slower-gaining ones on full-feed. Many of the same genes that cause fast gains also cause more efficient gains, so that the improvement of one trait through selection will cause an improvement in the other. The **correlation** between the two traits is high but not perfect.

When all are full-fed, faster-gaining animals make more efficient gains than slower-gaining ones, because (1) they use a smaller percentage of their total feed intake for maintenance; (2) they are usually healthier; (3) they are growing rather than depositing fat; and (4) they have a more efficient metabolic system, which allows better utilization of the food they consume for body-weight increases.

15.12.5 Other Factors

Many other factors may affect the efficiency of feed conversion in livestock. These include the diet quality and its **palatability** to the animal. Feeding certain **hormones** and drugs may also increase the economy of gains under some conditions.

15.13 SUMMARY

All animals depend directly or indirectly on plants for their supply of nutrients. The digestive system of each species is adapted to the kind of feeds normally consumed. Ruminants possess a complex digestive system consisting of four stomach compartments, in which microorganisms help in digesting cellulose and other compounds that cannot be broken down by the animal's own digestive enzymes. This enables them to utilize large quantities of forages. Other species, such as swine and poultry, consume a larger percentage of grains and other concentrates and depend almost entirely on digestive enzymes to chemically break down these compounds.

Foods must be broken down into simpler compounds before they can be absorbed into the bloodstream, where they are used for maintenance energy; body repair; growth of new tissues; reproduction; work; and the production of meat, milk, eggs, and wool. In the processes of digestion, absorption, and utilization, many plant nutrients are lost to the animal for productive purposes. Although animals are somewhat inefficient converters of plant nutrients into animal products, they possess digestive systems designed to enable them to serve as "middlemen" in converting many plants (especially those containing large amounts of cellulose, which is not utilized by humans) into animal products useful to the health and nutrition of humanity.

STUDY QUESTIONS

1 Define the following: herbivores, carnivores, omnivores, proventriculus, ventriculus, monogastric, and polygastric.
2 Compare the capacities of the digestive tracts of several farm animals.
3 Name the major parts of the digestive systems of farm animals.
4 How do monogastric (nonruminant) and polygastric animals differ in their ability to digest forages and foods high in fiber? Why do they differ?
5 Why do monogastric (nonruminant) animals not bloat as readily as polygastric animals? What causes bloat?
6 What affects and/or controls appetite in farm animals?
7 Distinguish between prehension and mastication of food.
8 Outline the processes and enzymes that digest carbohydrates in farm animals.
9 Discuss the process of protein digestion in farm animals.
10 What are the major enzymes involved in the digestion of fats?

11 What is an enzyme and what produces it?
12 Where are most food nutrients absorbed into the body in farm animals?
13 How does the digestive system of birds differ from that of mammals?
14 Discuss some factors that may affect the digestibility of feeds.
15 List and discuss some factors that affect the efficiency of feed conversion into animal products.

CHAPTER

THE NUTRITIONAL APPLICATION OF VITAMINS TO HUMAN AND ANIMAL HEALTH[1]

There is still an unknown substance in milk, which, even in very small quantities, is of paramount importance to nutrition. If this substance is absent the appetite is lost and with apparent abundance the animals die of want.

C. A. Pekelharing (1905)

16.1 INTRODUCTION

Prior to the twentieth century, **carbohydrates, fats, proteins,** and a few minerals were generally considered to be the only dietary requirements for normal body functions. Centuries before, however, certain observations suggested that other organic compounds are essential to the maintenance of good health. For example, it has been known for some 300 years that eating fresh fruits and vegetables is effective both **prophylactically** and **therapeutically** against scurvy. It has also been appreciated for a long time that rickets can be cured by the oral consumption of cod-liver oil. Also, in 1897 the Dutch physician Eijkman found that the disease beriberi, which afflicted those consuming diets of polished rice, could be cured by eating rice polishings.

These observations suggested that there were other naturally occurring food components that were indispensable for health but that were not carbohydrate, fat, or protein in nature. In 1912 Funk, a Polish biochemist working in London, first introduced the term *vitamine* (a vital amine), which later became **vitamin**

---

[1]The authors acknowledge with appreciation the contributions to this chapter of Dr. R. M. Forbes, Department of Animal Science, University of Illinois, Urbana.

478

(from the Latin *vita,* meaning *"life"),* to denote this group of organic compounds.

16.2   VITAMINS DEFINED

Vitamins are organic compounds essential for normal growth and maintenance of animal life. They are effective in relatively small amounts. Some vitamins are required for metabolic reactions essential for the transformation of energy, but they do not themselves contribute energy. Some are metabolic essentials but not dietary essentials, because they are synthesized in the bodies of certain species.

16.2.1   Nomenclature

Vitamins were originally categorized as (1) *fat soluble* and (2) *water soluble* because fat solvents could extract the former from foods, and water could be used to extract the latter. The fat-soluble vitamins include A, D, E, and K and contain only carbon, hydrogen, and oxygen. The water-soluble ones include ascorbic acid (C) and the B complex (B_1 to B_{12}); they also contain carbon, hydrogen, and oxygen and may contain nitrogen, sulfur, or cobalt. Pertinent information about the fat-soluble vitamins is summarized in Table 16.1 and about the water-soluble vitamins in Table 16.2.

16.3   THE FAT-SOLUBLE VITAMINS

Prior to the discovery of the fat-soluble vitamins, it was assumed that fats served only as a source of energy for animals. The fat-soluble vitamins are usually stored in the body and therefore need not be consumed on a daily basis.

16.3.1   Vitamin A

The earliest descriptions of symptoms now attributable to vitamin deficiency relate to vitamin A. Night blindness and its cure, feeding liver, is mentioned in the Ebers papyrus (ca. 1600 B.C.), in Chinese writings of the same period, and later by Hippocrates and Roman writers. Celsus (25 B.C.–A.D. 50) is believed to have been the first to use the term *xerophthalmia,* which literally means "dry eye," a condition resulting from a vitamin A deficiency.

While feeding a purified diet (containing all the known essential nutrients) to white rats at the University of Wisconsin in 1913, Dr. E. V. McCollum observed that the rats became sick and ceased to grow, their eyeballs became dry, their tear glands did not act, their eyelids became swollen and inflamed, and finally blindness occurred. When Dr. McCollum added milk fat to the diet, the rats soon became healthier, and their eye condition improved. The milk fat was a good source of fat-soluble vitamin A.

Vitamin A is essential to growth and efficient food utilization. It is important in the resistance of the body to **infection,** keeps epithelial tissue normal, and

Table 16.1
THE FAT-SOLUBLE VITAMINS

Nomenclature	Function	Clinical deficiency symptoms	Common sources
Vitamin A (carotene)	Important in cellular metabolism. Regeneration of visual purple in eyes. Essential for normal epithelial tissue lining the digestive, respiratory, and reproductive tracts.	Night blindness, **keratinization** of epithelium, retarded growth and appetite, xerophthalmia, muscle incoordination, rough hair coat or plumage, reduced fertility.	Yellow corn (beta-carotene), alfalfa, and grasses; egg yolk, liver, fish-liver oils, milk fat, green vegetables; commercial preparations.
Vitamin D	Role in calcium and phosphorus absorption and metabolism. Normal **calcification** of bones.	Rickets, **osteomalacia,** decreased egg laying and hatchability in poultry.	Fish-liver oils, irradiated yeast, egg yolks, milk fat, field-cured hays, commercial preparations.
Vitamin E (tocopherols)	Normal reproduction in some species, antioxidant, hatchability of eggs, activities of cell nucleus.	Infertility in rats, mice, guinea pigs, and possibly swine. Skeletal muscular dystrophy. **exudative diathesis, encephalomalacia,** liver **necrosis,** cardiac muscle abnormalities, dental depigmentation.	Cereal grains (mostly in germ); egg yolk; oils of soybean, peanuts, and cottonseed; alfalfa; beef liver; wheat germ oil; commercial preparations.
Vitamin K	Required for prothrombin formation, which is essential for normal blood clotting.	**Subcutaneous** and **intramuscular** hemorrhages, especially in poultry.	Green, leafy plants; liver, egg yolk, fish meal; synthetic form (menadione).

Table 16.2
THE WATER-SOLUBLE VITAMINS AND RELATED COMPOUNDS

Nomenclature	Function	Clinical deficiency symptoms	Common sources
Vitamin C (ascorbic acid)	Formation and maintenance of **intercellular** material in bones and in soft tissues. Acts as a tissue **catalyst** (aids in healing).	Scurvy, hemorrhages throughout body, swollen and bleeding gums, **anemia**. Loss of teeth.	Citrus fruits; tomatoes; green, leafy vegetables; potatoes. Synthetic preparations.
Thiamine (B_1)	Component of two **coenzymes**, essential in CHO metabolism and energy transfer. Promotes normal appetite and digestion. Helps keep the nervous system healthy and prevents irritability.	Beriberi in humans, polyneuritis in birds, lack of appetite, hyperirritability, incoordination, Chastek paralysis in foxes. Reproductive failure in horses.	Milk and milk products, brewers' yeast, wheat germ, unmilled cereals, grain by-products, lean pork, liver, kidney, egg yolk. Good-quality hay. Synthetic preparations.
Riboflavin (B_2 or G)	Forms a part of two flavoprotein coenzymes, role in energy transfer (helps cells use oxygen), function in protein metabolism, component of xanthine oxidase. Helps keep skin healthy.	Curled-toe paralysis in chick; **lesions** of skin, eye, and nervous system; depressed appetite; retarded growth	Milk, cheese, liver, kidney, eggs, fish, green forage, oil meals, fermentation products. Commercial preparations.
Pantothenic acid	Component of coenzyme A, which is involved in many metabolic reactions.	Retarded growth, skin lesions, **gastrointestinal** troubles, lesions of nervous system and adrenal gland, skin and hair depigmentation. Goose-stepping in pigs.	Liver, egg yolk, milk, alfalfa hay, peanut meal, cane molasses, yeast, rice, wheat brans. Cereal grain and by-products. Royal jelly of bees.
Nicotinic acid (nicotinamide or niacin)	Component of two coenzymes; energy transfer. Can be spared by the amino acid tryptophan. Health of digestive and nervous systems.	Pellagra or black tongue, **dermatitis**, diarrhea, **dementia**, loss of appetite and weight, vomiting, anemia. Enlarged hocks in turkey poults.	Milk, meat, eggs, green vegetables, peanut butter, animal and fish by-products, distillers' grains and yeast, fermentation solubles, oil meals. Hays and grains are fair sources.

(continued)

Table 16.2
THE WATER-SOLUBLE VITAMINS AND RELATED COMPOUNDS (CONT.)

Nomenclature	Function	Clinical deficiency symptoms	Common sources
Pyridoxine (B_6)	Part of enzyme concerned with protein metabolism. Essential for normal metabolism of tryptophan.	Dermatitis and convulsions in rats, pigs, and poultry. Microcytic hypochromic anemia in pups, pigs, and rats.	Yeast, liver, muscle, meat, egg, yolk, milk, cereal grains, vegetables.
Biotin	Functions in enzyme systems. Fat synthesis, **deamination** of certain amino acids.	**Perosis** in chick, **dermatitis**, loss of hair, disturbances of nervous system.	Yeast and organ meats, whole grains, molasses, milk.
Folic acid	Role in transfer of single-carbon units. Synthesis of purines and certain methyl groups. **Erythropoiesis.**	Retarded growth and **macrocytic,** hyperchromic anemia. Poor feathering and pigmentation of feathers.	Green, leafy materials; organ meats; cereals; soybeans; animal by-products.
Cyanocobalamin (B_{12})	Methyl-group synthesis, purine synthesis, CHO and fat metabolism. Synthesis of nucleic acids. Known as animal protein factor **(APF).**	Pernicious anemia in humans. Retarded growth, low hatchability of eggs, posterior incoordination, unsteadiness of gait.	Milk, meat, fish meal, animal by-products. Cow manure, built-up litter. Commercial preparations.
Related Compounds			
Choline	Component of **phospholipids,** essential in building and maintenance of cell structure. Transmission of nerve impulses. Fat metabolism in the liver	Fatty livers **(cirrhosis),** renal tubular degeneration, enlarged spleen and hemorrhagic condition of kidneys. Perosis in chicks.	Milk, meat, eggs, fish, all naturally occurring fats.
Inositol	Lipotropic action in certain rat diets, where other vitamins are deficient.	**Alopecia** (dropping of hair).	Occurs in plant products in the organic phosphorus substance phytin.
Para-aminobenzoic acid (PABA)	Anti-gray-hair factor in mice and rats. A growth stimulant in chicks.	Graying of hair **(achromotrichia)** in animals other than humans.	Synthetic preparations.

482

prevents night blindness. Vitamin A occurs in plants as carotene (**precursor** of vitamin A).

The value of grass was recognized in the time of Jeremiah, who recorded

Yea, the hind [a female deer] also calved in the field, and forsook it, because there was no grass. And the wild asses [four-footed, hoofed mammals related to the horse] did stand in the high places, they snuffed up the wind like dragons; their eyes did fail, because there was no grass. . . .

Jeremiah 14:5–6

Although Jeremiah associated good eyesight with the availability of grass, it was several thousand years later before scientists found the scientific explanation that grass contains carotene. Vitamin A is essential in the eye's regeneration of visual purple (rhodopsin), which is instrumental in preventing night blindness. Vitamin A and carotene are absorbed from the small intestine and are stored primarily (approximately 80 to 90 percent) in the liver.

Recent medical research demonstrated that high levels of vitamin A aid in the healing of wounds. Physicians noted slow healing of skin grafts among patients burned severely. This was especially marked among those receiving steroids. When high levels of vitamin A were administered, however, the speed of healing among the burn victims was near normal.

Clinical Deficiency Symptoms

1 Impaired growth results because vitamin A is essential in the building of new cells.

2 Keratinization of epithelial tissue occurs. Vitamin A is essential to the health and integrity of the epithelial tissue. A deficiency results in its keratinization. (Tissue changes into **keratin,** which is an insoluble protein and the chief structural constituent of horns, nails, hair, and feathers.) Keratinized tissues in the various tracts of the body (digestive, genitourinary, reproductive, and respiratory) lower the resistance of the affected epithelial tissues to infective **organisms.** Pneumonia, diarrhea, kidney stones, and bladder stones are more prevalent, and reproductive efficiency is greatly reduced.

3 Night blindness (xerophthalmia) occurs.

4 Rough hair coat and skin conditions (e.g., acne vulgaris, which is a chronic **inflammatory** disease of the sebaceous glands) are often closely associated with a vitamin A deficiency. Studies at the University of Missouri indicate that vitamin A may be beneficial in the prophylaxis and therapy of ringworm in cattle (Figure 16.1).

Sources   Sources of vitamin A include fish-liver oils, butter (milk fat), egg yolk, cheese, liver, green vegetables (in most of which it exists as carotene), and synthetic preparations. The green color of plants and dry forage is a good indicator of the carotene content. (Dark-green color indicates high carotene

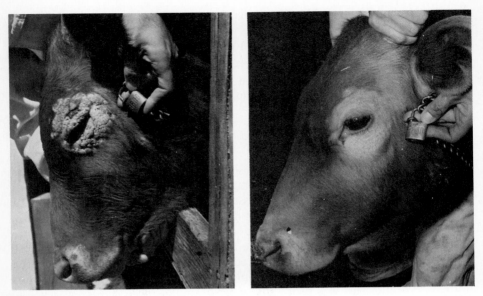

FIGURE 16.1
Effect of vitamin A on the health and integrity of the skin and on the fineness of the hair coat. Note the favorable effect of vitamin A injections on ringworm infection in this 6-month-old Guernsey bull. After four weekly injections (300,000 IU each), note the nearly complete recovery (right) and especially the change in skin and hair-coat appearance. *(Missouri Agr. Expt. Station.)*

content.) However, this visual evaluation of the carotene-containing qualities of forages is limited because the green pigment (chlorophyll) is not destroyed at the same rate as carotene.

16.3.2 Vitamin D

Prior to the twentieth century, humans resembled the living plant in being dependent on sunshine for health and well-being. Why the Egyptians worshipped sun gods is unknown, but their religion may have been a result of their observing the favorable effects of sunshine on the rachitic children of the royal families.[2] Rickets is a condition (caused by a vitamin D deficiency) that results in a disturbance of bone development. The disease is characterized by bending and distortion of the bones under muscular action and by the formation of nodular enlargements on the ends and sides of bones.[3] An early concept of the cause of rickets was a lack of exercise. One of the early (about 1650) remedies for rickets

[2]The role of vitamin D in the proper development of bones has a long history. The Greek historian Herodotus visited a battlefield where the Persians had defeated the Egyptians in 526 B.C. He observed that the skulls of the slain Persians were fragile whereas those of the Egyptians were strong. The Egyptians attributed the difference to the effect of sunlight; they were bareheaded from childhood, whereas the Persians wore turbans to protect them from the sun.

[3]The discovery of vitamin D was one of the greatest scientific and medical discoveries of all time. An estimated 90 percent of the young children of Europe had rickets as late as 1900.

was to "take roots of Smallage, Parsley, Fennel, and Angelica [all herbs], slice and boil them in distilled water of Angelica, unset Hyssop, and Collsfoot, of one part each, until they are tender; then strain and boil it to a syrup with white honey."

Later, cod-liver oil was observed to be beneficial in the therapy of rickets. In 1922 Dr. E. V. McCollum and his associates at Johns Hopkins University in Baltimore, Maryland, identified the antirachitic element in cod-liver oil and named it vitamin D. In 1924 Dr. A. F. Hess (physician) and Dr. H. Steenbock (biochemist at the University of Wisconsin) discovered that rickets could also be prevented by **irradiating** certain foods for animal consumption with ultraviolet light.

Although nature provided little vitamin D in common foods, it gave humans a way of forming it in their bodies. The ultraviolet radiation in sunlight (wavelength 280 to 320 nanometers) serves as a source of the radiant energy necessary to convert 7-dehydrocholesterol (an animal sterol stored beneath the skin surface) into biologically active vitamin D_3 (cholecalciferol). *Ergosterol* (a plant sterol), on irradiation, yields ergocalciferol (vitamin D_2). There are some 10 *sterol* derivatives having vitamin D activity; however, ergosterol and 7-dehydrocholesterol are the chief provitamins D, which yield D_2 and D_3, respectively.

Apparently, vitamin D_2 (the plant form of the vitamin) and vitamin D_3 (the animal form) have the same antirachitic value for the dog, pig, rat, **ruminant,** and human, whereas D_3 is more beneficial than D_2 for poultry. Recent evidence indicates that D_3 is more active than D_2 in promoting calcium absorption in Cebus monkeys.

Rickets seldom occurs in the tropics, where exposure to sunshine is maximal; it is principally confined to the temperate zone, where **radiation** may be markedly reduced, particularly in winter. (Arctic residents consume substantial amounts of vitamin D–rich fish.) Darkly pigmented skin filters out much (as much as 65 percent) of irradiation; therefore blacks are very susceptible to rickets. Could the need to regulate absorption of ultraviolet radiation partially explain why Caucasians are white in the winter (white skins allow maximum photoactivation of 7-dehydrocholesterol into vitamin D at low intensities of ultraviolet radiation) but pigmented in the summer, when more sunlight is available? In almost all races, the skin is lighter in the **neonate** and gradually darkens as the individual matures, a change that parallels the declining need for vitamin D. Is this phenomenon an inherent safety feature provided by nature to ensure ample vitamin D for the growth, development, and well-being of the young?

Examples of rickets are shown in Figure 16.2*a* and *b*.

The Role of Vitamin D　This vitamin is required for normal **calcification** of growing bones. It serves an important role in the metabolism of calcium and phosphorus in the body throughout life, but especially in growth, reproduction, and **lactation.** Vitamin D enhances the absorption of calcium and phosphorus from the small intestine and helps maintain normal blood levels of these

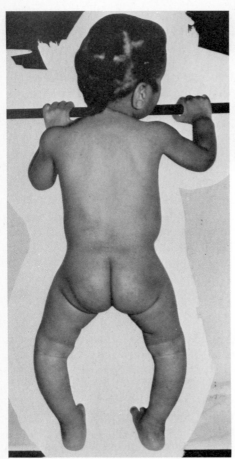

(a)

(b)

FIGURE 16.2
(a) Rickets in an infant. Note the bowlegs that
curve laterally, indicating that the weakened
bones have bent as a result of standing
*(Courtesy Dr. R. L. Nemir, New York
University, and the Upjohn Company,
Kalamazoo, Mich.) (b)* Advanced stage of
rickets caused by a lack of dietary vitamin D
when a pig was fed indoors. Note the leg
abnormalities. *(Courtesy Dr. J. M. Bell,
University of Saskatchewan, Canada.)*

minerals. Being fat-soluble, some vitamin D is stored in the liver; however, this
amount is limited, and the body needs a more regular and consistent **exogenous**
source of vitamin D than of vitamin A.

Excessive vitamin D may be harmful in that it mobilizes calcium and
phosphorus from the bony tissues. These minerals are redeposited in soft
tissues, principally in the walls of blood vessels, but also in kidney tubules,
bronchi, and the heart.

Clinical Deficiency Symptoms

1 Rickets, a bone deformity that results from the lack of calcium and/or
phosphorus deposition. It occurs primarily in young animals or infants (Figure
16.2a). Only mammals and birds are subject to rickets.

2 Osteomalacia, a condition characterized by a partial decalcification of normal bones that leads to a softening and brittleness of bones. It is commoner in older persons (senile osteomalacia) and animals whose bones are fully grown but whose diets are deficient in vitamin D or in calcium and phosphorus.

3 Low concentration of blood **plasma** phosphorus.

4 Thickening and swelling of joints.

Sources   Sources of vitamin D include liver oils of various fish, fortified milks, butter, egg yolk, irradiated animal and plant sterols, and commercial preparations. The supplementation of diets with vitamin D is recommended for animals spending the winter months in northern latitudes, when there is less sunlight and its efficiency in converting vitamin D precursors into vitamin D is greatly reduced, and for animals confined indoors during any season.

16.3.3   Vitamin E

This vitamin was discovered in 1924 and was called *tocopherol* (from the Greek *tocos,* meaning "childbirth," and *phero,* meaning "to bring forth").

A deficiency of vitamin E leads to sterility in the rat and to muscular dystrophy in the dog, guinea pig, rabbit, and certain other **species.** An early sign of deficiency in male rats is the loss of **spermatozoan** motility. Pregnancy may occur in the female, but embryonic development is retarded and often results in fetal resorption.

Tocopherol isomers are strong *antioxidants.* (They hinder **oxidation,** which chemically consists of an increase in positive charges on an **atom** or the loss of negative charges.) Vitamin A and carotene are susceptible to oxidative destruction in the presence of unsaturated fats (Chapter 14), both in the cell and in the digestive tract, and are protected against this change by vitamin E.

The vitamin is distributed throughout the body, is stored principally in fat, and is therefore not a daily requirement in the diet. Vitamin E appears to be relatively nontoxic, although some evidence indicates that large dosages may mobilize phosphorus and cause bone decalcification.

Vitamin E passes through the placental membranes and also into the mammary gland. Therefore the mother's dietary level of vitamin E is reflected in body stores of the young at birth and, similarly, by the amount it obtains from its mother's milk.

It is now known that selenium can perform some functions of vitamin E. An example is the protection afforded rats by selenium against **necrotic** liver degeneration, which normally results from a vitamin E–deficient diet.

Clinical Deficiency Symptoms   The most dramatic characteristics of a vitamin E deficiency are the changes in various body tissues. Those related to this discussion are as follows:

1 Infertility in guinea pigs, hamsters, mice, rats, and possibly swine.

2 A brown discoloration of the rat's uterus and **adipose** tissue.

3 Skeletal muscular dystrophy in guinea pigs, hamsters, **lambs,** rabbits, rats, and swine. (It is also called "stiff lamb disease," a form of muscular dystrophy in sheep, which is similar in appearance to "white muscle disease," caused by a deficiency of vitamin E and/or selenium in cattle (Figure 16.3a.).

4 Cardiac muscle abnormalities in cattle, lambs, monkeys, **poultry,** rabbits, and rats (often characterized by "brown fat" in the heart of pigs and sometimes referred to as "mulberry heart").

5 Nutritional **encephalomalacia** (edema of the brain) in **chicks,** which is also called "crazy chick" disease. Symptoms include lack of coordination, head retraction, convulsions, and twitching of the limbs (Figure 16.3b).

6 Liver **necrosis** (death of liver cells) in rats and liver and muscular degeneration in pigs.

Sources   Sources of vitamin E include wheat-germ oil, cereals, egg yolk, beef liver, and good-quality **forage.** It is also prepared synthetically.

16.3.4 Vitamin K

Vitamin K is essential to the formation of prothrombin by the liver. Blood coagulation consists of two major steps: (1) prothrombin (in the presence of thromboplastin, calcium, and other factors) is converted into thrombin, and (2) **fibrinogen** (acted on by thrombin) is converted into the fibrin clot (Figure 16.4).

Vitamin K was discovered in Denmark (1934) when chicks fed an ether-extracted (fat-free) diet developed a hemorrhagic condition. A deficiency may occur in humans, mice, rabbits, and rats in the presence of abnormal fat **absorption.**

Vitamin K is stored in appreciable amounts in the liver. Massive doses of vitamin K are apparently nontoxic. The bacteria of the intestinal tract of humans and most **domesticated** animals are capable of synthesizing vitamin K; hence a hemorrhagic tendency induced solely by the dietary lack of vitamin K rarely occurs. (Poultry have a short intestinal tract and host so few microorganisms that they must be provided with a dietary vitamin K source.) Infants, however, are an exception because (1) the prenatal stores of the vitamin are small, (2) the intestinal bacterial **flora** necessary to synthesize vitamin K is not established at birth, and (3) the usual diet of the neonate does not supply it. The result is that a mild vitamin K deficiency is common in the human neonate. Occasionally, the blood prothrombin concentration reaches such a low level that hemorrhages occur, usually within the first week of life, commonly starting with a hemorrhage

---

FIGURE 16.3
(a) White muscle disease of a 12-week-old calf caused by a vitamin E deficiency. Weak muscles, lameness, and abnormal locomotion characterize the calf. The other photographs show abnormal white areas in the cardiac muscles of calves 5 to 6 weeks old. *(Courtesy Dr. O. H. Muth, Oregon State University.)* *(b)* α-Tocopherol deficiency in the chick. Note the loss of control of legs and the head retraction. *(Cornell University, Poultry Science Dept.)*

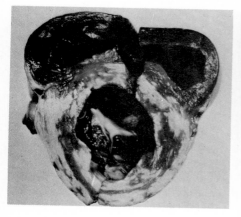
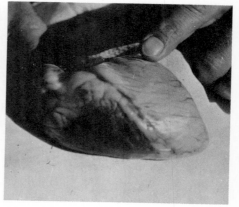

(a)

(b)

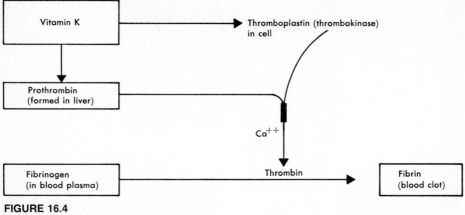

FIGURE 16.4
The role of vitamin K in blood clotting.

from the **gastrointestinal** (GI) tract. Hemorrhagic disease of the neonate can usually be prevented by the administration of vitamin K to the expectant mother **prepartum** or to the infant at birth. Oral administration of sulfonamides and **antibiotics** may destroy vitamin K–synthesizing organisms in the GI tract. Therefore the prophylactic administration of vitamin K is recommended to avoid the prothrombin-reducing effect of certain drugs.

Missouri studies (unpublished) indicate that female rats are less susceptible than males to a vitamin K deficiency. Apparently, this is related to the sex **hormones,** because when the male sex hormone testosterone is administered to females, they soon develop a vitamin K deficiency, whereas the males are less susceptible to vitamin K deficiency when they receive the female sex hormone, estrogen.

Dicoumarol   This compound was isolated orginally from spoiled sweet clover (commonly called *hemorrhagic sweet clover disease*) and later made synthetically. It is used clinically as an anti**coagulant** in thrombotic states and acts by depressing the factors concerned with the formation of thrombin. Since vitamin K will overcome the effect of **dicoumarol,** the action of the latter is similar to that of any **antimetabolite.**

Clinical Deficiency Symptoms

1 Hypoprothrombinemia, a deficiency of prothrombin in the blood (Figure 16.4).

2 Subcutaneous and intramuscular hemorrhages, especially in poultry (Figure 16.5).

FIGURE 16.5
Vitamin K deficiency in chicks. Note the spontaneous hemorrhages under the skin of the chick at left, which was fed a vitamin K–deficient diet for the first 15 days. A normal chick the same age but fed an adequate diet is shown at right. (Upjohn Company Vitamin Manual, *1965.*)

Sources Sources of vitamin K include alfalfa, spinach, cabbage, egg yolk, fish meal, and hog liver fat. Its synthetic **analogues** are prepared commercially. Naturally occurring vitamin K is fat soluble, but some of the synthetic preparations are water soluble.

16.4 THE WATER-SOLUBLE VITAMINS AND RELATED COMPOUNDS

The vitamins that are soluble in water include ascorbic acid (vitamin C), which is apparently required only in the diets of the human, monkey, and guinea pig, and the B complex, which is required only in the diets of **monogastric** animals. The ability of the **ruminant** to use microbiologically synthesized B-complex vitamins was discussed in Chapter 14.

The B vitamins may be subdivided into two groups: (1) those involved in the release of energy from food (thiamine, B_1; riboflavin, B_2; nicotinamide; pantothenic acid; and biotin) and (2) the **hematopoietic** vitamins, or those related to the formation of red blood cells (folic acid and B_{12}, sometimes called *cobalamine*). Pyridoxine, B_6, functions so that it fits both the energy-releasing and hematopoietic categories of vitamins.

B vitamins are involved in the absorption of food from the intestine and are therefore concerned in the first stage of the nutritional process. Oxidation of food is accomplished enzymatically, and certain B vitamins are involved in the process as essential components of **coenzymes.** Cellular energy is supplied by this oxidation of food within the cell.

16.4.1 Vitamin C (Ascorbic Acid)

> This is a wonderful secret of power and wisdom of God, that he has hidden so great an unknown virtue in this fruit—sour oranges and lemons—to be a certain remedy for this infirmity [scurvy].
>
> *Anonymous (sixteenth century)*

A tour of the *Mayflower* in Plymouth or of *Old Ironsides* in Boston reminds one of the tales recorded about the perils of mutiny and famine. Another peril, less recorded but equally dangerous on early ocean voyages, was disease among sailors. Of the 160 men who sailed with Vasco da Gama around the Cape of Good Hope in 1497, 100 died of scurvy.

Records of early voyages reveal that transportation was slow and there was no way to keep fresh fruits and vegetables for extended periods of time. Early symptoms of scurvy were loss of appetite and laziness. Later, the victim's mouth became sore, teeth became loose, bones broke easily, and small blood vessels burst **subcutaneously;** the majority of those afflicted eventually died of the disease.

After a voyage from France to Newfoundland in 1536, Jacques Cartier's men became sick with scurvy, and 26 men died of the disease. One day Cartier noticed that a native Indian who had had scurvy earlier appeared cured. He asked about the Indian's therapy and discovered that the Indian had made a drink by boiling the bark and leaves of a selected tree (probably an American spruce). When Cartier's men drank this "tea," they too were cured of scurvy.

In 1735 four ships (*Dragon, Hector, Susan,* and *Ascension*) sailed for the West Indies. On all ships except the *Dragon,* men developed scurvy and died. There were no casualties on the *Dragon,* whose men had fruit juice in their meals. Why were these observations of the benefits of fruit juices in the prevention of scurvy not quickly applied in human nutrition? Science advances slowly, with each experience and discovery contributing to previous knowledge. Furthermore, people are often slow to accept and apply new findings.

About two centuries after Cartier's curious observation of the benefits of "Indian tea" on scurvy victims, Dr. J. Lind, a British navy surgeon and perhaps the first "experimental nutritionist," applied Cartier's findings experimentally, in 1747. On one ship (the warship *Salisbury)* he used six pairs of sailors who were ill with scurvy. He gave cider, cream of tartar, oranges and lemons, and other special doses to pairs 1, 2, 3, and 4 to 6, respectively. In only 6 days, experimental pair 3 (who had received oranges and lemons) were greatly improved whereas the others remained ill. The results of this simple experiment prompted Dr. Lind to recommend that all British sailors receive orange and lime juice daily. By 1795 this recommendation was adopted by the British navy, and from that day to this, sailors have often been called "limeys," since limes were chiefly used at that time. Thus Dr. Lind laid the foundation for the concept that deficiency diseases result from the lack of essential food constituents.

The famous voyages and adventures of the hardy mariners played a very real role in establishing the basis of vitamin C. Perhaps history will record equally important contributions of our space travelers to the science of **nutrition** and **physiology** as humans continue to triumph over distance and disease.

Vitamin C is involved in the absorption of iron from food. Recent experiments with radioiron incorporated into food show that **ingestion** of ascorbic acid and food containing ascorbic acid enhanced the **absorption** of iron (Figure 16.6). Thus ascorbic acid is indirectly effective in preventing iron-deficiency anemia. Fever increases the need for vitamin C. Vitamin C seems to help to maintain the metabolic and functional properties of phagocytes, as consumption of too little of the vitamin or too much in supplemental form may impair phagocytic function.

Clinical Deficiency Symptoms   Scurvy occurs only in the human, monkey, and guinea pig;[4] other mammals are apparently capable of synthesizing their

---

[4]Certain other animals apparently need dietary vitamin C. For example, catfish develop broken-back syndrome when kept for extended periods in water devoid of vitamin C. In streams, they apparently obtain vitamin C from algae.

FIGURE 16.6

Effect of ascorbic acid on absorption of radioactive iron (^{59}Fe) from food in humans. This man, whose blood values were normal, was first given eggs containing 3.5 mg ^{59}Fe. Less than 10 percent of this radioactive iron could be subsequently measured in the blood. When 1 g of ascorbic acid was added to a comparable amount of radioactive food iron, absorption was increased to more than 70 percent. *(Courtesy Dr. C. V. Moore and Dr. R. Dubach, Washington University, St. Louis, Mo.)*

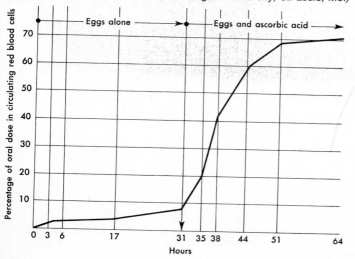

dietary needs of ascorbic acid. Scurvy is characterized by hemorrhages through-out the body (result of increased fragility of blood capillaries), swollen and bleeding gums, and **anemia.**

Sources   Sources of vitamin C include lemons, oranges, grapefruit, and tomatoes. Other fruits and vegetables also furnish appreciable amounts of vitamin C. In addition, it is manufactured synthetically (was the first vitamin to be synthesized in the laboratory). It is apparently not stored in appreciable amounts in the body since blood plasma concentrations of vitamin C reflect dietary intake very closely. Ascorbic acid is absorbed from the small intestine and excreted via the urine.

16.4.2   Thiamine (Vitamin B₁)

For many centuries people believed that persons who ate meat of certain animals would act like those animals. The person who ate snake's meat would be as sly and cunning as a snake. Because the heart was regarded as the center of courage, those wishing to be courageous considered the hearts of animals as choice meat. Similarly, those who wanted to run fast ate meat from the legs of deer.

Today advertising, promotion, and habit largely govern what people eat. Humans have eaten bread for centuries. During the Middle Ages, white bread was rare and expensive and, probably for those reasons, was preferred. A person's wealth often determined the kind of bread he or she ate, and black bread was generally considered a badge of poverty. This prompted people to attempt faster and easier ways to remove the **germ** and **bran** to produce a white flour. However, they were unknowingly throwing away valuable vitamin B₁ with the germ and bran. (Some 80 percent of this vitamin is lost from the flour in the milling process.) Today most white flour is *enriched*[5] with B₁ and other nutrients that are lost in milling.

People of those nations whose diets consisted largely of rice developed a serious nerve disease, called *beriberi,* when they switched to white (polished) rice. (Vitamin B₁ was discarded when the germ and skin were thrown away.) Beriberi in Singhalese means "I cannot," signifying that the person is too ill to do anything. The disease existed as early as 2600 B.C. in the orient, where it was called the "scourge of the orient." It is characterized by spasmodic rigidity of the lower limbs, with muscular **atrophy,** paralysis, and **anemia.** There is a change in

---

[5]*Enriched* refers to the practice of adding certain dietary nutrients to food products. For example, most white flour is *enriched* with the B vitamins thiamine, niacin, and riboflavin, and with iron. Research has shown, however, that "enrichment" of white flour with these vitamins and iron does not compensate for all losses in milling. The "enriched" bread is deficient in the amino acids lysine and valine, as well as in certain vitamins and minerals present in whole wheat flour. Incidentally, oatmeal (rolled oats) is a whole grain (rather than a milled one like decorticated wheat) and, nutritionally considered, it is richer in vitamins, minerals, and protein than even whole wheat. Nutritionally fortunate, then, are children whose parents prepare warm oatmeal and milk for breakfast, rather than serving the more convenient (and more costly) sweet rolls made with white flour, or the highly advertised dry cereals.

the nerves that control the body (especially the lower limbs). Frequently, the heart becomes enlarged.

The early experiments that eventually led to the association of thiamine (B_1) with beriberi are of interest. From 1878 to 1882, about one-third of the enlisted Japanese naval force of 5000 men were victims of beriberi. In 1882 Dr. Takaki, a naval medical officer, initiated an experiment, using two crews of 270 men each going on a 9-month voyage over the same route. One crew (control group) was given the regular diet, consisting largely of polished (**decorticated**) rice, and the diet of the other crew consisted of less rice and more whole-grain barley. In addition, the second crew had meat, milk, and vegetables added to their diet. Among the 270 control men receiving the regular navy rations, there were 169 cases of beriberi, resulting in 25 deaths. On the second vessel there were only 14 cases of beriberi and no deaths. In each case of beriberi on the second ship, the sailor had not eaten his allowance of new foods.

The next important observation about beriberi was made in 1897 by Dr. Christian Eijkman, a Dutch physician and medical officer for a prison in the Dutch East Indies, who later received the Nobel prize in medicine. While relaxing on a walk through the prison yard one day, Dr. Eijkman noticed that the chickens seemed paralyzed, exactly as did the patients he was attending. Could these **fowls** have beriberi as his patients in the prison had? The chickens were fed leftovers from the table, and so they were being fed the same food as the men. Prisoners had beriberi and so did the prison chickens, he concluded. He could not recall having observed these symptoms of beriberi in chickens outside the prison yard, which prompted him to postulate that the *food* the prisoners were eating was causing the disease.

Dr. Eijkman began an experiment in his laboratory by feeding different foods to chickens and then keeping careful records of their growth and condition. He made two important observations: (1) chickens fed diets composed entirely of polished rice developed a disease very similar to beriberi, and (2) diseased chickens fed rice polishings (the outer coating of the rice, which was normally discarded) recovered. Moreover, he proved that chickens fed unpolished rice never developed the disease. From these observations, Dr. Eijkman concluded that beriberi must be caused by some poisonous substance in rice that rice polishings could destroy. Conversely, Dr. Funk, who originally named vitamins, postulated that rice polishings contained a substance—namely, vitamins— essential to the good health of chickens. Somewhat later, Dr. R. R. Williams, an American chemist working in the Philippines, isolated vitamin B_1 (thiamine).

Thiamine is essential to carbohydrate **metabolism.** The principal role of thiamine is as a part of a coenzyme in the oxidative **decarboxylation** of alpha-keto acids. The deficiency state, as it occurs **spontaneously** in humans, is called *beriberi*. It is characterized by the accumulation of pyruvic and lactic acids, particularly in the blood and brain, and an impairment of the cardiovascular, nervous, and GI systems.

Raw fish apparently contain the **enzyme** thiaminase, which inactivates thiamine. A thiamine deficiency may exist in foxes that have eaten raw fish, causing a condition called *Chastek paralysis* (a thiamine-deficiency **syndrome** resembling Wernicke's disease in humans, due to alcoholism).

Thiamine is synthesized by bacteria in the intestinal tract of ruminants, but biosynthesis is not a dependable source for humans and other monogastric animals. The liver stores a limited amount of thiamine; however, it is generally considered to be a daily dietary essential for humans. Considerable thiamine is stored in the bodies of pigs. Excess dietary thiamine is excreted via the urine.

Bacteria synthesize some thiamine in the cecum of the horse, but apparently not enough to meet daily requirements.

Clinical Deficiency Symptoms

1 Classical beriberi.

2 Edema, especially in the legs (also called *wet beriberi).*

3 Polyneuritis in rats and birds (Figure 16.7*a* and *b*).

4 Loss of appetite, decreased growth (Figure 16.7*c*), muscle weakness, incoordination.

Sources   Sources of thiamine include brewers' yeast, wheat germ, and milk; egg yolk; meat (especially pork and organ meats, such as liver and kidney); whole-grain, enriched cereals and breads; nuts; and dried legumes, such as peas and beans.

In Europe and the United States, where diets usually provide an abundance of thiamine, the classical picture of beriberi is rarely seen except in alcoholism (Wernicke's disease).

16.4.3   Riboflavin (Vitamin B$_2$)

This vitamin was originally labeled vitamin G. It is a component of the greenish-yellow fluorescent pigments of milk (lactoflavin), eggs (ovoflavin), liver (hepatoflavin), and grass (verdoflavin). It was first reported in milk (1879) by Blyth, who called it *lactochrome.*

Riboflavin functions as a coenzyme and is essential in normal energy transfer in the body. It is also important in protein metabolism.

FIGURE 16.7
(a) Polyneuritis in the rat. This thiamine-deficient rat shows the typical arched back and hyperextended hind legs (left). Such rats have a spastic gait, turn awkwardly, and lose balance. At right, the same rat (8 h after receiving thiamine hydrochloride) has normal use of its hind legs. (Upjohn Company Vitamin Manual, *1965.) (b)* Thiamine deficiency (polyneuritis) in the chick. Note the retraction of the head. *(Courtesy Dr. H. R. Bird, University of Wisconsin.) (c)* Thiamine deficiency. Contrast the growth of these littermates. The pig on the left received the equivalent of 2 mg of thiamine per 100 lb liveweight, while the one on the right received none. *(Courtesy Dr. N. R. Ellis and the USDA.)*

(a)

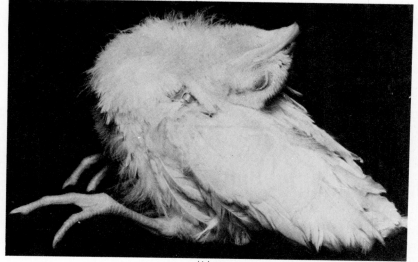

(b)

(c)

Clinical Deficiency Symptoms

1 Leg paralysis in poultry, "curled-toe paralysis" (Figure 16.8b).
2 Stiff and crooked legs in swine.
3 Impaired growth.
4 In humans there are eye, lip, skin, and tongue **lesions,** seborrhea (functional disturbance of the sebaceous glands), **dermatitis** (Figure 16.8a), and scrotal skin lesions.

Sources   Sources of riboflavin include milk and milk products; eggs; lean meats; legumes; distillers' solubles; and green, leafy vegetables. It was first synthesized in 1935. Being light-sensitive, riboflavin is destroyed by blue or violet rays. It is stored in limited amounts in the liver and kidneys; however, it is needed in the diet on a daily basis.

Information pertinent to the water-soluble vitamins and related compounds is presented in Table 16.2.

(a)

(b)

FIGURE 16.8
(a) Generalized dermatitis and growth failure in a riboflavin-deficient rat. Note the marked keratitis of the cornea (left). Same animal after 2 months of treatment (right). *(Upjohn Company Vitamin Manual, 1965.)*
(b) Riboflavin deficiency (curled-toe paralysis) in a young chick. Note the curled toe and tendency to squat on hocks. *(Cornell University, Poultry Science Dept.)*

16.4.4   Pantothenic Acid

The root word in the derivation of this term is Greek and means "from everywhere," since it is found so widely in living organisms. The universal distribution of pantothenic acid in all living material suggests that it performs a vital function in cellular metabolism. Like many other B vitamins, pantothenic acid is a component of an enzyme, coenzyme A, an important factor in the regulation of carbohydrate and fat metabolism. A dietary deficiency of pantothenic acid is rare in humans. It is a component of almost all foods but in varying concentrations.

Pantothenic acid is required in the chick, dog, pig, and rat. It is synthesized by **rumen** microorganisms in cattle, goats, and sheep and to an appreciable extent in the intestines of humans, horses, and rabbits.

Clinical Deficiency Symptoms

1 A pantothenic acid–deficient rat shows lesions of the adrenal cortex **(hypertrophy),** depletion of lipid materials, and ultimately hemorrhage and necrosis. Other symptoms in deficient rats include impaired growth, graying of the hair, testicular degeneration, impaired antibody formation, duodenal ulcers, and fetal abnormalities.

Adrenalectomy restores the hair and skin pigment in black rats made gray by pantothenic acid deficiency. This phenomenon apparently relates to the adrenal hormones and pantothenic acid in the production of **melanin.**

2 In chicks the principal deficiency symptom is dermatitis (especially of the eyelids, vent, corners of the mouth, and feet); however, spinal cord lesions, **involution** of the thymus, and fatty degeneration of the liver may also occur. Feathering is retarded and rough in appearance (Figure 16.9*a*).

3 Pantothenic acid–deficient swine develop an abnormal gait, called *goose-stepping* (Figure 16.9*b*). They also may develop gastrointestinal ulcers.

Sources   Sources of pantothenic acid include liver, egg yolk, milk, potatoes, cabbage, and peas. The royal jelly of bees and the ovaries of codfish are exceptionally rich sources of pantothenic acid.

16.4.5   Niacin (Nicotinamide)

This vitamin earlier was called vitamin B_5. It is associated with *pellagra,* which the Italian scientist Frapoli appropriately named (from the Italian *pella agra,* meaning "rough skin") in 1771.

The story of the search for the cause and cure of pellagra numbers among the great stories of modern medicine and is as exciting as reports of FBI person-hunts. Because the common stable fly *(Stomoxys calcitrans)* displayed certain salient characteristics that seemed to qualify it for the role of a transmitter of

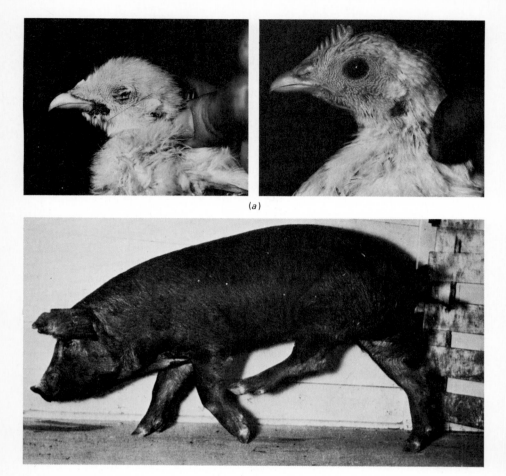

(a)

(b)

FIGURE 16.9
(a) Pantothenic acid deficiency in the chick. The eyelids, corners of the mouth, and adjacent skin areas are involved. Feathering is retarded and rough (left). The same bird (right) after receiving calcium pantothenate for 3 weeks. (Upjohn Company Vitamin Manual, *1965.) (b)* Pantothenic acid deficiency in the pig. The abnormal gait (goose-stepping) develops on a low pantothenic acid diet. *(Courtesy Dr. R. W. Luecke, Michigan State University.)*

pellagra, it was indicated as a **vector** of the disease. It was later cleared of this indictment. In 1735, Don Gasper Casal, the brilliant young physician of King Philip V of Spain, wrote about pellagra. In 1740, Jujati, an Italian scientist, confirmed the symptoms described by Casal: brown or red scaly patches appeared on the skin (especially on those parts of the body exposed to the sun), the tongue became sore and irritated, diarrhea and indigestion developed, a loss in body weight occurred, and finally the mind became affected.

Pellagra was first reported in Europe just before the American Revolution and was common in the United States about 100 years later. After 1907 pellagra spread with speed throughout the southern United States. It became common among people who had no milk or fresh meat but consumed corn bread and fat salt pork. Pellagra was commonly referred to as the disease of the "three D's"—*dermatitis, diarrhea,* and **dementia** (mental deterioration)—and was associated with a diet of "three M's"—*maize meal* (cornmeal), *molasses,* and *meat* (salt pork).

Changes in societies often result in nutritional problems. As populations move, economic, environmental, and social conditions may evolve that do not always allow people to consume diets that their ancestors found to support good health.

For two years, Dr. Joseph Goldberger of the U. S. Public Health Service attempted to isolate the **"germ"** that caused pellagra. Then he began to study the reports of southern hospitals for clues that might suggest that pellagra resulted from something other than a germ. He noted that although a high percentage of the patients in three southern hospitals had pellagra, the doctors, nurses, and orderlies who handled and even slept in the same ward with those ill with the disease were not afflicted. From this fact, Dr. Goldberger concluded that pellagra was not **contagious.** He searched for further clues to explain why patients had pellagra and hospital personnel did not. He found that the hospital staff did not eat the same food as the patients. Could diet be related to this disease, which was causing thousands of deaths annually? That answer appeared too simple; however, he initiated an experiment to see.

Dr. Goldberger selected an orphanage in Missouri where the diets of children consisted of corn pone (corn bread made without milk or eggs), salt pork, hominy, and molasses. To these foods he added milk and eggs. The results were indeed pleasing, for in less than a year, pellagra was eliminated among children in the orphanage. In other orphanages in which the dietary improvement had not been made, the rate of pellagra continued to be high.

Dr. Goldberger conducted a second experiment. This time his objective was to see if a poor diet per se would *cause* pellagra in healthy men. But where does a scientist get healthy men to volunteer for such an experiment? On promise of pardon, 12 healthy convicts on the Rankin Prison Farm in Georgia volunteered to serve as experimental subjects. (Eleven were selected for the experiment.) They were fed maize (corn) meal, white bread, flour, potatoes, salt pork, and syrup. After 6 months, seven had developed pellagra.

After experimenting in state hospitals, orphanages, and prisons, Dr. Goldberger then established the fact that pellagra resulted from an inadequate diet. He and 15 of his associates (in 1916) tried in a number of ways to transmit pellagra to themselves by a series of inoculations with blood, nasopharyngeal secretion, **feces,** urine, and **desquamating** epithelium. The results of this heroic experiment were negative. In 1926 Dr. Goldberger and his associates demonstrated similarities between pellagra in humans and blacktongue in dogs, and in 1937, Dr. C. A. Elvehjem and his coworkers at the University of Wisconsin found niacin to be the active antiblacktongue factor. It was soon found to be the specific cure and preventive of pellagra.

By 1947 it had become well established experimentally that the amino acid tryptophan was a precursor for niacin synthesis in the body (mostly in the intestine). Corn protein is low in tryptophan; thus individuals consuming a predominantly corn diet are predisposed to pellagra. Such a dietary regime increases the nicotinamide requirement. Thus pellagra might be considered as a dual deficiency of both nicotinamide and tryptophan. Nicotinamide is a component of two coenzyme systems and is important in cell respiration and energy transfer.

The horse, rat, and neonatal calf apparently have little dietary need for niacin, provided adequate amounts of tryptophan are available in their diets.

Recent research at Purdue University demonstrated that the feeding of 100 ppm of niacin significantly increases nitrogen retention in growing lambs.

Clinical Deficiency Symptoms

1 Dermatitis, dementia, diarrhea, loss of appetite and weight, vomiting, and anemia are commonly associated with a niacin deficiency. Usually, the first lesions are those of the mucous membranes of the mouth, tongue, and vagina. The dermatitis is bilaterally symmetrical and is especially prominent on the exposed areas of the face, neck, hands, forearms, and feet. It also occurs on the elbows and knees and under the breasts. Precipitating factors are sunlight, heat, and friction. For these reasons, most pellagra occurs in the warm months (Figure 16.10*a*).

2 As is true of other B vitamin deficiencies, pellagra often accompanies poverty, chronic alcoholism, fever, hyperthyroidism, pregnancy, and the **stress** of injury or surgical procedures.

3 Reduced growth occurs (Figure 16.10*b* and *c*).

Sources Sources of niacin include milk (see page 75), lean meat, eggs, green vegetables, cereals (except corn), and peanut-oil meals. It is not stored in appreciable amounts in animal tissues.

16.4.6 Pyridoxine (Vitamin B$_6$)

There are three related compounds with vitamin B$_6$ activity, namely, pyridoxine, pyridoxal, and pyridoxamine. Pyridoxal is a component of several enzymes. Pyridoxine functions in the conversion of tryptophan into niacin and in the metabolism of amino acids, and thus affects protein metabolism.

In laboratory animals fed diets deficient in pyridoxine (vitamin B$_6$), the lymphoid organs show signs of marked immune incompetence. For example, the thymus, spleen, and lymph nodes are smaller and not as well developed as organs of animals fed adequate levels of pyridoxine. Because of underdeveloped lymphoid tissue, pyridoxine-deficient animals are less able to mount an immune

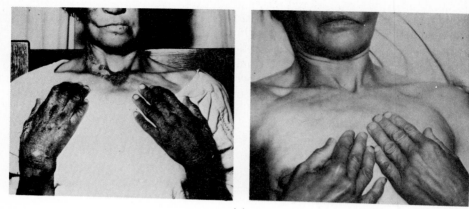

(a)

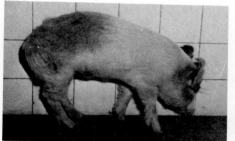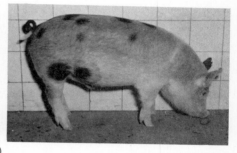

(b)

(c)

FIGURE 16.10
(a) Characteristic lesion of pellagra (left) on the back of the hands and dermatitis outlining the exposed area of the neck. The same patient (right) after nicotinamide therapy. (Upjohn Company Vitamin Manual, *1965.*) (b) Niacin deficiency in a pig (left) that received a diet containing 80 percent yellow corn. The growthier and normal-appearing pig (right) had niacin added to the diet. *(Courtesy Dr. D. E. Becker, University of Illinois.)* (c) Effect of niacin deficiency on chick growth. *(Courtesy Dr. H. R. Bird, University of Wisconsin.)*

response to infecting stimuli. Pyroxidine-deficient animals have decreased cell-mediated immunity and are slow in producing antibodies.

Severe deficiency is uncommon in humans, but pyridoxine status may be less than optimal in certain groups. For example, pregnant women and women using oral contraceptives need additional pyridoxine. Recent research showed that women who use oral contraceptives over extended periods of time secrete significantly lower levels of vitamin B_6 in their milk. Alcoholics risk a deficiency of the vitamin because of poor diets.

The diet of people residing in the United States commonly provides ample amounts of vitamin B_6, and deficiencies are seldom seen.

Clinical Deficiency Symptoms

1 Convulsive seizures (rats, poultry, dogs, infants, and swine).
2 Arterial lesions (monkeys).
3 Anemia (dogs and swine).
4 Dermatitis and edema of paws and nose (rats), as shown in Figure 16.11.
5 Impaired growth (all young animals).

Sources Sources of pyridoxine include milk, muscle meats, liver, green vegetables, whole-grain cereals, and yeast.

16.4.7 Biotin

Originally called vitamin H, this is a component of a coenzyme system and is concerned with CO_2 fixation, **decarboxylation,** and **deamination.**

A biotin deficiency can be induced in rats by feeding them raw egg white, which contains a biotin antagonist, **avidin,** that inhibits biotin absorption from the GI tract.[6] The deficiency state in the rat is characterized by dermatitis around

---

[6]A dramatic illustration of balance between vitamins is the interrelation between biotin and avidin. Many types of necessary balance between foods or nutrients are not entirely understood. For example, dogs digest raw starch and raw egg they are fed together but cannot digest raw starch or raw egg when each is fed alone.

FIGURE 16.11
Pyridoxine deficiency in the rat. The condition is characterized by edema; acanthosis; and denuding of the ears, paws, and snout (left). The same rat (right) after 3 weeks of treatment with pyridoxine hydrochloride. (Upjohn Company Vitamin Manual, *1965.*)

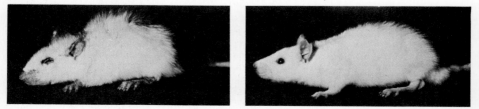

the eye. This "egg-white injury" has been produced in dogs, humans, and monkeys. Temperatures sufficient to coagulate egg albumen inactivate avidin.

Biotin is widely distributed in foodstuffs, and therefore a deficiency of it is rare. Biotin is synthesized in the intestines of humans, chickens, dogs, rats, and ruminants (rumen). It has been shown that humans will often excrete (via the feces and urine) more biotin than was ingested.

Clinical Deficiency Symptoms

1 Symptoms of an induced biotin deficiency include growth retardation, dermatitis (Figure 16.12a and b), loss of hair, and disturbances of the nervous system. In rats and dogs an ascending paralysis may be observed, with accompanying growth cessation and a spectacled condition.

FIGURE 16.12
(a) Biotin deficiency in a rat. Note the dermatitis, which begins around the eye (left). The same rat (right) 3 weeks after adequate biotin was added to the diet. (Upjohn Company Vitamin Manual, *1965*.) *(b)* Biotin deficiency in the chick. Note the severe lesions on the bottom of the feet. *(Courtesy Dr. H. R. Bird, University of Wisconsin.)*

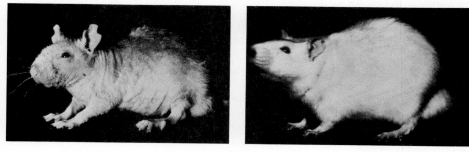

(a)

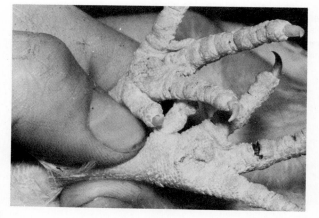

(b)

2 In chicks biotin is a good prophylactic measure for **perosis,** as are manganese, choline, and folic acid.

Sources Sources of biotin include milk, meats, yeast, and whole grains.

16.4.8 Folic Acid

This is needed to build red blood cells **(erythropoiesis).** The primary role of folic acid is apparently in the synthesis of nucleoprotein. It is especially important to mammalian cells during **mitosis** (required to carry **metaphase** to **anaphase**).

Folic acid–deficient chicks grow slowly, are incompletely feathered, have a reduced number of red blood cells and **leukocytes,** and have lower **hemoglobin** and hematocrit values.

Folic acid is interrelated with ascorbic acid and vitamin B_{12}. Apparently the chick, because of its short digestive tract, is the only domestic animal that requires a dietary source of folic acid. Ruminants synthesize the vitamin in their rumen, and sufficient intestinal synthesis apparently occurs in other species to meet requirements.

Clinical Deficiency Symptoms

1 Growth retardation (Figure 16.13).
2 Abnormal blood cells (red and white).
3 Poor feathering in chicks.
4 Poor pigmentation of colored feathers.

Sources Sources of folic acid include liver; dark green leafy vegetables; whole grains; and cereals.

FIGURE 16.13
Folic acid–deficient chick (left) is stunted, poorly feathered, and severely anemic. The healthy chick (right) received the same ration supplemented with 100 μg of folic acid per 100 g of ration. Both are 4 weeks old. (Upjohn Company Vitamin Manual, *1965.)*

16.4.9   Cyanocobalamin (Vitamin B$_{12}$)

This vitamin is also called *cobalamin* and, like folic acid, participates in nucleic acid synthesis.

Vitamin B$_{12}$ is closely related in function to folic acid. Apparently, three vitamins in particular (B$_{12}$, folic acid, and ascorbic acid) are involved in blood cell formation. The element cobalt is essential to the formation of vitamin B$_{12}$.

Pernicious anemia, resulting from a vitamin B$_{12}$ deficiency, has not been observed in species other than humans. Rumen bacteria synthesize the vitamin if sufficient cobalt is present in the diet. Intestinal synthesis also occurs in humans, the rat, and the pig, but the synthesized vitamin is not absorbed and is excreted in the feces.

Clinical Deficiency Symptoms

1 Pernicious anemia is the most severe form of vitamin B$_{12}$ deficiency. *Pernicious* stems from the Latin *perniciousus,* meaning "fatal," and *anemia* means *a reduction below normal in the number of erythrocytes.* Thus the name *pernicious anemia* was originally selected because the disease was inevitably fatal before the discovery of liver therapy. It is sometimes called *Addisonian pernicious anemia,* after Thomas Addison, who first described its clinical characteristics in 1855. It usually occurs after the age of 30 years. Addison's description included the following: "The countenance gets rather pale, the whites of the eyes become pearly, the general frame flabby rather than wasted, the pulse perhaps large but remarkably soft and compressible, and occasionally with a slight jerk, especially under the slightest excitement—the lips, gums, and tongue seem bloodless."

Failure of gastric mucosal secretion of "intrinsic factor" leads to reduced ability to absorb vitamin B$_{12}$ from the GI tract. This precipitates pernicious anemia. **Desiccated** hog stomach is a good source of intrinsic factor and is **orally** effective in the **therapy** of pernicious anemia. Also effective therapeutically are liver extracts and the **parenteral** injection of crystalline vitamin B$_{12}$.

2 Vitamin B$_{12}$ deficiency causes low **hatchability** of eggs and low vitality and retarded growth of chicks.

Sources   Sources of vitamin B$_{12}$ include milk, meat (especially organ meats), and fish products.

16.4.10   Choline

This organic compound is synthesized in the body from methionine and is a structural component of fat and nerve tissue. It is now known to participate as a **catalyst** in metabolic functions. It occurs along with the B-complex vitamins in milk, meat, eggs, fish, and most cereals. Choline is present in all fat-containing foods. As a component of **phospholipids,** choline is essential to the building and maintenance of cell structure.

No disease attributable to a choline deficiency has ever been demonstrated in humans. Choline prevents fat deposition in the liver. It has been called the "lipotropic factor," which means that it enhances the deposition of body fat (but not in the liver).

Choline is closely related to the amino acid methionine and may yield methyl groups (CH_3) to combine with homocysteine to form methionine. Choline is also closely associated with biotin and folic acid, since a deficiency of any one of the three will cause perosis in chicks.

Clinical Deficiency Symptoms

1 Impaired growth, fatty livers, enlarged spleen, and kidney hemorrhage.
2 Perosis in chicks (Figure 16.14) and reduced egg laying in hens.

Sources   Sources of choline include milk, meat, eggs, fish, and naturally occurring fats.

16.4.11 Inositol

This compound is closely related to the B-complex vitamins. It is a cell component in almost all animal tissues and is found in especially high concentrations in many organ tissues (heart, kidneys, spleen, **thyroid,** and testicles). It is

FIGURE 16.14
Perosis (slipped tendon), which may result from a deficiency of choline, biotin, or manganese. *(Cornell University, Poultry Science Dept.)*

apparently not a dietary requirement for humans and most domestic animals. (It possibly is for the chick.)

Clinical Deficiency Symptoms   Retarded growth and a loss of hair in mice and rats.

Sources   Sources of inositol include meats, nuts, fruit, and whole grain.

16.4.12   Para-Aminobenzoic Acid (PABA)

When rescued at Cape Sabine my hair was entirely white, probably due to semi-starvation, and darkened again within a year.

From Adolphus Greely's account of his 1881 Arctic expedition

PABA holds tentative vitamin status since it acts as an anti-gray-hair factor in mice and rats and as a growth-stimulating factor in chicks. It is a molecular component of folic acid.

Para-aminobenzoic acid counteracts the **bacteriostatic** effect of sulfonamides. Present evidence indicates that PABA increases the physiological potency of insulin and penicillin and that it may inhibit the production of thyroid hormones. There are no established dietary requirements for PABA in humans or domestic animals.

16.5   VITAMIN ASSAYS

Biological Assay   The oldest and still most commonly used technique to determine the vitamin potency of a food is the **bioassay.** Usually, animals are depleted of the vitamin by withholding its intake. Known amounts of the purified vitamin are then fed or administered (in a series of levels) to groups of the depleted animals. From the growth rates a standard response curve is prepared. The vitamin potency of the food in question is then determined in a similar way, comparing the animal response with the standard response curve. This technique is somewhat limited by the expense of labor, equipment, animals, and feed.

Microbiological Assay   The second assay technique for vitamin potency, **microbiological assay,** involves the use of microorganisms as test subjects. This method is faster and less costly; however, it has the disadvantage that the vitamin must first be extracted from the foodstuff before it is added to the assay medium.

Chemical Assay   A third test method for the concentration of a vitamin is based on the chemical characteristics of the particular vitamin. This technique is fast but should be confirmed with **bioassays** to make certain the substance has vitamin activity in the body.

16.6 EXPRESSING VITAMINS A AND D QUANTITATIVELY

The vitamin A potency of food is expressed in International Units (IU) or U. S. Pharmocopeia units (USP). One IU is defined as 0.344 μg of crystalline all *trans*–vitamin A acetate, equivalent to 0.3μg of crystalline vitamin A alcohol. Beta-carotene is used as the standard for provitamin A, 0.6 μg being equivalent in activity to 0.3 μg of vitamin A, using the rat as the assay animal. Beta-carotene is split enzymatically to yield vitamin A. In humans 1 IU of carotene is only about one-half as valuable as 1 IU of vitamin A, depending on its food source.

An IU (USP) of vitamin D is defined as the antirachitic activity, measured in a bioassay with rats, of 0.025 μg of crystalline vitamin D_3. Vitamin D potency is expressed in rat units per gram, except in poultry, where the International Chick Unit (ICU) is used to express the activity produced in chicks by 0.025 μg of crystalline vitamin D_3. Vitamin D_2 and D_3 have equal value for mammals, but D_2 has only 1/30 the activity of D_3 for poultry.

16.7 SUPPLYING VITAMINS TO FARM MAMMALS AND POULTRY

Dietary vitamin requirements of farm animals[7] can be met largely through the consumption of the common feedstuffs grown under normal conditions. Vitamins are needed in very small amounts (e.g., the addition of only 4 to 8 mg per ton of **ration** provides sufficient vitamin B_{12} for poultry). Most vitamins are now available commercially in a stabilized form and at a very nominal cost.

16.8 SUMMARY

A brief review of the principal roles, sources, and clinical deficiency symptoms of the fat- and water-soluble vitamins has been presented. Research leading to the identification and importance of vitamins is primarily a contribution of this century.

Some essential vitamins are stored in considerable quantities within the body during periods of adequate intake. These stores are drawn on later to meet requirements during periods of shortage. Others must be supplied regularly because storage in the body is limited. Moreover, there is a considerable variation of vitamin needs among species. Some species can synthesize adequate quantities of certain vitamins within their bodies. A considerable amount of

---

[7]Information pertaining to the specific vitamin requirements of the various species of farm mammals and poultry is available in "Nutrient Requirements of Domestic Animals," National Academy of Sciences–**National Research Council**, Washington, D.C. 20418.

research is needed to determine the exact requirements for certain vitamins and vitamin-related compounds in many species.

STUDY QUESTIONS

1 How can certain vitamins be metabolic essentials and yet not be required in the diet?
2 Name the fat-soluble vitamins. Name the water-soluble ones. Which group contains the most elements?
3 Why are fat-soluble vitamins not required in the diet on a daily basis?
4 Give at least one important function, one or more **clinical** deficiency symptoms, and two or more dietary sources of vitamins A, D, E, and K. Construct a table, and refer to Table 16.1 only as needed.
5 Differentiate between 7-dehydrocholesterol and ergosterol.
6 What mineral has recently been shown to perform some functions of vitamin E?
7 Which species have a dietary need for ascorbic acid? Does this mean that animals of other species have no need for vitamin C? Explain.
8 What significant contribution did Dr. Christian Eijkman make to the field of vitamins?
9 Give at least one important function, one or more clinical deficiency symptoms, and two or more dietary sources of ascorbic acid, the B-complex vitamins, and the related compounds. Construct a table, and refer to Table 16.2 only as needed.
10 Why do healthy ruminants have no apparent dietary requirement for the B-complex vitamins?
11 Write a short paragraph on the story of the search for the cause and cure of pellagra.
12 Which amino acid serves as a precursor for the synthesis of niacin in the body? Relate this to the antipellagric properties of milk. See Section 3.2.3.
13 What methods are used to determine the vitamin potency of foods for humans and animals?
14 How is the vitamin potency of food expressed?
15 What is a good reference for obtaining information pertaining to vitamin needs of farm animals? *(Hint:* See **NRC** in the Glossary.)

CHAPTER 17

THE NUTRITIONAL CONTRIBUTIONS OF MINERALS TO HUMANS AND ANIMALS[1]

In all science, error precedes the truth, and it is better it should go first than last.

Horace Walpole (1717–1797)

17.1 INTRODUCTION

Advances in the science of mineral **nutrition** during the twentieth century have been greater than in all the previous time of humans on earth. Nutritional studies have clearly shown that complex interrelations exist among mineral **elements.** Thus the mineral requirements for a given feeding regime vary, depending on the presence of certain **inorganic** and **organic** compounds. A classic example is that high levels of calcium and/or of phytic acid increase the zinc requirement of swine. Therefore the requirement for a given mineral element must be considered in relation to the natural food components of the diet. Feeding an excess of iron ties up phosphorus and results in iron rickets. Feeding a high level of copper results in a significant decrease of liver zinc stores. These are only a few illustrations of the delicate relation among minerals. However, it should be readily appreciated that fortifying animal diets with excessive amounts of either **micro-** or **macro**mineral elements may prove to be more detrimental than beneficial.

There seems to be little **species** difference with respect to the essential

[1]The authors acknowledge with appreciation the contributions to this chapter of Dr. R. M. Forbes, Department of Animal Science, University of Illinois, Urbana.

minerals. An apparent exception is cobalt, an essential dietary requirement for **herbivorous** animals only.

To understand better the nutritional contributions of minerals to humans and animals, it is necessary to consider them individually, as will be done in this chapter.

17.2 THE MACROELEMENTS

The macrominerals (from the Greek *macro,* meaning "major") include calcium, magnesium, sodium, and potassium as the principal **cations** and phosphorus, chlorine, and sulfur as the principal **anions.** Pertinent information related to these elements is summarized in Table 17.1.

17.2.1 Calcium and Phosphorus

More than 70 percent of the total body ash is calcium and phosphorus. About 99 percent of the calcium and 80 percent of the body phosphorus are located in the bones and teeth. It is readily apparent, then, that calcium and phosphorus are very important in the formation and maintenance of the skeleton of humans and animals. The Ca/P ratio by weight in bone is about 2:1.

Radioisotope studies have shown that there is a continuous exchange of calcium and phosphorus between the bones and soft tissues (about 1 percent daily). If the dietary intake of calcium is inadequate, the animal can draw on its soft-bone (vertebrae, skull, mandibles, and ribs) reserves. This is especially important during pregnancy, **lactation,** and egg laying. Parathyroid hormone (Chapter 7) and vitamin D (Chapter 16) are closely related to this calcium-mobilization mechanism. However, if an individual continues indefinitely to consume a diet with a negative mineral balance, **osteomalacia** may develop.

Subnormal **calcification** of bones in growing animals may cause *rickets* (Figure 17.1*a* and *b*). This condition may result from an inadequate dietary level of calcium and/or phosphorus or from decreased **absorption** of these inorganic elements, as in vitamin D deficiency.

Increasing the proportion of cereal (especially of oatmeal) in the diet of humans has been observed to increase the tendency to develop rickets. Much of the phosphorus in cereals is present as *phytin,* which is poorly absorbed by humans and poultry. Phytin combines with dietary calcium in the intestine and lowers its availability. The customary use of milk with porridge provides extra calcium needed under these conditions. Excessive intakes of aluminum, iron, and magnesium interfere with phosphorus absorption by forming insoluble phosphates.

Calcium is a normal and essential constituent of all living body cells, but it is more concentrated in blood. Parathyroid hormone regulates blood calcium and phosphorus levels. Calcium is essential to normal muscle contractions; therefore, when the blood calcium concentration drops markedly, **tetany** will result

TABLE 17.1
THE MACROMINERALS

Nomenclature	Function	Clinical deficiency symptoms	Major sources
Calcium (Ca)	Bone and tooth formation, blood clotting, enzyme activation, muscle contraction	Rickets, slow growth and bone development, **osteomalacia**	Milk, legumes, steamed bone meal, calcium phosphates, ground limestone, ground oyster shells
Phosphorus (P)	Bone and tooth formation, component of many enzyme systems, release of body energy, part of **DNA** and **RNA**	Rough hair coat, **pica**, lowered appetite, slow growth and low utilization of feed, lowered blood **plasma** phosphorus	Milk and eggs, oilseeds and hulls of cereals, steamed bone meal, dicalcium phosphate, tripolyphosphate, defluorinated phosphate
Magnesium (Mg)	Enzyme activator, constituent of skeletal tissue	**Anorexia** (lowered appetite), hyperirritability, muscular twitching, **tetany** (convulsions), profuse salivation, **opisthotonos** (muscle spasms)	All feeds, especially plant products (especially leafy vegetables and cereal grains)
Sodium (Na)	Muscle contraction; maintenance of **osmotic pressure** of body fluids; component of bile, which aids in fat digestion	Loss of weight, craving for salt, eating of soil, reduced appetite	Common salt; cured meats, cheese, many canned vegetables, soups
Potassium (K)	Maintenance of **electrolyte** balance, **enzyme** activator, muscle function	Heart **lesions,** loss of weight, reduced appetite, poor wool growth	Normal rations, widely distributed
Chlorine (Cl)	Acid-base relations, maintenance of osmotic pressure of body fluids, used to make hydrochloric acid, necessary for digestion	Craving for salt, reduced appetite, decreased blood chlorine level	Common salt
Sulfur (S)	Synthesis of amino acids in ruminants (component of sulfur-containing amino acids)	Slow growth, low **feed efficiency,** slow wool growth in sheep	Protein supplements, forages, cereals

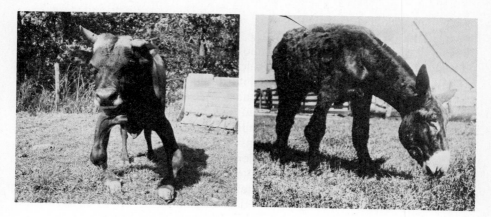

(a)

(b)

FIGURE 17.1
(a) Rickets in a young Jersey bull (left) and a Missouri mule (right). *(Courtesy Dr. W. A. Albrecht, University of Missouri.)* *(b)* Calcium deficiency in a pig. Note the abnormal bone development and rachitic condition. A lack of calcium retards normal skeletal development but does not usually depress total weight gain. *(Courtesy Dr. N. R. Ellis and the USDA.)*

(as in **parturient paresis**). As noted in Figure 16.4, calcium is essential for blood clotting.

Phosphorus is a component of the energy transfer system of the body, and calcium activates the **enzyme** ATPase. Also, phosphorus is a part of the genetic materials **DNA** and **RNA**. Hence it is concerned with the **metabolism** of almost all **nutrients** through its vital role in both vitamin and enzyme activity.

Calcium/Phosphorus Ratio   Since calcium and phosphorus are present in the body on a 2:1 basis, it would seem reasonable to assume that their intakes should be in about the same ratio. (Ca/P ratio range of 1:1 to 3:1 is satisfactory.) If much more calcium than phosphorus is consumed, the excess calcium is not absorbed in the proximal part of the small intestine, but accompanies the phosphorus to the lower portion of the small intestine (the point where most phosphorus is absorbed). This excess calcium combines with phosphorus to form insoluble tricalcium phosphate, thus interfering with phosphorus absorption. Conversely, an excess of dietary phosphorus over calcium will in the same way decrease the absorption of both calcium and phosphorus.

In general, **forages** (especially **legumes**) are high in calcium and low in phosphorus, whereas the cereal grains are high in phosphorus and low in calcium. Therefore the **livestock** feeder must consider the Ca/P ratio when making major shifts in the feeding regime.

Studies at Cornell University indicate that **atrophic rhinitis** can be produced in swine by feeding diets low in calcium or imbalanced in calcium and phosphorus.

Milk is an excellent source of both calcium and phosphorus (Chapter 3). Eggs are rich in phosphorus but low in calcium. Meats are low in calcium but provide significant amounts of phosphorus. **Bone meal** and dicalcium phosphate are good **supplemental** sources of calcium and phosphorus for farm mammals, and ground limestone and oyster shells are good supplemental calcium sources for poultry.

In countries where milk is not available, bones are often consumed as such and in bone soups (cooked with vinegar, which disintegrates the hard bone and renders its calcium available).

Clinical Deficiency Symptoms

1 Rickets.

2 Osteomalacia.

3 **Pica** (depraved appetite), especially noted in phosphorus deficiency. The animals may chew wood, bones (Figure 17.2), and rocks and develop a rough hair coat.

4 Impaired appetite **(anorexia),** slow growth, rough hair coat.

FIGURE 17.2
Phosphorus-deficient cow (left) showing unhealthy appearance and bone-chewing tendency. Same cow (right) after phosphorus was added to the diet. *(Courtesy Dr. W. E. Petersen, University of Minnesota.)*

5 Easily fractured bones.

6 Lowered blood **plasma** phosphorus.

17.2.2 Magnesium

About three-fourths of the body's magnesium is found in the skeleton. The balance is distributed throughout the body fluids. A close biological relation between calcium and magnesium has recently been demonstrated experimentally by the substitution of magnesium for calcium in bone formation without significantly changing the **x-ray** differentiation pattern. However, the body apparently exhibits "selective absorption" for calcium over magnesium when both are present in the diet. An increase of calcium or phosphorus (or both) in the ration apparently increases the minimum magnesium requirement and may result in symptoms of magnesium deficiency.

Magnesium is essential as an activator of many enzyme systems, especially those involved in carbohydrate metabolism. It is an essential constituent of bones and teeth. Magnesium is a component of chlorophyll (as iron is of **hemoglobin**). It is also vital in the proper functioning of the nervous system.

In general, leafy vegetables and cereal grains are rich sources of magnesium, whereas most animal products contain lesser amounts. Magnesium tetany has been observed in calves and children fed milk alone for extended periods without magnesium supplementation. Most of the commonly fed rations of farm animals supply ample magnesium without supplementation.

Clinical Deficiency Symptoms

1 Tetany, a disease of cattle called "grass tetany" or "grass staggers" (frequently observed in the Netherlands and New Zealand), associated with a magnesium deficiency. It is most commonly observed among dairy cattle subsisting on lush spring pastures that have had high nitrogen applications. Figure 17.3*a* depicts a lamb in tetany resulting from a magnesium deficiency.

2 Hyperirritability; magnesium, like calcium, depresses nervous **irritability.**

3 Muscular twitching, convulsions, weak pasterns, excessive salivation.

4 **Opisthotonos** (a form of spasm in which the head and the heels are bent backward and the body bowed forward).

5 Retarded growth and reduced feed efficiency.

6 Calcification of soft tissue in certain species (Figure 17.3*b*).

17.2.3 Sodium, Potassium, and Chlorine

These three essential dietary minerals are present largely in the body fluids and soft tissues. They function (1) to maintain **osmotic pressure** and acid-base equilibrium, (2) to control the movement of food nutrients into cells, and (3) to regulate water metabolism.

Since these minerals are not stored to any appreciable extent in the body, there is a regular dietary need. A deficiency in any of these minerals results in a

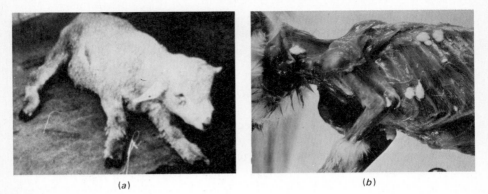

FIGURE 17.3
(a) Magnesium deficiency in a lamb. Note stiff legs. *(Courtesy Dr. U. S. Garrigus, University of Illinois.)* *(b)* Large calcium phosphate deposits on the rib cage of a guinea pig fed a diet deficient in magnesium. Frequently, such deposits form in and around joints, giving rise to an arthritis-like condition. The condition is entirely prevented by feeding adequate levels of magnesium. *(Courtesy Dr. B. L. O'Dell, University of Missouri.)*

reduced appetite, a decline of growth, a loss of weight, decreased production (meat, milk, eggs, and wool), and decreased blood concentrations. A rough hair coat often characterizes a salt (NaCl) deficiency. If allowed free access to salt, farm animals will not suffer a salt deficiency. However, **ruminants** fed little or no **roughage** may exhibit a potassium deficiency (Figure 17.4).

Sodium This element is essential for normal muscle contraction. It is also essential for maximal utilization of dietary energy and protein and for efficient reproduction. Deficiency of it in laying hens results in a loss of weight, decreased egg laying, and **cannibalism.**

FIGURE 17.4
The lamb at left received a potassium-deficient ration (0.1 percent K), whereas the lamb at right received sufficient dietary potassium (0.6 percent K). *(Courtesy Dr. R. L. Preston and the University of Missouri.)*

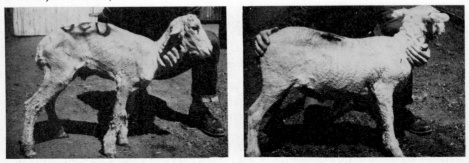

Potassium This mineral plays a vital role in muscle function. A potassium deficiency causes heart **lesions** and degeneration of the kidney tubules. Potassium chloride has an epinephrine-like effect (Chapter 7) and has been used to relieve the allergic effects of hay fever.

Recent studies have shown an interrelation between sodium and potassium in metabolism. (At inadequate levels of either, the deficiency symptoms are aggravated by a large excess of the other.) Recent evidence indicates that potassium may be required by one or more enzyme systems.

Chlorine The primary role of chlorine is in control of acid-base equilibrium and osmotic-pressure regulation. However, it is also the important component of gastric secretions. A limited amount of chlorine is stored in the skin and **subcutaneous** tissues.

Salt (NaCl) The inclusion of "common salt" in the diets of humans and animals has been practiced for a long time. Early humans used salt as a condiment (primarily as a seasoning) and probably were unconcerned about its nutritional and physiological support. Salt stimulates salivary secretion and promotes the action of diastatic enzymes. It is the mineral compound most often needed as a dietary supplement for most farm animals.

When the dietary intake of salt is low, the human kidney may excrete as little as 1 g of sodium chloride per day. Conversely, it may excrete a large quantity (as much as 40 g daily) when the salt intake is high. The latter situation requires an increased water intake.

When the rigors of exercise and/or heat result in profuse perspiration, humans may consume large quantities of water in relatively short periods of time. In such situations, cramps may result unless salt is taken concurrently with the water.

An excessive salt intake may result in water retention, which causes **edema.** A salt deficiency is more prevalent in herbivores than in other animals, because forages and grains contain small amounts of salt.

Salt-deficiency symptoms include (1) lack of appetite, (2) rough hair coat, (3) eating of soil, (4) unhealthy appearance, (5) decrease in production, and (6) loss of weight (Figure 17.5).

17.2.4 Sulfur

Most body sulfur is found in proteins and, more specifically, in the amino acids *cystine* and *methionine*. Sulfur is present in insulin and glutathione, as well as in wool. Thus most dietary sulfur is derived from an organic source (protein). Only a small fraction is **ingested** in the inorganic form, which is ineffective in satisfying body requirements for sulfur.[2] However, small amounts of inorganic sulfur supplements have been observed to increase the utilization of **urea** as a source of nitrogen for ruminants.

---

[2]Recent research indicates that inorganic sulfur may be useful in poultry diets.

FIGURE 17.5
Extreme salt deficiency in a cow. *(Courtesy Dr. S. E. Smith, Cornell University.)*

Sulfur functions in the synthesis of sulfur-containing amino acids in the **rumen** and certain other sulfur compounds of the body. A sulfur deficiency may result in reduced growth, lower **feed efficiency,** and slow wool growth in sheep (Figure 17.6).

FIGURE 17.6
Lambs fed a low-sulfur diet. Lamb 6 received 3 g of sulfur per pound of diet, whereas lamb 5 received none. Note the excessive salivation, lacrimation, and shedding of wool by lamb 5. *(Courtesy Dr. U. S. Garrigus, University of Illinois.)*

17.3   THE MICROELEMENTS (TRACE ELEMENTS)

The microminerals (from the Greek *micro,* meaning "minor") that have been shown to be essential for cattle, sheep, and swine include cobalt, copper, iodine, iron, manganese, molybdenum, selenium, and zinc. In recent experiments in which animals were fed highly **purified diets,** one or more of the following elements have been classified as essential trace minerals for some species: arsenic, chromium, nickel, tin, and vanadium. Pertinent information related to the microelements is summarized in Table 17.2.

17.3.1   Iron

Although needed in small quantities, iron plays an essential role in the life of humans and animals. As a component of the respiratory pigment hemoglobin, iron is essential for the normal functioning of every organ and tissue throughout the body. It is present as an iron-porphyrin nucleus (heme) in the hemoglobin molecule, in the protein fractions of cytochrome C, and in other important enzymes. Hence iron is a component of oxygen carriers and of oxidizing **catalysts** (enzymes), which are essential for cellular **oxidation.**

More than 60 percent of the body's iron is in the form of hemoglobin. Iron is stored in the liver, spleen, kidneys, and, to some extent, the bone marrow. However, because the destruction and formation of red blood cells is continuous (the average life of an RBC is about 4 months), there is a constant metabolism of iron in the body.

Iron-Deficiency Anemia   The **anemia** resulting from an iron deficiency may occur anytime the iron intake becomes deficient relative to the needs for hemoglobin formation. It is most likely to occur during the **suckling** period of mammals, since milk is deficient in iron. In Missouri studies, calves receiving iron dextran injections gained 22 percent faster to 12 weeks of age than noninjected controls. In the seventeenth century, physician Thomas Sydenham made a tonic of iron and wine and used it in the treatment of anemia.

In baby **pigs,** a condition commonly called **thumps** soon develops (within 3 to 4 weeks) if the **piglets** are confined without access to soil and do not receive supplemental iron during the suckling period. (Since many pigs are now raised in confinement, iron injections are routine.) Human babies also soon deplete their body stores (within 6 months) unless supplemental iron is provided in the high-milk diet. **Calves, foals,** and **lambs** seldom demonstrate iron deficiency because they usually have access to forage early in life.

Both pregnancy and egg laying increase the need for additional dietary iron. Studies utilizing **radioactive** iron (^{59}Fe) have shown that there is an increased **absorption** of iron from the small intestine during pregnancy to meet the needs for fetal growth. Feeding supplemental iron to a **lactating** female does not increase the iron content of the milk.

TABLE 17.2
THE MICROMINERALS

Nomenclature	Function	Clinical deficiency (or excess) symptoms	Major sources
Iron (Fe)	Component of **hemoglobin**, component of many enzyme systems	Nutritional anemia, **thumps** in pigs, diarrhea, loss of appetite	Eggs, soil, forages and grains, iron injections; liver, pork; ferrous sulfate
Copper (Cu)	**Erythropoiesis, coenzyme** system, hair pigmentation, reproduction, collagen and elastin synthesis, iron utilization	Depraved appetite, stunted growth, diarrhea, **osteomalacia** (in mature cattle), bleached hair and wool, **ataxic** gait, anemia, loss of **condition**, aortic rupture in swine and poultry	Feedstuffs and $CuSO_4$ (0.25 to 0.5% $CuSO_4$ added to salt fed free choice)
Iodine (I)	Synthesis of thyroxine	Enlarged necks in calves and lambs, goiter, hairless pigs and woolless newborn lambs, dead or nonviable calves	Iodized salt (KI in salt), cod-liver oil
Cobalt (Co)	Component of vitamin B_{12}, RBC formation, proper function of rumen microorganisms	Loss of appetite, weakness, **emaciation**, rough hair coat, anemia, reproductive failure	Cobalt pellets (for ruminants); 0.5 ppm of cobalt salt added to ration (vitamin B_{12} injection to relieve cobalt deficiency)
Zinc (Zn)	Carbonic anhydrase, enzyme activator	Retarded growth, **anorexia, parakeratosis** in swine, hyperkeratosis in chicks, poor feathering, poor **hatchability**	ZnO or $ZnCO_3$ added to ration; forages
Manganese (Mn)	Growth and bone formation, enzyme activator	**Perosis** (slipped tendons) in poultry, lowered hatchability and eggshell strength, lameness, stiffness	$MnSO_4$ at 100 g/ton of feed; widely distributed in feeds, nuts and seeds, milk, legumes, and cereals
Selenium (Se)	Destroys peroxides; related to vitamin E, which prevents peroxide formation	**Necrosis** of liver, white muscle disease in sheep (deficiency); "alkali disease" or "blind staggers" (excess, above 5 ppm)	Oil meals and grains
Molybdenum (Mo)	Enzyme systems. Affects copper absorption and availability to tissues	Excess: **teart**, diarrhea, loss of weight, emaciation	Widely distributed; rarely a problem
Fluorine (F)	1 to 2 ppm in water added to aid in preventing tooth decay	Excess: chalky and mottled teeth (fluorosis), decreased appetite, and slow growth	Water

It should be noted that there are many kinds of anemia not caused by iron deficiency. Hereditary anemia (sickle-cell anemia) was mentioned in Chapter 4. Deficiencies of copper, protein, and certain vitamins can cause anemia. Certain pathological conditions may result in anemia. It may result from an interference with or cessation of hemoglobin production or from an excessive loss of blood.

Relative to the amount needed by the body, most animal foods (except milk) contain liberal amounts of iron. Eggs, meat, and many leafy vegetables provide good sources of iron for humans. However, the iron of many leafy plants is utilized poorly, whereas iron from inorganic sources is utilized more completely. Hemoglobin iron is the best-utilized source.

Excessive dietary iron interferes with phosphorus absorption by forming an insoluble phosphate, and this may cause rickets. Iron toxicosis is uncommon in farm animals but has been reported in children as a result of accidental excessive intake of iron pills; in Bantus of South Africa due to cooking in iron pots and drinking Kaffir beer, which contains a high level of iron; and in certain genetic conditions.

An iron deficiency in cattle and sheep may cause pica, similar to that caused by a phosphorus deficiency. It is characterized by diarrhea, loss of appetite for the usual foods, and anemia.

17.3.2 Copper

Like iron, copper is important in hemoglobin formation. It, too, is stored in the liver and, to a lesser extent, in the spleen, kidneys, heart, lungs, and bone marrow. Since milk is low in both copper and iron, nature provides liver stores at birth to supply the mammal with copper throughout the suckling period. Copper is not an essential component of the hemoglobin molecule, but it is apparently needed as a catalyst in its formation. Therefore anemia may result from a copper or an iron deficiency or both. Animals suffering from inadequate copper intake appear to be unable to utilize iron at a normal rate, and a deficiency in hemoglobin synthesis exists. Copper, like iron, is vital to certain cellular enzyme systems involved with oxidation-**reduction** reactions.

A copper deficiency in sheep is reflected in changes in the **fleece.** The fibers become progressively less crimped (the wool becomes steely or stringy), wool growth is slowed, and black wool turns white. Depigmented feathers or a bleached hair coat is common in poultry and other animals (Figure 17.7). A copper deficiency interferes with synthesis of **keratin,** the principal constituent of hair and wool. These changes in hair and wool reflect a failure of certain enzyme activities for which copper is essential. Diarrhea, a loss of appetite, and swelling about the pasterns may occur in copper-deficient animals. Furthermore, in an extreme copper deficiency, the bones become fragile and the animal exhibits an **ataxic** (uncoordinated) gait. A disease of lambs, called "swayback," characterized by nervous symptoms is caused by a copper deficiency and results from damage to nerve tissue during embryonic development.

FIGURE 17.7
A typical copper-deficient poult at left. Note the depigmentation of the feathers. Age is 4 weeks. *(Courtesy Dr. J. E. Savage and Dr. B. L. O'Dell, University of Missouri.)*

Although essential in small amounts, excessive copper is toxic. Copper toxicity is accentuated by low molybdenum diets. The addition of molybdenum to diets (currently not approved by the **FDA**) containing toxic levels of copper will counteract possible copper poisoning. Thus it is apparent that copper and molybdenum are biologically antagonistic. Recent evidence indicates that high levels of dietary copper tend to deplete the liver stores of zinc.

Missouri studies have shown that when chickens are fed a purified diet deficient in copper during their early growth and development, they may die of aortic rupture. This results from defective elastin (scleroprotein) in the walls of the aorta. Copper is essential to the enzyme system that aids in the use of the amino acids needed to synthesize the elastin fibers (Figure 17.8).

17.3.3 Iodine

The animal body contains a very minute amount of iodine, approximately 60 percent of which is found in the **thyroid** gland. The foremost need for dietary iodine is in the synthesis of thyroxine. The discovery of iodine in the thyroid gland was made by Baumann in 1896.

Activity of the thyroid gland is regulated by thyrotropin (TSH, or thyroid-stimulating hormone) from the **anterior** pituitary (Chapter 7). When the blood concentration of thyroxine declines, the anterior pituitary increases its output of TSH, which in turn causes the thyroid to increase its thyroxine output. The thyroxine then enters the bloodstream and performs its physiological function of regulating the metabolic rate of the body. Thus iodine is indirectly involved in control of the rate at which the body uses energy.

When dietary iodine is insufficient, the thyroid cannot synthesize sufficient thyroxine, and the blood concentration declines, causing the release of thyrotropin. This hormone of the anterior pituitary gland in turn increases the

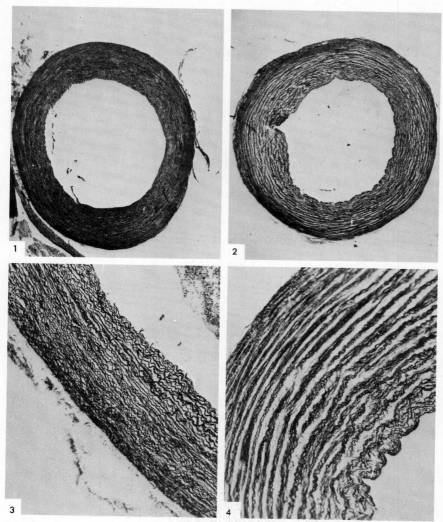

FIGURE 17.8
Cross section through arch of the aorta from control turkey poult fed a milk diet with 50 ppm of copper (upper left, × 24). Cross section through arch of the aorta from copper-deficient poult (upper right, × 24). Lower left and right show, respectively, control and copper-deficient aorta magnified 120 times. Note thicker aortal wall and the accumulation of nonelastin material and focal breaks in the copper-deficient poult. *(Courtesy Dr. J. E. Savage and Dr. B. L. O'Dell, University of Missouri.)*

activity of the thyroid gland, causing it to enlarge (Figure 17.9). This condition is called *goiter* (compensatory **hypertrophy,** or an enlargement involving the formation of more tissue in an effort to secrete more thyroxine). Cleopatra and the model for the Mona Lisa reportedly each had a small goiter.

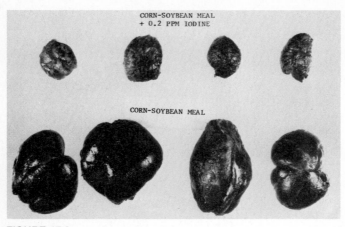

FIGURE 17.9
Thyroid glands of pigs fed a corn–soybean meal basal diet with or without 0.2 ppm iodine for 51 days. The goitrous thyroid glands of pigs fed the basal diet (bottom row) weighed approximately 6 times more on the average than those of pigs fed supplemental iodine (top row). *(Courtesy Dr. Gary L. Cromwell, University of Kentucky, Lexington.)*

    Goiters are more likely to occur during times of increased metabolic rate, such as **puberty** and pregnancy. In farm animals, the newborn may exhibit a goiter resulting from an iodine deficiency in the ration of its mother during pregnancy. Typical iodine-deficiency symptoms are hairlessness and abnormal growth and development in young pigs (Figure 17.10); enlarged necks in calves, lambs, and **kids;** and weak foals. Sheep showing a typical goiter are depicted in Figure 17.11. Iodine deficiency is a geographic problem,[3] and people living in goitrous areas observed the benefits of sea salt in preventing goiter hundreds of years before the discovery of iodine.[4] Crops reflect the level of iodine in the soil on which they are grown. When farm animals drink only rainwater, they are likely to be deficient in iodine unless they are fed iodized salt or an inorganic form such as either potassium or sodium iodide.

[3]For example, the Great Lakes region of the United States and Canada, the Pacific northwestern states, and Switzerland are areas in which an iodine deficiency is likely to occur unless supplemental dietary iodine is provided.
    How to provide iodine for the general population was at first a problem. Three proposals were advanced: (1) Since goiter is prevalent among iodine-deficient children, why not add it to candy? This method was tried in some schools and failed because not all children liked the same kind of candy. (2) Add iodine to drinking water. But then what should be done about the rural people? (3) Add iodine to salt. This method was and continues to be successful. At least a dozen countries now iodize their table salt. The amount added varies from 1 part in 10,000 in the United States and Canada to 1 part in 200,000 in Poland. In at least two countries, Canada and Switzerland, all salt for home use must be iodized. This requirement reduced the incidence of goiter by more than 85 percent in Switzerland.
    [4]Seaweed provides a reliable source of iodine and has been used in the treatment of goiter for more than 3000 years.

FIGURE 17.10
The pig in front was fed a basal diet plus 0.5 percent potassium thiocyanate for 51 days. The littermate in back was fed the basal diet plus 0.2 ppm iodine for the same period. The pig fed thiocyanate showed symptoms of hypothyroidism, including shortened legs and extreme lethargy. *(Courtesy Dr. Gary L. Cromwell, University of Kentucky, Lexington.)*

17.3.4   Cobalt

This element was not recognized to be essential for growth and health until 1935. As with iodine, cobalt deficiency is a regional problem.

Cobalt is an integral component of the vitamin B_{12} **molecule.** It is essential in the synthesis of this vitamin by the rumen **microflora.** Thus it is actually a deficiency of vitamin B_{12}, not of cobalt per se, that is responsible for the metabolic failure observed in cobalt-deficient ruminants. This explains why no essential role for cobalt has been demonstrated in swine and poultry. (They require dietary B_{12}.) It is believed that cobalt stimulates the appetite of

FIGURE 17.11
Ewe showing a typical goiter due to an iodine deficiency (left) and woolless neonatal lamb (right) born of an iodine-deficient ewe. *(Courtesy Dr. J. E. Catlin, Montana State University.)*

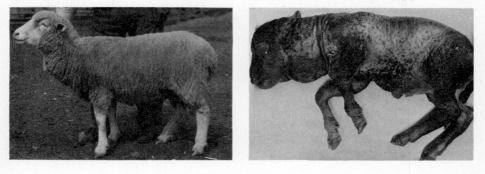

ruminants through its action on the rumen **flora.** Recent evidence suggests that cobalt may be associated with the synthesis of pyridoxine, niacin, and riboflavin by the rumen microorganisms.

Clinical symptoms of a cobalt deficiency include loss of appetite and body weight; **emaciation;** long, rough hair coat; retarded wool growth (and weak fibers); scaliness of the skin; abortion; reduced milk secretion; and anemia. Cobalt deficiencies in sheep and cattle are shown in Figure 17.12*a* and *b,* respectively.

17.3.5 Zinc

Most of the body's zinc is found in the liver, bones, and **epidermal** tissues (skin, hair, and wool). Since zinc absorption from grains and legume seeds by swine and poultry is very poor, diets based on these feedstuffs frequently must be supplemented with zinc salts.

FIGURE 17.12
(a) Comparison of a sheep fed a cobalt-adequate diet (left) with one fed a cobalt-deficient diet (right). *(Courtesy Dr. S. E. Smith, Cornell University.) (b)* The calf at left shows effects of a cobalt-deficient diet. The same calf is pictured at right a few weeks after cobalt was added to the diet. *(Courtesy Dr. R. C. Carter, Virginia Polytechnic Institute.)*

(a)

(b)

The primary physiological role of zinc is related to enzymatic activity. The most fundamental physiological action of zinc is in its role in protein synthesis. Zinc is required for the formation of DNA and RNA, the compounds that control cellular multiplication and growth through their influence on protein synthesis (see Chapter 4).

Clinical symptoms of a zinc deficiency include (1) **parakeratosis (dermatitis)** in swine (Figure 17.13a) and calves, (2) retarded growth, (3) decreased **feed efficiency.** Parakeratosis is apparently predisposed by rations high in calcium

FIGURE 17.13
(a) Parakeratosis in swine. *(Courtesy Dr. R. W. Luecke, Michigan State University.) (b)* Calf showing loss of hair on legs and severe scaliness, cracking, and thickening of the skin as a result of zinc deficiency at age 15 weeks (left). Same calf (right) 5 weeks after zinc was added to the diet. *(Courtesy Dr. W. J. Miller, University of Georgia.)*

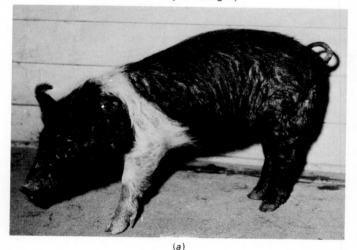

(a)

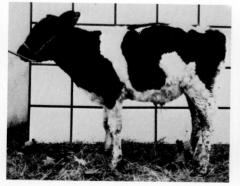

(b)

(especially noted when animals receive rations containing primarily plant proteins).

Calves fed a zinc-deficient diet develop parakeratosis. They are characterized by an unhealthy appearance; rough hair coat; stiffness of the joints; dry, scaly skin on the ears; and a thickening and cracking of skin around the nostrils (Figure 17.13b).

Zinc-deficiency symptoms in the chick include slow growth, shortened and thickened long bones, poor feathering, reduced **hatchability,** and embryonic **anomalies.** In severe deficiency, keratosis occurs. Recent Missouri studies, using purified diets, have shown zinc to be especially important to proper structural bone development in chickens (Figure 17.14). Zinc is closely related to the **assimilation** and proper use of other minerals.

17.3.6 Manganese

Manganese is found in the liver, bone, muscle, and skin. Probably the most important function of manganese in the body is to activate several enzymes concerned with carbohydrate, fat, and protein metabolism. Manganese has been shown experimentally to be essential for normal reproduction in several laboratory animals. Manganese-deficient diets delay sexual maturity in females,

FIGURE 17.14
A malformed chick embryo (note missing appendages) caused by a zinc-deficient diet (left). Both control (right) and zinc-deficient embryos were incubated 18 days. *(Courtesy Dr. J. E. Savage and Dr. B. L. O'Dell, University of Missouri.)*

cause irregular **ovulations** and weak young at birth, and may cause sterility in males.

In poultry a manganese-deficient diet causes **perosis** (slipped tendons), a malformation of the leg bones of growing chicks (Figure 17.15), and lowered hatchability. In rabbits, bone malformations may occur. Perosis is not caused solely by a manganese deficiency, because a vitamin deficiency of biotin, folic acid, or choline will also result in perosis (Chapter 16). Diets high in calcium and phosphorus apparently predispose poultry to perosis, probably by interfering with manganese absorption.

17.3.7 Selenium

Only recently has selenium been considered an essential micronutrient for animals. It is now well established that selenium can perform some of the functions of vitamin E. In vitamin E deficiency, hydroperoxides are formed during metabolism of unsaturated fatty acids. A selenium-containing enzyme, glutathione peroxidase, can destroy these hydroperoxides, thereby preventing them from damaging tissues. A low dietary selenium level (<0.05 ppm) may cause white muscle disease in sheep.

Basic research studies in the bull have shown that selenium is associated with the reproductive system. However, its physiological role in reproduction has not been clearly defined. Recent research at the University of Illinois indicates that low levels of selenium are anticarcinogenic in mice. The mode of action has not been resolved.

FIGURE 17.15
Manganese deficiency caused the "slipped tendons" in the chick at right. A normal chick is pictured at left. *(Courtesy Dr. J. E. Savage and Dr. B. L. O'Dell, University of Missouri.)*

Studies have shown that crops grown on soils high in selenium (>5.0 ppm) may contain toxic levels of that element. In some regions (South Dakota, for example), livestock consuming such feeds may develop "alkali disease," or "blind staggers," due to the destruction of certain oxidation catalysts. There is commonly a loss of hair in cattle, horses, and swine; the hoofs slough off, lameness occurs, appetite diminishes, and growth (in young animals) is retarded (Figure 17.16). Recent studies at the Ohio Agricultural Research and Development Center in Wooster indicate that dietary selenium levels are the most critical during early stages of growth and when cattle are fed diets marginal or deficient in protein.

Selenium is a cumulative poison, so that toxic symptoms may be observed only after an extended period of its consumption in low quantities. Small amounts of arsenilic acid or arsenic compounds are effective in reducing the toxicity of selenium.

A selenium deficiency may be reflected in a condition known as *nutritional myopathy*, or white muscle disease, in lambs and cattle, and as heart and skeletal muscular dystrophy in mink. Large doses of vitamin E will normally correct these.

Approval was gained from the FDA in 1974 to add selenium to chicken, swine, and turkey feeds in a premix as either sodium selenite or sodium selenate. Similar approval for the addition of selenium to ruminant feeds was obtained in 1979.

17.3.8 Molybdenum

This mineral is known to be a component of one or more enzymes.

When cattle graze on vegetation grown on soils high in molybdenum (especially noted in Canada and England), they may develop a condition called

FIGURE 17.16
Ewe (left) suffering from selenium toxicity. Loose wool is typical of sheep afflicted with chronic selenium poisoning. The lambs (right) show congenital deformity traceable to selenium injury during fetal development. *(Courtesy Dr. P. O. Stratton, University of Wyoming.)*

teart, which is caused by molybdenum poisoning. It is also recognized as "peat scours" in New Zealand. The chief symptoms of molybdenum toxicity are diarrhea, loss of weight, **emaciation,** anemia, and stiffness. These symptoms may be cured by the administration of copper sulfate. This fact points again to the interrelation between molybdenum and copper discussed previously (Section 17.3.2). Toxic levels of molybdenum interfere with copper metabolism, thus increasing the copper requirement. (One gram $CuSO_4$ per animal daily will prevent and cure symptoms of molybdenum toxicity.) Tolerance to molybdenum is believed to be affected by the intake of methionine and inorganic sulfate, as well as by the copper content of the diet.

17.3.9 Fluorine

The essentiality of this mineral has not been clearly established. It is found in very minute amounts throughout the body, but notably in the hair, bones, and teeth. It is known that fluorine, at appropriate intake levels, aids in preventing dental caries.[5] Conversely, higher levels will cause *fluorosis* (chalky and mottled teeth), as shown in Figure 17.17. Mottled teeth are structurally weak.

The effects of fluorine are cumulative, so that the intake of small quantities over an extended period of time may produce toxic effects. Rock phosphates

[5]The rate of dental caries in certain areas of the United States and Canada was reduced by as much as 65 percent following water fluoridation.

FIGURE 17.17
Effects of fluorine on teeth in cattle. All animals received a diet containing 7 ppm fluorine. In addition, animals *b* through *d* had, respectively, 30, 50, and 100 ppm fluorine added to their diets. Note the increasing discoloration of the teeth with increasing dietary fluorine levels in comparison with the teeth of the control cow, *a*. (*C. S. Hobbs and G. M. Merriman, University of Tennessee,* Agr. Expt. Sta. Bull. *351, 1962.*)

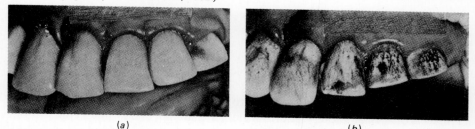

(a)   (b)

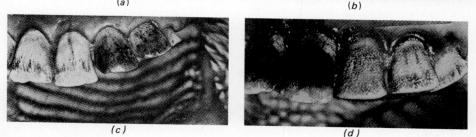

(c)   (d)

generally contain 3 to 4 percent fluorine (toxic amounts). Proper defluorination procedures are necessary to render this mineral safe for supplemental mineral purposes. Fluorine inhibits several enzyme systems, which may explain, in part, its toxic effects in animals.

17.3.10 Vanadium

Vanadium has recently been described by USDA scientists as an essential element for the health of chicks and rats. Their research showed that certain physiological functions of test animals become impaired when they consume a diet low in vanadium. Effects include reduced feather and body growth, impaired reproduction and survival of the young, altered red blood cell numbers and iron metabolism, impaired hard tissue metabolism, and altered blood lipid levels.

Researchers at Colorado State University reported that the required vanadium level may be between 50 and 500 parts per billion (ppb) when the element is consumed in a purified diet.

17.3.11 Arsenic, Nickel, and Tin

When extreme care in ration preparation is taken and mineral contamination of caging, bedding, and air supply are avoided, it can be shown that arsenic, nickel, and tin are essential for normal animal development, although natural deficiencies have not been reported. In addition to reduced growth, lowered reproductive performance and changes in hemoglobin concentration and liver function have been observed in experimental studies. University of Illinois researchers reported that a specific effect of nickel deficiency on urease activity in the rumen may explain the improved growth response of sheep and steers on low protein diets that are given nickel supplements.

17.4 SUMMARY

A review of the principal roles, sources, and clinical deficiency symptoms of the macro- and microminerals has been presented. Research is rapidly accumulating regarding the significance of minerals in animal nutrition. The need of an adequate and balanced supply to ensure good health and performance of farm animals is of special interest.[6] Certain elements are needed in small amounts but are toxic to animals in larger quantities.

Some essential mineral elements are stored in considerable quantities within the body during periods of adequate intake. These stores are drawn on later to meet requirements during periods of shortage. Other minerals must be supplied regularly because storage in the body is limited. Life cannot exist without

---

[6]Information pertaining to the specific mineral needs of the various species of farm mammals and poultry is available in "Nutrient Requirements of Domestic Animals," National Academy of Sciences–**National Research Council**, Washington, D.C. 20418.

mineral matter because many of the important body functions depend on mineral compounds.

Study Questions

1 Cite one or more examples of the interrelation among minerals.
2 What are the macrominerals? The microminerals?
3 Name a hormone and a vitamin that are closely related to calcium mobilization.
4 Why is the dietary Ca/P ratio important?
5 Are appreciable amounts of sodium, potassium, and chlorine stored in the body? Of what practical significance is this?
6 Which mineral compound is most often needed as a dietary supplement for most farm animals?
7 Give at least one major function, one or more clinical deficiency symptoms, and one or more dietary sources of each macromineral. Construct a table, and refer to Table 17.1 only as necessary.
8 Which *two* minerals are important in hemoglobin formation?
9 What is the relation of iodine and thyroxine?
10 Give at least one major function, one or more clinical deficiency symptoms, and one or more dietary sources of each micromineral. Construct a table, and refer to Table 17.2 only as necessary.
11 Cobalt is an integral component of vitamin B_{12}. Cobalt is now believed to be associated with the synthesis of what other B vitamins by the rumen microorganisms?
12 Is a zinc deficiency likely to occur among farm animals under normal feeding conditions? Why?
13 Selenium can perform certain functions of which vitamin?
14 Cite two examples of minerals whose toxic effects may not be readily apparent at first, but rather only after extended periods of their consumption in low quantities.
15 What are the effects of feeding high levels of fluorine to cattle?
16 Where can one obtain more information pertaining to the mineral needs of farm animals? See **NRC** in the Glossary.

CHAPTER 18

ANIMAL DISEASE AND THE HEALTH OF HUMANS[1]

He who has health, has hope; and he who has hope, has everything.

Arabian proverb

18.1 INTRODUCTION

The health, fitness, and ingenuity of people depend first on an adequate diet of wholesome and nutritious foods. In the development of modern civilizations, humans have come to rely heavily on meat, milk, and eggs as major sources of food nutrients. In the United States, animal products are the foundation of a diet that has helped to make a strong nation. Animal health is perhaps the most significant factor in the production of wholesome meat, milk, and eggs. Moreover, for aesthetic reasons, pleasure, sports, and other purposes, people frequently associate closely with animals. Because many **diseases** of animals may be transmitted to humans, the relation between animal health and **public health** becomes readily apparent. The prevention, control, and elimination of animal diseases contribute important safeguards to the health of humans. Crowding increases the prevalence of animal disease by increasing the animals' exposure to infectious agents (Figure 18.1). The same principle applies to humans. Thus, as animal and human populations of the world increase, it becomes imperative for

[1]The authors acknowledge with sincere appreciation the helpful suggestions and materials pertaining to this chapter of Dr. D. C. Blenden, Professor of Veterinary Microbiology and Community Health and Medical Practice, School of Medicine and College of Veterinary Medicine, University of Missouri-Columbia; Dr. K. L. Campbell, Assistant Professor of Small Animal Medicine, College of Veterinary Medicine, University of Illinois, Urbana; Dr. T. M. Curtin, Professor and Dean, School of Veterinary Medicine, North Carolina State University, Raleigh; and Dr. P. L. Nicoletti, Professor of Preventive Medicine, College of Veterinary Medicine, University of Florida, Gainesville.

students and the public to become more familiar with and knowledgeable about the relation and importance of animal disease and public health.

Many **insects** and other **arthropods** are **parasitic** on humans and farm animals. Some of these **ectoparasites** visit the **host** only for a blood meal. Several **species** of mites enter the skin and may cause **dermatitis** in humans, dogs, and other animals. Some mites are **vectors** of transmissible diseases. A large and important group of diseases are transmitted to humans by vectors such as mosquitoes; these diseases include malaria, yellow fever, and forms of **encephalitis.** Other rickettsial, bacterial, and spirochetal diseases may be transmitted to humans by arthropod parasites; e.g., rodent fleas transmit plague and murine typhus; ticks transmit Rocky Mountain spotted fever, relapsing fever, and tularemia; and the mouse mite transmits rickettsial pox.

It is often said that the United States is the healthiest place in the world in which to raise **livestock.** However, according to **USDA** surveys, there are still many animal losses; e.g., probably 25 percent of all young **pigs** die between **farrowing** and weaning time, an estimated 10 to 15 percent of all calves and **lambs** die before marketing age, and the annual losses of chickens and turkeys average 10 to 20 percent. Moreover, death losses are only part of the picture. **Morbidity** losses from many diseases are often greater than those from **mortality.** For example, **bovine mastitis** seldom leads to death, yet it is the most costly disease of dairy cattle in the United States.[2] **Intestinal** worms kill relatively few animals but render thousands unthrifty and unprofitable. A recent governmental report estimated the total annual losses of livestock and their products to be $6.9 billion (about 10 percent of the total annual income from livestock). There is no way to accurately determine the losses caused by parasites or digestive disorders.[3]

[2]J. R. Campbell and R. T. Marshall, *The Science of Providing Milk for Man,* McGraw-Hill, New York, 1975.
[3]An estimated 1.6 billion pounds of animal protein, which is needed immensely to improve human nutrition, are lost annually worldwide as a result of animal disease.

FIGURE 18.1
The spread of infectious diseases is enhanced by certain management practices, such as close confinement and crowding of animals (left). The photograph at right depicts an aborted fetus in a brucellosis-infected Florida dairy herd. The natural curious tendency of cows to lick a *Brucella*-laden fetus contributes to the spread of brucellosis. *(Courtesy Dr. P. L. Nicoletti, College of Veterinary Medicine, University of Florida, Gainesville.)*

Provisions for supplying people with safe animal products were discussed in Chapter 3. In this chapter, animals (**domestic** and wild) that are sources of **infections** for humans and those that harbor (serve as **reservoirs**) **organisms pathogenic** to humans are discussed. Certain insect-borne diseases are included, although most **life cycles** and other pertinent information related to insects and other arthropod vectors are presented in Chapter 19.

18.2   DISEASE AND HEALTH

However secure and well-regulated civilized life may become, bacteria, **protozoa, viruses,** and infected fleas, lice, ticks, mosquitoes, and bedbugs will always lurk in the shadows ready to pounce when neglect, poverty, famine, or war lets down the defenses. About the only genuine sporting proposition that remains is the war against these ferocious fellow creatures, which stalk us in the bodies of rats, mice, and all kinds of domestic animals; which waylay us in our food and drink and even in our love.

Hans Zinsser

The World Health Organization of the United Nations defines *health* as *a state of complete physical, mental, and social well-being, and not merely the absence of disease and infirmity.* In this chapter, *health* refers to a state in which all parts are functioning normally, whereas *disease* refers to a disturbance in function or structure of any organ or body part. *Public health* refers to the health of human populations, especially on a community basis.

18.2.1   Types of Disease

There are two general types of disease. *Infectious* diseases are caused by pathogens or **germs** (i.e., disease-producing microorganisms: viruses, bacteria, protozoa, and **fungi**). *Noninfectious* diseases may result from one or more of the following conditions: mechanical ailments such as flesh wounds and **ruptures;** digestive disturbances resulting from **bloat, ingestion** of hardware, or dental failure; intoxications of either chemical (lead, arsenic, nitrates, **insecticides**) or plant origin (black nightshade, hemlock, double-leaf stage of cocklebur, larkspur, lily of the valley, and others, see Section 18.6); nutritional deficiencies (the lack of fat, proteins, vitamins, minerals; Chapters 14, 16, and 17); or excesses; cell growth, whether malignant or nonmalignant; genetic disorders (e.g., sickle-cell **anemia**); or metabolic disorders such as diabetes mellitus or **acetonemia.**

18.2.2   Modes of Spreading Disease

There are a number of important means by which infectious diseases are spread. These include diseased animals, polluted streams or other bodies of water, vehicles used to transport animals, **carrier** animals (e.g., swine may carry leptospirosis to cattle), **carrion** feeders (e.g., dogs, foxes, or birds may carry bits of infected **carcasses** to clean farms), insects (especially flies, mosquitoes, and ticks), airborne pathways, and contaminated facilities and handling equipment (**cattle** chutes, surgical equipment, **poultry** crates, etc.).

18.2.3   Modes of Pathogen Entry

Certain pathogens affecting people and animals are present throughout the environment. Some important means by which they gain entrance into the body are through the respiratory tract; the digestive tract; wound contamination; the mucous membranes of the eye, e.g., pinkeye and leptospirosis (the latter may be acquired when the urine of an infected animal is introduced into the eye); the **genital** tract (especially during mating or **parturition**); the teat canal (especially in **lactating** females); the navel cord (in the **neonate**); contaminated instruments (syringes and/or surgical); and insect bites.

18.2.4   Body Defenses against Disease

Fortunately, nature provided a series of mechanisms whereby animal and human life is sustained in the presence of pathogens. These defensive mechanisms include the skin and mucous membranes as the first line of defense (certain body secretions—e.g., tears—contain **lysozyme,**[4] which has **antiseptic** properties); and the digestive tract (gastric juices and stomach acids depress bacterial growth), tissue fluids (**lymph** contains **leukocytes** such as **macrophages,** which either neutralize pathogens chemically or physically engulf them), and lymph organs and liver, which mechanically trap pathogens until the macrophages can destroy them. The defensive mechanisms also include the reactive defenses. The latter include *inflammatory reaction,* characterized by the following four cardinal signs: (1) an increased blood supply (redness), (2) increased temperature of the part (heat), (3) swelling of the part (**edema**), and (4) increased sensitivity (tenderness or pain). The reactive defenses also include **febrile** *reaction,* characterized by an overall increase in body temperature (caused by effects of **toxins** of microorganisms on the heat-regulating mechanism of the hypothalamus) and an increase in **metabolic activity**, and *immune reaction* (such as development of **antibodies**).

   Inflammation   One type of body reaction to injury (mechanical, chemical, or infectious) is called **inflammation.** It is an attempt by the body cells to destroy the injurious agents. Inflammation begins with an accumulation of fluid and white blood cells (**exudate**) around the area of injury or infection. There is often an accumulation of **pus,** which consists of a fluid containing dead cells, fibrin from the exudate, leukocytes, and the causative organism.

---

[4]Lysozyme has an interesting history. In 1922 Alexander Fleming discovered a substance in his own nasal mucus capable of dissolving, or *lysing,* certain bacteria. The substance was identified as an enzyme and named *lysozyme.* Fleming believed that some organisms produced antibacterial substances, and he went on to discover penicillin, the first true antibiotic.

18.2.5   The Resistance of Animals and People to Pathogens

Complete and lasting freedom from disease is but a dream remembered from imaginings of a Garden of Eden.

René J. Dubos (1901–   )

Immunity   Immunity is the ability an individual or animal acquires to resist and/or overcome an infection. Following an infectious disease or artificial immunization, an animal develops an increased resistance to the disease. This results largely from the production of antibodies that aid in the defense of the host by reacting or uniting with the causative microorganism or its toxin. Production of antibodies is stimulated by a variety of substances, known collectively as *antigens*. An **antigen** (from the Greek *anti,* meaning "against," and *geneo,* meaning "produce") is a substance that stimulates the formation of specific antibodies when it is introduced into an animal. Antigens are usually proteins, are foreign to the animal body, are soluble in body fluids (either **in vivo** or **in vitro**), and react or unite with a specific antibody. Each pure antigen is specific and stimulates the production of antibodies against itself but not against other antigens.

Introduction of an antigen into an animal stimulates the production of immune substances called *antibodies,* which may be demonstrated in both the tissues and blood of the recipient 7 to 14 days postinoculation. Antibodies are usually obtained from the blood **serum** of hyperimmunized animals or humans, known as *immune serum* or **antiserum.** Antibodies are quite antigen-specific; i.e., they usually react with only one antigen. **Serum** antibodies are proteins closely related to normal serum globulins. It is chiefly the *gamma* globulin that is increased in immune animals and humans.

Natural (Inherent) Resistance   Natural (inherent) resistance is the normal, or innate, resistance of animals to infection. This includes both *mechanical* (such as the skin) and *physiological* barriers to aid the host in resisting microorganisms. The latter include the unfavorable acidity of certain body secretions and unfavorable body temperature. Moreover, blood is **bactericidal** and may contain certain specific antibodies. A few natural antibodies are inherited, e.g., those responsible for the human blood groups. *Species immunity* is the natural resistance of a given animal species to microorganisms and/or parasites that may infect another species. An example is the malarial parasite, which produces disease chiefly in humans (also in certain birds and primates). People are naturally immune to Texas cattle fever and certain other animal diseases, whereas animals such as cats, cattle, and horses possess natural resistance or immunity to measles and other diseases of humans.

Acquired Resistance or Immunity   Acquired resistance or immunity is that gained by having had the disease (actively acquired immunity) or through artificial immunization (actively or passively acquired immunity). The latter involves receiving **vaccine** that causes the production of antibodies (actively acquired artificial immunity) or being injected with immune serum or serum containing antibodies produced in another host (passively acquired artificial immunity).

Actively acquired immunity is produced in response to the entrance of a particular antigen into the host. Following recovery from an infectious disease, the individual is commonly resistant to that disease for varying lengths of time. The body of the immune animal produces its own antibodies against the disease.

Actively acquired artificial immunity results from the injection of an immunologically active form of the infectious, or specifically related, agent (vaccine). There are several types of vaccines: suspensions of killed bacteria **(bacterin),** such as blackleg and typhoid vaccines; microorganisms that are **attenuated** or of reduced **virulence,** such as the viruses of smallpox, rabies, and yellow fever; the products of bacterial growth, such as the toxins associated with diphtheria and tetanus; and polyvalent vaccines, such as the triple vaccine of typhoid, paratyphoid A, and paratyphoid B bacteria. Use of the latter results in the concurrent production of antibodies against antigens of all three organisms. Live bacteria and viruses usually give stronger immunity than killed **cultures.** They are attenuated (of reduced virulence) when used in vaccines.

Passively acquired immunity occurs in the neonate as a result of antibody transfer from the mother to the offspring through either the placenta, as in humans, or the **colostral** milk, as in cattle, horses, sheep, and swine (see Section 11.12).

Passively acquired artificial immunity is possible through the injection of *immune serums* (antibacterial or antiviral serums or **antitoxins**), but the resistance lasts only a short while. In passive immunity, antibodies produced in one animal are transferred in serum to a recipient animal or person who does not participate in production of the antibodies. Antitoxins against tetanus and scarlet fever result in passively acquired artificial immunity. Injection of antiserum affords immediate increased resistance (passive immunity) to such diseases as swine erysipelas and **canine** distemper. The major differences between active and passive immunity are presented in Table 18.1.

18.2.6 Recent Application of Technology to Protection from Disease

Viral Vaccine Controls Cancer   The concept of protecting one animal with a viral vaccine from another originated with Jenner in 1798, when he observed that milkmaids survived epidemics of smallpox. He demonstrated that the cowpox virus produced little or no disease in humans yet protected humans from smallpox. This concept served as the basis for the recent development of a commercially produced, federally licensed vaccine for cancer in chickens, the first for cancer of any farm animal.

TABLE 18.1
MAJOR DIFFERENCES BETWEEN ACTIVE AND PASSIVE IMMUNITY

	Active immunity	Passive immunity
Body participation	Forms own antibodies	Recipient of antibodies produced in another animal
Material introduced into body	Antigens	Antibodies
How immunity is acquired	Naturally, in response to disease or **subclinical** infection	Naturally, by placental transfer of antibodies and/or from colostral milk
	Artificial exposure by inoculation with vaccines (immunity is produced naturally)	Artificially, by injection of antiserum (antitoxin; antibacterial serum)
Duration of immunity	Months and years	Relatively short (often only a few weeks)
Usefulness	Primarily **prophylactic** (preventive)	Primarily **therapeutic;** temporarily prophylactic

A cancer affecting chickens called Marek's disease has been the most commonly occurring cancer in the world. Before the vaccine was developed, almost all chicken flocks were affected, and in some instances over half the flock would die. The cancer can now be prevented by a highly effective vaccine, which is administered to most chickens in developed countries.

Marek's disease is named after the Hungarian veterinarian who first described it in 1907. The disease has been recognized in most countries of the world. In the 1950s a severe form of the disease appeared on the east coast of the United States and spread gradually over the country, causing many deaths and production losses in layers and broilers. Also, affected broilers are unfit for human consumption because of the presence of tumors, causing them to be condemned at processing plants even though contact with chickens or poultry products from birds with Marek's disease has never been shown to cause cancer in humans.

The virus of Marek's disease is spread from an infected chicken to a noninfected one through the air. A similar virus isolated from turkeys was found to protect chickens inoculated with it from developing Marek's disease. This virus, called the herpesvirus of turkeys, is highly effective as a vaccine and was first licensed for use in the United States in 1971. The vaccine is now used on a worldwide basis and has reduced losses from Marek's disease in the developed countries by over 95 percent.

In 1974, the first year of full adoption of the vaccine, it is estimated that the vaccine saved $628 million. Each year thereafter the saving has been put at over $168 million. This breakthrough of agricultural research alone represents a saving to the consumer of more than 2¢ per dozen eggs marketed, and nearly 6¢ per pound of broiler at the supermarket.

Monoclonal Antibodies Poultry producers lose an estimated $200 million annually to a disease called coccidiosis, because affected birds fail to grow as they should. They spend another $90 million on medicine to control the disease. An effective vaccine against coccidiosis would reduce costs of poultry production, resulting in more profits to producers and less costs to consumers.

Vaccines in general could greatly reduce the billions of dollars spent annually for drugs and medication to control a number of human and animal diseases. But even when vaccines are available, they are of limited or no value to some people and animals, because their bodies cannot produce the right kind of antibodies, which in turn control diseases. Thus, the problem is to provide humans and animals with effective antibodies.

One possible solution may lie in hybridomas, cells that produce one, and only one, very specific antibody. Hybridomas are made in the laboratory by fusing an antibody-producing cell from the spleen of a mouse with a cell from a mouse tumor (myeloma), which will multiply fast and indefinitely outside the mouse's body in tissue culture but will not grow in any other species of animal. Although the spleen manufactures a wide range of antibodies, individual spleen cells produce only one, specific antibody. Thus in a hybridoma, the spleen cell provides the design of the product and the tumor cell provides for mass production of the product, which is called a **monoclonal antibody.** Because of their purity these antibodies can be used in producing new vaccines, in improving the diagnosis of disease, and in fighting disease, including cancer.

The organism that causes coccidiosis consists of hundreds of antigens, each of which stimulates a specific antibody (multiple antigens are common in most disease organisms). Because all these antibodies—together with thousands of antibodies to other organisms—are circulating in the chicken's blood, it is difficult to study the antigens that could be used to develop a vaccine. But **hybridomas** can produce a pure antibody for each antigen on the coccidia organism. These antibodies will hook up with their specific antigen, thus enabling scientists to isolate and study them.

The same principle makes monoclonal antibodies very valuable for diagnosing disease agents. When a disease agent is isolated from a sick person or animal, it is often identified by mixing it with antibodies against several germs to determine which antibody it reacts with. These antibodies are now prepared in animals, but because the preparations contain antibodies to other germs the animal has contacted in nature, the tests are not always reliable. Monoclonal antibodies not only can pinpoint the cause of disease but also can be used to measure the degree of immunity that the person or animal had developed. In addition, they can be given to humans or animals to help fight infection on a temporary basis. Since they are pure, they do not produce side effects. Monoclonal antibodies developed in the past decade are now entering the market and will be studied extensively in the years ahead.

Genetic Resistance to Disease Many noninfectious diseases result from hereditary weaknesses. For example, it has been shown that in poultry and in laboratory animals, some inbred lines are more susceptible than others to certain nutritional diseases. Differences among inbred lines are largely genetic. In addition, many hereditary metabolic and nervous disorders have been reported in animals and humans. Genetic resistance to disease, however, usually refers to resistance to infectious diseases. **Zebu** cattle are more resistant to certain infectious diseases and parasites than are British breeds.

Genetic resistance to disease may result from the ability of an individual to prevent the entrance of disease organisms or from the ability to destroy these organisms effectively if they gain entrance into the body. Failure to combat organisms properly after they have entered the body is illustrated in humans by the genetic defect known as agammaglobulinemia. This defect results from the body's failure to produce gamma globulins (antibodies) in normal quantities. It is a sex-linked recessive trait. Agammaglobulinemia is observed in boys who fail to resist bacterial infections. Before the use of antibiotics, most, if not all, affected boys died of infectious diseases at an early age. The Chédiak-Higashi **syndrome** in humans, similar conditions (probably the same) in the blue frost mink and albino Hereford, and possibly certain types of gray coat color in collies are inherited. In addition to being partial albinos, people and animals with these conditions have leukocytes (neutrophils) that are also defective: they do not have the ability to phagocytize and kill microorganisms. Such individuals are therefore very susceptible to bacterial infections.

Attempts have been made to develop genetically disease-resistant strains of animals. Genetic resistance to disease appears to be specific rather than general. Therefore it may be possible to develop a strain of animals that is resistant to a specific disease but whose resistance to other diseases does not necessarily increase. To be of practical value, it would be necessary to develop strains or breeds of animals that are genetically resistant to several diseases. This would be extremely difficult, expensive, and time-consuming, although through the application of techniques associated with genetic engineering, many such exciting possibilities may become commonplace.

18.2.7 Antigen-Antibody Reactions

The commonest antigen-antibody reaction is a highly specific chemical union in which **molecules** are held together by strong attracting forces. The antibody molecules are large, complex proteins with receptor sites on their surfaces that unite chemically with reciprocal reactive sites on the antigen. The building of a large mass, or latticework, of these interlocking antigen and antibody units results in **agglutination,** or clumping together of cells or molecules. Agglutination tests are used in blood-typing and in diagnosing such diseases as typhoid fever, tularemia, and brucellosis.

Precipitation Reaction   A precipitin reaction results from the precipitation of a **soluble** antigen with its specific antibody (antiserum). The precipitin reaction is used in the laboratory identification of different species of meats, as in adulteration control, or in the identification of bloodstains in medicolegal situations. The *precipitation reactions* are similar to *agglutination reactions*, except that they involve the precipitation of molecules from a solution rather than the clumping together of cells.

Lytic Reaction   A lytic reaction requires the presence not only of an antigen and antibody (**lysin**) but also of a third substance, called *complement.* Complement is protein in nature and is found in normal and immune blood. It is responsible for the destruction of cells previously sensitized by lysin. A **bactericidal** *reaction* differs from bacteriolysis (the lytic reaction) only in that the cells are not disrupted. Complement is usually required, along with immune serum, to cause death of the bacteria (antigen).

Complement Fixation Reaction   A complement fixation reaction is based on the fact that complement that is bound in one reaction cannot react in a second (Figure 18.2). In the complement fixation test, complement is allowed to react

FIGURE 18.2
Principles of complement fixation. In the first stage, antigen (Ag) and antibody (Ab) are reacted in the presence of complement (●). The interaction of antigen and antibody fixes some, but not all, of the complement available. In the second stage, the residual or unfixed complement is measured by adding antibody-sensitized red blood cells, which are lysed by residual complement. Thus, a reciprocal relation exists between the amounts of lysis in the second stage and the amount of antigen present in the first stage.

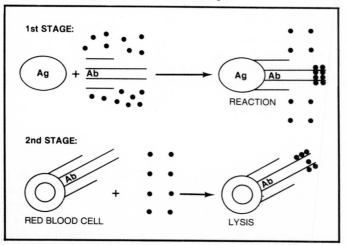

with one antigen-antibody system, and then a second indicator system (sensitized red blood cells) is added. If **hemolysis** of the red blood cells occurs, complement has not been fixed by the first antigen-antibody mixture, and the test is negative. Conversely, in a positive test, complement is fixed in the first reaction and therefore is unavailable to produce hemolysis of the red blood cells. The Wassermann test for syphilis is a widely known complement fixation reaction.

Toxin-Antitoxin Reaction A toxin-antitoxin reaction helps explain the remarkable protective action of immune serum against the toxins of diphtheria and tetanus. **Antitoxin** (from the Greek *anti,* meaning "against," and *toxin,* meaning "poison"), like other antibodies, is found in the blood serum of immune animals and reacts specifically with its antigen (toxin). The protective effect of antitoxin results from its direct neutralization of toxin.

18.2.8 Testing for Disease

The veterinarian has several means of diagnosing diseases. These include physical examination, antigen-antibody reactions of the blood (see Section 18.2.7), skin testing with antigen (e.g., tuberculosis of cattle), microscopic examination of blood (used to identify the type of infectious organism and also to make red and white blood cell counts), skin scrapings (especially useful for identifying fungus or parasitic mite infections), microscopic and/or visual **fecal** examination (primarily for parasites), chemical tests of blood and urine, body-temperature records, **biopsy,** and/or necropsy.

18.3 SELECTED ANIMAL DISEASES TRANSMISSIBLE TO HUMANS

Unhealthy farmers are poor producers. In many parts of the world ill health due to infection and ill health due to bad diet are inextricably interconnected. Acute protein **malnutrition** is so often precipitated by infectious diseases that there is a case for saying that an attack on these diseases should be the first step toward preventing it. Certainly public health measures against infectious diseases and measures to produce more protein should be undertaken simultaneously.

W. R. Aykroyd

18.3.1 Viral Infections

These infections are caused by a group of minute agents called *viruses.* A virus is characterized by a lack of independent metabolism and can therefore replicate only within living cells. The individual particles consist in part of either **DNA** or **RNA** (but not both).

Cells can be grown or cultivated under artificial conditions, which is called *tissue culture.* Such cultures can be used to support viral growth in the production of vaccines. These cultures are also used in diagnostic tests in which viruses produce cellular changes.

Rabies (Hydrophobia)   This is one of the oldest diseases known to humans. It was described in the fourth century B.C. by Aristotle, who wrote, "Dogs suffer from a madness which puts them in a state of fury, and all of the animals that they bite when in this condition become also attacked by rabies." *Rabies* is derived from the Latin word *(rabere)* meaning "rave," or "fury." It probably received its name because infected animals often become excited and attack any object or animal in their way.

All **mammals** are believed to be susceptible to this viral disease, which may be transmitted to humans through the saliva of a rabid animal (via a bite or skin laceration). The commonest vectors of rabies are the **carnivorous** or biting animals, such as cats, coyotes, dogs, foxes, raccoons, skunks, and wolves. Seventy-four percent of the animal rabies in the United States in 1981 occurred in skunks and bats. Bats are important vectors in certain other countries also. The incidence of rabies in the United States decreased from 8837 cases in 1953 to 7211 in 1981. The number of cases traced to selected animals and humans in 1953 as compared with those in 1981 (in parentheses) were dogs, 5688 (216); cats, 538 (285); farm animals, 1118 (465); foxes, 1033 (195); skunks, 319 (4480); bats, 8 (858); other animals, 119 (718); and humans, 14 (2). Six cows in 1978 and eight in 1979 were diagnosed as having rabies in Wisconsin. Recent trends in the number of cases of rabies in selected wildlife hosts are given in Figure 18.3.

The causative virus multiplies and causes degenerative changes in brain tissue, which results in convulsions, excessive salivation, madness, paralysis, and finally death of the infected animal.

As concern mounts over the increase of rabies among wild animals and household pets, researchers from the Michigan Department of Public Health, Ohio State University, and Texas A & M University recently reported a new vaccine for humans, developed from the cells of rhesus monkeys, that may be a useful alternative to the rabies vaccine from human diploid cells, which is scarce and expensive and can lead to allergic reactions and neurological problems. In tests on 60 subjects, the new vaccine gave as much protection as vaccine of human origin and had no serious side effects.

Severed chicken heads laced with rabies vaccine have been scattered as bait for foxes throughout valleys in Switzerland in a highly successful campaign to impede the spread of rabies. The effort marks one of the first times that an attempt has been made to immunize wild animals. One problem has been that immunization required the use of live virus, a potentially dangerous practice. But zoologist Alexander Wandeler and his coworkers at the University of Bern in Switzerland have used a weakened virus that is not infectious and has proved safe and effective. An estimated 60 percent of the foxes in the test areas ate the bait.

The average incubation period of rabies in humans is from 30 to 60 days. (The range is 10 days to more than 8 months.) Humans infected with rabies virus

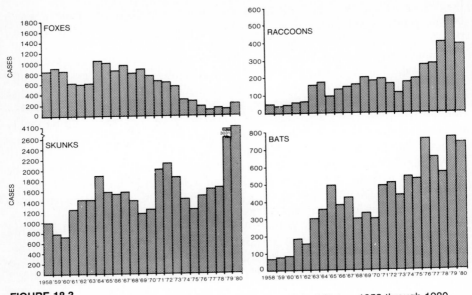

FIGURE 18.3
Number of cases of rabies in selected wildlife hosts in the United States, 1958 through 1980.
(Courtesy Centers for Disease Control, Atlanta, Ga.)

commonly develop hydrophobia, or the "fear of water" phenomenon. This results from painful spasms of the **pharyngeal** muscles caused by the act of drinking. The condition causes the victim to avoid swallowing saliva, which then drools from the mouth. Rabies almost always terminates in the death of infected animals. Identification of the disease may be made by examination of histologic sections of brain (cerebellum) tissue of the suspected animal for the presence of *Negri bodies* (inclusion bodies). A second and more accurate laboratory diagnosis of rabies can be made through the intracerebral inoculation of varying dilutions of suspected tissue suspensions into test mice. (Dogs, guinea pigs, hamsters, and rabbits may also be used as laboratory test animals to isolate the virus.) Recently, a third, and even more accurate, diagnostic test based on a specific antigen-antibody reaction has been developed using direct fluorescent microscopy procedures. The latter test is becoming the most widely used means of diagnosing rabies.

Rabies virus has been isolated from the milk of infected cows. However, the disease rarely follows the consumption of such milk. Rabies is an infection usually requiring broken skin for entrance of the virus. Thus ingestion of the virus is not dangerous, unless there are lesions in the mouth and throat, because the virus is destroyed by the stomach's gastric juices. If a person who has been bitten has the wound properly treated and quickly accepts the Pasteur treatment (developed by Louis Pasteur in 1883), he or she rarely succumbs to rabies. **Immunization** in humans is accomplished through a series of daily injections.

People working in a high-risk occupation should be vaccinated against rabies before exposure. Estimates are that not more than 15 to 20 percent of the persons bitten by rabid animals develop the disease even when no antirabies treatment is given. However, because of the high mortality associated with rabies, a person exposed to the disease should be immediately placed under the care of a physician.

A practical rabies immunization of dogs (after 3 months of age) can be accomplished through the administration of a modified, living-virus vaccine produced in chicken embryos or tissue culture. A current common control in dogs is the injection of modified live or inactivated rabies vaccine at 3 and 12 months of age, followed by booster vaccinations every 2 to 3 years.

A major factor in the increased incidence of rabies in recent years is the small number of cats vaccinated against it (only 4 percent of the total population). The Centers for Disease Control reported in 1983 that cats now pose a greater risk of exposing humans to rabies than dogs. This is a particular problem with cats permitted to roam outdoors. Cats should be vaccinated with an approved vaccine after 3 months of age and given booster shots annually thereafter.

Psittacosis (Ornithosis) This chlamydial[5] infection, found in many species of birds, may be transmitted to humans. Psittacosis is most common in parakeets, pigeons, parrots, and other related pet birds. However, other species of domestic poultry (turkeys) and nonpsittacine wild birds may become infected. In the latter species, the infection is generally termed *ornithosis*.

The disease may be transmitted to humans in two principal ways: (1) by the airborne route, in which a person inhales dust containing contaminated **desiccated** fecal materials, feathers, urine, and droplets of nasal secretions; or (2) by direct contact through bite wounds. In people the disease is manifested by influenza-like symptoms 7 to 15 days after contact with the infected birds.

In 1974, a total of 154 human cases of psittacosis were reported among 560 employees of six turkey processing plants in Missouri, Nebraska, and Texas. The sources of infection were traced to turkey flocks in east central Texas. There were 124 cases reported in humans in 1980.

Newcastle Disease This is a contagious, and often fatal, viral infection of poultry and many wild birds, characterized by neurologic and pneumonic disturbances. It is also called *avian pneumoencephalitis* and *avian distemper*. Newcastle disease represents a mild potential hazard to persons exposed to infected carcasses and to those involved in the production of vaccines against it. The disease is characterized in humans by a severe conjunctivitis (inflammation of the membranes lining the eyelids) and a syndrome of fever, chills, headache, and general malaise (weakness and/or discomfort). **Killed virus** vaccine and

---

[5]Members of the genus *Chlamydia* are agents having a cell structure intermediate between that of viruses and bacteria.

modified live, or attenuated, virus of avian embryo origin are capable of stimulating immunity in poultry and wild birds. One form of the disease (velogenic) is considered an exotic disease in the United States, and all exposed birds are slaughtered by order of the USDA.

Yellow Fever   This acute febrile disease, characterized by jaundice and liver **dysfunction** in humans, is naturally transferred by the bite of the infected female *Aedes aegypti* mosquito. The mosquito takes up the virus when it bites an infected person or monkey. The virus develops rapidly in the mosquito's body. After about 12 days, the mere bite of such a mosquito is sufficient to introduce enough virus into people to cause yellow fever. Once a mosquito is infected, it may transmit yellow fever for life. An attenuated vaccine is prophylactically effective against the disease in humans. However, the best controls are avoiding being bitten and eradicating mosquitoes. The Nobel prize in **physiology** and medicine for 1951 was awarded to Dr. Max Theiler for his discoveries concerning yellow fever and especially for contributions in the development of a vaccine that protects persons exposed to the disease. The U.S. Public Health Service has a section of its National Communicable Disease Center working on the eradication of *A. aegypti* in the United States and neighboring countries.

Dengue   This infectious, eruptive, febrile disease is caused by a **filterable virus** and is transmitted by the bite of *A. aegypti* and *A. albopictus* mosquitoes. Symptoms of the disease (high fever; extreme prostration; severe pains in the head, eyes, muscles, and joints; and sore throat) appear 3 to 6 days after a virus-bearing mosquito bites the victim. The rash (eruption) breaks out on the fourth or fifth day of the illness. The disease is seldom fatal. Dengue occurs chiefly in Egypt, India, Iran, the West Indies, and Hawaii and other islands of the Pacific. Like yellow fever, dengue is a disease in which monkeys serve as an inter**epidemic reservoir host.** More than 500,000 persons in the Galveston and Houston, Texas, areas had dengue in the 1922 outbreak.

The last identification of indigenous dengue infection reported in the United States was that of a 5-year-old girl in Brownsville, Texas, in 1980. Dengue virus was isolated from 11 patients of the Brownsville, Harlingen, and Laredo area. Several thousand cases of dengue have been reported in Puerto Rico during the past decade.

Encephalomyelitis   This viral disease found in humans and horses appears primarily in two forms, a western form (WE) and a more severe eastern form (EE). Its principal reservoir is wild birds. The disease is communicable to humans (only from a bite by the vector), in whom an inflammation of the brain and spinal cord develops. It is characterized by fever, incoordination, and, in advanced stages, convulsions and coma. The St. Louis encephalomyelitis strain is found chiefly in humans. The virus is transmitted by the *Culex pipiens* and *C. tarsalis* mosquitoes. Selected arthropod-borne diseases transmissible from animals to humans are presented in Table 18.2.

TABLE 18.2
SELECTED ARTHROPOD-BORNE DISEASES TRANSMITTED FROM ANIMALS TO HUMANS*

Name of disease	Vector or means of transmission	Animal affected	Animal hosts (reservoirs)
Western encephalitis	Mosquitoes to birds to mosquitoes	Mosquitoes to humans or horses (terminal hosts)	Birds
Eastern encephalitis	Mosquitoes to birds to mosquitoes	Mosquitoes to humans or horses (terminal hosts)	Birds
St. Louis encephalitis	Mosquitoes	Chiefly humans	Fowl, wild birds
Venezuelan encephalitis	Mosquitoes	Horses and other equines, humans, fowl	Vampire bat
Russian encephalitis	Ticks, milk from infected animals	Sheep, goats, birds, rodents, humans	Cattle, elk, goats, sheep, fox
Colorado tick fever	Ticks	Humans	Ground squirrels, mice, hamsters
Yellow fever	Mosquitoes	Humans, monkeys, marmosets, lemurs	Primates, rodents
Korean hemorrhagic fever	Chiggers (larvae of mites)	Humans	Rodents

*More than 200 arthropod-borne viruses (**arboviruses**) have been identified and associated with an animal host. More than one-third of these cause a febrile or more severe disease in humans.

Venezuelan equine encephalomyelitis, a sleeping sickness virus spread by marsh mosquitoes, killed more than 1000 horses in Texas and 1500 horses in other southwestern states in 1971. The cooperative action of veterinarians and government agencies was commendable. Suspected cases had to be confirmed, an experimental vaccine brought into use, nearly 2 million horses vaccinated, and millions of acres of land and marshes sprayed to arrest the disease-carrying mosquitoes—all in a few weeks' time.

Bluetongue   This viral disease is transmitted by insects (species of *Culicoides*, biting gnats or midges) and affects sheep, cattle, goats, and wild ruminants. The clinical signs are most severe in sheep. Symptoms may include a swollen and cyanotic ("blue") tongue, oral ulcers, lameness, hair loss, pneumonia, and deformed offspring. Cattle and wild ruminants are a reservoir, and infection may not be apparent. Bluetongue disease can be transmitted through fresh or frozen bovine semen. Only five of the 22 known serotypes of the virus have been isolated in the United States. There is no effective treatment for this difficult-to-control disease. The vaccine available in the United States protects only against one of the strains. The major impact of bluetongue on the United States cattle industry has been its constraint on exports of breeding animals. The ban on bovine semen exports from the United States to the United Kingdom, Australia, and New Zealand has resulted in an estimated annual loss of approximately $24 million. Ruminants imported from countries with bluetongue must be tested and found negative for the disease prior to entry into breeding units. Hopefully, these measures will prevent the dissemination of bluetongue virus.

The Pox Diseases of Humans and Animals

Human **ecology** and animal ecology have developed in a curious contrast to one another. Human ecology has been concerned almost entirely with . . . the effects of man upon man, disregarding often enough the other animals amongst which we live.

Charles Elton

Smallpox (Variola)   Smallpox (variola) is caused by a virus and is usually transmitted from person to person. It may be transmitted by inoculation into several animal species and thereby lose much of its virulence for humans. When carried back to humans (by vaccination), a mild disease, known as *vaccinia,* confers immunity to smallpox. Until a satisfactory prophylactic vaccination for smallpox was developed, the disease was feared by all. The only time George Washington left the continental United States was to visit the British West Indies, where he became infected with smallpox.

A disease known as *cowpox* is characterized by lesions on the teats and skin of the cow's udder. The viruses of smallpox and cowpox are generally considered to be immunologically similar. In 1798, Jenner reported that persons who milked cows infected with cowpox (vaccinia) had resistance to smallpox. Conversely, it was later observed that healthy lactating cows became infected with cowpox from milkers who had been recently vaccinated against smallpox. This prompted the recommendation that recently vaccinated persons should not milk cows until their vaccine reactions are well healed. Smallpox vaccine was commonly produced by propagating the virus on the scarified skins of **calves.** The resulting vaccine virus has proved to be effective in the prophylaxis of smallpox in people for more than a century. The word *vaccination* is derived from the Latin *vacca,* meaning "cow." (The term originally referred only to the use of cowpox virus to prevent smallpox.)

On December 9, 1979, the World Health Organization's Global Commission for the Certification of Smallpox Eradication declared that smallpox eradication had been achieved throughout the world. This has resulted in discontinuation of the practice of routine smallpox vaccination by many countries.

Foot-and-Mouth Disease (FMD)   Loffler and Frosch found in 1899 that this highly infectious disease of cattle and other **cloven-footed** animals is caused by a filterable virus. It is only one-millionth of an inch in diameter and is believed to be the smallest of the viruses affecting humans or animals (even smaller than the polio virus). Cattle and swine are more susceptible than sheep and goats. Wild ruminants (buffalo, camel, deer, llama, antelope) are also susceptible, and humans may become infected with a mild form of the disease. The disease is transmissible from infected animals to people, presumably by contact, through contaminated meat and milk products, and/or through the air. The virus can be carried in meat or milk products, by birds or other animals, on vehicles, on the clothing or shoes of humans, and for short distances through the air. The virus is concentrated in lymph nodes, blood, and bone marrow. A dog traveling with a bone containing infected bone marrow could spread the virus. Studies have shown that the FMD virus can survive for about 3 months in the refuse on boots that become contaminated on infected premises, on hay for about 4 months, on hair for a month, and in a refrigerated carcass for up to 4 months. The common cool, damp climate of Britain provides conditions in which it flourishes. Spreading like a plague, foot-and-mouth disease caused 422,500 animals (208,700 cattle, 113,500 hogs, and 100,300 sheep) to be killed (then burned or buried) in England and Wales during a 5-week period in late 1967. Direct costs attributable to eradication were estimated at $250 million.

An effective vaccine against foot-and-mouth disease has been developed using recombinant DNA technology. Research on the new vaccine was done under a cooperative agreement between the USDA and Genentech, Inc., a San Francisco–based research firm. Developmental work and testing were done in the USDA's high-containment facility at the Plum Island Animal Disease Center off the coast of Long Island. The nonhazardous aspects of the research were carried out at Genentech's California facilities.

Scientists used gene cloning to reproduce a fraction of the FMD virus coat. The fraction, called VP3, is one of four major polypeptides in the virus coat. In the recombinant DNA production method, scientists use *Escherichia coli,* strain K-12, as the host for reproducing the VP3 polypeptide. A plasmid (small ring of DNA) is removed from the *E. coli* bacterium with a cutting enzyme. Workers then isolate the VP3 DNA fragment, splice it into the *E. coli* plasmid, and insert the recombined plasmid back into the bacterium. The bioengineered plasmid can then be cloned in the bacteria to produce FMD vaccine. The new technique makes production of FMD vaccine possible on a commercial scale. The vaccine can be stored without refrigeration.

Vesicular Stomatitus (VS) This disease is caused by a virus. In 1982 the USDA reported 393 laboratory confirmations of VS, which causes blisterlike lesions in cattle, horses, sheep, swine, and humans. The clinical symptoms of VS closely resemble those of foot-and-mouth disease, but VS is usually short-lived and not fatal. Although FMD last invaded the United States in 1929, occasional cases of VS are seen in this country. VS generally occurs at 10- to 15-year intervals and is most commonly diagnosed in animals that have been near low-lying marshes, swamps, and similar areas following periods of heavy rainfall and high humidity. These conditions also favor increased populations of mosquitoes and gnats that may spread the disease.

Humans affected by the virus commonly have blisters on the lips, tongue, and foot and have flulike symptoms in the respiratory tract. There were 30 clinical cases reported in humans during the 1982 outbreak.

Contagious Ecthyma of Goats and Sheep More commonly known as *sore mouth*, this is a communicable disease of goats and sheep caused by a filterable virus. In people it causes vesicles on the gums, tongue, cheeks, lips, or areas where the skin is scratched, but the symptoms are mild and usually last about 3 weeks. Lambs and **kids** may be protected by **vaccination.**

18.3.2 Rickettsial Infections

Soldiers have rarely won wars. They more often mop up after the barrage of epidemics. Typhus with its brothers and sisters—plague, cholera, typhoid, dysentery —has decided more campaigns than Caesar, Hannibal, Napoleon and all the inspector generals of history. The epidemics get the blame for defeat, the generals get the credit for victory.

Hans Zinsser

First described in 1909 by Ricketts, the organisms that cause rickettsial diseases are neither true bacteria nor filterable viruses. **Rickettsiae** are generally considered to cause animal diseases that are transmitted from animal to animal (including humans) by insects, ticks, mites, and various other arthropods. Recent evidence suggests that rickettsiae are parasites of arthropods (commonly **intracellular**), and humans and animals become infected when infected arthropods feed on them or when a wound is contaminated with the **feces** of an infected arthropod.

Typhus Fevers Two types of typhus affect humans: **epidemic** *typhus,* which is transmitted from person to person by lice, and **endemic** *(murine) typhus,* which is transmitted by fleas from rodent hosts (especially rats) to people. Epidemic typhus may be prevented through vaccination and the eradication of body lice. The vaccine is prepared from rickettsiae grown in the yolk sac of a developing chick embryo. Destruction of rats and their breeding places is the most effective prophylactic means of controlling murine typhus. There were 62 cases of *murine typhus* (flea-transmitted) reported in the United States in 1980.

Rocky Mountain Spotted Fever In 1909, Ricketts discovered that this rickettsial disease is transmitted to humans by two or more tick species, which included the Rocky Mountain wood tick (*Dermacentor andersoni,* which also transmits Colorado tick fever) and the American dog tick *(D. variabilis);* see Table 18.3. The disease is most common among persons whose occupation exposes them to tick bites. When in areas infested with ticks, one should check oneself frequently for their presence (the tick must be attached for about 2 hours to transmit the organism), even if insect repellents are used.

The two ticks indicated above as the most common vectors (the brown dog tick *Rhipicephalus sanguineus* can also transmit the disease) of this disease are the "three-host" type and spend the periods between feedings away from the

TABLE 18.3
SELECTED RICKETTSIAL DISEASES TRANSMITTED FROM ANIMALS TO HUMANS

Disease	Responsible organism	Common vector	Carrier animal(s)
Epidemic typhus	*R. prowazekii*	Body louse	Humans
Murine typhus fever	*R. mooseri*	Rat flea	Wild rat and other rodents
Q fever	*Coxiella burnetii*	Ticks	Cattle, goats, sheep, rats
Rocky Mountain spotted fever	*R. rickettsii*	Rocky Mountain wood tick	Mice, rats, rabbits, opossum, goats, and other animals
		American dog tick	Dog
Tsutsugamushi disease	*R. tsutsugamushi* (*R. orientalis*)	Chigger mite	Field mice and rats

hosts. The larval and nymphal stages are found on smaller animals (especially **rodents**), whereas the adults prefer larger animals (both wild and domestic). It is the adult tick that transmits the infection to people. (The two immature stages are rarely found on humans.) Dogs may carry ticks into the house and should therefore be inspected frequently.

The life cycle and feeding habits of these ticks are interesting and important in understanding the transmission of the disease. The adult attaches itself to one animal and feeds 7 to 10 days, or until engorgement occurs. After feeding, the female drops to the ground, lays 2000 to 6000 eggs, and promptly dies without feeding again. Six-legged **larvae** hatch from the eggs, select a host (usually a small rodent such as a rabbit or ground squirrel), and feed 4 to 6 days, or until engorgement occurs. They then detach, fall to the ground, and molt, from which they emerge as eight-legged **nymphs.** These nymphs find a host and feed 6 to 8 days, or until they are engorged. The engorged eight-legged nymphs drop off, molt, and become eight-legged adult males and females. These ticks probably use 2 to 3 years to complete the cycle from the first-**generation** to second-generation adults. (Unfed adult ticks may survive up to 4 years.) Infection is maintained from generation to generation by transmission through the eggs. The best prophylactic measure for humans is to avoid ticks. When this is impossible, the use of inactivated vaccine offers some immunity.

Trench Fever   This is a relapsing fever characterized by headache, dizziness, and pain in the back and legs. It is an infectious rickettsial disease (the agent is *R. quintana*) transmitted by the body louse, *Pediculus humanus.*

Tsutsugamushi Disease (Scrub Typhus)   This disease is an infection of rodents of the orient and South Pacific caused by *R. tsutsugamushi.* Humans may become infected from the bite of the **chigger** mite. Only the larval mite forms are parasitic and transmit the disease. The nymph and adult forms live on the ground, where they feed on plants. Prevention is dependent on protection against the bite of infected mites. No effective vaccines have been developed. It is a dangerous disease of humans, with about a 35 percent mortality rate.

Q Fever   First recognized in Australia in 1935, this disease is classified as a rickettsial disease, although the mode of infection to humans differs from that of other infections in this group. The causative organism of Q fever is *Coxiella burnetii* (originally named *R. burnetii*). Ticks are the most important vector. People may acquire Q fever through the inhalation of contaminated dust (including tick feces). However, most people become infected through exposure to livestock (cattle, goats, and sheep) or through the **ingestion** of their products (raw milk or meat of infected animals). Since milk-borne transmission of Q fever has occurred, pasteurization temperatures have been elevated slightly to ensure killing of the causative organism (Chapter 3). The disease has been identified in cattle in some 35 states within the United States and therefore is recognized as endemic. A recent Ohio study indicates that the disease is more prevalent among large than among small dairy **herds.**

Preferred prophylactic measures are avoidance of ticks, care in aiding animals through **parturition,** and proper pasteurization of milk. A vaccine against Q fever is now available. The armed forces have studied this disease from the standpoint of biological warfare. The atomic bomb dropped on Nagasaki weighed about 5 tons and killed an estimated 80,000 people, but only a fraction of an ounce of **chick** embryo tissue inoculated with *C. burnetii* could, if properly placed, infect a billion people and would likely be fatal to as many as 10 million.

18.3.3 Bacterial Infections

The discussion of infectious diseases of bacterial origin will be confined to the more common ones affecting the health of humans and animals.

Tuberculosis   The ease of transmissibility of pathogenic species of the genus *Mycobacterium* between animals and humans makes it significant to public health. There are three common species of tubercle bacilli responsible for tuberculosis in **homeotherms.** *Mycobacterium bovis* (cause of bovine tuberculosis) is capable of infecting several species, including cats, dogs, swine, and humans. Similarly, *M. tuberculosis* (human types of tubercle bacilli) may be transmitted from humans to caged primates, cattle, dogs, parrots, and swine. *Mycobacteruim avium* (usual avian tuberculosis organism) may also infect swine.

Tuberculosis has been observed widely in most homeotherms and in such **poikilotherms** as alligators, fish, frogs, snakes, and turtles. Hence it is doubtful that any species of animal has an absolute resistance to tuberculosis. Fortunately, the occurrence of bovine tuberculosis in the United States has been greatly reduced through the joint cooperative efforts of farmers, veterinarians, and government agencies to eliminate tuberculin-reacting cows. The tuberculosis eradication program in cattle was initiated in the United States in 1917, when about 5 percent of the cattle were infected. By 1936, over 3.5 million infected cattle had been slaughtered. Today, less than 0.1 percent of cattle in the United States are tuberculin-positive. It is estimated, however, that 12 percent of the cattle in South America have bovine tuberculosis.

Methods of Controlling Infection from Cattle to People These methods include the maintenance of healthy cows through testing programs and sanitation and the pasteurization of milk (Chapter 3). The **USDA** meat-inspection service also aids in locating farms that may be sources of infection. Federal legislation requires **intra**state as well as **inter**state meat inspection. In countries having a high occurrence of tuberculosis, the use of a vaccine to immunize humans and animals against the disease is practical. However, in the United States, where human infection of bovine tubercle bacillus is rare and the disease has been virtually eliminated in cattle, since vaccination renders the diagnostic test with tuberculin invalid, vaccination in any species is not advisable.

Tuberculosis in Mammals Other Than Cattle Dogs are resistant to *M. avium* but susceptible to *M. bovis* and *M. tuberculosis*. Cats are susceptible to *M. bovis* but are resistant to infection with *M. tuberculosis* and *M. avium*. They probably become infected by drinking raw milk on farms. Although the transmission of tuberculous infection of cats to humans is very rare, children may become infected by fondling infected cats that have open skin lesions or by being bitten. Swine are susceptible to infection by any of the three common species. However, most tuberculosis of swine is of the avian type.

The federal meat-inspection records indicate that about 15 percent of the swine slaughtered 60 years ago had tuberculosis lesions (primarily found in lymph nodes of the head, neck, and **mesentery**), whereas today less than 1 percent have the characteristic lesions. This striking reduction is attributed largely to increased confinement of poultry and restriction of their freedom to roam farms and hog lots, where they deposited large amounts of fecal material containing tubercle bacilli. Requirements that garbage be cooked have aided in reducing the prevalence of swine tuberculosis.

Monkeys are very popular mammals at the zoo, both for children and the organisms causing tuberculosis. Monkeys rarely contract tuberculosis in the wild; however, when they are exposed to infected humans or other **primates** in captivity, their susceptibility to the disease is greatly enhanced and presents a hazard to public health. This is another example of humans giving animals their disease. (Monkeys are especially susceptible to *M. tuberculosis*.)

Apparently sheep possess considerable resistance to both *M. bovis* and *M. avium*. They have some degree of natural resistance to the human type of tubercle bacillus, but a few cases of such infections have been reported. Goats possess considerable resistance to tuberculosis. However, if allowed to associate with infected cattle, they may become infected with *M. bovis*. Therefore goats that produce milk for human use should be tested for tuberculosis, and their milk should be pasteurized. Chickens are very susceptible to *M. avium*, but through confined housing and sanitary precautions, its prevalence has been greatly reduced in recent years. The relatively short time that poultry are now kept on farms has also helped to reduce the occurrence of avian tuberculosis in the United States. However, tubercle bacilli may remain **viable** and pathogenic in chicken litter and/or infected soil for several years. *M. avium* is virulent in varying degrees but may cause tuberculosis in cattle, deer, mink, rabbits, rats, sheep, and swine. Humans are relatively resistant to this pathogen, but *M. avium* may be transmitted to them. Badgers are known carriers of tuberculosis and therefore many are killed in England.

Brucellosis This bacterial infection derives its name from a British Army surgeon, Sir David Bruce. In 1887, he discovered the bacteria that later was named *Brucella melitensis*. It is also called "Bang's disease" in cattle (after a Danish veterinarian, Dr. Bernard Bang, who isolated *B. abortus* in 1897). Bang established that the organism commonly causes abortion in pregnant cows infected with the disease.

Brucellosis is found throughout the world, but especially in certain countries having large populations of cattle, goats, sheep, and swine. There are three species of the genus *Brucella* of major public health significance: *B. abortus* (of bovine, or cattle, origin), *B. melitensis* (of **caprine,** or goat, and of sheep origin), and *B. suis* (or **porcine,** or swine, origin). The latter was isolated in 1914. The correlation of Malta fever in humans with infected goats and of undulant fever in humans with infected cattle was not proved until the twentieth century. The *brucellae* enjoy a variety of hosts, including cattle, goats, horses, humans, sheep, and swine. They have also been isolated from buffalo, reindeer, caribou, elk, camel, and yak. Hence there is a large potential animal reservoir of infection to people. The disease has been experimentally transmitted to chickens, guinea pigs, hamsters, mice, monkeys, rabbits, and rats. Infected caribou or reindeer sometimes transmit *Brucella* infection to humans. (This has occurred in Eskimos.)

Occurrence of Brucellosis Data obtained by the Animal and Plant Health Inspection Service of the USDA indicate that the incidence of brucellosis in cattle (based on blood agglutination tests) in the United States decreased from 11.5 percent in 1935 to less than 0.3 percent in 1981. Much of this decrease is attributable to the surveillance programs of the USDA (such as the brucellosis ring test and market-cattle testing program), which involve testing for *Brucella* agglutinins (types of antibody). Those animals found to be **reactors** are sold for slaughter. The market-cattle test program is a system of applying back tags to market beef cattle and testing the blood at slaughter or at auction markets. Cases of infection are traced to the farm of origin, where further testing is done.

Agglutination Test Specific agglutinins for *Brucella* may be observed in the blood serum of infected animals. To determine the presence of *Brucella* antibodies, multiple dilutions of serum are tested against a prepared *Brucella* antigen. No agglutination indicates that the animal is negative, whereas a positive diagnosis is indicated by an agglutination at 1:100 or greater dilution. (Interpretation of the reaction depends on species, previous vaccination, and other factors.) Infected cattle sometimes fail to have **titers** that classify them as reactors, especially just following calving or an abortion.

   Brucellosis Milk Ring Test   A modification of the blood serum agglutination test, called the brucellosis ring test (BRT), is used to detect *Brucella* agglutinins in milk. It serves as a screening test for brucellosis in cattle by detecting antibodies in bulk milk (a pooled sample of herd milk). The BRT test is performed by mixing nonhomogenized milk in a test tube with stained (by means of a purple dye) *Brucella* organisms. If the milk contains antibodies, agglutination occurs, and as the cream layer rises, the agglutinated *Brucella* cells are carried with the milk fat globules. A purple cream layer with white milk below indicates a positive test. Conversely, in a negative test, the cream layer is white, and the milk is diffusely purple, since the stained bacterial cells remain in suspension. In all 50 states, milk samples are tested by the BRT 2 to 4 times annually. Reduction in the number of herds positive to the BRT reflects the efforts of the cooperative state-federal brucellosis-eradication program (Figure 18.4).

   Brucellosis Card Test (BCT)   This test is based on the use of a disposable card on which blood serum or **plasma** is mixed with a buffered whole-cell suspension of *B. abortus* (antigen), which reacts (agglutinates) with antibodies in the blood serum of animals infected with brucellosis. The test may be conducted in the laboratory or on the farm or ranch and is read as either negative or positive (there are no suspect titers). The BCT is especially useful as a diagnostic test in cattle populations in which the prevalence of brucellosis is high due to its

FIGURE 18.4
Brucellosis milk ring testing (herd test). Note decline in the number of suspicious herds.
(Courtesy Centers for Disease Control, Atlanta, Ga.)

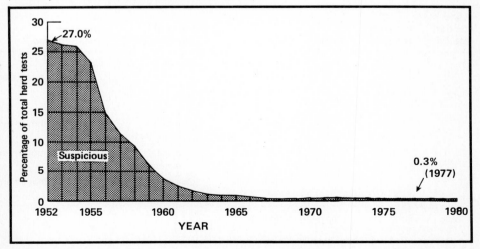

simplicity. However, it is too sensitive to use in vaccinated cattle or in herds or areas where the prevalence of brucellosis is low. It is also useful in identifying infected swine.

Brucellosis in Cattle, Swine, and Goats   The *Brucella* organism is an extracellular and intracellular parasite. In pregnant animals, the organism often localizes in the placenta. The **chorionic** tissue of the placenta reacts to the infection with inflammation, becomes edematous, and finally decreased circulation in the **fetus** results in an abortion. (Infected goats do not have an interrupted pregnancy as frequently as do cattle and swine.) At the time of abortion, the uterine exudate is heavily contaminated with organisms, which may contaminate food, bedding, and the surroundings. (Aborted material may be carried by birds, dogs, or wild animals.) Animals ingesting contaminated materials readily acquire brucellosis. Water contaminated with *Brucella* is also a potential source of infection. The fact that the *B. abortus, B. suis,* and *B. melitensis* organisms usually become localized in the mammary **glands** and may be secreted into milk is an important potential health hazard to humans. (Once infected, the mammary glands of most cows remain infected throughout their lives.) The three species of *Brucella* are relatively species-specific, but there can be cross infections. Routes of natural *Brucella* infection to humans are depicted in Figure 18.5.

The new USDA classifications for brucellosis control became effective in 1982 and are summarized in Table 18.4; see also Figure 18.6.

Brucellosis in Sheep   Sheep are susceptible to infection by *B. melitensis* and occasionally by *B. abortus*. Infections in people have been traced to sheep's milk and to cheese manufactured from raw sheep's milk.

TABLE 18.4
SUMMARY OF STATE BRUCELLOSIS CLASSIFICATIONS FOR CATTLE

Herd infection rate, %	MCI reactor rates,* %	Class	Status	Intrastate tests required	Interstate tests required
0	<0.05	Free	Vaccinates and nonvaccinates	None	None, but certification required[†]
<0.25	<0.10	A	Vaccinates and nonvaccinates	None[†]	One before movement[†]
<1.5	<0.30	B	Vaccinates and nonvaccinates	One before movement[†]	One before movement; one after
>1.5	>0.30	C	Vaccinates	One before movement[†]	One before movement; one after
			Nonvaccinates	One before movement[†]	Two before movement; one after

*Adjusted market cattle infection (MCI) reactor rate.
[†]Postmovement test recommended.
Source: Animal and Plant Health Inspection Service of the USDA.

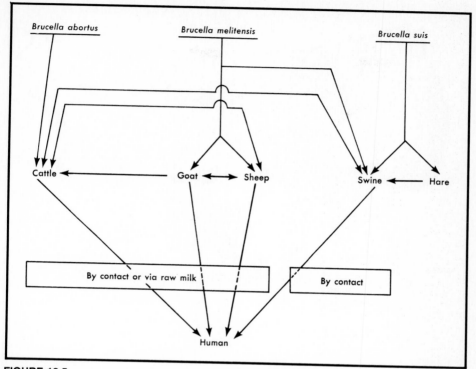

FIGURE 18.5
Routes of natural *Brucella* infection to humans.

Brucellosis in Humans   The virus may be transmitted to humans, in whom it causes brucellosis, or undulant fever (also called *Malta fever, goat fever,* and *Mediterranean fever*). Undulant fever is so named because the body temperature of infected persons varies, or *undulates.* Acute brucellosis in people is characterized by fever, chills, headache, night sweats, and weakness. These symptoms may extend from 10 to 14 days in mild cases and from 4 to 6 months in acute cases. The fever normally ranges from 100 to 103°F but may reach 104 to 105°F. Recurring fevers are common. The occurrence of brucellosis in humans continues to be high throughout the world, except in the Scandinavian countries, the Netherlands, Switzerland, Canada, Australia, New Zealand, West Germany, and the United States. The reported incidence of human brucellosis has been reduced appreciably in the United States during the past 25 years (Figures 18.7 and 18.8), paralleling the increased efforts of the brucellosis-eradication program and increased pasteurization of milk. **Abattoir** workers accounted for about half of the 185 cases reported in 1981. Brucellosis is most prevalent among adult males (especially veterinarians and others handling infected animals).

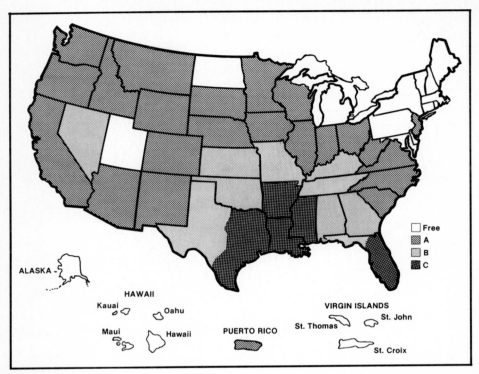

FIGURE 18.6
State classifications of brucellosis that were implemented by the USDA on May 1, 1982. See also Table 18.4 *(Courtesy USDA-APHIS, Hyattsville, Md.)*

Other potential sources of infection include drinking unpasteurized milk from infected animals, inhaling contaminated air (especially with *B. suis*), handling infected animals (especially swine) in an abattoir or meat-packing plant, and being accidentally exposed to *Brucella,* as may occur when laboratory personnel work with *Brucella* cultures and when veterinarians accidentally inoculate themselves with *B. abortus* vaccine (Strain 19). *B. melitensis* and *B. suis* are more pathogenic for humans than is *B. abortus.*

Prophylaxis and Control The two basic control measures responsible for the marked reduction of brucellosis among farm animals during the past 35 years have been (1) the slaughter of infected animals (as detected by the blood agglutination test) and (2) the immunization with living **cultures** of *B. abortus* (Strain 19 vaccine in bovine). The vaccine (a lyophilized agglutinogenic strain of *B. abortus*) was developed by Buck in 1930 and remains the vaccine recommended for calfhood immunization (2 through 6 months of age in dairy and 2 to 10 months in beef calves). Also, it is now permissible to vaccinate cattle over 1 year of age with a reduced dose of Strain 19.

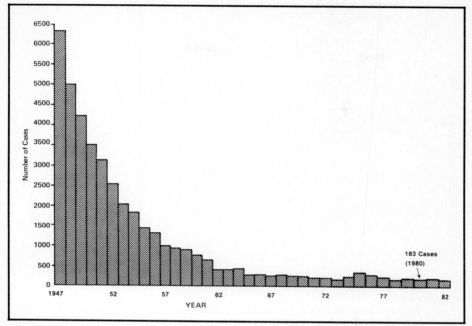

FIGURE 18.7
Reported human brucellosis in the United States from 1947 to 1982. *(Courtesy of Centers for Disease Control, Atlanta, Ga.)*

The nationwide cooperative state-federal brucellosis-eradication program has been successful in its effort to reduce the incidence of bovine brucellosis in the United States.

There are several plans for eradication of brucellosis from swine herds. An example is the "blood-testing and slaughter" plan. Except for **purebred** herds, the disposal of infected swine herds does not present as great an economic loss as does the disposal of cattle. To date, vaccination of swine has not been successful and is not recommended. The control of brucellosis in goats has been impeded because of limitations of the blood agglutination test. A good vaccine, called Rev. 1 (*B. melitensis*–attenuated), is now available for goats.

Therapy of Brucellosis No single dependable and effective therapeutic agent has been developed for animals because long-term **therapy** is economically unfeasible. Also, therapy is relatively ineffective because the bacteria are intracellular. In humans, however, several drugs and combinations of antibiotics have therapeutic value in the treatment of brucellosis.

Leptospirosis This infectious disease is produced by *Leptospira* (from the Latin *lepto*, meaning "thin," and *spira*, meaning "spiral") bacteria. Domestic animals susceptible to leptospirosis include cattle, dogs, goats, horses, sheep,

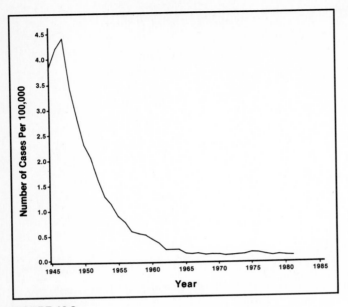

FIGURE 18.8
Cases of human brucellosis reported in the United States per 100,000 population from 1945 to 1981. The incidence of brucellosis in the United States decreased sharply from 1947 to 1965 as a result of the widespread adoption of the pasteurization of dairy products and the favorable results achieved through the cooperative efforts of the bovine brucellosis eradication program. The downward trend has continued, but at a slower rate. There were 185 cases of human brucellosis reported to the Centers for Disease Control in 1981. *(Courtesy Centers for Disease Control, Atlanta, Ga.)*

and swine. Wild animals that are affected include bats, deer, foxes, jackals, mice, muskrats, opossums, skunks, water birds, reptiles, turtles, and various others, including ticks. The chief ports of entry for leptospirae are through mucosal surfaces (conjunctival, nasal, oral, and vaginal) and broken skin. Modes of infection of humans include occupational assignments as abattoir, fish-cleaning, or sewer worker; exposure to infected cattle, goats, horses, rodents, sheep, and swine; swimming in ponds or slow-moving streams contaminated by diseased domestic or wild animals; and close contact with infected dogs.

Leptospirae thrive in a moist environment but do not survive in dry **environments** or under acid or highly alkaline conditions. *L. pomona* is the chief causative organism in cattle and swine. In many areas of the United States, *L. hardjo* also affects cattle. These organisms may be shed in the urine, which may be a "port wine" color (due to hemoglobinuria from destruction of red blood cells). Renal dysfunction is common in infected animals and in humans. Abortions (commonest in late pregnancy) may occur in cows and **sows.**

Leptospirosis in humans often resembles influenza. The number of cases of leptospirosis reported in the United States increased from 14 in 1947 to 85 in 1980. Vaccination with multivalent bacterins at 6- to 8-month intervals is a good prophylactic measure and is especially desirable in domestic animals. Vaccination in people is not usually recommended.

Anthrax Some historians believe that anthrax was the fifth plague of Egypt mentioned in the Bible. Several authors have noted that Moses threatened Pharaoh with a disease that would infect his livestock:

> The hand of the Lord will fall with a very severe plague upon your cattle which are in the field, upon the horses, the asses, the camels, the oxen, and upon the sheep: *there shall be* a very grievous murrain [a pestilence, or plague].
>
> *Exodus 9:3*

Although anthrax is chiefly a disease of animals, it was known at the end of the sixteenth century that it could be transmitted to humans. It was the first animal and human disease to be traced to a microorganismic source. Koch identified the anthrax bacillus in 1876. Moreover, it was the first disease against which a bacterial vaccine was found to be effective. Pasteur produced a preventive vaccine in 1880. Anthrax is caused by the organism *Bacillus anthracis.* Its **spores** are very resistant to cold, heat, and drying and may even be viable and virulent for 50 years or longer.

Humans and swine possess considerable natural resistance to anthrax, whereas cattle, horses, goats, sheep, and wild **herbivores** are quite susceptible. Cats, dogs, and wild animals in captivity may also become infected with the disease (especially from eating infected animals). Birds are highly resistant naturally but may be infected experimentally. Insects may carry and mechanically transmit the disease. Farm animals may become infected with anthrax by grazing on contaminated pasture, ingesting contaminated feed (especially **bone meal,** fish meal, and **tankage** that have not been subjected to high temperatures sufficient to kill the spores) or water, or being bitten by infected flies. The disease may also be spread by birds and small animals that have fed on infected **cadavers.**

Anthrax infection in humans is rare in the United States. A person may become infected through a skin lesion or a small wound on an exposed surface or by eating improperly cooked meat (especially of animals slaughtered in uninspected abattoirs in certain countries). Moreover, studies indicate that human anthrax may be acquired from direct contact with animal excretions, secretions, or infected tissues; by contact with animal skin, hair, or hide; or from bites of infected flies. In the United States, it is primarily an occupational hazard. During World War I, there were several cases of human anthrax traced to the use of horsehair shaving brushes that were imported from China and Siberia, where anthrax was prevalent. Apparently, the last case of anthrax in humans in the

United States was reported in 1980. This case involved a 30-year-old man who skinned infected carcasses in a Colorado rendering plant. In one Australian factory, goat hair is now **irradiated** to destroy the harmful anthrax bacterium before the hair is sold for use in making cloth and rugs. Certain areas of the Americas, Africa, and Asia still have dreaded animal losses resulting from anthrax.

The following account of the tracking down of a recent case of anthrax is of interest. The story began when a packhorse died suddenly on a pack trip in the Olympic Mountains of the state of Washington. Because there was no apparent cause of death, meat from the dead horse was fed to leopards, cougars, and other large cats on a private game farm near Sequim, Washington. Three days later, 38 of the big cats died. Cultures were made and anthrax was diagnosed. The packhorse was suspected, and cultures were made of the remaining horse meat. It, too, was positive for anthrax. But why did that particular horse die of anthrax and none of the others in the pack train? Where did the disease organisms originate?

Knowing that saddle pads are often made of imported goat hair and that new saddle pads had been used for the first time on the trip during which the horse had died, the state epidemiologist traced the pads to a Rhode Island saddle pad company that had used coarse cashmere scraps that originated in Afghanistan in manufacturing the pads. The cashmere scraps were positive for anthrax. The company quickly recalled that lot number of saddle pads. Another recent case of anthrax was traced to imported Haitian drums that were made in part from anthrax-infected goatskin.

In 1976, 160 cattle and horses died of anthrax in Texas.

Infected animals should be isolated, and if they die, the carcasses should be cremated. Pasteur produced vaccines composed of cultures of living anthrax organisms of reduced virulence (attenuated by growing them at high temperatures). When he injected these weakened bacteria into healthy sheep, they became slightly ill, but thereafter they exhibited immunity to further anthrax infection. Many of Pasteur's contemporaries questioned his work, and so he arranged a demonstration. With his influential skeptics as witnesses, he injected weakened bacteria into 25 sheep and selected 25 uninjected others as controls. Several weeks later, his doubting contemporaries assembled to observe Pasteur give a massive injection of fully active bacteria (more than enough to kill any normal healthy sheep) to all 50 sheep. A few days later, all 25 controls were dead, whereas the 25 treated sheep were alive and healthy. Pasteur's sheep reacted to the weakened anthrax bacteria by producing antibodies against them (Sections 18.2.5 to 18.2.7). Today an attenuated vaccine (called Sterne's) is used to control anthrax in farm animals (in areas where climatic conditions are favorable for the spores to survive).

Salmonellosis   There are more than 1600 **serotypes** of *Salmonella* organisms, and many cause gastroenteritis in humans. Five serotypes cause about 60 percent of the reported isolations from humans. Of these, *S. typhimurium* is responsible for approximately 30 percent. In the United States the reported cases of salmonellosis increased from 723 in 1942 to 33,715 in 1980. Salmonellosis is now estimated to affect more than 2 million persons (approximately 1 percent of the United States population) annually. The high occurrence of the disease is attributed to poultry and egg products and the centralization of food-processing operations. *S. pullorum,* once an important disease of poultry that can be transmitted through the egg (commoner in duck eggs), causing high death losses in chicks, has been largely eliminated by blood testing of parent stock. It was rarely a cause of disease in humans. Sheep and swine may also host *Salmonella* organisms, which may infect people when they eat improperly cooked meat.

Septicemia and certain uterine infections in cows may be caused by food-borne salmonellae. They may be transmitted to farm animals in contaminated feed or water. To prevent the possible cycling of *Salmonella* through swine and poultry to people, raw garbage intended for use as food by swine or poultry should be cooked. As a further precaution, eating of raw meat and eggs (either fresh, dried, or frozen) and drinking of unpasteurized milk should be avoided. Rats, other rodents, cockroaches, houseflies, and many other animals may be mechanical carriers of salmonellae.

Salmonella dublin is an increasingly common cause of diarrhea and death in young calves.

Typhoid Fever   Milk and milk products may serve as vehicles of typhoid fever transmission. Such milk has usually been contaminated by a dairy-farm handler and then consumed **raw.** Pasteurization destroys the causative organism, *S. typhi,* and since almost all milk consumed in the United States (except on dairy farms) is pasteurized, milk-borne outbreaks are rare. The housefly, *Musca domestica,* and other fly species are capable of transferring *S. typhi* from feces to food or to water.

Paratyphoid Fever   This disease resembles typhoid fever and is caused by *S. paratyphi.* It affects humans, domestic **fowl,** and other animals. The infectious organisms usually enter the body via improperly cleaned and/or improperly cooked food or through contaminated water.

Cholera   This is an infectious intestinal disease of humans common in India and other Asiatic countries. It is caused by a bacterium known as *Vibrio comma,* or *V. cholerae.* It is very **contagious.** The pathogen is usually transmitted in food and water that have been contaminated by people who have the disease. Flies also spread the bacteria. Cholera vaccine provides some immunity. People traveling to Asiatic countries usually receive two injections of such vaccine. These are followed by stimulatory injections at intervals of 4 to 6 months to maintain partial protection. Cholera in people should not be confused with hog cholera, which is a highly infectious viral disease of swine but has no apparent public health significance.

In 1980, cholera was diagnosed in a 46-year-old woman in Florida who had
ingested raw oysters, which were believed to be the source of infection. Ten
cases of cholera were reported in the United States in 1980.

Tularemia   This disease of mammals was first described in 1911 in Tulare
County, California (from which the name is derived), as a "plague-like disease
of rodents." In 1919 it was described in Utah as "deerfly fever." Today it is
known that tularemia is caused by *Francisella tularensis* (closely related to the
plague bacillus). The chief vectors of tularemia are ticks (especially the dog tick
and the wood tick) and bloodsucking flies (especially the deerfly and horsefly),
but it may also be spread by fleas, certain mosquitoes, and lice. Infected animals
that die may contaminate water and thereby spread the infection to sheep and
perhaps other species. Susceptible animals include cattle, cats, chipmunks, dogs,
horses, hamsters, muskrats, sheep, swine, beavers, coyotes, foxes, mice, rats,
rabbits, deer, opossums, squirrels, guinea pigs, and many other wild mammals.
Primary **modes** of transmission to humans include bites from infected ticks or
other arthropods, the handling of infected animals, and the inhalation of dust or
vapor containing *F. tularensis*. In people the disease is characterized by irregular
fever, which lasts for several weeks. Cooking readily renders the infected tissues
safe for human consumption.

A recent case of tularemia was reported in a 33-year-old man in Colorado
who had been bitten on the hand by an 8-week-old kitten. At the time of the
bite, the patient had observed the mother cat bringing wild cottontail rabbits to
her litter for food. There were 234 cases of tularemia reported in the United
States in 1980.

Plague   This is an **acute** infectious disease divided primarily into the two
types *bubonic plague,* in which the causative organism, *Pasteurella pestis,* is
transmitted chiefly from the rat reservoir of infection to humans by the rat flea,
and *pneumonic plague,* in which transmission is direct through airborne infection
from person to person. Since the dawn of history, bubonic plague has been
regarded in some regions of the world as one of the major pestilences of
humanity. The ill effects of the disease were first recorded in the Bible (I Samuel
5,6) and have been a serious threat to humans ever since. When Charlemagne
led his army against the Danes in 810, he was followed back into France by
plagues. In the 1300s, a form of bubonic plague called the *black death* (so called
because it caused formation of spots of blood or hemorrhages that turned black
under the skin) killed a fourth of the population of Europe. From 1347 to 1350,
about 20 million people in Asia, Africa, and Europe died from plague. In 1658,
Athanasius Kircher, a Jesuit priest, reported that the blood of plague patients
"was filled with a countless brood of worms, not perceptible to the naked eye,
but to be seen through the microscope."

Plague is primarily a disease of rodents. Bubonic plague occurs more
frequently in tropical or semitropical climates, whereas pneumonic plague in

epidemic form is a disease of cold and temperate regions. The control of fleas and rats (which are responsible for more than 75 percent of the plague in humans) controls epidemics of bubonic plague. In the United States, squirrels (11 dead squirrels in the Denver, Colorado, area showed positive signs of bubonic plague in 1968), chipmunks, prairie dogs, marmots, mice, rats, and rabbits may be naturally infected with the plague. Sylvatic plague, or plague of the woods, is widely spread among wild rodents of the western United States. Sporadic cases of plague in humans occur in the United States.

Eighteen cases of bubonic plague were reported in humans in the United States during 1980. All were in west and southwestern states. During the Vietnam war, pneumonic plague was one of the most dangerous hazards that nurses and physicians had to face.

The Endemic Relapsing Fevers   These infectious diseases are caused by spirochetes of the genera *Borrelia* and *Spirochaeta,* which are transmitted primarily by ticks of the genus *Ornithodoros* and to a lesser extent by lice. People and many domestic animals (including cattle and horses), wild animals (including opossums and squirrels), and laboratory animals (including mice, rats, and guinea pigs) are affected by the disease. The most effective means of prophylaxis is to avoid ticks and lice.

Bartonellosis   This is an infectious disease caused by *Bartonella bacilliformis,* which is transmitted by the sand fly. It occurs chiefly in the valleys of the Andes Mountains in Peru, Chili, Bolivia, and Colombia. It is also called *Carrión's disease* (after D. A. Carrión, a student in Peru who inoculated himself and died of the disease).

Vibriosis   This is an infectious disease of ruminants caused by several subspecies of *Campylobacter fetus* (formerly named *Vibrio fetus).* In cattle vibriosis is transmitted through the semen of infected **bulls,** both in natural service **(coitus)** and, although rare, through artificial insemination. The infection results in infertility, irregular estrous cycles (due to early embryonic deaths), abortions (in the fifth or sixth month of gestation), and retained placentas. The disease is diagnosed by culturing the organism from cervical samples, a technique requiring specialized equipment and **aseptic** collection methods. Other methods used for diagnosis are the vaginal mucus agglutination test (infected animals have agglutinating antibodies in their vaginal mucus) and fluorescent antibody stains of sheath washings from bulls and cervicovaginal mucus from cows. The best method of bringing the disease under control is to use artificial insemination. The semen from infected bulls can be used safely if it is diluted and treated with penicillin and streptomycin (Chapter 10). Intrauterine infusions of streptomycin and penicillin can be used to accelerate the elimination of infection in females. Vaccination is effective in controlling the disease in herds where natural service must be continued.

Vibriosis in sheep results in significant economic losses by causing abortions, stillbirths, and weak lambs that die soon after birth. The infection is transmitted by ingestion of organisms from feces, uterine discharges, and aborted fetuses and membranes. The disease is diagnosed by finding the organism in the stomach contents of aborted lambs or through culturing methods. The **intramuscular** administration of penicillin and dihydrostreptomycin and the feeding of low levels of chlortetracycline have been effective in reducing abortions in infected sheep. Vaccination is recommended for flocks with a history of vibriosis.

In humans, *Campylobacter* may be associated with obscure febrile illnesses, diarrhea, and perhaps abortion. Transmission is via ingestion. The origin of most cases in humans is unknown.

Glanders This is a highly communicable disease of horses caused by *Pseudomonas mallei*. It may be transmitted to humans through contact with diseased animals or their infected meat. Hippocrates recorded its clinical signs in about 450 B.C. A century later, Aristotle associated it with the Greek word *malleus,* meaning "malignant disease" or "epidemic." Glanders is characterized by lesions in the lungs and upper respiratory tract. Abscesses, nodules, and ulcers of the skin may also accompany the disease. Prophylaxis of glanders in humans is concomitant with its control in horses. (For 5000 years the horse has been close to people; see Chapter 21.) Because the disease has been eradicated in horses, it has not been reported in people in the United States for several decades.

Listeriosis This disease is caused by the bacteria *Listeria monocytogenes* and affects cattle, chickens, foxes, sheep, swine, rodents, and humans. Its most common effect is inflammation of the central nervous system (**encephalitis**). A diagnosis of listeriosis is very uncommon in people. The disease may cause abortion in women and meningitis in newborn infants.

Erysipelas This is a communicable disease of swine and poultry caused by *Erysipelothrix rhusiopathiae*. Sheep, mice, pigeons, and humans are also susceptible. People have considerable resistance to the disease (especially to infection through the **gastrointestinal** tract). The causative organism is resistant to drying and may even survive exposure to sunlight for 10 or more days. It may retain viability and virulence for several months in salted or pickled meats. The organism is widespread in soil. Animals that have had the disease may serve as carriers and thereby expose other animals. Vaccination offers a prophylactic measure against swine erysipelas. Humans may become infected through skin abrasions, especially among those handling infected swine (in abattoirs), and by handling infected fish. The skin form of the disease is called *erysipeloid* in humans. Penicillin is effective therapeutically in humans, whereas antiserum is also used to treat infected swine.

Actinomycosis   This chronic bacterial disease often localizes in the jaw, causing swelling, abscesses, fistulous (draining) tracts, and bony proliferation. The condition is referred to as "lumpy jaw" in cattle and may also affect horses, swine, and humans.

18.3.4   Protozoal Infections

Malaria   Malaria was apparently recognized by Hippocrates, who wrote: "Should there be rivers in the land, which drain off from the ground the stagnant water and rain water, these [the people] will be healthy and bright. But if there be no rivers, and the water that the people drink be marshy, stagnant, and fenny, the physique of the people must show protruding bellies and enlarged spleens."

According to an old Vietnamese legend, an ungrateful wife was transformed into a mosquito. To return to her original form, she had to find a special drop of human blood. Since that time, the story goes, she has never ceased biting people, seeking in vain the *special* drop of blood that would terminate her imprisonment. More than a mere nuisance, the dapple-winged female *Anopheles* mosquito spreads malaria, one of the world's greatest killers of humans. (It causes an estimated 1 million deaths annually.) In 1897 the English scientist Sir Ronald Ross observed malarial protozoa in a mosquito, thus indicting the mosquito as the vector of malaria. For this discovery, Ross received the Nobel prize in physiology and medicine in 1902.

Mosquitoes are among the deadliest enemies of humans and domestic animals. The spread of disease by mosquitoes was responsible for the downfall of certain ancient civilizations around the Mediterranean Sea. Mosquitoes once made it impossible for people to inhabit certain warm parts of the world. Before the Panama Canal could be built, it was necessary to kill the disease-carrying (yellow fever) mosquitoes of that area. *Aedes aegypti* and other mosquitoes carry heartworms, which are infectious to dogs, and the organism that causes encephalomyelitis of horses and humans. Malarial parasites also infect lizards, monkeys, and birds (the latter via the *Culex* mosquito). There are some 200 species of *Anopheles* mosquitoes (about 40 of which carry malaria) and several hundred species of *Aedes* mosquitoes (apparently only one of which carries the causative organisms of yellow and dengue fevers).

When the *Anopheles* mosquito bites a person who has malaria, it becomes infected by consuming blood cells that contain the parasite. These parasites develop and multiply (**maturation** of the sporozoites) in the mosquito's stomach, then migrate to its salivary glands (mouthparts). When the female mosquito bites another person, it injects saliva containing the malarial parasites into the victim. The tiny parasites enter the bloodstream and travel to the liver. They then migrate to the red blood cells, in which they multiply rapidly and cause the blood cells to burst, resulting in varying degrees of **anemia.** As the cells are destroyed, materials are liberated, and the host reacts with chills and fever.

Plasmodium is the parasitic *protozoan* (one-celled) organism that causes malaria. It spends part of its life cycle in the red blood cells of humans and part of it in the body of a female *Anopheles* mosquito. The best way to break the parasite's life cycle and prevent malaria, then, is to eradicate *Anopheles* mosquitoes.

Malaria has been virtually eradicated in Europe, North America, and the U.S.S.R. It remains, however, a leading cause of death in Africa, India, Pakistan, southeast Asia, and South America, where, collectively, malaria claims the lives of a million or more people annually. Through the use of the pesticide DDT, and more recently malathion in certain countries, malaria has been eradicated in more than 40 countries since World War II. There were 262 cases of malaria reported in the United States in 1980.

African Trypanosomiasis (African Sleeping Sickness)   This **chronic** infectious disease is caused by flagellates (protozoa) of the genus *Trypanosoma*. *Trypanosoma gambiense* is the causative agent of African sleeping sickness (not to be confused with ordinary sleeping sickness, or encephalitis, which is caused by a virus). The trypanosomes live in the blood of their host, where they multiply and release poisonous by-products of metabolism. In humans or domestic animals, they invade the nervous system, causing lethargy and finally death. The trypanosomes are spread from host to host by bloodsucking tsetse flies. When a tsetse fly withdraws blood from an infected animal, trypanosomes are sucked into its intestine, where they multiply and undergo developmental changes. They migrate to the fly's salivary glands, in which they further develop and multiply. The fly is then capable of transferring the trypanosomes via saliva into a vertebrate host. This disease makes large areas of Africa uninhabitable for humans and is difficult to control because many wild animals serve as a reservoir of trypanosomes. The organism causing sleeping sickness has a life cycle similar to that of the malarial parasite. Like malaria, it can be spread from person to person, but unlike malaria, it can also be carried from animals to people.

Anaplasmosis   This disease of ruminants is caused by a parasite that lives inside the victim and destroys red blood cells, resulting in anemia. It is transmitted from sick or carrier cattle (untreated animals that recover are generally considered to be carriers for life) to healthy animals by bloodsucking insects (especially horseflies, stable flies, horn flies, mosquitoes, and ticks). Improperly **sterilized** surgical instruments, vaccinating needles, **dehorning** equipment, etc., may also spread the disease.

Deer and antelope may serve as reservoirs. The disease is diagnosed by finding antibodies in the blood or by visualizing the causative organism in blood smears. A killed vaccine is useful in prophylaxis of the disease. Anaplasmosis presents no known public health significance.

Cryptosporidiosis   This protozoan parasite produces enterocolitis and diarrhea in several species including cattle, sheep, deer, goats, and humans. The majority of human cases have been observed in farm workers having close contact with infected calves.

18.3.5 Fungal Infections

Two groups of fungal diseases may be transmitted from animals to people: (1) those affecting humans *externally* (superficial, or dermatomycoses) and (2) those affecting humans *internally* (localized, or systemic granulomatous mycoses). External fungal infections include the *Trichophyton* organisms, which cause dermatophytosis (ringworm) in cattle, cats, dogs, horses, and other domestic animals. It is estimated that approximately 80 percent of ringworm infections of persons residing in rural areas are acquired from farm animals.

There are probably 10 or more fungal diseases that affect humans internally. *Coccidiomycosis* is an internal fungal infection that begins as a respiratory infection and may progress to produce abscesses throughout the body, especially in the subcutaneous tissues, skin, bone, peritoneum, testis, thyroid, and central nervous system. The disease may infect cats, cattle, dogs, horses, sheep, swine, rodents, humans, and other species. These species may also serve as reservoirs. The infection is not directly transmitted from animal to animal, but rather is acquired through inhalation of spores. Humans, too, acquire infection through the inhalation of spores. *Histoplasmosis* is a fungal disease resembling miliary tuberculosis (characterized by the formation of lung lesions resembling millet seeds). A survey of 185,000 naval recruits with lifetime residence in a single county revealed that the disease is most prevalent in the central and south-central United States. The fungus has been isolated from a number of mammalian species, including mice, rats, cats, cattle, dogs, bats, horses, foxes, opossums, sheep, skunks, swine, and woodchucks. People usually become infected by inhalation of spore-laden dust existing in nature in a saprophytic state. Direct transmission from person to person or from animal to humans has not been demonstrated.

18.3.6 Helminthic (Worm) Infections

Humans may become infected from several species of parasites of domestic and wild animals by several means. Important ones include passive introduction of parasites into the body by the oral (food or drink) route (e.g., the eggs or larvae of certain tapeworms), active penetration of the infective stage of the parasite into the skin (e.g., certain hookworms from moist soil or blood flukes from water), introduction of the parasite into the skin via an alternate host (e.g., malarial parasites by mosquitoes and tsetse flies), and skin contagion from the animal reservoir (e.g., the eggs or larvae causing cattle and/or sheep bots). These cause **helminthic** infections.

Trichinosis The parasitic nematode *Trichinella spiralis* has a cosmopolitan distribution but a fairly simple life cycle. Its usual hosts are meat-eating mammals. Infection occurs when the host **ingests** meat containing the living larvae. (A person becomes infected from eating inadequately cooked pork, bear, or walrus meats.) A few hours postingestion, the young larvae are liberated by gastric digestion from the cysts (tiny muscle sheaths, or sacs) in which they live. The worms grow rapidly, and some travel from the stomach to the small intestine, where they attach themselves to the inner lining, grow, and become sexually mature in about 2 days. The adult worms mate and produce larvae, which are released **viviparously** from the maternal body into the small intestine. About 8 days after mating, each new female worm begins to produce small but active larvae that first penetrate the walls of the host's intestine. They then enter the lymphatic and blood systems and are transported throughout the body. Although most body organs and tissues are subject to invasion, the worms largely infect the voluntary muscles. An inflammatory response to the worm results in the degeneration of the muscle fibers surrounding the parasite and in the formation of a connective tissue capsule (cyst) around it. The encapsulated larvae ultimately become **calcified** and eventually degenerate and die unless the meat is eaten by a person (Figure 18.9) or another suitable host. (The larvae may live in these cysts for a year or more.)

Swine, rats, cats, dogs, and other mammals are subject to infection. Swine become infected mainly by feeding on garbage containing raw pork scraps. It has been estimated that the collection and use of garbage for swine production constitutes a $100 million enterprise annually. Therefore it is not economically practical to prohibit the use of garbage for feeding swine, but it is practical to legislate a requirement that garbage be cooked and thereby rendered safe for swine. (All states now require that garbage intended for swine food be cooked.) USDA and **U.S. Public Health Service** reports indicate that since 1906 the occurrence of trichinosis has been reduced by more than 80 percent in humans and by 94 percent in swine. Human trichinosis is no longer a serious public health problem. Only 130 cases were reported in the United States in 1980. The disease probably served as the basis for the first food sanitation code. The Mosaic code forbade the children of Israel to eat pork, as recorded in Leviticus 11:7–8: "And the swine, though he divide the hoof, and be clovenfooted, yet he cheweth not the **cud;** he *is* unclean to you. Of their flesh shall ye not eat, and their carcase shall ye not touch; they *are* unclean to you." Methods of providing safe pork for humans were discussed in Chapter 3.

Taeniasis   Domestic animals play an important role in most of the tapeworm infections of humans. The pork tapeworm, *Taenia solium,* may be viable in improperly cooked pork. The adult tapeworm (3 to 6 ft in length) inhabits a person's small intestine. (Tapeworms have no mouth or digestive system, but rather attach themselves to the intestinal lining by means of a small head that has suckers, thus allowing them to absorb already digested food **nutrients**.) Its larval stage (bladder worm) develops in the **musculature** of swine. Infected pork containing these larvae is commonly called "measly pork."

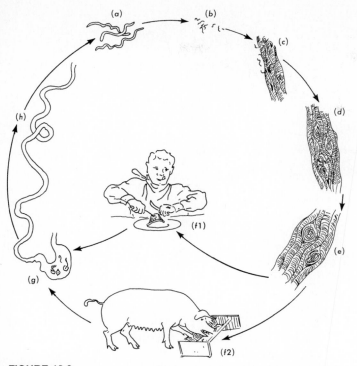

FIGURE 18.9
Life cycle of *Trichinella spiralis* (trichina worm). *(a)* Adult worms in
mucus of small intestine of swine; females produce many larvae;
(b) larvae enter bloodstream and are distributed throughout body; *(c)*
larvae enter voluntary muscle tissues and develop; *(d)* larvae grow to
infective stage and encyst; *(e)*, *(f*1), *(f*2) human (swine, rats, and other
animals) becomes infected by eating improperly cooked pork containing
T. spiralis; *(g)* cysts dissolved in stomach pass to small intestines *(h)*,
in which they reach sexual maturity (adults); and the cycle begins
again. *(Courtesy Dr. P. D. Garrett and Dr. D. E. Rodabaugh.)*

The life cycle of the 12- to 25-ft-long beef tapeworm, *T. saginata,* is similar to
that of the pork tapeworm, except that cattle and other herbivores rather than
swine serve as intermediate hosts. (The ox is the essential host of the larval, or
cysticercus, stage.) Cattle acquire this parasite by ingesting the tapeworm eggs
voided in human feces (humans are the only known host of the mature worm), as
shown in Figure 18.10. Therefore the disease is more prevalent in countries
where pastures are irrigated with sewage effluent. The larvae of *T. solium* and *T.
saginata* are killed when heated to 55°C. Quick-freezing and thorough curing or
salting of meat are also destructive to these larvae.

The commonest tapeworm in humans is *Hymenolepis nana,* which is about 2
in long. It is also common in murine rodents (mice and rats), which serve as a

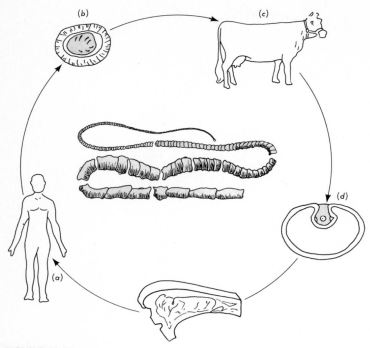

FIGURE 18.10
Life cycle of *Taenia saginata* (beef tapeworm). *(a)* Humans become
infected with adult worm by eating improperly cooked beef containing the
cysticerci; *(b)* ova voided in feces; *(c)* calves and cows become infected
with the cysticerci via infested soil or forage; *(d)* cysticerci in beef tissue
(muscle).

continuous source of human infection, due to contamination of humans' food
and drink with their feces. However, human infections may also be perpetrated
by ingestion of foods contaminated with feces from other infected humans.

The broad, or fish, tapeworm, *Diphyllobothrium latum,* is the largest cestode
(tapeworm) found in humans. It may be 30 or more feet in length. The infection
is acquired through the consumption of raw or insufficiently cooked fish that is
infected with the larvae. The life cycle of *D. latum* is rather complex. The adult
worm inhabits the small intestine. Eggs passed in the feces hatch and, when
deposited in fresh water, release larvae. The free-swimming larva that moves
about in the water must be swallowed by a suitable *first* **intermediate host**
(usually a water flea) to continue its development. The life cycle is dependent on
the infected first intermediate host being eaten by a suitable *second intermediate
host* (usually freshwater fish, such as the pike, salmon, trout, and perch). In this
second host the larva penetrates the muscles and continues to develop. When
this host is ingested by a fish-eating mammal, the parasite attaches itself to the

intestinal mucosa and grows into an adult tapeworm. Wild reservoir hosts include the bear, fox, mink, and wolf.

Echinococcosis (Hydatid Disease)   The small tapeworm associated with this disease is found most frequently in dogs and wolves. It is also found in cats. Humans may acquire the disease from these infected carnivores. The larva, known as the *hydatid,* may develop in nearly all mammals, where it forms hydatid tumors, or cysts, in the **viscera** (liver, lungs, kidneys, and other organs). In many countries, these infections are responsible for huge losses of animal protein through condemnation of internal organs, especially liver. Dogs may become infected by feeding on infected animal viscera. Dogs harboring adult worms constitute the foremost hazard to humans. Children playing among infected dogs are endangered by worm eggs adhering to the dog's hair. Humans sometimes serve as an intermediate host for the huge cysts (hydatids) of these tiny dog tapeworms.

Trematodiasis   The intestinal trematodes, or flukes, are parasitic in humans and animals. A person becomes infected from the ingestion of raw or insufficiently cooked infected freshwater fish, crustaceans, and vegetation, which are the intermediate hosts of trematodes. Humans are parasitized by only those trematodes that require some mollusk (usually a snail) as an intermediate host. Swarms of tailed larvae (called **cercariae**) emerge from the mollusk and either invade a person via the skin or first encyst in the tissues or on the surface of a second intermediate host. When infected tissues of this host are consumed, the trematode then proceeds with its life cycle. The life cycles of these human parasites are quite interesting. Eggs excreted in the feces of the **definitive** (final) **host** hatch in water, in which they release motile larvae **(miracidia),** which in turn invade snails. Here they undergo **asexual** reproduction and development to the larval stage. The larvae then leave the snail and, depending on the species of trematode, encyst on vegetation or in fish, other snails, or clams. Several piscivorous (fish-eating) mammals, such as the cat, dog, and fox, serve as reservoir hosts for trematodes.

Blood flukes (schistosomes) produce cercariae, which directly penetrate the host's skin. However, in the United States, these are of minor importance to humans. Also, certain blood flukes parasitize cattle, sheep, and goats in tropical countries.

Fascioliasis (Liver Fluke Disease)   This disease of cattle, sheep, horses, goats, and other herbivorous animals is caused by *Fasciola hepatica.* People may acquire the disease by ingesting raw watercress (found especially in certain sheep- and cattle-raising areas) to which the infective cysts of the fluke are attached. The flukes may obstruct the biliary passages in the human and animal liver. An enlarged liver with degeneration and cyst formation ensues. The eggs of *F. hepatica* are excreted in the feces of infected mammals. If they reach fresh water, they mature in about 2 weeks. These miracidia then develop within snails to cercariae, which in turn encyst on water vegetation. Ingestion of viable encysted metacercariae attached to the raw grass or water plants causes infection. In the infected mammal, the fluke migrates to the liver and bile ducts, where it lays eggs that are shed in the feces, and the cycle repeats itself. This disease causes large losses among cattle in the southern parts of the United States owing to condemnation of liver at slaughter.

Clonorchiasis (Oriental Liver Fluke Disease)   This chronic disease of the liver and bile ducts is caused by the trematode *Clonorchis sinensis*. Humans become infected (especially in China and Japan) through the consumption of raw freshwater fish containing the encysted stages (metacercariae) of *C. sinensis*. Adult flukes shed eggs into the bile ducts, from which they are excreted via the feces. The causative parasitic trematodes have two intermediate hosts. Certain snails (mollusks) act as the first intermediate host, in which development to the cercaria stage occurs. The cercariae then escape from the snails and, on entry into a suitable freshwater fish (second intermediate host), encyst in the meat as metacercariae. When infected fish are ingested, the metacercariae are released, and the life cycle is complete. Dogs, cats, and other fish-eating mammals may serve as reservoir hosts for *C. sinensis*.

Paragonimiasis (Oriental Lung Fluke Disease)   Human infection by the lung fluke, *Paragonimus westermani*, is acquired by the ingestion of infected crabs and crayfish that contain the encysted stages of the parasite. Its life cycle requires two intermediate invertebrate hosts. The first is an operculate snail (mollusk), in which development from the miracidium stage to that of cercaria occurs. The second intermediate host is either a freshwater crab or a crayfish (crustacean). Paragonimiasis is also found in mammals other than humans, including the dog, cat, tiger, leopard, wolf, wildcat, and swine.

When viable cysts are ingested by the final (definitive) host, **excystation** occurs in the duodenum. The young worm then migrates through the **peritoneal** cavity, penetrates the diaphragm, and enters the lungs. Eggs are excreted by **expectoration** or, if swallowed, in the feces. When they reach fresh water, they embryonate and are then ready to enter a suitable snail vector, and the life cycle is thus completed. Proper cooking of crabs and crayfish is the best prophylactic measure.

Hookworm (Ancylostomiasis)   There are several hookworms that may parasitize humans. Some writers believe that nearly a quarter of the world's population is infected with hookworms. It has been estimated that 500 human hookworms can remove from 50 to 250 cc of blood (about 4 percent of a person's total blood volume) daily. Hookworm infection is common where people ordinarily walk barefoot, since the small animal parasites usually enter the body through the skin, but the parasite may also be acquired by ingestion. It may be controlled by proper disposal of human and animal (cat, dog) feces (to prevent soil pollution), as noted in Deuteronomy 23:13:

> Thou shalt have a paddle . . . and . . . when thou will ease thyself . . . thou shalt dig therewith, and shalt turn back and cover that which cometh from thee. [Commentarial consensus adds "so that your camp be kept clean and free of disease."]

Eggs produced by the female are excreted with the feces. When the temperature and moisture are suitable, the larvae hatch and, after attaining a certain stage of development, enter into the body of a new host. In this new host, they migrate to the lymph and blood vessels, which carry them to the lungs. From there they migrate up the trachea to the throat, are swallowed, and enter the intestines. Hookworms infect all common animals. Life cycles of the several species are similar; that of the dog hookworm is depicted in Figure 18.11.

Larvae of the cat and dog hookworm, *Ancylostoma braziliense*, burrow beneath the skin of a person and cause a thin red line of eruption, commonly referred to as *cutaneous larva migrans*. Actually, there are several parasites specific to animals that may accidentally infect people that are categorized as larva migrans. They migrate in vain to different human body tissues in search of a suitable site for development to maturity. Those that migrate in the **subcutane-**

FIGURE 18.11
Life cycle of *Ancylostoma caninum* (dog hookworm). *(a1)* Larvae penetrate skin (or foot pads), then *(b1)* burrow to blood vessels and migrate to the lungs; or *(a2)* larvae may be ingested and then *(b2)* migrate to the lungs; *(c)* larvae move to air spaces of the lungs and then to the bronchial tubes and trachea; *(d)* ciliary motion of epithelium and coughing carry larvae to the throat, from which they may be swallowed; *(e)* larvae pass to the small intestine, in which they mature and become adults; the adult worms attach, suck blood (up to 0.8 cc per worm daily), and the females lay eggs; *(f)* eggs (ova) are passed in the feces, undergo development, and hatch into larvae in the soil, thus completing the cycle. *(Courtesy Dr. P. D. Garrett and Dr. D. E. Rodabaugh.)*

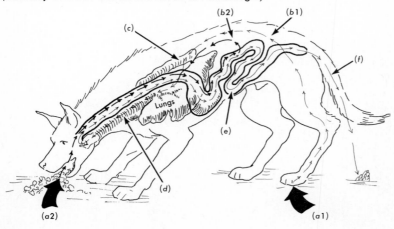

ous tissue are called *cutaneous* larva migrans (creeping eruption), whereas nematode larvae penetrating the internal organs are referred to as *visceral* larva migrans.

Roundworms   People may become infected by several parasitic roundworms commonly found in domestic and wild animals. Nearly two-thirds of the world's population is or has been infected by these. The ascaris of swine morphologically resembles the roundworm of humans. Humans may be infected occasionally from ingesting infective-stage *Ascaris* eggs from **porcine** sources. The adult cat ascarid, *Toxocara cati*, has been found in the human intestine. The dog ascarid, *Toxocara canis*, is an important cause of human visceral larva migrans and may cause blindness in children. *Ascaris* is a relatively large parasitic roundworm (Figure 18.12) that causes ascariasis and lives in the intestines of humans and sometimes in those of swine and horses. The female (longer than the male) may grow as long as 16 in. It feeds on already digested food in the host's intestine. Roundworms reproduce by laying eggs, which may be found in improperly cleaned vegetables or in unsanitary places. The eggs are then swallowed and **hatch** inside the **lumen** of the host's intestines. Roundworms may travel to other body parts, such as the liver and appendix. *A. lumbricoides* of swine *(A. suis)*

FIGURE 18.12
Life cycle of *Ascaris suis* (swine roundworm). *(a)* Eggs (ova) ingested with soil; *(b)* eggs pass through the stomach; *(c)* larvae hatch from eggs in the intestine; *(d–f)* larvae burrow through the intestinal wall and via portal circulation pass to the liver *(d)*, via hepatic veins to the heart *(e)*, and ultimately to the lungs *(f)*; in the lungs growing larvae migrate to the bronchi, ascend the trachea to the pharynx *(g)*, and pass down the esophagus, eventually reaching the intestine; *(h)* in the small intestine the larvae develop into adults, which lay eggs that are passed with feces *(i)*, contaminating the soil, thus completing the cycle. *(Courtesy Dr. P. D. Garrett and Dr. D. E. Rodabaugh.)*

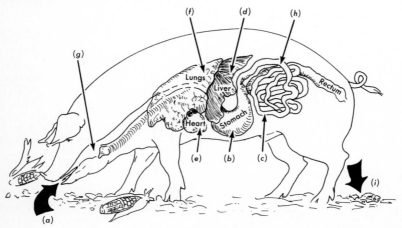

may inhabit the lumen of the small intestine in children and cause **colicky** pains and diarrhea.

Elephantiasis   This skin disease is most common in the tropics. It is so named because the affected skin becomes rough like that of an elephant. The disease is usually caused by a tiny worm, the *filaria,* which is carried by mosquitoes. These parasitic filarial worms block the flow of **lymph.**

18.4   SELECTED HUMAN DISEASES TRANSMISSIBLE TO ANIMALS

Certain diseases of human origin may be acquired by animals and transmitted back to humans.

18.4.1   Viral Infections

Certain of the great apes may acquire *chicken pox* (varicella) from children and serve as a source of infection to other children. Present evidence indicates a close relationship between the viruses of measles and canine distemper. (Measles virus may be used to immunize dogs against canine distemper.)

Chimpanzees are susceptible to the viruses that cause the *common cold* and are capable of transmitting the infections back to people. The viruses that cause *influenza* in humans (e.g., type A influenza) are apparently closely related to those isolated from avians (chickens, ducks, and turkeys), horses, swine, and possibly cattle and sheep. There has been much speculation with respect to the role of animals in the **epidemiology** of human influenza, particularly as a source of new major antigenic variants. It is now generally accepted that the swine strain represents the prototype of the influenza strain responsible for the 1917–1918 human **pandemic** and that this strain was introduced to swine by people. An unanswered question that often arises is whether a human influenza strain has the potential to establish itself in animals and then at some later date be reintroduced into people when the latter's antibody status permits.

It has been shown that swine may receive their initial influenza infection either from infected swine or through an intermediate host, such as the swine lungworm. The complete cycle involves **latent** infection of lungworm ova with influenza virus. These lungworm eggs carry the virus as they pass through a developmental cycle in the earthworm and reenter the primary swine host when the earthworm is ingested. The migrating lungworm larvae then set up an **occult** viral infection in the lung tissues of the susceptible host. The cycle is perpetuated with the passage of newly infected lungworm ova.

18.4.2   Bacterial Infections

The streptococcus (*Streptococcus pyogenes*) that causes *scarlet fever* in humans may infect the cow and be shed into milk, resulting in milk-borne epidemics.

However, no such epidemics have been reported in the United States since 1944. Staphylococcal infections of humans may be transmitted to cows. However, the staphylococci of people are usually less virulent in cows than in humans. Humans are often carriers of staphylococci without exhibiting symptoms of infection.

Yersinia enterocolitica in contaminated milk has been responsible for gastroenteritis in humans. The organism has been isolated from swine. The role of domestic animals in the epidemiology of the disease in humans remains unclear.

18.4.3 Protozoal Infections

Several parasitic infections, e.g., *amoebic dysentery,* may be passed from a person to an animal and back to a person (by means of the cat, dog, monkey, and rat).

18.5 DISEASES TRANSMISSIBLE BY ANIMALS AS PASSIVE CARRIERS

Animals may host the spores of several pathogenic organisms in their intestinal tracts, usually with no apparent harm to themselves. We have included information concerning a few such diseases.

Tetanus The tetanus organism, *Clostridium tetani,* is commonly found in the intestine of herbivorous animals (primarily horses) and to a lesser extent in that of humans. **Manured** soil is a prime source of these spores. People working around horses and stables are more likely to be carriers than those engaged in other occupations and activities. In humans and farm animals (especially horses, sheep, and goats) tetanus results from a deep-puncture wound. (The deep puncture provides an entry and **anaerobic** conditions for growth of the organism.) Invasion of the organisms requires broken skin. Gunshot and firecracker wounds are especially susceptible to infection. *Passive immunization* with tetanus antitoxin after injury is very effective. However, *active immunization* using tetanus **toxoid** is preferred in tetanus prophylaxis and is recommended for all persons.

Gas Gangrene Like tetanus, gas gangrene infection in humans results from the introduction of **spores** through broken skin. The clostridia organisms (especially *C. perfringens)* are frequently found in the intestinal tract of most farm animals and of humans. The disease is more prevalent in combat zones where people are apt to be wounded and to be in contact with soil contaminated with animal or human **excreta.**

Botulism (From the Latin *botulus,* meaning "sausage.") Spores of the causative bacterium inhabit the intestinal tract of many animals. Botulism in humans and animals is caused by the ingestion of food in which the organisms *C. botulinum* have produced **toxins.** Among farm animals, horses are the most susceptible to the disease. They develop botulism from eating moldy or spoiled forages or grains in which botulism toxin has been produced. Poultry are also susceptible to the disease.

18.6 TOXIC PLANTS[6]

> Who taught the natives of the field and wood
> To shun their poison and to choose their food;
> Prescient, the tides and tempests to withstand;
> Build on the wave or arch beneath the sand?
>
> *Alexander Pope (1688–1774)*

The **lethal** effects of certain plants have been known for many centuries. Socrates (469–399 B.C.) was sentenced by the rulers of ancient Athens to die by drinking brew made from the poisonous hemlock plant. Since the diet of most farm animals is largely of plant origin, selected plants that may adversely affect the health of animals providing useful products to humanity will be reviewed.

The toxicity of a plant varies with the animal's species, age, sex, nutritional status, state of health, and **stress** factors and with the formulation of the compound and the route of its administration. A common way of expressing the toxicity of a compound is by means of an LD_{50} value. This is a statistical estimate of the dosage necessary to kill 50 percent of a large population of test species under stated conditions.

There are several poisonous plants that may cause extensive economic losses to the producer of animal products. (Cattle and sheep are the most frequently affected farm animals.) Two commonly accumulated poisons are cyanide (hydrocyanic acid, or HCN) and nitrate (potassium nitrate, or KNO_3). Plants that may accumulate HCN include arrow grass, chokecherries, and sorghums. Oats, Sudan grass, and corn (especially in hot, dry weather, when nitrogen fertilizer application is high) are common plants that may accumulate KNO_3. Many wild plants, including the Russian thistle, white ragweed, and lamb's-quarter, are known to accumulate KNO_3.

Some ingested plants interfere with the metabolism of certain nutrients. An example is the bracken fern, which on ingestion apparently causes a thiamine deficiency in horses, cattle, and swine. The deficiency develops because this fern contains an **enzyme** called *thiaminase,* which adheres to the thiamine molecule and thereby renders it inactive. The administration of thiamine apparently corrects the deficiency in most animals, except cattle. Ingestion of certain plants may cause congenital malformations (Figure 18.13).

Horses and, occasionally, cattle and sheep may become poisoned by eating the horsetail plant, *Equisetum.* Cocklebur seedlings *(Xanthium strumarium)*

[6]The authors are grateful to Dr. A. A. Case, Professor Emeritus of Veterinary Medicine and Surgery, University of Missouri-Columbia, for his contributions to this section.

(a)   (b)

FIGURE 18.13
Congenital malformation of the above calf *(a)* was induced by maternal ingestion of the lupine plant, *Lupinus caudatus (b),* during the period of pregnancy between 40 and 70 days. The deformity may occur in the front legs, hind legs, back, or palate (cleft), depending on the stage of fetal development at the time of ingestion. *(Courtesy Dr. Wayne Binns, Director, Poisonous Plant Research Laboratory, USDA, Logan, Utah.)*

contain a toxic compound, hydroquinone. One-fourth to one-half pound of the seedlings will kill a 50-lb pig if the plants are consumed in a short period of time. Sheep, cattle, and poultry are also susceptible to poisoning by the plant. Jimson weed and larkspur plants may poison cattle. Milkweeds are quite toxic to sheep (1.5 oz of green leaves may be fatal) and to a lesser degree to cattle, horses, and poultry.

Milk sickness in humans and *trembles* in certain farm mammals are due to a poison in the white snakeroot *(Eupatorium rugosum)* plant, in Jimmy fern *(Notholaena sinuata)* or in the rayless goldenrod *(Hapalopappus heterophyllus).* This poison is of danger chiefly to cattle, milk goats, horses, and sheep, in which it causes trembles. Humans and milk-fed animals are poisoned by ingesting milk from affected mammals that contains tremetol. Many people (including Abraham Lincoln's mother) died of this **insidious** disease during the nineteenth century. Because of improved pastures, cattle today are rarely exposed to these poisonous plants. This, coupled with the dilution of milk from exposed cows with that from other sources, renders milk sickness (vomiting sickness) a rarity today.

Pets and pleasure horses have increased in relative numbers in large-population centers. Owners are generally unfamiliar with toxic plants and their possible harmful effects. This has tended to shift the danger of toxic plants from an agricultural to a metropolitan situation. *Native* toxic plants have largely disappeared in the intensively cultivated croplands of the eastern, central, and southern United States and Canada. However, substitute plants, shrubs, or trees that are often poisonous have taken their place.

In subtropical North America, the kinds and numbers of toxic plants often differ from those of other North American areas. In much of Florida and the other southeastern states, an increasingly important beef cattle industry makes pasture management very important from the standpoint of improving the kind of pasturage and of avoiding native poisonous plants and shrubs. When such soil-improvement cover crops as the **legume** *Crotalaria* are used, care must be exercised to keep the seeds from contaminating hay or grain crops destined for use in the manufacture of animal feeds. The same is true of castor beans, tung nuts, avocados, and many ornamentals, such as oleander, jequirity beans *(Abrus* spp.), *Lantana* spp., various lilies, the laurels, and many others.

For example, the generally accepted lethal amounts of the common Japanese yew shrub *(Taxus* spp.) for most animals have been observed to be as little as 0.1 percent of an animal's body weight. The poisonous element includes an alkaloid that depresses heart function. An increasing number of cattle and horses die annually as a result of poisoning from ornamental shrubbery, especially the Japanese yew. Dr. A. A. Case, based on his experiences with poisoning from the fresh leaves and twigs of the ornamental yew, suggests that the lethal dose for domestic animals and pets varies from 300 g for a mature cow to 10 g for a mature German shepherd dog. Ingestion of less than 100 g is usually fatal in swine, and only 30 g will kill most fowls. Dr. Case advises further that the average sheep is poisoned by about the same amount of yew twigs, leaves, and immature fruit as a small horse. The green growth and living bark of the black locust *(Robinia pseudoacacia),* as in oleander and ornamental yew, are extremely toxic for horses and other farm animals as well as for humans. Domestic animals should not have access to pruned new growth nor to dried materials of such plants.

There is considerable variation in the plant species common to each region of the United States. From Key West, Florida, to Kodiak, Alaska (a distance of over 5000 miles), every kind of climate from humid suptropical, semiarid, and arid to frigid subarctic may be found. It would be impractical to attempt an enumeration of all the toxic plants and shrubs that may be found in such vast areas in a brief survey such as this. For example, a specific area such as Missouri has approximately 100 genera and several hundred species of plants, shrubs, and trees that are described as toxic for one or more species of animals under given circumstances. Florida workers have described at least 50 genera of poisonous plants. The Wyoming Agricultural Experiment Station considered over 50 genera and 80 species important enough to list as poisonous,[7] especially to cattle, sheep, goats, horses, and swine. Texas workers, representing the southwestern area, have reported over 50 genera as having toxic species. Kansas recognizes 58 genera and 79 species of toxic plants. USDA workers have selected 22 genera as

[7] *Wyoming Agr. Expt. Sta. Bull.* 324, 1953

the most troublesome toxic plants, principally in the west and far west, although at least a dozen more may be found in most areas of the United States and Canada. At least 25 state agricultural experiment station bulletins include the subject of poisonous plants as the whole or a part of their subject matter. Toxic and noxious weeds and their seeds are listed and/or described in numerous monographs and research papers.

The kind and intensity of the animal industry in a region determine whether a given toxic plant is considered important. It has been observed that quite often the toxic vegetation is also the hardiest and persists when wholesome forage plants have largely disappeared. In the mountainous areas of the United States and Canada, much of the original native plant population continues to exist. However, over**grazing** has almost eliminated wholesome pasturage vegetation in many areas. Such practices have allowed the more undesirable and often poisonous plants, such as *Halogeton glomeratus,* to encroach on and almost to destroy the ranges for grazing cattle and sheep. Overgrazing has also seriously damaged good pasturelands in many other areas (e.g., the Osage Hills of Oklahoma and the Flint Hills of Kansas). The populations of less desirable weeds and shrubs almost always increase, usually because such plants are not eaten by cattle, sheep, goats, and other animals unless nothing else is available. Proper range and pasture management is the best and most practical way to control or avoid losses from animals grazing areas with toxic plants under the conditions mentioned.

Anticoagulant Action of Clovers   White and yellow sweet clover and, to a lesser extent, lespedeza hays contain a compound called *coumarin,* which, under conditions of spoilage, may be converted into **dicoumarin.** Dicoumarin interferes with prothrombin formation, which is essential for proper blood clotting (Figure 16.4). This principle was used to develop the **rodenticides** *pindone* and *warfarin.*

18.6.1   Mycotoxins[8]

Mycotoxins are chemical substances produced by fungi (molds) that may result in illness and death of animals and humans when feed or food containing them is eaten. Certain mycotoxins are nerve poisons, others are liver and kidney poisons, and some affect other body organs. Mycotoxins may cause birth defects, abortions, tremors, cancers, or other adverse effects.

Fungi that form mycotoxins occur worldwide. Foods and feeds may be contaminated with mycotoxins produced by fungi that have grown on crops in the field, in storage, or during processing. Crops frequently contaminated include corn, peanuts, cottonseed, tree nuts, wheat, rye, barley, and rice.

[8]An especially good reference on this subject is "Aflatoxin and Other Mycotoxins: An Agricultural Perspective;" Council for Agricultural Science and Technology Report No. 80, December 1980.

Poisonings of humans and animals by mycotoxins have been recorded for more than 5000 years. Documentation in the early 1960s that mycotoxins in feed killed over 100,000 turkeys and 14,000 ducks in Great Britain stimulated substantial recent scientific investigation in this important area.

Of the more than 100 mycotoxins, the aflatoxins are of the greatest concern. They are produced by two fungi that occur worldwide in many agricultural commodities. Fungus growth and aflatoxin production are favored by the warm temperatures and high moistures typical of tropical and subtropical areas, including the southern United States. Aflatoxins are found frequently in corn, peanuts, cottonseed, and tree nuts and also in animal products, such as milk and milk products and certain meat products of cows and swine that have ingested contaminated feed. Significant losses from aflatoxin contamination in recent years have occurred in the peanut, corn, cottonseed, poultry, cattle, and swine industries. Considerable progress has been made in controlling the contamination of foods and feeds and in detoxifying contaminated products. Ammonia is particularly valuable as a detoxifying agent.

Aflatoxin B_1 has adverse biological effects, including liver damage, reduced growth rate, and immunosuppression, when the mycotoxin is present at levels approximating 1 ppm in the diets of most domesticated and experimental animals; liver damage has been observed in humans. Aflatoxins seem to produce both acute and delayed effects in humans. Acute toxicity effects of aflatoxin poisoning occurred in one area of India in 1974 when 106 persons died and 297 others became ill from eating corn contaminated with aflatoxins (6 to 15 ppm).

Aflatoxins have been established as a cause of liver cancer in certain animals. Aflatoxin B_1 is the most potent naturally occurring cancer-producing substance known. Less than 1 ppb in the diet is sufficient to produce significant incidence of liver cancer in rainbow trout, the most sensitive cancer test animal known. Although evidence from some developing countries showing an increase in the incidence of liver cancer with an increase in the aflatoxin content of the diet tends to implicate aflatoxins as a cause of human liver cancer, there is no evidence that aflatoxins are a significant cause of human liver cancer in the United States. For example, the per capita aflatoxin intake in the southeast is approximately 9 times greater than that in the United States as a whole, but the proportion of total deaths due to liver cancer is lower in the southeast than in the country as a whole.

Several mycotoxins are formed by *Fusarium* fungi. Their production is favored by high moisture and temperatures. Although corn is the crop most affected in the United States, *Fusarium* mycotoxins are produced on several other cereal grains. One of these compounds, zearalenone, is hormonally active (estrogenic) in swine and certain other animals; it is an important cause of infertility in swine. Other *Fusarium* mycotoxins cause necrosis and ulceration of the mouth, stomach, and small intestine, accompanied by diarrhea and hemorrhages. These mycotoxins may cause death in cattle and swine. Vomit toxins cause feed rejection and vomiting in swine. Prompt drying of cereal grains after harvest and storage at low moisture contents are good control measures.

Ochratoxin A is produced by several species of fungi throughout the world. This toxin has been associated with kidney disease of swine in Scandinavia and is believed to cause a kidney disease in humans in the Balkans. Annual losses in swine production from this toxin in Denmark are estimated at several million dollars.

Staggers syndromes are nerve diseases of domestic animals, especially cattle, that are of known or suspected mycotoxin origin. They are characterized by muscle tremors; hyperexcitability; unsteady, staggering movements; and inability to walk or stand. Leukoencephalomacia of horses results from a specific brain lesion caused by consuming *Fusarium moniliformen* contaminated corn. Affected horses stagger, act sleepy, and often die. Slaframine is a mycotoxin produced by the fungus that causes black patch disease of red clover. Animals ingesting slaframine-contaminated red clover produce excessive amounts of saliva; hence the designation "slobbers" syndrome.

Ergot and its effects are mentioned in the Bible. It is produced by several species of fungi that infect the blooms and replace the seed with ergots in more than 1000 species of the grass family, especially the cereal grains. Ergotism caused thousands of human deaths in the Middle Ages and five deaths in France as recently as 1951. The occurrence of ergotism in humans is rare today, but it is observed occasionally in food-producing farm animals. Manifestations of ergotism include nervous symptoms and convulsions in most animals and gangrene in cattle and humans. The incidence of ergot is greatly reduced and/or controlled by a number of agricultural practices, including seed treatment and selection and breeding for ergot resistance.

18.7 GOVERNMENTAL SAFEGUARDS FOR ANIMAL AND HUMAN HEALTH

18.7.1 U.S. Food and Drug Administration (FDA)

From 1880 to 1900, numerous articles and books were published alerting the public to the hazards of unsanitary food production practices. This prompted the United States Congress to enact, in 1906, the Pure Food and Drug Law. Dr. H. W. Wiley led the crusade for this law and was made head of the Bureau of Chemistry (USDA), which was charged with enforcement. In 1956 a United States postage stamp bearing his picture was issued, commemorating the golden anniversary of this important event. Concurrently, the Federal Meat Inspection Act was passed, and both laws became effective in 1907. The U.S. Food and Drug Administration (FDA) was established as a separate unit of the USDA in 1927. Then, in 1940, the **FDA** was transferred from the USDA to the Federal Security Agency, which later became the Department of Health, Education and Welfare and today is the Department of Health and Human Services. The foremost mission of the FDA is to safeguard American consumers against injury, unsanitary foods, and fraud. It also protects industry against unscrupulous competition. In addition to inspection and sample analyses, it conducts independent research on such things as toxicity in laboratory animals, disappearance curves for **pesticides,** and long-range effects of drugs.

18.7.2 U.S. Department of Agriculture (USDA)

The Food Safety and Inspection Service of the USDA is charged with maintaining the wholesomeness and safety of meats being processed in packing plants that are involved in the **inter**state shipment of meat and meat products (including inspection of poultry and poultry products). Veterinarians and trained lay inspectors examine each animal carcass and its internal organs before it can be sold for human consumption. Rejected carcasses are consigned to rendering vats and made into tankage (an animal protein **supplement**) or fertilizer. Federal legislation now requires that states inaugurate inspection programs for meat intended for *intrastate* as well as *interstate* commerce.

The Food and Drug Administration requires pesticide manufacturers to present new products with their proposed labels for approval before they are authorized to sell them. The label must indicate, as a minimum, the name of the product; its active and inert ingredients, together with the percentage of each in the formulation; the directions for use; the pest(s) controlled; the method and rate of application; and any restrictions to be observed in application and handling. If an antidote is known, it must be given. The FDA also sets legal tolerances for pesticides on or in raw agricultural commodities. Moreover, it sets the "safe" intervals between last application of the **insecticide** and the time of harvest or slaughter of the crop or animal. Through the cooperative supervision of the FDA and USDA, then, the pesticide user and the consumer of the product are safeguarded (Sections 19.6.1 and 19.6.2).

The Veterinary Services Division (VSD) of the Animal and Plant Health Inspection Service (APHIS) of the USDA is responsible for programs to control and eradicate, if possible, certain diseases of livestock (e.g., brucellosis, tuberculosis, scabies, and screwworms). The VSD conducts nationwide state–federal cooperative programs for control and eradication of animal diseases, suppresses spread of disease through control of interstate and international movements of livestock, keeps informed of the overall animal disease situation nationally and internationally, administers laws to ensure humane treatment of transported livestock and certain laboratory animals, collects and disseminates information on morbidity and mortality, and provides training for USDA employees and others in related government agencies.

18.7.3 U.S. Department of Health and Human Services (USDHHS)

The U.S. Public Health Service (USPHS) section of the USDHHS is largely concerned with the prevention and treatment of disease in humans. It functions in vector control, prevention of water pollution and pollution abatement, and control of communicable diseases. It traces the causes and sources of epidemics.

A part of this important complex is the National Institutes of Health (NIH), organized in 1930 and composed of eleven sister institutes: the National Cancer Institute; the National Heart, Lung, and Blood Institute; the National Institute of Allergy and Infectious Diseases; the National Institute of Arthritis, Diabetes, Digestive, and Kidney Diseases; the National Institute for Dental Research; the National Institute of Aging; the National Institute of Environmental Health Sciences; the National Eye Institute; the National Institute of Neurological and Communicative Disorders and Stroke; the National Institute of Child Health and Human Development; and the National Institute of General Medical Sciences. Aside from its own research endeavors, the USPHS provides funds for health-related research at many universities and research institutes in the United States.

18.7.4   Local and State Agencies

These agencies cooperate with the foregoing federal agencies in an effort to safeguard animal and human health.

18.8   PROTECTING UNITED STATES LIVESTOCK FROM FOREIGN DISEASES

Today, animal disease can move like lightning. Distance no longer provides a buffer against invasion. A jet air transport can outpace the development of **clinical** signs in an animal that has been exposed to a disease just prior to shipment. (More than 90 percent of animals imported into the United States from overseas countries enter by air.) **Epizootic** diseases capable of debilitating or destroying livestock populations still exist in Asia, Africa, and Latin America, where animal diseases and pests constitute the most important limitation to animal production. The most important epizootic diseases are rinderpest, contagious bovine pleuropneumonia, foot-and-mouth disease, African horse sickness, African swine fever, Newcastle disease, fowl plague, trypansomiasis, East Coast fever, and piroplasmosis. These, as well as Rift Valley fever, are potential hazards to the health of United States livestock.

The movement of foreign livestock into the United States was free and easy until 1875, at which time the United States prohibited the importing of cattle and hides from Spain, where foot-and-mouth disease was rampant. By the 1880s, the situation had reversed itself. European countries were refusing to buy United States cattle or beef for fear of spreading contagious bovine pleuropneumonia. In 1884, Congress established the Bureau of Animal Industry in the Department of Agriculture and gave the Secretary of Agriculture authority to enforce **quarantine** laws. Today there are stations at 68 entry points, where inspectors of the USDA's Veterinary Services Division inspect all animals and poultry to be imported into the United States (Figure 18.14). If no evidence of communicable disease is found, the animals may then be quarantined for a period of time, and

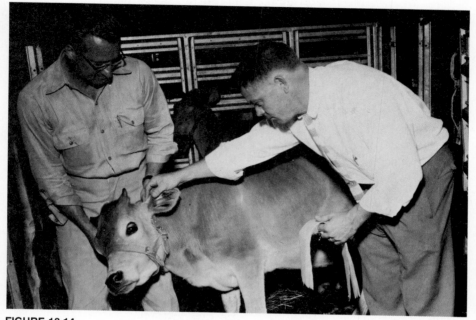

FIGURE 18.14
USDA veterinarian checks ear tags of calves from Channel Islands. *(Courtesy Veterinary Services Division, USDA.)*

if no communicable diseases appear during this period, they are released to the purchaser. Animals under quarantine are held in strict isolation. USDA veterinarians give them precautionary treatment against external parasites and subject them to various tests (e.g., blood test for glanders in horses and tests for tuberculosis and brucellosis in cattle).

A new quarantine center, named the Harry S. Truman Animal Import Center, located on an island abutting Key West, Florida, was opened in 1980. The center is used to quarantine and clear breeding stock imported from countries otherwise prevented from shipping cloven-hooved animals to the United States because of the presence of foot-and-mouth disease. It is administered by the USDA's Animal and Plant Health Inspection Service (APHIS).

The Veterinary Services Division is also responsible for maintaining very stringent safeguards on the entry of zoo animals into this country (Figure 18.15). Wild animals may be brought into the United States only after extensive quarantine periods abroad and a further quarantine period at the animal quarantine station at Newburgh, New York. Moreover, they are allowed to go only to approved zoos, in which animals are isolated from **domestic** livestock and where proper measures are taken to dispose of waste to prevent the spread of diseases.

FIGURE 18.15
Veterinary Services Division checks health of zoo animals; giraffe is unloaded from ship (left); after quarantine, white swans from the Netherlands are recrated for release to a bird farm (below); quarantine zebras are sprayed (right). *(Courtesy Veterinary Services Division, USDA.)*

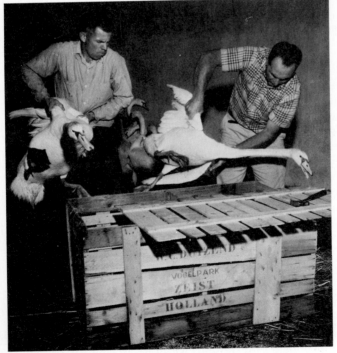

18.9 SUMMARY

Even before recorded history, people probably compared diseases of humans with those of animals. However, it has been only relatively recently that **epidemiological** links between diseases of humans and animals have been revealed. In this chapter, an attempt has been made to give a panoramic view of the relation of animal health to public health. A discussion of selected arthropod-borne diseases was also included.

It is possible that most infectious and parasitic diseases of humans originated in animals. The constant challenge of nature to animal health often is also a challenge to the health of people. Several **zoonoses** (anthrax, plague, rabies, and bovine tuberculosis) are nearly equally harmful to humans and animals. Others (brucellosis, Q fever, and hydatidosis) cause more serious illness in people than in animals. A third group (foot-and-mouth disease, pasteurellosis, and pseudo-rabies) is less harmful to humans but serious in animals. More than 200 diseases can be transmitted from animals to people. An understanding of their modes of transmission and control enables humans to associate safely with animals and to utilize the products and services that they provide for humanity.

The universal insidiousness of the **infections** and **infestations** in the wide and somber spectrum of animal parasitism alone constitutes a major concern and a challenge to modern agriculture and veterinary medicine. The diseases of humans and animals and those common to both constitute a serious impediment to the increases in production that are so vitally necessary to the world of today and tomorrow. An estimated 1.6 billion pounds of animal protein, which might be used to improve human **nutrition,** are lost each year as a result of animal diseases. The value of this meat in the marketplace is unrealized, and the labor and capital invested in herds and flocks later lost to disease are wasted. As world food needs mount, it becomes imperative for humans to exercise improved control over animal health, which will greatly reduce morbidity and mortality losses of farm animals and greatly increase quantities of animal foods useful to humanity. Based on FAO estimates, "a 50 percent reduction of losses from animal diseases in the developing countries, which is a realistic goal, would result in a 25 percent increase in animal protein."

STUDY QUESTIONS

1 What is meant by public health?
2 Differentiate between infectious and noninfectious diseases.
3 How are infectious diseases of farm animals spread?
4 How do pathogens gain entrance into an animal's body?
5 What is the body's first line of defense against pathogens?
6 What are the four cardinal signs of an inflammatory reaction?
7 What stimulates the production of antibodies in an animal?
8 Differentiate between natural and acquired resistance to disease. What is meant by species immunity? Give two or more examples of genetic resistance to disease.

9 What is a vaccine? Give one or more examples of the different types of vaccines.
10 Define prophylactic. Define therapeutic.
11 Which mammals are susceptible to rabies?
12 How can ornithosis be transmitted to humans?
13 What significant contribution did Jenner make to the health of humanity?
14 Can foot-and-mouth disease of livestock be transmitted to humans?
15 Cite examples of some common vectors and some carrier animals of rickettsial diseases that may be transmitted from animals to humans.
16 Why is vaccination of humans and animals against bovine tuberculosis not advisable in the United States?
17 Why has the cooking of garbage intended for swine food aided in reducing the prevalence of swine tuberculosis in the United States?
18 Name three species of *Mycobacterium* and three species of *Brucella* that are of public health significance.
19 For what are the following used: *(a)* market-cattle test program, *(b)* brucellosis ring test, and *(c)* brucellosis card test?
20 Why is *undulant fever* an appropriate name for brucellosis in humans?
21 How may a person acquire leptospirosis?
22 What was the first disease of animals and humans to be traced to a microorganismic source?
23 What was the first disease against which a bacterial vaccine was found to be effective?
24 Is the reported occurrence of salmonellosis increasing or decreasing in the United States?
25 Does hog cholera have public health significance?
26 Is glanders a serious public health threat in the United States? Why?
27 Is malaria a current threat to the health and well-being of humans? What is the best way to prevent malaria?
28 What important disease of humans is spread from host to host by the bloodsucking tsetse fly?
29 How may anaplasmosis be spread?
30 What are *(a)* actinomycosis and *(b)* histoplasmosis? Do these diseases affect people?
31 Trace the life cycle of the trichina worm *(Trichinella spiralis).*
32 Trace the life cycle of the beef tapeworm *(Taenia saginata).*
33 What is the effect of quick-freezing and thorough curing or salting of meat on pork and beef tapeworm larvae?
34 Trace the life cycle of sheep liver fluke disease (fascioliasis).
35 Name three intestinal parasites that humans may acquire from eating improperly cooked fish.
36 Trace the life cycle of the dog hookworm *(Ancylostoma caninum).*
37 Differentiate between cutaneous and visceral larva migrans.
38 Trace the life cycle of the swine roundworm *(Ascaris suis).*
39 Name three or more viral and one or more bacterial infections of humans that may be transmitted to animals.
40 How may animals be passive carriers of diseases? Cite three or more examples of public health importance.
41 What is meant by an LD_{50} value?
42 Cite several examples of plants that may accumulate cyanide or nitrate. Are all of these wild plants? Of what significance are mycotoxins to animal and human health? Cite examples.

43 Cite an example of a plant that interferes with the metabolism of certain nutrients.
44 Of what public health significance are the poisons of the white snakeroot plant, jimmyweed, or rayless goldenrod when consumed by lactating cows? Cite a historical example.
45 Where would one write to obtain information pertaining to the poisonous plants of a given state? *(Hint:* See Appendix D.)
46 What effect does overgrazing of pastures often have on the populations of undesirable plants?
47 What is coumarin? Is dicoumarin of importance to livestock feeders? Why?
48 What are the primary missions of the U.S. Food and Drug Administration?
49 What role does the USDA play in providing wholesome and safe meats for humans?
50 What governmental agency is responsible for the labeling of pesticides? For setting legal tolerances for pesticides on or in raw agricultural commodities?
51 What is the function of the U.S. Public Health Service in the prevention and treatment of disease? Name the sister institutes of the National Institutes of Health.
52 What is meant by epizootic diseases? Cite some important examples.
53 Briefly discuss controls and safeguards on the movement of livestock and zoo animals into the United States.
54 Cite several zoonoses that are nearly equally harmful to humans and animals; cite some that cause more serious illness in humans than animals; and finally, cite examples that are less harmful to humans than animals.
55 Why is it imperative for humans to exercise improved control over animal health throughout the world?

SELECTED INSECTS AND PARASITES OF SIGNIFICANCE TO HUMANS AND ANIMALS[1]

Insects are our dangerous rivals for the good supplies of the world, and they are important rivals and enemies in many other ways.

Leland Ossian Howard (1857–1950)

19.1 INTRODUCTION

In no branch of the animal kingdom did nature provide such extremes of color, shape, and size as in **insects.** They range in size from tiny beetles that can literally creep through the eye of a needle to the well-camouflaged tropical walkingsticks, which are often more than a foot in length. Their colors represent all tints of the rainbow. Some crawl slowly, whereas others can fly very fast (dragonflies have been clocked at 60 mph).

There are more species of insects than of any other form of animal life. However, according to a survey conducted by the **USDA,** less than 0.04 percent of the known insects are harmful to humans. There are approximately 600 injurious **species** of insects of primary importance in North America.[2] Even this relatively small number cause considerable damage, and even death (directly or indirectly), to humans. During the Civil War, more soldiers died of typhoid fever (transmitted by houseflies) than were killed in combat. During World War II,

[1]The authors acknowledge with appreciation the contributions to this chapter of Dr. W. R. Enns and Dr. R. D. Hall, Department of Entomology, University of Missouri-Columbia.
[2]"Insects," *USDA Yearbook,* 1952.

596

Allied armies in the Pacific theater had 5 times as many casualties from malaria (transmitted by mosquitoes) as from combat. Each American pays several dollars more for cotton goods because of boll weevils. (Estimated annual damage is $200 million, or about 1 of every 7 bales produced.) The annual cost of controlling insects in the United States, when added to their damage to crops and **livestock,** is estimated by the USDA to be more than $1600 million.

Because insects have six legs, they are classed as *Insecta* (some authorities also use the term *Hexapoda,* meaning "six-legged"). True insects have three body segments: (1) the *head,* which contains the brain, antennae (feelers), eyes, and mouthparts; (2) the *thorax,* or "motor room," which holds the chief muscles used in flying, swimming, or walking (the legs and wings are always attached to the middle body segment); and (3) the *abdomen,* which contains the digestive, reproductive, and excretory organs. Because insects have no internal skeleton, their shell or outside covering is composed of a tough or horny material that usually contains some percentage (0 to 60 percent) of a substance called *chitin* (an insoluble polysaccharide). In the growing process, insects shed this outer skeleton several times, each time forming a new one. This process is called *molting.* The young insects between **molts** are called **instars,** and the periods of time between molts are called *stadia.*

Insects respond to the environment as cold-blooded animals. Therefore their activity is governed largely by air temperature. Scientists have demonstrated that one can determine the room temperature by counting the number of chirps certain insects make per minute and applying a simple mathematical formula. A scientist at Harvard University observed that he could determine the **ambient temperature** by timing the speed of ants running along a mountain trail. Since insects are invertebrates, they possess no spinal cord. Instead, they have *ganglia* (knots of nerves) spaced along paired nerve cords on the **ventral** side of their bodies. Each ganglion controls certain phases of the insect's activity and is often capable of functioning independently. Hence a **decapitated** ant may survive for several weeks.

Not all winged insects have the same number of wings. True flies have two wings; bees, butterflies, dragonflies, moths, and wasps have four. Insects can taste and smell. The sense of taste of a monarch butterfly for certain sweets is 1200 times as sensitive as that of humans. The *antennae* are used to detect odors and sounds. Some male moths can fly for miles in the dark by following scent trails through the air, thereby to discover females. Many insects have effective sound-transmitting devices. (Some have been known to transmit sounds for a distance of a mile.) Those that undergo a *complete* **metamorphosis** have four stages: (1) egg, (2) **larva,** (3) **pupa,** and (4) adult. The branch of biological science that deals with the study of insects is called *entomology.*

Dogs were among the first animals **domesticated** by humans (Chapter 1). The first domesticated dog probably brought with it a familiar **parasite,** the flea, which was recognized then as a nuisance and later as a **vector** of **disease.** Fossils indicate that fleas of the same form, size, and structure as those of today were living on ancestral dogs some 30 million or more years ago in the Oligocene epoch of the Tertiary period. From minute mice to enormous elephants, all animals are plagued by parasites. A *parasite* is defined as an organism living in or on, and at the expense of, another **organism** (the host) while rendering nothing in return. Parasites "steal" food and shelter from other animals and thereby sap the strength and health of their **hosts.** The study of parasites (both **microscopic** and **macroscopic**), including their life histories, modes of transfer from one host to another, and effects on the host, constitutes the discipline known as *parasitology.* The fields of entomology and parasitology often overlap. Entomologists may specialize in the study of insect species that are parasites of humans and domestic animals, or they may study the parasites that affect (and in some instances offer control of) these pest insects.

19.2 TAXONOMY

Before reviewing selected insects and other parasites of importance to the health and well-being of humans and animals, it seems desirable to classify animals briefly according to their presumed natural relationships (Table 19.1). The seven basic levels in classification are

Kingdom
Phylum
Class
Order
Family
Genus
Species

Genus and species names are italicized when they appear in print. When typed or handwritten, they should be underlined. The combination of a genus and species name is termed a *binomen,* and it provides a unique description of each animal species. No two genera in the animal kingdom may have the same name; however, this rule does not apply to specific names in separate genera. Therefore, to properly designate an animal by its scientific name, both the genus and species must be given. Generic names always begin with a capital letter and specific names never do. The binomen is often followed by the name of the person who first described that particular species. If the describer's name appears within parentheses, it means that the species was originally described in a genus different from the one in which it is currently placed.

The description and naming of animal species forms an important part of the work done by *taxonomists.* Many entomologists are engaged almost exclusively in this type of endeavor. The application of names to animal species is governed by an extensive set of regulations known as the International Code of Zoological Nomenclature, which is used by scientists throughout the world. Irregardless of native language, nationality, or ideology, zoological nomenclature is one of the few areas where worldwide cooperation exists.

TABLE 19.1
THE ANIMAL KINGDOM

Phylum*(selected classes of the subphylum Vertebrata)	Examples (common names)	Estimated no. of living species described†
Chordata		
Amphibia	Frogs, salamanders, toads	2,000
Aves	Birds, fowls	15,000
Mammalia	Humans, bats, cats, cattle, horses, whales	10,000
Pisces (Osteichthyes)	Bony fishes	25,000
Reptilia	Alligators, lizards, snakes, turtles	6,000
Subtotal		58,000
Arthropoda	See Table 19.2 for classes and examples	973,000
Mollusca	Clams, mollusks, oysters, slugs, snails	100,000
Echinodermata	Feather stars, sand dollars, sea urchins, starfish	5,800
Annelida (Annulata)	Earthworms, leeches (segmented worms)	8,000
Bryozoa (Polyzoa)	Moss animals, sea mats	4,000
Brachiopoda	Lampshells	500
Nemertinea	Nemertines, ribbon worms	600
Nemathelminthes (Aschelminthes)	Filariae, roundworms, trichinae, wheel animalcules	12,000
Platyhelminthes	Flatworms, flukes, tapeworms	15,000
Ctenophora	Comb jellies, sea walnuts	100
Coelenterata	Coral animals, hydras, jellyfishes	10,000
Porifera	Sponges	5,000
Protozoa	Amoebae, euglenas, malarial organisms, paramecia, trypanosomes	30,000
Grand total		1,222,000

*Not all phyla are listed (only the most important ones).
†These estimates represent the mean of several references.

19.2.1   The Five Phyla of Foremost Importance in Animal Science

1 Protozoa (amoebae, ciliates, and flagellates) can be disease agents. Internal parasites impair or destroy tissues, organs, and/or systems. Certain **protozoa** cause malaria, the loss of red blood cells, and damage to the **reticuloendothelial** system. Others cause anaplasmosis in cattle, which results in the destruction of red blood cells and consequently in **anemia.** Still another protozoan causes trypanosomiasis in humans, which is characterized by **fever,** anemia, **erythema,** and often damage to the central nervous system (Chapter 18).

2, 3 Platyhelminthes (flukes and tapeworms) and Nemathelminthes (nematodes and filariae) may cause disease (Chapter 18). These internal parasites impair the function of organs and/or systems. Essentially, all tissues and organs are subject to their damage by blood loss, mechanical attack, toxic effects, **allergic** reactions, and **nutrient** losses; from egg laying in tissues; and by predisposing the mucosa to bacterial invasion.

4 Arthropoda (insects, mites, and **ticks**) are important as internal and external parasites causing *direct* effects such as irritation and worry, blood loss, **envenomization, dermatosis,** allergy, **myiasis,** and other related **infestations** or having *indirect* effects as vectors (mechanical or biological) or as **intermediate hosts** for bacteria, protozoa, **rickettsiae, viruses,** and **helminthes.**

The phylum Arthropoda includes the invertebrates that are bilaterally symmetrical (i.e., when the body is cut exactly in half lengthwise, the right half

TABLE 19.2
THE PHYLUM ARTHROPODA*

Classes	Examples (common names)
Insecta	All true insects. The class Insecta includes members of the phylum Arthropoda that, as adults, have three body regions (head, thorax, and abdomen), one pair of antennae, three pairs of jointed legs, and usually two pairs of wings.
Selected orders	
Anoplura	Bloodsucking lice (e.g., cattle lice and swine lice)
Coleoptera	Beetles, weevils
Diptera	Flies, gnats, midges, mosquitoes
Hemiptera	True bugs (e.g., bedbugs and stinkbugs)
Hymenoptera	Ants, bees, wasps
Isoptera	Termites
Mallophaga	Biting lice (e.g., bird lice)
Orthoptera	Cockroaches, crickets, grasshoppers, katydids
Siphonaptera	Fleas
Chilopoda	Centipedes
Diplopoda	Millipedes
Arachnida	
Selected orders	
Acarina	Mites and ticks
Araneida	Spiders
Phalangida	Harvestmen, or daddy longlegs
Scorpionida	Scorpions
Minor orders	Pseudoscorpions, whip scorpions
Crustacea	Barnacles, crabs, crayfish, cyclops, lobsters, sow bugs, water fleas
Minor classes	Bear animalcules, king crabs, pauropods, peripatuses, sea spiders, symphylans

*Adapted from "Insects," *USDA Yearbook,* 1952, which defines arthropods as jointed-legged invertebrate animals.

forms a mirror image of the left, and vice versa), with segmented bodies and appendages. Selected examples within the phylum Arthropoda are given in Table 19.2.

5 The phylum Chordata, which includes humans and domestic animals, is important in animal science because its constituents may serve as hosts, intermediate hosts, or **reservoirs** for many parasites that cause diseases of humans and animals. Selected external and internal parasites and their significance to humans and/or animals are presented in Tables 19.3 and 19.4.

19.3 CONTRIBUTIONS OF INSECTS TO HUMANITY[3]

The class Insecta is the largest of all classes of animal life. In fact, of every five described (known) kinds (species) of animals in the world, approximately four are true insects. Fortunately, only a relatively small number of these known insects are "harmful" to humans. The others are either "beneficial" or have no known economic or health importance. It would be remiss not to mention some ways in which insects are useful to humans.

[3]Portions of this section were adapted from C. L. Metcalf, W. P. Flint, and R. L. Metcalf, *Destructive and Useful Insects,* 4th ed., McGraw-Hill Book Co., New York, 1962.

TABLE 19.3
SELECTED EXTERNAL PARASITES OF HUMANS AND/OR ANIMALS AND THE DISEASE OR CONDITION EACH MAY CAUSE

Class	Disease or condition	Host(s)
Insecta		
Chewing lice	Irritation	Domestic animals, including poultry
	Intermediate host	Dog (dog tapeworm)
Sucking lice	Blood loss, irritation	Humans, domestic animals
	Typhus, trench fever, relapsing fever	Humans
Mosquitoes	Malaria, dengue, filariasis, yellow fever, **encephalitis**	Humans
Biting flies: stable flies, horn flies, horseflies, blackflies	Blood loss, irritation, vectors, intermediate hosts	Humans, domestic animals
Arachnida		
Mites	Itch or scab, mange, blood loss, irritation	Humans, cattle, horses, sheep, poultry
Ticks	Blood loss, irritation	Humans, domestic animals
	Anaplasmosis	Cattle
	Rocky Mountain spotted fever and tularemia	Humans
	Equine piroplasmosis	Horses

TABLE 19.4
SELECTED INTERNAL PARASITES OF HUMANS AND/OR ANIMALS AND THE DISEASE OR
CONDITION EACH MAY CAUSE

Phylum	Disease	Host(s)	Infection
Protozoa	Malaria	Humans	*Anopheles* mosquitoes
	African sleeping sickness	Humans, cattle	Tsetse flies
	Anaplasmosis	Cattle	Humans, ticks, insects
	Trichomoniasis	Cattle, humans, swine	By contact
Platyhelminthes	Schistosomiasis (lung and liver flukes)	Humans (snails are intermediate hosts)	Skin contact with contaminated water
	Tapeworms	Humans, domestic animals (arthropods often intermediate hosts)	Ingestion of intermediate host or (rarely) the eggs
Nemathelminthes	Hookworms	Humans, dogs	Through the skin
	Ascaridiasis	Humans, swine	Ingestion
	Trichinosis	Humans, rats, swine	Ingestion of raw pork
Arthropoda	Internal myiasis, bots, warbles	Humans, cattle, horses	Direct (egg laying)
	Traumatic myiasis (screwworms)	Humans, cattle, horses, sheep	Direct (egg laying) deposition of larvae

Insects produce and collect many products or articles of commercial value. Many *secretions* of insects are valuable; e.g., a silkworm secretion is the true natural silk. The silk industry began in China, where the source of silk was kept secret for more than 2000 years. Attempts to take silkworm eggs out of the country were punishable by death. Beeswax is secreted as thin scales, or flakes, by the hypodermal glands on the underside of the worker bee's abdomen. Bees use wax to make the comb in which honey is stored and the young are reared. Shellac is secreted from the hypodermal glands found on the back of a scale insect of India. The light-producing secretions of giant fireflies in the tropics are used in minor ways for illumination and could lead to the synthesis of a substance that would give a brilliant light with almost no accompanying heat.

The *bodies* of some insects are useful or may contain certain useful compounds. Cochineal and crimson lake are pigments made by drying the bodies of a cactus scale insect found in the tropics. These pigments are used in cosmetics, as a coloring for beverages, in decorating cakes and pastries, and as a dye for textiles. Cantharidin is a dangerously irritating substance obtained from the dried bodies of a European blister beetle, commonly called *Spanish fly*. Insects such as the dobson, or hellgrammite, are widely used as fish bait, and the best artificial flies used for fishing are modeled after insects. The bodies of insects serve as food for many animals that are valuable to humans. Many fish used for food subsist largely on aquatic insects. Many highly prized song and game birds depend on insects for a large portion of their food. Chickens and turkeys feed on insects naturally and, under certain conditions, can be raised almost exclusively on a diet of insects. Swine may feed and fatten on white grubs rooted from the soil. Several wild animals, e.g., moles, raccoons, shrews, and skunks, eat many insects. In many regions of the world, from ancient times to the present, insects have been and are eaten extensively by humans. Grasshoppers (called *locusts* in the Bible), crickets, walkingsticks, certain beetles, caterpillars, and pupae of moths and butterflies, termites, large ants, aquatic bugs, cicadas, and honeybee larvae and pupae are foods that were highly prized by many of the more primitive races of humans. The Australian aborigines are cited as an example of a living population whose diet includes insects.

Insects *collect, elaborate,* and *store* valuable plant products. Honey is nectar that is collected from plant blossoms, concentrated, modified chemically, and sealed in waxen "bottles" by the honeybee (Chapter 3). Insects cause plants to produce galls, some of which are valuable. They are especially useful as tanning materials and medicines. The Aleppo gall produced by a wasplike insect on several species of oaks in Asia and Europe has been used for centuries as a tonic, astringent, and antidote for certain poisons. The early Greeks used it for dyeing hides, mohair, and wool. Other galls have been used for dyeing fabrics and as tattoo dyes. The Aleppo dye is used in preparing the finest and most permanent inks.

Insects aid in the production of flowers, fruits, seeds, and vegetables by *pollinating* the blossoms as they search for nectar or gather pollen. It is estimated that if all bees were eliminated, some 100,000 species of flowering plants would concurrently cease to exist. The Smyrna fig produces only female flowers and no pollen. Therefore the growing fig is dependent on small wasplike insects from another fig (the caprifig) which crawl into its flower cluster and thereby serve as pollinators. Most varieties of apples, sweet cherries, and plums would be barren without insect pollinators such as bees, butterflies, flies, moths, and wasps. Clover seed does not form without the visit of an insect (usually some kind of a bee, such as the long-tongued bumblebee) to each blossom. When the plant was first introduced in New Zealand, it failed to produce seed until bumblebees were introduced to transport the pollen from flower to flower. Certain beans, melons, squash, and many other vegetables require insect visits before the blossoms set (form fruit in the blossom). Many ornamental plants, e.g., chrysanthemums, iris, orchids, and yucca (both in and out of greenhouses), are pollinated by insects.

Many insects *destroy* other injurious insects. This is accomplished by their action as *parasites*—living on or in their bodies and/or their eggs, or as **predators**—capturing and devouring other insects. The praying mantis (so named because of the position it assumes as it rests on twigs or stalks its prey) is a good example of the latter. It eats a large number of insects, ranging in size from tiny aphids to large caterpillars. Some insects have even been imported from other countries to halt the spread of a serious insect pest. A classic example is the importation of the Australian ladybird, a beetle that was imported to save the citrus industry of California from the cottony-cushion scale insect. When 140 of the beetles reached San Francisco, they devoured the insects at a fast rate and thereby preserved the lemon and orange industry. Further discussion of this subject as it relates to **biological control** of insects is presented in Section 19.6.2.

Many insects *destroy* harmful weeds in the same way that they injure crop plants. Control of prickly pear by insects in Australia is an example of what can be accomplished in this way. In about 1787, cactus plants were taken to Australia for culturing cochineal insects for dye. Various species of cacti took root beyond the gardens, and in the absence of natural enemies, the prickly pears spread rapidly. Within a century, about 60 million acres were affected (one-half so densely covered that the land was rendered useless). Then more than 500,000 insects of 50 different species were dispatched from North and South America to Australia. Several were successfully established. The insects checked the new growth of cactus and reduced the density of the plants, thereby allowing grass to return. Recently, certain beetles have been released in the United States in a similar program against thistle plants.

Insects *improve* the physical condition of the soil by burrowing throughout the surface layer (allowing air to penetrate the soil) and *enhance* its fertility through the fertilizing properties of their droppings and dead bodies. Insects perform a valuable scavenger service by devouring dead plants and the bodies of animals and by burying **carcasses** and **dung.**

Certain insects are *invaluable* in scientific endeavors. The ease of handling, rapidity of multiplication (short **generation interval**), great variability, and low cost of rearing and maintaining make certain insects valuable in the study of biochemistry, **ecology,** genetics, and **physiology.** Studies of variation, geographical distribution, and the relation of color and pattern to surroundings have been greatly advanced by studying insects. Many genetic fundamentals were derived from studies of *Drosophila* (pomace flies, or fruit flies). Principles of regeneration and **parthenogenesis** have been established through the study of insect physiology. The behavior and psychology of higher animals have been illuminated by studies of the reactions of insects such as the honeybee, whose behavior can be analyzed into very simple **tropisms.** Valuable information in sociology has been deduced from a consideration of the economy of social insects. Insects have been used in the **bioassay** of certain crops (Figure 19.1).

Insects have *aesthetic* and *entertainment* value. Their shapes, colors, and patterns serve as models for artists, decorators, florists, and milliners. The more highly colored and striking forms of insects are much used as ornaments in dishes and trays and in necklaces, pins, rings, and other jewelry. Butterflies and moths are admired universally, whereas those using the microscope find much to admire in the colors and patterns of many smaller insects. The songs of insects have been found highly interesting for centuries. Insects have served as subject matter for many poems and songs. The inimitable variety and the curious habits found in insects afford entertainment and diversion for thousands who collect and study them. Many persons have participated in the oriental game of gambling on crickets trained for fighting. Fleas have frequently been used for circus stunts.

Insects are often *useful* in the practice of medicine and surgery. Although not

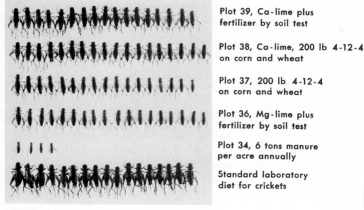

Bioassay of red clover, Sanborn Field
Crickets from 25 eggs per plot
Four-year rotation: corn, oats, wheat, clover
All produce removed, save plot 39 with straws,
stovers, and second clover crop turned under

Plot 39, Ca-lime plus
fertilizer by soil test

Plot 38, Ca-lime, 200 lb 4-12-4
on corn and wheat

Plot 37, 200 lb 4-12-4
on corn and wheat

Plot 36, Mg-lime plus
fertilizer by soil test

Plot 34, 6 tons manure
per acre annually

Standard laboratory
diet for crickets

FIGURE 19.1
Crickets from 25 eggs per plot were used as test animals in the bioassay of red clover on Sanborn Field at the University of Missouri. *(Courtesy Dr. W. A. Albrecht and Missouri Agr. Expt. Station.)*

a particularly pleasant thought, it is a fact that the larvae (**maggots**) of certain flies (reared **aseptically**) have unique value in the treatment of certain wounds. Such maggot therapy is currently being practiced in cases in which sophisticated medical science has failed. In brief, such maggots tend to consume only necrotic (dead) tissue, leaving healthy tissue alone. They also secrete natural antibiotics that speed and enhance wound healing. The stings of honeybees may have remedial value for disorders such as arthritis and rheumatism. Extracts from the bodies of honeybees and certain other insects are used to a limited extent as medicines.

Bees are now being bred with a preference for particular plants. Because bees are responsible for 80 percent of insect pollinating activity, this is an especially important agricultural development. Pollen preferences among bees can determine success or failure for crops such as alfalfa and clover.

19.4   HARMFUL EFFECTS OF INSECTS[4]

The harmful effects of insects have been appreciated since biblical times. The Bible records in Exodus that God commanded Moses to stretch out his hand to make locusts (some commentaries say grasshoppers) come over the whole land

[4]Portions of this section were adapted from C. L. Metcalf, W. P. Flint, and R. L. Metcalf, *Destructive and Useful Insects,* 4th ed., McGraw-Hill Book Co., New York, 1962.

of Egypt. This infestation was designed to intimidate Pharaoh, who had rebelled against God. According to the scripture,

> The locusts went up over all the land of Egypt . . . so that the land was darkened; and they did eat every herb of the land, and all the fruit of the trees . . . and there remained not any green thing in the trees, or in the herbs of the field, through all the land of Egypt.

> *Exodus 10:14–15*

Insects *damage* all kinds of growing crops and other valuable plants. Destruction is accomplished in a number of ways, which include chewing bark, buds, stems, or fruits of the plants; sucking sap from buds, leaves, stems, or fruits; and boring or tunneling into the bark, stems, or twigs, into fruits, nuts, or seeds ("worms" and/or "weevils") or between leaf surfaces (leaf miners). Their damage takes the form of making cancerous growths on the plants within which they live and obtain nourishment (gall insects); attacking roots and underground stems in any of the foregoing ways (subterranean or soil insects); laying eggs in plants; taking plant components for use in the construction of their nests or shelters; carrying other insects to a plant where they become established; disseminating organisms (bacteria, **fungi,** and viruses) that are pathogenic to plants and injecting them into the plant tissues in the feeding process by carrying them into insect tunnels or by making wounds through which such disease organisms may gain entrance; and causing cross-fertilization of certain disease-producing rusts, whose aecia **(spores)** would not otherwise have developed.

Insects annoy and injure humans and animals in a number of ways. They cause *annoyances* by their mere objectionable presence in places; the sound of their flying or "buzzing" about; the unpleasant odor of their secretions and/or decomposing bodies (the increasing concentration of **organic** materials in the Great Lakes near Chicago has caused a reduction in the oxygen content of the water, resulting in fewer insects making their home in the Great Lakes); or the offensive taste of their secretions and/or excretions left on food or tableware. They irritate by crawling and moving about the skin; chewing, nibbling, or pinching the skin; entering the ears, eyes, nostrils, or alimentary canal, causing **myiasis;** and laying their eggs on feathers, hair, or skin. Some insects *injure* humans and animals by applying **venoms.** They may apply their venoms by means of a stinger and/or pincers; by penetrating with nettling hairs; by leaving caustic or corrosive body fluids directly on the skin when the insect is handled or crushed (e.g., blister beetles); or by poisoning the animal on being swallowed (e.g., blister beetles cause irritation to the digestive tract when swallowed by chickens and may poison other animals when present in stored forage).

Certain insects make their homes on or in the body as external or internal *parasites,* thereby injuring the host animal by causing nervous irritation when they crawl about; causing **inflammation** by chewing or piercing the skin; contaminating feathers or fur with their eggs and **excreta;** sucking blood; tunneling into auricular, muscular, nasal, ocular, or **urogenital** passages, causing mechanical injury and/or promoting infections; and attaching to the stomach and intestinal lining, thus mechanically blocking food passages, disturbing the nutritional processes, causing an ulcerous condition, or secreting **toxins.**

As noted in Table 19.5, insects and other arthropods may disseminate disease agents (bacteria, fungi, parasitic worms, protozoa, or viruses) from diseased to healthy animals, from certain wild animals (the reservoir) to humans or domestic animals, and from a diseased parent or antecedent life stage in one or more of the following ways: accidentally conveying **pathogens** from filth to food; transporting pathogens from filth or diseased animals to the eyes, lips, or wounds of healthy animals; infecting a susceptible large animal when it swallows the insect host of a pathogen; inoculating the pathogen hypodermically as the insect bites the animal; depositing the pathogen directly on the skin or in its **feces** either via its proboscis or dead crushed body, allowing entrance of the pathogen through an insect bite, a scratch, or even the unbroken skin; and finally, by serving as a host in which the pathogen may merely cling to the insect's body, multiply in the insect's internal organs, and undergo within the insect an essential part of its **life cycle** that cannot occur elsewhere.

Insects may *destroy* or depreciate the value of stored products and possessions, including clothing, drugs, food, animal and plant collections, books, papers, bridges, buildings, furniture, mine timbers, telephone poles, telegraph lines, railroad ties, and trestles. Insects accomplish these destructive processes in one or more of the following ways: devouring matter as food; contaminating it with their secretions, excretions, eggs, or even their own bodies (even though the product may not be eaten); seeking protection or building nests on or tunnels

TABLE 19.5
SELECTED DISEASES OF HUMANS TRANSMITTED BY ARTHROPODS

Vector	Disease
Mosquitoes	
Anopheles	Malaria
Culex	Filariasis (elephantiasis)
Aedes	Yellow fever, dengue fever, equine encephalomyelitis
Flies	
Houseflies	Typhoid fever, paratyphoid fever, dysentery, Asiatic cholera
Deerflies	Tularemia
Tsetse flies	African sleeping sickness
Sand flies	Pappataci fever
Lice, *Pediculus humanus*	Epidemic typhus fever, relapsing fever
Fleas	
Rat flea	Bubonic plague
Other fleas of wild rodents	Endemic (murine) typhus fever
Ticks: wood ticks, dog ticks, rabbit ticks	Spotted fever, tularemia, equine encephalomyelitis, Colorado tick fever, relapsing fever
Mites, trombiculids (chiggers)	Tsutsugamushi disease, rickettsial pox

within these various things; and finally, causing increased expenditures for labor in sorting, packing, and preserving certain items, such as food and clothing. Livestock pests lower the production of meat and milk and the value of hides.

19.5 SELECTED ARTHROPODS AFFECTING DOMESTIC ANIMALS AND/OR HUMANS

Among the most important and widespread infectious diseases of humans and the animals that serve humanity are those spread by insects such as fleas in **regurgitated** food and in their feces. Studies have shown that these vectors may prove important in providing knowledge for a means of their control.

19.5.1 Flies

Houseflies   If there is anything "as harmless as a fly," it is not the common housefly. Although dark-gray, dull-looking houseflies seldom attack farm **mammals,** they do annoy them and spread filth. The tiny hairs on the legs of houseflies usually carry millions of bacteria, some of which may be pathogenic and cause such diseases as dysentery and diarrhea. They can carry microorganisms on their footpads, in regurgitated food, and in their feces. Studies have shown that these flies may transport disease-producing organisms as far as 15 to 20 miles. They may place microorganisms inside the human body (through contaminated food) that cause such diseases as cholera, dysentery, tuberculosis, and typhoid.

Whereas most insects have four wings, flies have only two. They thus belong to the order Diptera, which means "two wings." (These two wings are almost as transparent as window glass.) The wing joints vibrate and cause a buzzing sound when the fly beats its wings. However, this sound is partially muffled by the rush of air created by the beating wings. Do you remember how difficult it is to catch or hit a housefly? This is because the fly is equipped with two large *compound eyes* that cover most of its head. These eyes are composed of thousands of tiny jewel-like parts, called *facets,* which enable the fly to see in different directions simultaneously and to become quickly aware of movement. On top of the housefly's head (between the large eyes) are three very small eyes, called *ocelli.* The *antennae* (feelers) are also between the compound eyes.

The sucking mouth tube, called the *proboscis,* is shaped like a funnel turned upside down. The fly uses this proboscis as one might use a straw in a glass of milk. Small solid particles of food (such as sugar) may be ingested by the fly's first dissolving them with saliva. The mouthparts function as a sponge. Since the proboscis is soft, the housefly cannot bite. Tiny pads, called *pulvilli,* are attached to each of the fly's six feet. Tiny hairs cover these pads and secrete a sticky fluid that enables the fly to cling to smooth surfaces and walk upside down on ceilings. The housefly breathes through minute openings, called *spiracles,* which lie in a row on each side of the thorax and abdomen.

A female may lay from 2 to 25 batches of approximately 300 to 500 eggs each during her lifetime of 2 to 4 weeks. Female houseflies prefer to lay their eggs in fresh **manure,** garbage, or fermenting vegetable waste. Tiny, white, legless larvae (maggots) hatch 12 to 48 h after the eggs are laid. Before developing into adults, the larvae mature by feeding for 4 to 14 days on the material about them and then pupate. The control of houseflies is difficult because they breed in so many places. Sanitation, therefore, is the key to their control. Manure must be kept in flyproof areas or spread thinly over fields. Applications of **insecticides** as space or residual treatments are a necessary adjunct to sanitation in controlling the housefly.

Control of houseflies has become increasingly important to livestock producers during recent years. Animal confinement facilities that originally may have existed in remote rural areas have often been surrounded by new residential developments. In such cases, flies originating at the agricultural operation and subsequently annoying homeowners nearby may be sufficient grounds for legal action. Decisions frequently instruct the livestock producer to control such fly populations or risk being driven out of business.

Horn Flies   These small black flies are about one-half the size of houseflies. They feed chiefly on cattle (as many as 4000 horn flies may settle on one animal) but may also attack other mammals. Because they feed only on blood, a stabbing or sucking process must be employed. In cases of heavy infestation, this may cause cattle to lose weight and lactating animals to secrete less milk. According to USDA estimates, horn flies reduce the potential value of cattle and milk production in the United States by $150 million annually. Beef cattle subjected to heavy attacks of horn flies during the fly season were observed to gain 50 lb less than cattle treated with insecticides. Research at the University of Illinois revealed that dairy cows protected from horn flies produced 10 to 20 percent more milk than unprotected cows. The entire adult life of this fly is spent on the host, except for the brief time in which it leaves to deposit its eggs underneath fresh, warm cattle dung. It takes about 1 week for eggs to develop into adults. An application of residual insecticide to the host is an effective way to control horn flies. New insecticide-treated ear tags and various self-applicating devices (back rubbers and dust bags) afford excellent control.

Stable Flies   These flies, often called *biting houseflies,* resemble houseflies in appearance but are actually bloodsuckers; this ability is made possible by a hard, piercing proboscis. The stable fly is known as the *dog fly* in some areas of the United States. Stable flies are annoying to cattle and horses because they frequently bite the legs and lower parts. (They may also feed on humans.) They lay eggs in moist, fermenting straw and manure, in decaying vegetable matter (especially at the base of silos); and frequently on the moist bottom of large, round hay bales stored outside. About 3 weeks is required to develop from egg to adult. Unlike the horn fly, the stable fly stays on an animal only long enough to engorge itself with blood. During warm weather, stable flies may feed several times daily. After feeding, the adult rests on nearby walls, trees, or other objects to digest the blood meal. Stable flies are controlled by frequent cleaning of premises, manure disposal, and application of insecticides to facilities (resting places) and animals.

Screwworm Flies   These blowflies are bluish green, with three dark stripes down their backs. They are somewhat larger than the housefly. Screwworm larvae resemble a ringed spinal column and have the appearance of a wood screw; hence the name screwworm. Eggs laid in a human or an animal wound (or the navel of the neonate) hatch in 6 to 12 h. The larvae **ingest** living tissue for 3 to 10 days and then drop to the ground, where further development occurs. (Depending on the temperature, it may take a few days to 2 months for the larvae to pass through the pupal stage and emerge as adults.) The pupae are killed by hard freezes. Thus screwworm flies are more troublesome in the warm southern United States, where, until recently, they caused an estimated annual loss of $20 million to livestock producers. Treating the wounds with smears that contain insecticide is a good **prophylactic** measure. The release of **sterilized** males (because the female usually mates only once) is a current method of control (Section 19.6.2). The flies spread northward in the summer with livestock shipments. The **dehorning** of cattle or goats during the fly season may expose animals to screwworm infestations. Odor from the infected wound attracts more screwworm flies, and if not treated, the animal will die.

Severe infestations of the Gulf Coast tick in the ears of cattle may predispose the animals to screwworm infestation. It is interesting that insecticidal ear tags, now widely used for fly control on cattle, were developed originally as an aid to control these ear ticks and hence to reduce screwworm populations.

Fleece worms are larvae of blowflies that are similar to the screwworm fly. They deposit eggs on the soiled wool of sheep. After **hatching,** the larvae feed on the skin surface and cause severe annoyance. Most infestations in sheep occur in early spring, when the wool gets soiled by feces, urine, or rain. Infestations of fleece worms are often associated with screwworm infestations in sheep.

True screwworms are presently confined to the far southern United States, ranging into Mexico. However, throughout the country other species of blow-flies may produce similar maggot infestations and are often mistaken for screwworms. For example, larvae of certain species of greenbottle and bluebottle flies that resemble larvae of screwworms annually infest fistulated cattle at the Missouri Agricultural Experiment Station. These same species often infest pet animals (especially cats and rabbits) when a female pet gives birth during the summer months. Approved screwworm insecticide smears may be valuable in such cases.

Heel Flies   These hairy black-and-yellow-striped flies look like small bumblebees. The common heel fly, which appears during the first warm days of spring, is nearly 3 times larger than the housefly. It has no mouthparts and does not bite or sting. Heel flies cause no pain, but rather annoy cattle while depositing their eggs. Animals attempt to escape from them by running with their tails held high. The female heel fly (also called *warble fly,* or *cattle warble*) attaches eggs to hairs on the heels or legs if the animal is standing and, if lying down, to the flanks of cattle and certain wild **ruminants.** These eggs hatch in 3 to 5 days. The larvae (called *cattle grubs*) penetrate through the skin and live in the **viscera** for about six months. They then migrate to the back, cause painful swellings, called *warbles,* and finally cut breathing holes through the skin. The grubs develop under the skin for about 7 weeks and then drop to the ground to pupate. Pupation requires 2 to 11 weeks and yields winged adult flies. Because millions of tons of infested meat containing these grubs must be discarded and hides intended for leather uses are damaged, the economic losses to livestock producers are significant. (USDA estimates indicate losses of approximately $140 million annually in the United States.) Certain wild animals (e.g., deer) have learned to partially avoid heel flies by standing in water. Certain systemic insecticides (materials that are absorbed through the skin) can produce excellent control of cattle grubs. Care must be taken to apply such compounds at the prescribed time of the year.

Botflies   The sheep botfly (sheep gadfly, or nasal fly) carried by sheep and goats may cause conjunctivitis in humans. The eggs hatch within the female sheep botfly. The females then deposit the larvae enclosed in a drop of sticky fluid into the sheep's nostrils. These larvae migrate to the nasal sinuses, where they feed on mucus secreted by the tissues and mature for 6 to 8 months before dropping to the ground to pupate. They cause irritation of nasal passages and difficult breathing. The adult sheep botfly does not feed. Goats may become infested with the sheep botfly when they occupy ranges with infested sheep. Related species often parasitize deer and reindeer.

The ox warble (botfly) of cattle, goats, and other animals lays its eggs on hairs of the legs. The larvae hatch in about 4 days and then migrate up the hair and eventually penetrate the skin. These larvae pass through the **subcutaneous** connective tissue toward the diaphragm and finally migrate to the animal's back. They then escape from their cysts and fall to the ground. This parasite causes considerable loss of meat and milk and damage to hides. The larvae migrate about in the subcutaneous tissues of humans, causing so-called larva migrans, or "creeping eruption" (swelling in the various affected body parts). This species is closely related to the heel fly.

The human botfly may attack humans and certain domestic animals. It is found largely in the tropics (especially in Central and South America and in Mexico). The presence of botfly larvae is very painful to humans, in whom they migrate subdermally. In infested areas humans should have a well-screened house, use bed nets, and avoid bloodsucking flies and mosquitoes.

Horse botflies harass the horse by darting around its head in an attempt to lay their eggs. (Females lay their eggs on hair of the horse's legs, throat, or mouth.) Larvae hatch on contact with the horse's tongue. The larvae, or *bots,* burrow into the mucous membrane of the tongue, mouth, and throat, where they live for about 2 weeks before passing to the stomach. They then live for about 9 months in the horse's stomach and intestines before passing out with feces. They pupate in the soil. There is one generation annually. Because adult botflies do not eat, their control depends on treatment of the host to eliminate bots. Carbon disulfide is one drug used to treat horses for botflies; it may be administered **orally** in a capsule, or via a stomach tube by a veterinarian. Occasionally, eggs of the horse botfly may accidentally be transferred to a human mouth. The small larvae may subsequently cause discomfort along the gum line.

Farm animals seem to realize that botflies are dangerous to them. There is no pain associated with egg laying; the animals merely have an instinctive fear of the fly. This is probably nature's way of helping them avoid excessive parasitism. When these flies are buzzing among the **herds,** cattle and horses become greatly excited. This excitement reduces meat and milk production.

Horseflies and Deerflies These beautifully colored (brown, black, orange, or metallic green) flies attack cattle, deer, horses, and other **homeotherms,** including humans. Only the adult females are bloodsuckers; the males feed on plant juices and nectar. The lancelike mouthparts of the female are developed for cutting skin and sucking blood that oozes from the wound. (They are vicious biters.) The big, bothersome, bloodthirsty horsefly (gadfly) is familiar to, and despised by, all farm people. Its less conspicuous and smaller cousin, the deerfly, is equally despised in many areas, where it causes pain by biting and may also spread tularemia (Chapter 18). Horseflies and deerflies breed in moist places. Their development requires several months, and control is difficult. (However, pyrethrum sprays kill horseflies on contact.) Female horseflies lay their eggs in clusters, commonly on plants that overhang water or grow in wet soil. The larvae live in debris along ditches, swamps, and rice fields. Adult horseflies may carry infections such as anaplasmosis, anthrax, and tularemia. Horseflies and deerflies will move several miles from their breeding places in search of nourishment. Local control of horseflies is often facilitated by using canopy traps. A black object, usually a painted beach ball, attracts female flies to the trap.

Face Flies These annoying pests of cattle and other livestock are closely related to the housefly. The larvae are found only in fresh bovine feces, in which they feed; then they pupate in the adjacent soil. There are commonly several generations during the summer. The pupa may be distinguished from pupae of other species of flies by its color, white; those of other species are dark brown. Face flies do not bite or suck blood, but their feeding is known to cause irritation to the eyes of cattle. They stay near or on cattle, where they feed on animal secretions from the body **orifices** (they cluster around eyes, mouth, nose, and wounds). Horses bothered with face flies often stand head-to-tail (i.e., in reverse directions), flicking the flies off each other's face. These flies are **carriers** of pinkeye disorders. Face flies **hibernate** in attics and the walls of houses. Insecticidal ear tags, dust bags, and sprays all offer some degree of control.

19.5.2 Mosquitoes

The female of the species is more deadly than the male.

Rudyard Kipling (1865–1936)

In order to control mosquitoes, one must learn to think like a mosquito.

Samuel Taylor Darling

All mosquitoes start life as eggs and must have water in which to develop. Eggs hatch into larvae, which are often called *wigglers* because of their activity in water. The larva (after four molts) is covered with a shell and then passes through the pupal stage. Finally, it leaves the shell as an adult. Female mosquitoes are stronger than males, and they usually live longer. Moreover, they (only females bite) are the world's number one vector (carrier) of human diseases (Chapter 18). Male mosquitoes are quite gentlemanly; they feed on plant juices such as the nectar of flowers and decomposing fruits. Female mosquitoes exhibit a preference for specific host animals. Some species (especially in the genus *Culex*) feed on birds, whereas *Aedes* and *Anopheles* prefer domestic animals and humans.

19.5.3 Ticks

All ticks are parasitic on animals. Adult ticks have four pairs of legs. Many enjoy long life spans and can survive months or perhaps years between blood meals, which are required at the time of egg production. Some have several successive hosts; e.g., the Gulf Coast tick may pass its first stage on a quail, the second on a rat, and the third on a cow. The life history of a tick involves four stages: (1) the egg (almost all eggs hatch); (2) larva, or **seed tick;** (3) **nymph;** and (4) adult. Eggs are not deposited until the engorged female (body size may be increased 10-fold or more with blood) has left the host. Some ticks (e.g., the **fowl** tick) lay a few hundred eggs, return to the host for another blood meal, and then lay more eggs; they may repeat this process several times. However, most ticks lay only a single batch of eggs and die when **oviposition** is completed. The six-legged larva, or seed tick, must find a suitable host for engorgement when it hatches if the life cycle is to be continued. Larval metamorphosis terminates in an eight-legged nymph that engorges and molts into an adult. Many individuals in each of the developmental stages (larvae, nymphs, and adults) die without finding suitable hosts. Ticks are blood feeders and, because of this, transmit disease-producing organisms to humans and domestic animals. (They rank number one in transmitting animal diseases.) In range animals and certain wild mammals (especially the elk and moose) death may result from heavy infestations of the winter tick, *Dermacentor albipictus*. They are particularly severe in late winter and early spring, when the animal's food supply is often short.

For many years cattle ranchers feared "Texas cattle fever" (also called *cattle tick fever* and *red water fever*), which is characterized by high fever, red corpuscle destruction, enlarged spleen, and engorged liver and which frequently terminates in the animal's death. In 1890, Drs. S. Theobald Smith and F. L. Kilborne of the USDA linked the transmission of this disease with infestations of cattle ticks, which were later found to transmit the causative organism (*Babesia bigemina*). Their observations, coupled with studies and descriptions of the life history and characteristics of the cattle tick by Dr. Cooper Curtice (a USDA veterinarian), pointed the way for studies that later resolved similar problems with respect to parasitic vectors that spread such human diseases as malaria, yellow fever, tularemia, typhus, and Rocky Mountain spotted fever.

Ticks may be classified as *scutate,* or hard ticks, which are vectors of rickettsia, tularemia, certain viruses, and many animal diseases, and *nonscutate,* or soft ticks, which are responsible for **endemic** relapsing fever in humans. Examples of these groups include (1) the *one-host* scutate tick (i.e., it passes through all stages on one host), *Boophilus annulatus,* the cattle tick; (2) the *two-host* nonscutate tick. *Otobius megnini,* the ear tick; (3) the *three-host* scutate tick, *Dermacentor variabilis,* the American dog tick; and (4) the *many-host* nonscutate tick, *Argas persicus,* the fowl tick. Certain ticks may pass disease-producing agents to their young. This type of transmission is called *transovarial* and explains how a one-host tick can be a vector of disease-producing pathogens.

19.5.4 Mites

These tiny tick cousins are about 1/25 in or less in length and have four pairs of legs. Most **poultry** mites engorge themselves with blood and then hide in cracks in the walls and floor of the poultry house. However, a second group spends its entire life cycle on birds. (Chicken mites, *Macronyssids,* are known to kill chickens by **exsanguination**.)

The northern fowl mite, *Ornithonyssus sylviarum,* is currently the most important external parasite of caged laying hens in the United States. Infestations are usually centered near the cloaca (or vent) of the chicken and may reduce its egg output. Roosters are often severely affected and frequently die unless treated. Persons engaged in handling infested poultry or eggs may be bitten. Treatments for control of the northern fowl mite, consisting of *acaricides* applied to the poultry as sprays or dusts, are not generally effective unless the birds' vent regions are adequately covered. Modern poultry facilities that house many thousands of hens in limited-access cages compound the treatment problem. Acaricide-treated plastic strips, similar to insecticide ear tags for cattle, have given excellent experimental control of northern fowl mites when the strips were hung in poultry cages. Such slow-release plastic formulations may be useful in future poultry pest-management programs.

Mange and scab mites burrow into their host's skin and reside there for life. Common sheep scab is caused by *psoroptic* mites,[5] which reside on the host's

---

[5]Psoroptic sheep scabies has been virtually eradicated in the United States. The same mite species occurs on cattle, and the number of cases reported during recent years has increased.

skin. The mites live on blood that oozes from skin punctures made with their sharp mouthparts. The skin reacts by **exuding serum** that forms itching crusts, and patches of wool soon fall out. Scab is readily transmitted to other sheep, to goats, and probably to rabbits. *Sarcoptic* mites burrow into the skin of the face and head. *Chorioptic* mites reside and cause sores on the skin surface of the legs. *Demodectic* mites burrow into hair follicles and skin glands. Regular systematic dipping controls the scab-causing psoroptic, chorioptic, or sarcoptic mites; demodectic mange is more difficult to control. These four types of mites also attack cattle, dogs, other animals, and humans. Horses are subject to their own psoroptic mange. Much leather is rendered of poor quality because of damage caused by demodectic mange.

Two kinds of mange in swine are caused by mites. The commonest is known as sarcoptic mange; demodectic (follicular) mange is the least common. The causative mites spend their entire lives on infested swine, in which they produce skin wounds, or **lesions,** as they feed on the host's tissues and blood. The feeding and burrowing of mites cause irritation, itching, inflammation, and swelling of the affected tissues. Nodules and vesicles form, which break and discharge serum that dries into large granules (scales) that characterize mange. Sarcoptic mange spreads primarily by direct contact with infested swine. (Swine have a habit of resting and sleeping in close contact with one another.) The disease can be transmitted to people and to certain other animals, although the mites live only a limited time on such new hosts. Demodectic (follicular) mange of swine is caused by microscopic (the adult female is about 0.01 in long), parasitic, wormlike mites, called *Demodex folliculorum suis.* They penetrate the hair follicles and oil glands of the skin, where they complete their life cycle. A completely effective **therapy** of demodectic mange has not been developed.

The so-called 7-year itch is caused by mites. Actually, it does not (and never did) last 7 years, but rather was permanent prior to effective therapy. In certain areas humans are harassed by the six-legged red larva of the mite *Eutrombicula alfreddugesi,* better known as the **chigger** (also called the *harvest mite* and *red bug*). Chiggers (first active stage of these mites) chase about rapidly on the shoes and clothing of humans but settle down to feed in areas where the clothing is rather tight against the skin. Chiggers do not burrow into the skin, but rather grasp the host's skin with their papal claws, insert their mouthparts into the skin at a hair follicle, and inject a salivary solution (toxin) into humans as they feed. This secretion dissolves tissue and sets up a reaction that causes a **wheal** (an **edematous** elevation) on the skin and is accompanied by intense itching. (Chiggers cause considerable loss in business in certain resort areas.) Chiggers transmit important diseases of humans, e.g., scrub typhus. Turkeys, as well as birds, frogs, rabbits, and snakes, are also susceptible to chiggers.

The unfertilized adult female mite *Ornithonyssus bacoti* produces eggs by parthenogenesis, which gives rise only to males; these males are capable of fertilizing. The mother mites of chiggers are fertilized by wandering about the ground, on the surface of which males have carefully placed *spermatophores* (small capsules situated on the end of a fine stalk).

19.5.5   Lice

Most lice are considered permanent parasites (i.e., they spend their entire life on the host). The human body louse transmits three dangerous diseases: epidemic typhus, relapsing fever, and trench fever (Chapter 18). All farm mammals and poultry may become infested with their own peculiar kind (often species-specific) of either *biting* or *bloodsucking* lice. Lice have three pairs of legs and are true insects. Large numbers of eggs (**nits**) are laid on hair or feathers. Most lice require from 2 to 4 weeks (depending on the temperature) to develop from an egg into an adult. Whereas only one species attacks swine, at least seven species of biting lice attack chickens. Lice of cattle, goats, sheep, and other animals are spread by contact of one animal with another. Cattle infested with lice rub and scratch themselves, causing patches of hair to be lost (usually associated with the presence of the biting, or red, louse). Large numbers of lice retard the growth of **calves,** reduce weight gains and the efficient utilization of feed in cattle and poultry, impair the lactation of dairy cows, and significantly reduce egg laying in poultry. Because cattle lice live nearly all their lives on the host, they are easier to control through residual insecticides than are many insect pests. Lice of sheep and goats are best controlled by dipping the animal in one of several effective insecticide solutions.

The swine louse, *Haematopinus suis,* is a bloodsucking parasite and is the largest louse that preys on domestic animals (females often attain a length of 1/4 in). These brownish-gray lice are parasitic on swine only and pass their entire life cycle on this host. They are easily seen on swine. They feed frequently by puncturing the skin and sucking blood in a new site at each feeding. Each puncture produces irritation and itching, which cause infested swine to rub themselves vigorously against any available object. The frequent scratching and rubbing destroy patches of hair and often wound the skin. Oils and medicated liquids are usually effective against swine lice.

It is interesting that deer often paw the ground, making a shallow mud bath to lie in, and coat their bodies with mud, which dries on their skin and kills body parasites such as lice and ticks.

19.5.6   Gnats

Many **gnats** (small flies of the order Diptera) suck blood from animals, whereas others do not even bite. Most gnats lay their eggs on water, where they float 1 to 5 days and then hatch. Others lay eggs in decaying plant material. Some of the many species of gnats are vectors of a serious viral disease of sheep called *bluetongue.* An almost invisible gnat, called the *punkie,* is a nuisance in woods of the west. Indians called it the "no-see-um." Related gnats that bite humans in the southeastern states are called *sand flies.* One family of gnats attacks plants and makes galls on them. The most harmful of these is the Hessian fly (so named because people once believed that it was brought to America in the bedding of Hessian troops during the Revolutionary War).

19.5.7 Fleas

These insects are extremely important vectors of human diseases. Perhaps the most deadly of these diseases is bubonic plague, which fleas transmit from rats to humans (Chapter 18). Fleas also transmit endemic typhus from rats to humans. Fleas serve as intermediate hosts of certain tapeworms. These insects may infest poultry houses. They may be controlled by thorough cleaning, followed by a treatment with insecticide. Fleas commonly infest cats and dogs. All fleas pass through four stages: the (1) egg, (2) larva (two molts), (3) pupa, and (4) adult. Eggs are laid while the female is on the host (they are not attached, however) and then drop to the ground, where they hatch in a few days into wormlike larvae. The larvae are nonparasitic and live on organic matter of the soil. In about 2 weeks the larva becomes fully grown and spins a tiny cocoon in which it develops into a pupa. The pupa usually emerges as an adult flea in a week or so. Most fleas are rather host-specific (i.e., they commonly attack only one host); however, some of the commoner fleas may attack humans, cats, dogs, rodents, swine, and many other animals (including certain birds). Cat and dog fleas are controlled by dusting with appropriate insecticides.

19.5.8 Bedbugs

These insects attack humans, mice, rabbits, guinea pigs, cattle, horses, and poultry. Fortunately, they are not known to be vectors of pathogenic organisms. Bedbugs pass through three stages: (1) egg, (2) nymph (five molts), and (3) adult. Females lay from 75 to 500 eggs at the rate of three to four per day. There may be one to four generations annually. The nymphs feed before each molt. The wingless adult lives about 1 year. Unlike many insect species, both males and females are avid bloodsuckers and if present in beds and mattresses make short appearances to obtain blood meals from humans at night. Nymphs require only 6 to 9 and adults 10 to 15 min for engorgement, after which they drop from their victim and retreat quickly to their hiding places (they are especially active at night). It is possible for humans to carry bedbugs into their houses from infested poultry houses or equipment. However, the use of DDT after World War II essentially eliminated bedbug infestations, which had annoyed human beings for many years. The EPA now prohibits the use of **DDT,** and bedbug and lice populations are making a strong comeback.[6] Related bugs, parasites of bats or birds, are often mistaken for true bedbugs.

19.6 ARTHROPOD CONTROL—ESSENTIAL FOR HUMANITY

Insects are humans' greatest competitors for food. In many parts of the world, insects reduce a potential agricultural abundance to such an extent that people

---

[6]It is somewhat ironic that a Swiss chemist, Paul Mueller, was awarded the Nobel prize in medicine in 1948 for developing DDT as an insecticide, and only a generation later, its use is banned in several countries.

die of malnutrition. Scientists are probing the biological, biochemical and behavioral differences that set insects apart from other animal life, in an effort to learn ways to control their populations (especially their capacity for reproduction) and thereby increase the food and fiber supply for humanity.

19.6.1 Chemical Control

Civilization advances proportionately to the ability of humanity to overcome problems associated with securing the essentials of survival: food, water, and shelter. The biological checks and balances of nature are too slow and inadequate to ensure plentiful food for humanity. In a natural setting, humans would compete with a host of natural enemies seeking the same food supply. Today, they compete favorably by protecting their food from ravages of insects, rodents, and disease organisms with agricultural chemicals. The use of chemicals to fight pests dates back to the ancient Greeks, who employed brimstone (sulfur) as an insecticide. Cave dwellers observed that food treated with the salty residue of seawater is safe from insects. Settlers in the Great Plains of the United States used Paris green (a crude arsenical) to save their potato crops from the Colorado potato beetle. They also treated their grain seeds with copper sulfate to protect them from plant disease. Today over $1 billion is spent annually for pesticides in the United States.

It would be difficult to appraise accurately the benefits to humanity that have accrued from the use of agricultural chemicals in the control of pests. Similarly, it would be difficult to imagine the adverse impact on humanity's future welfare and economy if suddenly they were made unavailable. Chemicals have been used to tip the so-called biological balance in nature in favor of humans.

Pesticides are poisons used to destroy pests of any sort. They include **fungicides, herbicides,** insecticides, acaricides, and rodenticides. Pesticides are often the most effective weapon available to fight pests that damage or destroy crops, livestock, and forests. These chemicals also help protect humanity's health and well-being from insects that transmit such diseases as malaria, yellow fever, and typhoid. The effectiveness of pesticides in controlling agricultural pests helps keep food costs down and quality high. According to USDA estimates, if pesticides were not used, crop and livestock production in the United States would decrease by approximately 30 percent. Without pesticides, many vegetables and fruits would be destroyed by insects and disappear almost completely from food markets. These losses in animal and vegetable food production would result in the starvation of many people.

Insecticides for the control of insects are effective in two ways: as *stomach* (ingestant) poisons (e.g., lead arsenate) and **contact poisons** (e.g., pyrethrum). Stomach poisons are used to combat such pests as beetles, which chew and swallow leaves and concurrently swallow the insecticide. Pyrethrum kills insects by causing paralysis of the central nervous system. DDT, which was first synthesized in Austria in 1873, and chlordane function as both a stomach and a contact poison. They belong to the class of insecticides known as chlorinated hydrocarbons. Other examples include dieldrin, methoxychlor, and toxaphene.

These broad-spectrum, inexpensive residual insecticides revolutionized agricultural technology. Today, because of known or suspected adverse effects, many of their uses have been cancelled. At present, most insecticides commonly employed are organophosphorus or carbamate compounds. These agents are nerve poisons that inhibit the enzyme acetylcholinesterase. This action halts nerve transmission at the synapse, or junction, of nerve cells. Many organophosphorus and carbamate insecticides are extremely poisonous to humans and domestic animals, and they should be used only with adequate protective equipment. Several new classes of insecticidal materials have been investigated in recent years. Formamidine compounds affect insects' biogenic amines, thus offering a different *mode of action* with possible utility against insecticide-resistant strains. Synthetic pyrethroid compounds (similar to the pyrethrins obtained from certain flowers) have found wide use in the livestock industry as fly and mite control agents. Insect growth regulators (IGRs) may mimic the action of insect hormones or inhibit formation of the chitin that hardens the insect skeleton. Exciting as such research may be, the fact is that many corporations are turning away from insecticide development. The huge capital investment required and the small probability of successfully contending with governmental regulations make such development a poor risk financially.

Pesticide Residues in Animal Products Agricultural chemicals are not used without some risk. It has been established experimentally that cows fed foods contaminated with certain chemical compounds (e.g., DDT) store them in their body fat and secrete them in their milk fat. Many pesticides used to treat plants and/or animals are fat soluble and may be present in various levels in eggs, meat, milk, and their by-products. However, at the low levels observed, no adverse physiological effects have been noted. There are no medically documented records of death in humans resulting from the presence of pesticide residues in properly treated foods. This is because the safety margins between permitted levels of use by the EPA or FDA and the minimum toxic doses are so great that violations of legal tolerances are extremely unlikely to involve significant health hazards. Tolerance limits for pesticide residues in foods are established by the EPA and FDA. The permissible level is generally determined by considering the acceptable safe intake level of that chemical for humans and the amount of the food that will likely be consumed during a lifetime. The safe level of intake is calculated by extrapolating from data derived from small-animal experimentation (usually data of the most sensitive laboratory animal are used), after which an additional safety factor is applied. The FDA monitors pesticide concentration in animal products. If foods are found to contain residues in excess of the minumum safe tolerances as established by the FDA, they are withdrawn from the market.[7] Extensive studies of the pathways and possible retention of pesticides in animals and their excretion patterns are being made (Figure 19.2).

---

[7]An example of food contamination involving a pesticide that resulted in significant economic losses (estimated at $10 million) to poultry producers and nutritional losses to consumers was the high levels of dieldrin that were found during routine inspection of slaughtered poultry and declared unsafe for human consumption. The dieldrin was later traced to contaminated feed.

FIGURE 19.2
Pesticide metabolism is studied using catheterized lactating cows. *(Courtesy Dr. D. J. List, Pesticide Residue Laboratory, Cornell University.)*

19.6.2 Biological Control

The man of science ought to realize the factors which have given him the vantage which he holds.

Carl von Voit (1831–1908)

Insects constitute at least 90 percent of all the world's animal life and can produce astronomical numbers of offspring. For example, if all the descendants of a pair of houseflies lived and mated fully, the family would number approximately 190,000,000,000,000,000,000 flies after only 4 months.[8] A queen termite is equally **fertile.** She lays thousands of eggs daily during a life span of up to 50 years. Vast numbers of insects compete with humans for fruits, grains, and other food plants and transmit diseases to animals and humans. However, significant and important progress has been made in biological control of these arthropods.

Scientists have investigated a number of methods and techniques, from the use of the back of one's hand to special chemicals, in attempts to kill these destructive and disease-bearing insects. Chemicals are being used to sterilize the

[8]"Insects," *USDA Yearbook,* 1952.

pink boll worm in Texas. They also appear valuable in controlling the gypsy moth and boll weevil. Some chemists predicted that DDT would free the world from insects. However, the more vigorous ones lived, and, apparently, subsequent generations were somewhat resistant to this insecticide. After contact with an insecticide, the insect population that survives is generally resistant, and a larger proportion of the population becomes resistant through genetic **selection.** The rapidity with which the population increases its resistance depends on the frequency of the specific genes for resistance and the intensity of selection. If the **gene** is **recessive,** a longer period is required for acquiring insecticide resistance than when the gene is **dominant.** Besides this resistance to the insecticides, their residues on plants present a possible health hazard to humans, who ingest these food products of animals that consume the plants.

Public concern over the possible dangers of chemical pesticides has prompted scientists to emphasize the use of natural enemies of insects against them. Thousands of insects prey on other insects. For example, a wasp, *Dendrosoter protuberans,* is a natural enemy of the European bark beetle, which carries Dutch elm disease and spreads it by depositing its eggs through the tree bark. In Europe the beetle has been held in check by this wasp, which destroys the beetle larvae. Much interest has been shown in the potential for biological control of insects affecting livestock. It is known that major fly pests such as the housefly, face fly, stable fly, and horn fly suffer considerable natural mortality because of insect parasites and predators. In some areas, tiny parasitic wasps (particularly *Spalangia endius*) are deliberately reared using houseflies and then released by the millions in poultry or dairy facilities. The released wasp females seek new fly pupae in which to deposit their eggs. The developing wasp larvae kill their fly host, emerge as adults, and in turn seek other fly pupae. The cycle thus becomes cumulative, and effective control of fly populations has been demonstrated in the southern United States. This concept of *inundative* or *mass release* has been likened to a "living insecticide" and offers great possibilities in other areas if parasitic wasp species can be found that are suited to local climatic conditions. Predatory arthropods can be very effective biological fly control agents; however, techniques are not yet available for their commercial use.

Japanese beetles, which are enemies of more than 200 crops and ornamental plants, may be controlled by infecting them with a bacteria-caused milky disease.

> So, naturalists observe, a flea
> Hath smaller fleas that on him prey;
> And these have smaller still to bite 'em
> And so proceed *ad infinitum.*
> *Jonathan Swift (1667–1745)*

This quotation emphasizes the principle of natural insect elimination, as shown in the above examples. A further variation of this principle is found in nature's biological control of insects by means of the parasite-host relationship.

One of the most interesting episodes in the history of biological control was the use of myxomatosis virus to control the rabbit population of Australia. The virus exists naturally and causes tumors in the Brazilian wild rabbit. In 1950, myxomatosis virus was introduced into several experimental groups of rabbits, which were then released in the countryside of Australia. Within 10 months after their release, other rabbits throughout a 500,000-square-mile area were found to be infected. Within 3 years the estimated original rabbit population of 500 million had been reduced to an estimated 50 to 100 million, thus allowing sufficient regeneration of vegetation to permit a 10 to 15 percent increase in wool production of sheep.

The application of this biological relation is also important in insect control. There are more than 1000 viruses, fungi, nematodes, protozoa, and rickettsiae that infect and kill insects naturally. Researchers can even inject these pathogenic organisms into the larvae of some insects. A poison secreted by the bacterium *Bacillus thuringiensis* is **lethal** to caterpillars of more than 100 butterflies and moths. Dried spores of this bacterium are being used in the United States to control larvae of the alfalfa caterpillar, cabbage looper, imported cabbage worm, hornworm, tent caterpillar, and gypsy moth. Fortunately, this substance is harmless to humans, domestic animals, wildlife, and most beneficial insects. Other agents, such as the polyhedral virus, can be prepared commercially and therefore offer a means of controlling insects. This virus is grown on insect larvae because viruses require cellular material to multiply. Each 100 infected larvae produce enough virus particles to treat an acre of cropland. Thus, through science, researchers have applied the natural principle of biological control and adapted it to rid humanity of insect pests.

An experimental laboratory (Figure 19.3) is useful in studying various ways of controlling insects biologically. A means of rearing host insects is essential to such investigations. Laboratory rearing of horn flies is depicted in Figure 19.4 *a* and *b*.

Entomologists throughout the world are currently collecting and assembling research data on the anatomy and behavior of harmful insects. This research has led to a number of ways for their elimination. Knowing that certain insects commonly mate only once in a lifetime, scientists released millions of artificially sterilized males to disrupt the insect's reproductive cycle. The direct application of this principle is found in the screwworm eradication program. It may be possible to treat insects with chemosterilants that will induce sterility without affecting mating.

The technique of **radiation sterilization** using cobalt 60 was developed by Dr. E. F. Knipling of the Entomology Research Division of the USDA. His early work was designed to control the screwworm fly. Many female insects (e.g., screwworm flies) usually mate only once (they carry the **sperm** in a special sac, the spermatheca, from which the sperm emerge to fertilize the eggs as they pass down the oviducts). Knipling reasoned that if the females mated with sterile males, they would lay only unfertilized eggs, and the reproductive cycle would be broken. A dose of cobalt 60 sufficient to render the males sterile, but low enough to ensure that the treated flies were still strong enough to compete with

FIGURE 19.3
Biological Control of Insects Research Laboratory. *(USDA, Agr. Res. Service, Columbia, Mo.)*

nonirradiated ones, was employed. A strong advantage of using sterilized males rather than powerful insecticides is that the males seek out *all* females for mating and leave none to form a reservoir for reproduction.

Tests in 1954 rendered a 170-square-mile island off the coast of Venezuela free of screwworm flies. In 1958, researchers used an air-terminal hanger in Florida to raise massive numbers of sterile male flies. To produce 50 million flies weekly,

FIGURE 19.4
(a) Rearing cages for obtaining horn fly eggs. Their food source (blood meal) is at the top, and pads for oviposition (obtaining eggs) are located at the bottom of the traps. *(b)* Rearing horn flies (larvae to pupae) in the laboratory. These supply host materials for rearing various species of parasitic insects that attack horn flies. They are also used for insecticide and other investigations. *(USDA, Biological Control of Insects Research Laboratory, Columbia, Mo.)*

(a)   (b)

40 tons of horse meat (or whale meat) and 4500 gal of beef blood were used. After being sterilized by exposure to radioactive cobalt, the flies were air-dropped and spread over infested areas in the southeastern United States. Through the use of 200 to 1000 flies per square mile, the southern United States was made essentially free of screwworm flies.[9] A weekly release of 4000 sterile male flies per square mile was effective in complete eradication of screwworms in Alabama, Florida, and Georgia. This represents an important landmark in insect-eradication attempts. In 1965, sterilized males were used to eradicate the oriental fruit fly on Guam. More than 20 million sterilized Mexican fruit flies were released along the United States–Mexican border in 1966 to protect the citrus orchards of Arizona and California.

Australian scientists are attempting to breed flies out of existence. Their plan, as reported by *Science Service,* calls for breeding flies "whose only **offspring** will be males." Eliminating female flies would mean the end of flies. Male flies are being **irradiated** in the experiment to give the **X chromosomes** (male) in males a better chance than the Y chromosomes (female). After enough male-breeding flies have been produced to dominate a fly community, they will be released in an effort to increase the proportion of male to female flies. The island of Capri (off the coast of Italy) was rendered free of fruit flies in 1967 through the use of flies that had been radiation-sterilized in Israeli and Austrian laboratories and flown to Capri for release.

Research indicates that attractants can be synthesized and used to lure unsuspecting insects into traps instead of attracting them to mates. More than 200 insect species have been found to possess sex attractants. Although odors released by female insects are usually for the purpose of attracting males, they may also serve to excite the male sexually before **copulation.** The production of female sex attractant in most insects is depressed appreciably within a few hours after mating. Moreover, it is interesting that sex attractants are produced by insects only just before or during the period of the day in which mating normally occurs. In some insects the male *and* female exude attractants (assembling scents). For example, the male boll weevil can attract females from a distance of 30 ft. The sexual odors released by males are primarily for the purpose of sexually exciting the female, making her more receptive to the male's advances. There are more than 50 insects in which males lure or excite the females. Certain female beetles emit an odor that will attract male beetles of that species from 1800 ft away. The female gypsy moth also emits a sex attractant that makes it easy for the male to find her. Scientists can now synthesize this attractant, or lure, called *gyplure.* Possibly scientists could synthesize a large amount of this attractant and spread it over a large acreage of timberland that the moth has infested. (Gypsy moth caterpillars are colorful but destructive insects that kill

---

[9]The southeastern United States is free of screwworms. However, the southwestern states cannot be free because of reinfestations from Mexico. Therefore, a joint United States–Mexico program is under way to eliminate screwworm infestations. The USDA facilities for this project were recently moved from Texas to southern Mexico.

trees by devouring their leaves.) With the sex odor literally everywhere, the males should become confused in their attempt to find females, quickly become exhausted, and give up in their mating effort or, possibly because of overexposure, become insensitive to the sex attractant. A chemical is classed as a sex attractant if it brings to it an insect that then assumes a mating position.

Research entomologists have discovered that a pair of glands, called the *corpora allata,* found in the female cockroach brain secrete a lure on signal. Scientists are currently attempting to develop a way to interrupt this message, which would result in failure of females to secrete the sex attractant. This would then make it difficult for males to find females and consequently would reduce reproduction. A recent study identified valeric acid as the sex lure produced by adult female sugar beet wireworms, which attack sugar beets, potatoes, corn, lettuce, and certain other crops. Diluted solutions of valeric acid attracted male beetles from a distance of 12 m. This chemical is economical and readily available for use as a bait in traps for population control.

Some plants possess attractants. Cotton, for example, contains chemicals that can attract or repel the boll weevil or cause it to feed uncontrollably. When these chemical compounds are synthesized, the boll weevil may well be retired to insect museums.

Further research has indicated that lights and sounds sensed only by insects can also be used to ward them off or to lure them to their deaths. Many insects possess eyes and ears of a sort. Black-light traps in combination with the female sex lure are being used to control cabbage looper moths. Some insects are attracted by ultraviolet lights and can be lured by them into traps. Other insects possess mechanisms that are sensitive to sound. Several years ago scientists discovered that bats use high-pitched squeaks that serve as a type of radar to zero in on their prey at night. One of the bat's favorite prey is *Heliothis zea,* a moth with many names (the larva is known as the *cotton bollworm, corn earworm,* and *tomato fruit worm*). Nature protects this moth by equipping it with sensitive hearing organs, which hear the bat's cry as a warning. This information has enabled researchers to install rotating loudspeakers in cotton fields to transmit a recording of the bat cry as an "electronic scarecrow."

Naturally occurring **hormones** are being applied to prevent insect eggs from hatching or the pupae from maturing. This, too, is a part of the research data being established. This process involves natural elimination of insects. *Juvenile* hormones are naturally occurring chemicals that are essential to insect development during the early stages of life. They are secreted by the same tiny glands located at the base of the brain that secrete the sex attractant in cockroaches. Juvenile-hormone secretion apparently ceases at maturity. Researchers reasoned, therefore, that if juvenile hormone were administered to insects late in the pupal stage, normal metamorphosis would be prevented. A juvenile hormone was isolated in 1956 by Carroll Williams at Harvard University, and several such hormones have been synthesized since then. A new type of hormone was synthesized in 1966 by researchers of the USDA. The chemical is derived from farnesol, an oil commonly present in many plants and animals. A drop of this compound applied to the pupa's skin penetrates rapidly. This new synthetic juvenile hormone is apparently nontoxic to noninsect species and has been observed to affect the early stages of development. When sprayed on an insect egg, it effectively blocks development of the embryo. Scientists theorize that the hormone interferes with the insect's genes.

It has long been observed that certain individuals or varieties of plants and animals possess unusual ability to resist attack or damage by arthropod pests. This phenomenon is termed *host resistance*. It generally is dependent on genotype and is an inherited trait. It is obvious that if such a trait can be combined with adequate production performance, it may prove a superior, economical, and ecologically sound method of reducing losses to pests.

In plants, such host resistance may take the form of extra-thick stems through which pests cannot easily penetrate or a characteristic known as *vigor tolerance*. The latter indicates that the host organism is capable of suffering without loss to production a level of damage that would severely reduce the productivity of nonresistant cultivars.

Host resistance to external parasites has also been noted in many species of domestic animals. In some cases, antibodies are formed by the host's immune system in response to arthropod feeding. These antibodies may act to prevent subsequent infestations of the same parasite species. It has long been observed that male animals frequently suffer more severe infestations of external parasites than do females. Sex hormones may be involved in this type of resistance. Chickens genetically selected for a high level of plasma corticosterone (secreted by the adrenal gland; Chapter 7) response to social stress are more resistant to northern fowl mite infestation than chickens selected for a low level of corticosterone response. Feeding chickens this hormone can produce similar results. Breeding resistant varieties of domestic animals and manipulating the animals' own hormone systems to promote resistance offer entomologists exciting prospects for effective control of external parasite populations. In addition, certain of these approaches may apply to internal parasites as well.

19.6.3 Living Insecticides

Insecticides and other pest controls formulated from living microorganisms rather than chemicals are being developed to control, for example, tapeworms, nematodes, ticks, and other parasites and diseases of humans, domestic animals, and plants.

Insect pests, like other forms of life, are susceptible to diseases caused by bacteria, fungi, protozoa, and viruses. These microorganisms are not new; they are already present in the environment, and they often help to control naturally populations of insects that harm humans and cause extensive damage to food and fiber crops. They have been screened naturally for centuries to fit the environment. Now scientists are finding how to use them effectively for the benefit of humanity.

Commercial formulations of these microbes are called *living insecticides*.

There are more than 1500 natural, safe, biodegradable microorganisms that might be formulated into living insecticides. Moreover, new microbes are being discovered each year.

The potential impact of substituting microbial insecticides for chemical insecticides is indeed substantial. In the United States alone, use of a living viral insecticide to control bollworms and budworms on cotton, a living fungus against the citrus rust mite, a bacterium against cabbageworms, and a living protozoan against grasshoppers could replace an estimated 20 to 30 million pounds of chemical insecticides annually.

This new but now formalized science has blossomed during the past two decades. It holds great promise as a major means of biologically controlling, on a safe and effective basis, damaging insects such as caterpillars, grubs, flies, mosquitoes, mites, weevils, and even the difficult-to-kill grasshoppers that are perpetual pests of humans and the foods, fibers, and forests important to humanity.

19.7 SUMMARY

Since the beginning of time, human beings have shared the planet with insects and parasites. Some are beneficial, whereas others may transmit disease and inflict significant economic losses. Historically, diseases transmitted by insects have been notoriously effective in determining the outcome of military conflicts (Napoleon's army is said to have suffered greatly from scabies), some say even more effective than the decisions of military commanders (epidemic typhus and malaria are among the most telling).

Many protozoans are parasites. One type of amoeba destroys the intestinal linings of humans and causes the painful disease amoebic dysentery. Other protozoans may invade the blood of mammals and cause diseases such as malaria and Texas cattle fever. Bloodsucking insects and ticks collect parasites from infected animals and convey them to humans and other animals. Parasitic insects, mites, and ticks usually attack the skin. Certain ticks transmit Rocky Mountain spotted fever to humans. One type of mosquito spreads yellow fever, and another carries malaria. The tsetse fly transmits African sleeping sickness. Humans may acquire typhus fever from a body louse.

The total mass of protoplasm produced annually by insects exceeds that produced by all other terrestrial (earthly) animal life combined. A knowledge concerning insects and how to circumvent their harmful effects is important if humans are to improve the health and productivity of animals that provide useful products to them. Great strides in parasite and insect control have been made possible through new and improved agricultural chemicals. (DDT has saved millions of lives from malaria and typhus.) Pesticides allow humans to control arthropod-transmitted diseases and also to increase their food supplies by providing a means of combating agricultural pests that formerly attacked and destroyed crops and food animals. However, because of possible injurious effects of certain chemicals on humans, new methods of insect control are being studied and exploited. These include biological, physical, genetic, ecological, and/or cultural approaches. It seems almost certain that future control of insects will be largely biological and depend less on the use of pesticides.

Evidence of this trend is found in the recent interest in *integrated pest management.* This technique stresses a thoughtful and balanced approach to pest control involving all appropriate technology available. It does not depend exclusively on biological, chemical, or cultural control but is based on maximum economic benefit within biological and ecological limits. Its most effective implementation requires a thorough understanding of the adverse effects of pest populations, and in some cases computer *models* are used to predict future trends. Management decisions involving *economic thresholds* and *injury levels* are expected to be commonplace in the years ahead.

STUDY QUESTIONS

1 Of what economic and public health significance are insects?
2 What are the three principal body segments of insects?
3 Define molting. Identify instars and stadia.
4 Can insects taste, smell, and listen?
5 What are their four metamorphic stages?
6 Define a parasite. An arthropod. What is entomology? Parasitology?
7 What are the seven basic levels in the classification of animal life?
8 Name and cite a few examples of the five most important phyla.
9 Briefly discuss some contributions of insects to humanity.
10 Discuss some harmful effects of insects.
11 Cite one or more examples of diseases of humans that may be transmitted by *(a)* mosquitoes, *(b)* flies, *(c)* lice, *(d)* fleas, *(e)* ticks, and *(f)* mites.
12 What is the proboscis? Of what use is it?
13 How is it possible for flies to walk upside down on ceilings?
14 Trace the life cycle of houseflies. What is the best way to break this cycle?
15 What is the source of food for horn flies? How may these flies be controlled?
16 Trace the life cycle of stable flies. How are their numbers held down?
17 Trace the life cycle of screwworms. What environmental factors aid in their control in certain areas?
18 What recent application of agricultural research aids in controlling screwworm populations?
19 Trace the life cycle of heel flies. In what ways do they cause economic losses to livestock producers?
20 Trace the life cycles of botflies of cattle, of sheep, and of horses. How does the human botfly affect people?
21 What diseases of humans are transmitted by horseflies?
22 How can one distinguish the pupae of face flies from those of other flies? Why do horses often stand head-to-tail in reverse directions?
23 Trace the life cycle of mosquitoes. How do female mosquitoes rank as a vector of human diseases?
24 Trace the typical life cycle of ticks. Cite some diseases transmitted by ticks.
25 What are chiggers? Are they linked with any diseases of humans?

26 Why are lice considered *permanent* parasites? What dangerous diseases of humans may be transmitted by the human body louse?

27 How can one control lice of *(a)* sheep and *(b)* swine?

28 Name a serious viral disease of sheep for which gnats serve as a vector.

29 Trace the life cycle of fleas. What important human diseases do they transmit? How are they controlled?

30 Trace the life cycle of bedbugs. What insecticide is effective in their control?

31 Define *(a)* fungicides, *(b)* herbicides, *(c)* insecticides, *(d)* pesticides, and *(e)* rodenticides.

32 Briefly discuss the use and importance of chemicals in the control of insects and parasites.

33 Which governmental agencies monitor pesticide concentrations in animal products?

34 What influences the rapidity with which an insect population increases its resistance to an insecticide?

35 What is meant by biological control? Cite examples of biological control of animal life.

36 Of what practical significance is the use of radiation to sterilize male screwworm flies?

37 What are sex attractants? How may they be used in controlling insects?

38 What other means of insect control are being investigated?

39 What is meant by the term "integrated pest management"?

40 What are living insecticides?

CHAPTER **20**

ANIMAL BEHAVIOR[1]

The future trends of human behavior are not predictable because human behavior is modified by intelligence, by education, by laws of human making in contrast to the immutable natural laws which govern the social trends of subhuman populations.

K. L. Frank

20.1 INTRODUCTION

Behavior is the reaction of an animal to a certain stimulus or the manner in which it interacts with its **environment.** Since animals encounter many different stimuli during their lifetime, it necessarily follows that they react in many different ways. Moreover, animals are capable of **spontaneous** actions that are independent of stimuli. Thus animal behavior becomes a complex subject involving many scientific aspects. One **species** may react one way in response to a certain stimulus, whereas another may react in another way. The same is true of individuals within a species. In general, the behavior of animals as a species determines their ability to survive in the wild state. This suggests that natural **selection** has played a part in the development of a certain type of behavior. Perhaps species that have become extinct did not develop the type of behavioral response to their environment that would have enabled them to survive in the face of adversity.

[1]The authors acknowledge with appreciation the contributions to this chapter of Dr. H. W. Gonyou, Department of Animal Science, University of Illinois, Urbana.

Some wild animals seem to teach their young to fear humans. A fawn may pay no attention when a human comes near. However, when its mother shows fear, the fawn will flee with her. Similarly, mother wolves warn their cubs about traps. When the mother wolf and her cubs come upon a trap for the first time together, the mother shows fear. Thereafter the cubs stay away from traps when with their mother or when alone.

Nature equipped many animals with an inherent means of avoiding the potentially harmful. However, their survival may depend on recognizing a particular danger in time to avoid it. There are exceptions. For example, nature was not so kind to the frog as it was to other animals, because a flaw exists in the frog's early warning system that may prove fatal. If a frog is placed in a pan of warm water to which additional heat is being applied gradually, it will typically show no inclination to escape. Being a cold-blooded creature, its body temperature approximates that of the surrounding water, and it does not detect the slow change in temperature. As the temperature continues to rise, the frog remains oblivious to its ensuing danger. And, although it could easily hop to safety, the frog is content to stay put, even as the steam begins to fill its nostrils. Eventually, the frog succumbs to an unnecessary misfortune that could have been avoided if it had simply been alert to the danger at hand.

A knowledge of the reaction of farm animals to certain stimuli and of the forces responsible for a certain type of behavior is of considerable practical importance. With this knowledge, the **livestock** producer will know with some degree of accuracy how an individual will react in a given set of circumstances, what makes it act in that particular way, and how the expected reaction might be changed or controlled for the best results. This knowledge can be of considerable practical value in a program designed to produce more efficient breeding, feeding, and management of farm animals.

Research at Southern Illinois University demonstrated that the average sucking span of piglets can be reduced from 50 min to 40 min by playing tape recordings of the sounds of nursing piglets. This has practical implications, since piglets provided with the tape recordings nursed more frequently than those without, and had a greater intake of milk and heavier body weights at weaning.

The same basic principles that govern behavior in animals also govern the behavior of humans. Much remains to be learned in this area, however. The greatest single problem in the world today may well be the behavior of humans toward other humans. This problem is encountered in the family, in the school, and even in national and international associations. Humanity must learn the principles and methods as well as the application of peaceful coexistence between individuals and nations. It is evident that humans have not learned to use them as well as they should, because since the beginning of time, there have been wars and rumors of wars. Although humans have taught animals many things, animals have taught humans many more. When humans better understand animal behavior, perhaps they can apply certain aspects toward a more lasting peaceful coexistence in the world.

We suffer less from a want of science and technology than from lack of understanding of the aims of life and of society.

R. W. Hutchins

20.2 CAUSES OF BEHAVIORAL RESPONSES IN ANIMALS

Behavioral responses in all animals are determined by **heredity,** or *internal factors,* and by learning experiences, or *external factors.* Some responses appear to be controlled by an interaction between these factors. For example, certain strains or **breeds** within a species have been selected on a genetic basis and then trained for a specific behavioral response. Breeds of horses have been developed that are excellent for running, pacing, or **draft** purposes. A single breed excels in one but not all three. Certain cow **ponies** have been developed that make outstanding roping horses when well trained. Others excel as "cutting horses," which are capable of sorting (cutting out) a specific individual from a large **herd** of cattle (Figure 20.1). A good cutting horse does not necessarily make a good roping horse, and a draft horse could not become a fine pacer, even with the best training. This shows that certain basic inherited internal patterns must be present before the animal can learn from experience to excel in a given area.

20.2.1 Basic Hereditary Patterns

There is considerable evidence linking behavioral responses to hereditary factors. Selection has developed dogs specifically adapted to fighting, herding cattle, or hunting various wild animals. **Crossbreeding** and raising the young on foster mothers is another method used to demonstrate inherited basic behavioral patterns. Some traits have been observed to persist in the young even though they are raised by foster mothers and have not had the opportunity to learn the traits from their own mothers. It has been shown that some traits in dogs, mice, and rats **segregate** according to the laws of heredity when breeds within these species are crossed. Nevertheless, the traits are often modified to a certain extent by environment.

Genetic differences appear to involve **threshold** responses. Some thresholds have a high value and others a low one, which means that heredity can increase or decrease the ease with which an animal can be motivated to show a particular response. This is especially true of the fighting trait. Terriers developed for fighting will fight at the slightest provocation. Beagles have been developed for hunting and run in packs with little or no fighting. They will fight, however, if properly stimulated.

Heredity produces behavioral responses in several ways. It can affect the growth and development of many body parts, such as the **sensory** and/or the motor organs. Genetic defects of the nervous system have been reported in many species ranging in size from mice to cattle. All animals have brains, and farm mammals have the same brain components as humans. However, only a few species (e.g., the parrot) have ever been taught to speak a human language.

FIGURE 20.1
A cutting horse in action. Some horses can be trained to cut an animal from a herd of cattle with great skill. Others cannot be trained to do this. *(Courtesy Robert Sibbitt.)*

This seems to be a particular genetic learning ability limited almost exclusively to humans. It appears that an animal can show only those behavioral responses for which it has the appropriate neural mechanisms.

Genes are responsible for the production of various **enzymes, hormones,** and other chemical substances in the body. These can affect the behavioral response of animals. Chickens can be made to show male or female behavioral responses, depending on whether they have male or female hormones in their bloodstream. Many examples of sex reversals (changes from a female to a male and vice versa) have been reported in the **avian** species (Chapter 12). Strains of mice develop convulsions when exposed to high-pitched sounds. They appear to lack one or more of the chemical substances necessary to supply energy for normal brain or muscle action under such conditions.

20.2.2 Responses to Learning and Experience

Behavior includes all the responses of animals and their ways of acting. The general behavior of most animals depends on the particular reaction patterns with which they have been born. These are called *instincts* and **reflexes.** They are unlearned forms of behavior. Animals often show fixed behavioral responses when they are subjected to certain environmental conditions. The behavior of an animal may be modified as a result of experience, which is called learning. *Insects* are guided almost entirely by instinct. Their life span is short and allows little time for them to learn from experience. Although birds possess an innate nest-building capacity, studies have shown that the quality of nest construction improves with succeeding attempts. This is the result of learning. However, birds can improve their nest building only because they possess the appropriate inherited neural pathways. Several types of learned behavior are known.

Habituation   This is the means by which an animal learns to ignore certain stimuli. It is the simplest type of learning and is necessary to prevent animals from responding continuously to their environments. Animals habituate quickly to common environmental sounds (e.g., ventilation fans) and no longer react to them.

Conditioning   There are two basic types of conditioning, or *associative* learning. *Classical* conditioning occurs when an animal learns to respond to an artificial stimulus (conditioned stimulus) in the same way (conditioned response) as it would to a normal or natural stimulus (unconditioned stimulus). An example of classical conditioning is the simple experiment conducted by the Russian scientist Pavlov, who rang a bell each time he fed meat to a group of dogs. Eventually, the dogs produced saliva when the bell was sounded, even though they did not see, smell, or taste the meat. Animals may be conditioned to perform several reflex activities. This kind of learning is the basis of training farm animals. For example, **lactating** cows often "let down" their milk on entering the milking parlor (Chapter 11).

The second type of conditioning is called *operant conditioning* or *trial and error*. If animals perform a certain behavior and are rewarded when they perform it correctly (positive reinforcement) or are punished when they do it incorrectly, they soon associate the behavior with its results. Thus, if a horse does a certain trick and is rewarded each time with a sugar cube, it soon does the trick because of the reward. Trial and error helps it to associate the reward with the trick and vice versa. Similarly, animals soon learn to stay away from an electric fence when they touch it and receive a shock. In fact, many times when an electric fence is removed, it is almost impossible to drive pigs over the spot where it once was. Recently, domestic animals such as sheep and swine have learned to control the light and temperature of their environment by operating switches.

A behavior that is learned through reinforcement and reward can be eliminated if the reward is withheld long enough. This process is called extinction, and it may be useful to people who want to alter the characteristics of those with whom they associate. The animal world provides many interesting examples of extinction. For example, the walleyed pike is a large fish with a pronounced appetite for minnows. If it is placed in a tank of water with its small aquatic associates, it will soon consume the minnows and be alone in the tank. However, an interesting phenomenon occurs if a piece of clear glass is placed down the middle of the water tank, thereby separating the pike from the minnows. The pike cannot see the glass and strikes it solidly in pursuit of food.

Again and again it will swim into the glass, bumping its head repeatedly. Since this behavior is not reinforced, it is eventually extinguished. Finally, the pike becomes discouraged and gives up—it has learned that it is impossible to get at the minnows. If the glass is then removed from the tank, the minnows can swim freely about their mortal enemy in perfect safety. The pike will no longer try to eat the minnows because it has learned that they are unreachable. The pike will actually starve to death while one of its favorite foods is nearby.

Insight Learning (Reasoning)   This is a third type of learning that is most prevalent in higher mammals. It is the ability to respond correctly the first time the animal encounters a certain situation. It enables the animal to solve a new problem by reasoning without trial and error. Animal studies indicate that severe **malnutrition** in very early life can result in long-lasting behavioral changes, including what has been interpreted as retarded ability to solve relatively complex problems.

Imprinting   This is a fourth type of learning (a form of early social learning) that has been observed in some species, such as chickens, ducks, geese, and turkeys. The Austrian zoologist Lorenz did the pioneering work in this field. When a baby duckling is exposed immediately after **hatching** to some moving object (especially if the object emits a sound), it adopts that object as its parent. Ducklings may adopt a human, dog, cat, or any other moving object as a parent. Imprinting can usually be accomplished only within the first 36 h following hatching of the duckling. Apparently, inheritance controls the length of the sensitive period when the individual can be imprinted, the object to which it can be imprinted, the tendency to respond to the first object to which it is exposed, and the permanence of attachment to the object once imprinting occurs. Although imprinting does not occur as distinctly in higher mammals, early association of mammalian young with their species and humans are important for relations later in life (socialization).

20.2.3 Intelligence

This is the ability of animals to learn to adjust successfully to certain situations. It is sometimes defined as *the organization of behavior* (the ability to learn from experience and to solve problems). The degree of intelligence varies greatly among individuals within a species as well as among species.

Animals learn to do some things, but they inherit the ability to do others (often called *instinct*). It is evident that birds do not have to learn how to build a nest, because young that are hatched and reared away from their parents know instinctively how to build one. The spider spins a web peculiar to its own species without learning from its parents how to do it. There are many other examples of inherited behavior in animals that allows them to do certain things spontaneously (without previous learning or thinking). Intelligence, however, is not an instinctive type of behavior.

Much has been written in popular articles about the high intelligence of coyotes, crows, dogs, horses, and many other wild and **domesticated** animals. Arguments have raged for years as to which species is the most intelligent. Instances of unusual displays of intelligence by individual animals were formerly used to decide the issue. These methods have now been replaced by more precise experimental means of measurement. It should be emphasized that because species differ in many innate tendencies, tests for intelligence may be biased accordingly.

In most tests, mammals have been found to be the most intelligent of all animals. Heading mammals in intelligence are the **primates,** which include humans, monkeys, and apes. Ocean mammals rank second. Next are the **carnivores,** which include the dog, cat, fox, coyote, wolf, lion, tiger, bear, etc. Following the carnivores in intelligence are the **ungulates,** or **grazing** animals, which include cattle, elephants, horses, and swine. The most intelligent farm animal, contrary to popular belief, is probably the pig and not the horse. Both the elephant and the pig seem to rank above the horse in intelligence. However, these rankings are based on only a few studies and may not be completely accurate. Recent studies with swine show that their ability to learn varies significantly among individuals, sexes, and breeds.

Much recent study has been directed toward the mechanisms involved in memory, which plays an important role in determining intelligence. These studies show rather clearly that memory is of two types, *short term* and *long term.* They also indicate that memory is related to the synthesis of protein by the brain. For example, if protein synthesis is blocked in various ways, an animal forgets what it has been taught. This means that **DNA** and **RNA** within the cells are also involved, since they play an important role in protein synthesis by the cellular **cytoplasm.**

20.3   MOTIVATION

This is the internal state of an animal that causes the immediate behavioral response. Scientists often refer to motivation in terms of *drives* or *tendencies* to respond in a particular manner. Hence animals experience hunger, thirst, elimination, sex, and pain drives.

The hypothalamus is the part of the central nervous system that controls many types of behavior. Apparently, both inhibitory and stimulatory centers are located in this gland and function in behavioral control. For example, destroying a certain area of the hypothalamus that is related to appetite may result in an animal starving to death even when feed is available. If another portion of this area is destroyed, the animal may overeat until it becomes extremely fat. The hypothalamus is involved in many behavioral responses other than eating.

Endocrine gland secretions are also involved in certain behavioral responses, especially sexual activities. Substances from the hypothalamus cause the release of gonadotropic hormones from the **anterior** pituitary gland, which in turn cause

ovulation, the onset of **puberty,** and occurrence of the normal estrous cycle in females (Chapter 9).

Stimuli that are effective in triggering a certain behavioral response are called *releasing stimuli.* A certain part of an animal's environment may act as a source of the stimulus (releaser) to which a releasing mechanism in the animal is sensitive. If the stimulus is sufficiently high, activation through neural pathways causes a certain behavioral response. For example, a record of the vocalizations of a **boar** played over a loudspeaker will cause a **sow** or **gilt** in **heat** to assume the mating **stance,** from which she can be moved only by much force and effort. Releasing stimuli can often be designed that are more effective than those found in nature.

20.4 METHODS OF ANIMAL COMMUNICATION

Animals are such agreeable friends—they ask no questions; they pass no criticism.
George Eliot

Although animals cannot speak in the same way as humans, they do communicate very effectively with one another. The following are some ways in which they do this.

20.4.1 Sound

'Tis sweet to hear the honest watch dog's bark.
Lord Byron (1788–1824)

Sound is an important means of communication, especially among farm mammals (Figure 20.2). Most female farm animals respond to the distress calls of their young. Cattle of all ages also respond to a distress call of another of their species, regardless of its age. Research has shown that bats use a type of radar to fly in dark caves or at night. They transmit high-pitched vocalizations that hit an object and bounce back, thus helping them to avoid destructive contact with various objects. This phenomenon of echolocation by bats was first reported by the great Italian naturalist Lazzaro Spallanzani, in 1793. He removed the eyes of several bats and found that they could fly about a room without hitting the walls or furniture. However, when he plugged their ears, the eyeless bats could not navigate without hitting the walls. He concluded, therefore, that bats use sound cues. Dolphins use a similar method of navigation to swim in the ocean depths. Farm animals do not appear to possess such a system of navigation. Birds are happy animals that sing with joy as they welcome the morning and the spring and as they express love for their mate. In a sense, they also communicate with humans by sound, as suggested by the following quotation:

FIGURE 20.2
This bellowing bull exemplifies the many animals that communicate by sound. *(Missouri Agr. Expt. Station.)*

I value my garden more for being full of blackbirds than cherries, and very frankly give them fruit for their song.

Joseph Addison (1672–1719)

20.4.2 Chemicals

Many female insects secrete chemical substances (social hormones) that attract males. These compounds are called **pheromones** (Chapter 19). In mammals such as dogs, females in estrus apparently secrete a substance that attracts males from miles around. Other female farm mammals also secrete chemical compounds when in **heat,** and males appear to locate them by sense of smell. However, this does not appear to be as effective over long distances as in the case of the **bitch.** Dogs use urine as a marker of their presence and possibly of their home territory. **Stallions** are reported to do the same thing by depositing their **feces** at particular locations. Compounds present in saliva and the preputial secretions of boars are now used commercially to stimulate estrous sows and gilts to stand for mating.

20.4.3 Visual Displays

Birds are noted for their visual displays in the act of courtship. Visual displays during courtship are less evident among farm mammals but do occur to some extent. Dogs, when they strike a hostile stance, cause the hair to rise on top of their necks (raise their **hackles**). This serves to make them look larger and probably more formidable. Honeybees communicate in an interesting way. When a worker bee finds a good source of food, she returns to the **hive** and performs a certain kind of dance that directs other workers to the site of the food. (The type of dance apparently indicates the distance to the food.) Scientists believe that the location of the sun is used by worker bees to give exact directions to other workers. The flashing lights male fireflies use to attract females are an example of another type of visual signaling system.

20.5 ORIENTATION (NAVIGATION OR HOMING) BEHAVIOR

Many stories have been told about how cats, dogs, cattle, and horses often find their way back home when moved to distant places. They apparently can return home by observing certain landmarks and/or by smell.

The best-known example of returning to home range after several years' absence is that of salmon. These fish are hatched (spawned) in freshwater streams. They then swim to the ocean, where they spend several years growing to maturity. When the time comes for them to spawn, they unerringly return to the same stream in which they hatched. Research shows that they do this largely by their sense of smell, because salmon are unable to locate their home stream when this sense is destroyed. Certain migratory birds, homing pigeons, and turtles are thought to use the sun and stars as compasses to guide them.

20.6 TYPES OF ANIMAL BEHAVIOR

Animals display several different types of behavior. Those discussed here will be restricted mostly to farm animals.

20.6.1 Ingestive Behavior

This type of behavior includes eating and drinking and is characteristic of animals of all ages. It is very important because it is closely related to the rate of growth and well-being of an individual. The first **ingestive** behavior trait common to all **neonatal** mammals is **suckling.**

Farm mammals vary in the way they ingest food. Swine and horses possess teeth in the upper and lower jaws, so that they can bite off grass or take a mouthful of grain, chew, and then swallow. The juices of the digestive tract complete the digestive process. Cows have no upper incisors, and so they normally ingest food during grazing by wrapping their tongues around a bunch of grass and then jerking the bunch forward so that it is cut off by the lower teeth. Sheep graze similarly to cattle but have a **cleft** upper lip that allows them to graze more closely to the ground. Goats graze similarly to cattle and sheep, but they are excellent browsers, often feeding on shoots of shrubs and trees. Once the food is taken into the mouth by these **ruminants** (cow, sheep, and goat), it is swallowed. It is later **regurgitated,** chewed more thoroughly, and then swallowed once more. This is known as the process of **rumination** (Chapter 15).

The grazing habits of these farm animals give practical importance to proper pasture and range management. Grazing animals eat the more tender parts of plants and reject the coarser ones. They also tend to eat first those species of plants more to their liking. When the more acceptable species are completely consumed, however, animals will graze the less desirable ones. When overgrazing is practiced over a period of years, the more **palatable** plants may be gradually eliminated, being replaced by those of lesser palatability and productivity (Section 18.6). Thus overgrazing over a period of years may cause a considerable change in the species of plants on a pasture or range and thereby reduce its quality and carrying capacity. Most grazing animals do not consume poisonous weeds when the supply of desirable **forages** is adequate; however, when overgrazing occurs, animals may consume poisonous weeds, become sick, and possibly die.

Cattle prefer some species of forages to others, perhaps because some may have a bitter or undesirable taste. In the United States, a pure stand of fescue is grazed by animals, and they remain in good condition. However, if the fescue is present in a pasture mixed along with more acceptable forages, the fescue will likely be the last grazed.

Cows on the range generally spend from about one-third to one-half of their time grazing. This time varies with forage availability and density. Cows spend a little less time, on the average, in ruminating than in grazing, but this also varies with the abundance of the forage. Cattle graze at all hours of the day and night, but peak activity usually occurs (when **ambient** temperature and other environmental factors are favorable) just after daybreak, in the afternoon, and before dark. Nursing activities may occur both during the day and at night, but they seem to be more frequent early in the morning, at midday, and about dusk.

Swine have very distinctive feeding and drinking habits. **Pigs** are born with the inherent tendency to root (dig and turn the soil). This is probably a characteristic carried over from their wild ancestors, who rooted in the soil to find worms and insects that would provide them with protein, energy, and, possibly, certain essential mineral elements. Domesticated swine will root less if fed a well-balanced **diet** and a good mineral **supplement.** Swine can balance their diet if fed a grain such as shelled corn in one compartment of a feeder and a protein supplement in another. Their choice of a balanced diet might be termed "nutritional wisdom." They seem to prefer some foods to others, possibly because of a difference in taste. When the protein supplement fed free choice with a grain is largely soybean meal, pigs tend to eat more protein than needed to balance their diet. When fed a diet rich in meat scraps or **tankage,** they are less likely to eat more protein than needed. It is of interest that pigs seldom, if ever, become sick or die from overeating. This is a problem encountered in other species of farm mammals.

Grazing in **poultry** is limited because they do not have a digestive system designed to handle forages (Chapter 15). However, they can utilize a limited amount of tender, low-fiber forage. Poultry subsist largely on grain diets and, like swine, do not commonly suffer from overeating.

20.6.2   Eliminative Behavior

Some farm animals deposit their feces at random; others do not. Eliminative behavior in farm animals tends to follow the general pattern of their wild ancestors but can be influenced by the method of management.

Cattle deposit their feces in a random way. They also do not appear to avoid contacting these deposits. When cattle on pasture lie down to rest in a group, many animals may have **defecated** by the time they arise. Cows can defecate while walking, so that their feces are scattered, but in general, feces are deposited in a neat pile while the cow is in a standing position. It is evident that the American *bison* must have had a similar habit of defecation, because the small piles of **manure** (called *buffalo chips*) were collected by the early settlers as a source of fuel for cooking. Most cows and bulls urinate in a random way while standing, although some variations in posture have been observed. Eliminative behavior in sheep is similar to that in cattle. **Ewes,** however, assume more of a squat position when they urinate than do cows.

Although swine are usually thought to be unclean in their habits, they are among the cleanest of all farm animals when given the opportunity. Normally, pigs keep their beds and nests clean, dry, and free of feces and urine. They usually deposit their feces in a corner of the pen, away from the sleeping quarters. When maintained on pastures with shelters, they keep the shelters free of **excreta.** Pigs usually deposit their excreta near the source of their drinking water; however, modern methods of rearing pigs place them in crowded pens, where their natural eliminative patterns are thwarted. It has been suggested that when pigs are first placed in the confines of a pen having a concrete floor, the act of watering one corner for a few days will induce the pigs to deposit most of their feces and urine at this place.

Horses tend to deposit their feces in a certain place and will often return to this place. Some writers report that on the range where several stallions are in the same pasture with a group of mares, each stallion has his own band of mares and a territorial range in which key border positions are marked with his feces. Chickens seem to deposit their excreta at random, except for the usual heavy deposition under **roosting** places at night. Cats bury their feces and urine, whereas dogs tend to deposit them at particular places (scent posts). The sniffing, selection of upright vertical targets, and leg lifting by adult male dogs at the time of urination have been shown to be a secondary sex characteristic under hormonal influence.

20.6.3   Shelter-Seeking Behavior

Shelter seeking by animals is an attempt to avoid injury from the sun, wind, rain, **predators,** and/or insects (Figure 20.3).

FIGURE 20.3
Pigs seeking shelter from the heat of midday.

Cattle in the range country of the southwestern United States have been known to descend from the high mountain country to the lower ranges just ahead of a violent storm. Hundreds of cattle are often seen making such a descent in single file. They seem to have the ability to sense and to avoid violent storms. Cattle may also run and "act up" prior to storms. Cattle usually seek shade in the heat of the summer day and do not come out into the open to graze until dusk or sunset. Cows on an open range devoid of trees often congregate around a water source beginning about 10:30 A.M. Usually, they remain quiet and rest when the weather is hot. They leave the water source later in the afternoon.

It is absolutely necessary for swine to seek shade in hot weather and to avoid direct rays of the sun, because they do not possess an efficient internal cooling system (Chapter 13). If water is available, when the ambient temperature is high, they wallow in it to keep cool. If forced to remain in the hot sun, they pant rapidly and often utter grunts or sounds of distress. In summer pigs sleep stretched out full length to expose maximal body surface, whereas in cold weather they sleep curled up to expose minimal body surface.

During a rain- or snowstorm, cattle, horses, and sheep turn their backs to the storm and often drift in the opposite direction. This is probably done to protect their eyes and face. The bison is one of the few animals that will face a storm head on.

20.6.4 Agonistic (Combat) Behavior

This type of behavior involves defense (submission), offense (aggression), escape, and passive activity. Among species of farm mammals, males (Figure

20.4) are more likely to fight than females, but females may exhibit fighting behavior under certain conditions. **Castrated** males are usually quite passive, which indicates that hormones (especially testosterone) are involved in this type of behavior.

Boars, bulls, **rams,** and stallions that run together from a very young age seldom fight. Perhaps they have already settled their social rank among themselves, when they are not being observed. Mature stallions can run together with a group of mares, although each stallion usually has his own band. Much has been written about fights between stallions over a band of mares, and no doubt this does occur, the victor either retaining the band or taking it from the vanquished. This is nature's way of guaranteeing that the strongest stallion will be the protector of the band and will leave more and better **progeny** to reproduce the next **generation.**

Under range conditions there are instances in which dozens of bulls are run together with a large group of cows. Even though the gathering includes many different bulls of all ages, fighting among them is seldom observed. In the fall, on an Arizona range, it is not uncommon to see a group of mature Hereford bulls grazing together in a group without a cow in sight. They seem to prefer the company of their own sex at that particular time. This behavior was observed on

FIGURE 20.4
Bulls demonstrating agonistic (combat) behavior. *(Photograph by J. F. Lasley.)*

an extremely large range on which the bulls were allowed to be with cows all year.

Although young males raised together will seldom fight, often in a large group of young bulls many will **ride** a single individual to death if he is not removed from the group. In feedlot steers, this phenomenon is called the *buller syndrome,* and it can have serious economic consequences. Similar riding behavior is often noted in a group of young boars.

The bringing together of sexually mature strange males of the same species almost always results in a fight. The intensity of fighting depends on the tenacity of the two combatants. Sometimes one becomes submissive after a brief skirmish. At other times the fighting is intense, and although fighting rarely results in death among farm animals, a serious injury to one or both often results. Fighting among two young boars on a hot summer day may end in the death of one or both from heat exhaustion. If the combatants are allowed to "fight it out," one usually becomes submissive and the fight ends. In other words, one animal learns that the other is boss. When a group of **roosters** run with a **flock** of **hens,** it is not unusual to see two roosters fight at intervals for several days. Sometimes the fight is ferocious, both combatants being covered with blood.

Some **strains** and breeds of cattle are born with the instinct to fight. Bulls in Spain and Mexico are bred especially for their fighting ability. Dogs and game chickens are also bred for this purpose. Many cows on the range will attack a human on foot or on horseback if they are roped or aggravated and become hot and disturbed. Many wild range cows have sent cowboys to the top of a corral fence. If these same animals had been raised in closer contact with humans, they would be more docile. Early experiences greatly affect an animal's reactions later in life. Some of the fighting behavior of domesticated animals is probably a carryover from their wild ancestors. In the wild, the ability to fight was necessary for survival of the individual and the species. Interspecies fighting is not commonly observed.

Most animals possess the trait of *play behavior,* but in varying degrees. This type of behavior is especially noted in young vertebrate animals (e.g., cats, dogs, and primates). The novice observer may confuse *play* behavior with *combat* behavior in some species.

20.6.5 Sexual Behavior

Oh, Love's but a dance, where time plays the fiddle.

Austin Dobson (1840–1921)

This type of behavior involves courtship and mating. It is largely controlled by hormones (Chapter 7), although bulls and stallions that are castrated after sexual maturity (**stags**) may retain a large amount of sex drive and exhibit all types of sexual behavior. This suggests that psychological (learned) as well as hormonal factors are involved. Sexual behavior is of importance because it is responsible for the reproduction and continuation of the species. The number of young produced and the vitality of the young are major factors determining profits in livestock production. It is interesting that sheep and goats will readily mate with each other when confined together, although interspecies matings are rarely **fertile.**

Males of the domesticated species of farm mammals are usually kept separate from females except during the breeding season. This is particularly true of registered cattle, sheep, swine, and horses. The detection of **estrus** in females often depends on the observation of the farmer or livestock producer. At times this is difficult for humans, because the signs are rather faint in some females. Male animals are much more efficient in detecting estrus; moreover, they maintain a 24-h vigil.

Because of the restlessness exhibited during estrus, the female is often said to be in *heat*. Females in heat seem to be burning with an intense sexual desire. If the herdsperson places his or her hand on top of the back or hips of a sow or gilt in heat, she will frequently stand absolutely still, with her ears erect or partially erect. **Mares** in heat will squeal and urinate profusely in the presence of other horses, but the efforts of mares and ewes to find a male while they are in heat are not so obvious as are those of females of other species of farm mammals. When in heat, cows and sows will stand when **mounted** by other females, whereas mares and ewes seldom exhibit this type of behavior. Some cows, when kept alone, will become restless, walk the fence, and bawl when they are in heat. Some may even go through fences, which they would otherwise not attempt to do, looking for a male. When kept alone, sows often break out of pens when they come into heat and range far and wide in search of a male. This is called *appetitive behavior.*

Males in most species of farm mammals detect females in heat by sight or smell (Figure 20.5). Courtship is more intense on the open range than under restricted farm conditions. A bull can often detect the proestrous cow 24 to 48 h before estrus and frequently remains in her general vicinity. The stallion bites or **teases** the mare in a kind of courtship that is responsible many times for getting the mare into the frame of mind to accept the male in **copulation.** The boar often nudges the sow or gilt around the head with his head and nose and emits varied grunts *(chant de coeur)*. He will then attempt to mount her. A rooster will spread one of his wings toward the ground and scratch the wing feathers with one leg, performing a sort of dance (waltz) around the hen. This is also a form of courtship. Sometimes the hen runs away, but at other times she squats, allowing the rooster to mount and copulate.

20.6.6 Mother-Young Behavior

The various forms of maternal (parental) behavior in farm animals begin shortly before birth and end after the young are weaned. This behavior, however, varies widely among the different species of farm animals.

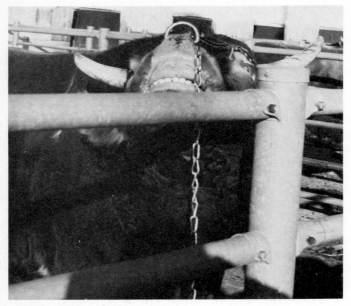

FIGURE 20.5
The upturned, or curled, upper lip indicates that bulls can detect an
odor signifying estrus in the cow. *(Missouri Agr. Expt. Station.)*

When **parturition** approaches, a cow in a large pasture or on the range seeks
seclusion from the herd. She usually chooses a small depression or ditch, a patch
of shrubs, or a clump of trees within which she is hidden from view. After the
young is born, the mother rises to her feet and closely inspects the **calf,** licking it
so that the hair coat is clean and shiny. Within a few minutes after birth, the calf
stands and, although somewhat wobbly, attempts to nurse. The attempts are
awkward at first but usually prove successful. The mother cow stands patiently,
often emitting a soft moo, until the young one has nursed. After nursing, all
seems well, and the calf exhibits a sudden burst of vigor. A little later, the
mother hides the calf and may return to the herd. The mother continues to cast a
wary eye in the direction of her hidden young, but she is careful not to reveal its
location, especially to humans. While hiding, the calf remains motionless in a
sort of stupor, and if picked up, it may appear ill or dead to the novice. The calf
is so quiet that often one can walk within a few feet of where it is hiding without
seeing it. After 2 to 4 days of intermittently nursing and then hiding her young,
the cow will usually bring her baby to join the herd.

If a calf is **stillborn** or dies shortly after birth, some cows soon leave the place
where the young lies, never to return. Others may return to their dead calves at
frequent intervals during the day and for several days, smelling it and mooing
gently to it.

The mother cow is very possessive with her young throughout the nursing period. If her calf encounters trouble from a human or beast, she quickly comes to its rescue. Some cows become quite belligerent when their calves are disturbed and may physically harm a person who comes too close. The tendency of the mother to give protective care must have an inherited base, since it varies widely among individuals and breeds. It has been reported that the F_1 female from the American *bison*–European cattle cross does not possess a normal maternal instinct, and the young must be reared by a foster mother. This finding also suggests that inheritance is involved in maternal care.

Cows apparently locate their young by smell, although sight may also be involved. Regardless of the way in which a cow and her calf recognize each other, it is accurately done even when a large herd of cows and calves are crowded together. At roundup time on the range, where many different brands are represented in the herd, the cowboys must wait until the cows and calves "mother up" in order to put the same brand on the calf that is on the cow (Figure 20.6). A mistaken pairing of a cow and calf means that one owner loses a calf while another gains one that does not belong to him or her.

Hereford cows on an Arizona range exhibit an interesting type of protective behavior toward their calves. Often, in a pasture of thousands of acres, there are only one or two watering places. It is not unusual to see 12 to 15 baby calves in an area accompanied by only one or two mother cows. Usually, the other mothers have gone to water, leaving a nurse or two to protect their young from

FIGURE 20.6
Waiting for cows and calves to "mother up" on an Arizona range so that the mother's brand can be placed on her calf. *(Photograph by J. F. Lasley.)*

predators. The giraffe and elephant also employ a "baby-sitter" to look after their young while they seek food and water.

One of the noisiest times on the farm or ranch is weaning time, when cows in a herd are cut away from their calves. The calves and their mothers bawl (often in unison) almost continuously for 2 or 3 days before they again become quiet. After the cows and calves have been kept separated for several days, the bond between them is completely severed, and they show no mutual signs of recognition. Since the dairy calf is removed from its mother within a few days after birth and reared separately, the tie between the mother and her offspring is soon severed. This is not true, however, with range cattle.

One to two days before parturition, the sow, as a general rule, will build a nest. (However, many sows today are **farrowed** in confinement where there are no materials available for nest building.) In building a nest the sow carries leaves, straw, and other litter to the site and hollows out an area in the ground that fits her body contours. She leaves enough space in the nest for the pigs and lies in such a way that soon after birth the pigs will go to the side containing the mammary glands. The time of farrowing in sows spans a period of 2 to 4 h, although cases are known in which pigs were born on two consecutive days. Usually, however, a delay of several hours in farrowing due to a difficult birth (**dystocia**) results in the remainder of the pigs being stillborn. Some sows are highly nervous during the farrowing process and eat their pigs. The practice

FIGURE 20.7
Nursing behavior in pigs. *(Courtesy Howard Cowden.)*

commonly used to prevent this is to remove all dead pigs and placental membranes from the pen or nest as soon as possible and before the sow has had an opportunity to eat them. Once the sow has acquired a taste for flesh, she may develop a permanent pig-eating habit.

Sows are very protective of their young and will defend them when they squeal. The sow approaches the intruder with her mouth open and emits a series of sharp, barking grunts in rapid succession. This defensive behavior is much more evident in some sows than in others. The sow continues to care for her pigs until they are weaned, but after 2 to 3 days' separation from the young, she appears to lose interest in them. If pigs are left with the sow for several weeks past the usual weaning age, she may cease lactating and wean them herself. Sows will accept pigs from another **litter** if the foster pigs are placed with them during the first day or two after they have farrowed. Exchanging of pigs among sows is a common practice in herds in which many sows are farrowing at approximately the same time.

Baby pigs nurse at intervals of 1 or 2 h. After the first few days, each **piglet** usually has located a teat that it will nurse for the balance of the nursing period (Figure 20.7). A few pigs may nurse two teats if the litter is small. The rear teats of the sow are less likely to be functional at weaning time than are the front ones (Table 20.1). An explanation for this may be that milk begins to flow first in the front teats and last in the rear ones (see Chapter 11). The stronger pigs tend to get the nipples that are first to give milk.

Many times the **runt** pigs get the rear teats, probably because they are not successful in competing with the larger and more vigorous pigs for the front ones. The extreme rear teats usually yield less milk than those toward the front,

TABLE 20.1
LOCATION OF TEATS THAT WERE
FUNCTIONAL IN 147 SOWS WHEN
THEIR PIGS WERE WEANED AT 56
DAYS

Location of teats*	Percent of teats functional at weaning
First pair	81
Second pair	79
Third pair	77
Fourth pair	69
Fifth pair	43
Sixth pair	27
Seventh pair	6

*The first pair in all sows was the pair nearest the front legs. Some sows, of course, did not possess seven pairs of teats. The percentage figures are based only on total teats present in that particular location.
Source: Missouri Agr. Expt. Station.

which may partially explain why pigs nursing the rear teats are commonly the runts in the litter.

After parturition, the ewe licks the newborn **lamb,** removing the moisture and placental membranes. The neonatal lamb soon staggers to its feet and makes awkward efforts to find a teat to nurse. The ewe stands quietly during this process. After a few days, the lamb goes directly to the ewe's teat and immediately begins nursing. Some time is required for the letdown of milk, however (Chapter 11). The suckling position of the lamb is usually a standing one. If the lamb's tail has not been **docked** (cut off), it wiggles from side to side during nursing. The ewe also shows a strong protective behavior toward her young. Although normally timid and easily frightened, she will defend her young even if the attacker is formidable. It is interesting that sheep will accept and suckle orphan goats **(kids)** and vice versa (interspecies rearing).

Mares show the same maternal behavior toward their **foals** that is shown by females in other species of farm mammals. A mare calls her foal with a **neigh** or **whinny** and exhibits considerable nervousness and distress when her young is disturbed. When horses were used for pulling farm implements, foals were often left in the barn while their mothers worked in the fields. When the teams were brought in at midday or nighttime, there was usually a noisy exchange of whinnies between mother and foal until they were put together and the foal was allowed to nurse. It is of interest that a mare will give as much attention and affection to a **mule foal** as she will to a horse foal.

Broodiness (an inclination to sit on eggs) of hens is inherited. For example, few hens in the Leghorn breed show this tendency, whereas many Cornish breed

FIGURE 20.8
Hen demonstrates the mother-young behavioral trait. *(Courtesy Dorothy Tompkins.)*

hens are broody. Broodiness is related to the secretion rate of and/or the response to prolactin (Chapter 12). Broody hens will sit on a nest full of eggs or on one that contains no eggs. They are determined (intense) in their nesting behavior and will spend very little time obtaining food or water during the nesting, or incubation, period.

Hens that are allowed to run with their **chicks** show an intense maternal behavior (Figure 20.8). While foraging for food, they will cluck to their chicks and call them each time a choice morsel is found. At the cluck of the hen, all the chicks will come running. Hens hover over their chicks by covering them with their wings and nestling them close to their bodies during the night and during periods of the day when they rest or need protection from the surrounding environment. Hens with chicks acquire a definite antagonistic attitude and will attack any enemy that bothers their young. The hen emits a loud, shrill cry to warn her chicks and others of the flock of imminent danger. The chicks respond quickly to such a warning. Today, few hens are allowed to incubate their eggs and care for their young. Incubation is accomplished by mechanical means because the efforts of hens are much more profitable when directed toward laying maximum numbers of eggs.

20.6.7   Investigative (Exploratory) Behavior

This type of behavior involves exploration (through seeing, hearing, smelling, tasting, and touching) of new quarters or pastures and the investigation of strangers that come close to the herd. Cattle investigate an object that they do not fear by approaching it closely, with their ears pointed forward and their eyes focused directly on it (Figure 20.9). Calves are likely to be more curious and investigative than older cows. When placed in a new pasture, cows graze quickly away from the place of entrance and soon have investigated every nook and cranny. As every stock producer knows, if there is an open gate or a hole in the

FIGURE 20.9
Investigative behavior in heifers.

fence around the new pasture, the cows will soon find it and enter another pasture or range.

Sheep react similarly to cattle in their investigation of strange objects in their quarters. They also approach the object with their heads up, ears forward, and eyes fixed on the object with a steady stare. However, they usually are more timid than cattle and will bunch and/or run if the object makes a strange and unexpected move or noise.

Swine also are curious. When approached by a person, they give a "woof," scatter as if badly frightened, and run as fast as they can for a short distance. If the person stands still and makes no motion to scare them, they will usually return and may approach the person, sniffing about him or her and nibbling or biting shoes and clothes. Perhaps their inquisitive nature is responsible for some pigs chewing and biting off the tails of penmates.

The young **foal** is quite curious and spends much time looking at and smelling the objects in its pasture or pen. As it grows older, it becomes less curious. The investigative habits of the young foal are a source of continual worry to its nervous mother, who at times seems to doubt the judgment of her offspring in making investigations.

20.6.8 Gregarious Behavior

Gregarious behavior in animals (observed only in certain species) means that they prefer to live in flocks or herds. In the wild state this is an advantage in detecting the enemy and protecting themselves. It may also be advantageous for obtaining food. It may be that some animals simply prefer a crowd. A method of trapping groups of wild cattle is depicted in Figure 20.10.

Cattle tend to roam in groups of various sizes, depending on the size of the pasture or range. However, there is usually considerable space between individuals when they are grazing. In a farm herd in which the number of cattle can be easily counted, one or two individuals missing from the group signifies that something is wrong. The missing animal(s) may be sick or dead or may have escaped into another pasture.

Cattle breed promiscuously, and there is not a tendency for a male and female to pair for life, as is usual in some species of mammals (e.g., foxes and wolves). However, if several cows in a herd are concurrently in heat and only one bull is present, the bull may show a preference for a certain female and not mate with one or more of the others.

Sheep are notorious for their tendency to run in a flock. They also have a very strong tendency to follow a leader. The leader is usually an older ewe, since rams usually run with the flock only during the breeding season. Lambs in a slaughterhouse are often led from one place to another by a ram or goat that is halterbroken and led by an attendant. Sheep raising was a very important occupation in biblical times. The writers of the New Testament certainly were aware of the gregarious nature of sheep and the ease with which they could be led. They were also aware of the strong tie between the shepherd and his flock,

FIGURE 20.10
A wild cow trap on a rough mountain range in Arizona. The country is often so rough and brushy that it is not possible to round up cattle, and they become wild. Thus they must be trapped during roundup time. Note that the triggers are set so that animals can go to water and salt but cannot return. When it is not roundup time, the triggers are opened and the animals become used to going in and out of the opened trap.

as noted in John 10:14: "I am the good shepherd, and know my sheep, and am known of mine."

Swine are gregarious in nature but now have little opportunity to show this because of modern methods of husbandry. They are usually confined to a limited area during the growing-fattening period. Domesticated pigs that have grown up in the wild have been known to roam in herds of fewer than 10 and are usually under the leadership of a mature boar or sow. A mature wild boar, with his long, sharp tusks, is a good match for almost any enemy. Thus he protects and guards his small group very efficiently. Under domestication the pig has lost much of its ferocity and usually is a gentle and easily handled animal, although there are exceptions. A part of the ease of gentling pigs under domestication is the result of **selection** for this trait. Selection progress in swine may be rapid because of their high reproductive rate, as compared with other farm mammals, and the relatively short **generation interval** (Chapter 5).

Horses in the wild were gregarious in nature and tended to run in herds of varying sizes. During the breeding season, each stallion had his own small band of mares, which he carefully guarded from other stallions and predators. Some domesticated horses show a definite preference for one horse and will avoid others. There seems to be an especially strong bond between the members of a team of draft horses that have combined their efforts for many days or years in pulling heavy loads.

20.7 SOCIAL DOMINANCE

Within most groups of gregarious animals of the same species there is usually a well-organized social rank order. One individual is dominant over all others, one ranks second, one third, and on down to the one that is subordinate to all. This order exists in all species of farm animals, although it is not always a clear-cut linear rank order. Since mature male farm animals seldom are run together in groups, the social rank order is usually important only in females. It has been said that much more attention is paid to social rank in a group of female farm animals in which social dominance exists than in any human social gathering.

The existence of a social rank order is most noticeable when the species has an instinct to fight or when resources such as food or mates are limited. At most other times, the social order is evident only in subtle ways. Social rank order is called the *peck order* in chickens, in which it was first discovered. When two strange hens meet for the first time, a vigorous fight usually follows. One hen will finally run away, thus ending the fight. The next time the two meet there is often another fight; however, it is usually less intense than the first but has the same result. At each meeting thereafter the hens fight again, but less and less intensely, until finally, only the threat of a peck from the dominant hen sends the subordinate one scurrying. It has been postulated that the establishment of a social rank order within a group is important because it prevents fighting and thereby allows more time to be spent eating and for productive purposes.

The rooster assumes the indisputable command of the flock of hens during the laying season, and the hens seem to accept this unanimously. The hens, however, establish their social rank order (hierarchy) as usual within the flock. Some observations indicate that if the rooster takes a fancy to a certain hen and pays her a lot of attention, her social rank rises in the flock. It goes down again, however, if the rooster drops her and transfers his attentions to another hen.

One of the principal advantages of social rank order in a group of wild animals is that it gives a mating priority to the top-ranked males. Hence they leave more progeny than the less dominant ones, and their progeny probably have a better chance of survival. The **offspring** of dominant females of some species grow faster, probably because they get more food. They may also have more of a particular hormone or a better balance of two or more hormones.

Social rank order among farm animals on pasture may be of little economic importance, but it may become very important when a group of animals are fed in confinement. This is especially true if the animals are fed a limited amount of feed. The dominant individuals under such conditions often get more than their share of feed, whereas the subordinate individuals may practically starve, become very thin, and perform inefficiently.

Several factors influence social rank in animals. When a male chicken is castrated (**caponized**), it tends to go to the bottom in rank. The injection of hens and/or roosters with the male hormone testosterone increases their social rank. Age is also an important factor. Young animals have a low rank and seldom try to establish a rank over their elders. Early experience also seems important, because if individuals are placed in a subordinate rank early, they usually remain there. Other young animals that do not rank as low as these may eventually assume a higher rank. In cattle, weight or size and aggressiveness or timidity are related to social rank. To rank high in the social order, the individual must have the will and ability to fight.

Social rank in chickens and dairy cattle appears to be highly **heritable,** which indicates that by selection and mating of subordinate animals, genetic progress can be made in achieving timidity and reduced aggressiveness in the progeny (Chapter 5). Some animal breeders believe that under domestication, humans have selected for the least aggressive animals, whereas in the wild state, selection favors those animals that are the strongest and most aggressive.

A social rank order similar to that found in animals is observed in humans. However, such rank order is related to social activity. Of two individuals, one might have a dominant rank over the other in one activity but be in a subordinate rank to the other in some other activity.

20.8 POPULATION DENSITY AND ANIMAL BEHAVIOR

Most animal populations increase in size until they reach a certain level and then remain more or less constant. Within such a population there are limited fluctuations from season to season and from year to year. Human populations differ from this by showing a long-term upward trend in numbers (Chapter 1).

The principal factors regulating population size or density of animals include predation, starvation, accidents, **parasites,** and **disease** (Chapter 1). However, present evidence indicates that these factors alone do not altogether account for this regulation. It seems that there are certain forms of social behavior that limit the rate of reproduction and prevent population size from exceeding the food supply.

The behavior of animals is related to control of their population density. Examples of this kind of control are found among certain species that limit the size of their territorial habitat during the breeding season. In some wild species, those individuals low in social rank may be driven from a particular area and not be allowed to reproduce there. The population of a particular area thus can be limited.

The question of population density and social **pathology** of rats has been examined.[2] A population of wild Norwegian rats was confined in an enclosure of about 1/4 acre. Plenty of food was supplied, and predation and disease were minimized. This left social behavior as the only important force that could

[2]*Sci. Am.*, **206**:139 (1962).

prevent an increasing population density. Within 27 months, the population size had become stabilized at about 150 adults. This lack of young was due to the fact that there was so much **stress** from social interaction that maternal behavior was upset and few young survived. Later, a similar experiment was conducted with domesticated albino rats, and similar results were obtained. Females were more affected by this treatment. Some did not carry pregnancies to full term, and **litters** of others failed to survive birth when pregnancies were of normal length. Some other females built inadequate nests or none at all. Infant **mortality** increased to 96 percent in these crowded females. The males showed disturbances ranging from certain sexual deviations to **cannibalism.** In fact, the entire social organization of the group was disrupted.

Although the relation of population density and social behavior of farm mammals has not been studied extensively, some livestock producers believe that a high density of animals affects the fertility rate. Whether this is an important factor is not known, but purchasing a pregnant female or one near breeding age and moving her to a new farm often results subsequently in poor reproductive performance. The production of specific **pathogen-**free pigs, in which the young are taken from their mothers before birth by a cesarean operation, is sometimes followed by poor reproductive performance of females. It is not known, however, if this is related to emotional stresses and disturbances encountered early in life. This is a fertile field for study in farm mammals and may be of considerable practical importance.

20.9 SUMMARY

Behavior of animals may be defined as the reaction to a certain stimulus. It includes all responses and ways of acting of animals. It is a complex subject and involves many scientific aspects. Many types of behavior in animals can be traced to their ancestors' ability to survive in the wild state. Domesticated animals have certain behavioral patterns similar to those found in their wild ancestors. An understanding of the basic principles of animal behavior is necessary in successful and efficient livestock production.

Behavioral responses in all animals are determined by heredity, learning, or experience. Hereditary responses are internal and can be illustrated by an animal behaving in a certain way without a need to learn the behavior. This is sometimes called *instinctive behavior.* Certain types of behavior, however, must be learned through experience. Most animals tend to act and respond in accordance with past experiences. The rapidity with which an individual learns in this way is closely related to its intelligence. Animals communicate with one another by sight, sound, and smell. Communication has an important influence on their behavior.

Typical kinds of behavior include ingestive, eliminative, shelter-seeking, agonistic (combat), sexual, mother-young, investigative, and gregarious. The importance of social rank order in domestic farm animals and the relation of population density to abnormal behavior need further study if people are to better understand animal behavior and its role in the production of foods useful to humanity.

STUDY QUESTIONS

1 Define behavior.
2 Of what practical value is a knowledge of animal behavior to the livestock producer?
3 What factors cause behavioral responses in animals?
4 In what ways may heredity produce behavioral responses?
5 What evidence is there that certain basic hereditary patterns affecting behavior exist?
6 List and describe the several types of learned behavior in animals.
7 What is intelligence in animals?
8 Which farm animal probably has the highest intelligence?
9 In what way can chemistry affect the ability to remember?
10 Define the term *motivation* (as it relates to animal behavior).
11 How may endocrines affect the behavior of animals?
12 In what general ways do animals communicate?
13 What are pheromones?
14 Name and define some of the most important types of animal behavior.
15 Describe some interesting examples of mother-young behavior.
16 What is meant by gregarious behavior?
17 What advantages are there in gregarious behavior?
18 Name one major difference between goats and sheep in their behavior.
19 What is meant by social dominance? How can it be of importance in livestock production?
20 What is the relation between population density and animal behavior in some species?

HORSES IN THE SERVICE OF HUMANITY[1]

The horse raises what the farmer eats, and eats what the farmer raises. But you cannot plow in the ground and get gasoline. You do not have to pay a finance company 10 to 15 percent to own a horse. Brood mares work on the farm and raise colts as well.

Horse power was much safer when only the horse had it.

William Penn Adair (Will) Rogers (1879–1935)

21.1 INTRODUCTION

The pathways of history are paved with the bones of horses. Archeological records indicate that the horse was first **domesticated** approximately 5000 years ago (Chapter 1). The domesticated horse immediately took a leading role in the destiny of the human race. No other animal has so inspired the human imagination. History reports demon horses, angel horses, ghost horses, sea horses, headless horses, etc. The **centaur** of the early Greeks (a weird combination of a horse's lower body with a man's upper body and head) probably resulted from impressions formed by the first encounter of Greek warriors with the enemy mounted on horses. It is said that the American Indians also believed that horse and rider were one being when they first met mounted Spanish soldiers.

One of the first uses made of horses was that of helping conduct wars. As early as 2000 B.C., the Assyrians used horses to draw bronze chariots that carried

[1]The authors acknowledge with appreciation the contributions to this chapter of Dr. Laurie M. Lawrence, Department of Animal Science, University of Illinois, Urbana.

armored soldiers. Soldiers of ancient Egypt and Greece rode horses in their conquests. Ghenghis Khan led his army of over 700,000 horsemen in the conquest of China. In addition to riding horses in battle, his soldiers used them to round up large **herds** of goats and sheep. England had its own breeds of horses before the Roman invasion, but the Romans also brought horses with them, so that interbreeding with British breeds likely occurred. In the age of knighthood in the British Isles, each knight used horses in defending his king, his religion, and his women. Horses that carried weapons, armor, and baggage during war were used as packhorses and were not ridden by the knights.

Columbus brought five **mares** and two **stallions** to the West Indies. Spanish colonists later brought additional horses to the new world. Cortés took horses with him when he invaded Mexico in 1519. These were the first horses on the American Continent since prehistoric days. Cortés later imported hundreds more to use in fighting the Aztecs. Some of these horses escaped, as did some of those belonging to early explorers and missionaries who journeyed into the southwestern United States. The horses that escaped reproduced rapidly because of almost ideal conditions, and by 1580 wild horses were so plentiful that Indians were capturing, breaking, and riding them. The Indian **ponies** were called **cayuses** (named after the Cayuse Indians, now living in Oregon). Later, when cowboys used these wild ponies to **work** (cut out, round up, etc.) cattle on the range, they were called **mustangs** (from the Spanish *mustengo,* meaning "wild"). A **bronco** was the name given a wild mustang that had not been tamed and broken.

Horses changed the lives of American Indians, who learned to depend on the speed of horses to help hunt and kill wild buffalo, their main source of meat. They made ropes of twisted horsehair and used horsehides to make clothing, beds, tents, saddles, and moccasins. The wealth of an Indian man came to be judged by how many horses he owned. When a chief died, his favorite horse was killed and placed in his grave with many of his other belongings. Indians believed the chief's horse would go with him to the Happy Hunting Grounds.

The cattle industry in the southwestern United States probably would not have thrived as it did without the tough, durable cow pony. Even today there are parts of the range where cattle cannot be worked without horses. The cowboy and his pony are as close as were the Indian and his pony. The horse also proved to be a hero of the Pony Express.

The development of farming in the grain belt of the United States depended on **draft** horses to pull farm implements. Later, horses were replaced by tractors. Horses of all kinds gradually decreased in numbers during the next few decades. Recently, however, horse numbers have been increasing because of the popularity of light horses for shows, sports, racing, and pleasure riding. Horses are still used to advantage for working cattle on the range and farm and in feedlots.

Additionally, horses are used for the production of certain **vaccines**; for example, Jumbo, a **bay gelding**, provided enough blood over an 11-year period to produce tetanus **antitoxin** and pneumonia **antiserum** inoculations for 210,000 people. Moreover, physicians frequently prescribe PMS (pregnant mare serum) because of its follicle-stimulating properties. This **hormone** is obtained from the blood serum of mares between 40 and 80 days of pregnancy (Chapter 7).

Many horse owners do not live on farms; nevertheless, they are careful to maintain the health and well-being of their prized pets. It is to this group that this chapter is dedicated, and many aspects of management are therefore considered, a departure from the biological and scientific orientation of the other chapters.

21.2 CHARACTERISTICS AND TYPES OF HORSES

The horse is a member of the family Equidae, which includes all living horses, donkeys, zebras, and **onagers**, as well as their many extinct ancestors. All living Equidae are members of the genus *Equus,* which includes six **species:** domestic horses and their closest wild relatives *(E. caballus);* the asses, domestic and wild; the onagers, wild asses of southwestern Asia; and three kinds of zebras.

The major external parts of the domestic horse were shown in Chapter 6. The skeleton of the horse is presented in Figure 21.1. A knowledge of the skeletal parts is useful in describing the location of certain abnormalities in horses.

21.2.1 Size and Type

Domesticated horses vary greatly in height and weight. Ponies, the smallest, weigh 300 to 800 lb; draft horses, the largest, may weigh a ton (2000 lb) or more at maturity.

FIGURE 21.1
The skeletal system of a horse. *(Courtesy American Museum of Natural History, New York.)*

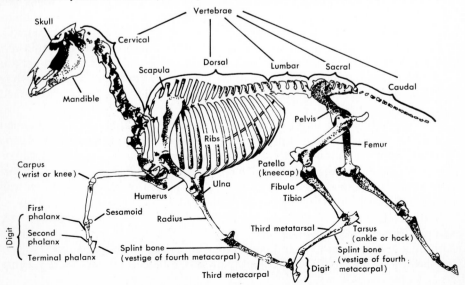

The height of horses is measured in **hands** from the ground to the highest point of the *withers,* the ridge between the shoulder bones. A hand is the average width of a man's hand, *4 in.* Mature horses measuring less than 58 in. (14.2 hands) in height are usually called *ponies.* Certain draft horses attain a height of 20 hands, whereas most horses of the riding and racing breeds are from 15 to 16.2 hands high. Racing horses weigh from 800 to 1100 lb and most riding horses from 1000 to 1400 lb.

Humans have developed the natural qualities of horses by **selective** matings. For example, speed and power have been vastly increased. **Breeds** for many other purposes, such as riding and working livestock, have been developed.

Horses are generally divided into three types: *light horses,* which have small bones, thin legs, and weigh approximately 900 to 1200 lb at maturity; *heavy horses,* which have large bones, thick and sturdy legs, and weigh 1600 lb or more at maturity; and *ponies,* which usually weigh less than 800 lb at maturity. Each of these types includes several breeds, and there may be more than one type within the same breed (e.g., there are Hackney *ponies* and *heavy* Hackneys). Names of selected breeds of horses were presented in Table 2.1. The names and general usages of various breeds are given in Table 21.1.

21.2.2 Gait

The way a horse walks or runs is referred to as its **gait.** In describing gaits, consideration is given to speed, beat, and leg movement. Gaits overlap somewhat, so that it is difficult to describe them clearly.

TABLE 21.1
SELECTED BREEDS OF HORSES CATEGORIZED BY FORM AND FUNCTION

Saddle horses	Heavy harness horses (coach horses)
American Saddlebred Horse	Cleveland Bay
Appaloosa	French Coach (Normand)
Arabian	German Coach (Oldenburger)
Lippizzan	Hackney
Morgan	Ponies
Quarter Horse (most registrations recently)	Connemara (good jumper)
Tennessee Walking Horse	Dartmoor (strong and surefooted)
Thoroughbred	Hackney
Light harness horses (roadsters)	Shetland
Morgan	Welsh
Standardbred (American trotting horse)	Wildhorses
Draft horses	Ass
Belgian	Forest horse
Clydesdale	Przhevalski's horse
Percheron	Zebra
Shire	
Suffolk Punch	

Horses have four natural gaits: *walk, trot,* **canter,** and *gallop. Walk* is the slowest gait (approximately four miles per hour). The *trot* is a two-beat gait in which each diagonal pair of legs (near hind and off fore, followed by off hind and near fore) hits the ground concurrently. The trotting speed is approximately nine miles per hour. The *canter* is a comfortable, three-beat, rhythmic riding gait at 10 to 12 miles per hour. There are three hoofbeats to each stride. Many people consider the canter to be the same as the gallop. However, a galloping horse may move at speeds of up to 40 miles per hour. The *gallop* is a four-beat gait in which the two hind feet give a two-beat count (left hind, then right hind), followed by a two-beat landing of the forefeet (left fore, then right fore).

Horses may be trained to use four artificial gaits: pace, slow gait, fox trot, and rack. When a horse *paces,* it moves the legs on the same side of the body at the same time, and with some pacers there is a movement from side to side as well as up and down. The natural pacing ability of a horse is influenced greatly by **heredity.** The pace is an uncomfortable riding gait and is avoided by those using cow ponies on the range. The **slow gait** is a high, methodical, showy stepping pace done very slowly and with restrained speed. If done properly, it is the most showy and thrilling of the five gaits (found in the American Saddlebred **five-gaited horse).** A slight variation of the slow gait is the *running walk.* This is a slow four-beat gait of the Tennessee Walking Horse, in which each foot hits the ground separately and never in combination with one of the others. It is quite comfortable for the rider. The **fox trot** is a broken trot in which the hind foot reaches the ground slightly ahead of the diagonal forefoot. This gives the appearance of walking in front and trotting behind. It is a medium-speed, four-beat gait that is very easy on both horse and rider. The **rack** is a rapid four-beat gait, free from any lateral or pacing motion, in which the legs move in pairs, but not quite simultaneously, so that each foot is lifted and put down separately. The rear foot strikes the ground a fraction of a second before the front on the same side. An experienced person can recognize this gait by ear if blindfolded. It is normally found only in the American Saddlebred five-gaited horse.

21.2.3 Eyes

Horses have larger eyes than any other land animal. Their eyesight is keen in both daylight and darkness. Horses are believed to be capable of distinguishing between their masters and other people at a distance of 1/2 mile. Their oval-shaped eyes are set at the sides of their head and move independently. Each eye moves in a half circle from front to back. Thus a horse can concurrently look forward with one eye and backward with the other. The horse's eyes make objects far to one side or to the back of the head seem to move faster than they actually do. This explains why a horse may *shy* (suddenly jump sideways or forward) at the slightest movement to its side or rear. Some horsepersons use *blinkers* (blinders that fit alongside the eyes) to block a horse's side and rear vision.

21.2.4 Coat Color

The **coats** of horses are many colors and combinations of colors. Solid body colors include black, brown, **bay** (reddish brown), liver chestnut (dark reddish brown), light chestnut or **sorrel** (red shades of chestnut or yellowish brown), **palomino** (golden), gray, and white. Many gray horses are born black but gradually turn lighter with age. Some gray horses (e.g., the **Lippizzans** of Austria) are white by maturity. Bays have a reddish coat and black *points* (legs, mane, and tail). Chestnuts vary the most in color.

Some horses have mixed colors, and these are given various names. **Piebald** indicates a black coat with white spots; **skewbald** is any color, except black, spotted with white (any piebald or skewbald horse may also be called a **pinto**); *pied* is a solid-colored coat with a few small spots; **roan** is a solid color (red or black) with white hairs growing throughout the coat.

Horses may be grouped according to *color type*. The terms *pinto* and *paint* are often used synonymously. However, pinto is now used more appropriately to refer only to coat color pattern, whereas the paint horse has additional characteristics and qualifications for purposes of registration. Paint horses are now generally recognized as pinto horses with Quarter Horse heritage. **Palominos** have golden coats and light blond or silvery manes and tails. **Appaloosas** have dark-brown or black leopard spots on a roan background. They have also been called "raindrop" horses because of their spots. The Appaloosa color pattern in horses is quite variable, with at least six basic patterns: (1) frost, (2) leopard, (3) marble, (4) snowflake, (5) spotted blanket, and (6) white blanket. The skin is mottled around the **genitalia** and sometimes around the eyes and nostrils. The **sclera** of the eye is white, and the hooves are striped. *Albinos* have a white or pale-colored coat. In nearly all mammals except the horse, the gene for albinism is recessive. However, in the horse the gene for albinism is believed to be lethal, so that apparently no horses are *true albinos*. True albinos have pink eyes and skin and no color in the hair. Pseudo (false) albinos may have a light skin and hair coat, but their eyes are colored.

Markings differ considerably among horses. Selected markings are depicted in Figure 21.2. Short descriptions of these markings are as follows:

Snip is any marking, usually vertical, between the two nostrils.

Star is any marking on the forehead.

Strip is a long vertical marking running down the entire length of the face from the forehead to the nasal peak.

Blaze is a broader, more open strip.

Star and strip is a marking on the forehead with a strip to the nasal peak; the strip does not have to be an extension of the star.

Star, strip, and snip is a marking on the forehead with a narrow extension to the nasal peak and opening up again between the nostrils.

Bald is a very broad blaze; it can extend out and around the eyes and down to the upper lip and around the nostrils.

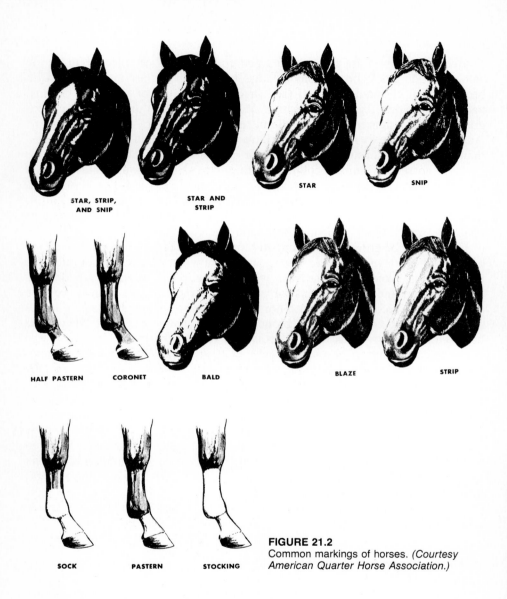

STAR, STRIP, AND SNIP

STAR AND STRIP

STAR

SNIP

HALF PASTERN

CORONET

BALD

BLAZE

STRIP

SOCK

PASTERN

STOCKING

FIGURE 21.2
Common markings of horses. *(Courtesy American Quarter Horse Association.)*

Coronet is any narrow marking around the coronet above the hoof.

Half pastern is a marking that includes only half the pastern above the coronet.

Pastern is a marking that includes the entire pastern.

Sock is a marking that extends around the leg from the coronet halfway up the **cannon bone,** halfway to the knee on the foreleg, or halfway to the hock on the back leg.

Stocking is a full marking almost to the knee on the foreleg and almost to the hock on the hind leg; it is an extended sock.

Suggested modes of inheritance of the many coat colors are presented in Tables 21.2 and 21.3. Geneticists do not agree on these, and changes will be made as more knowledge is gained from research.

21.3 SELECTION OF HORSES

Because humans have been able to develop many types and breeds of horses, it is probable that their form, structure, and function are strongly inherited, even though detailed research into the genetics of traits has not been carried out. Methods successfully used for the improvement of other species of livestock,

TABLE 21.2
GENES THOUGHT TO BE INVOLVED IN DETERMINING COAT COLORS IN HORSES

Gene(s)*	Coat color(s)
AA, Aa	A is a pattern determiner, making color appear in the mane and tail, causing black mane and tail, or white or light mane and tail of the **sorrel** and **palomino.**
aa	Uniform color, mane and tail same color as body.
BB, Bb	Black pigment in hair, skin, and eyes.
bb	Brown pigment in hair, skin, and eyes.
c^{cr}	Sometimes called the palomino gene. It dilutes bay to buckskin, chestnut or sorrel to palomino when heterozygous and to glass-eyed whites when homozygous, regardless of the background genotype for color. It has no effect on black or mouse colors.
D	D is a dominant dilution gene separate and apart from the c^{cr} gene. It converts black into mouse, bay into yellow dun with dark mane and tail, and chestnut or sorrel into yellow dun with dun mane and tail.
DD	Extremely dilute color (almost white); D is **incompletely dominant** to d.
Dd	Moderately dilute color (palomino).
dd	Allows color with no dilution.
EE, Ee	Full extension of black or brown pigment throughout the coat except as modified by the A gene.
E^d	Dominant black color, nonfading, Shetland type, completely covers the effect of the A gene. Thus E^d is **epistatic** to A.
ee	Red and yellow range in the body coat.
RR, Rr	**Roan.** (Also referred to as the Rn gene.)
rr	Nonroan.
GG, Gg	Progressive graying; any color at birth but grays with age.
gg	Does not cause a progressive graying of the hair.
SS, Ss	**Piebald** and **skewbald** spotting.
ss	Nonspotted, solid color.
FF, Ff	Face and legs solid color.
ff	White face and legs; ranges from extreme to star in forehead.
Ww	White (WW not seen—probably **lethal**); W is epistatic to all other colors.
ww	Allows pigmentation with other color genes.
W^{ap}	Appaloosa with blanket.

*None of these genes appears to be linked on the same **chromosome.**

TABLE 21.3
SUGGESTED GENOTYPES OF HORSES OF DIFFERENT COAT COLORS

Color	A, a, a^t	B, b	W, w	C, c^{cr}	D, d	E^d E, e	R, r	G, g	S, s	Comments
Bay										
Normal	AA or Aa	BB or Bb	ww	CC	dd	EE or Ee	rr	gg	ss	Bay, black mane and tail
Red	AA or Aa	BB or Bb	ww	CC	dd	ee	rr	gg	ss	Red, black mane and tail
Black										
Normal	AA or Aa	BB or Bb	ww	CC	dd	E^dE^d or E^de	rr	gg	ss	Dominant black, will not fade
Normal	aa	BB or Bb	ww	CC	dd	EE or Ee	rr	gg	ss	Black (fades brown in sun)
Smoky	aa	BB or Bb	ww	CC	dd	ee	rr	gg	ss	Black (fades red in sun)
Dilute	aa	BB or bb	ww	CC	Dd	Ee	rr	gg	ss	Dirty black; sometimes called good black
Chestnut	AA or Aa	bb	ww	CC	dd	EE or Ee	rr	gg	ss	Dark-brown mane and tail
Liver chestnut	aa	bb	ww	CC	dd	EE	rr	gg	ss	Uniform brown body, mane and tail
Seal brown	$a^t a^t$	BB or Bb	ww	CC	dd	EE or Ee	rr	gg	ss	Very dark brown
Sorrel	AA or Aa	bb	ww	CC	dd	ee	rr	gg	ss	Light mane and tail
Sorrel, uniform	aa	bb	ww	CC	dd	ee	rr	gg	ss	Uniform color body, mane, and tail
White			WW or Ww							WW is lethal, not seen
Gray			ww					GG or Gg		Foal colored at birth Becomes gray with age
Piebald and skewbald (see Sec 2124)			ww						SS or Ss	Spotted any color with white, depending on color of parents

Type	A	B	W	C	D	E	R	G	S	Phenotype
Roan										
Red	AA or Aa	BB or Bb	ww	CC	dd	EE or Ee	Rr	gg	ss	Bay and white with dark mane and tail
Blue	aa	BB or Bb	ww	CC	dd	EE or Ee	Rr	gg	ss	Blue (black and white hairs)
Strawberry	AA or Aa	bb	ww	CC	dd	ee	Rr	gg	ss	Chestnut or sorrel and white with light mane and tail
Buckskin	AA or Aa	BB or Bb	ww	$c^{cr}C$	dd	EE or Ee	rr	gg	ss	Sooty cream, black mane and tail
	AA or Aa	BB or Bb	ww	$c^{cr}C$	dd	ee	rr	gg	ss	Clear cream, black mane and tail
Palomino	AA or Aa	bb	ww	$c^{cr}C$	dd	EE or Ee	rr	gg	ss	Sooty type, white mane and tail
	AA or Aa	bb	ww	$c^{cr}C$	dd	ee	rr	gg	ss	Clear type, white mane and tail
	aa	bb	ww	$c^{cr}C$	dd	EE or Ee	rr	gg	ss	Sooty type, mane and tail same color
	aa	bb	ww	$c^{cr}C$	dd	ee	rr	gg	ss	Clear type, mane and tail same color
Pseudo albino	AA or Aa	bb	ww	$c^{cr}c^{cr}$	dd	Ee	rr	gg	ss	Pink skin, blue eyes, sooty pale cream
	AA or Aa	bb	ww	$c^{cr}c^{cr}$	dd	EE or Ee	rr	gg	ss	Pink skin, blue eyes, clear pale cream
	aa	bb	ww	$c^{cr}c^{cr}$	dd	ee	rr	gg	ss	Cremello
	aa	bb	ww	$c^{cr}c^{cr}$	dd	EE or Ee	rr	gg	ss	Cremello
	aa	bb	ww	$c^{cr}c^{cr}$	dd	ee	rr	gg	ss	Cremello
Mouse, grulla*	AA or Aa	BB or Bb	ww	CC	Dd	EE, Ee, or ee	rr	gg	ss	Sooty black
Dun*	AA , Aa	BB or Bb	ww	CC	Dd	EE, Ee, or ee	rr	gg	ss	Yellow dun with dark mane and tail
*Dun**	aa	bb	ww	CC	Dd	ee	rr	gg	ss	Yellow dun with dark mane and tail

*The actual genotype and phenotypes are controversial at present.

such as cattle, sheep, and swine, most likely would be effective in horse breeding. These methods and principles of mating were discussed in Chapters 4 and 5.

The improvement of horses through the application of proved breeding methods involves finding or identifying superior individuals and then mating them with the expectation that they will produce superior **offspring.** Superior individuals may be located and identified in several ways. These include close observation of the type and performance of an individual and of its close relatives. **Type** refers to the physical form and structure that should best fit individuals for a particular function. To know what is the most desirable form and structure, one must become familiar with the parts of a horse (Figure 6.5) and the characteristics of strength for specific purposes. One must also be able to evaluate and compare body conformation (type). For this reason a guide for evaluating conformation is presented in Section 21.3.1. Performance involves proper functioning of the horse. **Thoroughbreds** are **selected** for speed, saddle horses for various gaits, draft horses for the ability to pull heavy loads, and cow ponies for ruggedness and the ability to perform certain jobs in working cattle. To be successful in finding superior breeding stock, one must have a definite usable ideal in mind.

The selection of an individual can be based on the animal's own type and performance, often called *individuality.* Three kinds of records of relatives may aid in this selection: those of an individual's ancestors in its **pedigree,** those of its **progeny** or descendants (if any), and those of its **collateral relatives** (brothers, sisters, uncles, aunts, and cousins). The latter are not related to the individual as ancestors or descendants (Table 5.5).

A rule of thumb for improving horses through breeding methods is to select superior individuals that have superior relatives. The closer the relationship between an individual and its relatives, the more attention should be given to the type and performance of those relatives. Most of the major horse-breed associations now publish this information.

21.3.1 Evaluating Body Conformation of Horses

The usefulness and, to a large extent, the longevity of a horse are dependent on its conformation. Sound, strong feet and legs, for example, are essentials in all types of horses that serve humanity. Since conformation tends to denote service and utility, a few selected desirable and undesirable traits are presented in Figure 21.3.

21.4 CARE AND MANAGEMENT OF BREEDING ANIMALS

The reproductive rate in horses is low compared with that of other farm mammals. This results from a comparatively long **gestation** period, an older age when they first produce young, and a low level of **fertility.** (Fertility averages about 50 percent for mares in confinement and 60 to 75 percent for mares on

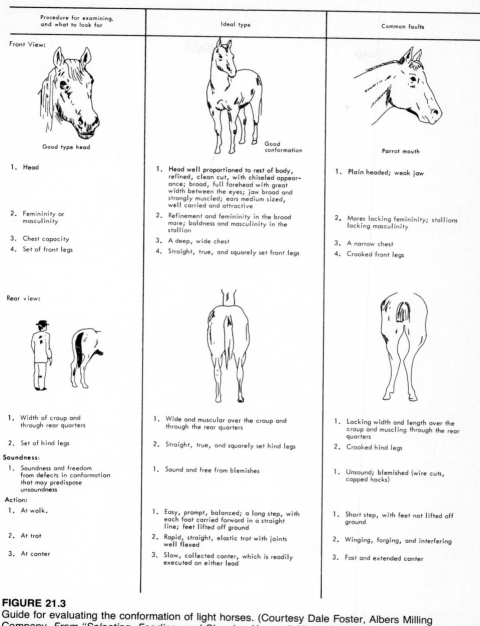

Procedure for examining, and what to look for	Ideal type	Common faults
Front View: Good type head	Good conformation	Parrot mouth
1. Head	1. Head well proportioned to rest of body, refined, clean cut, with chiseled appearance; broad, full forehead with great width between the eyes; jaw broad and strongly muscled; ears medium sized, well carried and attractive	1. Plain headed; weak jaw
2. Femininity or masculinity	2. Refinement and femininity in the brood mare; boldness and masculinity in the stallion	2. Mares lacking femininity; stallions lacking masculinity
3. Chest capacity	3. A deep, wide chest	3. A narrow chest
4. Set of front legs	4. Straight, true, and squarely set front legs	4. Crooked front legs
Rear view:		
1. Width of croup and through rear quarters	1. Wide and muscular over the croup and through the rear quarters	1. Lacking width and length over the croup and muscling through the rear quarters
2. Set of hind legs	2. Straight, true, and squarely set hind legs	2. Crooked hind legs
Soundness:		
1. Soundness and freedom from defects in conformation that may predispose unsoundness	1. Sound and free from blemishes	1. Unsound; blemished (wire cuts, capped hocks)
Action:		
1. At walk.	1. Easy, prompt, balanced; a long step, with each foot carried forward in a straight line; feet lifted off ground	1. Short step, with feet not lifted off ground
2. At trot	2. Rapid, straight, elastic trot with joints well flexed	2. Winging, forging, and interfering
3. At canter	3. Slow, collected canter, which is readily executed on either lead	3. Fast and extended canter

FIGURE 21.3
Guide for evaluating the conformation of light horses. (Courtesy Dale Foster, Albers Milling Company. *From "Selecting, Feeding, and Showing Horses,"* Albers Research Bull., *1966.)*
Figure continues on page 670.

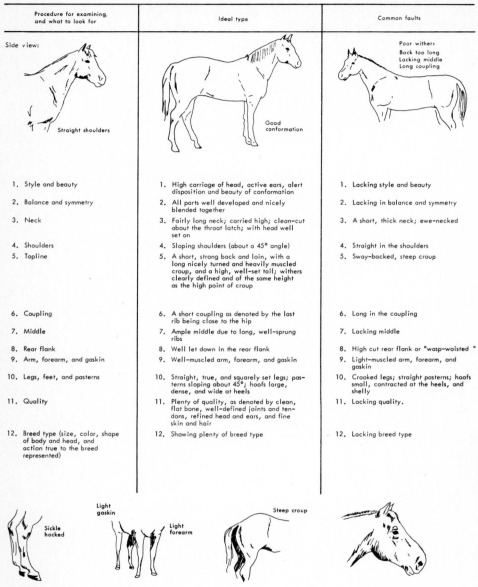

Procedure for examining, and what to look for	Ideal type	Common faults
Side view: Straight shoulders	Good conformation	Poor withers Back too long Lacking middle Long coupling
1. Style and beauty	1. High carriage of head, active ears, alert disposition and beauty of conformation	1. Lacking style and beauty
2. Balance and symmetry	2. All parts well developed and nicely blended together	2. Lacking in balance and symmetry
3. Neck	3. Fairly long neck; carried high; clean-cut about the throat latch; with head well set on	3. A short, thick neck; ewe-necked
4. Shoulders	4. Sloping shoulders (about a 45° angle)	4. Straight in the shoulders
5. Topline	5. A short, strong back and loin, with a long nicely turned and heavily muscled croup, and a high, well-set tail; withers clearly defined and of the same height as the high point of croup	5. Sway-backed, steep croup
6. Coupling	6. A short coupling as denoted by the last rib being close to the hip	6. Long in the coupling
7. Middle	7. Ample middle due to long, well-sprung ribs	7. Lacking middle
8. Rear flank	8. Well let down in the rear flank	8. High cut rear flank or "wasp-waisted"
9. Arm, forearm, and gaskin	9. Well-muscled arm, forearm, and gaskin	9. Light-muscled arm, forearm, and gaskin
10. Legs, feet, and pasterns	10. Straight, true, and squarely set legs; pasterns sloping about 45°; hoofs large, dense, and wide at heels	10. Crooked legs; straight pasterns; hoofs small, contracted at the heels, and shelly
11. Quality	11. Plenty of quality, as denoted by clean, flat bone, well-defined joints and tendons, refined head and ears, and fine skin and hair	11. Lacking quality.
12. Breed type (size, color, shape of body and head, and action true to the breed represented)	12. Showing plenty of breed type	12. Lacking breed type

Light gaskin · Light forearm · Steep croup

Sickle hocked

Coon footed · Splay footed · Tail set to low · Roman nose

FIGURE 21.3 (Cont.)

pasture.) Fertility is influenced by a low **conception** rate. Mares seldom deliver more than one **foal** at **parturition;** however, they do have a long reproductive life, often reproducing until they are 20 or more years of age.

The usual breeding and **foaling** season for horses is the spring, although some foals are produced during other seasons. Foaling in the spring is advantageous because of the warm weather and the abundance of green grass for both mother and foal. These induce fast growth and development of the young.

21.4.1 The Stallion

Although the stallion attains **puberty** in about 1 year, he is seldom used for breeding before 2 years of age. Even then, he is used on a limited number of mares with 10 to 12 matings well spaced throughout the breeding season. Most breeders do not use a stallion for breeding purposes until he is 3 years of age, at which time he can service 25 to 30 mares during the breeding season. Older, mature, highly fertile stallions may be hand-mated (closely scheduled and supervised matings) with 75 to 100 mares during a long breeding season. Usually breeding is limited to once daily; however, if the occasion demands, two or three well-spaced **services** per day are possible for a short period. Proper **nutrition** and exercise are important in maintaining good physical condition in stallions to be used for breeding purposes.

Artificial insemination of horses has been practiced for many years (Chapter 10). It has not been widely used, however, because the life of collected **sperm** is limited to a few hours. A method of successfully freezing stallion semen was developed recently, and if this method proves to be practical, artificial insemination of horses will likely become more widespread, although artificial insemination of horses is restricted or prohibited by some breed registries.

The volume of semen per **ejaculate** from the stallion is from 50 to 150 cc (the average for Thoroughbreds is 50 to 60 cc), and the normal sperm concentration is 30 to 800 million per cubic centimeter. Thus the total number of sperm per ejaculate is in the billions. However, sperm production varies with an individual, its age, and **environmental** conditions.

21.4.2 The Mare

Fillies (young females) reach sexual maturity between 10 and 18 months of age, but they do not reach physical maturity (mature body size) until 3 to 5 years of age. If kept as **broodmares,** they are not usually bred until they are at least 3 years old. Saddle or racing mares often are not bred until they are much older.

The length of the estrous cycle in normal mature mares averages 19 to 21 days. The length of **estrus** varies from 2 to 9 days, with an average of 5 to 6. **Ovulation** most frequently occurs 1 or 2 days before the end of estrus (Chapter 9). In some mares, ovulation occurs even though they do not show outward signs of estrus (quiet estrus). Only one egg is usually released from the **ovary,** although twin ovulations occur 2 to 5 percent of the time, depending on breed and season. Twin ovulations seldom result in delivery of two live foals.

Mares in healthy physical condition at breeding time are more likely to conceive. Proper feeding and plenty of exercise are vital. Extremes in **condition** should be avoided. If mares are too thin, their feed should be increased so that they are gaining weight at the time of breeding. (Thin mares often need worming.) Overfeeding and extreme fatness should be avoided. Mares usually get sufficient exercise if allowed to run in large pastures. When confined, however, they should be exercised (ridden or driven) regularly.

One can usually determine if a mare is in **heat** by trying her in the presence of a stallion. The stallion is restrained by a halter or bridle. He is allowed to tease the mare with a barrier such as a solid board fence, tease bar, or breeding stall between them. If the mare is in heat and receptive to the stallion, she may be bred. The optimum time to breed mares, if only a single service is given, is late in the heat period. This is when most ovulations occur. To increase conception, one may breed some mares on alternate days during estrus if the stallion is not being used too frequently on other mares.

During estrus, quiet mares will stand for service, whereas nervous ones may have to be **twitched** or **hobbled** to prevent movement and the possibility of their injuring (kicking) the stallion. The use of strong breeding hobbles or a properly constructed breeding stall is recommended. Occasionally mares may be tranquilized. Most mares come into heat 5 to 10 days after foaling, and mares are sometimes bred on the ninth day. However, breeding at this time is recommended only for mares free of **infection.** Uterine infections are indicated by the birth of a diseased or dead foal and a **retained placenta.** Infected mares should be treated and not bred until they return to a healthy condition.

Pregnancy is indicated by the cessation of the usual estrous periods. If a stallion is kept on the same farm, it is desirable to check mares for estrus on alternate days, beginning the fifteenth day after breeding. Pregnancy may be diagnosed by means of rectal palpation or by hormonal methods. A high concentration of gonadotropic hormone (mostly FSH) is present in mare's blood **serum** between 40 and 80 days of pregnancy. (Peak concentration is reached at about the seventieth day.) Blood serum from mares is injected into immature female mice or rats. A positive pregnancy test is indicated if the ovaries of the injected females show newly ruptured follicles or newly developed corpora lutea. A high level of estrogenic hormone is present in the urine of mares between the seventh and ninth months of pregnancy. Biological (an injection into mature, **castrate** female rats) or chemical tests may be used to detect the presence of these hormones.

The gestation periods of mares average approximately 340 days but vary between 335 and 345 days. However, a mare may carry her foal a year or more. At birth such young are usually extra large, weak, and difficult to raise. A mare shows several signs of approaching parturition. Her **udder,** often enlarging for several days, becomes distended with milk, which may drip from the teats 24 to 48 h before foaling. Beads of wax form on the teat canal in some mares just before foaling. The muscles around the tail head relax, and the lips of the vulva become enlarged. Restlessness is indicated by the successive lying down and rising of the mare.

Foaling usually occurs while the mare is lying down. Shortly before parturition, the fetal membranes (water bag) appear. They rupture, and soon the front feet and nose of the foal appear at the vulva. The normal and usual presentation of the foal is front legs first and outstretched, with the head and chin resting between the legs. The back of the young should be near the top of the mare's pelvis. Any other presentation is abnormal and increases the probability that the mare will require assistance for completion of the birth process. If labor is too difficult and prolonged (normal delivery time is 10 to 15 min), a veterinarian should be called. It is interesting that sizable breeding fees (often several thousand dollars in **purebred** racehorses) charged for selective matings are not paid until a veterinarian has observed parturition and declared the foal alive and normal. When foaling appears normal, it is better not to disturb the mare, although assistance may be given after the withers of the foal have passed the rim of the pelvis. Pulling the legs downward and outward toward the mare's hocks when labor pains occur is often helpful. Once the fetal membranes have ruptured, normal delivery requires only a few minutes, and delays (especially more than 45 min) may result in the death of the foal.

One should make certain that the foal is breathing after delivery and that all fetal membranes are removed from the mouth and nostrils. If the foal is not breathing, blowing into its mouth, tickling the nostril with a straw, working the ribs in artificial respiration, or lifting the foal and dropping it gently to the ground will often initiate the breathing process.

The **afterbirth** is usually expelled within 2 to 6 h postfoaling. If it is retained longer than 6 h, a veterinarian should be called. Even when the afterbirth is expelled normally, the hindquarters and tail of the mare should be washed and **disinfected** and stall litter or other debris removed. This aids in preventing uterine infections.

After foaling, the mare should be given small quantities of warm water at frequent intervals. She should be fed lightly with a laxative feed. (**Bran** mash mixed with oats is good.) The quantity of the first feed should be about one-half the normal **ration,** and this can be gradually increased for a week at which time the mare is switched to her regular diet. Feed should be increased more rapidly for thin mares or those with an inadequate milk supply.

21.4.3 The Foal

A disease caused by navel infection, *navel ill* or *joint ill,* is serious in new foals. It is characterized by lameness and swollen joints. The navel cord should not be cut. Rather, it should be allowed to dry up and drop off naturally. The navel cord may be disinfected with iodine or dusted with a drying powder recommended by a veterinarian 2 or 3 times daily until it dries and drops off.

It is imperative that within 24 h of birth the foal get **colostrum,** which is rich in vitamins and **antibodies** and is a natural laxative. Antibodies do not pass in large amounts from the blood of the mare to that of the foal. Thus the newborn foal without colostrum is likely to succumb to infections. Because of the importance of colostrum to the foal, the mare should not be milked before foaling.

A strong, healthy foal should be standing 1/2 to 2 h after birth. The foal should not be encouraged to stand right away but rather should be given plenty of time to stand on its own. Before the foal is allowed to nurse, the mare's udder should be washed with a mild disinfectant and rinsed with clean water. Some foals nurse almost immediately, whereas others may need assistance. A weak foal may be fed colostrum with a bottle and nipple. Owners should watch for constipation and diarrhea in the foal the first few days after birth and apply treatment if necessary.

The healthiest place for a mare and foal is a green pasture free of other animals. Green grass is rich in **nutrients** that stimulate rapid growth. Many foals begin nibbling hay and grain by 2 to 3 weeks of age. The foal should be encouraged to eat grain as soon as possible so that the dietary transition changes at weaning will be easier. Foals are usually weaned at 4 to 6 months of age by being separated from their mothers. If practical, the mother should be out of sight and sound range. Many farms now follow the practice of removing one mare at a time and leaving the weaned foal with other mares and foals. When all the mares have been removed, the foals can be turned out to pasture together. They should be fed hay and grain to keep them growing. When possible, foals are kept separate from older horses.

Horse **colts** may be castrated when only a few days of age, but most horse owners prefer to wait until they are approximately 1 year or older. When horses are on the range some owners prefer not to castrate colts until they are 2 or 3 years of age because they believe this allows better body development and greater vigor and strength. However, the older the horse at castration, the greater the danger of complications. Castration is best performed by a veterinarian. Springtime, before extremely hot weather, flies, and danger of screwworms (these have been eradicated from parts of the United States), is an excellent time for castration. Colts can then be turned out on pasture where there is maximum freedom from infectious microorganisms.

21.4.4 The Young Horse

Yearlings and 2 year olds will develop rapidly and possess sounder feet and legs if they are on pasture rather than confined in close quarters. Periodic handling and grooming, together with leading and training, are important. If properly trained during the nursing period, there will be very little training or breaking, except riding, to do when the young horses are older. It is desirable to teach the colt to lead as soon as possible, at latest by 1 month of age. Likewise, handling and care of the feet at 1 month is a good practice. It is also particularly important to control **parasites** (external and internal) in young horses.

21.4.5 The Working Horse

Mature horses perform many jobs, so that it is difficult to make specific recommendations for their care and management. They should be fed adequately, be worked frequently, have their feet properly trimmed and shod, and have their health protected.

21.5 NUTRITION OF HORSES

Feed is an expensive item in raising and maintaining horses. Rations should be adequate in both quantity and quality, yet formulated as economically as possible. The best diet for horses depends on the availability and relative costs of feeds.

Horses require the same food nutrients as other **livestock.** These include carbohydrates, fats, proteins, minerals, vitamins, and water. Grain, dry **forage,** and pasture supply most of these nutrients. Nutrient requirements of horses depend on age, size, and amount of work performed.[2] Horses used for breeding have additional requirements.

21.5.1 Concentrated Feeds

Oats are the most common grain fed to horses in many parts of the United States, especially in the middle west, where they are readily available. They are **palatable** and provide sufficient bulk to prevent packing in the stomach (which causes **colic**). Good-quality oats have a high protein content and, when fed with timothy hay, make a good balanced diet for mature horses. However, when grown on improperly fertilized soil, timothy hay and oats may give a high-phosphorus/low-calcium ratio, and hence horses require calcium **supplementation.** Oats may be fed in whole, crimped, or crushed forms. Rolled or crimped barley is substituted for oats where it is available in large quantities and at an economical price. Barley is the principal grain fed to horses in the southwestern United States, where horses are used to work cattle.

Corn is often fed to horses in the middle west, or corn belt, because it is available in large supply and is often cheaper than other concentrated feeds. When horses and mules were widely used for pulling farm implements, a combination of ear corn, oats, and limited **hay** was fed twice daily, in the morning and at night. Corn is a heavy concentrated feed that lacks fiber and bulk. However, corn has a high caloric density and is an excellent source of energy for hard-working horses. It is low in protein, which must be supplied by good-quality hay or a small amount of protein supplement. It may be fed as ear or as shelled corn but should not be ground. Finely ground corn forms a solid mass in the horse's stomach and often causes colic. It is also eaten more quickly than whole corn.

---

[2]For additional information, see *Nutrient Requirements of Horses,* 4th ed., National Academy of Sciences–National Research Council, Washington, D.C., 1978.

Grain sorghums can be fed to horses and are best when fed after crushing or coarse grinding. Whole wheat is not a commonly recommended feed for horses because it often gums when chewed and causes digestive disturbances. When competitively priced, however, it may be fed in limited quantities with other grains. Wheat bran can be fed to horses mixed with oats and hot water to form a bran mash. Molasses is frequently added to horse rations because it increases palatability and decreases dustiness. Wheat bran is a desirable feed for horses because of its bulk, palatability, nutrient content, and laxative properties. When economically priced, it may replace a portion of the oats in diets for breeding stock.

The traditional protein supplement for horses has been linseed meal. Linseed meal that is produced by mechanical means has a high lipid content and thus contributes to an excellent finish or sheen of the hair coat. However, the mechanical process has been replaced almost entirely by a chemical process that removes most of the oil from the meal. Consequently, linseed meal has been replaced in many horse diets by soybean meal, which is a higher-quality protein. Protein supplementation is usually only necessary for growing horses and pregnant or lactating mares.

21.5.2   Forages

Bright-green grass hay is the most common dry forage fed to horses. Timothy, brome, and orchard grass hays are popular because they are usually free of dusts and molds. Since grass hay is low in protein, it should be fed with a **concentrate** or protein supplement so that a balanced ration is provided. Prairie and well-cured oat hays are approximately equal to timothy hay as forage for horses. Bright-green and properly cured **legume** hays such as alfalfa, red clover, and lespedeza are also excellent for horses. Legume hays may be fed alone or in a mixture with grass hays, but it is important that they be practically free of dust and molds.

Good-quality **silage** that is finely chopped and free of decay and molds may be used to replace about one-half of the hay in the diet. Corn silage is preferred, but grass silage, legume **haylage,** or milo silage can be fed. Silage should be added to the ration gradually. Most silages are too bulky for working horses, which require a high-energy diet, but they are satisfactory for less active ones.

Pelleted feeds for horses are very popular because they are convenient to handle and feed wastage is minimized. They are particularly useful as a feed for horses with heaves (difficult breathing). Pellets may be slightly more expensive than a diet consisting of farm-produced hay and grain.

21.5.3   Pasture

This is the natural feed for horses. It more nearly provides the nutritional needs of horses than does any other feed. Mature horses doing little or no work can obtain all their nutrients from grass. Good pasture (improved and properly fertilized) is almost indispensable for broodmares, foals, and young horses. The

Bluegrass region of Kentucky is noted as a center of light-horse (especially Thoroughbred) breeding, largely because of its excellent pastures.

21.5.4 Salt and Minerals

The working horse has an additional dietary need for salt (NaCl) because it may lose 75 g or more daily through sweat and urine. A horse's need for salt may be met by having block or granular salt available at all times. Granular salt may be included in the diet at the rate of 1 to 2 oz daily.

Pregnant mares, **lactating** mares, foals, and young growing horses require additional calcium, phosphorus, and vitamin D. Milk production and the growth of bones make these requirements high. Mature horses, other than broodmares, require no supplemental minerals other than salt while on good-quality pasture or when fed high-quality hay, a portion of which is **legume.** For added protection, however, horses may be fed (**ad lib**) a mineral mixture (equal parts of steamed **bone meal,** ground limestone, and salt). Iodine should be provided to pregnant mares as iodized salt in iodine-deficient areas so that goiter is prevented in newborn foals. The need for other trace minerals has not been established, except in certain limited areas. A common practice is to make trace mineralized salt available free choice. However, the forced feeding of trace (micro-) minerals may even be detrimental (Chapter 17).

21.5.5 Vitamins

Horses receiving good pasture or high-quality hay (including legumes) do not need supplemental vitamins. Those confined for long periods and fed low-quality roughages are more likely to develop vitamin deficiencies. The vitamin requirements for such horses may be satisfied with suitable commercial supplements. Excess vitamins contribute nothing to the horse's health but are merely an added expense.

B-complex vitamins are synthesized in the large intestine and cecum of healthy horses. However, a need for supplemental dietary thiamine has been demonstrated when horses consume certain plants that contain thiaminase, which impairs the utilization of thiamine (Section 18.6), and a low riboflavin intake appears to contribute to periodic ophthalmia (moon blindness). Green plants and high-quality legume hays are good sources of all B-complex vitamins and also of vitamin A. The latter is essential for good health of the epithelial tissues, especially those of the eyes. Vitamin A is also necessary for proper bone growth and maintenance (Chapter 16). Vitamin D requirements of horses are usually satisfied when horses are exposed to sunlight. Claims that vitamin E supplements improve fertility in horses have not been verified by research.

21.5.6 Water

Horses should be supplied with fresh, clean drinking water at all times. Generally, horses drink 10 to 12 gal of water daily. Additional water is needed when they are working or when the weather is hot. Horses not having access to water free choice should be watered at least twice daily at regular intervals. An inadequate water intake may cause **impaction.** Horses should be permitted to drink only small amounts while they are hot, or **founder** may result.

21.5.7 Feeding Horses

Horses vary widely in the amount of feed required to maintain weight and condition. Some are **easy keepers,** whereas others are **hard keepers.** Defective teeth and/or parasite **infestations** increase feed requirements.

A rule of thumb for feeding a horse doing light work is 1 lb of grain and 1 lb of hay daily per 100 lb body weight. An idle horse usually requires only one-half this amount of grain. If a horse is ridden or driven hard, the amount of hay may remain constant, but the grain should be increased to 1¼ to 1⅓ lb per 100 lb body weight. A stallion used heavily for breeding purposes should be fed approximately the same amount as a horse doing heavy work. Broodmares, even when nursing a foal, do well on high-quality pasture alone. However, a good-quality feed should be available in a creep (a structure barring entrance of larger horses) to foals 2 or more months of age. In winter broodmares should be fed small amounts of grain in addition to dry forage.

Young horses (**weanlings** to 2 years of age) should be fed sufficient amounts to maintain good condition. They may gain satisfactorily if fed 1/2 lb of grain per 100 lb body weight daily when on pasture or when fed good-quality hay. This amount should be increased to 1 lb if pasture or hay is of inferior quality.

21.6 TRAINING AND GROOMING HORSES

Proper training of all horses is very important. Early training of a foal usually results in a gentler and better-dispositioned individual. Lessons should be given one at a time and in the proper sequence.

Place a halter on the foal at 2 to 3 weeks of age, and allow the foal 2 or 3 days to become accustomed to it. The foal may then be tied near its mother for 30 to 40 min daily. A good time to brush and groom a foal is when it is tied in this way. One can train the foal to lead by first leading it with its mother and then by itself. At first it should be led at a walk, and then at a trot. This is a good time to teach the foal to start and stop on command and to stand in show position. Patience and gentleness accompanied by firmness are of great importance in training a foal.

Breaking a young horse to a saddle or harness may be accomplished during the winter before it is a 2 year old. (Racehorses have the same official birthday, January 1.) Many expert "bronc riders" break cow ponies that have not been handled as colts and are wild when they are brought in from the range. Each horsebreaker has methods he or she prefers, and it is surprising how quickly young horses can be broken and trained. One cowboy who could scarcely write his name was outstanding in breaking and training cow ponies. When asked how he got so much out of a horse, he replied, "You just have to be smarter than the horse."

Horses that are to be shown must be well-mannered and well-groomed. The coat can be conditioned by proper feeding and grooming. A balanced diet fed in adequate amounts helps ensure a pleasing appearance. The best grooming is useless if a horse is underfed and unthrifty.

A currycomb or rubber brush should be used to remove mud and dirt from the coat. Otherwise, vigorous rubbing with a stiff brush followed by rubbing with a soft one is recommended. Rubbing the coat briskly with a cloth will make the hair shine. It has been said that the best way to make a horse's coat shine is to apply plenty of "elbow grease." The groom should attend to the entire body (including the face, belly, and legs). It is recommended that a brush be used with care on the mane and tail of show horses because it may remove considerable hair. Many suitable mane and tail combs are available.

Daily washing is not usually recommended, because it removes the oils that give the coat its natural sheen and glossy appearance. However, it is sometimes necessary to wash certain parts of the body if the hair is stained. White legs and ankles and light manes of show horses are usually washed before showing. Most people wash a horse after a hard workout. (Racing horses are usually washed following a race.) In cold weather, the recommended procedure is to scrape the coat with a sweat scraper, wipe it with a damp sponge, and rub briskly with cloths or towels. Following this, the horse is covered with a blanket, and its legs are rubbed with liniment (especially if the horse's legs are sore) and wrapped in bandages to ensure gradual cooling. The horse is then walked and given swallows of water every few minutes until it has cooled. This detailed procedure is not necessary for pleasure horses, but they should be dried, cooled slowly, and given a rubdown following a ride.

Blanketing of show and pleasure horses that are kept in a stable helps keep their coats clean, smooth, and short. This is especially beneficial in cold weather. Horses look neater if the long hair on the ears, jaws, and back of the fetlocks is kept trimmed. Some of the mane should also be removed at the point where the headstall of the bridle is normally located (commonly called the "bridle path," at the poll). Some owners of pleasure horses prefer to roach (clip) the mane completely. It is recommended that those contemplating the showing of horses investigate fully the methods used by the most successful show winners.

21.7 COMMON DEFECTS AND UNSOUNDNESS IN HORSES

Defects and/or unsoundness in horses may involve both anatomy and physiology. They may be due to environment or heredity or the interaction of the two.[3]

---

[3]The discussion that follows includes information adapted largely from "Light Horses," *USDA Farmers' Bull.* 2127, 1962.

21.7.1   Defects of Movement

The normal movement of a horse going forward is for the legs to move parallel to an imaginary centerline drawn in the direction of travel. A defect in movement is one that deviates from this in any way.

Cross-firing is "scuffing" of the inside of the diagonal forefeet and hindfeet and is usually found only in pacers and trotters.

Dwelling is a noticeable pause in the foot's flight, making it appear that the stride was completed before the foot reached the ground. This is most noticeable in trick-trained horses.

Interfering is a condition in which the fetlock or **cannon** is struck by the opposite foot. It is observed most frequently in base-narrow, toe-wide, or splayfooted horses.

Lameness is an indication of a structural or functional disorder in which the individual favors one or more feet while standing or walking. It is also called *claudication*. Pressure between the foot and ground is eased, and the head bobs up as the affected foot contacts the ground.

Paddling involves throwing the fore feet outward from the body as they are lifted and before they reach the ground again. This is most common in toe-narrow or **pigeon-toed** horses.

Pointing is a more than usual extension of the stride, with little or no flexion (joint movement).

Pounding is a heavier-than-normal contact of the feet with the ground (observed especially in Thoroughbreds, hackneys, and some draft horses).

Rolling is an excessive **lateral** motion of the shoulder (usually observed in horses with protruding shoulders).

Scalping is a condition in which the toe of a forefoot hits a hind foot at the top of the hairline.

Speedy cutting is contact between the inside of the diagonal fore and hind pastern. It may be seen in fast-trotting horses.

Stringhalt is excessive and involuntary flexing (bending) of the hocks during progression (moving ahead) and may affect one or both rear limbs. It is more noticeable when the horse is turning.

Trappy is the term for a short, quick, and choppy **stride,** which is most common in horses with short, straight pasterns and straight shoulders.

Winding, or *rope-walking,* is a twisting of the striding leg around and in front of the supporting leg, like a person walking a rope.

Winging is an exaggerated paddling, particularly noticeable in high-stepping horses.

21.7.2   Blemishes and Unsoundness

Blemishes include abnormalities, such as cuts and splints, that usually do not affect the performance or service of horses. However, *unsoundness* includes defects that may impair performance or service. Although blemishes seldom disqualify horses from show-ring competition, unsoundness (particularly hereditary conditions) does. Points of the body where some of these abnormalities occur are shown in Figure 21.4. Those interested in buying and breeding horses should become familiar with these defects.

Definitions for selected examples of blemishes and unsoundness include the following:

Blindness is either impaired eyesight or complete blindness.

FIGURE 21.4
Location of common points of unsoundness in horses. (*Adapted from* USDA Farmers' Bull. 2127, 1962.)

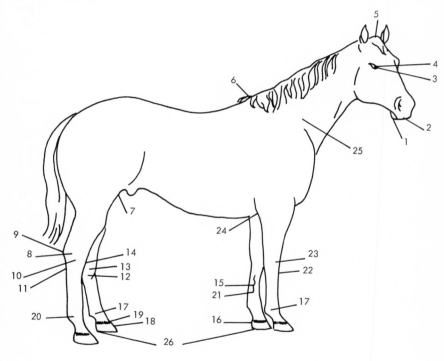

1. Undershot jaw	10. Stringhalt	18. Quittor	26. Contracted feet,
2. Parrot mouth	11. Curb	19. Ring bone	corns, founder,
3. Blindness	12. Bone spavin or	20. Windpuffs	thrush, quarter
4. Moon blindness	jack	21. Splints	or sand crack,
5. Poll evil	13. Bog spavin	22. Knee sprung	scratches or
6. Fistulous withers	14. Blood spavin	23. Calf kneed	grease heel
7. Stifled	15. Bowed tendons	24. Capped elbow	General: heaves,
8. Thoroughpin	16. Sidebones	25. Sweeney	hernia, roaring, thick
9. Capped hock	17. Cocked ankles		wind

Blood spavin is enlargement of the hock, due to an enlargement of the superficial saphenous vein on the anteromedian aspect (inside) of the hock.

Bog spavin is the enlargement of the hock joint capsule, due to an accumulation of fluid.

Bone spavin (or jack spavin) is a bony growth on the hock.

Bowed tendons are shortened, enlarged tendons behind the cannon bones of the legs.

Calf-kneed is the opposite of knee-sprung, so that the horse stands with its knees too far backward.

Capped elbow is the enlargement of the elbow (bursa).

Capped hock is a swelling on the point of the hock.

Cocked ankles is a condition in which the legs (usually rear) are bent forward at the fetlocks in a cocked position.

Contracted heels is a contraction or drawing in of the hoof at the heel.

Corns are bruises of soft tissue under the horny part of the hoof's sole.

Curb is an enlargement at the plantar aspect of the fibular tarsal bone, due to an inflammation and thickening of the plantar **ligament;** in acute cases lameness results.

Fistulas are inflammations or infections in the withers, accompanied by passages or sinuses through the tissues to the skin's surface.

Founder (laminitis) is an inflammation of the fleshy part of a hoof.

Knee-sprung (bucked-kneed) is a condition in which the knees protrude too far forward.

Moon blindness (periodic ophthalmia) is a periodic cloudy or inflamed condition of the eye that appears and disappears at intervals of approximately 1 month; it was given this name because people thought it was related to moon cycles.

Parrot mouth is a condition in which the lower jaw is shorter than the upper jaw.

Poll evil is an acute swelling on the top (poll) of a horse's head.

Quarter, or sand, crack is a vertical split in the inside of the horny wall of a hoof that extends from the coronet downward.

Quittor is an open draining sore at the coronet (hoof's head).

Ringbone is a bony growth on the pastern bone (usually of a forefoot).

Scratches, or grease heel, is a mangelike inflammation of the **posterior** surfaces of the fetlocks; it occurs more frequently in the hind legs.

Sidebones is the hardening, or ossification, of the lateral **cartilage** of the third phalanx.

Splints is a condition most frequently affecting the forelimbs of young horses. Splints are most commonly found on the medial aspect of the limb, between the second and third metacarpal bones.

Stifled is a stifle joint stiffened and/or extended.

Sweeney is the **atrophy,** or shrinking, of the shoulder muscles because of a nerve injury; it results in a depression of the shoulder.

Thoroughpin is a swelling just above the hock on both sides of the leg; it involves the tarsal sheath that encloses the deep digital flexor **tendon.**

Thrush is a disease of the horny layers of the **frog** of a foot; it is usually accompanied by a discharge and an unpleasant odor.

Undershot jaw is a condition in which the lower jaw is longer than the upper jaw.

Wind puffs are soft enlargements on the fetlock, due to excessive production of **synovial** fluid.

21.7.3 Stable Vices and Other Bad Habits

Horses in confinement often develop habits that may be annoying to their owners and even harmful to themselves. Most stable vices (bad habits) result from sheer boredom.

Bolting is a horse's habit of eating its concentrate feed very rapidly and without chewing it properly.

Cribbing refers to a condition in which the horse bites the manger or another object while sucking in air. It sometimes causes a **bloated** condition, and the horse may be more likely to develop **colic.** This condition may be remedied by buckling a strap snugly around a horse's neck in such a way that it compresses the **larynx** when the neck is flexed or extended. If properly placed, such a strap will cause no discomfort at other times.

Halter pulling refers to the habit of pulling back on the halter rope or bridle reins when tied. The halter and/or strap may be broken, allowing the horse to run away.

Some horses develop a habit of *kicking* or striking the stall or other objects with their rear feet. They apparently do this for the satisfaction derived from striking something.

Tail rubbing is a habit of persistently rubbing the tail against the side of a stall or on another object. This may result from parasites.

Weaving is a condition in which a horse weaves, or sways from side to side, while standing in a stall.

Draft horses may develop the habit of *balking* when hitched to heavy loads. Thus they refuse to push their shoulder against the collar. Such a habit sometimes develops because of a poorly fitting collar, which causes a sore on their necks or shoulders.

A horse may develop the habit of *rearing* on its hind legs when an attempt is made to lead it. Other horses may rear when mounted. Some horses develop the habit of *striking out* with the forefeet when they are being led and/or when they rear up on their hind legs. Other horses develop a habit of *shying* when ridden and may even unseat their rider. Draft horses sometimes develop the dangerous habit of *running away* when hitched to a wagon or other moving vehicle. Riding horses, especially cow ponies, may "cold-jaw," a habit of taking the bit in their teeth and running at top speed. Pulling on the reins, and other methods commonly used in stopping them, are often unsuccessful. Once these runaway habits develop, little can be done to correct them.

Another undesirable habit horses develop is that of playing "hard to catch" when their owners try to drive them into a stable or catch and bridle them in a pasture or pen. Other horses object to being harnessed or saddled. Gentle horses may develop the habit of expanding their chest cavity (breathing deeply, commonly called *swelling up*) when the saddle cinch is tied or buckled. They exhale air after the cinch is fastened and thereby cause the saddle to be loosened. Some cow ponies, probably because of the way they are first broken and handled, develop the habit of *bucking*. This habit marks some horses as unsuitable for riding; others seem gentle but may "break in two" and throw their riders when least expected.

Most vices of horses are difficult to correct once they are developed. Preventing the habit is much more desirable than trying to correct it. Older horses, however, can be broken of many undesirable habits and taught new tricks. It should be remembered that horses are not generally considered "pals," as dogs often are. Rather, horses must be taught to serve and respect their master.

21.7.4 Care and Trimming of the Feet

For want of a nail the shoe was lost,
For want of a shoe the horse was lost,
For want of a horse the rider was lost,
For want of a rider the battle was lost.

George Herbert (1593–1633)

Most horses are now kept for either riding or driving purposes. These uses involve movements of various kinds, and proper movement is dependent on healthy and sound normal feet. In fact, many movement defects among horses are related directly or indirectly to improper care and trimming of the feet. Therefore, horse owners should care for the hooves of their horses properly and use the services of a professional when appropriate. Tools most often used to trim the feet include hoof nippers, a hoof knife, a hoof pick, and a rasp. Properly and improperly trimmed hooves are shown in Figure 21.5.

Proper care of feet starts with the young foal. An unshapely hoof causes uneven wear and may result in unsound legs. Faulty legs can often be corrected by proper trimming of the hooves; however, it is more desirable to keep hooves properly trimmed as a **prophylactic** measure against such an unsoundness. The foal's feet should be observed while it is standing on a hard surface and then when in action, both at a walk and at a trot.

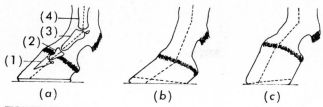

FIGURE 21.5
Trimming horse feet. (a) Properly trimmed hoof with normal foot axis. (1) Coffin bone, (2) short pastern bone, (3) long pastern bone, (4) cannon bone. (b) Toe too long, which breaks the foot axis backward. Horizontal dotted line shows how hoof should be trimmed to restore normal posture. (c) Heel too long, which breaks the foot axis forward. Horizontal dotted line shows how trimming will restore the correct posture. (USDA.)

Most necessary foot trimming of young foals can be accomplished with a rasp because the feet seldom need more than a little smoothing and rounding. Even if little shaping is necessary, handling the foal's feet makes shoeing easier at maturity. Regular hoof care in the foal and young horse will favor proper development of the legs and correct movement. It is seldom necessary to put shoes on young foals, although light plates are sometimes used on young show horses. Some horsepersons who raise cow ponies on the range prefer to run them in a rocky area or where the ground is hard and firm. They believe that horses raised under such conditions have harder and more durable feet in later life. Mares and foals are never shod under such conditions.

The hoof of a normal healthy horse grows about 1/2 in monthly. If the ground is mostly soft and the hoof is not trimmed, the wall will break off and wear unevenly. Under such conditions the hoof should be trimmed monthly. Nippers may be used to trim the horn, and the wall may be leveled with a rasp. If the hoof is uneven, it may be necessary to correct this over a period of time with several trimmings. The **frog** of a foot should also be carefully trimmed by removing the ragged edges so that dirt or other extraneous matter will not accumulate in the crevices (Figure 21.6). Since it serves as a cushion for the hoof, the frog should not be trimmed excessively. The sole should be trimmed little, if at all. It is recommended that the outside walls of the hoof not be rasped, since this removes the outer protective layer that normally prevents **evaporation.** The bars (either of the recurved ends of the wall of a horse's hoof where they curve in to the sole) should never be trimmed below the level of the walls. Frequently, the novice trims these excessively, which allows the wall to collapse inwardly and thereby pinches the lateral **cartilage.**

Shoes are a necessity when horses pull loads or are ridden on hard surfaces. Otherwise, the hooves wear down, the feet become tender, and lameness results. Cow ponies worked on the range, especially on rocky ground, must be

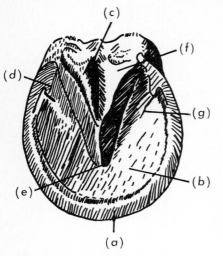

FIGURE 21.6
Parts of a horse's foot. (*a*) Bearing edge and hoof wall; (*b*) sole; (*c*) median furrow of the frog; (*d*) lateral furrow of the frog; (*e*) apex of the frog; (*f*) branch of the frog; (*g*) bar. *(USDA.)*

shod. Shoes should be the proper size for each horse. They are shaped to fit the contour of the hooves. When nailing on a shoe, one should take care not to drive the nail into the "quick," or sensitive, part of the hoof. When nailed through the hoof, the sharp ends of nails should be twisted off, and the remaining blunt ends tapped lightly to bend them downward and thereby hold the shoe firmly in place. Shoes are commonly replaced or reset at intervals of 4 to 8 weeks. Those worn too long may allow the hoof to grow out of proportion and thereby throw a walking or running horse off balance. Corrective shoeing to relieve an abnormality requires skill.

21.8 DETERMINING THE AGE OF HORSES

A mature male horse has 40 teeth, whereas a mature female has only 36. Foals of both sexes have 24. Frequently, a small pointed tooth, sometimes called a "wolf tooth," may appear in front of each first molar in the upper jaw, thus increasing tooth numbers by two. Since these teeth may interfere with the bit, they are often removed at a young age. The mature male (usually not the female) commonly has **tushes,** or pointed teeth, located between the incisors and molars. The appearance of teeth is used to determine a horse's age, which is important when one is purchasing horses. Age is determined by noting the time of appearance, the shape, and the degree of wear of temporary and permanent teeth. Temporary (milk) teeth are easily distinguished from permanent teeth because they are smaller and whiter. Even the novice can recognize a **smooth-mouthed** horse (normally considered a horse 10 or more years old). A description of the teeth of horses at various ages is given in Figure 21.7.

Temporary incisors to 10 days
of age; first or central upper and
lower temporary incisors appear.

Temporary incisors at 4 to 6
weeks of age; second or
intermediate upper and lower
temporary incisors appear.

Temporary incisors at 6 to 10
months; third or corner upper
and lower temporary
incisors appear.

Temporary incisors at 1 year;
crowns of central temporary
incisors show wear.

Temporary incisors at 1½ years;
intermediate temporary incisors
show wear.

Temporary incisors at 2 years;
all show wear.

Incisors at 4 years;
permanent incisors replace
temporary centrals and inter-
mediates; temporary corner
incisors remain.

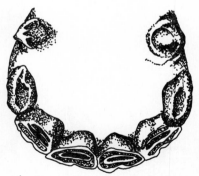

Incisors at 5 years;
all permanent; cups in all incisors.

FIGURE 21.7
The teeth of horses at various ages. *(Redrawn from "Breeding and Raising Horses,"*
USDA Agr. Handbook No. 394, *1972.)* Figure continues on page 688.

Incisors at 6 years; cups worn
out of lower central incisors.

Incisors at 9 years;
cups worn out of upper central
incisors; dental star on upper
central and intermediate pairs.

Incisors at 7 years; cups also worn
out of lower intermediate incisors.

Incisors at 10 years;
cups also worn out of upper intermediate incisors, and dental star
is present in all incisors.

Incisors at 11 or 12 years;
cups worn in all incisors
(smooth mouthed), and dental
star approaches center of cups.

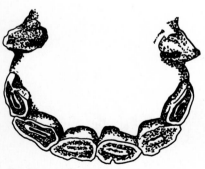

Incisors at 8 years;
cups worn out of all lower incisors, and dental star (dark line
in front of cup) appears on lower
central and intermediate pairs.

FIGURE 21.7 (Cont.)

Characteristic shape of lower
incisors at 18 years.

21.9 DISEASE AND PARASITE CONTROL

The best possible breeding, feeding, and training will be of no avail if a horse is sick and unhealthy. Important selected characteristic body functions of healthy horses are described in Section 21.9.1. Variations from these, indicating illness or disease, are discussed briefly in Sections 21.9.2 to 21.9.5.

21.9.1 Selected Normal Body Functions

The pulse rate of a horse at rest varies between 28 and 42 beats per min. During extreme exertion it may increase to more than 200 beats per min, but when the horse is rested, the pulse rate usually returns to normal in a few minutes. Pulse rate is higher in young than in mature horses. It may be determined by placing the fingertips over the external maxillary artery that comes around the **ventral** surface of the **mandible** (jawbone) immediately in front of the large masseter (cheek) muscle.

The respiration rate of horses at rest varies between 9 and 12 per min. It may increase to 100 or more per min during running or vigorous work. It usually returns to normal within a few minutes after the physical exertion stops. Breathing should be free, easy, and noiseless, without a whistling or roaring sound. One may determine respiration rate by observing movements of the flanks and/or nostrils. (It is especially important to monitor respiration rate during summer months.)

The normal body temperature of a horse at rest is about 100°F. High **ambient temperatures,** excitement, and exercise increase body temperature, which may be determined by inserting a thermometer in the rectum and allowing it to remain for at least 3 min.

21.9.2 Disease Symptoms

The frequency of defecation and appearance of its feces are indicative of a horse's condition. Normally, horses **defecate** 8 to 10 times daily. Hard feces may be caused by consumption of feed that is too dry, an inadequate water intake (especially in winter), or inadequate exercise. Soft, watery feces often result from the consumption of large quantities of green, watery grass, too much bran, or an intestinal irritation. Slimy, strongly odoriferous droppings suggest that the feed is too concentrated, that the intestines are irritated, or that something has caused a shift from normal in the intestinal bacterial **flora.** If the droppings contain large amounts of whole grain, this indicates either that the horse eats too fast, without properly chewing, or that it cannot chew grain properly because of defective teeth.

Indications of disease are a loss of appetite, high body temperature (above 101°F), fast respiration and heart rates, profuse sweating, stiffness, nasal discharge, coughing, diarrhea, constipation, pawing, rolling, groaning, lameness, etc. When such symptoms are observed, a veterinarian should be consulted.

Studies of horses treated in the Veterinary Clinic at the University of Missouri-Columbia indicate that wire cuts are the number one reason for veterinary treatment. Thus barbed-wire fences are undesirable for restricting horses, especially in small lots. Board fences are preferred.

21.9.3 Nutritional Diseases

Selected nutritional **diseases** of horses and their probable causes, symptoms, and methods of prevention and treatment are given in Table 21.4. These data show that poor management is the general cause of most noncommunicable horse diseases. Since horse owners can prevent many of these diseases, it is well for them to become acquainted with such information.

21.9.4 Communicable Diseases

Most diseases of horses are caused by bacteria, **viruses, fungi,** or **parasites.** As with noncommunicable diseases, the application of preventive measures is more desirable than trying to treat a disease once symptoms appear. Preventive measures include daily observation and care, cleanliness, disinfection of premises, isolation and/or **quarantine** of sick animals, and **vaccination** when appropriate. Close cooperation with veterinarians is important in planning preventive measures and is essential for proper treatment of diseased animals.

21.9.5 Common Horse Parasites

Parasitic infestations may cause the appearance of **acute** symptoms, but they are more likely to result in a gradual and progressive **insidious** unthriftiness. Parasitic infestations are associated with general unthriftiness or weakness, **emaciation,** rough hair coat, and frequent attacks of **colic.** They are more damaging to foals and young horses and often result in slow growth and development. Heavily parasitized older horses perform inefficiently and have increased nutrient requirements.

Most parasite eggs laid internally pass in the feces and are deposited on pasture or in stalls and lots. They develop to the infective stage on the ground and may reenter a horse's body with grass, other feed, or water. Preventive measures include breaking the **life cycle** at some vulnerable point. (Horses are commonly wormed at 6-month intervals.)

The female botfly attaches her eggs to hairs on a horse's chin, nose, or legs. Eggs hatch in 7 to 10 days and are taken into the mouth when the horse licks or bites the hair. The **larvae** then burrow into the mucous membranes of the tongue, where they remain approximately 2 weeks before emerging and passing to the stomach. They remain attached to the stomach walls for 8 to 10 months or until they fully develop. The larvae are then passed with the feces, **pupate** in the ground for 20 to 70 days, and finally emerge as adult flies (Chapter 19). Large numbers of bots cause irritation and injury to the stomach walls, and sometimes severe colic develops. A common treatment is the administration of carbon disulfide or an organophosphate (it should be administered by a veterinarian).

Ascarids, or large roundworms, infect horses and, when fully grown, may attain a length of 12 in. They locate mostly in the upper portion of the small intestine but sometimes locate in the stomach and cecum. Eggs pass from the body in feces and, in the infective stage, may reenter the body when swallowed by the horse. Mature horses apparently build up a resistance to ascarids; consequently, these parasites are primarily a threat to the health of young horses. Heavy infestations with ascarids may even be fatal to young horses, although they more frequently cause unthriftiness, rough hair coat, and slow, inefficient gains. Cleanliness of quarters and pasture rotation are the recommended preventive practices. Several available drugs are active against ascarids.

Large strongyles (bloodworms, or palisade worms) are the most important parasites of horses. These bloodsucking worms attach themselves to the colon and cecum. Their greatest damage is caused by the loss of blood, intestinal injury, and the migration of larvae through many body organs. Migrating larvae cause many problems, depending on where they move and lodge. Many "colics" in adult horses are caused by restriction of the blood supply to the large bowels, in which parasites damage the arteries. Small strongyles are also an important internal parasite in horses. Serious infestations cause general unthriftiness, **anemia,** and weakness. The life cycle includes the discharge of eggs into the feces; larvae enter the body when the horse **grazes.** Prevention includes keeping quarters clean and rotating pastures. Drugs are available for control of large and small strongyles but should be chosen after consultation with a veterinarian.

Various species of lice also parasitize horses and can be effectively controlled through spraying with recommended **insecticides.**

21.10 SUMMARY

Where in this wide world can one find nobility without pride, friendship without envy, or beauty without vanity? Here, where grace is laced with muscle, and strength by gentleness confined. He serves without servility; he has fought without enmity. There is nothing so powerful, nothing less violent, there is nothing so quick, nothing more patient. England's past has been borne on his back. All our history is his industry. We are his heirs, he is our inheritance. The HORSE.

Sir Ronald Duncan (1914–1982)

Since horses were first domesticated approximately 5000 years ago, they have been of great value and service to humanity. Horses probably were first used to carry soldiers into battle, but they soon found a role in transportation and farming (especially for draft purposes). The introduction of horses into America had a significant effect on the lives of American Indians. They rode horses to capture and kill buffalo for food and made clothing and other items from horsehair and horsehide.

TABLE 21.4
SELECTED NUTRITIONAL DISEASES OF HORSES

Disease	Cause	Symptoms (and age or group most affected)	Distribution and losses
Anemia, nutritional	Commonly an iron deficiency but may be caused by a deficiency of copper, cobalt, and/or certain vitamins.	Loss of appetite, progressive emaciation, and death.	Worldwide. Losses consist of retarded growth.
Azoturia (Monday morning disease, blackwater)	Associated with faulty carbohydrate metabolism and with work following a period of idleness in the stall on full diets.	Profuse sweating, abdominal distress, wine-colored urine, stiff gait, reluctance to move, and lameness.	Worldwide, but the disease is seldom seen in horses at pasture and rarely in horses at constant work.
Colic	Improper feeding, working, or watering; internal parasites	Excruciating pain; depending on the type of colic, other symptoms are distended abdomen, increased intestinal rumbling, violent rolling and kicking.	Worldwide.
Founder	Overeating, overdrinking, or inflammation of the uterus following parturition. Also intestinal inflammation.	Extreme pain, fever (103 to 106° F), and reluctance to move. If neglected, chronic laminitis develops.	Worldwide. Actual death losses from founder are not very great.
Heaves	Often associated with the feeding of damaged, dusty, or moldy hay. Often follows severe respiratory infections.	Difficulty in expiring air, resulting in a jerking of flanks (double flank action) and coughing. The nostrils are often slightly dilated.	Worldwide. Losses are negligible.
Iodine deficiency (goiter)	A failure of the body to obtain sufficient iodine from which the thyroid glands can form thyroxine.	Foals may be weak.	Northwestern United States and the Great Lakes region.
Periodic ophthalmia (moon blindness)	Probably due to dietary deficiency of riboflavin. Leptospiral microorganisms may also be involved.	Periods of cloudy vision, in one or both eyes, which may last for a few days to a week or two and then clear (recur at intervals).	In many parts of the world. In the United States, it occurs most frequently in the states east of the Mississippi River.

Treatment	Control and eradication	Prevention
Provide dietary sources of the nutrient or nutrients lacking.	Balanced diet.	Feed adequate levels of iron, copper, cobalt, and vitamins.
Absolute rest and quiet. While awaiting the veterinarian, apply heated cloths or blankets or hot-water bottles to the swollen and hardened muscles.	Azoturia is noncontagious. When trouble is encountered, decrease the diet and increase the exercise on idle days.	Restrict the diet and provide daily exercise when animals are idle, or turn idle horses out to pasture.
Call a veterinarian. To avoid danger of inflicting self-injury, place the animal in a large, well-bedded stable or take it for a slow walk.	Proper feeding, working, and watering.	Adequate diet with minimal dust. Control internal parasites.
Pending arrival of the veterinarian, the attendant should stand the animal's feet in a cold-water bath.	Alleviate the causes, namely, overeating, overdrinking, and/or inflammation of the uterus following parturition.	Avoid overeating and overdrinking (especially when hot).
Affected animals are less bothered if used only at light work, if the hay is sprinkled lightly with water at feeding, or if the entire ration is pelleted.	See Prevention.	Avoid the use of damaged and/or dusty feeds.
Once the iodine-deficiency symptoms appear in farm animals, no treatment is entirely satisfactory.	At the first signs of iodine deficiency, an iodized salt should be fed to all farm animals.	In iodine-deficient areas, feed iodized salt containing 0.01% potassium iodine to all farm animals.
Anitibiotics administered promptly are helpful.	If symptoms of moon blindness are observed, immediately change to greener hay or grass or add riboflavin to the diet at the rate of 40 mg per horse daily.	Feed high-riboflavin green grass or well-cured green leafy hays, or add riboflavin to the diet at the rate of 40 mg per horse daily.

TABLE 21.4
SELECTED NUTRITIONAL DISEASES OF HORSES (Cont.)

Disease	Cause	Symptoms (and age or group most affected)	Distribution and losses
Rickets	Lack of either calcium, phosphorus, or vitamin D and/or incorrect ratio o the two minerals.	Enlargement of the knee and hock joints, and the animal may exhibit great pain when moving about. Irregular bulges (beaded ribs) at juncture of ribs with breastbone, and bowed legs.	Worldwide. It is seldom fatal.
Salt deficiency	Lack of salt (sodium chloride).	Loss of appetite, retarded growth, loss of weight, a rough coat, lowered production of milk.	Worldwide.
Urinary calculi (gravel, stones, water belly)	Unknown, but incidence is higher when there is a high potassium intake, an incorrect Ca/P ratio, or a high proportion of beet pulp or grain sorghum.	Frequent attempts to urinate, dribbling or stoppage of the urine. Usually only males affected.	Affected animals seldom recover completely.
Vitamin A deficiency (night blindness)	Vitamin A deficiency.	Night blindness, the first symptom of vitamin A deficiency, is characterized by impaired vision.	Worldwide.
Osteomalacia	Lack of vitamin D. Inadequate intake of calcium and phosphorus. Incorrect ratio of calcium and phosphorus.	Phosphorus deficiency symptoms are depraved appetite (gnawing on bones, wood, or other objects or eating dirt), lack of appetite, stiffness of joints, failure to breed regularly, decreased milk production.	Southwestern United States is classified as a phosphorus-deficient area.
Calcium deficiency	Inadequate dietary calcium intake.	Calcium deficiency symptoms are fragile bones, reproductive failures, and lowered lactations.	Calcium-deficient areas have been reported in parts of Florida, Louisiana, Nebraska, Virginia, and West Virginia.

Source: Adapted from "Selecting, Feeding, and Showing Horses," *Albers Res. Bull.,* 1966

Treatment	Control and eradication	Prevention
If the disease has not advanced too far, treatment may be successful by supplying adequate amounts of vitamin D, calcium, and phosphorus, and/or adjusting the ratio of calcium to phosphorus.	See Prevention.	Provide sufficient calcium, phosphorus, and vitamin D and a correct ratio of the two minerals.
Salt-starved animals should be gradually accustomed to salt, slowly increasing the hand-fed allowance.	See Prevention and Treatment.	Provide salt free choice.
Once calculi develop, dietary treatment appears to be of little value. Smooth-muscle relaxants may allow passage of calculi if used before rupture.		Good feed and management appear to lessen the incidence. 1 to 3% salt in the concentrate diet may help (using the higher levels in the winter).
Treatment consists of correcting the dietary deficiencies.	See Prevention and Treatment.	Provide carotene (vitamin A) through green, leafy hays; silage; or lush, green pastures. Synthetic preparaions of vitamin A are recommended for winter feeding.
Select natural feeds that contain sufficient quantities of calcium and phosphorus.	Increase the calcium and phosphorus content of feeds through fertilizing the soils.	Feed balanced diets and allow animals free access to a phosphorus and calcium supplement.
Select natural feeds that contain sufficient quantities of calcium and phosphorus.	Increase the calcium and phosphorus content of feeds through fertilizing the soils.	Feed balanced diets and allow animals free access to a phosphorus and calcium supplement.

Tractors, trucks, and other power-driven vehicles have almost replaced horses for draft purposes in the United States, and for a period of time the future of horses appeared bleak. However, the number of horses has increased decidedly in recent years. This is the result of an increased demand for horses for pleasure and recreation.[4]

Machines will never replace horses for many purposes. The beauty of horses and the wide variety of sizes, colors, and gaits make them very popular for riding, driving, and showing. The horse project is this nation's largest 4-H youth project. Horse racing is the world's most popular sport (Chapter 1). In some areas, range cows are still worked with horses, and this will probably always be true.

The breeding, feeding, managing, training, and showing of horses require patience and skill, but in this day of rockets and spaceships, the horse offers a pleasurable and desirable means of recreation and relaxation to those seeking a change of pace.

There is nothing as good for the inside of a man as the outside of a horse.[5]

Anonymous

STUDY QUESTIONS

1 What was probably the first major use of the horse by humans?
2 How and why did the horse gain a foothold in the new world?
3 Why did the wealth of an Indian chief come to be measured in terms of the number of horses he owned?
4 Why did horse numbers decline on farms a few decades ago? Why have they increased in recent years?
5 What are some of the most important uses of horses at the present time?
6 List and describe the three major types of horses.
7 What are the four natural gaits of a horse? The four artificial gaits?
8 How does the trot differ from the pace?
9 What are some distinguishing features of the eyes of horses?
10 How many different pairs of genes are known to affect coat color?
11 Why does mating a palomino mare to a palomino stallion fail to always produce a palomino foal?
12 List some of the common faults of conformation in the light horse.
13 Define filly, gelding, stallion, foal, colt, and mare.
14 Outline the recommended care that should be given a mare at foaling.
15 Outline the recommended care that should be given a foal at parturition and shortly thereafter.
16 What are the major nutrient requirements of horses?
17 Describe some important factors to consider in training and grooming horses.

[4]Melvin Bradley, *Horses—A Practical and Scientific Approach,* McGraw-Hill Book Company, New York, 1981.

[5]This quotation was kindly provided by Dr. J. O. Swink, Jr., who specialized in the practice of equine veterinary medicine in Union, Missouri, for many years.

18 Name and describe some of the common defects and types of unsoundness often found in horses.
19 What are some stable vices and other bad habits of horses? How may they be corrected?
20 Describe some important points to consider in trimming horses' hooves.
21 How many teeth does a horse have? What are wolf teeth?
22 Name the major parts of a horse's foot.
23 What is moon blindness in horses?
24 What are some important communicable diseases of horses?
25 What are some common horse parasites? Describe a treatment for each.

ANIMAL RESEARCH IN RETROSPECT AND PROSPECT[1]

And he gave it for his opinion, that whoever could make two ears of corn, or two blades of grass, to grow upon a spot of ground where only one grew before, would deserve better of mankind, and do more essential service to his country, than the whole race of politicians put together.

Jonathan Swift (1667–1745), in *Gulliver's Travels*

22.1 INTRODUCTION

The first European settlers to enter this country encountered many hardships and dangers. They found a different climate and brought little experience that fitted them for the struggle with nature that followed. Instead, experience was to be gained with axe in hand for forest clearing and gun at side for self-defense. It is not surprising, then, that progress in **agriculture** was slow early in American history.

For nearly three centuries the nation's agriculture expanded *horizontally* across the continent. Thus new land was the frontier of agriculture in the United States prior to the twentieth century, but farsighted pioneers of agriculture in Congress realized that population pressures and soil depletion would eventually lead to the need for more productive plants and animals. Realizing that research would show the way, they enacted the Hatch Act in 1887, thereby authorizing the establishment of agricultural experiment stations (Chapter 1). During the

[1]The authors acknowledge with appreciation the contributions to this chapter of Dr. D. H. Baker, Department of Animal Science, University of Illinois, Urbana.

twentieth century, research has become the frontier of agriculture, and increased production no longer depends on new acres but rather on greater productivity per acre. This might be termed *vertical* expansion. When de Tocqueville of France toured America in 1831, he could not have understood what changes agricultural research could bring about in a democratic society or he would not have recorded the following statement in his book *Democracy in America:*

> Agriculture is perhaps, of all the useful arts, that which improves the most slowly among democratic nations.
>
> *Alexis de Tocqueville (1805–1859)*

Animal research (including biochemistry, applied and molecular biology, entomology, genetics, immunology, microbiology, nematology, nutrition, parasitology, pathology, pharmacology, physiology, and virology) is a systematic method of efficiently obtaining and applying knowledge to the production of animals and their products for humans. It shows the cause and effect relation between relevant variables by subjecting them to critically designed treatments or constraints. The results are evaluated by the use of precise measurements and by probability analysis techniques.

In the early days of agricultural experimentation, two or three scientists attempted to research the entire field, including animals, plants, soils, economics, engineering, etc. The rapid accumulation of knowledge and the increasing complexity of many agricultural problems make it essential that solutions now be derived by teamwork among highly trained personnel of widely different disciplines.

> Art is I; science is we.
>
> *Claude Bernard (1813–1878)*

Looking Backward[2] was a best seller when it appeared in 1888. It foretold that in the year 2000 there would be a very good thing in Boston—an ideal society. Although realization of this society is not likely, agricultural research has contributed substantially to the United States' becoming the first major nation in the history of the world to achieve an era of food abundance.[3] Today, food for more than 235 million people is produced in the United States, where just two centuries ago only 1 million American Indians and a small white population (probably less than 3 million) were fed. American consumers can buy a greater variety of pure and wholesome foods at less cost, in hours of work,

[2]E. Bellamy, *Looking Backward: 2000–1887*, Boston, 1888; New American Library, New York, 1960.

[3]Except for one year, the United States was a net importer of food from 1922 to 1941. Agricultural research and the application of technology has changed agricultural productivity enough that the United States exported approximately $44 billion of agricultural products in 1981.

than any other consumers at any time in history. Thus, through research in genetics, nutrition, physiology, and management, the **USDA,** colleges of agriculture, and state agricultural experiment stations (SAES) have been very influential in changing animals and their productivity. For example, in 1983, it took less than 6 lb of feed and only about 7 weeks to raise a 3-lb **broiler,** whereas in 1940, it took about 12 lb of feed and 12 or more weeks.[4] The average annual egg production per hen increased from about 140 eggs in 1935 to 245 annually in 1983, and the average annual amount of milk produced per cow in the United States increased from about 4600 lb in 1940 to about 15,000 lb in 1983. The application of science to animal breeding has resulted in cattle that produce more beef and resist disease, hogs that produce leaner meat (the average hog yielded 16.5 lb more lean meat in 1983 than in 1956), and bees that make more honey. These contributions of animal scientists have benefited consumers by increasing availability and decreasing *real* costs of foods (as a percentage of total income).

Science is never static. New truths are learned, old ideas are modified or discarded, and the body of knowledge continues to grow. Humanity is at a time when history is ahead of schedule—when events people once thought might happen in the next century have already occurred. Today, more people are engaged in animal research than ever before, yet urgent problems remain unsolved. If animal agriculture is to meet these needs and challenges, it must develop more efficient livestock; producers must raise more animals on less land; there must be dramatic advances in preparing better feedstuffs and feed additives; and finally, there must be constant vigilance to reduce disease and health-related losses in livestock and poultry.

The search for knowledge is as old as humanity. Today, *equipment*[5] aids animal scientists in their unending search for new knowledge. The decades to follow will bring more tools and techniques to animal research. In this chapter, selected research tools, equipment, and techniques available to animal scientists are presented. We hope the information presented will serve as the "spark" needed to ignite research interests in students and guide many into the challenging and rewarding field of animal-science research in the service of humanity. This is in keeping with the following quotation.

> Any number of young men are groping about without any vision of where they are going—but it does not necessarily follow that such a man is doomed to failure—in every man's life there lies latent energy only waiting for a spark and if it strikes will set the whole being afire, and he will become a human dynamo, capable of accomplishing anything to which he aspires.[6]

---

[4]It has been calculated that if the average normal human infant grew at the same rate as a broiler chick, it would weigh 256 lb at 10 weeks of age (day-old chicks increase their body weight from 1 oz at hatching to 3¾ lb at 8 weeks, a 60-fold increase).

[5]It should be noted that although expensive equipment is a major contributor to the spiraling cost of animal/biology–related research, these new pieces of equipment and the accompanying technologies are facilitating great strides in the acquisition of new knowledge that will benefit humanity.

[6]James Cash Penney, *The Man with a Thousand Partners,* Harper & Row, 1931.

22.2 KINDS OF ANIMAL RESEARCH

Science started with the organization of ordinary experience.

Alfred North Whitehead (1861–1947)

Science is the attempt to make the chaotic diversity of our sense-experience correspond to a logically uniform system of thought.

Albert Einstein (1879–1955)

Research is the use of systematic methods to evaluate ideas or to discover new knowledge. There are two principal kinds of agricultural research: (1) basic, or fundamental, and (2) applied, or directed. Scientists often conduct basic research to satisfy their own curiosity. They need not have any practical goal in mind but rather attempt to discover or learn more about the basic laws of nature. Basic research is important because it supplies the fundamental knowledge for applied research.[7] When farm animals are used in basic-research studies, the information gained can often be readily applied to practical farm conditions. Fundamental research is difficult to plan and direct, and its results are often unpredictable. It consists in exploring the unknown, moving forward the frontiers of knowledge. Although each new step is planned and chosen from the results of the previous one, the scientist is often reminded of the following quotation:

We don't know where we are going, but we're on our way.

Stephen Vincent Benét (1898–1943)

Since the half-life of scientific information in the animal sciences is relatively short, greater emphasis is being given to basic research than ever before.

Seek to understand things; their utility will appear later. First of all it is knowledge which matters.

Charles Richet (1850–1935)

Early agricultural research was of an applied nature, as recorded by J. W. Sanborn in the following statement:

It is intended that the experiments carried on shall cover those practical problems on the farm that affect the economy of its operations, and thereby aid in increasing its revenues. The public rightly demands that Agricultural Colleges shall do work, aside from that of the schoolroom, that shall be of public utility.[8]

Applied research aims at a specific objective, such as the development of a new crop or method of selecting and mating animals. This type of research is an

---

[7]In science, one discovery often must wait for another in order to have practical application.
[8]*Missouri State Agr. College Farm Bull.* 1, 1883.

application of basic knowledge directed toward a specific end result. It is usually possible to plan and organize applied research. Scientists can often predict the probability of success. However, applied research findings and subsequent recommendations, e.g., husbandry practices, are more likely to become obsolete than are those of basic research.

22.2.1 Goals of Animal Research

The essence of knowledge is, having it, to apply it.

Confucius (551–479 B.C.)

Certain goals provide the framework for evaluating the contributions that animal research has made and also provide a basis for developing future projected research needs. These research goals include the production of adequate amounts of animal products for humans at decreasing "real" costs; protecting and ensuring consumer nutrition, health, and well-being; protecting farm animals from **insects, diseases,** and other hazards; ensuring a stable and productive animal agriculture[9]; expanding the demand for animal products by developing new and improved products and processes and by enhancing product quality; expanding export markets and assisting developing nations with their animal agriculture; and enhancing the national capacity to develop and disseminate knowledge and improved methodology for solving problems.

Agricultural research is directed toward benefiting the agricultural industries and the people affected by those industries and toward the advancement of science.

More than ever before, it is appropriate that attention be focused on the contributions of agricultural research to the betterment of humanity. The results of agricultural research have been of benefit not only to farm production. Many findings have been directly applicable to (1) human health (e.g., the discovery, development, and use of vitamins and studies of iodination of salt, anticoagulants of blood, rodenticides, vaccines, antibiotics, and nutritional anemias due to iron and copper deficiencies[10]), (2) the quality of the environment (e.g. soil and water conservation, eutrophication, reforestation, and game management), and (3) people's cultural and aesthetic needs (e.g., the release of millions of acres from agricultural use to public parks and recreational facilities).

It is probable that future federal budgets will channel an increasing amount of money into research focusing on life-supporting systems. Indeed, a resolve should be made to put priorities in proper sequence, and increasing the supply and quality of people's food must be emphasized as science continues to serve humanity.

---

[9]Forces outside the animal sciences (e.g., energy, transporation, marketing) could have a greater effect on the future of animal agriculture than the research conducted by those of us in the animal science field.

[10]Moreover, fatherless turkeys are being used in skin-grafting research in cooperation with workers in human medicine.

22.3 THE SCIENTIFIC METHOD

Sit down before fact as a little child, be prepared to give up every preconceived notion, follow humbly wherever and to whatever abysses nature leads, or you will learn nothing.

Thomas Henry Huxley (1825–1895)

I maintain this as an incontestable fact. The results of a properly conducted and properly appreciated experiment can never be annulled, whereas a theory can change with the progress of science.

Carl von Voit (1831–1908)

Investigation of a problem must seek to enlarge the boundaries of knowledge. Mere repetition does not bring progress.

Max Rubner (1854–1932)

Scientists have a lively curiosity concerning natural phenomena; they attempt to understand and explain the world in which they live. Through observation and experimentation, scientists formulate principles describing patterns of action in the branch of knowledge called *science*. This consists of the broad field of human knowledge of acts and laws (principles or rules) arranged in an orderly system. *Science* is verified knowledge; i.e., it can be validated and communicated to other people.

The early philosophers often speculated regarding natural phenomena without recourse to experimentation. They accepted their power of reason without documentation. Some of the ancients believed that the earth was flat, that the stars revolved about the earth, and that the sun sank into the sea. Scientists of today realize that mere thinking and reasoning about natural phenomena are inadequate; they must have facts on which to base reason or to make a sound theory. Scientists discover and test facts and principles by the *scientific method*. This scientific approach to solving problems employs the following basic orderly steps.

Stating the Problem

Perfect knowledge alone can give certainty, and in Nature perfect knowledge would be infinite knowledge, which is clearly beyond our capacity. We have, therefore, to content ourselves with partial knowledge—knowledge mingled with ignorance, producing doubt.

W. S. Jevons (1835–1882)

The basis of an experimental investigation is curiosity and a desire to find answers to a problem or an unknown. Such a basis includes any limiting conditions, the ultimate objective of the research, and the proposed method(s) of attacking the problem. Planning for basic research is dependent largely on the ideas, the imagination, and often the speculation of the scientist.

Forming the Hypothesis

It is a matter of secondary importance whether our theories and assumptions are correct, if only they guide us to results in accord with facts—by their aid we can foresee the results of combinations of causes which could otherwise elude us.

A. W. Rücker (1901)

In forming a possible explanation, or hypothesis, scientists consider what is known pertaining to the problem. They then advance a hypothesis (an untested theory), which is something assumed (speculated), because it seems a likely explanation based on observations and reasoning. Often, an alternative hypothesis is also advanced. Some people call a hypothesis "a reasonable guess."

Experimenting and Observing

The greatest joy of those who are steeped in work and who have succeeded in finding new truths and in understanding the relations of things to each other, lies in work itself.

Carl von Voit (1831–1908)

An **experiment** is designed and conducted to test the hypothesis. Close supervision and detailed observations by the scientist are important in data collection.

Interpreting the Data

A science can grow only by the observation of individual facts.

Sir Francis Bacon (1561–1626)

An important step in any scientific investigation is to arrange properly, summarize, and analyze the data. This allows the researcher to develop possible solutions to the problem. High-speed computers have greatly reduced long computations in exploring solutions (Section 22.6).

Drawing Conclusions

Theory always tends to become more abstract as it emerges successfully from the chaos of facts by processes of differentiation and elimination, whereby the essentials and their connections become recognized, while minor effects are seen to be secondary or unessential, and are ignored temporarily, to be explained by additional means.

Oliver Heaviside (1850–1925)

There is something fascinating about science. One gets such wholesome returns of conjecture out of such a trifling investment of fact.

Mark Twain (1835–1910)

By carefully classifying the data, scientists can draw certain conclusions. Statistical methods (Section 22.5) are employed to estimate the probability that the experimental results are really due to the treatment or simply to chance.

It should be mentioned that pilot studies to develop techniques and refine certain problems often precede a full-scale experiment. A simplified version of selected essentials in organized research is presented in Figure 22.1.[11]

[11]When Dr. C. W. Turner, a world-renowned endocrinologist, reviewed Figure 22.1, he added the following adage in a footnote: "If at first you don't succeed, try, try, again."

FIGURE 22.1
A schematic depicting a typical pathway of organized research and selected qualitites of a scientist.

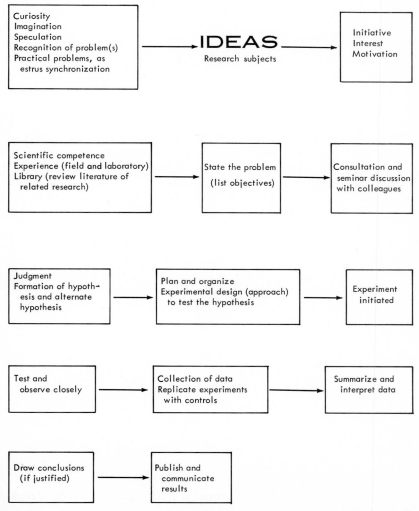

Publishing the Results

Even the most significant research findings are of minimal use until they are published.

R. T. Marshall (1932–   )

Research is conducted to gain new information, and published results are a scientist's tangible output. Agricultural research is responsive to the immediate needs of its clients. An example is the fast and effective response to the corn blight in 1970. Within a year the blight was controlled and corn and meat production were saved. However, this was accomplished only by scientists sharing their experimental data and expertise with plant and animal producers and processors.

22.4 RESEARCH CONTROL

Art and science have their meeting place in Method.

Sir Henry Bulwer (1801–1872)

Because minimizing uncontrolled variables is so important in research, it would be remiss not to emphasize the importance of research control. For example, a nutrition study was designed and initiated at the University of Missouri to determine the effects of a zinc-deficient diet on the growth and well-being of **chicks.** However, scientists found the performance of chicks fed a diet low in zinc to be almost equal to that of controls fed a diet high in zinc. On examination, it was found that the chicks fed the experimental (test) diet were getting zinc from the wire cages in which they were kept. When plastic-coated or stainless-steel cages were used, animals fed the test diet soon developed characteristic zinc-deficiency symptoms (Section 17.3.5).

Experimental animals often eat bedding (straw, shavings, etc.) and chew on materials, especially wood, from which the pens are constructed. Therefore, when certain experiments (e.g., mineral **metabolism** studies) are being conducted, it is imperative that research control be of the highest priority (Figure 22.2).

22.5 STATISTICS AS A RESEARCH TOOL

A scientist suffered confusions
Brought on by statistical delusions
Though he worked 'till he lathered
The data he gathered
Refused to support his conclusions.

W. W. Armistead (1916–   )

Those working in biological research are coming more and more to appreciate the statistical measures or tests employed to evaluate research *data* or facts.

FIGURE 22.2
Stalls of stainless-steel construction aid in research control of digestion and mineral-balance studies. *(Courtesy Dr. W. H. Pfander, Dr. R. L. Preston, and the University of Missouri.)*

Statistical methods provide ways of simplifying masses of numbers and facts and of presenting them in an understandable, condensed, and consistent form.

Estimation and Tests of Significance   Scientists wish to do more than describe the particular observations that have resulted from their experiments. They wish to generalize their results as a part of the scientific method, so that the results of similar treatments in future experiments may be predicted. Suppose nutritionists are looking for a diet **supplement** that will increase weight gains of **weanling** pigs. They believe that they have found such a supplement and proceed to feed it to four **litters** selected at random from their herd of 40 litters. They compare the weight gains of the selected litters with the gains that would have occurred had the litters received the usual diet. They set up the *null hypothesis* (assume) that the supplement is of no value. This hypothesis is formulated for the express purpose of being rejected; i.e., it states that there is no difference in the growth-promoting effects of the two diets, and therefore the feeding of either diet will yield equal weight gains. The hypothesis is based on the assumption that the four litters to receive the test diet were *selected at random* from all the litters available for the experiment. *A random sample* is one in which each litter has an equal probability (chance) of being selected for use as a test litter or as a control litter.

At the end of 10 weeks, the nutritionists find that the average pig weight of the four test litters exceeds that of the controls by 40 lb. By considering the variation among pigs of the same litter, applying appropriate statistical techniques, and consulting published probability tables, the researchers find that the probability of a difference of this magnitude occurring under the null hypothesis is 0.02. This means that if the null hypothesis were really true (i.e., that the supplement has no additional value), random sampling would result in differences as large or larger than the one observed in only 2 percent of a large number of samples (analogous to the chance of guessing *one* particular number correctly from 50). The researchers would then conclude that one of two possibilities occurred. *Either* they obtained data from a rare sample and the diets are equally effective *or* their data are from a more likely sample and the supplement has in fact increased weight gains. The experimenters therefore would likely reject the null hypothesis. The probability of being wrong in concluding that the difference in gains was really due to diet supplement is 0.02. Treatment differences that are not likely to be due to chance are called *statistically significant* differences. Conventionally, differences that occur with probabilities of less than 0.05 (<0.05) are usually considered to be statistically significant.

> The statistical method, like other methods, is not a substitute for, but a humble aid to, the formation of a scientific judgment.
>
> *E. Bidwell Wilson*

Statistical methods should be used by the animal scientist as tools in designing research (this allows an accurate and valid statistical analysis to be made on completion of an experiment), in analyzing data, and in drawing inferences or conclusions therefrom. No amount of statistical manipulation will overcome a poorly designed experiment. Moreover, statistics are no substitute for good judgment.

Statistical evaluation of quantitative traits was discussed in Section 5.2.

22.6 COMPUTERS: AID TO RESEARCH

> Operations involving intense mental effort may frequently be replaced by the aid of other operations of a routine character, with a great saving of both time and energy.
>
> *Ernst Mach (1838–1916)*

People's first conscious mathematical operations probably involved only simple counting: the number of faces in the tribe, the number of **cattle** in the **herd,** and the number of wives owned. When the numbers involved exceeded the number of fingers and toes, a new form of numbering was invented. Piles of sticks and stones would have been logical calculating aids for primitive people. (The English words calculate and calculus are based on the Latin *calculus,* a small pebble used in calculating and counting.) People's mathematics and technical ingenuity have progressed to the availability of computers to aid in the advancement of science.

Computers are currently used for balancing animal diets, calculating "least-cost" diets (i.e., which combination of feeds will provide the desired **nutrients** at the least cost), evaluating production records that can be used as a basis for the selecting and mating of farm animals (also to determine the degree of **inbreeding,** the intensity of **selection,** the degree of **dominance** of one type of gene over another, and the physical linkages of **genes** on **chromosomes),** and analyzing research data. Computers will aid the animal scientist of the future by storing scientific information that can be recalled quickly. This will greatly reduce library time required to review published literature related to a given research subject or problem. Physicians will use computers to aid in the diagnosis of blood diseases and other disorders. For example, there are more than 80 blood diseases on file, including the anemias and leukemias. A patient completes a self-administered questionnaire pertaining to his or her personal and family medical history, and this information is processed electronically. He or she is then given a battery of automated tests, including 14 chemical determinations of blood constituents, **x-rays,** and an electrocardiogram. A computer processes the results immediately and provides the doctor with a prediagnostic information package.

Scientists are utilizing computers to help design certain experiments that enable them to estimate the sample size needed to show statistically significant differences with varying treatment levels. In short, computers are as useful in the planning of experiments and in the summarization and appraisal of data collected from them as human ingenuity can instruct them. Moreover, much of animal science teaching in the future will utilize the computer. For example, in studying genetic progress in breeding for certain production traits, the student could include in the computer program values for such variables as selection superiority, **heritability estimates, progeny-testing** records, environmental factors, interactions, etc. On the basis of these assigned values, one may predict genetic gain in each generation and compare this gain with that expected with another set of values. If the assigned **genetic values** are correct, the student could give proper weight to each of several traits for greatest overall gain. A series of such studies could be made in a short span of time through simulation and the aid of a computer.[12]

22.7 LITERATURE AND THE LIBRARY

Reading maketh a full man, conference a ready man, and writing an exact man. Read not to contradict, nor to believe, but to weigh and to consider.

Sir Francis Bacon (1561–1626)

---

[12]J. R. Campbell and R. T. Marshall, *The Science of Providing Milk for Man,* McGraw-Hill Book Co., New York, 1975, pp. 396–397.

> To be perfectly original one should think much and read little, and this is impossible, for one must have read before one has learnt to think.
>
> *Lord Byron (1788–1824)*

It should be clear from materials presented in the preceding chapters that the various aspects of animal science are continuously being explored and written about. Only by extensive library effort can one keep abreast of the published literature in a given field. At the beginning of the nineteenth century there were about 100 scientific journals. In 1850 there were 1000, and by 1900 the number had reached 10,000. Some estimates of the number at present go as high as 100,000. *The World List of Scientific Periodicals* indexes approximately 50,000 different publications in the sciences.

> It might be suggested that, in time, the amount of knowledge needed before a new discovery could be made would become so great as to absorb all the best years of a scientist's life, so that by the time he reached the frontier of knowledge he would be senile. I suppose this may happen some day, but that day is certainly very distant. In the first place, methods of teaching improve. Plato thought that students in his academy would have to spend ten years learning what was then known of mathematics; nowadays any mathematically minded schoolboy learns much more mathematics in a year.
>
> *Bertrand Russell (1872–1970)*

Scientists gather information by studying technical journals and books to determine what is already known relative to an idea or a problem. A number of services provide *abstracts* (e.g., *Dairy Science Abstracts, Nutrition Abstracts,* and *Chemical Abstracts*), or summaries of major facts, from long, often complicated articles. Examples of scientific journals in the animal sciences published in the United States are the *Journal of Animal Science,* the *Journal of Dairy Science,* and *Poultry Science.* Scientists can now utilize elaborate information-retrieval systems to help search the literature. The researcher selects appropriate key-word buttons on the computer retrieval program to obtain a bibliography of papers, reports, conference proceedings, and personal communications pertinent to his or her area of interest.

22.8 RESEARCH ORGANIZATIONS

The U.S. Department of Agriculture (USDA) and the agricultural experiment stations operated by state colleges and universities are the most important scientific animal science research agencies in the United States. Federal government research organizations include the regional laboratories of the USDA, the Public Health Service, and the Atomic Energy Commission. The **National Research Council** was organized in 1916.

The Agricultural Research Service (ARS)   The Agricultural Research Service is an agency within the USDA that conducts research pertaining to the production and use of farm products. It has programs designed to lead to the control or destruction of diseases and pests that harm crops and farm animals. The largest ARS research center, an 11,000-acre tract near Beltsville, Maryland, serves as headquarters for the ARS. The largest solely animal-oriented USDA research facility, the Roman L. Hruska U.S. Meat Animal Research Center, ARS, is located at Clay Center, Nebraska.

State Agricultural Experiment Stations (SAES)   These are research centers in which scientific investigations pertaining to the agricultural sciences are conducted. The United States has 53 stations, employing approximately 10,000 workers. Each state has at least one main station, usually at a state college or university (Appendix D). In addition to conducting research, the SAES help train future animal scientists, as noted in the following quotation:

> In its overseas activities as well as at home, the university will function as a university and not merely as a pool of technical talent or an employment broker. It will remember that its unique role is not only to apply present knowledge but to *advance the state of knowledge,* not only to supply experts today but to *train the next generation of experts.*
>
> *John W. Gardner (1912–   )*

Associations and Societies   Professional societies, such as the American Dairy Science Association, the Poultry Science Association, and the American Society of Animal Science, are composed of members interested in the advancement of animal agriculture through scientific endeavors. Members of these organizations sponsor scientific meetings at which researchers present their latest experimental findings. These societies also contribute to research by publishing journals with worldwide distribution that report research results.

> The Greeks had no classical education, but they had the two essentials of true education: first, the ability to express themselves correctly in words, and second, to appreciate their own relation to their surroundings, which latter is science.
>
> *E. H. Starling (1866–1927)*

Information about scientific research reaches farmers in many ways. Scientists publish their findings in bulletins and scientific papers. County agents and specialists from agricultural colleges report on these findings to farmers at local meetings and by field demonstrations. Farm newspapers, magazines, and radio and television programs bring information pertaining to agricultural research to farmers.

22.9 ATOMS IN ANIMAL RESEARCH

To know what questions to put to Nature—that is 95 percent of scientific research.
Alfred North Whitehead (1861–1947)

Each atom, though it is quite incommensurable, has in it the power of a thousand horses—infinite power is at our fingertips awaiting release.
Ernest Rutherford (1871–1937)

In animal research, the number of questions to be asked of nature seems infinite. Future animal scientists seeking to answer these questions will probably rely more on techniques using radioactive isotopes than on any other method known today. **Radioactive tracers** and **radiation** sources have beome indispensable to most phases of agricultural research. They permit scientists to study biological pathways; e.g., **tracer** techniques have shown that more of the phosphorus in milk comes from cow's bones than from their feed, whereas most of the phosphorus in an egg comes directly from the **hen's** feed and not from her body stores. **Hormones,** in microgram amounts (a microgram has the same relation to a 1000-lb **steer** as a penny, or one cent, does to 4.5 billion dollars), accelerate rate and efficiency of weight gain in cattle and sheep. However, it is necessary to know if meat from treated animals is suitable for human consumption. When an **isotope** of hydrogen (^{3}H) or carbon (^{14}C) was linked to hormones and then administered in normal amounts to cattle, subsequent tests for the radioactive hormones showed less than one part per billion of the hormone in the meat. This level of hormone is not detrimental to humans, and the meat of such animals is therefore safe for human consumption.

Radioisotopes have been used to study **insects** and their **life cycles,** dispersion patterns, mating and feeding habits, **parasites,** and **predators.** All these are important factors in the **biological control** of insects (Chapter 19). The biosynthetic efficiency of penicillin production has been increased 1000-fold by the initiation of repeated **mutations** in the microorganisms that produce this **antibiotic. Ionizing radiation** is being used to destroy the bacteria (or slow their **metabolism** or reproductive rate) and **enzymes** that normally cause food deterioration. **Irradiation** of food products can virtually eliminate problems of food poisoning, e.g., with salmonella, and the transfer of parasites, e.g., the trichina worm, to humans. Radiation preservation of food is accomplished in two ways: (1) *pasteurization,* which is accomplished with low dosages of radiation, and (2) **sterilization,** which requires higher levels (about 10 times more than pasteurization). Bacon and other pork products, chicken, fish, and beef can be packaged and then irradiated, resulting in preservation that permits storage for 1 year at room temperature. An ingenious application of atomic energy to animal science was the sterilization (with cobalt 60) and release of male screwworm flies (Figure 22.3). The female mates only once. If mated to a **sterile** male, she lays only

FIGURE 22.3
Cobalt-60 gamma cell used to expose small animals to varying doses of ionizing radiation. *(Courtesy G. W. Leddicotte and the University of Missouri.)*

infertile eggs. Through this medium, the insect was completely eliminated in parts of the southern United States (Chapter 19). Brain tumors have been located using radioactive arsenic and eye tumors with radioactive phosphorus.

22.9.1 Low-Level-Radiation Research Laboratory

There is no longer margin for doubt that whatever the mind of man visualizes, the genius of modern science can turn into fact.

David Sarnoff (1891–1971)

Research facilities of this type (Figure 22.4) provide a nondestructive **in vivo** method (called **whole-body counting**) of studying nutritional influences on the change of muscle, fat, and bone during growth, reproduction, **senescence,** and various metabolic disorders. The effects of pregnancy, starvation, and major dietary changes on body fluids can be determined quickly and accurately. Scientists have substituted cobalt 60 atoms for stable cobalt in molecules of vitamin B_{12} to study the metabolism of this vitamin within the body.

The laboratory can serve as a research tool in studying the interrelations among **macro-** and **micro**minerals; e.g., the interaction of copper with iron in the etiology of **anemia** can be studied. One can also study iron metabolism by injecting small doses of radioactive iron (^{59}Fe). The **absorption** of iron from the intestine and its half-life (Section 22.9.2) during various **hemolytic** anemias can

FIGURE 22.4
A typical low-level-radiation laboratory. *(Courtesy W. J. Coffman and the University of Missouri.)*

be determined in humans and animals. These data can then be correlated with various blood and tissue hematologic parameters to assess iron status in the body.

Whole-body counters provide a technique for studying the transfer of nutrients and other substances from mother to **offspring** while **in utero.** Scientists have utilized the whole-body counter to study and compare animals **infested** with internal parasites and parasite-free animals. Researchers at Cornell University reported a positive relation between the level of parasite infestation and loss of ^{59}Fe-labeled blood from the digestive tract. This technique has potential uses in evaluating parasite-killing drugs.

A whole-body counter utilizing potassium 40 (^{40}K) is useful to investigators interested in identifying breeding animals possessing superior genetic traits, e.g., meatiness in swine, which consumers prefer (Chapter 3). (^{40}K is a naturally occurring radioisotope of potassium.) Utilizing this facility, researchers can determine the heritability of body composition at different ages to find the youngest age at which animals possessing the most desirable lean/fat ratios can be selected effectively. This will allow scientists to correlate the lean/fat ratio (since fatty tissues have a low potassium concentration and muscle tissues a higher level) with performance and longevity and thereby to determine if the same genes affect two or more of these traits. This is made possible by determining body composition and performance traits in the progeny of several **sires.** Moreover, **carcass** composition can be correlated with blood composition, hormonal levels, and **feed efficiency.** Using radioactive iodine (^{131}I), scientists at the University of Missouri have found that **thyroid** activity increases with the onset of **lactation** and egg laying. It may be that dairy-cattle breeders can soon select animals as young **calves** for potential milk production by measuring the activity of their thyroid glands (Figure 22.5). This would eliminate the risk and expense of waiting until they begin lactating to determine production characteristics. The test involves administration of a minute amount of ^{131}I, which soon concentrates in the thyroid glands. Generally, a high concentration of ^{131}I indicates an increased thyroid activity, which is related to high milk production.

Studies of medical interest are possible in a low-level-radiation laboratory. It is known that the radioactivity of ^{40}K in a person's body decreases with age. (Under normal conditions ^{40}K is the most abundant radioactive isotope in the human body.) Is this rate of decrease correlated with longevity? Does it reflect the nutritional status and/or environmental conditions of humans? The ^{40}K content of a person's body in normal and diseased states can be determined (Figure 22.6). Are potassium metabolism and body stores important in muscular dystrophy (it has been indicated that there is a gradual and progressive decrease of body potassium during the unrelenting course of this disease), congestive heart failure, **chronic** renal **dysfunction,** diabetes (a sharp drop in potassium content accompanies the profound muscle weakness that follows diabetic coma), malabsorption **syndrome, hyper-** or **hypo**thyroidism, various **gastrointestinal**

FIGURE 22.5
A young dairy animal positioned for counting in a low-level-radiation laboratory. *(Courtesy W. J. Coffman and the University of Missouri.)*

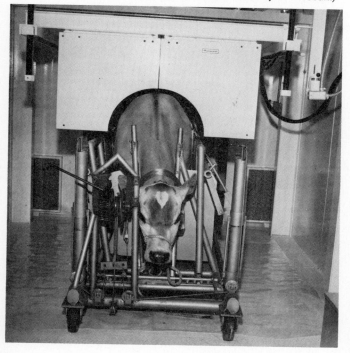

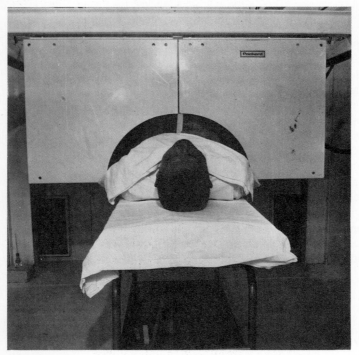

FIGURE 22.6
Using the low-level-radiation laboratory to make ^{40}K and ^{137}Cs counts in an adult human. *(Courtesy W. J. Coffman and the University of Missouri.)*

disorders, hematological disorders such as leukemia and lymphomas, and other disorders? Medical research teams can utilize low-level-radiation laboratories to help answer these and other related questions.

Whole-Body Counting   Whole-body counting provides a means by which researchers can look inside the body and observe interacting biological systems. Whole-body counters detect **gamma rays** emitted by radioactive atoms disintegrating within the body. Whole-body counting is quick, accurate, and painless except for possible psychological discomfort associated with being locked in a closed environment of steel (to shield detectors from stray gamma radiation), and it provides scientists with a useful technique for examining these systems. In short, whole-body counters can be used to add to scientific knowledge and increase the ability to manage and control conditions and processes important to people's health and well-being. Each fact uncovered by using whole-body counters seems to trigger more penetrating research. This is how science advances. With each advance in science comes knowledge on which humanity can build and find solutions to problems affecting our lives.

How It Works The whole-body counter (gamma counter) is a liquid **scintillation counter** (detector), which is housed in a steel room. The walls are often made of 6- to 8-in-thick steel from a pre-World War II naval ship. It has detector tanks (filled with special chemicals) that are attached to photomultiplier tubes (photoelectric cells) and preamplifiers (Figure 22.7). A gamma ray **(photon)** passing through the liquid scintillator loses energy as it collides with **atoms** or **molecules** of the liquid. The absorbed energy produces *light photons,* or scintillations, in the liquid. The light photons strike the light-sensitive surfaces of the photomultiplier tubes and emerge as a volley of **electrons** (tiny pulses of electricity). These electrons then pass through a preamplifier that amplifies the electronic signal prior to counting and recording. These photomultiplier devices are similar to the equipment in the familiar "electric-eye" door openers.

Whole-body counters reveal the kinds and amounts (count emissions) of radioactive substances that have accumulated in the body from natural sources (every person is slightly radioactive); as fallout from human activities, e.g., ^{137}Cs or ^{131}I; or from tracer isotopes given for medical purposes. Each radioactive substance emits gamma rays (certain isotopes, for example, ^{32}P and ^{45}Ca, emit beta and/or other rays) with a characteristic energy level. Whole-body counters can measure a *specific* energy spectrum, or "fingerprint," and thereby identify the kind of atom producing the **radiation.** The number of light photons produced

FIGURE 22.7
Low-level-radiation detectors of a whole-body counter. *(Missouri Agr. Expt. Station.)*

in the scintillation fluid is proportional to the energy transferred by the incoming gamma ray. For example, gamma rays emitted by ^{40}K emit about twice as much energy as those of ^{137}Cs. When both these radionuclides are producing flashes of light in the scintillation fluid at once, the photomultiplier tubes produce two different strengths of electric pulses, which are proportional to the energy of the gamma photons from the two radioisotopes. Electronic devices, called *multichannel pulse-height analyzers,* sort and record the number of each (Figure 22.8).

22.9.2   Radioactive Half-life

Out of monuments, names, words, proverbs, traditions, private records, and evidences, fragments of stories, passages of books, and the like, do we save and recover somewhat from the deluge of time.

Sir Francis Bacon (1561–1626)

FIGURE 22.8
Counting and printout equipment of a low-level-radiation laboratory.
(Missouri Agr. Expt. Station.)

In general, **half-life** is the amount of time it takes any amount of a radioactive substance to decrease to half its original value. For radioactive isotopes this means the length of time required for one-half of a particular isotope to disintegrate (decay) into a stable form (Figure 22.9a). If a graph of the activity of

FIGURE 22.9
(a) Half-life pattern of ^{90}Sr. *(From "Radioisotopes in Medicine," U.S. Atomic Energy Commission, 1967.)* (b) The decay of an isotope having a half-life of 3 h, such as ^{134}Cs. Each radioisotope is characterized by its own half-life, i.e., by the time necessary for half of any mass of this element to decay. *(From AEC document, in "Radioisotopes in the Service of Man," UNESCO.)*

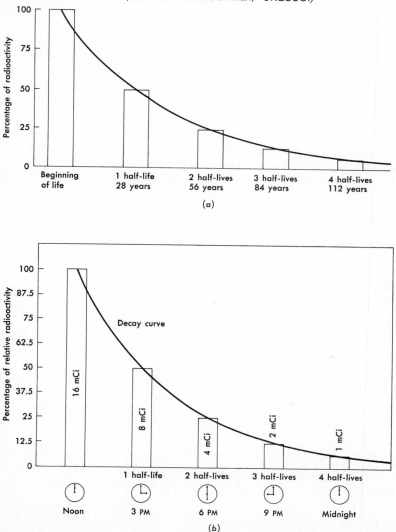

a radioactive substance is plotted against time, the characteristically shaped curve known to mathematicians as an exponential curve is obtained (Figure 22.9b). In animal experimentation, three half-lives of radioisotopes must be considered: (1) *physical half-life (T_P)* of a radioisotope, defined as the natural decay of radioactivity and its corresponding half-life; e.g., half of the ^{40}K atoms will disintegrate over a period of 1.3 billion years; (2) **biological half-life** *(T_B)*, defined as that time required for the body to eliminate one-half of any substance by the regular processes of elimination (biological half-life varies among species[13]; e.g., the half-life of radioactive ^{131}I-thyroxine is only about 16 h in rats but 48 to 60 h in cows); and (3) *effective half-life (T_E)*, defined as the time required for a radioactive element, in an animal's tissues, to diminish by 50 percent. The effective half-life results from the combined action of radioactive decay and biological elimination.

Radioisotopes are being used extensively in animal research. Important information on the study of **senescence** has resulted from the use of radiation as a **stress.** Studies at the U.S. Atomic Energy Commission's Division of Isotope Development at Oak Ridge, Tennessee, suggest that aging is primarily associated with damage to chromosomes. If **DNA** is damaged, animals grow older because of the basic instability of the DNA (Section 8.9).

Controlled radiation often benefits humanity. Its uses include the bombardment of plants from a radioactive cobalt source to induce genetic changes, the irradiation of deep-seated tumors with a beam from a **particle accelerator,** and **therapy** of thyroid cancer with radioactive iodine. It is also being used in treating brain tumors.

Australian researchers used radioactive copper to determine the amount of copper needed in the diet of Merino sheep to ensure high-quality fleece (copper helps govern the value of fleece; Chapter 17).

22.9.3   Nuclear Research Reactor

It is the avowed purpose of scientific thought to reduce the number of mysteries, and its success has been marvelous.

J. H. Robinson (1863–1936)

The story of atomic energy evolves from the curiosity of people concerning the nature and structure of matter, the stuff of which all material things are made.

C. Jackson Craven (1908–   )

Throughout the history of humanity, only a few single events have materially altered the course of civilization. Among these was the development of a successful **nuclear reactor,** an accomplishment that has been compared to the

---

[13]Body size affects the loss of radioisotopes, e.g., the mouse loses ^{14}C much faster than the cow.

invention of the steam engine or the manufacture of the first automobile in its enormous impact on the future and its significance for social change.

The unleashed power of the atom has changed everything except our ways of thinking.
Albert Einstein

An atomic reactor is a controlled nuclear device used to produce and release atomic energy without an explosion. As the energy is produced, **neutrons** and radioisotopes are generated, and, just as people may be sunburned from overexposure to sunlight, they may also be burned by overexposure to these radiations. Furthermore, this energetic radiation is capable of killing bacteria and insects. A research reactor is designed to provide a source of neutrons (uncharged atomic particles) and/or gamma radiation for research into basic or applied biology, chemistry, or physics.

In an atomic reactor, energy is released from the **nucleus** (core) of the atom. Scientists usually call such machines *nuclear reactors;* however, they may also be called *atomic piles* (so named because the first reactors were essentially piles of **graphite** blocks and uranium fuel) or *atomic machines.* The nuclear reactor produces power through **fission** (splitting) of the nuclei of **atoms** of uranium or plutonium. The atoms split when they are bombarded by neutrons. Each atom splits into two smaller *fission fragments* (products). Each of these is approximately one-half the mass of the original atom. Fission also releases energy and two or three neutrons from each atom. (Heat is generated as they fly apart at great speeds and collide with surrounding matter.) These neutrons fly out and may hit more uranium atoms, thus causing them to split. In this manner, a *chain reaction* occurs. The neutrons generated in the reaction become available to provide neutron environments, in which materials can be bombarded to produce radioisotopes.

Nuclear reactors are used to supply intense fields or beams of neutrons for scientific experiments; to produce new elements or materials by neutron irradiation; and to furnish heat for electric power generation, propulsion, industrial processes, or other applications (Figures 22.10 and 22.11). Some people predict that devices employing nuclear radiations (for example, ^{60}Co) will replace the autoclaves now used in hospitals and microbiological laboratories to sterilize surgical instruments, glassware, and equipment.

A relatively new, sensitive, versatile analytical tool employing nuclear energy is *neutron activation analysis.* In **activation analysis,** a sample of unknown material is first irradiated (*activated*) with nuclear particles (almost always neutrons) and becomes radioactive. The radioactive atoms disintegrate with the emission of energetic **beta** and gamma radiations. When these radiations are counted (*analyzed*), the half-lives of the radioactive nuclei and the radiation energies are determined. These nuclear fingerprints are used to identify the artificially created nuclei and, by inference, the stable elements in the original nonradioactive sample as well.

FIGURE 22.10
Research Reactor Facility of the University of Missouri-Columbia. It houses animals, laboratories, counting rooms, and a 5-MW nuclear research reactor. It produces neutron flux environments of up to 5×10^{14} thermal neutrons per square centimeter per second. It is used to assist research in many disciplines. *(Courtesy G. W. Leddicotte and the University of Missouri.)*

Activation analysis permits agricultural scientists to detect **pesticides** on crops and in prepared foods by monitoring traces of bromine and chlorine that remain on foods even after manufacturing. (The technique allows detection of extremely minute quantities, as little as one part per billion.) It may be used to study mineral metabolism of humans and animals. Another biological application is the identification of the geographical source of opium.

Human hair contains small traces of metallic elements such as copper, gold, and sodium. Activation analysis has shown that the quantities of these elements present in each hair are relatively constant for an individual but vary from person to person. English scientists found an unusual amount of arsenic in a relic of hair from Napoleon's head. This has given rise to the suspicion that he was slowly poisoned to death. The technique is also used to identify gunpowder residues. Scientists involved in space-related research have determined the compositions of lunar and planetary surfaces using activation analysis.

Activation analysis is being used to analyze animal feeds for nutrient elements and to study metabolic pathways of various minerals in animals (e.g., Sr in bone chemistry). The relation of selenium to white muscle disease (Chapter 17) was originally established using neutron activation analysis techniques. Scientists are analyzing hair to study the possible relation of trace (micro-) mineral changes in the hair of animals having certain diseases and/or metabolic disorders. Activation analysis is also used to determine the trace mineral content of water. Since the usefulness of activation analysis has just begun to be demonstrated to science and industry, many new applications can be expected in the future. Selected

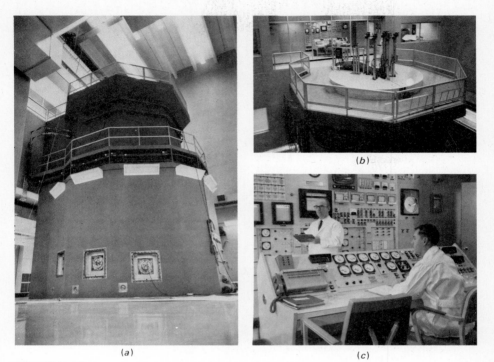

(a)

(b)

(c)

FIGURE 22.11
(a) A nuclear reactor showing beam port facilities (near bottom); (b) top view of a reactor showing control rods and operating console room; (c) console room of a nuclear reactor. *(Courtesy G. W. Leddicotte and the University of Missouri.)*

items of equipment involved in neutron activation analysis are depicted in Figure 22.12a to c.

22.9.4 Nonradioactive Isotopes in Animal Research

Man . . . has found ways to amplify his senses . . . and, with a variety of instruments and techniques, has added kinds of perception that were missing from his original endowment.

Glenn T. Seaborg (1912–   )

Most compounds of biological importance contain the elements carbon, hydrogen, nitrogen, and oxygen. However, of these, only carbon and hydrogen have radioactive isotopes with half-lives long enough to be useful in following their fate in the various metabolic processes of plants and animals.

(a)   (b)

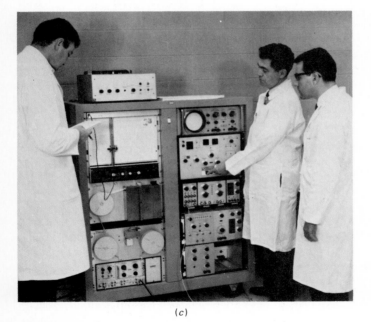

(c)

FIGURE 22.12
(a) Placing a test sample in an irradiation facility. This pneumatic system conveys the small
plastic container, or **rabbit**, into a neutron environment to activate trace elements in the
experimental sample. (b) After irradiation is accomplished, the activated material is usually
processed by a radiochemical method. (c) The atomic fingerprints produced in an experimental
sample can be detected, identified, and measured by an appropriate counting system. The
system shown is a gamma-ray spectrometer. (*Courtesy G. W. Leddicotte and the University of
Missouri.*)

Most elements occur normally as mixtures of two or more isotopes. Although some are nonradioactive, they can be as stable as the isotope form present in the greatest abundance. For example, nitrogen contains 99.63 percent ^{14}N and 0.37 percent ^{15}N. Oxygen contains 99.759 percent ^{16}O, 0.037 percent ^{17}O, and 0.204 percent ^{18}O. These isotopes are stable and emit no radioactive rays or atomic particles. Fortunately, their **stable isotope** forms may be utilized in studying the fate of nitrogen or oxygen in metabolic pathways by using a **mass spectrometer** (Figure 22.13). This instrument is very sensitive and will accurately measure the relative concentrations of different stable isotope forms. Sample preparation is usually quite tedious and somewhat more difficult than is that for radioactive-isotope detection. However, since there are no usable radioactive isotopes of nitrogen and oxygen available to animal scientists, they must rely on the use of their stable forms.

One distinctive feature characterizing **proteins** and amino acids is that they contain *nitrogen* in addition to carbon, hydrogen, and oxygen. When these nutrients are utilized by animals, the nitrogen may be metabolized in one pathway and the other elements in another. A researcher can use ^{15}N as a **tag** for the more abundant ^{14}N, since the former follows the same metabolic pathways as the latter. Thus an amino acid that has been **enriched** with ^{15}N can be administered to an animal, and by isolating and analyzing various compounds

FIGURE 22.13
A scientist determines the stable isotope ^{15}N concentration of a metabolite with a mass spectrometer. *(Courtesy Dr. R. A. Bloomfield and the University of Missouri.)*

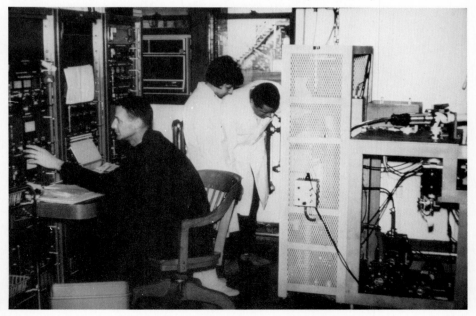

for ^{15}N enrichment, the researcher can determine the tentative pathway of its metabolism. This technique, using the mass spectrometer, is especially useful in nutrition studies with farm animals.

22.10   RESEARCH IN ENDOCRINOLOGY

In Chapter 7 the roles of hormones secreted by the **endocrine** glands were discussed. These body regulators of various chemical makeups are intimately involved in such vital life processes as reproduction, growth, fattening, lactation, and egg laying.

Experimentation involving these endocrine-controlled processes is being expanded and intensified immensely, so that the challenging area of science, known as *endocrinology,* has moved to front stage in the field of animal science. One important development at the Missouri Agricultural Experiment Station has been a technique for measuring the thyroxine secretion rate of cows (Figure 22.14) by injecting radioactive iodine (^{131}I) and varying amounts of the hormone *thyroxine.* This enables researchers to study the relation between thyroid secretion rate and one or several of the vital physiological processes mentioned above.

FIGURE 22.14
Measuring the thyroxine secretion rate of a cow. *(Courtesy Dr. C. W. Turner and the University of Missouri.)*

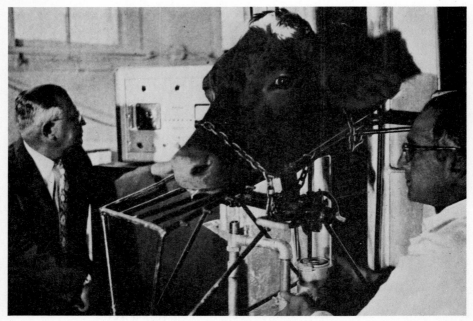

Because of similarities in the effects of hormones on mammals, including small laboratory animals, farm animals, and humans, most preliminary endocrinology experimentation involves the use of mice and rats, for reasons of efficiency and economy. An important operative procedure to be mastered by the endocrinologist is removal of the pituitary gland at the base of the brain, since it is considered to be the master endocrine gland of the body (Chapter 7). A unique method of performing a **hypophysectomy** is depicted in Figure 22.15.

22.11   GAS CHROMATOGRAPHY (GC) IN ANIMAL RESEARCH

The day of the genius in the garret has passed, if it ever existed. A host of men, great equipment, and long, patient, scientific experimentation to build up the structure of knowledge, not stone by stone, but grain by grain, are now our only sure road of discovery and invention.

Herbert Hoover (1874–1964)

Because of its accuracy, small sample sizes required, simplicity of sample preparation, speed of analysis, versatility, relative sensitivity, and automatic recording of data, GC has become an increasingly important tool in biological research.

How GC Works   A liquid, gaseous, or solid sample is injected or placed into port *A* (Figure 22.16), where it is instantly volatilized, and the vapors are swept onto a column with a carrier gas. This column is packed with an inert solid support that is coated with a thin film of a liquid (similar to the varnish on a floor). The liquid has a boiling point significantly higher than the column temperature. As the carrier gas sweeps the volatilized molecules through the

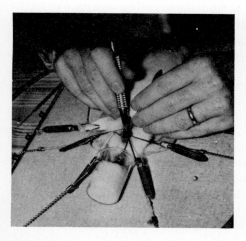

FIGURE 22.15
A scientist uses a dental drill to penetrate the skull of a rat. He then uses vacuum to evacuate the pituitary gland from the body. Replacement therapy with hormones normally secreted by this gland can then be studied. *(Courtesy Dr. R. R. Anderson, University of Missouri.)*

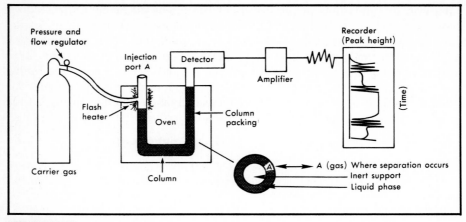

FIGURE 22.16
Essential components of a gas or solid chromatography unit.

column, the various components separate because of their differences in vapor pressures (volatility) and solubilities in the liquid film. The most volatile components will move through the column the fastest and the least volatile ones the slowest. As a consequence, the various chemical compounds reach the detector at different times. The signal from the detector is transmitted to the recorder, where a permanent record of the analysis is made. The size (peak height or area) of the recorded response is related to the quantity of each compound. Injecting known compounds and noting the time required for each to move through the column is a means of identifying the components of an unknown mixture.

Gas chromatography is an important tool in **ruminant** nutrition research. Since the volatile fatty acids **(VFA),** acetate, propionate, and butyrate, are the primary energy-yielding nutrients resulting from the microbial digestion of feeds in the **rumen,** much research has been devoted to a study of their production and metabolism. The use of gas chromatography in analyzing blood and rumen contents for VFA has allowed great strides to be made in ruminant nutrition in recent years.

Studying the effects of concentrations of VFA in the rumen and blood of animals on milk-fat synthesis and composition is possible using GC. Cows fed low-**forage**–high-grain diets have a different proportion of VFA in the rumen than those fed a high-forage–low-grain diet. Cows fed the former diet secrete milk with a lowered fat content (Chapter 11). Feedlot steers and heifers fed either monensin or lasalocid, both anticoccidial antibiotics, produce more propionate relative to acetate in the rumen (less methane gas is also produced). The net effect of these processes is an improvement in **feed efficiency** (see Chapter 14).

Another area in which GC is being used is in the analysis of respiratory gases resulting from metabolic studies. This research may aid in detecting certain metabolic disorders in farm animals. The composition of rumen gas may also be determined by this technique. Gas chromatography is used to determine the purity of organic chemicals and to identify unknown components resulting from the metabolism of various organic compounds by animal tissues and micro-organisms.

Pesticide residues in feeds and in animal tissues and their products may be determined by this technique. Gas chromatographic studies have aided scientists in synthesizing and stabilizing many artificial flavors now used in food processing. More than 100 components of strawberry flavor have been identified. Gas chromatography is used in human medicine to study and compare compositional differences in the blood of diseased and healthy persons. It is an important tool in research involving drugs, enzymes, food nutrients, hormones, and other biologically important molecules at the cellular level. Internal revenue officials have even used gas chromatography to differentiate between "white lightning" and legitimate bourbon.

Quantitative gas chromatographic techniques can also be used to analyze the 20 amino acids of protein and more than 40 other nonprotein amino acids (Figure 22.17). Moreover, methods for the analysis of purines, pyrimidines, **nucleosides,** and **nucleotides** have made possible determination of base ratios in DNA and **RNA.** These scientific breakthroughs have broad implications in advancing studies in genetics (plant and animal, e.g., in the development of high-lysine–high-tryptophan corn), protein biochemistry, nutrition, medical research, and space-related research.

22.12   HIGH-PRESSURE LIQUID CHROMATOGRAPHY (HPLC)

In gas chromatography, the components to be analyzed are gaseous or converted into a gas, and they are swept onto and through a column by a gas. In liquid chromatography, the components are dissolved in a liquid and are moved through the column by a liquid. Because some compounds are difficult to volatilize, liquid chromatography makes possible the analysis of more compounds. The high pressures necessary to push the liquid phase through the column gave rise to the name for this new methodology: *high-pressure liquid chromatography* (HPLC). It is sometimes referred to as high-performance liquid chromatography because of the excellent separation and reproducibility achieved with the method.

HPLC requires a very reliable and accurate pump to move the liquid phase at the desired flow rate (Figure 22.18). If the components are eluted from the column with a gradient (increasing amounts of one liquid), two pumps are required, and the control of both pumps must be highly reproducible. To effect this kind of control and to make automation possible, computers with microprocessors are being used to drive these instruments. They can also direct the analysis of the resultant peaks. HPLC is highly sensitive and can detect some compounds at concentrations of less than one part per million. The detectors can measure the absorbance of ultraviolet or visible light by the components of interest. Alternately, refractive index or ionization detectors can be used. The

FIGURE 22.17
Scientists at the University of Missouri observe data being recorded
by the gas chromatograph. *(Courtesy Dr. C. W. Gehrke, University of
Missouri.)*

separation of components depends on the compositions of the liquid phase and
the column material, which may separate the components on the basis of their
hydrophobicity, ionic charge, or size.

The HPLC methodology is being used to analyze the 20 amino acids of
proteins; the purine and pyrimidine components of RNA and DNA; and
protein, RNA, and DNA molecules themselves. Molecular engineers make use
of this and related technologies to harness genetic components in the service of
humanity. These methods also have broad implications for protein biochemistry, nutrition, medical research, and space-related research.

22.13   IN PRAISE OF PIGS AS RESEARCH ANIMALS

The sciences are of a sociable disposition, and flourish best in the neighborhood of
each other; nor is there any branch of learning but may be helped and improved by
assistance drawn from other arts.

Sir William Blackstone (1723–1780)

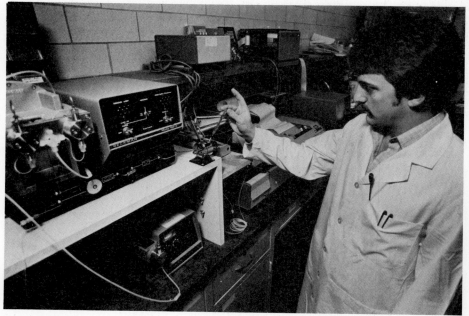

FIGURE 22.18
Injecting a sample into the high-pressure liquid chromatograph (HPLC). The computer-driven HPLC unit separates, identifies, and quantifies components of biological samples. *(Courtesy Dr. J. L. Robinson and the University of Illinois.)*

These little pigs go to research! Swine have embarked on a new career as subjects for medical research. Their anatomy approximates that of humans more closely than that of almost any other animal. **Physiologically,** swine are similar to humans; heart, circulatory system, digestive tract, and even teeth possess striking similarities. Like humans, they suffer from peptic ulcers, brucellosis, tuberculosis, influenza, and cardiovascular disease. Their skin gives easy readings to **allergy** reactions. Swine have about the same nutritional requirements as humans. Moreover, they digest their food in a manner much the same as humans. However, until recently, the usefulness of swine as laboratory animals was limited because they required too much food and space and too large a dosage of expensive experimental drugs. They were also difficult to handle. Then scientists developed *miniature pigs* by first crossing four kinds of wild pigs and then saving the smallest offspring for new crosses. Selecting and mating the smallest to the smallest resulted in pigs that weigh about 40 lb at 5 months[14] and about 180 lb at maturity (the approximate weight of a mature man). These miniature pigs are now being used to test mechanical hearts, the effects of

[14]Compare this with recent experiments in which pigs reached a market weight of over 200 lb in 120 days with about 3 lb of feed per pound of gain.

 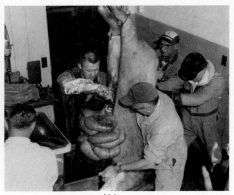

(a)   (b)

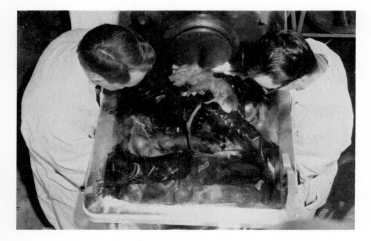

(c)

FIGURE 22.19

(a) The "transfer" isolator (left) and "surgical" isolator (right) are connected by a plastic "sleeve" (barely visible). These isolators are in position for a hysterectomy operation. (b) The uterus (containing the fetuses) of the anesthetized sow is being severed. The uterus drops into a germicidal trap and is brought up into the sterile environment of the surgical isolator. (c) A top view of the transfer isolator showing the pigs that have just been passed from the surgical isolator via the attached sleeve. The umbilical cords are being clamped and severed. (d) The "rearing" isolator (left) has been attached to the transfer isolator (right) via a plastic sleeve, and two pigs are passed from the latter to the former. The tub portion of the rearing isolator is divided by a partition, so that one pig can be placed on each side (separated so that they will not nurse each other). Pigs are reared in these isolators for several weeks. They are fed sterile milk (see cans below the pigs). (e) Top view of the transfer isolator. A piglet is being passed through a sleeve to the adjoining rearing isolator. *(Courtesy Dr. E. H. Bohl, Department of Veterinary Science, Ohio Agricultural Research and Development Center, Wooster, Ohio.)*

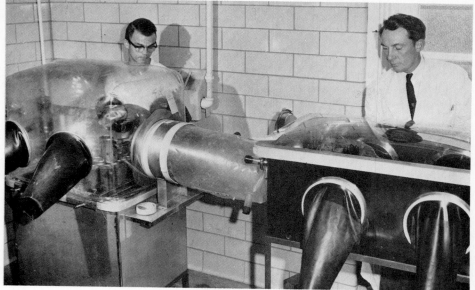

(d)

(e)

radiation, and the genesis and development of **atherosclerosis** and to study the physiological processes involved in senescence.

Skins of young pigs are used as covering materials for persons with severe third-degree burns. Pigs were also used as test animals in researching and developing the type of flight suits that would give maximal protection in the flash fires frequently encountered in aircraft mishaps.

22.13.1   Germ-Free Pigs

There are many possible applications of **germ-free** pigs in biological research. Some examples of their use at the Ohio Agricultural Research and Development Center are in the production of highly specific **antisera** against certain bacteria and **viruses.** [Since the pig does not acquire **antibodies** through the placenta (Section 11.12), it is a good research animal for such studies.] Such antisera are of great value in comparisons of the **antigenic** relationships among various microorganisms, in the study of the pathogenesis of microbial **infections** (the influence of extraneous or unknown microorganisms can be avoided), and in the study of the significance and interrelation of various microorganisms of the **gastrointestinal** tract.

Photographs of equipment and isolators used in procuring (by cesarean operation) and rearing germ-free pigs at the Ohio station are depicted in Figure 22.19*a* to *e*. Germ-free mice, rats, rabbits, chickens, guinea pigs, dogs, and lambs have also been used in experimental studies.

22.14   FISTULATED ANIMALS: AID TO RESEARCH

The rumen of a cow is the darkest place on earth.
William Dempster Hoard (1836–1918)

An absolutely clear and exhaustive understanding of any single thing in the world would imply a perfect comprehension of everything else.
Arthur Schopenhauer (1788–1860)

Animal scientists (especially ruminant nutritionists) have used fistulated animals in various microbiological, nutritional, and physiological studies since 1928, when Dr. A. F. Schalk of the North Dakota Agricultural Experiment Station published his first article describing fistula surgery in cattle. The opening, or "window," leading from the rumen to the outside is called a **fistula.** The metal, rubber, glass, or plastic tube connecting the body cavity with the outside is called a **cannula.** This surgical alteration and installation gives researchers an opportunity to study rumen temperatures, to introduce liquid or solid materials into the digestive tract, or to withdraw rumen contents for microbiological and chemical determinations. (Figure 22.20).

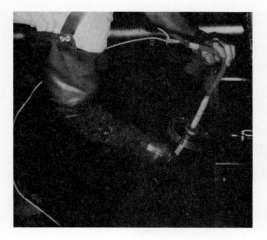

FIGURE 22.20
An animal scientist inserts probes through a rumen cannula to monitor rumen temperatures when the animal drinks water of varying temperatures. Weighed nylon bags containing ground forages are introduced into the rumen and, after certain periods of time, are withdrawn, dried, and weighed to determine the amount of hay digested when the animal drinks cold or warm water. Microbial populations and the concentration of volatile fatty acids are also studied concurrently. *(Courtesy Dr. M. D. Cunningham, Purdue University, and Dr. F. A. Martz, University of Missouri.)*

Fistulating animals apparently causes them very little discomfort or difficulty. They live a quite normal life and provide a ready source of microorganisms for laboratory studies (Figure 22.21*a* and *b*). Mixed **cultures** taken from fistulated animals are being used under standardized laboratory conditions to digest forage samples in studies designed to develop an in vitro technique to evaluate forages.

An **abomasal** fistula may be used in certain experiments in which the researcher wishes to study the effects of bypassing the rumen.

22.15 ANIMAL-DISEASE RESEARCH

Let knowledge grow from more to more.

Lord Tennyson (1809–1892)

There is a close relation between the health of animals and that of humans (Chapter 18); between the health of farm animals and the profits realized by those involved in animal production; and finally, between animal health and the availability and cost of animal products (meat, milk, eggs, and wool). One of the world's most modern animal-disease research centers is the National Animal Disease Center (NADC), located near Ames, Iowa (Figure 22.22).

Approximately 50 different diseases can be investigated in separate work areas. Scientists study the cause, transmission, mechanism of infection, diagnosis, prevention, treatment, and control of animal diseases. Air-locked doors, dressing room–shower entrances, washed and filtered air, and autoclaves help prevent cross-contamination of research areas (modules). When scientists enter and leave research areas in which studies of highly infectious biological agents, capable of infecting animals and humans, are being conducted, they must shower

(a)

(b)

FIGURE 22.21
(a)Sheep are one of the smallest domesticated animals used in rumen
studies. (b) This set of twin lambs was born after their mother was
fistulated.(Courtesy Dr. W. H. Pfander, University of Missouri.)

and change clothes. So that organisms cannot escape from a research area,
modules (laboratories) are kept at a lower pressure than the corridors. When the
air-locked doors are opened, the difference in pressure forces air into the
module. Special hoods afford protection to those studying pathogenic microorganisms and viruses (Figure 22.23). Plastic isolation cages are used for study of
highly infectious diseases in natural **hosts** (Figure 22.24).

The electron microscope is used to study structural characteristics of microorganisms and tissue **cells** (Figure 22.25). This searchlight of science helps
scientists to see disease-causing viruses in complex with natural antibodies. In
the electron microscope a beam of electrons is substituted for the light rays of a
light microscope. Special electronic lenses bend the beam. This instrument
enables scientists to observe objects too small to be seen through conventional
microscopes. A research tool recently developed by the United Kingdom
Atomic Energy Authority is the **proton**-scattering microscope. This microscope

FIGURE 22.22
USDA research center for the study of livestock and poultry diseases.*(Courtesy NADC.)*

FIGURE 22.23
Protective biological hood for research with pathogenic organisms.
(Courtesy NADC.)

FIGURE 22.24
Plastic large-animal isolation unit with its own feed, water ventilation,
and waste-disposal system. *(Courtesy NADC.)*

can be used to examine single crystals and crystalline layers that are only tens or
hundreds of **atoms** thick.

Like many experimental studies, animal-disease research is best approached
as a team effort by scientists having diversified training. Research veterinarians,
bacteriologists, biochemists, nutritionists, pathologists, physiologists, virolo-
gists, and other biological scientists join the attack on the disease being
investigated. Each team working on a disease has its own research area or

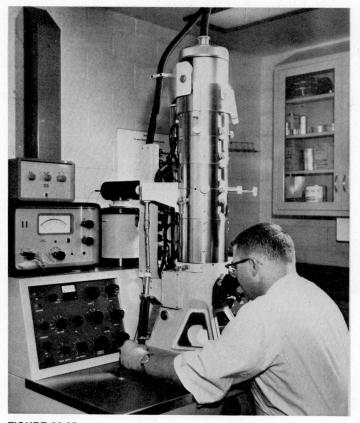

FIGURE 22.25
NADC scientist adjusts an electron microscope. Magnifications of up to
1 million are possible. *(Courtesy NADC.)*

module. Excellent but expensive equipment aids in collecting physiological data
(Figure 22.26).

The NADC is part of the Agricultural Research Service of the USDA. Other
USDA animal-disease centers include the Plum Island, New York, Animal
Disease Center, which is used for the research and study of foreign diseases (not
now established in the United States), and the Parasitological Research Center,
in Beltsville, Maryland, in which investigations of parasites that infest **livestock**
and **poultry** are made. Complementing the USDA's research on animal diseases
is that conducted by schools of veterinary medicine and the state agricultural
experiment stations throughout the United States.

The Veterinary Services Division of **APHIS** (USDA) directs diagnostic
services for **enzootic** or **epizootic** national disease programs. It serves as a

FIGURE 22.26
Oscilloscope monitors and makes a permanent recording of heart and respiratory rates, blood pressure, and other physiological functions by direct-wire telemetry. *(Courtesy NADC.)*

reference unit in helping state diagnostic laboratories solve unusual problems, such as sickness or death caused by chemicals. The NADC conducts research pertaining to **mycotoxins** in feedstuffs (Figure 22.27). The National Veterinary Services laboratory of APHIS tests new biologicals intended for veterinary use for purity, **innocuity,** and efficiency. The latter research is accomplished under the auspices of the Veterinary Services Program of APHIS, which is located near the NADC in Ames, Iowa.

22.16 FINANCING AGRICULTURAL RESEARCH

The marshaling of research and researchers on a mass scale is a relatively new development in the history of humanity. Research is power, and agricultural research is equated with the power to feed people. With the establishment of agricultural experiment stations came Hatch funds from the federal government. Later, other federal funds became available (Chapter 1). The percentage of federal research and development funds going to agricultural research decreased from 40 percent in 1940 to less than 0.8 percent in 1982. However, federal support to agricultural research increased more than tenfold during that period

FIGURE 22.27
Extraction of toxigenic materials from moldy animal feeds. *(Courtesy NADC.)*

of time.[15] (The total federal research budget has increased more than 4000 percent since 1940.)

The expenditures for animal agricultural research from all public and private sources in the United States amount to less than 1 percent of the total gross sales at the farm gate.[16] Few important industries can survive at such a low outlay for research purposes. Animal agriculture contributes 52 percent of the cash farm receipts from farm marketing in the United States and is the most important source of marketed protein in *quantity* and *quality* (Chapter 1). Yet it receives one-fifth of the funds expended for agricultural research. Sales of cattle and calves for meat represent 23 percent of cash farm receipts, yet funds for beef cattle research represent only about 7 percent of the amount spent for farm research. This does not, however, include other related and supporting research, such as pasture and forage production. To continue to produce at a profit and thereby remain a competitive industry, United States livestock and dairy producers must continually become more efficient. Only greatly expanded animal research designed to solve their problems will make this possible.

In 1982, the ARS allocated 20 percent ($82 million) of its $413 million research budget in support of animal productivity (Figure 22.28*a*). A 1983 USDA report of planned distribution of research funds among major objectives of the ARS program for the year 1990 showed a reduction of the percentage of

---

[15]In 1900, approximately one-half of the federal dollars appropriated for agricultural research went to the states and about one-half was appropriated to USDA agencies. By 1983, only about one-fifth of the total federal agricultural budget went to the states and approximately four-fifths went to USDA agencies.

[16]According to the International Food Policy Research Institute, the developed countries spend about 1.0 percent of their income from agriculture in support of agricultural research, compared with only 0.3 percent in the developing countries.

FIGURE 22.28
Distribution of ARS funds by six major program objectives in 1982. (a) Total ARS funding for these six program areas in 1982 was $413 million. (b) Planned distribution of funds among the six major objectives of ARS research programs for 1990 is shown. *(Agricultural Research Service Program Plan—6-Year Implementation Plan, 1984–1990, ARS, USDA, February 1983.)*

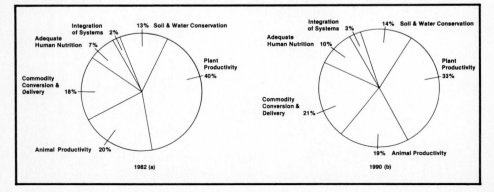

research monies allocated to animal productivity to 19 percent (Figure 22.28*b*). The relative emphasis given by the agricultural experiment stations to major categories of research is depicted in Figure 22.29.

Agricultural Research by Private Organizations   The private sector of the economy conducts and funds about 53 percent of the total national food and agricultural research (Figure 22.30). Agricultural research conducted by industry complements that conducted by federal and state governmental agencies.

Grants   Grants of money are important sources of support for agricultural research. These may be *free grants,* with few or no guidelines attached, or *contracts* to perform specific studies. These may be financed by individuals, industry, government, the military, or foundations.

22.17   TRENDS AFFECTING FUTURE ANIMAL RESEARCH

What is past is prologue.

William Shakespeare (1564–1616)

For I dipt into the future, far as human eye could see,
Saw the Vision of the world, and all the wonders that would be . . .

Lord Tennyson (1809–1892)

FIGURE 22.29
Major categories of research efforts of agricultural experiment stations of the land-grant colleges of agriculture. *(USDA.)*

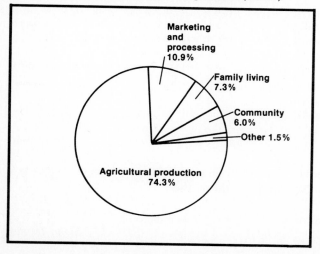

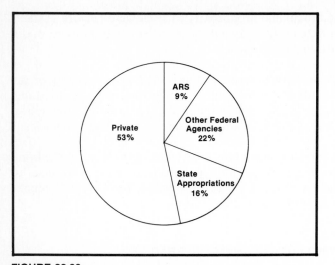

FIGURE 22.30
Public and private funds, by percentages, available for United
States food and agricultural research. *(Agricultural Research
Service Misc. Pub. No. 1429, USDA, January 1983.)*

Before future trends in animal research can be discussed, anticipated problems
must first be defined. Unless the right questions are asked, it may not be possible
to even come close to answers that are pertinent to problems humanity wishes to
solve in the future.

> There is one quality more important than "know-how"—this is "know-what," by
> which we determine not only how to accomplish our purposes, but what our purposes
> are to be. . . . Whether we entrust our decisions to machines of metal, or to those
> machines of flesh and blood, bureaus and vast laboratories and armies and corpora-
> tions, we shall never receive the right answer to our questions unless we ask the right
> questions.
>
> *Norbert Wiener (1894–1964)*

> The primary aim of research must not just be more facts and more facts, but more
> facts of strategic value.
>
> *Paul Weiss*

If from a high tower above the world, one could view in perspective the state
of animal research day by day, what would one see?

> To look at his picture as a whole, a painter requires distance.
>
> *John Tyndall*

The authors possess no crystal ball. However, if current trends are projected, one can get some idea of future problems that agriculture will be facing. Population, for example, is expected to increase both in the United States and in the world (Chapter 1). This means that production (plant and animal) per acre must increase if mass starvation is to be avoided. Hungry children seldom develop into healthy, creative human beings. Starving people create riots and revolutions. The dead do not speak. Thus maximum use of land and animals is essential for peace and prosperity and for the sustenance of life itself.

What research is needed to provide information basic to increased productivity of food for humans? We believe that research will be expanded in the areas of animal diseases; pests and parasite control[17]; reproductive physiology,[18] e.g., in estrus synchronization and in artificial insemination; genetics[19]; **feed efficiency,** especially for swine and ruminants (to include calculating least-cost feeds and the formulation of complete diets); production influences on animal product quality; and environmental control of production practices. More emphasis will likely be given to research pertaining to the genetics and nutrition of fish, rabbits, and companion animals. Fish are tremendously efficient converters of feed into food for human use, and rabbits are prolific and can subsist, reproduce, and grow on diets (e.g., forage and even many weeds) not directly competitive with foods of humans. Companion animals are certain to continue to frequent the premises and lives of humans (there were approximately 49 million dogs, 39 million cats, and 8 to 10 million horses in the United States in 1983). Although pets do not contribute to food or fiber production, they provide substantial companionship and psychological benefits to humanity. Moreover, regardless of the view one takes of the overall merits and contributions of companion animals to society, the fact remains that people will continue to use substantial food of food of edible quality for people to provide sustenance for their pets. Therefore,

---

[17]An estimated 25 percent of the food grown in many developing countries is lost to insects, birds, rodents, monkeys, rusts, molds, mildews, bacteria, and the combined effects of high temperatures and humidity.

Marek's disease is a highly contagious disease affecting the peripheral nerves and the visceral organs of chickens. It has resulted in an annual loss to the poultry industry of more than $200 million in the United States alone. USDA scientists studied the disease for more than 30 years before showing that it is caused by a herpesvirus. A similar virus was isolated in turkeys, and within 2½ years a vaccine had been developed that reduced the incidence of Marek's disease in vaccinated chickens by 90 percent, reduced condemnations in broilers by a similar amount, and increased egg production in layers by 4 percent. Further examples of the economic significance of animal disease–related research will surface in the decade ahead.

[18]Low reproduction rates are the most important single factor in reducing production efficiency in beef cattle. The percentage of mated cows in United States beef herds that raise calves to maturity is approximately 80 percent. The figure is about 90 percent in the Netherlands and Denmark but only 30 to 50 percent for many developing countries. This economically important area of animal production will be researched intensively in the years ahead.

Studies are under way to discover how to make ewes breed and **lamb** the year around and thereby extend the restricted breeding season of sheep. This could result in large gains in lamb production.

[19]The curious question arises, Is it possible to alter the genetic codes of animals so that they can be more readily adapted to the needs of increased animal production as scientists aim their research for the survival and betterment of humanity?

it is incumbent on animal scientists to research ways of improving the health, nutrition, and breeding of the animals that provide enjoyment and companionship to humanity.

Future research with farm animals will include increased emphasis on the efficient utilization by ruminants of materials not suitable for human consumption. This means greater utilization of increased amounts of fiber (**cellulose,** hemicellulose, pectin, and **lignin**) by ruminants and possibly other **ungulates.** Because of present and projected protein needs of humans, additional emphasis will be given to utilizing more waste and by-product materials and to providing nonprotein nitrogenous compounds to ruminants, which can utilize them in the synthesis of microbial protein. This will spare increasing quantities of vegetable protein for humans, but allow maintenance of their meat and milk supply. There will be more interdisciplinary research. This cooperative effort among scientists will help to achieve their common objective of improving the quality of life for humans.

Expansion of animal-related research efforts will be an important part of the overall commitment to increasing the efficiency and productivity of American agriculture. A brief backward look gives an interesting perspective on productivity increases. United States agricultural productivity remained virtually unchanged from 1910 to 1930. It increased an average of 1.3 percent annually in the 1930s and 1.7 percent in the 1940s. Total farm productivity increased 2.7 percent annually from 1950 to 1965 but increased only 1.7 percent annually during the 1965–1980 period.[20] This recent decline in the overall productivity of American agriculture is ample evidence of the need to expand the investment of public and private monies in agricultural research.

> The direction in which one moves and the pace at which one moves are more important than the place at which one stands at the moment.
>
> *Wilbur J. Larson (1921–   )*

Tomorrow's agricultural research is therefore going to be influenced by many factors, only some of which can be predicted. In general, research will become more refined. More emphasis will be placed on research in depth. It is likely that the SAES in themselves will develop areas of competence and that more research programs will be coordinated among states. Public and private research will work more closely together. This coordination and cooperation will help to reduce duplication of research efforts. Certainly, the challenge is clear for both public and private research teams. If in the United States future research needs are not accepted and met, the present standard of living may become but a memory. Public demands for research related to health, nutrition, social behavior, environmental quality, new energy sources, and agricultural production will likely increase in the decades that follow.

[20]B. R. Eddleman, Research and Education Support for Food and Agriculture in the National Interest, Miss. Agr. and Forestry Expt. Sta. Pub. No. 108, 1980.

22.17.1 Waste-Management Research

Domestic animals contribute an estimated 2 billion tons of waste annually in the United States. When animal agriculture was a family operation, animal waste was not considered a problem. Indeed, it was regarded as a valuable agricultural asset, and most of it was returned to the land and plowed under. With the advent of large-scale confinement operations for beef, dairy, swine, and poultry production, waste disposal came to be regarded as an important problem. For example, drainage from feedlots is an important consideration in lake and stream pollution.

Thus, increased size and concentration of many animal production operations have presented important problems in waste disposal and management. Maintenance of environmental quality necessitates development of disposal practices that minimize health, odor, and conservation problems without introducing additional ecological hazards. However, environmentalists must be careful to not condemn the cow while overlooking the coyote, which also urinates in rural America.

It is important that animal agriculture enlist the interest, expertise, and activity of microbiologists, chemists, ecologists, engineers, and others competent to recommend and implement research projects related to principles of waste disposal and management.

Through the application of research and technology, scientists at the University of Illinois are using animal waste to produce combustible gas (methane) as an energy source for the farm (Figures 22.31 and 22.32). Residues from the manure used for gas production are useful as fertilizer. Some 60 to 90 percent of the quantities of nitrogen, phosphorus, potassium, and other mineral elements fed to livestock is retained in the manure.

22.17.2 Researching the Use of Nonprotein Nitrogen (NPN) and the Recycling of Wastes

Increasing world demands for protein in human nutrition and an increasing ability in many countries to purchase plant proteins have caused sharp increases in the cost of protein for ruminants. Through microbial synthesis of protein, ruminants can utilize NPN and thereby provide humanity with the high-quality proteins of meat and milk without directly competing with people for plant proteins. In Finland, cows have produced up to 9000 lb of milk and 360 lb of protein in a lactation without any feed source of protein. In USDA studies, beef animals have been raised on protein-free diets from 7 months of age and bred and then have given birth to normal calves (Figure 22.33).

Ruminants have digestive systems that are enormous in proportion to their body size. Thus, they are able to subsist and produce meat and milk on feeds too bulky and too poor to nourish humans. Millions of tons of feed are available from crop and livestock-processing **offals** and roughages unfit for human consumption, which can continue to support substantial livestock production throughout the world. Future research studies will investigate the expanded use

FIGURE 22.31
This anaerobic digester uses gas-producing bacteria to create methane and carbon dioxide from the organic matter of livestock manure. The resultant biogas can then be used as a replacement for propane or natural gas in many energy applications, such as heating buildings and/or the powering of a fuel production unit that creates ethanol (ethyl alcohol) from corn, or for fueling an engine-generator to produce electricity at the rate of 300 kW per day. The above digester is a horizontal tank 16 ft in diameter and 90 ft long, and it consists of four sections: the main tank for gas production, and sections for gas processing, gas storage, and sludge storage. It is banked with soil on three sides with a solar collector on its south face to help preheat the manure it receives from 1200 hogs. It produces enough methane to heat 10 average-size homes. An odorless by-product of the biogas process can be utilized as fertilizer. See also Figure 22.32. *(Courtesy R. A. Easter, Department of Animal Science and the University of Illinois Agricultural Experiment Station.)*

of animal manures, garbage, wastepapers and wastes from canneries, cheese factories, and packing plants in feeding ruminants. Moreover, future agricultural research will include greater emphasis on improved ways and means of producing, harvesting, and preserving high-quality forages for livestock.

22.17.3 Food Protein Research

The mounting world needs for food, especially protein, were discussed in Chapter 1. Never before has the need been as great for more food and agricultural engineers and scientists capable of solving the world's food shortages. Numerous colleges and universities have accepted the challenges of greatly expanded efforts in protein research. For example, Texas A&M University recently opened the new Food Protein Research and Development Center, which is well-positioned to work in concert with private industry and individuals in solving the problems of converting crops into edible food products for

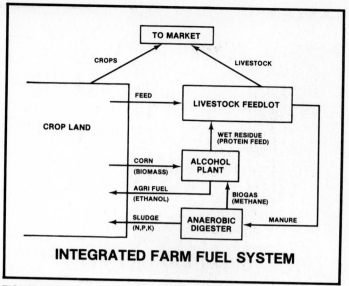

FIGURE 22.32

This schematic depicts the concept of an integrated biomass energy system for agriculture using renewable biomass. The system is aimed at alleviating the problem of on-farm production of fuel alcohol with the aid of methane from animal manure. In this system, corn is used as the biomass. Stillage residue from producing alcohol is an excellent source of protein for feeding livestock. In an integrated alcohol-producing system, manure from the feedlot goes to an anaerobic digester (Figure 22.31), where it is converted into a biogas that is about 60 percent methane. The methane is then used to fuel the alcohol plant. Sludge from the digester has fertilizer value when spread on cropland, thereby completing the energy and ecological cycle for a farming unit. *(Courtesy Dr. D. L. Day, Department of Agricultural Engineering, and Dr. M. P. Steinberg, Department of Food Science, University of Illinois, Urbana.)*

houses many state-of-the-art research tools, such as ultrafiltration and reverse osmosis equipment in its protein-isolate pilot plant, a modern extruder (for processing formed protein, cereals, and snack foods, for example) in its food texturization laboratory, a continuous-chain solvent extractor, and multiphase centrifuges for aqueous extraction.

22.17.4 Biotechnology in Animal Research

Many new technologies are being established that are having a profound impact on animal research. Two technologies of particular importance are recombinant DNA and the manufacture of hybridomas. Recombinant DNA technology

FIGURE 22.33
This cow is believed to be the first in the world to successfully have
grown, conceived, and given birth to a healthy calf when fed since
weaning (7 months) on a protein-free **purified diet** that contained
urea as the sole source of nitrogen. The cow weighed 930 lb and the
calf 61 lb at time of birth. In other USDA studies, five calves have
been obtained from a Shorthorn cow under similar feeding and
management conditions. Additionally, one Angus bull has not received
protein since 84 days of age and was 12½ years old when the
authors communicated with Dr. R. R. Oltjen, then Chief, Ruminant
Nutrition Laboratory, Nutrition Institute, USDA. *(Courtesy USDA.)*

involves isolating and manipulating DNA, the physical matter that makes up a
gene (Chapter 4). Recombinant DNA techniques form the core methodology of
gene transfer experiments and genetic engineering. Hybridoma techniques
provide the basis for the production of monoclonal antibodies.

Recombinant DNA Recombinant DNA technology can be defined as the
technique of joining pieces of DNA together in vitro and then introducing them

into living cells, where they can replicate. The purpose is to separate a piece of DNA of particular interest from other DNA and to use the DNA replicating systems that nature devised to produce enough of the chosen DNA to use and study (Figure 22.34).

This mass-produced DNA (or gene) can be used to genetically manipulate bacteria to produce the protein encoded by that DNA. The approach has been used to synthesize a growing number of biologically useful substances such as hormones, enzymes, and nutrients. Availability of large quantities of these substances will greatly facilitate research on their function in animals. Proteins from disease pathogens can also be synthesized in this way, forming the basis for highly effective vaccines, for animal health research.

Recombinant DNA technology also yields highly specific DNA probes for genes of interest. These probes are being used at the University of Illinois and other institutions to determine differences in the structure of genes in individual animals. Such differences in gene structure can then be correlated with **phenotype.** For instance, if there are several forms of a gene involved in growth rate, and one form is found most often in the fastest-growing animals, then the DNA probes can be used to screen breeding stock to identify carriers of that gene form. It should be cautioned that there are numerous genes involved in functions such as growth, egg production, and milk production, so that at least several genes must be identified and screened before this method reaches its full potential. The procedure can be very effective in detecting deleterious genes in the animal population. For example, researchers at the University of Illinois

FIGURE 22.34
A University of Illinois researcher loads DNA from individual animals onto a gel to identify the presence of gene forms. *(Courtesy Dr. W. L. Hurley, Department of Dairy Science, University of Illinois.)*

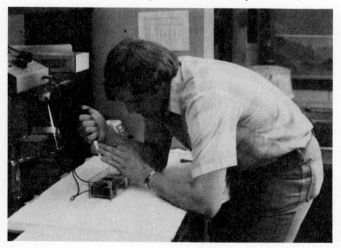

have identified a recessive gene that, in the homozygote, may result in increased calf mortality. These researchers are developing DNA probes to detect the presence of this mutant gene so that it can be eliminated from the breeding population (Figure 22.35).

The most spectacular use of recombinant DNA would be to supply genetic material for gene transfer in genetic engineering attempts with animals. The DNA would be injected into a one-cell embryo so that every cell of the resulting animal contained that gene. This suggests that any gene could be transferred to an animal, bestowing on it the functions that gene encodes. Cattle and swine could be made to grow larger and at a faster rate, and cows could be made to produce more milk. However, many problems must be confronted before these examples become reality. There are numerous technical problems in transferring genes, and the knowledge of how to control genes once they are transferred is limited. Perhaps most important, the knowledge of which genes·are the most desirable to transfer is only beginning to be achieved. The technological ability to transfer genes has greatly surpassed the basic understanding of how the animal works biologically. Closing this gap between technological capability and fundamental biological knowledge offers some of the most challenging research aims for animal researchers in the next decade.

Hybridoma Technology and Monoclonal Antibodies Antibodies are produced by lymphocytes in response to an antigen (a protein, carbohydrate, or

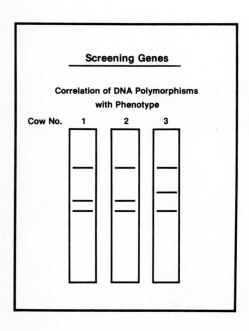

FIGURE 22.35
Total DNA from cows is digested with an enzyme and the fragments separated by size on agarose gels. Fragments of specific genes are detected by molecular hybridization to DNA probes. Cows 1 and 2 have the same gene fragment pattern and therefore have the same gene form (DNA polymorphism), whereas cow 3 has a different gene form. These different DNA morphisms will be correlated with a particular phenotype and the DNA probes used to screen cattle for the presence of that gene form most highly correlated with the phenotype. *(Courtesy Dr. W. L. Hurley, Department of Dairy Science, University of Illinois.)*

another organic compound) that is foreign to that animal (see Chapter 18). Each lymphocyte produces only one type of antibody, which recognizes and binds to only one site on the antigen. Thus, whole-serum antibodies are a mixture of many different antibodies. To make a highly specific antibody to an antigen, one simply needs to find that lymph cell secreting that particular antibody and grow that cell in culture. The concept is simple, but lymphocytes die in tissue culture. The highly innovative means developed to circumvent this problem is to fuse lymphocytes (spleen cells) with a myeloma cell line. A myeloma cell line is derived from cells of bone marrow tumors that do not secrete antibody. Fusion of these cells yields hybrid cells (hybridomas) that secrete an antibody and will grow in culture. Each will grow sufficiently in culture to permit characterization of its antibody, called a **monoclonal antibody,** and the **hybridoma** cells can be used to produce large quantities of the antibody.

With these monoclonal antibodies, researchers have tools that make possible the detection of the original antigen with high specificity. Such antibodies are used in measuring hormones or other blood and tissue compounds; localizing an antigen within a cell; and in disease detection, since they can indicate the presence of parasites, bacteria, or viruses. Monoclonal antibodies are having a significant impact on animal research, and this impact will be important in both livestock production and animal health industries in the future.

22.17.5 Applying Electronic Technology in Animal Research

Recent developments in engineering and electronics have had a major impact on the mechanization and automation of dairy production practices. Several of these developments have now become widespread among modern dairy producers. Electronic animal identification (ID) was the major technological advance that facilitated much of the research and development of dairy automation (Figure 22.36).

Automatic grain dispensers became widely accepted by dairy producers in the early 1980s following the commercialization of this application for electronic animal ID. As cows came to a feed dispenser they were identified electronically via a transponder around the cow's neck. A small computer, usually in the farm office and connected to the feeder stall by electronic cables, would check its data bank for that cow and send a signal to operate the feed dispenser if the cow was scheduled to receive additional grain. The dispensing was controlled to spread the cow's daily allowance over several periods (usually at least four). Feed was dispensed at just the right speed so that each cow would consume her feed and not leave a large quantity for another, more aggressive cow that might force the first one out of the stall. This system provides a way to feed grain to cows individually that are housed in groups. Thus it solved the vexing problems of how to feed enough grain to high producers to meet their needs without overfeeding low producers on the same lot, and without slowing down the milking process to wait for high producers to eat grain in the milking parlor. It

FIGURE 22.36
Examples of electronic animal identification units. The top center unit is battery-powered, with a magnetic switch to turn it on. The other units are passive (no battery) and obtain their energy when they are placed in an electronic field of a defined wavelength, which is emitted from a transmitter-receiver located at a feed-dispensing stall or a milking parlor stall. The elongated unit at lower left is for permanent subdermal implantation. The small square at lower right is a miniaturized unit suitable for use as an ear tag. *(Courtesy Dr. S. L. Spahr, University of Illinois.)*

also was an automatic system that allowed many dairy producers to reduce the labor necessary for feeding grain.

Automation also had an impact in the milking parlor. The first commercial electronic milk flowmeter appeared in the late 1970s. The unit is coupled with an automatic detacher and provides a computer listing of each cow, her milking time, and her milk yield. Advanced versions of the system provide an electronic record system storing reproduction information on each cow in the herd.

Headed by research at the University of Illinois, complete electronic record systems with many automatic functions have started to appear commercially. One such system detects automatically subclinical mastitis, relying on the electrical conductivity of milk (Figure 22.37). The system uses high-speed electronic measurement and recording to measure the electrical conductivity of milk at various stages of milking. An on-line computer assimilates the data and identifies quarters that have milk with high electrical conductivity. High conductivity is indicative of a high content of ions, mainly chloride, which have leaked through the secretory cell membrane from blood (see Chapter 11). The system, like the one for automatic feed dispensing, operates automatically and provides the user with additional information for management.

FIGURE 22.37
Modified lid assembly for milking machine claw to measure milk temperature and electrical conductivity of milk for subclinical mastitis detection (bottom). Probes, such as the one shown unattached, are inserted into each conductivity cell to allow in-line measurement of milk conductivity by quarters. Data collection is controlled by a microprocessor located at each stall in the milking parlor. An on-farm computer identifies cows with subclinical mastitis, notes the infected quarter(s), and maintains a record of the udder health by quarters. A milking claw with mastitis-detection probes is shown in top photo. *(Courtesy Dr. S. L. Spahr, University of Illinois.)*

Electronic cow ID in the milking parlor, electronic measurement of milk yield at every milking, and on-farm electronic cow records promise to improve the level of management and reduce the repetitive tasks of the modern dairy producer.

22.18 OPPORTUNITIES IN ANIMAL RESEARCH

The sole meaning of life is to serve humanity.

Count Leo N. Tolstoy (1828–1910)

No endeavor is in vain;
Its reward is in the doing,
And the rapture of pursuing
Is the prize the vanquished gain.
Henry Wadsworth Longfellow (1807–1882)

Students are often only dimly aware of the scientific opportunities in animal research in this age of missilery and space exploration. Although there are no **mustangs** or moon shots to attract young scientists in the field of animal research, there is the challenging problem of keeping the world alive and healthy. There is currently before humanity the specter of mass starvation—a problem larger than any previous one in history.

The animal science research team includes persons trained in all areas of the biological and physical sciences and in many areas of the social sciences. All have a common objective: the betterment of humanity.

The USDA has more than 4000 scientists working on agricultural research problems; the 53 state agricultural experiment stations employ about 10,000 scientists.

Identifying an Animal Scientist Traits characterizing excellence in an animal researcher include originality (novel ideas), the ability to recognize problems, the ability to integrate remotely associated concepts (the ability to discover a feature common to two or more apparently unrelated concepts), the ability to detect implications of a problem or phenomenon, energetic work habits, self-sufficiency (independence), the ability to analyze problems and data thoroughly, and a liking for difficult and challenging problems. Moreover, it is important that an animal researcher possess a broad understanding of the biological sciences together with a working knowledge of the husbandry practices of the animals (farm, laboratory, and companion types) that are used as research models. Still another helpful characteristic of the animal scientist is the innate ability to take advantage of serendipity, or chance discoveries. Animal scientists must also possess skills in communication (verbal and written), have perseverance, be technically and theoretically competent, and have the ability to get along with their associates and fellow staff members. Scientists rarely are geniuses, but they are dedicated to excellence and productivity, and they often do not adhere to merely an 8-hour day.

22.19 SUMMARY

If a man will begin with certainties he shall end in doubts; but if he will be content to begin with doubts he shall end in certainties.

Sir Francis Bacon (1561–1626)

Thus, the task is, not so much to see what no one has seen yet; but to think what nobody has thought yet, about what everybody sees.

Arthur Schopenhauer (1788–1860)

If a little knowledge is dangerous, where is the man who has so much as to be out of danger?

Thomas Henry Huxley (1825–1895)

Research is the process of gaining new knowledge and uncovering new relations and truths. Research is as essential to the expanded production of food and fiber as oxygen is to the marathon runner.

Agricultural research has been fanning the flames of our "agricultural revolution" during the past several decades. American agriculture has provided the basis for a standard of living unprecedented in the history of humanity. The early decision to put American agriculture on a scientific basis was instrumental in this achievement. Establishment of the land-grant colleges of agriculture and state agricultural experiment stations, with the energetic support of American industry, has played a vital role. In retrospect, after the operation of land-grant colleges for more than a century, of agricultural experiment stations for over 90 years, and of the agricultural extension service for more than half a century, it is readily apparent that the decision to put faith in the triad of teaching, research, and extension for the betterment of humanity was indeed wise.

Animal production is vital to a healthy U.S. economy; it generates approximately 52 percent ($70 billion in 1980) of total farm income. Through a combination of research, development, and education, the efficiency of many phases of animal production has increased substantially during the past 20 years. However, significant further progress requires additional research. Especially needed is basic knowledge in areas related to improving the efficiency with which foods and other products from animals are produced and utilized.

Expanded animal research is providing and will continue to provide American consumers with meat, milk, eggs, and fibrous products, which contribute to a more abundant, healthful, and vigorous life. In the future those involved in the production of such animal products will be expected to be better informed on the basic factors of both genetics and environment in producing more milk with fewer cows, more beef with fewer cattle, more pork with fewer sows, more lamb and wool with fewer sheep, and more eggs with fewer hens. The age of genetic engineering and molecular biology is upon humanity. Tremendous progress can be expected during the balance of this century in the development and production of new products in animal agriculture and in more efficient production of existing products. Fermentative production of vaccines, vitamins, and antibiotics, of growth promoters such as growth hormone, and of other hormones to control or enhance reproduction and lactation are examples among many of how genetic engineering promises to improve the efficiency of animal production. Embryo transfer techniques and applications (discussed in Chapter 9) will enable animal breeders to concentrate and expand superior genetic materials. The use of **monoclonal antibodies** will aid in the diagnosis, control, and eradication of many diseases detrimental to food-producing and companion animals.

Genetic engineering research holds great promise for yielding new techniques for identifying and transferring genetic materials among animals and between animals and microorganisms, especially bacteria. Possibly through cloning and related technologies, genetic copies of especially superior animals can be produced. Other research is needed on genetic manipulation to produce animals that are more resistant to diseases and parasites and have greater reproductive efficiency, greater efficiency of **feed conversion,** and greater overall production of meat, milk, eggs, and wool.

Genetic merit will be more precisely detemined by progeny evaluation and other approaches (Chapter 5). A greater utilization of genetic excellence is expected through more use of artificial insemination, **estrous** synchronization, and possibly even sex control. Crossbreeding and other means of obtaining **heterosis** will become more prevalent. The feeding to ruminants of more **NPN** and of **cellulose** materials, such as cornstalks, silage, corncobs, hulls, by-products, and even sawdust and wastepaper, can be expected. (Ruminants provide the vital link between forages and nonconventional feedstuffs inedible to humans and the human foods meat and milk.)

> It is difficult to say what is impossible, for the dream of yesterday is the hope of today and the reality of tomorrow.
>
> *Robert H. Goddard (1882–1945)*

Precision instruments will be further developed that will expose many of the secrets of biochemistry, genetics, physiology, nutrition, etc. The future of humanity rests largely in its ingenuity in developing new techniques that can make it possible for the first time to achieve a numerical balance between people and their food supply without the dismal but repeated validation of the Malthusian doctrine.

Since people live longer than most other animals, they should study animals and attempt to determine what causes them to age and become old. Animals are chattels, and researchers, who are limited only by humane laws, may do what they please with them—even to the extent of taking their lives. There are, of course, those who believe that substantial advances in biology and the biomedical sciences can be made without resorting to animal experimentation. Others hold the view that it is better to impair the growth of, or even kill, animals with diets deficient in the essential nutrients or with toxic doses of nutrients or environmental contaminants in controlled research studies than to impair the growth of or kill humans because of the lack of baseline data. The acquisition of new knowledge pertaining to adequate and inadequate levels of dietary nutrients and doses of various drugs and pharmaceutical agents is imperative to the health, productivity, and well-being of humans, as well as to the food- and fiber-producing and companion animals that serve humanity.

The contributions of science to the progress and productivity of American agriculture are immense and of tremendous consequence to the health and well-being of humanity. But the task is not finished: millions of people are malnourished worldwide. The need is indeed great for additional research that will eventually lead to a well-fed human populace throughout the world.

Because of "research freedom" in many universities and institutes, human ingenuity is the principal limiting factor in research progress. Since it is estimated that 90 percent of all scientists who have ever lived are still living, the concentration of ability has become very great. The increased impetus for animal research has been due not only to more interest but also to greater financial support (public and private).

Research often results in change; however, change usually means more opportunity and a better way of life for people and the animals that serve humanity. There are times, however, when progress seems to advance so rapidly that new problems are created at nearly the same rate that old ones are being solved. This thought is exemplified in the following quotations:

Science never solves a problem without creating ten more.

George Bernard Shaw (1856–1950)

Man sometimes speaks as though the progress of science must necessarily be a boon to mankind, but that, I fear, is one of the comfortable nineteenth century delusions which our more disillusioned age must discard.

Bertrand Russell (1872–1970)

Perhaps the most characteristic realization of the scientist today is that nature and the universe are too complex to be fully understood and described, that concepts must change repeatedly to absorb new findings, and that the recurring miracle of life is more majestic than any formula, any computer, or any rocket that humans can devise.

Nature keeps much on her shelf, but more in her closet.

Science corrects the old creeds, sweeps away, with every new perception, our infantile catechisms, and necessitates a faith commensurate with the grander orbits and universal laws which it discloses.

Ralph Waldo Emerson (1803–1882)

Ever since I have been inquiring into the works of nature, I have always loved and admired the simplicity of her ways.

George Martine (1747)

Science cannot determine origin, and so cannot determine destiny. As it presents only a sectional view of creation, it gives only a sectional view of everything in creation.

Theodore T. Munger (1830–1910)

The authors have a deep faith in the future of animal scientists. As they develop their capabilities, they extend even further their expectations. Whatever and whenever demands and needs arise for special projects and developments, the adaptable animal researcher will accept the challenge, spring to meet the new crises, and demonstrate that there are extraordinary potentials in ordinary people.

STUDY QUESTIONS

1 What areas of study does animal research encompass?
2 Define research. Differentiate between basic and applied research.
3 What is science?
4 What is the scientific method? Briefly discuss the basic orderly steps it embraces.
5 Why is good experimental control imperative to sound research?
6 What is the null hypothesis? What is a random sample?
7 What is meant by the term statistically significant?
8 Of what value are computers to animal research?
9 Where do scientists obtain scientific information?
10 What is the role of the USDA and the SAES in animal research?
11 How is the practical application and importance of scientific information conveyed to farmers?
12 Cite several examples of the use of atoms in animal research.
13 Cite several uses of a low-level-radiation research laboratory.
14 What is meant by the half-life of radioactive isotopes?
15 What is an atomic reactor? Give several examples of its use in animal research.
16 Of what use is activation analysis to animal researchers?
17 How can nonradioactive isotopes be used to follow the metabolic pathways of nutrients?
18 Cite one or more examples of the use of radioactive isotopes in endocrine research.
19 What are some potential uses of gas chromatography in animal research? Of high-pressure liquid chromatography?
20 What are miniature pigs? Why are they valuable research animals? Cite one or more uses of germ-free pigs in biological research.
21 Differentiate between a fistula and a cannula.
22 Of what use are fistulated animals?
23 Where is the National Animal Disease Center? What governmental agency is responsible for its operation? Is animal-disease research its only function? Explain.
24 What are the relative financial contributions of industry, state, and the USDA to agricultural research?
25 What are some areas of animal research that will likely be expanded during the next decade?
26 What are some characteristics of animal researchers? Does there seem to be a bright future for such persons?

COMMON TERMS OR NAMES APPLIED TO SELECTED FARM ANIMALS

	Cats	Cattle	Dogs	Goats	Horses	Poultry	Sheep	Swine
Male of breeding age	Tom	Bull	Dog	Buck	Stallion	Cock*	Ram	Boar
Mature female	Pussy	Cow	Bitch	Doe	Mare	Hen	Ewe	Sow
Young male	. . . . .	Bullock	Puppy dog	Buck kid	Colt	Chick†	Ram lamb	Shoat‡
Young female	. . . . .	Heifer	Puppy bitch	Doe kid	Filly	Chick†	Ewe lamb	Gilt
Newborn	Kitten	Calf	Pup	Kid	Foal	Chick†	Lamb	Pig
Unsexed male	Gib	Steer	Castrate	Wether	Gelding	Capon	Lamb	Pig
Unsexed female	Neuter	Spayed	Spayed	Spayed	Spayed	. . . . .	Wether	Barrow
Groups	Bevy	Herd	Pack	Band	Herd	Flock	Spayed	Spayed
Genus	*Felis domestica*	*Bos*	*Canis*	*Capra*	*Equus*	*Gallus*§	*Ovis*	*Sus*
Act of breeding	Mating	Serving	Copulating	Serving	Covering	Mating	Tupping	Coupling
Act of parturition	Littering	Calving	Whelping	Kidding	Foaling	. . . . .	Lambing	Farrowing

*Called a *tom* in turkeys.
†Called a *poult* in turkeys, a *gosling* in geese, and a *duckling* in ducks.
‡Applied to a young pig of either sex under 1 year of age.
§Genus for chickens (Aves is the class of birds).

CONVENIENT CONVERSION DATA

1 milliliter (ml or cc) = 12 drops
1 teaspoon = 60 drops (5 cc)
1 tablespoon = 3 teaspoons

1 pint = 32 tablespoons
1 cup = 0.5 pint
1 quart = 2 pints (946.4 cc) and 1/8 peck

1 gallon = 4 quarts (3.785 liters, 8.345 pounds)
1 kilogram (kg) = 2.2046 pounds
1 liter = 1.0567 quarts (1 kg water)

1 foot = 0.3048 meter
1 meter = 3.2809 feet (39.37 inches)
1 kilometer = 0.6214 mile

1 acre = 0.4047 hectare (43,560 square feet)
1 hectare = 2.471 acres
1 barrel = 31.5 gallons (119 liters)

Grains, pounds per bushel
Oats = 32
Corn and rye = 56 (ear corn = 70)

Barley = 48
Beans and wheat = 60

1 bushel = 2150 cubic inches (32 quarts)
1 gallon = 231 cubic inches

Temperature conversion

$$°C = \frac{°F - 32}{1.8}$$

$$°F = (°C \times 1.8) + 32$$

MEASURE OF MASS

Grains	Drams	Ounces	Pounds	Metric equivalents, grams
1	0.0366	0.0023	0.00014	0.0647989
27.34	1	0.0625	0.0039	1.772
437.50	16	1	0.0625	28.350
7000	256	16	1	453.5924277

TABLES OF WEIGHTS AND MEASURES

METRIC WEIGHT

Micro-gram	Milli-gram	Centi-gram	Deci-gram	Gram	Deka-gram	Hecto-gram	Kilo-gram	Metric ton
1								
10^3	1							
10^4	10	1						
10^5	100	10	1					
10^6	1000	100	10	1				
10^7	10^4	1000	100	10	1			
10^8	10^5	10^4	1000	100	10	1		
10^9	10^6	10^5	10^4	1000	100	10	1	
10^{12}	10^9	10^8	10^7	10^6	10^5	10^4	1000	1

METRIC MEASURE (VOLUME)

Micro-liter	Milli-liter	Centi-liter	Deci-liter	Liter	Deka-liter	Hecto-liter	Kilo-liter	Myria-liter
1								
10^3	1							
10^4	10	1						
10^5	100	10	1					
10^6	10^3	100	10	1				
10^7	10^4	10^3	100	10	1			
10^8	10^5	10^4	10^3	100	10	1		
10^9	10^6	10^5	10^4	10^3	100	10	1	
10^{10}	10^7	10^6	10^5	10^4	10^3	100	10	1

METRIC MEASURE (LENGTH)

Micro-meter*	Milli-meter	Centi-meter	Deci-meter	Meter	Deka-meter	Hecto-meter	Kilo-meter	Myria-meter	Mega-meter	Equivalents	
1	.001	10^4	....	....	....	....	....	....	....	0.000039	in
10^3	1	10^1	....	....	....	....	....	....	....	0.03937	in
10^4	10	1	....	....	....	....	....	....	....	0.3937	in
10^5	100	10	1	....	....	....	....	....	....	3.937	in
10^6	1000	100	10	1	....	....	....	....	....	39.37	in
10^7	10^4	1000	100	10	1	....	....	....	....	10.9361	yards
10^8	10^5	10^4	1000	100	10	1	....	....	....	109.3612	yards
10^9	10^6	10^5	10^4	1000	100	10	1	....	....	1093.6121	yards
10^{10}	10^7	10^6	10^5	10^4	1000	100	10	1	....	6.2137	miles
10^{12}	10^9	10^8	10^7	10^6	10^5	10^4	1000	100	1	621.370	miles

*Formerly called the *micron* and abbreviated μ.

AGRICULTURAL COLLEGES AND EXPERIMENT STATIONS IN THE UNITED STATES

State	University name	Address
Alabama*,†	Auburn University	Auburn University, AL 36849
Alabama†	Alabama Agricultural & Mechanical College	Normal, AL 35762
Alaska*	University of Alaska Agr. Expt. Station	Palmer, AK 99645
Alaska†	University of Alaska	Fairbanks, AK 99701
Arizona*,†	University of Arizona	Tucson, AZ 85721
Arkansas*,†	University of Arkansas	Fayetteville, AR 72701
Arkansas†	Agricultural, Mechanical & Normal College	Pine Bluff, AR 71601
California*,†	University of California	Berkeley, CA 94720
	University of California	Davis, CA 95615
	University of California	Riverside, CA 92521
California*	University of California	Los Angeles, CA 90024
Colorado*,†	Colorado State University	Fort Collins, CO 80523
Connecticut*,†	University of Connecticut	P.O. Box 1106, Storrs, CT 06268
Delaware*,†	University of Delaware	Newark, DE 19711
Delaware†	Delaware State College	Dover, DE 19901
Florida*,†	University of Florida	Gainesville, FL 32611
Florida†	Florida A&M University	Tallahassee, FL 32307
Georgia*,†	University of Georgia	Athens, GA 30602
Georgia*	Georgia Coastal Plain Expt. Station	Tifton, GA 31794
Georgia†	Fort Valley State College	Fort Valley, GA 31030
Guam*,†	University of Guam	Mangilao, Guam 96913

*Agriculture experiment station.
†Land-grant college of agriculture.
Note: Agricultural colleges and experiment stations will send, on request, a list of available bulletins relating to their published research and activities.

766

State	University name	Address
Hawaii*,†	University of Hawaii	Honolulu, HI 96822
Idaho*,†	University of Idaho	Moscow, ID 83843
Illinois*,†	University of Illinois	Urbana, IL 61801
Indiana*,†	Purdue University	Lafayette, IN 47907
Iowa*,†	Iowa State University of Science & Technology	Ames, IA 50011
Kansas*,†	Kansas State University	Manhattan, KS 66506
Kentucky*,†	University of Kentucky	Lexington, KY 40506
Kentucky†	Kentucky State College	Frankfort, KY 40601
Louisiana*,†	Louisiana State University	Baton Rouge, LA 70893
Louisiana†	Southern University & Agricultural & Mechanical College	Baton Rouge, LA 70813
Maine*,†	University of Maine	Orono, ME 04469
Maryland*,†	University of Maryland	College Park, MD 20742
Maryland†	Maryland State College	Princess Anne, MD 21853
Massachusetts*,†	University of Massachusetts	Amherst, MA 01003
Michigan*,†	Michigan State University	East Lansing, MI 48824
Minnesota*,†	University of Minnesota	St. Paul, MN 55108
Mississippi*,†	Mississippi State University	Mississippi State, MS 39762
Mississippi†	Alcorn State University	Lorman, MS 39096
Missouri*,†	University of Missouri	Columbia, MO 65211
Missouri†	Lincoln University	Jefferson City, MO 65101
Montana*,†	Montana State University	Bozeman, MT 59717
Nebraska*,†	University of Nebraska	Lincoln, NB 68583
Nevada*,†	University of Nevada	Reno, NV 89557
New Hampshire*,†	University of New Hampshire	Durham, NH 03824
New Jersey*,†	Rutgers—The State University of New Jersey	New Brunswick, NJ 08903
New Mexico*,†	New Mexico State University	Las Cruces, NM 88003
New York*,†	Cornell University	Ithaca, NY 14853
New York*	New York State Agr. Expt. Station	Geneva, NY 14456
North Carolina*,†	North Carolina State University	Box 5847, Raleigh, NC 27650
North Carolina†	North Carolina Agricultural & Technical State University	Greensboro, NC 27411
North Dakota*,†	North Dakota State University of Agriculture & Applied Science	Fargo, ND 58105
Ohio†	Ohio State University	Columbus, OH 43210
Ohio*	Ohio Agricultural Research & Development Center	Wooster, OH 44691
Oklahoma*,†	Oklahoma State University	Stillwater, OK 74074
Oklahoma†	Langston University	Langston, OK 73050
Oregon*,†	Oregon State University	Corvallis, OR 97331
Pennsylvania*,†	The Pennsylvania State University	106 Armsby Bldg., University Park, PA 16802
Puerto Rico*,†	University of Puerto Rico	Mayaquez, PR 00708
Rhode Island*,†	University of Rhode Island	Kingston, RI 02881
South Carolina*,†	Clemson University	Clemson, SC 29631
South Carolina†	South Carolina State College	Orangeburg, SC 29117
South Dakota*,†	South Dakota State University	Brookings, SD 57007
Tennessee*,†	University of Tennessee	P.O. Box 1071, Knoxville, TN 37901
Tennessee†	Tennessee State University	Nashville, TN 37203

*Agriculture experiment station.
†Land-grant college of agriculture.

State	University name	Address
Texas*,†	Texas A&M University	College Station, TX 77843
	Prairie View Agricultural & Mechanical University	Prairie View, TX 77445
Utah*,†	Utah State University of Agriculture & Applied Science	Logan, UT 84322
Vermont*,†	University of Vermont	Burlington, VT 05405
Virgin Islands*	College of the Virgin Islands	St. Croix, VI 00850
Virginia*,†	Virginia Polytechnic Institute and State University	Blacksburg, VA 24061
Virginia†	Virginia State College	Petersburg, VA 23803
Washington*,†	Washington State University	Pullman, WA 99164
West Virginia*,†	West Virginia University	Morgantown, WV 26506
Wisconsin*,†	University of Wisconsin	Madison, WI 53706
Wyoming*,†	University of Wyoming	Box 3354, Laramie, WY 82071
	Other colleges granting degrees in agriculture	
Arizona	Arizona State University	Tempe, AZ 85281
Illinois	Southern Illinois University	Carbondale, IL 62901
Texas	Texas Technological College	Lubbock, TX 79409

*Agriculture experiment station.
†Land-grant college of agriculture.

ALPHABETICAL LIST OF ELEMENTS AND SYMBOLS

Element	Symbol	Atomic number	Atomic weight*	Element	Symbol	Atomic number	Atomic weight*
Actinium	Ac	89	227	Curium	Cm	96	247
Aluminum	Al	13	26.98	Dysprosium	Dy	66	162.50
Americium	Am	95	243	Einsteinium	Es	99	254
Antimony	Sb	51	121.75	Erbium	Er	68	167.26
Argon	Ar	18	39.942	Europium	Eu	63	152.0
Arsenic	As	33	74.91	Fermium	Fm	100	253
Astatine	At	85	210	Fluorine	F	9	19.00
Barium	Ba	56	137.35	Francium	Fr	87	223
Berkelium	Bk	97	249	Gadolinium	Gd	64	157.25
Beryllium	Be	4	9.013	Gallium	Ga	31	69.72
Bismuth	Bi	83	208.99	Germanium	Ge	32	72.60
Boron	B	5	10.82	Gold	Au	79	197.0
Bromine	Br	35	79.913	Hafnium	Hf	72	178.49
Cadmium	Cd	48	112.40	Helium	He	2	4.003
Calcium	Ca	20	40.08	Holmium	Ho	67	164.93
Californium	Cf	98	251	Hydrogen	H	1	1.0079
Carbon	C	6	12.010	Indium	In	49	114.81
Cerium	Ce	58	140.12	Iodine	I	53	126.90
Cesium	Cs	55	132.90	Iridium	Ir	77	192.2
Chlorine	Cl	17	35.455	Iron	Fe	26	55.85
Chromium	Cr	24	52.01	Krypton	Kr	36	83.80
Cobalt	Co	27	58.94	Lanthanum	La	57	138.91
Copper	Cu	29	63.54	Lawrencium	Lr	103	257

*Atomic weight of the most abundant or best-known isotope or (in the case of radioactive isotopes) of the isotope with the longest half-life, relative to atomic weight of carbon 12 = 12.

Element	Symbol	Atomic number	Atomic weight*	Element	Symbol	Atomic number	Atomic weight*
Lead	Pb	82	207.20	Rhodium	Rh	45	102.90
Lithium	Li	3	6.940	Rubidium	Rb	37	85.48
Lutetium	Lu	71	174.98	Ruthenium	Ru	44	101.1
Magnesium	Mg	12	24.32	Samarium	Sm	62	150.34
Manganese	Mn	25	54.94	Scandium	Sc	21	44.96
Mendelevium	Md	101	256	Selenium	Se	34	78.96
Mercury	Hg	80	200.60	Silicon	Si	14	28.09
Molybdenum	Mo	42	95.95	Silver	Ag	47	107.875
Neodymium	Nd	60	144.26	Sodium	Na	11	22.990
Neon	Ne	10	20.182	Strontium	Sr	38	87.63
Neptunium	Np	93	237	Sulfur	S	16	32.064
Nickel	Ni	28	58.71	Tantalum	Ta	73	180.94
Niobium				Technetium	Tc	43	99
(columbium)	Nb	41	92.91	Tellurium	Te	52	127.60
Nitrogen	N	7	14.007	Terbium	Tb	65	158.92
Nobelium	No	102	254	Thallium	Tl	81	204.38
Osmium	Os	76	190.2	Thorium	Th	90	232.04
Oxygen	O	8	15.999	Thulium	Tm	69	168.93
Palladium	Pd	46	106.4	Tin	Sn	50	118.69
Phosphorus	P	15	30.973	Titanium	Ti	22	47.90
Platinum	Pt	78	195.08	Tungsten	W	74	183.85
Plutonium	Pu	94	242	(wolfram)			
Polonium	Po	84	210	Uranium	U	92	238.06
Potassium	K	19	39.098	Vanadium	V	23	50.95
Praseodymium	Pr	59	140.91	Xenon	Xe	54	131.29
Promethium	Pm	61	147	Ytterbium	Yb	70	173.03
Protactinium	Pa	91	231	Yttrium	Y	39	88.92
Radium	Ra	88	226	Zinc	Zn	30	65.38
Radon	Rn	86	222	Zirconium	Zr	40	91.22
Rhenium	Re	75	186.21				

*Atomic weight of the most abundant or best-known isotope or (in the case of radioactive isotopes) of the isotope with the longest half-life, relative to atomic weight of carbon 12 = 12.

GLOSSARY

abattoir A slaughterhouse.

ablactate To wean.

abomasum The fourth compartment (true stomach) of a ruminant (cow, deer, goat, sheep).

absorption The process of taking in food through the intestinal wall.

acanthosis Hypertrophy, or thickening, of the prickle cell layer of skin.

acariasis An infestation of humans or animals with mites or ticks.

Acarina An order that includes mites and ticks.

accelerator A device for increasing the velocity and energy of charged *elementary particles,* e.g., electrons or protons, through application of electric and/or magnetic forces.

acclimatization Acclimation, or the complex of processes of becoming accustomed to a new climate or other environmental conditions. A series of compensatory alterations in an animal.

acetate See **acetic acid.**

acetic acid A weak organic acid (CH_3COOH) with a characteristic pungent odor. It is a clear colorless liquid. Vinegar is a dilute, impure acetic acid.

acetonemia (ketosis) A condition characterized by an abnormally elevated concentration of ketone (acetone) bodies in the body tissues and fluids.

achromotrichia Graying of the hair.

acromegaly A chronic disease caused by an overfunction of the anterior pituitary and characterized by an enlargement of the bones and soft parts of the hands, face, and feet. Also called *Marie's disease.*

activation The process of making a material radioactive by bombardment with neutrons, protons, or other nuclear particles. Also called *radioactivation.*

activation analysis   A method for identifying and measuring chemical elements in a sample of material. The sample is first made radioactive by bombardment with neutrons, charged particles, or gamma rays. The newly formed radioactive atoms in the sample give off characteristic nuclear **radiations** (such as gamma rays), which can be instrumentally determined. Activation analysis is usually more sensitive than chemical analysis. It is especially useful in animal research, industry, archeology, and criminology.

acute   Generally characterized by a short and often severe course.

adaptation   The adjustment of an organism to a new or changing environmental condition.

adaptation syndrome   The defensive response of the body through the endocrine system to systemic injury evoked by stresses and worked out by an initial stage of shock.

ADF (acid detergent fiber)   Fiber extracted with acidic detergent in a technique employed to help appraise the quality of forages.

ADG   See **average daily gain.**

adipose   Consisting of or resembling fat.

ad libitum **(ad lib)**   At pleasure. Commonly used to express the availability of feed on a free-choice basis.

AEC   The U.S. Atomic Energy Commission. An independent civilian agency of the federal government with statutory responsibility for atomic energy matters.

aerobic   A term usually applied to microorganisms that require oxygen to live and reproduce.

afebrile   Without fever.

afterbirth   The placenta and allied membranes with which the fetus is connected. It is expelled from the uterus following parturition.

agalactia   A failure to secrete milk following parturition.

agglutination   The clumping together of cells, especially bacteria or red blood corpuscles, distributed in a fluid. In biology, it is generally a result of antigen-antibody reactions. Called also *clumping.*

agglutination titer   The highest dilution of a serum that causes clumping together of cells, especially bacteria or red blood corpuscles.

agglutinin   Antibody formed against cells (as bacteria), which, when mixed with these cells, causes them to clump together, or agglutinate.

agriculture   The utilization of biological processes, on farms, to produce food and other products useful to humans.

AI daughters   Daughters of a sire that were sired after the bull was placed in artificial-insemination service.

albinism   Congenital absence of pigment in the skin, hair, and eyes. See also **true albino.**

algae   Chlorophyll-bearing miscroscopic plants that synthesize food by photosynthesis. Algae are mostly aquatic but lack true stems, roots, or leaves.

allelomorphs (alleles)   Genes are paired in each animal, and each kind of gene at a particular chromosome location is called an *allele.* They are alternative forms of genes. Alleles are two genes that occupy the same location (locus) on homologous chromosomes and affect the same trait phenotypically but in a different or alternative way. For example, the gene for brown eyes (B) in humans is an allelomorph of the gene for blue eyes (b).

allergen   Any substance that gives rise to the formation of antibodies and the resultant allergic reaction. Also called an **antigen.**

allergy A severe reaction that occurs in some individuals following introduction of antigens into their bodies.

alopecia The loss of hair (baldness).

alpha (α) particle A positively charged particle (two protons and two neutrons) emitted by certain radioactive materials. It is the least penetrating of the three common types of radiation (alpha, beta, gamma) emitted by radioactive material, being stopped by a sheet of paper.

ambient temperature The prevailing or surrounding temperature.

amnion A thin membrane forming a closed sac around the developing embryo. It contains the amniotic fluid, in which the embryo is immersed.

amplifier An electrical device for increasing the current of a pulse of electricity (strengthening electric impulses).

anabolic Productive.

anabolism Any constructive process by which simple substances are converted by living cells into more complex compounds (constructive metabolism).

anaerobe A microorganism that normally does not require molecular oxygen to live and reproduce.

analogue A chemical that resembles another in structure or function.

anaphase The third stage of mitosis. It is characterized by the separation of centromeres and identical chromatids.

anaphylactic shock (anaphylaxis) Increased sensitivity to the action of a normally nontoxic protein when injected with it for the second time. It may cause a severe or even fatal reaction.

anaphylaxis An acute allergic reaction. See **anaphylactic shock.**

anemia A condition in which the blood is deficient in the amount of needed hemoglobin or in the number of red blood corpuscles or in both. It is characterized by paleness of the skin and mucous membranes, loss of energy, and palpitation of the heart (unduly rapid action of the heart, which is felt by the individual).

anesthesia Loss of the feeling of pain, touch, cold, or other sensation, produced by ether, chloroform, morphine, and other compounds; by hypnotism; or as the result of hysteria, paralysis, or disease.

anestrous period That time when the female is not in estrus; the nonbreeding season.

aneuploidy Animals or plants that possess $2n + 1$ or $2n - 1$ number of chromosomes. Also includes two, or more or less than two, extra chromosomes.

animal protein factor See **APF.**

anion An ion carrying a negative charge of electricity.

anogenital Denotes both the anal and the genital regions.

anomaly Abnormal state; marked deviation from normal.

anorexia Lack or loss of the appetite for food.

anoxia Lack of oxygen. Unless given oxygen, pilots develop anoxia (altitude sickness) when flying at high altitudes.

anterior Denotes the front or forward part. It means the same as the *ventral* surface of the body in human anatomy.

antibiotic A product of a living organism (especially of a bacterium or a fungus) that, when present in very low concentrations, destroys or inhibits the growth or action of another microorganism. Penicillin, tetracycline, and streptomycin are antibiotics.

antibody A protein substance (modified type of blood-serum globulin) developed or synthesized by the lymphoid tissue of the body in response to an antigenic stimulus.

Each antigen elicits production of a specific antibody. In disease defense the animal must have had an encounter with the pathogen (antigen) before a specific antibody can be found in its blood.

antigen   A high-molecular-weight substance (usually protein) that, when foreign to the bloodstream of an animal, stimulates the formation of a specific antibody and reacts specifically in vivo or in vitro with its homologous antibody.

antimetabolite   A substance bearing a close structural resemblance to one required for normal physiological functioning, and exerting its effect, perhaps, by replacing or interfering with the utilization of the essential metabolite.

antiseptic   (from the Latin *anti,* meaning "against," and *sepsis,* meaning "putrefaction"). A substance that prevents growth and development of microorganisms either by destroying them (bactericidal action) or inhibiting their growth (bacteriostatic action).

antiserum   A serum that contains an antibody or antibodies. It gives temporary protection against certain specific infectious diseases.

antitoxin   Antibody formed against poisonous toxins, such as the bacterial exotoxins, which specifically neutralizes (counteracts) the effects of the toxin. Diphtheria antitoxin, obtained from the blood of horses infected with diphtheria, is injected into persons to make them immune to diphtheria or to treat them if they are already infected.

APF (animal protein factor)   The original label given to vitamin B_{12}.

APHIS   Animal and Plant Health Inspection Service (of USDA).

aphrodisia   Sexual excitement, especially if morbid or excessive; e.g., sexual odors released by male insects sexually excite female insects.

apiarist   A beekeeper.

apiary   Bee colonies, hives, and other bee equipment assembled in one location.

aplasia   Incomplete or defective development of a tissue or organ.

apocrine   Denotes that type of glandular secretion in which the secretory products become concentrated at the free end of the secreting cell and are thrown off, along with the portion of the cell in which they have accumulated, as in the mammary gland.

apodous   Legless.

Appaloosa   A breed of horses with dark-brown or black leopard spots on a roan background.

apparent digestible energy (DE)   The food-intake gross energy minus fecal energy. Also called *apparent absorbed energy,* or *apparent energy of digested food.*

apterous   Wingless.

aquaculture   The raising of plants or animals, as fish or shellfish, in or under a sea, a lake, a river, or another body of water.

arable   Suitable for cultivation; tillable.

arboreal   Living in or among trees.

arbovirus   Any one of several groups of small viruses transmitted by arthropods such as mosquitoes and ticks. Yellow fever, dengue, and equine encephalomyelitis are caused by arboviruses.

arteriosclerosis   A progressive thickening and hardening of the walls of the arteries, often associated with high blood pressure or with chronic disease of the kidneys.

arthropoda   The phylum of the animal kingdom that includes insects, spiders, and *Crustacea;* characterized by a coating that serves as an external skeleton and by legs with distinct movable segments or joints.

Artiodactyla   Cattle, swine, goats, deer.

asepsis   Aseptic condition, methods, or treatment.

aseptic   Free from living germs that cause disease, putrefaction, or fermentation.

asexual   Sexless.

assay   The determination of the purity of a substance or the amount of any particular constituent of a mixture.

assimilation   The process of transforming food into living tissue (constructive metabolism).

asthenia   Lack or loss of strength.

atavism   Reappearance of a character after a lapse of one or more generations.

ataxia   Failure of muscle coordination.

atheroma   Fatty degeneration of the walls of the arteries.

atherosclerosis   A fatty degeneration of the connective tissue of the arterial walls. A form of **arteriosclerosis** characterized by **atheroma.** Lesions within the arteries with plaques containing cholesterol and other lipid materials.

atom   A particle of matter indivisible by chemical means. It is the fundamental building block of the chemical elements. The elements, such as iron and sulfur, differ from each other because they contain different kinds of atoms. There are approximately six sextillion (6 followed by 21 zeros, or 6×10^{21}) atoms in an ordinary drop of water. An atom consists of a dense inner core (the nucleus) and a much less dense outer domain of **electrons** in motion around the nucleus. Atoms are electrically neutral.

atomic number   The number of protons in the nucleus of an atom and also its positive charge. (See Appendix E.)

atomic reactor   A nuclear reactor.

atomic weight   The mass of an atom relative to other atoms; the atomic weight of any element is approximately equal to the total number of protons and neutrons in its **nucleus.** (See Appendix E.)

atrophic rhinitis   A chronic inflammation of the mucous membranes and turbinate bones of the nose. It often results in distortion of the snout in swine. Growth rate is retarded. Secondary bacterial infection and pneumonia are common complications.

atrophy   A defect or failure of nutrition or physiological function manifested as a wasting away or diminution in the size of a cell, tissue, organ, or part.

attenuate   To make (microorganisms or viruses) less virulent.

auditory   Of or pertaining to hearing or the organs of hearing.

autogamous   Self-fertilizing.

autopsy   Examination, including dissection, of a carcass to learn the cause and nature of a disease or cause of death. Also called **postmortem.**

autoradiograph   A photographic record of *radiation* from radioactive material in an object, made by placing the object very close to a photographic film or emulsion.

autosomal genes   Genes having their loci on chromosomes other than the sex chromosomes.

autosomes   All chromosomes except the sex chromosomes.

average daily gain (ADG)   Measurement used by scientists to indicate the daily change in body weight when experimental animals are fed test diets.

avian   Pertaining to all species of birds, including domestic fowls.

avidin   A specific protein component of egg albumen that interacts (combines) with biotin to render it unavailable to an animal, thus producing the syndrome known as egg-white injury. The phenomenon is observed only when rats, for example, consume uncooked egg, because the avidin is denatured by heat and thereby inactivated.

axilla   The small hollow beneath the arm where it joins the body at the shoulder (armpit).

axon   The central core forming the essential conducting part of a nerve fiber. The long extension of a nerve cell that carries impulses away from the body of the cell.

backcross   The crossing of an F_1 hybrid with one of the parental types (breeds). The offspring are referred to as the backcross generation, or backcross progeny.

background count   The number of impulses per unit time registered on a counting instrument when no sample is present.

bactericidal   (from the Latin *caedere,* meaning "to kill"). Capable of destroying bacteria.

bactericidins   Antibodies that cause the death of bacteria but not their disssolution or destruction.

bacterin   A suspension of killed bacteria (vaccine) used to increase disease resistance.

bacteriophage   An ultramicroscopic, filterable virus composed of deoxyribonucleic acid (DNA) and a protein coat. Infects particular strains or species of bacteria and usually causes lysis, or explosive dissolution, of the infected bacterium. Bacteriophages multiply at the expense of bacterial cells. They have no enzymes of their own.

bacteriostasis   Retardation of the life processes (growth or development) of bacteria without killing them. A *bacteriostatic* substance is a product that retards bacterial growth.

bag   See **udder.**

balanced ration   The daily food allowance of livestock or fowl, mixed to include suitable proportions of the nutrients required for normal health, growth, production, reproduction, and well-being.

barred   A term used to describe striped markings on fowl.

barren   Incapable of producing offspring, seed, fruit, or crops.

barrow   Young castrated male pig.

basal (diet)   A diet common to all groups of experimental animals to which the experimental substance(s) is (are) added.

basal metabolism (BM)   The chemical changes that occur in the cells of an animal in the fasting and resting state, when it uses just enough energy to maintain vital cellular activity, respiration, and circulation as measured by the basal metabolic rate (BMR). Basal conditions include thermoneutral environment, resting, postabsorptive state (digestive processes are quiescent), consciousness, quiescence, and sexual repose. It is determined in humans 14 to 18 h after eating and when at absolute rest. It is measured by means of a calorimeter and is expressed in calories per square meter of body surface.

bastard   A term applied to sheep with hair. It is reputed that "hair sheep" result from interbreeding of sheep and goats, but there is no biological verification of this. Instead, it is more probable that "hair sheep" are the result of domestication and selective breeding.

battery   A series of pens, cages, etc.

bay   Reddish coat color of horses, but with black points (mane, legs, and tail).

beam hole   An opening through a reactor shield that permits a beam (stream) of radioactive particles or electromagnetic radiation to be used for experiments outside the reactor.

beaver lamb   Sheep or lamb skin with short fine wool that has been dressed with the wool on, dyed, and finished by a process giving a weather-resistant straightness and brightness to the wool.

beebread   The flower pollen gathered by bees, which is mixed with a little nectar or honey and deposited in the cells of the comb.

beefy (beefiness)   A term used to designate the desirable physical conformation of a beef animal as contrasted with a dairy animal, which is lean (not beefy) and more angular.

beekeeping   Apiculture.

beget   To procreate, like a sire.

beta (β) particle   An *elementary particle* (an electron) emitted from a nucleus during radioactive decay. Beta radiation may cause skin burns, but beta particles are easily stopped by a thin sheet of metal.

bioassay   The use of animals to determine the active power of a compound as compared with the effect of a standard preparation.

bioclimatology   The science that studies the effect of physical environment upon living organisms.

biological control   The destruction or suppression of undesirable insects, other animals, or plants by the introduction or propagation, encouragement, artificial increase, and dissemination of their natural enemies, which include predaceous and parasitic insects, predatory vertebrates, nematode parasites, bacteria, protozoa, viruses, and parasitic fungi.

biological half-life   The time required for a biological system, such as a human or an animal, to eliminate by natural processes half the amount of a substance (such as radioactive material) that has entered it.

biologicals   Medicinal preparations made from living organisms and their products, used for the prevention or detection of disease; they include serums, vaccines, antigens, and antitoxins.

biological value   The percent utilization of protein within the animal body, expressed by the formula

$$\%\,\mathrm{BV} = \frac{N \text{ intake} - [(\text{fecal } N - \text{metabolic } N) + (\text{urinary } N - \text{endogenous } N)]}{N \text{ intake} - (\text{fecal } N - \text{metabolic } N)} \times 100$$

biological year   The time interval between the first and last egg laid before a hen enters a molt.

bionomics   The study of the relations of organisms to their environment; **ecology.**

biopsy   The microscopic or chemical analysis of tissue removed from a living body.

biotic   Pertains to life or living matter; biologicals.

bird egg   A very large reproductive cell.

bitch   A female dog.

blastema   A group of cells that give rise to an organ or structure either during regeneration or in normal embryogenesis.

bleeder   An animal that has **hemophilia.**

blind quarter   A quarter of an udder that does not secrete milk or one that has an obstruction in the teat that prevents the removal of milk.

bloat   A disorder of ruminants usually characterized by an accumulation of gas in the rumen.

blocky   Term commonly applied to meat-producing animals and draft horses, meaning a deep, wide, and often low-set animal.

bloom   A term commonly used to describe the beauty and freshness of a cow in early lactation. A dairy cow in bloom has a smooth hair coat and presents evidence of milking ability (dairy character).

BMR   Basal metabolism rate. See under **basal metabolism.**

boar Sexually mature uncastrated male pig. (See Appendix A.)

bolus Regurgitated food that has been chewed and is ready to be swallowed. A large pill for dosing animals.

bomb calorimeter An apparatus for measuring the heat of combustion, as of human foods, animal feeds, and fuel.

bone meal (steamed) A mash of ground animal bones that were previously steamed under pressure. It contains 1.5 to 2.5 percent nitrogen, 12 to 15 percent phosphorus, and 20 to 34 percent calcium. It is used as a fertilizer and as feed for farm animals.

bovine Pertaining to the ox or cow.

bran The seed coat of wheat and other cereal grains, which is separated from the flour and used as animal food.

branding Marking or printing with a hot iron, freeze process, caustic soda, or punch. Cattle are branded with markings that identify ownership. Branding, of course, damages the hide.

bray The cry of a donkey.

breed Animals having a common origin and characteristics that distinguish them from other groups within the same species.

breed average The average milk production of cows for a given dairy breed. (Usually, computations are based on **DHIA** records of the past 5 years.)

breed-average herdmates Daughters of a sire in the same herd producing at the current average production for that particular dairy breed. See **herdmates.**

breeding value (genetic value) The genetic ability of an animal to secrete milk, lay eggs, etc. One half of this genetic ability is transmitted to sons or daughters; hence the term *breeding value.*

breed out To eliminate undesirable characteristics through selective matings.

breed true To have the ability to transmit a characteristic uniformly to offspring.

breed type A particular type or form characterizing a breed. It includes special breed features in head, ear, color, or other traits common to a particular breed.

British breeds Those breeds, such as Hereford, Angus, and Shorthorn, native to Great Britain.

broad-spectrum antibiotic An antibiotic that is active against a large number of microbial species.

broilers (fryers) Chickens (meat type) that are 6 to 12 weeks of age.

bronco (bronc) Any wild or untamed western horse. A wild mustang that has not been broken.

brood A group of baby chickens.

brood animal An animal reserved for breeding and raising young.

broodiness Desire of birds to set in a nest on eggs for the purpose of hatching. Maternal behavior for hatching and rearing young.

buck A male sheep (ram), goat, rabbit, deer, or antelope.

buckskin General term applied to leather from deer- and elkskins; used for shoes and gloves and, to some extent, in clothing. Most buckskin is oil-dressed, which produces a soft, pliable leather having a buff color and resembling, before finishing, chamois leather. Also refers to coat color in horses (see Chapter 21).

buffer solution A substance in a solution that makes the degree of acidity (hydrogen-ion concentration) resistant to change when an acid or base is added. See **pH.**

bull A sexually mature uncastrated male. (See Appendix A.)

bulling A cow in heat (estrus).

bullock A castrated bull. English term for a finished, or fat, steer.

burro A donkey; an ass.

buttermilk Thick, smooth liquid. Usually made from skim milk, using bacteria to produce acid and flavor; contains at least 8.25 percent nonfat milk solids.

cadaver The body after death.

calciferol Commonly known as vitamin D_2.

calcify To deposit or secrete calcium salts that harden. An injured cartilage sometimes calcifies.

calf The sexually immature young of certain large mammals.

calf crop Calves produced by a herd of cattle in one season.

calorie (cal) A small calorie is the amount of heat required to raise the temperature of 1 g of water from 14.5 to 15.5°C. This is equivalent to 4.185 J.

calorimetry Measurements of the amount of heat absorbed or given out.

candling The illumination of the egg interior by holding the egg before a light. This is possible because of the translucency of the eggshell and the differences in the capacity of other egg constituents to transmit light. Candling ensures the elimination of practically all inedible eggs.

canine Pertaining to the dog family; includes dogs, wolves, jackals, and others.

cannibalism A habit of some fowls of pecking at or eating other fowls.

cannon bone Either the metacarpal or the metatarsal bone of the horse.

cannula A device used to connect the rumen with the outside environment. The opening leading from the rumen to the outside is called a *fistula*. A metal-, rubber-, or glass-tube cannula may be inserted into a body cavity to allow the escape of fluids or gas. Liquids and other materials may be introduced into the body through a cannula.

canter The three-beat gait of a horse in which the feet strike the ground one at a time; an easy gallop.

capacitation The physiological process that occurs naturally in the female reproductive tract, which gives mammalian spermatozoa the ability to initiate fertilization. In swine, for example, a population of capacitated spermatozoa first becomes available for fertilization 2 to 3 h after natural mating or artificial insemination with fresh semen.

caponettes Male chickens that have had their reproductive organs made useless by the injection of an estrogenic hormone (stilbestrol). The testes of these animals decrease in size, and the secretion of testosterone is inhibited, which in turn results in a regression of the secondary sex characteristics (comb, wattles, earlobes, mating instinct, and crowing).

capons Male chickens that have had their reproductive organs (testes) surgically removed. Caponized birds lose some of their male sex characteristics. (The comb loses its bright red color and shrinks in size, and libido is lost.)

caprine Pertaining to or derived from a goat.

captive bolt A pistol used as an alternative to the sole ax or knife for stunning while slaughtering cattle. When fired, a plunger or "bolt" in the barrel penetrates the animal's brain.

caput Head.

carbohydrates Materials consisting chemically of carbon, hydrogen, and oxygen. The most important carbohydrates are the starches, sugars, celluloses, and gums. They are so named because the hydrogen and oxygen are usually in the proportion to form water $(CH_2O)_n$. Formed in plants by photosynthesis, carbohydrates constitute a large part of animal food.

carcass The body of a dead animal. The whole trunk of a slaughtered animal.

carcinogen Any cancer-producing substance.

cardiac That which is related to the heart.

carnivorous Meat-eating. Carnivorous animals are dogs, cats, weasle-like mammals, etc.

carrier A heterozygote for any trait. A disease-carrying animal.

carrion Dead or decaying flesh of a human or animal.

carry To bear, as a pregnant cow carries a calf.

cartilage A firm but pliant type of tissue forming portions of the skeleton of vertebrates. The proportion of cartilage in the skeleton of young animals is greater than in mature animals. Also called *gristle*.

casein The major protein of milk.

castrate To remove the testicles or ovaries.

catabolic Destructive.

catabolism Any destructive process by which complex substances are converted by living cells into simpler compounds (destructive metabolism).

catalyst A substance that alters the speed of a chemical reaction without becoming a part of the end product. Enzymes are catalysts.

catgut Tough cord obtained from the intestines of cattle and sheep and used for strings of musical instruments, tennis rackets, and stitching in surgery.

cathartic A compound (medicine) that quickens and increases evacuation from the bowels. A laxative.

cation An ion carrying a positive charge of electricity. Cations include the metals and hydrogen.

cattalo (catalo) A hardy crossbreed of the American bison and domestic cattle.

cattle Collectively refers to mature bovine animals. In biblical times it referred to all livestock.

caudal Denotes a position toward the tail (rump) or posterior end (same as *inferior* in human anatomy).

cayuse Indian pony, named after the Cayuse Indians, who currently live in Oregon.

cecum (caecum) The first part of the large intestine.

cell A bit of protoplasm that usually contains a nucleus and cytoplasm. Cells are the building blocks of which the body is made. An unfertilized hen's egg is a single cell.

cellulose The principal carbohydrate constituent of plant cell membranes. It is made available to ruminants through the action of microorganisms that inhabit the rumen (rumen bacterial flora).

Celsius A centigrade temperature scale. Can be converted into Fahrenheit by using the formula ($°C \times 9/5$) + 32 = $°F$ (Appendix B).

centaur A legendary monster of the Greeks with a horse's lower body and a man's head, arms, and chest.

centriole A cell organ usually present as two small chromatic granules in the cytoplasm closely opposed to the nuclear wall. The mitotic center in many cells. It plays a role in cellular division.

centromere The point on the chromosome at which identical chromatids are joined and by which the chromosome is attached to a spindle fiber.

cercaria The final free-swimming larval stage of a trematode, consisting of a body and tail.

cervical Referring to the neck.

chamois In the United States, the term refers to the fresh split of sheepskin tanned solely with oils. In other countries, the term includes any one, or several, oil-tanned suede leathers made from sheep or lambskin, deer-, goat-, or kidskin, mountain antelope or chamois, or cattle-hide splits.

chelating agent An organic compound that can bind cations (metallic ions) by forming a stable, inert complex that is soluble in water; used in softening hard water, in purifying sewage, and in eliminating high concentrations of undesirable metallic or radioactive elements in the blood or tissues.

chemotherapy The treatment of disease by means of chemicals.

chevon Meat of a goat kid.

chick A baby chicken.

chigger A larva (mite) infesting humans, domestic animals, some birds, snakes, turtles, and rodents. Its bite results in inflamed spots, accompanied by intense itching.

cholesterol A white, fat-soluble substance found in animal fats and oils, bile, blood, brain tissue, nervous tissue, the liver, kidneys, and adrenal glands. It is important in metabolism and is a precursor of certain hormones.

chondrogenesis Formation of cartilage.

chorion The outermost membrane that encloses the unborn fetus in mammals.

chromatid A chromosome that appears doubled at the metaphase of cell division, connected at one point by a small beadlike body called the *centromere*.

chromosomes Dark-staining rodlike or rounded bodies visible under the microscope in the nucleus of the cell in the metaphase of cell division. Chromosomes occur in pairs in body cells, and the number is constant for a species. Chromosomes carry genes arranged linearly along their length. The backbone of a chromosome is the DNA molecule.

chronic Of long duration as opposed to acute.

chyme A thick liquid of partially digested food. It passes from the stomach into the small intestine.

cirrhosis A chronic disease of the liver characterized by a destruction of liver cells and an increase of connective tissue.

clean Often used to mean to be free of, as with disease, parasites, etc.; also a lay term used by cattle breeders to mean that a cow has shed her afterbirth.

clear egg An infertile egg.

cleft Split or divided to a certain depth. A cleft palate is a longitudinal opening in the roof of the mouth. A *cleft-footed* animal, also called *cloven-footed*, e.g., the cow or hog, has a divided hoof or foot.

cleidoic Closed or locked in, as the egg, which is cut off or isolated from free exchange with the environment by a more or less impervious shell.

climate The prevailing weather conditions that affect life.

clinical Referring to direct observation.

clonorchiasis Oriental liver fluke disease (especially in China and Japan). The causative parasitic trematodes have two intermediate hosts; the first is a molluscan, and the second is some edible fish from which humans become infected.

close breeding A form of inbreeding, e.g., the mating of brothers to sisters, sire to daughter, and son to dam.

closed herd (flock) A herd (flock) in which no outside blood is introduced.

cloven-footed See **cleft.**

clutch The eggs laid by a hen on consecutive days are referred to as a clutch. In domestic birds, the number of eggs laid successively are often referred to as a *cycle of laying* or the *laying rhythm*. The term *clutch* for wild birds refers to the set of eggs laid for one incubation. Some birds lay a single egg (e.g., penguin); others lay two (e.g., pigeon); others may lay 12 to 20 (partridges and most domestic fowls).

coagulant   A substance that acts on a liquid to coagulate it.

coat   Hair on the body of an animal.

cock   A male chicken; rooster. (See Appendix A.)

cockerel   A male chicken less than a year old.

cod   The part of the scrotum left after castration.

coefficient of digestibility   The percentage value of a food nutrient that has been absorbed. For example, if a food contains 10 g of nitrogen and it is found that 9.5 g has been absorbed, the digestibility is 95 percent.

coenzyme   A substance, usually a vitamin or mineral, that works with an enzyme to perform a certain function.

coitus   Coition; sexual intercourse between individuals of the opposite sex.

cold sterilization   The use of a cathode ray or an electron-beam gun in food processing to kill bacteria or insect life. Chemical sterilization of instruments.

colic   A digestive disturbance causing pains in the abdomen, as in horses.

colitis   An inflammation of the mucous membrane of the colon.

collagen   Protein contained in connective tissue, cartilage, and bones. It is the chief protein of raw hides and skins.

collateral relatives   Those related individuals that are not related as ancestors or descendants.

colostrum   The first milk secreted pre- and postpartum.

colt   A young horse; foal. Male horse under 4 years of age. (A young female horse is called a *filly.*)

comb   A piece of red flesh (tissue) on the top of a chicken's head.

comfort zone   The temperature interval during which no demands are made on the temperature-regulating mechanisms. The temperature at which humans and animals feel most comfortable. Also called **thermoneutral zone.**

commensal   Living on or within another organism and deriving benefit without injuring or benefiting the other individual. *Not* parasitic symbiosis.

compaction   Closely packed feed in the stomach and intestines of an animal causing constipation and/or digestive disturbances.

complementary genes   Genes that so interact that when both are present a new or novel trait appears.

complement fixation   A biological reaction that is the basis of many tests for infection. It utilizes a three-phase system: (1) antigen, (2) complement, and (3) antibody. When a specific antigen and antibody are mixed in the presence of complement, a reaction occurs that fixes or renders the complement inactive or fixed. Therefore, if the presence of a specific antigen or antibody in a mixture is known, the presence of the other can be determined by the mixture's reaction on the complement. Such a reaction is the basis of many tests for infection, including the Wassermann test for syphilis and reactions for gonococcus infections, glanders, typhoid fever, and tuberculosis.

complete ration   A blend of all feedstuffs (forages and grains) in one feed. A complete ration fits well into mechanized feeding and the use of computers to formulate least-cost rations (least-cost diets).

concentrate   A feed that is high in **NFE** and total digestible nutrients and low (less than 18 percent) in crude fiber. It includes the cereal grains, soybean-oil meal, cottonseed meal, and by-products of the milling industry, such as corn gluten and wheat bran. A concentrate may be either poor or rich in protein.

conception   The fecundation of the ovum. The action of conceiving or becoming pregnant.

condition   Refers to the amount of flesh (body weight), the quality of hair coat, and the general health of animals.

conduction A means of dissipating heat, especially on contact with cool water. The transfer of heat.

conformation The physical form or physical traits of an animal; its shape and arrangement of parts.

congenital That which is acquired during prenatal life. It exists at or dates from birth.

constitution The general strength and bodily vigor of an animal.

contact insecticide Any substance that kills insects by contact, in contrast to a stomach (ingestant) poison, which must be ingested.

contagious Transmissible by contact.

control rod A rod, plate, or tube containing a material that readily absorbs neutrons used to control the power of a nuclear reactor. By absorbing neutrons, a control rod prevents the neutrons from causing further fission.

convection Either cooling or warming of an animal by wind (breezes) according to whether the wind is cooler or warmer than the surface temperature of the animal.

coprophagy The ingestion of feces.

copulation The act of mating (sexual congress).

core The central portion of a **nuclear reactor,** containing the *fuel elements.*

cornicle An abortive spur on a hen's leg that hardens with age. It does not develop into a regular spur, as in the cock.

corn stover The dried cornstalk from which the ears have been removed.

corpora allata Paired glands of insects that secrete the juvenile hormones.

correlation A measure of how two traits vary together. A correlation of $+1.00$ means that as one trait increases the other also increases—a perfect *positive* relation. A correlation of -1.00 means that as one trait increases the other decreases—a perfect *negative,* or *inverse,* relation. A correlation of 0.00 means that as one trait increases, the other may increase or decrease—no relation. Thus a correlation coefficient may lie between $+1.00$ and -1.00

cosmic rays Radiations of many sorts, but mostly atomic **nuclei** (protons) with very high energies, originating outside the earth's atmosphere.

cosset A lamb raised without the help of its dam.

cotyledon The area of attachment of the fetal placenta to the maternal placenta (carunde) in certain types of ruminants.

counter A general designation applied to *radiation-detection instruments* or *survey meters* that detect and measure radiation.

covey A flock of birds; quail, partridge.

cow A mature female bovine.

cowhide The hide of any kind of cow.

cow-hocked A condition of a cow or horse in which the hocks are close together and the fetlocks (just above the hoof) are wide apart.

cranial Of or pertaining to the skull or the anterior (front) or superior end of the body.

crawler A newly hatched insect.

creep feeding A system of feeding young animals prior to weaning. It is designed to exclude mature animals.

crest The ridge of an animal's neck.

cribbing A condition in which the horse bites the manger or another object while sucking in air.

crimped Rolled with corrugated rollers. The grain to which this term refers may be tempered or conditioned before crimping and may be cooled afterward.

crossbreeding The mating of animals of different breeds.

crossing See **crossbreeding.**

crossing over Exchange of parts by homologous chromosomes during synapsis of meiosis prior to the formation of sex cells or gametes. Thus the homologous chromosomes exchange genes.

crude fiber (CF) That portion of feedstuffs composed of cellulose, hemicellulose, lignin, and other polysaccharides that serve as the structural and protective parts of plants. It is high in forages and low in grains. Poultry and swine are limited in the ability to digest fiber, whereas ruminants (cattle and sheep) can benefit from it through rumen bacterial activity. It is the least digestible part of a feed.

crude protein (CP) The total protein in a feed. In calculating the protein percentage, the feed is first chemically analyzed for its nitrogen content. Since proteins average about 16 percent (1/6.25) nitrogen, the amount of nitrogen in the analysis is multiplied by 6.25 to give the CP percentage.

crural Pertaining to the leg or to a leglike structure.

cryotherapy The therapeutic use of cold.

cryptorchidism A failure of the mammalian testes to descend into the scrotum.

cuboidal Shaped like a cube.

cud A bolus of regurgitated food (common only to ruminants).

culling The process of eliminating nonproductive or undesirable animals.

culture The growing of microorganisms (or cells) in a special medium.

curie (Ci) The basic unit to describe the intensity of **radioactivity** in a sample of material. The curie is equal to 37 billion (3.7×10^{10}) disintegrations per second. It is also a quantity of any nuclide having one curie of radioactivity.

customary host One in which a parasite commonly matures and reproduces.

cyclotron A particle **accelerator** in which charged particles receive repeated synchronized accelerations by electric fields as the particles spiral outward from their source.

cytokinesis The changes that occur in the cytoplasm of the cell during division and fertilization.

cytology The science relating to cells and their origin, structure, and function.

cytoplasm Protoplasm of a cell outside the nucleus.

dairy character The evidence of milking ability. See also **bloom.**

Dairy Herd Improvement Association (DHIA) An organization whose program is operated jointly by the USDA and the colleges of agriculture of the land-grant universities to aid dairy producers in keeping milk production and management records.

Dairy Herd Improvement Registry (DHIR) A modification of the DHIA program to make the records acceptable by the dairy breed associations. An **official production record** program.

dam The female parent.

dapple A circular pattern in an animal's coat color in which the outer portion is darker than the center.

dark chestnut A term used to describe a brownish black, mahogany, or liver-colored horse.

dark meat The legs and thighs of cooked fowls.

day length The length of daylight in hours during a 24-h period.

deacon A veal calf that is marketed before it is a week old. Also called *bob*, or *bob veal*.

deadborn Stillborn.

deamination Removal of the amino group ($-NH_2$) from an amino acid.

decapitate To behead; to cut off the head.

decarboxylation The removal of carboxyl group ($-COOH$) from an organic acid.

decay (radioactive) The spontaneous transformation of one nuclide into a different nuclide or into a different energy state of the same nuclide. The process results in a decrease, with time, in the number of the original radioactive atoms in a sample. It involves the emission from the **nucleus** of **alpha particles, beta particles** (or electrons), or **gamma rays.** Also called *radioactive disintegration.*

decortication Removal of the bark, hull, husk, or shell from a plant, seed, or root. Also, removal of portions of the cortical substance of a structure or organ, as in the brain, kidney, and lung.

defecation The evacuation of fecal material from the rectum.

definitive host (final host) The animal in which a parasite undergoes its adult and sexual life.

deglutition The act of swallowing.

dehorn To remove the horns from cattle, sheep, and goats or to treat young animals so that horns will not develop.

dehydrate To remove most or all moisture from a substance, primarily for the purpose of preservation.

dehydrogenase Enzyme that "activates" hydrogen.

deletion The absence of a portion of a chromosome, causing genes to be absent or lacking.

dementia A general designation for mental deterioration.

dental pad The very firm gums in the upper jaw of cattle, sheep, goats, and other ruminants.

dermatitis An inflammation of the skin.

dermatophytosis Any skin infection caused by a fungus, e.g., ringworm.

dermatosis Any skin disease.

desiccated Dried out; exhausted of water or moisture content.

desquamate To peel or come off in layers or scales, as the epidermis in certain diseases. The shedding of epithelial cells.

detector A radiation-detection instrument.

deutectomy The removal of the yolk sac from newly hatched chicks.

dewlap The pendulous skin fold hanging from the throat, particularly of the ox tribe and certain fowl.

DHIA See **Dairy Herd Improvement Association.**

DHIR See **Dairy Herd Improvement Registry.**

diaphysis The shaft of a long bone.

dicoumarol A chemical compound found in spoiled sweet clover and lespedeza hays. It is an anticoagulant and can cause internal hemorrhages when eaten by cattle. The trade name is Dicumarol.

diestrum (dioestrum) That portion of a female's cycle between periods of estrus (heat).

differentiation The process of acquiring individual characteristics, such as occurs in the progressive diversification of cells and tissues of the embryo. The transformation of mother cells into different kinds of daughter cells (brain, kidney, liver, etc.). This process is irreversible.

digestibility That percentage of food taken into the digestive tract which is absorbed into the body as opposed to that which is evacuated as feces.

digestible energy (DE) That portion of the energy in a feed which can be digested or absorbed into the body by an animal.

digestible protein (DP) That portion of the protein in a feed which can be digested or absorbed into the body by an animal.

digestion coefficient (coefficient of digestibility)   The difference between the nutrients consumed and the nutrients excreted expressed as a percentage.

digitalis   A valuable drug having diuretic properties, made from the dried leaves of the foxglove, used in the treatment of heart diseases.

dihybrid   A double heterozygote such as *AaBb*.

dihybrid cross.   A cross involving two pairs of alleles, each of which regulates different characteristics.

diploid   Refers to chromosomes paired (2*n*) in body cells; i.e., there are two of each kind of chromosome in the nucleus as compared with one of each pair (1*n*) in sex cells. Somatic cells are usually diploid, whereas gametic cells are usually haploid.

direct calorimetry   Measurement of the amount of heat produced within a small chamber (as in a **bomb calorimeter,** where the heat produced through combustion is absorbed by a known quantity of water in which the container is immersed). A means of determining the caloric value of foods.

disease   Any deviation from a normal state of health that temporarily impairs vital functions of animals.

disinfect   To destroy or render inert disease-producing germs (pathogens) and harmful microorganisms and to destroy parasites.

dislocating   A method of killing poultry in which the fowl's head is pulled and twisted until the neck separates from the skull and the spinal cord is severed.

dispermic   Refers to an ovum fertilized by two spermatozoa (rather than one).

disposition   The temperament, or spirit, of an animal.

dissect   To cut an animal into pieces for examination.

dissipate   To cause to disappear; to spread in different directions.

distal   Remote, as opposed to close or proximal; away from the main part of the body.

diuresis   Excessive discharge of urine.

diurnal variation   The amount of variation in one day.

dizygotic twins   Twins originating from two separate fertilized eggs. They are no more alike genetically than full brothers and sisters born at different times.

DNA   Deoxyribonucleic acid.

dock   To cut off the tail (especially in sheep).

doddie   A **polled** cow.

doe   An adult female rabbit, goat, or deer.

dogie   A motherless calf.

domesticate   To bring a wild animal or fowl under control and to improve it through careful selection, mating, and handling so that its products or services become more useful to humans. Domesticated animals breed under the control of humans.

dominant   Describes a gene that when paired with its allele covers up the phenotypic expression of that gene. For example, brown eyes (*B*) in humans is dominant to blue eyes (*b*). Thus *BB* and *Bb* individuals have brown eyes whereas *bb* individuals have blue eyes. Dominant genes affect the phenotype when present in either homozygous or heterozygous condition.

dormancy   Depressed metabolism. A state in which organisms are inactive, quiescent, or sleeping.

dorsal   Pertaining to the back, or more toward the back portion; opposite to *ventral.* It means the same as *posterior* in human anatomy.

double mating (double cover)   The mating of female livestock with a male on successive days of an estrous period to enhance the probability of conception. Also, mating a female to two different males during the same heat period.

draft animal An animal, e.g., a horse, mule, or ox, used to pull heavy loads.

drake An adult male duck.

drape A term applied to a cow or ewe incapable of bearing offspring.

drifting The moving of bees from one hive to another because of loss of direction caused by wind or other circumstances.

drone The male bee hatched from an unfertilized egg. It is larger than the workers, does not gather nectar for honey, and has no sting.

drone layer A queen bee that has exhausted her supply of spermatozoa stored after mating, so that her eggs produce only drones.

dropping A term commonly used to mean *parturition* (the act or process of giving birth).

drove A collection or mass of animals of one species, as a drove of cattle.

dry Nonlactating female. The dry period of cows is the time between lactations (when a female is not secreting milk).

dry ice Solid carbon dioxide that is purified, liquefied, expanded to form snow, and finally pressed into blocks. It is used as a refrigerant, as in the storage of semen.

drylot A relatively small area in which cattle are confined indefinitely as opposed to being allowed to have free access to pasture.

duct A canal (tube) that conveys fluids or secretions from a gland.

dung The feces (manure) or excrement of animals and birds.

duplication Process in which a chromosome is attached to parts from its own homologous chromosome and thus has duplicate genes.

dyad A unit of two sister chromatids that are synapsed as a tetrad.

dysfunction Partial abnormality, disturbance, or impairment in the function of an organ or a system.

dysgenesis A defect in breeding so that hybrids cannot mate between themselves but may produce offspring with members of either family of their parents. Such offspring are sterile.

dyspnea Difficult or labored breathing.

dystocia Abnormal or difficult labor, causing difficulty in delivering the fetus and placenta.

easy keeper An animal that does well on a minimum of food.

eccrine Glands or tissues that secrete a substance without a breakdown of their own cells, e.g., sweat glands. See **exocrine.**

ecology The study of the relation of organisms to their environment, habits, and modes of life; **bionomics.**

ectoblast An embryonic cell layer.

ectoderm The outermost of the three primary germ layers of the embryo; gives rise to skin, hair, nervous system, etc.

ectoparasites Parasites that inhabit the body surface.

eczema An inflammatory skin disease of humans and animals characterized by redness, itching, loss of hair, and the formation of scales or crusts.

edema The presence of abnormally large amounts of fluid in the intercellular tissue spaces of the body. Swelling.

efficacy Effectiveness. For example, the effectiveness of an antibiotic in the therapy of a certain disease.

e.g. For example, from the Latin *exempli gratia.*

eggbeater A horse that places his feet carefully and smoothly.

ejaculation A sudden ejection, or discharge, as of semen from the male.

elastin The yellow proteinaceous connective tissues of the skin that provide structural support for the bood vessels and thermostat mechanism. It is obtained when elastic tissue is boiled in water.

electrolyte Any solution that conducts electricity by means of its ions.

electromagnetic radiation Radiation consisting of electric and magnetic waves that travel at the speed of light (e.g., light waves, radio waves, gamma rays, x-rays).

electromagnetic spectrum (ES) A series of waves of varying length. At one end of the ES are gamma rays, whose waves measure only about one-billionth of an inch; at the other end are radio waves, which may be many miles long. Between these extremes, in order of increasing wavelength, are x-rays, ultraviolet light, visible light, infrared light, microwaves, and shorter radio waves. All waves have different characteristics, but all travel at the same speed (186,000 miles/s).

electron (symbole) An *elementary particle* with a unit negative electric charge. Electrons surround the positively charged *nucleus* and determine the chemical properties of an *atom.*

electron radiation Corpuscular radiation, consisting of streams or beams of electrons accelerated to energies of up to 10 MeV. There are a number of machine sources available for electron radiation.

element One of the 105 known chemical substances that cannot be divided into simpler substances by chemical means. A substance the atoms of which all have the same *atomic number,* e.g., hydrogen, lead, and uranium. (See Appendix E.)

emaciation A wasted condition of the body.

emasculation Excision of the penis. Castration.

embryology The science relating to the formation and development of the embryo.

emissivity (or absorptivity) Ratio of the rate of emission (or absorption) of radiant energy per unit area by the given body to the rate of emission (or absorption) from a blackbody at the same temperature.

encephalitis An inflammation of the brain that results in various central nervous system disorders.

encephalomalacia A condition characterized by softening of the brain. A disease causing lesions of the brain in young poultry, caused by a deficiency of vitamin E in the diet.

endemic (enzootic) Pertaining to a disease commonly found with regularity in a particular locality.

endocrine Pertaining to glands that produce secretions that pass directly into the blood or lymph instead of into a duct (secreting internally). Hormones are secreted by endocrine glands.

endogenous Internally produced in the body, e.g., hormones and enzymes.

endoparasites Parasites that inhabit the body tissues or cavities, e.g., the tapeworm.

endophily The feeding or resting of insects in houses (inside).

endotoxins Toxic substances (such as those causing typhoid) retained inside the bacterial cells until the cells disintegrate.

energy balance (EB) The relation of the gross energy consumed to the energy output (energy retention). It is calculated as follows: $EB = GE - FE - UE - GPD - HP.$

enriched material Material in which the percentage of a given **isotope** present has been artificially increased so that it is higher than the percentage of that isotope naturally found in the material.

ensilage A green crop (forage) preserved by fermentation in a silo, pit, or stack, usually in the chopped form. Also called *silage.*

enteritis Any inflammatory condition of the intestinal linings of animals or humans.

entozoa   Internal animal parasites, e.g., stomach worms.

entrails   The inner organs of animals, specifically the intestines.

envenomization   The injection of a poison into an animal (as by a wasp).

environment   The sum total of all external conditions that affect the life and performance of an animal.

enzootic   Occurring endemically among animals, i.e., continuously prevalent among animals in a certain region.

enzyme   A complex protein produced in living cells that causes changes in other substances within the body without being changed itself. An organic catalyst. Body enzymes are giant molecules.

epharmonic   Differing from the normal or usual because of environmental influences.

ephemer   Designating a very short-lived animal.

epidemic (epizootic)   Rapidly spreading (as a disease) so that many animals or people have it concurrently.

epidemiology   The field of science dealing with the relationship among various factors that determine the frequencies and distributions of infectious diseases.

epiderm   Epidermis (the outer layer of skin or tissue).

epigenesis   A theory of embryogenesis, stating that development consists in the successive formation of new parts that do not preexist in the fertilized embryo.

epinephrine   A drug used to arrest hemorrhage and to stimulate heart action. It is a slaughterhouse by-product obtained from the adrenal glands. Also called *adrenaline.*

epiphysis   A portion of bone separated from a long bone by cartilage in early life, but later becoming a part of the larger bone. It is at this cartilaginous joint that growth in length of bone occurs. It is also called the *head* of a long bone.

epistasis   Interaction between genes that are not alleles, causing the appearance of a different phenotype. For example, *BB* causes an animal to be black. However, *II* inhibits pigment production. *BBii* would be black, whereas *BBII* would be white. Epistasis can be due to two or more pairs of alleles on the same pair of homologous chromosomes or to two different pairs of genes on two different homologous chromosomes. Epistatic genes interact so that one pair of genes masks or suppresses the phenotypic expression of another pair of genes.

epistaxis   Bleeding from the nose.

epizootic   Designating a widely diffused disease of animals, that spreads rapidly and affects many individuals of a kind concurrently in any region, thus corresponding to an epidemic in humans.

equestrian   One who rides horseback.

equine   Pertaining to a horse.

eradication   The total elimination of the etiologic (disease-causing) agent from a region.

ergosterol   A plant sterol that, when activated by ultraviolet rays, becomes vitamin D_2. Also called *provitamin* D_2 and *ergosterin.*

eructation   The act of belching, or casting up gas from the stomach.

erythema   A severe redness of the skin associated with some local inflammation.

erythropoiesis   The production of erythrocytes (the red blood cells that transport oxygen). Occurs in the red bone marrow.

escutcheon   The part of a cow that extends upward just above and back of the udder, where the hair turns upward in contrast to its normal downward direction. Also called the *milk mirror.*

essential amino acid   An amino acid that cannot be synthesized in the body from other amino acids or substances or that cannot be made in sufficient quantities for the body's needs.

essential fatty acid A fatty acid that cannot be synthesized in the body or that cannot be made in sufficient quantities for the body's needs. Linoleic and linolenic acids are essential for humans.

estivation The adaptation, or modifications, in an animal that enable it to survive a hot, dry summer. Estivation is probably an evolutionary adaptation to periods of water scarcity. It is often called *summer hibernation.*

estrus (oestrus) The recurrent, restricted period of sexual receptivity (heat) in female mammals, marked by intense sexual urge.

ether extract The fatty substances of foods or other materials that are soluble in ether. Used in food analysis.

etiology The study or theory of the causation of diseases.

euploid An organism whose chromosome number is a whole-number multiple of the basic, or haploid, number.

European breeds Those, such as Charolais, Simmental, and Limousin, native to Great Britain or continental Europe.

evaporation (transpiration) As related to environmental physiology, the loss of moisture (sweating).

eviscerate To remove the entrails, lungs, heart, etc., from an animal or fowl when preparing the carcass for human consumption.

ewe A female sheep. (See Appendix A.)

excise To cut out or off.

excreta Waste materials discharged from the bowels.

excystation The escape of a cyst; especially, a stage in the life cycle of parasites occurring after the cystic form has been swallowed by the host.

exfoliate To remove the surface scale or layer.

exocrine (eccrine) Secreting outwardly, into or through a duct.

exogenous Externally provided, e.g., vitamins.

exomphalos A hernia that has escaped through the umbilicus (navel).

exophily The feeding or resting of insects out of doors.

exotoxins *Soluble* toxins that diffuse out of living bacteria into the environment (the culture medium of the living host). Toxins secreted by living microorganisms.

expectoration The act of coughing up and spitting out materials.

experiment From the Latin *experimentum,* meaning proof from experience. It is a procedure used to discover or to demonstrate a fact or general truth.

exsanguination The withdrawal of blood, as by the bloodsucking insects.

extirpation The complete removal or eradication of a part.

exudate An abnormal seeping of fluid through the walls of vessels into adjacent tissue or space.

exudative diathesis Symptom of vitamin-E deficiency in poultry. It is characterized by an accumulation of fluid in subcutaneous fatty tissue.

F_1 generation The first-filial-generation, or the first-generation progeny, following the parental, or P_1, generation.

F_2 generation The second-filial-generation progeny resulting from the mating of F_1-generation individuals. Produced by an inbreeding of the first filial generation.

facultative The ability of a microorganism to live and reproduce under either aerobic or anaerobic conditions.

fag Any tick or fly that attacks sheep.

Fahrenheit   The temperature expressed in degrees, where 32 and 212° are the freezing and boiling temperatures, respectively, of water at sea level. A more universal and scientific graduation of temperature is the centigrade Celsius scale. Fahrenheit is converted into Celsius by using the formula (°F − 32) 5/9 = °C (Appendix B).

fallout   Airborne particles containing radioactive material that fall to the ground following a nuclear explosion.

false heat   The display of estrus by a female animal when she is pregnant.

false (pseudo) albino   A solid-white horse with colored eyes.

family   A related group of animals.

FAO   Food and Agriculture Organization. A specialized international agency of the United Nations that collects and disseminates information on the production, consumption, and distribution of food throughout the world. It was organized in 1945 and has some 120 member nations; it is headquartered in Rome. (See also Section 1.4.7.)

farrow   To give birth to a litter of pigs. (See Appendix A.)

far side   The right side of a horse.

fascia   A thin sheet or band of fibrous (connective) tissue covering, supporting, or binding together a muscle, part, or organ.

fasting   Abstaining from food. Certain types of experiments involve withholding food from the test animal for varying periods of time.

fat   Adipose tissue; an ester of glycerol with fatty acids.

favor   To protect; to use carefully, as an animal favors a lame leg.

FCM (4 percent fat-corrected milk)   A means of evaluating milk production records of different animals and breeds on a common basis energywise. The following formula is used: FCM = 0.4 × milk production + (15 × lb fat produced).

FDA   Federal Food and Drug Administration.

febrile   Pertaining to a fever, or rise in body temperature.

fecal energy (FE)   The food energy lost through the feces.

feces (faeces)   Excrement discharged from the bowels.

fecundation   Impregnation or fertilization.

fecundity   The ability of an individual to produce eggs or sperms regularly.

feed conversion (feed efficiency)   The units of feed consumed per unit of weight increase. Also, the production (meat, milk, eggs) per unit of feed consumed.

feeder   A young animal that does not have a high finish (fatness) but shows evidence of ability to add weight economically.

femininity   Physical appearance resulting from well-developed secondary female sex characteristics.

femoral   Pertaining to the femur or to the thigh.

feral   Pertaining to animals in the wild, or untamed, state.

fertility   The ability to reproduce.

fertilization   The process in which two haploid gametes fuse, forming a diploid cell, the zygote.

fetus   The unborn young of animals (usually vertebrates) that give birth to a living offspring.

fever   Abnormally high body temperature.

fibrinogen   A soluble protein present in the blood and body fluids of animals that is essential to the coagulation of blood.

fibroblasts   Connective tissue cells whose function is one of repair. Also called *fibrocytes* and *desmocytes*.

fibrous carbohydrates Those feed components (e.g., cellulose and hemicellulose) not readily digested by animals.

filled milk Milk from which milk fat has been removed and replaced with other fats or oils.

filly A young female horse. A young mare. (See Appendix A.)

filterable virus A name commonly applied to a pathogenic agent capable of passing filters that retain bacteria.

find Lay term meaning *to give birth to young;* e.g., a cow finds a calf.

finish The degree of fatness or the distribution of fat in an animal.

first calf A term commonly used to indicate the first calf born to **bovine** females.

first meiotic division The first of two divisions occurring in reduction cell division and resulting in the production of two cells, each of which is haploid, the chromosomes occurring as paired chromatids joined at the centromere.

fission The splitting of a heavy **nucleus** into two approximately equal parts, accompanied by the release of a relatively large amount of energy and generally one or more neutrons.

fistula See **cannula.**

five-gaited saddle horse A horse trained to use the following gaits: walk, trot, rack, slow gait, and canter.

flaccid Limp; weak; flabby.

flatulence A digestive disturbance in which there is often a painful collection of gas in the stomach or bowels.

flay To remove the skin from a carcass.

fleece The wool from all parts of a sheep.

flesh The muscle and fat covering of an animal. See **condition.**

fleshing Removal of adipose tissue on the flesh side of skins or hides to be used in the leather industry.

flock A group of birds or sheep; called *band* in goats. (See Appendix A.)

flora The bacteria in or on an animal. Usually referred to as *bacterial flora,* as in the digestive tract.

fluid milk Milk commonly marketed as fresh liquid milks and creams. It is the most perishable form of milk and commands the highest price per unit volume. Also called *market milk.*

flux A measure of the intensity of neutron radiation. It is the number of neutrons passing through one square centimeter of a given *target* in one second.

foal Young (usually unweaned) horse or mule of either sex.

fodder Coarse food such as cornstalks, for cattle or horses.

fodder units (FU) This is a net-energy system of feed evaluation that is used commonly in Scandinavian countries. Values of feeds are measured and expressed in relation to a reference feed, barley. One kilogram of FU is equivalent to 1650 kcal NE.

football leather Leather for covering footballs. Traditionally of pigskin, but today it is generally made of embossed or printed cattlehide leather, and sometimes of sheepskin.

forage Roughage of high feeding value. Grasses and legumes cut at the proper stage of maturity and stored without damage are called *forage.*

forager A *worker* bee that goes out in search of food.

forceps A plierslike instrument used in medicine and certain research studies for grasping, pulling, and compressing.

forequarters The front two quarters of an animal. Also called *fore udder.*

founder Inflammation of the fleshy laminae within the hoof, from concussion, overfeeding, and many factors causing an oversupply of blood to this region of the hoof. It causes great pain to the affected animal. Also called *laminitis.*

fowl Any bird, but more commonly refers to the larger ones.

fox trot An easy, short, broken gait of a horse, between a walk and a trot.

freemartin Female born twin to a bull calf. (About 9 out of 10 will not conceive.)

freeze-drying See **lyophilization.**

fresh Designating a cow that has recently given birth to a calf.

Friedman test A test for pregnancy in which a small amount of urine of the tested animal is injected into the bloodstream of a virgin rabbit. Pregnancy is indicated by certain changes in the ovaries of the rabbit.

frog The elastic, horny, middle part of the sole of a horse's hoof.

full-feed The term commonly applied to fattening cattle being provided as much feed as they will consume safely without going **off feed.**

fumigant A liquid or solid substance that forms vapors that destroy pathogens, insects, and rodents.

fungi (singular, **fungus**) Plants that contain no chlorophyll, flowers, or leaves. They get their nourishment from dead or living organic matter.

fungicide An agent that destroys fungi.

fur Skins of wild animals, commonly covered with short fine hair, which are tanned or dressed for garments.

furrowing The process of cytokinesis in animal cells.

fusion The formation of a heavier **nucleus** from two lighter ones (such as hydrogen isotopes), with the attendant release of energy (as in a hydrogen bomb).

gait Any forward movement of a horse, such as walking or galloping.

galactophore A milk duct.

galactophorus Carrying or producing milk; conveying milk.

galactopoietic That which stimulates or increases the secretion of milk.

gamete A male or a female reproductive cell. A sperm or an ovum.

gametogenesis Cell-division process that forms the sex cells.

gamma rays (γ **rays**) High-energy, short-wavelength *electromagnetic radiation.* Gamma rays are very penetrating and are best stopped or shielded against by dense materials, such as lead. They are similar to x-rays but are usually more energetic (their penetrating power is greater) and are of nuclear origin. Gamma rays are emitted by isotopes of such elements as cobalt and cesium as they disintegrate spontaneously.

gander A mature male goose.

gaseous products of digestion (GPD) Include the combustible gases produced in the digestive tract incident to the fermentation of the diet. Methane constitutes the major proportion of the combustible gases produced by the ruminant; however, nonruminants also produce methane. Trace amounts of hydrogen, carbon monoxide, acetone, ethane, and hydrogen sulfide are also produced.

gastritis An inflammation of the stomach, especially of the lining or mucous membrane.

gastrointestinal Pertaining to the stomach and intestines.

gastrula An early stage in embryonic development; it follows the blastula stage.

GE See **gross energy.**

geiger counter Instrument that counts pulses produced by radioactivity, consisting of a counting tube with a central wire anode, usually filled with a mixture of argon and organic vapor.

gelatin An organic colloidal substance made from animal bones, skins, or hide fragments. Used in leather finishes to produce a tough film on the leather. Glue is an impure form of gelatin.

gelding A castrated male horse. To geld is to render sterile or to remove the testicles. Horses are usually castrated at about 2 years of age.

gene The smallest unit, or particle, of inheritance—a portion of a DNA molecule. Genes occur in pairs located on chromosomes in the nucleus of every cell.

generation See **F₁ generation** and **F₂ generation.**

generation interval The time from the birth of one generation to the birth of the next generation. In humans, it is calculated by subtracting the average age of the children from the average age of their parents.

genetic correlation Condition in which two or more quantitative traits are affected by many of the same genes.

genetic value The total value of all the genes of an animal for secreting milk or for some other trait. See **breeding value.**

genital Pertaining to the organs of reproduction.

genome The total genetic composition of an individual or population that is inherited with the chromosomes. A haploid set of chromosomes with their genes.

genotype The actual genetic constitution (makeup) of an individual as determined by its germ plasm. For example, there are two genotypes for brown eyes, *BB* and *Bb*. See **dominant.**

gentle Designating an even-tempered, docile, quiet animal.

germ A small organism, microbe, or bacterium that can cause disease in humans and/or animals. Early embryo; seed.

germ-free Designating an animal that is free of harmful organisms, as germ-free pigs used for experimental purposes. (See Chapter 22.)

germicide A substance that kills disease-causing microorganisms (pathogens).

gerontology The scientific study of the phenomena of aging. (See Chapter 8.)

gestation Pregnancy (gravidity). The period from conception to birth of young.

get The offspring of a male animal. A *get-of-sire* refers to a given number (commonly four) of progeny from a male or sire.

ghee A semifluid butter preparation from the milk of a buffalo, cow, sheep, or goat. It is nearly 100 percent milk fat and is used mostly in Asia and Africa.

giblets Any of the internal organs of a fowl used as food, particularly the heart, liver, and gizzard.

gigot A leg of mutton, venison, or veal that is trimmed and ready for consumption.

gilt A young female swine; commonly called *gilt* until the first litter of pigs is farrowed. (See Appendix A.)

girth The circumference of the body of an animal behind the shoulders. Also, a leather or canvas strap that fits under the horse's belly and holds a saddle in place.

giving milk Lactating, or the act of yielding milk by a mammal.

gland An organ that produces a specific secretion to be used in, or eliminated from, the body.

glasser Calf or kipskin taken from animals that are poorly fed and possess coarser grain hide or skin.

gluconeogenesis Formation of glucose from protein or fat.

glycogenesis Conversion of glucose into glycogen.

glycogenolysis Conversion of glycogen into glucose.

glycolysis   Conversion of carbohydrate into lactate by a series of catalysts. The breaking down of sugars into simpler compounds.

gnat   Any dipterous, biting insect.

gonad   The gland of a male or female that produces the reproductive cells; the testicle or ovary.

gosling   Any young goose before its sex can be determined. (See Appendix A.)

gossypol   A toxic yellow pigment found in cottonseed. Heat and pressure tend to bind it with protein and thereby render it safe for animal consumption. It may cause discoloration of egg yolks during cold storage.

GPD   See **gaseous products of digestion.**

grade   Animals showing the predominant characteristics of a given breed. They usually have at least one **purebred** parent, usually the male.

grading up   The continued use of purebred sires on grade dams.

grafting   A process of removing a worker bee larva from its cell and placing it in an artificial queen cup, for the purpose of having it reared as a queen.

gram-negative bacteria   Bacteria that are decolorized by acetone or alcohol and therefore have a pink appearance when counterstained with safranine, e.g., *Escherichia coli, Pseudomonas aeruginosa, Salmonella, Vibrio,* and *Pasteurella.*

gram-positive bacteria   Bacteria that are able to retain a crystal-violet dye even when exposed to alcohol or acetone, e.g., staphylococci, streptococci, and *Bacillus anthracis.*

graphite   A very pure form of carbon.

GRAS   An acronym for the phrase *generally recognized as safe.* This term is commonly used by the FDA and others when referring to food and feed additives.

grass tetany   A magnesium-deficiency disease of cattle characterized by hyperirritability, muscular spasms of the legs, and convulsions.

gravidity   See **gestation.**

Gray (GY)   The unit (or level) of energy absorbed by a food from ionizing radiation as it passes through in processing. One gray (Gy) equals 100 rad; 1000 Gy equals 1 kilogray (kGy).

grazing   Consumption of standing vegetation, as by livestock or wild animals.

green chop (fresh forage)   Forages harvested (cut and chopped) in the field and hauled to livestock. This minimizes the loss of moisture, color, nutrients, and wastage. Also called *zero grazing.*

gristle   Cartilage.

gross energy (GE)   The amount of heat, measured in calories, released when a substance is completely oxidized in a **bomb calorimeter** containing 25 to 30 atm of oxygen (heat of combustion).

grow out   To feed cattle so that the animals attain a certain desired amount of growth with little or no fattening.

growthy   Designating an animal that is large and well-developed for its age.

grub   Also called warble fly. These cause extensive and widespread damage to hides and skins of food-producing animals by puncturing holes along either side of the spinal line.

gruel   A food prepared by mixing a ground feed with hot or cold water.

gut   The digestive tract; sometimes referring only to the intestines of animals.

gut tie   A condition of animals in which the intestines become twisted, causing an obstruction.

habituation   The gradual adaptation to a stimulus or to the environment. The extinction of a conditioned reflex by repetition of the conditioned stimulus (also called *negative adaptation*).

hackamore   A bridle (with no bits) that controls a horse by pressure on its nose.

hackles   Erectile hairs on the backs of certain animals.

half-life   The time in which the radioactivity originally associated with an isotope will be reduced by one-half through radioactive decay. The time in which half the atoms of a particular radioactive substance disintegrate into another nuclear form. Measured half-lives vary from millionths of a second to billions of years. *Effective half-life* is the time required for a radionuclide contained in a biological system, such as a person or an animal, to reduce its activity by one-half as a result of radioactive decay and biological elimination.

half-sib   In genetics, a half-brother or half-sister.

halogens   A group of elements, including fluorine, chlorine, bromine, and iodine, which are strong oxidizing agents and have disinfectant properties.

ham shank   The hock end of a ham.

hamstring   The large tendon that is above and behind the hock in the rear leg of quadrupeds.

hand   Term used to describe the height (at the point of withers) of a horse. It is the average width of a person's hand (4 in).

handbag leather   Any leather used in making handbags. The most commonly used leathers for handbags are calf, patent (mostly made from cattle hides), goat, and sheep.

hand milking   The manual milking of an animal as opposed to machine or mechanical milking.

haploid   Refers to the 1*n* number of chromosomes in sex cells. Chromosomes are paired in body (somatic) cells. Sex cells contain one chromosome of each pair. *Haploid* means, then, having one of each kind of chromosome in the nucleus.

hard breeder (shy breeder)   A female animal that has difficulty or is slow in conceiving.

hard keeper   An animal that is unthrifty and grows or fattens slowly regardless of the quantity or quality of feed.

hatch   To bring forth young from the egg by natural or artificial incubation.

hay   Dried forage (e.g., grasses, alfalfa, clovers) used for feeding farm animals.

haylage   Low-moisture silage (35 to 55 percent moisture). Grass and legume crops are cut and wilted in the field to a lower moisture level than normal for grass silage, but the crop is not sufficiently dry for baling.

head-shy   Designating a horse on which it is difficult to put a bridle, to lead, or to work around its head.

health physics   The science concerned with recognition, evaluation, and control of health hazards from **ionizing radiation.**

heart girth   A measurement taken around the body just in back of the shoulders of an animal, used to estimate body weight.

heat increment (HI).   The increase in heat production following consumption of food when an animal is in a thermoneutral environment. It consists of increased heats of fermentation and of nutrient metabolism. There is also a slight expenditure of energy in masticating and digesting food. This heat is not wasted when the environmental temperature is below the critical temperature. This heat may then be used to keep the body warm. Also called *work of digestion*. See also **heat increment of feeding.**

heat increment of feeding (HIF)   The heat produced by an animal during fermentation in the gastrointestinal tract and during processing and use of food nutrients in the body. The HIF represents an inefficiency, unless the animal can use the heat to help keep the body warm in a cool or cold environment. It depends on species, diet quality, level of feed intake, and productive performance. In human physiology, the HIF is commonly referred to as **specific dynamic action (SDA).**

heat of fermentation (HF)   Heat produced anaerobically by microbes of the digestive tract. It is much greater in ruminants than in nonruminants.

heat of nutrient metabolism (HNM)   The heat produced as a result of the utilization of absorbed nutrients.

heat period   That period of time when the female will accept the male in the act of mating. In heat. Estrous period.

heat production (HP)   Results from (1) the heat increment (heat of fermentation plus heat of nutrient metabolism) plus (2) the heat used for maintenance (basal metabolism plus heat of voluntary activity) when an animal is consuming food in a thermoneutral environment. In direct calorimetry, heat production is measured by use of an animal calorimeter.

heat stroke   A condition caused by exposure to excessive heat; may occur in one of three forms: (1) sunstroke (thermic fever), (2) heat exhaustion, or (3) heat cramps.

heat tolerance   The ability of an animal to endure the impact of a hot environment without suffering ill effects.

hectare   A European unit of land measurement (2.47 acres). (See Appendix B.)

heel fly   A name applied to the common cattle grub because it enters the body through the feet.

heifer   A female of the cattle species less than 3 years of age that has not borne a calf. (See Appendix A.)

helminth   An intestinal worm or wormlike parasite.

hematopoietic   An agent that promotes the formation of blood cells.

hemoglobin   The red pigment in red blood cells of animals and humans that carries oxygen from the lungs to other parts of the body. It is made of iron, carbon, hydrogen, and oxygen and is essential to animal life.

hemolysis   The liberation of hemoglobin. Hemolysis consists of the separation of hemoglobin from the corpuscles and its appearance in the fluid in which the corpuscles are suspended. The freezing of blood will cause hemolysis. Many microorganisms are able to hemolyze red blood cells by production of hemolysins.

hemophilia   An inherited condition in which blood does not clot normally. An animal so affected may be called a *bleeder.*

hen   An adult avian female (usually refers to chicken or turkey). (See Appendix A.)

herbicide   A preparation for killing weeds.

herbivorous   Pertaining to animals that subsist on grasses and herbs.

herd   A group of animals (especially cattle, horses, and swine) collectively considered as a unit.

herdmates (stablemates or contemporaries)   A term used when comparing records of a sire's daughters with those of all their nonpaternal herdmates, both being milked at the same season of the year (under the same conditions). Each daughter is compared with her herdmates that calved over a 5-month period (2 months prior and 2 months succeeding, plus the month the daughter calves). This is done to adjust for effect of seasonal differences. The *herdmate comparison* removes, from the evaluation of breeding value, complications arising from herd, year, and season of calving production variations and assumes that the measured production differences are due to inheritance (Chapter 5).

heredity   The hereditary transmission of genetic or physical traits of parents to their offspring.

heritability   A technical term used by animal breeders to describe what fraction of the differences in a trait, such as milk production, is due to differences in genetic value rather than environmental factors; variation due to genetic effects divided by the total variation (genetic plus environmental variation).

heritability estimate   An estimate of the proportion of the total phenotypic variation between individuals for a certain trait that is due to heredity. More specifically, hereditary variation due to *additive* gene action.

hermaphrodite   An individual possessing both male and female reproductive organs. May be capable of producing both ova and sperm; however, this seldom is true.

hernia   The protrusion of an organ or part through some opening in the wall of its cavity. Also called *rupture*.

heterogametic   This term refers to the sex cells. In humans and livestock (except poultry) the male possesses an X and a Y chromosome and thus is *hetero-* (prefix meaning unlike) *gametic* in sex chromosomes. The female is XX, and so is *homo-* (alike) *gametic* for sex chromosomes.

heterosis (hybrid vigor)   Amount the F_1 generation exceeds the P_1 generation for a certain trait, or amount the crossbreds exceed the average of the two purebreeds that are crossed to produce the crossbreds. Also called *nicking*.

heterothermic mammals   Animals that have a fluctuating body temperature.

heterozygote   An individual possessing one or more pairs of allelic genes for a given trait, each pair being different genes. For example, *Aa* or *Bb* or *AaBb* are hetero- (unlike) zygotes. Each parent contributed unlike genes with respect to any given allelic pair governing contrasted characters.

HI   See **heat increment.**

hibernation   The ability of an animal to pass the winter in a dormant state in which the body temperature drops to slightly above freezing and metabolic activity is reduced nearly to zero. Hibernation is probably an adaptation to prevent starvation during periods of food scarcity.

hidrosis   Excessive sweating.

high-moisture silage   **Silage** containing 70 percent or more moisture.

hinny   The offspring of a stallion and a jennet (female jackass).

hip sweeney   A horse with atrophy of the hip muscles.

hive   A home for honeybees made by humans.

hobble   To tie the front legs of a horse together by means of a rope or straps so that it cannot run or kick.

hogget   A sheep from weaning until its first shearing.

homeostasis (homeokinesis)   The physiological regulatory mechanisms that maintain constant the "internal environment" of the organism in the face of changing conditions.

homeothermic (homoiothermic)   Having a relatively uniform body temperature maintained nearly independent of the environmental temperature. Warm-blooded animals (mammals and birds) are homeothermic.

homeotherms (endotherms)   The warm-blooded species. They maintain their *internal* temperature constant in the face of widely changing *external* temperature.

homogenized milk   Milk that has been treated to ensure breakup of fat globules to such an extent that after 48 h of quiescent storage at 7°C, no visible cream separation occurs on the milk. The reduced size of fat particles results in formation of a softer curd in the stomach.

homologous chromosomes   From the Greek *homo,* meaning "like," and *logous,* meaning "proportion." The pairs of chromosomes (twins) in body cells. *Homologous* refers to chromosomes that are structurally alike.

Homo sapiens   The species, including all existing races, of humans.

homozygote   An individual possessing like genes for a pair of allelomorphs, for example, *AA* or *aa.* Having only one type of allele for a given trait.

honey butter   A mixture of creamery butter and 20 to 30 percent liquid honey, used as a spread.

hood   A protective device, usually providing special ventilation to carry away gases, in which dangerous chemical, biological, or radioactive materials can be safely handled.

hoof oil   Pale yellow liquid obtained from skin, bones, and hooves of cattle. It is used in leather manufacture as a lubricating and waterproofing agent. Frequently called neat's-foot oil.

hormone   A chemical substance secreted by an endocrine gland that has a specific effect on the activities of other organs.

host   Any animal or plant on or in which another organism lives as a parasite. Infected invertebrates (which are actually hosts in the true sense) are usually referred to as *biological vectors.*

hot   Highly radioactive.

HP   See **heat production.**

humerus   The bone that extends from the shoulder to the elbow.

humidity   The mass of water vapor per unit volume. See **relative humidity.**

hutch   A boxlike cage or pen for small animals.

hybrid   A heterozygote, or progeny of genetically unlike parents.

hybridomas   Plasma cells that are specific for one set of antigenic determinants are fused to myeloma cells (malignant plasma cells) to produce a single clone of antibody-producing cells that can be grown indefinitely in tissue culture. These cells produce huge quantities of **monoclonal antibodies.**

hydrogenate   To combine with hydrogen (to reduce).

hydrolysis   Chemical decomposition in which a compound is broken down and changed into other compounds by taking up the elements of water. The two resulting products divide the water, the hydroxyl group (OH) being attached to one and the hydrogen atom (H) to the other.

hydroponics   Literally, water plants. These include grains grown by sprouting in chambers under conditions of controlled temperature and humidity to provide a source of green feed at times when it is not possible to produce it normally.

hygrometer   An instrument used for determining the relative humidity of the air.

hyper-   A prefix signifying above, beyond, or excessive.

hyperemia   Congestion, or an excess of blood in any body part.

hyperesthesia   Excessive sensitivity of nerves.

hyperplasia   The abnormal multiplication or increase in the *number* of normal cells in normal arrangement in a tissue.

hyperploid   A plant or animal whose chromosome number is greater than a whole-number multiple of the haploid number, e.g., $2n + 1$ or $2n + 2$.

hyperpnea   Panting (as in dogs); deep breathing.

hypertrophy The morbid (diseased) enlargement, or overgrowth, of an organ or part due to an increase in the *size* of its constituent cells.

hypo- A prefix signifying under, beneath, or deficient.

hypocalcemia A significant decrease in the concentration of ionic calcium, which results in convulsions, as in tetany or parturient paresis.

hypophysectomy Surgical removal of the pituitary gland (hypophysis).

hypothalamic releasing factors Substances secreted by the hypothalamus that regulate the release of hormones from the anterior pituitary gland.

hypothermia The lowering of body temperature, as in treatment of extremely high fever. The patient is given a sedative to inhibit shivering. Then, under minimal anesthesia, he or she is placed in ice water or may have ice packs applied. The rectal temperature may drop to about 86° F in 1 to 2 h. Subnormal body temperature is often induced artificially to facilitate heart surgery.

hypotrophy Degeneration, or loss of vitality, or an organ.

hysterectomy Partial or total removal of the uterus.

ice milk A frozen product resembling ice cream, except that it contains less fat (2 to 5 as opposed to 10 percent) and more **NMS** (12 as opposed to 10 percent) than ice cream. Both ice milk and ice cream contain a stabilizer or emulsifier and about 15 percent sugar.

identical twins Two individuals that developed from the same fertilized egg. The egg separated into two parts shortly after fertilization. If separation is late, Siamese (joined) twins or one individual with two heads or two bodies (etc.) may result. Identical twins are genetically alike. Therefore all differences between them represent environmental effects.

idiopathy A morbid (diseased) state of spontaneous origin.

i.e. That is, from the Latin *id est.*

ileocecal valve Valve at the junction of the lower end of the small and large intestines (junction of ileum and cecum).

imbibe To drink or inhale; to absorb moisture.

imitation milks Mixtures of nondairy ingredients (ingredients other than milk, milk fat, and nonfat milk solids) that are combined, forming a product similar to milk, low-fat milk, or skim milk. Sodium caseinate, although derived from milk, is commonly termed a nondairy ingredient and is often used as a source of protein in imitation milks. Vegetable oils are commonly used as the source of fat.

immunity The power that an animal has to resist and/or overcome an infection to which most or many of its species are susceptible. *Active* immunity is attributable to the presence of antibodies formed by an animal in response to an antigenic stimulus. *Passive* immunity is produced by the administration of preformed antibodies. (See also Section 18.2.5.)

immunize To render an animal resistant to disease by vaccination or inoculation.

immunoglobulins A family of proteins in body fluids that have the property of combining with **antigen** and, in the situation in which the antigen is pathogenic, sometimes inactivating it and producing a state of immunity. Also called *antibodies.*

immunology The study of resistance to disease; the science dealing with the nature and causes of immunity from diseases.

impaction Constipation. See **compaction.**

impregnate To fertilize a female animal.

inbreeding Production of offspring from parents more closely related than the average of a population. Genetically, inbreeding increases the proportion of homozygous genes in a population. For example, in a population of 100 individuals that are all heterozygous *(Aa),* none would be homozygous. However, in a population in which 50 were homozygous *(AA* or *aa)* and 50 were heterozygous *(Aa),* the percent homozygous pairs of genes at this locus would be 50.

incaparina   A so-called food substitute that includes corn, cottonseed meal, and torula yeast. First marketed in 1957, consumer acceptance has been slow.

incomplete dominance   A situation in which neither allele is dominant to the other, with the result that both are expressed in the phenotype, which is intermediate between the two traits.

independent assortment   A situation in which the separation or segregation of one pair of alleles has no effect on any other; occurs when the different pairs of alleles are on nonhomologous chromosomes.

indirect calorimetry   Measurement of the amount of heat produced by a subject by determination of the amount of oxygen consumed and the quantity of nitrogen and carbon dioxide eliminated. The determination of heat production from the respiratory exchange.

inert matter   Any material included in the diet that has no nutritive value. It may be used as a carrier for micronutrients.

infection   The invasion and presence of viable bacteria, viruses, parasites, etc., in a host that result in disease.

infestation   An invasion of the body by arthropods, including insects, mites, and ticks.

inflammation   The reaction of a tissue to injury, which tends to destroy (through increased white blood cells) or limit the spread of the injurious agent. It is characterized by pain, heat, redness, and swelling.

inflection   The point in the growth curve of plants and animals at which growth rate ceases to increase and from which it begins to decrease.

in foal   Designating a pregnant mare. Likewise, *in calf* refers to a pregnant cow.

ingest   To take in food for digestion via the mouth; to eat.

ingesta   Food or drink taken into the stomach.

inguinal   Pertaining to the groin.

inhibine   A substance found in human sputum that is active against a number of bacteria.

in milk   Designating a lactating female.

innervation   The distribution of nerves to an organ or body part.

innocuity   Harmlessness; some vaccines have more innocuity and efficacy than others.

inorganic   Pertaining to substances not of organic origin (not produced by animal or vegetable organisms).

insect   An air-breathing animal (phylum Arthropoda) that has a distinct head, thorax, and abdomen.

insecticide   A substance, such as a stomach poison, contact poison, or fumigant, that kills insects by chemical action.

insensible   Commonly used to denote insensible heat losses, e.g., water vapor and carbon dioxide. Heat dissipation by vaporization.

insidious   More dangerous than is apparent.

in situ   In the natural or normal place; the normal site of origin.

instar   Any one of the larval stages of an insect between molts.

integrated reproduction management (IRM)   Reproductive efficiency in livestock is affected by many factors such as nutrition, genetics, disease, and physiological conditions. IRM is a relatively new concept of integrating, on a "systems" or whole-animal-management basis, the various aspects of animal health, production, and genetics into an overall workable management system.

integument   A covering layer, as the skin of animals or the body wall of an insect.

inter-   A prefix meaning between.

intercellular   Situated between the cells.

intermediate host   A host in which a parasite develops only in part, before getting into its final (or definitive) host, where it develops to maturity.

intermuscular   Situated between muscles.

interphase   The period between successive mitotic divisions and the period of growth and usually duplication of chromosomes.

intestine   The lower portion of the alimentary canal from the stomach to the anus. Also called the *gut* or *bowels.*

intra-   A prefix meaning "within."

intracaudal   Situated or applied within the tissues of the tail.

intracellular   Situated or occurring within a cell or cells.

intracranial   Within the cranium.

intracutaneous   Into or within the layers of skin.

intradermal   Within the dermis. An injection into the layers of skin.

intramammary   Within the mammary gland, as injection into a mammary gland through a teat opening.

intramuscular   Within the substance of a muscle, e.g., an injection into muscle.

intraperitoneal   Within the cavity of the body that contains the stomach and intestines. Administration through the body wall into the peritoneal cavity.

intrasternal   Within the sternum.

intravenous   Within a vein or veins.

in utero   Within the uterus (intrauterine).

inversion   Rearrangement of genes on a chromosome in such a way that their order is reversed or different.

in vitro   Within an artificial environment, as within a glass or test tube.

in vivo   Within the living body.

involution   The return of an organ to its normal size or condition after enlargment, as of the uterus after childbirth. A decline in size or activity of other tissues; e.g., the mammary gland tissues normally involute with advancing lactation. The drying-off process of lactating cows.

iodinated casein   Milk protein (casein) that has been treated with iodine. It has the same physiological effect as thyroxine (hormone produced by the thyroid gland). It is commonly referred to as *thyroprotein* and is sometimes used to stimulate cows to secrete more milk.

iodine value   The degree of unsaturation of the fatty acids in a fat or oil can be quantitatively expressed as the iodine value, which refers to the number of grams of iodine absorbed by 100 grams of fat. Since the iodine reacts at the sites of unsaturation, much as would hydrogen in hydrogenation, the higher the iodine value the greater the degree of unsaturation that existed in the fat.

ion   An atom or a group of atoms (molecules) carrying an electric charge, which may be positive or negative. Ions are usually formed when salts, acids, or bases are dissolved in water.

ionization   The adding of one or more **electrons** to, or removing one or more electrons from, atoms or molecules, thereby creating ions.

ionizing radiation   Any radiation displacing electrons from atoms or molecules, thereby producing **ions**, e.g., alpha, beta, and gamma radiation and short-wave ultraviolet light. Ionizing radiation may produce severe skin or tissue damage.

IRM See **integrated reproduction management.**

irradiation The process of exposing (treating) materials (as in a nuclear reactor) to roentgen rays (x- or γ-radiation) or other forms of radioactivity. An example of practical significance to nutrition is the application of ultraviolet rays to a substance to increase its vitamin D efficiency.

irritability The ability to respond to stimuli.

is by An indication of the male parent, referring to progeny.

iso From the Greek word *isos,* meaning "equal." A prefix or combining form meaning "equal," "alike," or "the same"; e.g., "isocaloric" refers to the *same* caloric value.

isotonic Characterized by equal osmotic pressure; e.g., a solution containing just enough salt to prevent the destruction of the red corpuscles when added to the blood would be considered isotonic with blood. Describing a solution having the same concentration as the system or solution with which it is compared.

isotope An element of chemical character identical with that of another element occupying the same place in the periodic table (same atomic number), but differing from it in other characteristics, as in radioactivity or in the mass of its atoms (atomic weight). Isotopes of the same element have the same number of protons in their nucleus but different numbers of neutrons. A *radioactive* isotope is one with radioactive properties. Such isotopes may be produced by bombarding the element in a cyclotron.

isotope enrichment A process by which the relative abundances of the isotopes of a given element are altered by increasing (enriching) the concentration of one form.

-itis A word termination (suffix) denoting *inflammation* of a part indicated by the word stem to which it is attached.

IU International unit. A unit of measurement of a biological (e.g., a vitamin, hormone, antibiotic, antitoxin) as defined and adopted by the International Conference for Unification of Formulas. The potency is based on the bioassay that produces a particular biological effect agreed on internationally. See **USP.**

jack A male uncastrated donkey (ass).

jenny A female donkey (ass). Also called a *jennet.*

jerky Long thin strips of sun-dried beef or lean meat.

jowl Meat taken from the cheeks of a hog.

karyotype The chromosomes of a plant or animal as they appear at the metaphase of a somatic division. Arrangement by pairs, size, and centromere location when presented for a species.

katydid A long-horned grasshopper of the family Tettigoniidae.

keel bone The breastbone of birds; the sternum.

keratin An insoluble complex protein that constitutes hair, horn, claws, and feathers.

keratosis Any horny growth, such as a wart, causing the cornification, or hardening, of the epithelial skin layers.

ketosis See **acetonemia.**

kid A young goat or antelope up to the first birthday. (See Appendix A.)

killed virus A virus whose infectious capabilities have been destroyed by chemical or physical treatment.

kilo- A prefix that multiplies a basic unit by 1000.

kilocalorie (kcal) Equivalent to 1000 small calories.

kip or kipskin Light rawhide from a grass-fed, immature bovine animal (between the size of a calf and a mature animal).

Kjeldahl   Relating to a method of determining the amount of nitrogen in an organic compound. The quantity of nitrogen measured is then multiplied by 6.25 to calculate the protein content of the food or compound analyzed. The method was developed by the Danish chemist J. G. C. Kjeldahl in 1883.

kwashiorkor   A syndrome produced by a severe protein deficiency, with characteristic changes in pigmentation of the skin and hair, edema, anemia, and apathy.

label.   See **tracer isotope.**

labile   Changeable or unstable. Readily or continuously undergoing chemical, physical, or biological change or breakdown.

lactate   To secrete or produce milk.

lactation period   The number of days an animal secretes milk following each parturition.

lacteal   Pertaining to milk.

lactogenic   Stimulating the secretion of milk.

lactometer   An instrument used to determine the specific gravity of milk, providing information for calculating the percentage of solids in a sample of milk.

lamb   A sheep less than 12 months old. To give birth to a lamb.

lamb hog   A male lamb from weaning time until it is shorn.

larva (pl. larvae)   The immature form of insects and other small animals, which is unlike the parent or parents and which must undergo considerable change of form and growth before reaching the adult stage; e.g., white grubs in soil or decayed wood are larvae of beetles. Caterpillars, maggots, and screwworms are also larvae.

larvicide   A chemical used to kill the larval or preadult stages of parasites.

larviparous   Bearing and bringing forth young that are larvae (especially of insects).

larynx   The upper portion of the windpipe (trachea).

latent   Hidden or not apparent.

lateral   Of, at, from, or toward the side or flank.

laying   The expulsion of an egg. The term is commonly associated with hens in active egg production. See **oviposition.**

LD$_{50}$ The lethal dose for 50 percent of the animals tested.

legume   Refers to those crops that can absorb nitrogen directly from the atmosphere through bacteria that live in their roots. The clovers and alfalfa are common examples of legumes.

lesion   Injury or diseased condition reflected in discontinuity of tissues or organs, often causing loss of function of a part.

lethal   Deadly; causing death.

lethal gene   A gene, or genes, causing the death of an individual. Most lethal genes are recessive or partially dominant, requiring two genes of the same kind in a pair (homozygous) to cause death.

leukocytes (leucocytes)   The white blood cells. They have amoeboid movements and include the lymphocytes, monocytes, neutrophils, eosinophils, and basophils.

libido   Sexual desire or instinct.

lice   Small nonflying biting or sucking insects that are true parasites of humans, animals, and birds.

life cycle   The changes in form and mode of life that an organism goes through between recurrences of the same primary stage; life history.

ligaments   Tissues connecting bones and/or supporting organs.

ligate   To tie up, or bind, with a ligature.

lignin   A compound that in connection with cellulose forms the cell walls of plants and thus of wood. It is practically indigestible.

limited feeding   Feeding animals to maintain weight and growth but not enough to fatten or increase production. Feeding animals less than they would like to eat.

limiting amino acid   The essential amino acid of a protein that shows the greatest percentage deficit in comparison with the amino acids contained in the same quantity of another protein selected as a standard.

limnology   The study of inland waters (lakes and ponds), especially with reference to their biological and physical features.

linebreeding   A form of inbreeding in which an attempt is made to concentrate the inheritance of some ancestor in the pedigree.

linecross   A cross of two inbred lines.

linkage   Refers to two or more genes carried on the same chromosome.

linkage group   All the genes having loci on a particular chromosome.

linked genes   Sets of genes that tend to be inherited together and that are presumed to have their loci on the same chromosome.

lipid   Any one of a group of organic substances that are insoluble in water but are soluble in alcohol, ether, chloroform, and other fat solvents and have a greasy feel. They include fatty acids and soaps, neutral fats, waxes, steroids, and phosphatides (by United States terminology).

Lippizan   A famous breed of Austrian horses. They are milk white in color at maturity.

lipolysis   The hydrolysis of fats by enzymes, acids, alkalis, or other means to yield glycerol and fatty acids.

litter   The pigs farrowed by a sow or the pups whelped by a bitch at one delivery period. Such individuals are called *littermates.* Also, the accumulation of materials used for bedding farm animals.

livability   The inherited stamina, strength, and ability to live and grow.

livestock   Domestic farm animals kept for productive purposes (meat, milk, work, and wool); include beef and dairy cattle, sheep, swine, goats, and horses. Also called *stock.*

local infection   An infection restricted to a small area, as an ear infection.

locus   Region of a chromosome where a particular gene is carried or located. The segment or part of a chromosome concerned with regulating a particular trait.

loin   That portion of the back between the thorax and pelvis.

long feed   Coarse, or unchopped, feed for livestock, such as hay, as contrasted to *short,* or *chopped,* feed.

lope   A slow gallop of a horse.

low-fat milk   Milk that contains between 0.5 and 2 percent milk fat.

low-level counting (low-level analysis)   A procedure to measure the radioactive content of materials with very low levels of activity, using sensitive detecting instruments and good shielding to eliminate the effects of *background radiation* and cosmic rays.

low-moisture silage   Silage that contains 35 to 55 percent moisture. (See also **haylage.**)

low-set   Designates a short-legged animal.

lucerne   Alfalfa; a legume of high feeding value for ruminants.

lumbar   Pertaining to the loins.

lumen   The cavity on the inside of a tubular organ, e.g., the lumen of the stomach or intestine.

luteal phase   The stage of the estrous cycle at which the corpus luteum is active and progesterone influence predominates.

lymph   A transparent, slightly yellow liquid found in the lymphatic vessels. It may have a light-rose color due to the presence of red blood corpuscles.

lymphocyte A variety of white blood corpuscles that originate from the lymph glands.

lyophilization The evaporation of water from a frozen product with the aid of high vacuum. Also called *freeze-drying.*

lysin Antibody that causes the death and dissolution of bacteria, blood corpuscles, and other cellular elements.

lysozyme A substance present in human nasal secretions, tears, and certain mucus. It is also present in egg whites, in which it hydrolyzes polysaccharidic acids. It is bactericidal for only a few saprophytic bacteria and is inactive against pathogens and organisms of the normal flora.

macro Large or major. Abnormal size or length. Usually a prefix.

macrocyte An abnormally large erythrocyte.

macrophages Large phagocytes. Large scavenger cells, the function of which is primarily phagocytosis and the destruction of many kinds of foreign particles. These cells are strategically located in the spleen, liver, bone marrow, small blood vessels, and connective tissue. Collectively, they compose the reticuloendothelial system.

macroscopic Visible to the unaided, or naked, eye.

maggot The larva of a fly.

maiden An unbred animal.

malnutrition Any disorder of nutrition. Commonly used to indicate a state of inadequate nutrition.

mammal Any animal that suckles or provides milk for its young.

mammilla The nipple of the female breast.

mandible Bone of the lower jaw.

manufacturing milk Milk used for the manufacture of dairy products, such as cheese, butter, powdered milk, and ice cream.

manure Excreta of animals, dung and urine (commonly with some bedding).

marasmus From the Greek *marasmos,* meaning "a dying away." A progressive wasting and emaciation. *Enzootic* marasmus is a condition of malnutrition in domestic animals due to a deficiency of one or more trace elements, especially cobalt and copper.

marbling The distribution of fat in muscular tissue that gives meat a spotted appearance.

mare A mature female horse (usually more than 4 years of age).

market milk Milk that is consumed as fresh fluid milk. See **fluid milk.**

marsupial One of a class of mammals characterized by the possession of an abdominal pouch in which the young are carried for some time after birth.

marsupilia Pouched mammals, e.g., the kangaroo.

masculinity Physical appearance resulting from well-developed secondary male sex characteristics.

mass spectrometer An instrument for separation and measurement of isotopes by their mass.

mastication The chewing of food.

mastitis An inflammation of the mammary gland(s).

maternal Pertaining to the mother.

matrix The basic material from which a thing develops. A place or point of origin of growth.

matron A mare that has produced a foal.

maturation The process of becoming mature.

mature equivalency (ME) Age-conversion formulas (provided by the USDA and breed associations) applied to the milk production records of young cows to predict their expected *mature milk production* potential. Breed ME factors are used for comparative purposes in selecting and mating animals.

maverick   An unbranded animal, particularly a calf (also refers to a motherless calf in some areas). A **dogie.**

maxilla   The upper jawbone in vertebrates.

meat analogues   Material usually prepared from vegetable protein to resemble specific meats in texture, color, and flavor.

meat bird   A fowl raised for its meat as contrasted to one kept for egg laying.

meconium   The first excreta of a newborn animal.

median   Situated in the middle; mesial.

median plane   A line or plane (from cranial to caudal) dividing an animal into two equal halves.

megacalorie (Mcal)   Equivalent to 1000 kcal or 1,000,000 cal. A megacalorie is equivalent to a **therm.**

meiosis   A type of cell division that produces the sex cells, or gametes, possessing the $1n$ (haploid) number of chromosomes. Thus the chromosome number of daughter cells is reduced to one-half of the somatic number (chromosome number is reduced by one-half from diploid to haploid).

melanin   A dark-brown or black pigment found in hair and/or skin.

meninges   The membranes that cover the brain and spinal cord.

mesenchyme   Embryonic connective tissue that gives rise to the connective tissues of the body and blood vessels.

mesentery   A membrane that supports a visceral organ, particularly the intestine, and contains the vessels and nerves that supply that organ.

meta-   A prefix meaning "between" or "among"; indicating change, transformation, or exchange; after or next.

metabolic body size   The weight of the animal raised to the 0.75 power ($W^{0.75}$).

metabolism   The sum total of the chemical changes in the body, including the building up (anabolic, assimilation) and the breaking down (catabolic, dissimilation) processes. The transformation by which energy is made available for body uses.

metabolite   Any substance produced by metabolism or by a metabolic process.

metabolizable energy (ME)   The food-intake gross energy minus fecal energy minus energy in the gaseous products of digestion (largely methane) minus urinary energy. For birds and monogastric mammals, the gaseous products of digestion need not be considered. See also **physiological fuel value.**

metamorphosis   A change in shape or structure involving a transition from one developmental form to another, as in insects and frogs.

metaphase   The second stage of mitosis, characterized by alignment of chromosomes on the equator of spindle fibers.

metritis   An inflammation of the uterus.

MF   Milk fat.

micro   Small or minor. Usually a prefix, designating *tiny* or *microscopic in size.* Also, a prefix that divides a basic unit by 1 million.

microbiological assay   The use of microorganisms to assay. See **bioassay.**

microcurie   The amount of radioactive material that has the same intensity of radiation as one-millionth of a gram of radium.

microflora   The flora consisting of microorganisms. Commonly used in reference to the bacteria populating the rumen. (Protozoa are also present in the rumen.)

micromicro   See **pico.**

microphages Little phagocytes. Polymorphonuclear leukocytes that are formed primarily in bone marrow. They are particularly active during bacterial infections.

microscopic Invisible to the unaided eye.

milk equivalent The quantity of milk, as produced, required to furnish the milk solids in manufactured dairy products to be sold: e.g., approximately 10 and 20 lb of whole milk is required to manufacture 1 lb of cheddar cheese and butter, respectively.

milk fat The lipid components of milk.

milk serum The nonfat components of milk.

milli- A prefix that divides a basic unit by 1000.

miracidium The free-swimming larva of a trematode that penetrates the body of a snail host for further development into a **cercaria.**

miscible Designating two or more substances that when mixed together form a uniformly homogeneous solution.

miticide A compound that is destructive to mites. An acaricide.

mitosis A type of cell division in which cells with the $2n$ (diploid) number of chromosomes produce daughter cells that also possess the $2n$ (diploid) number of chromosomes. Thus the daughter cells receive the full complement of chromosomes existing in the original cell before division.

mitotic centers Two (usually) polarizing units of cells located opposite each other, like the spindle fibers oriented between the poles of a cell.

mode A statistical term referring to the value (number) that occurs the greatest number of times in a frequency distribution.

modified live virus A virus that has been changed by passage through an unnatural host, such as hog cholera virus passed through rabbits, so that it no longer possesses pathogenic characteristics but will stimulate antibody production and immunity when injected into susceptible animals.

molecule A group of atoms held together by chemical forces. A molecule is the smallest unit of matter that can exist by itself and retain all its chemical properties.

molt (molting) The shedding and replacing of feathers (usually in the fall). Snakes and certain arthropods also shed their outer covering and develop a new one.

mongrel Animal of mixed or unknown breeding.

monoclonal antibodies These are antibodies with specificity against only one set of antigenic determinants. They are produced in large quantities by hybridomas and have virtually revolutionized immunology. Specific *monoclonal antibody* has been used successfully for immunotherapy of cancer patients. Such antibody can also be used to develop immunodiagnostic techniques. See **hybridomas.**

monoestrous animal An animal that has only one estrous (heat) cycle annually.

monogastric Having only one stomach or stomach compartment, as do dogs, humans, and swine.

monohybrid A trait in an individual controlled by a single pair of genes with a genetic makeup of *Aa* or *Bb,* etc.

monohybrid cross A cross involving one pair of alleles, each of which controls an alternative form of the same characteristic.

monorchid A male animal that has only one testicle in the scrotum. Also called a *ridgling.*

monotreme Any member of the lowest order of mammals (Monotremata), comprising the duckbill platypus and the echidnas, which lay eggs and have a common opening for the genital and urinary organs and the digestive tract. They nourish their young by a mammary gland that has no nipple, in a shallow pouch developed only during lactation.

monoxenous parasite   A parasite that requires only one host for its complete development.

morbid   Diseased or unhealthy.

morphogenesis   The origin and evolution of morphological characters (form and structure). The establishment of shape and patterns.

morphology   The science of the forms and structures of animal and plant life without regard to function.

mortality   Death. Death rate.

mount   To copulate with, as certain male animals mount a female in the act of coitus.

mule   The cross resulting from mating a mare horse with a jack (male ass).

muley   A **polled** cow.

multiparous   Producing many (more than one) offspring at one time.

multiple alleles   Two or more alleles at the same locus on one pair of homologous chromosomes, affecting the same trait but in a different way. For example, in humans, gene *A* produces antigen A; gene *B*, antigen B; and gene *a*, neither antigen (or the O blood type). All three alleles may be at the same locus in a population, but only one is in a gamete and only two are in body cells. Thus, *multiple alleles* describes a condition in which three or more forms of the same gene exist, any one of which may occupy a specific locus at any given time.

mummified fetus   A shriveled or dried **fetus** that has remained in the uterus instead of being aborted or expelled.

musculature   The muscular system of any body part.

mustang   A wild horse, ridden by cowboys of the western plains, that descended from Spanish horses. The name is derived from the Spanish *mustengo*, meaning "wild."

mutation   A change in a gene, often resulting in a different phenotype. More specifically, a change in the code sent by a gene, causing the formation of a different protein by the cytoplasm of the cell. A permanent, transmissible change in the characteristics of an offspring from those of its parents.

mutton   The flesh of a grown sheep as opposed to that of a lamb. Goat meat is also called *mutton* in some countries.

myalgia   Muscular pain or rheumatism.

mycotoxins   Toxic metabolites produced by molds during growth on a suitable substrate.

myiasis   A disease due to the presence of fly larvae in warm-blooded animals.

myositis   Inflammation of a voluntary muscle.

nag   A horse or pony of nondescript breeding.

nanism   Dwarf growth.

nano-   A prefix that divides a basic unit by 1 billion (10^9).

nape   Back of the neck of an animal.

National Research Council   See **NRC.**

natural immunity   Immunity to a disease, infestation, etc., which results from qualities inherent in an animal. See **immunity.**

natural radiation   Background radiation.

natural service   In farm animals, it means to allow natural mating, as opposed to artificial insemination.

NDM   Nonfat dry milk. The product obtained by removing water from pasteurized skim milk. It contains not more than 5 percent moisture and not more than 1.5 percent MF unless otherwise indicated.

neat's-foot oil   A yellowish oil prepared by boiling the bones and joints of cattle (and sometimes of horses and sheep) and skimming off and clarifying the oil obtained.

necropsy An examination of the internal organs of a dead body to determine the apparent cause of death. Also called *autopsy, postmortem.*

necrosis Death of tissue, usually in individual cells, groups of cells, or small localized areas.

neigh The characteristic cry of a horse.

neonate A newborn infant.

net energy (NE) The difference between ME and the heat increment; includes the amount of energy used either for maintenance only or for maintenance plus production.

net energy for maintenance (NE$_m$) The portion of net energy expended to keep the animal in energy equilibrium. (There is no net loss or gain of energy in the body tissues.)

net energy for production (NE$_p$) The portion of net energy required in addition to that needed for body maintenance that is used for work or for tissue gain (growth and/or fat production) or for the synthesis of productive end products (a fetus, milk, eggs, wool, fur, or feathers).

neutron (symbol *n*) An uncharged elementary particle with a mass slightly greater than that of the **proton,** found in the **nucleus** of every atom heavier than hydrogen. Neutrons sustain the fission chain reaction in a **nuclear reactor.**

neutron activation analysis Activation analysis in which neutrons are the activating agent.

NFE See **nitrogen-free extract.**

NFS Nonfat solids of milk. They comprise proteins, lactose, minerals, and other water-soluble constituents of milk. Also called *SNF* (solids-not-fat) and *NMS*.

nicking Breeding of progeny that are superior to their parents. Also called **heterosis.**

nit The egg of a louse or similar insect.

nitrogen balance The nitrogen in the food intake minus the nitrogen in the feces minus the nitrogen in the urine. Nitrogen retention.

nitrogen-free extract (NFE) Comprises the carbohydrates, sugars, starches, and a major portion of the material classed as hemicellulose in feeds. When crude protein, fat, water, ash, and fiber are added and the sum is subtracted from 100, the difference is the NFE.

NMS Nonfat milk solids. See *NFS*.

nocturnal Of the night; a nocturnal parasite is one that is active at night.

nondisjunction Failure of a pair of homologous (sister) chromosomes (dyads) to separate during the reductional division of **meiosis**. Thus both chromosomes go into a gamete (usually only one of each pair goes into a gamete).

nonfat dry milk See **NDM.**

nonreturn The breeding efficiency of bulls expressed as the percentage of cows that conceive on the first service.

nonruminant An animal without a rumen, e.g., a chicken or a pig.

notching Cutting dents in the ears of animals for identification.

noxious Harmful, not wholesome.

NPN Nonprotein nitrogen (e.g., urea).

NRC National Research Council. A division of the National Academy of Sciences established in 1916 to promote effective utilization of the scientific and technical resources available. This private, nonprofit organization of scientists periodically publishes bulletins giving the nutrient requirements of domestic animals. Copies are available through the National Academy of Sciences–NRC, 2101 Constitution Avenue NW, Washington, D.C. 20418.

nuclear reactor A device in which a fission chain reaction can be initiated, maintained, and controlled. Its essential component is a core with fissionable fuel.

nucleoside A compound composed of a nitrogen base and a sugar.

nucleotide A compound composed of a nitrogen base, a sugar, and a phosphate.

nucleus A deep-staining body within a cell, usually near the center; the heart and brain of the cell, containing the chromosomes and genes. Also, the small, positively charged core of an **atom.** All nuclei contain both **protons** and **neutrons,** except the nucleus of ordinary hydrogen, which consists of a single proton.

nuclide A general term applicable to all atomic forms of the elements.

nulliparous Having never given birth to viable young.

nurse To suckle; to give milk to (e.g., a baby).

nurse cow A milk cow used to supply milk for nursing calves other than her own.

nutrient A substance (element or ingredient) that nourishes the metabolic processes of the body. It is one of the many end products of digestion.

nutrient-to-calorie ratio An expression of nutrients in weight per unit of energy needed. For example, the protein-to-calorie ratio is expressed as the grams of protein ($N \times 6.25$) per 1000 kcal metabolizable energy (grams of protein per 1000 kcal ME). This same dimension may be extended to other nutrients such as grams of calcium per 1000 kcal.

nutrilite A nutritional element.

nutriment That which is required by an animal as a building material and fuel (nourishment).

nutrition The science encompassing the sum total of processes that have as a terminal objective the provision of nutrients to the component cells of an animal.

nutritive ratio (NR) The ratio of digestible protein to other digestible nutrients in a foodstuff or diet. (The NR of shelled corn is about 1:10.)

nymph The immature stage of insects having only three stages (egg, nymph, and adult) in their development. Nymphs resemble adults in form and appearance (as contrasted with larvae, which do not resemble their adults) but do not have wings.

obligate parasite A parasite incapable of living without a host.

occult Obscure; concealed from observation; difficult to understand.

oestrus See **estrus.**

offal The internal fat of cattle. The viscera and trimmings of a slaughtered animal removed in dressing. Also, the by-products of milling used especially for livestock feeds.

off feed Having ceased eating with a healthy and normal appetite.

official production record Standard DHIA and DHIR records pertaining to milk production that are made under the supervision of an unbiased individual. Records are used for management purposes (e.g., feeding and culling), genetic evaluation (sire summaries), publicity, and sales. See **DHIR.**

offspring The sons and daughters of parents.

oil gland Gland at the base of the tail in chickens, ducks, turkeys (and most wild birds) that secretes an oil used by birds in preening their feathers. Also called *preen gland.*

olfactory Pertaining to the sense of smell.

omasum The third division of the stomach of ruminant animals. Also called *manifold, manyplies,* and *psalterium.*

omnivore An animal that subsists on feed of every kind (plant and animal), as with humans.

onager Wild ass of southwestern Asia.

on the hoof Designating a living meat animal.

opaque Not letting light through; neither transparent nor translucent.

open A term commonly used of farm mammals to indicate a nonpregnant status.

opisthotonos A form of tetanic spasm in which the head and heels are bent backward and the body bowed forward.

opsonins Antibodies of blood serum that sensitize (weaken) the cells of microorganisms so that they are readily ingested, or engulfed, by the phagocytic body cells (white blood cells).

oral Pertaining to the mouth. See **per oral** and **per os.**

orchidectomy Surgical removal of the testes.

orchitis An inflammation of a testis.

organic Pertaining to substances derived from living organisms. Referring to carbon-containing compounds.

organism Any complete living plant or animal.

organogenesis The origin or development of the organs of an animal.

orifice Entrance or outlet of a body cavity.

orthopnea Inability to breathe except in the upright position. This state is common during dehydration exhaustion.

osmosis The tendency of two fluids of different strengths that are separated by something porous (a semipermeable membrane) to diffuse or spread through a membrane or partition until they are mixed. Osmosis is the chief means by which the body absorbs food and is, specifically, the tendency of a fluid of lower concentration to pass through a semipermeable membrane into a solution of higher concentration.

osmotic pressure The force acting on a semipermeable membrane placed between a solution and its pure solvent, caused by the flow of solute molecules through the membrane toward the pure solvent.

osteofibrosis A loss of calcium salts from the bones that causes them to become fragile. It occurs chiefly in horses but may affect pigs, goats, and dogs.

osteogenesis Formation of bone.

osteomalacia A condition marked by softening of the bones, pain, tenderness, muscular weakness, and loss of weight. It results from a deficiency of vitamin D or of calcium and phosphorus. May also be caused by an overactive parathyroid gland.

osteoporosis A reduction in total bone mass. This disorder of bone metabolism occurs in middle life and older age in both men and women. The bone becomes porous and thin due to a failure of the osteoblasts (bone-forming cells) to lay down bone matrix. This disorder may result from (1) a dietary deficiency of calcium and/or protein, (2) a lowered calcium absorption, or (3) a hormonal disturbance.

osteosclerosis Abnormal hardening of bone.

outcross Mating of an individual to another in the same breed that is not closely related.

out of Refers to *mothered by* in animal breeding.

ovary Female reproductive gland in which the ova (eggs) are formed.

overdominance A type of gene expression in which the heterozygote *Aa* is superior to either homozygote *AA* or *aa*. Interaction between genes that are alleles.

ovicide Any substance that kills parasites in the egg stage.

ovine An animal of the subfamily Ovidae; sheep, goats.

oviparous Producing offspring from eggs that hatch outside the body.

oviposition The laying (expelling) of a fully developed egg.

ovoviviparous Producing eggs within the maternal body. The eggs hatch within or immediately after extrusion from the parent.

ovulation The shedding of a follicle by the ovary. The ovary of a hen contains a series of follicles (called the *follicular-size hierarchy*). After ovulation, the smaller follicles advance one position in size and reestablish the hierarchy as it existed just before ovulation.

ovum The female reproductive cell (gamete), which, after fertilization, develops into a new member of the same species. The male gamete is the *sperm*.

owner sampler (OS) An unofficial milk production record system, the records of which originate with the breeder and are summarized and supervised by the USDA and the colleges of agriculture throughout the United States.

ox (pl. oxen) Any species of the bovine family of ruminants. Specifically, a domesticated and castrated male bovine used for work purposes.

oxidase Enzyme that activates oxygen.

oxidation Chemically, the increase of positive charges on an atom or the loss of negative charges. There may be a loss of one electron (univalent O) or two electrons (divalent O). The combining of oxygen with another element to form one or more new substances. Burning is one kind of oxidation. Also called oxydation.

packer One who operates a slaughter and meat-processing business.

paint A coat color of horses (white patches interspersed with darker colors, usually black); also, a breed of horses.

paired feeding (food equalizing) A method of comparing nutritional effects at an arbitrary low level set by the animal that consumes the least food. Littermates or twins (especially monozygous ones) are considered best for paired-feeding studies.

palatability The relative relish with which feeds are consumed by animals.

paleontology The study of fossil remains.

palomino Color of horses: golden coat with light-blond or silvery mane and tail.

pandemic Prevalent (as a disease) throughout an entire country or continent or the world.

Papanicolaou stain A method of staining smears of various body secretions from the respiratory, digestive, or genitourinary tracts. It is used to diagnose cancer or the presence of a malignant process. Exfoliated cells of organs, such as the stomach or uterus, are obtained, smeared on a glass slide, and stained for microscopic examination. It was named for the Greek physician George Papanicolaou, who developed it. Also known as **Pap smear**.

parakeratosis Any abnormality of the stratum corneum (horny layer of epidermis) of the skin (especially a condition caused by edema between the cells, which prevents the formation of keratin).

parasite An organism that lives at least for a time on or in and at the expense of living animals.

parasiticide An agent or drug destructive to parasites.

parchment Tanned sheepskins. Vellum is essentially the same as parchment, except that it is made from calfskin. Parchment is used for diplomas, records, banjos, drumheads, lampshades, etc.

parental combinations Genotypes and phenotypes like those of the parents in a particular cross.

parental generation The first generation in a particular genetic experiment; frequently purebreeding lines.

parenteral Pertaining to administration by injection, not through the digestive (food) tract, i.e., such as subcutaneous, intramuscular, intrasternal, intravenous.

paresis Partial paralysis that affects the ability to move but not the ability to feel.

parrot mouth A malformed mouth of an animal (most common in horses) in which the upper jaw abnormally protrudes beyond the lower jaw.

parthenogenesis (parthogenesis)   Reproduction by the development of an egg without its being fertilized by a spermatozoon, e.g., drone bees. It occurs in certain lower animals and has been observed in turkeys. It does not occur in mammals.

partial dominance   A situation in which one gene of a pair of alleles is not completely dominant with respect to another. For example, in comprest Hereford cattle an individual that possesses no comprest genes or is of genotype *cc* is of normal size, those that possess two comprest genes *CC* are dwarfs, and those that possess one comprest gene *Cc* are midway in size between dwarfs and normals.

particle   A minute constituent of matter. The primary particles involved in radioactivity are **alpha particles, beta particles, neutrons,** and **protons.**

parturient paresis   A condition caused by a low blood-calcium concentration that results in partial to complete paralysis soon after parturition.

parturition   The act or process of giving birth to young.

passerine   Belonging to or having to do with the very large group of perching birds, including more than half of all birds, such as the warblers, sparrows, chickadees, wrens, thrushes, and swallows.

passive immunity   Disease immunity given to an animal by injecting the blood serum from an individual already immune to that disease. See **immunity.**

pasteurization   The process of heating milk to at least 62.8°C (145°F) and holding it at that temperature for not less than 30 min (holding method) or to 71.7°C (161°F) for 15 s (HTST).

pasture   Plants, such as grass, grown for feeding or grazing animals. To feed cattle and other livestock on pasture.

patency   The condition of being open or unobstructed.

patent leather   A term associated with the finish produced by the covering of the surface of leather with successive coats of daub and varnish. Most patent leather is made from cattle hides or kips, although horsehide, goatskin, kidskin, and calfskin are sometimes used.

paternal   Pertaining to the father or male parent.

pathogen   Any disease-producing microorganism or virus.

pathology   The branch of science dealing with disease, especially with structural and functional changes in tissues and organs of the body affected by disease.

paunch (rumen)   The first stomach compartment of a ruminant. See **rumen.**

PD   See **predicted difference.**

pectoral   Pertaining to the breast.

pedigree   A list of an animal's ancestors, usually only those of the five closest generations.

pelt   The natural, whole skin covering, including the hair, wool, or fur of smaller animals, such as sheep and foxes.

penetrance   A genetic term that refers to the percentage of times a phenotype actually shows up when it is expected.

percutaneous   Performed or introduced through the skin, as an injection.

perfusion   The act of pouring through or immersing in a physiological fluid, e.g., blood or saline.

pericardium   The membrane that encloses the cavity containing the heart.

perineum   The anatomical region of the body between the thighs, especially between the anus and the genitals.

periodic table (periodic chart)   A table or chart listing all the elements, arranged in order of increasing **atomic numbers** and grouped by similar physical and chemical characteristics into "periods." The table is based on the chemical law that physical or chemical properties of elements are periodic (regularly repeated) functions of their **atomic weights.**

periosteum The membrane that covers bone.

perissodactyl Having an uneven number of toes on each foot. A hooved animal with an uneven number of toes, such as a horse.

peristalsis The rhythmic contractions and movements of the alimentary canal.

peritoneum The membrane that lines the abdominal cavity and invests the contained viscera (digestive organs).

per oral Administration through the mouth.

per os Oral administration (by the mouth).

perosis A disease of chicks marked by bone deformities and associated with deficiency of certain dietary factors, such as biotin, choline, folic acid, or manganese. Also called *slipped tendon* or *hock disease.*

per se By, of, or in itself. As such.

Persians Crust leathers made from India-tanned hair (as opposed to wool growth) sheepskin. Leather from "bastard skins" (see **bastard**) is sometimes designated as Persian.

pesticide A compound used to control any plant or animal considered to be a pest.

PFV See **physiological fuel value.**

pH A symbol used (with a number) to express acidity or alkalinity in analyzing various body secretions, chemicals, and other compounds. It represents the logarithm of the reciprocal (or negative logarithm) of the hydrogen-ion concentration (in gram atoms per liter) in a given solution, usually determined by the use of a substance (indicator) known to change color at a certain concentration. The pH scale in common use ranges from 0 to 14, pH 7 (the hydrogen-ion concentration, 10^{-7} or 0.0000001, in pure water) being considered neutral; 6 to 0, increasing acid; and 8 to 14, increasing alkali.

phagocytes From the Greek *phago,* meaning "eat," and *kytos,* meaning "cell." Defensive cells (leukocytes, or white blood cells) of the body that ingest and destroy bacteria and other infectious agents. See also **macrophages** and **microphages**.

phagocytosis The engulfing of microorganisms, cells, or foreign particles by phagocytes (certain forms of leukocytes).

pharynx The tube, or cavity, that connects the mouth and nasal passages with the esophagus (throat).

phenocopy A phenotype determined by environment that mimics the same phenotype produced by heredity (genotype).

phenotype Expression of genes that can be measured by the human senses. What is seen in an animal for some trait. For simply inherited traits such as color in Holsteins, either black and white or red and white is seen.

pheromone A substance secreted externally by certain animal species (especially insects) to affect the behavior (especially sexual) or development of other members of the species. The queen substance of honeybees that inhibits ovary development in workers is a pheromone. Also called *assembling scent* and *sex pheromone* in insects.

phoresy That form of symbiosis in which one symbiont rests on or attaches to another for means of transportation.

phosphatide See **phospholipid.**

phospholipid A lipid containing phosphorus that on hydrolysis yields fatty acids, glycerin, and a nitrogenous compound. Lecithin, cephalin, and sphingomyelin are examples. Also called *phosphatide.*

photon The carrier of a quantum (unit quantity of energy) of electromagnetic energy. Photons have an effective momentum but no mass or electric charge.

photoperiodism The physiological response of animals and plants to variations of light and darkness.

physiological fuel value (PFV) Units (expressed in calories) of food energy in human nutrition. It corresponds to **metabolizable energy** as related to domestic animals.

physiological saline A salt solution (0.9 % NaCl) having the same osmotic pressure as the blood plasma.

physiology The science that pertains to the functions of organs, systems, and the whole living body.

phyto- A prefix, meaning "pertaining to plants."

pica A craving for unnatural articles of food, such as is seen in hysteria, pregnancy, and phosphorus deficiency. A depraved appetite.

pico- A prefix; divides a basic unit by 1 trillion (10^{12}). Same as *micromicro*.

piebald A horse having a black coat with white spots.

pig A young swine weighing less than 120 lb.

pigeon-toed Designating a horse or other animal whose feet (toes) turn inward.

piglet A young pig.

pincers The incisor teeth of a horse. Also called *nippers*.

pinocytosis The absorption of liquids by cells.

pinto Designating a horse that has a spotted or piebald coat color (white spots on any dark background).

pipped egg An egg through which the chick has forced its beak in the first step of breaking out of the shell during incubation.

pithing A method of animal slaughter in which the spinal cord is severed to cause death and/or to destroy sensibility.

Pituitrin Trademark for posterior pituitary injection (oxytocin).

placebo In Latin means "I shall please." An inactive substance or preparation given to please or gratify a patient. Also used in contolled studies to determine the efficacy (virtue) of medicinal substances.

placentitis Inflammation of the placenta.

plain A term suggesting general inferiority; coarse, lacking the desired quality.

plasma The liquid portion of blood or lymph in which the corpuscles or blood cells float.

plasmolysis A process causing water to leave the cell (contraction of protoplasm), as in the use of high concentrations of nontoxic salts and sugar for bacteriostatic purposes.

pleasure horse Horse used for riding, driving, or racing.

pleiotrophy The state in which one gene affects two or more traits.

pleiotropism The action of one gene on two or more traits.

PLM The acronym indicating protein, lactose, and minerals of milk.

poikilotherms (ectotherms) Cold-blooded animals. Ocean fish exemplify cold-blooded species.

poikilothermy The ability of animals to adapt themselves to variations in their environmental temperature. The quality of varying the body temperature with the environmental temperature. Most invertebrates, fish, amphibians, and reptiles are poikilotherms.

polled A naturally hornless animal.

polydipsia An excessive thirst.

polygastric Possessing more than one (many) stomach or stomach compartments; characterizing the cow and other ruminants.

polygenic (multiple-factor) inheritance   Inheritance involving a series of independent genes; characterized by the trait showing a continuous distribution pattern owing to the additive effect of genes, none of which show dominance.

polymer   A large molecule composed of repeating smaller units.

polyneuritis   Inflammation of many nerves concurrently.

polyp   A smooth, stalked, or projecting growth from a mucous membrane.

polyploidy   Complete duplication of all sets of chromosomes giving 3*n*, 4*n*, etc., numbers of homologous chromosomes in body (somatic) cells.

polypnea   A condition in which the respiration rate is increased; rapid, shallow breathing.

polyspermy   Entrance of many (poly) spermatozoa into the ovum at the time of fertilization.

pony   Any small horse of nondescript breeding, commonly less than 58 in tall at maturity.

porcine   Pertaining to swine.

porker   A young hog (pig).

portal system   The system of blood vessels conveying blood from the digestive organs and spleen to the liver.

posterior   Denotes the back or back portion (caudal). It means the same as dorsal surface of the body in human anatomy.

post-legged   Describing an animal (especially a horse) with too much set in the hocks, resulting in the hind legs being too straight.

postmortem   An examination of an animal carcass or human body after death. **Necropsy.**

postnatal   Occurring after birth.

postpartum   Occurring after birth of an offspring.

pot-bellied   Designating any animal that has developed an abnormally large abdomen.

poult   An immature turkey. After the sex can be determined, the turkey is called a *young tom* (male) or *young hen* (female). (See Appendix A.)

poultry   Birds raised for meat and eggs.

ppm   Parts per million (1 mg/liter).

precipitin   Antibody that forms a precipitate with its soluble antigen. An antibody formed in blood serum as a result of inoculating with a foreign protein.

precursor   A compound or substance from which another is formed.

predatism   Intermittent parasitism, such as the attacks of mosquitoes and bedbugs on humans.

predator   Any animal, including an insect, that preys on and devours other animals, e.g., a coyote or dog preying on sheep. Some predators, such as ladybugs, may be beneficial in that they kill and eat parasites.

predicted difference (PD)   A measure of a bull's ability to transmit milk-producing capacity to his daughters. The PD may be positive or negative, depending on whether the bull's daughters yield more or less milk than daughters of other bulls (**herdmates**) under the same conditions.

preen gland.   See **oil gland.**

prehension   The seizing (grasping) and conveying of food to the mouth.

premortal   Existing or occurring immediately before death.

prepartum   Occurring before birth of the offspring. Before parturition.

prepotent   Designating an animal that transmits its own characters to its progeny to a marked or highly uniform degree.

prick   To pierce or cut the tail of a horse so that it will be carried higher.

primates Humans, monkeys, and the great apes.

primiparous Bearing or having borne only one young or set of young.

prodome A symptom indicating the onset of a disease.

produce A female's offspring. The *produce of dam* commonly refers to two offspring of one dam.

progeny The offspring of animals.

progeny testing Evaluating the genotype of an individual by a study of its progeny.

progestational A phase of the estrous cycle (menstrual cycle) in which the corpus luteum is active and the endometrium is under its influence.

prolapse Abnormal protrusion of a part or organ; displacement of an organ from its normal location.

prolapsed uterus A condition in which the uterus is partially or completely turned inside out, usually following parturition.

proliferation Growth by rapid *multiplication* of new cells.

prolific Capable of producing abundant offspring.

prophase The first stage of mitosis, characterized by chromosome shortening and thickening and the disappearance of nuclear membrane as well as by the appearance and polarization of spindle fibers.

prophylactic A preventive, preservative, or precautionary measure that tends to ward off disease.

prophylaxis The prevention of disease.

prostaglandins A large group of chemically related 20-carbon hydroxy fatty acids with variable physiological effects in the body (Chapter 7).

prostate Gland in the male reproductive system that lies just below the bladder and surrounds part of the canal that empties the bladder.

protean Variable; readily assuming different shapes or forms; changeable.

protective antibodies Antibodies that when combined with pathogenic organisms render them noninfectious.

protein A substance composed of amino acids, containing about 16 percent (molecular weight) nitrogen. Thus protein content is computed by multiplying the chemically determined value for nitrogen by the factor 6.25 ($N \times 6.25$).

protein efficiency ratio (PER) The weight gained by growing experimental animals divided by the weight of the protein consumed, for example, the PER for casein is 2.5 (i.e., for each 1 g of casein ingested, test animals gain 2.5 g in body weight). The caloric intake must be adequate and the concentration of dietary protein must be adequate but not excessive, because gain is not proportional to intake at high levels of dietary protein.

Adjusted PER = (PER of test food $\times$ 2.5)/(PER of standard reference casein)

protein equivalent A term indicating the total nitrogenous contribution of a substance in comparison with the nitrogen content of protein (usually plant protein). For example, the nonprotein nitrogen (NPN) compound urea contains approximately 45 percent nitrogen and has a protein equivalent of 281 percent (6.25 $\times$ 45 percent).

protein-fortified Describing low-fat and skim milks that contain at least 10 percent nonfat milk solids *(NMS)*. When milk derivatives other than *NDM* are used to satisfy the 10 percent requirement, the protein added must be milk protein and must equal or exceed the quantity that would be added if the additive were NDM.

protein supplements Feed products that contain 20 percent or more protein.

prothoracic gland An endocrine gland of insects that secretes ecdysone, the growth and differentiation hormone.

proton An *elementary particle* with a single positive electric charge.

protozoan   A microscopic animal that consists of a single cell.

proved sire   A sire whose transmitting ability has been measured by comparing the production performance of his daughters with that of the daughter's dam and/or herdmates under similar conditions. See **herdmates.**

proximal   Nearest; closer to any point of reference; opposite to *distal.*

proximate analysis   Also known as the *Weende analysis* (developed in 1895 at the Weende Experiment Station in Germany); used to determine the gross composition of feed.

psychobiology   That branch of biology which considers the interactions between body and mind in the formation and functioning of personality; the scientific study of the personality function.

psychro-   From the Greek *psychros,* meaning "cold." The prefix denoting relations to cold.

psychroenergetics   Science dealing with the effect of ambient temperature and humidity on conversion of feed into bodily heat and energy.

psychrometer   An apparatus for measuring atmospheric moisture by the difference in readings of two thermometers (one dry bulb and one wet bulb).

puberty   The age at which the reproductive organs become functionally operative and secondary sex characteristics develop.

pubic   Pertaining to the pubes (hair growing over pubic area) or pubic bones. The lower part of the hypogastric region.

public health   An organized effort to prevent disease, prolong life, and promote physical and mental efficiency. Also, the health of the community taken as a whole.

pudic   Pertaining to the external genital parts, especially of the female.

pullet   A female chicken less than a year old.

puncher   One who herds cattle; a cowboy.

pupa   The quiescent or inactive stage during which an immature insect or larva transforms into an adult.

pupal stage   Period in the life history of insects between the caterpillar, or grub, stage and the mature, or adult, insect.

pupate   To change from an active immature insect into the inactive pupal stage.

purebred   An animal of a recognized breed that is eligible for registry in the official herdbook of that breed.

purebreeding (truebreeding)   Breeding a stock that is homozygous for one or more characteristics.

pure culture   A population of microorganisms that contains only a single species. Cultures are useful in the manufacture of many animal products, e.g., cheeses and yogurt.

purified diet   A mixture of the known essential dietary nutrients in a pure form that is fed to experimental (test) animals in nutrition studies.

pus   A liquid inflammatory product consisting of leukocytes, lymph, bacteria, dead-tissue cells, and fluid derived from their disintegration.

putrefaction   The bacterial decomposition of proteins.

pyrexia   A fever or febrile condition. An abnormal elevation of body temperature.

qualitative traits   Those traits, such as black and white or polled and horned, in which there is a sharp distinction between phenotypes. Usually, only one or two pairs of genes are involved.

quality   A term indicating fineness of texture as opposed to coarseness. It commonly is used to indicate relative merit, e.g., superior breeding or genetic merit.

quantitative traits   Those traits, such as skin color in humans, in which there is no sharp distinction between phenotypes, with a gradual variation from one phenotype to another. Usually, several genes, as well as environmental factors, are involved.

quarantine   Commonly thought of as the segregation of the active case of an infectious disease, but more technically, it includes compulsory segregation of exposed susceptible animals or individuals for a period of time equal to the longest usual incubation period of the disease to which they have been exposed. A regulation under police power for the exclusion or isolation of an animal to prevent the spread of an infectious disease.

rabbit   A device used to move a sample rapidly from one place to another, e.g., in a research reactor.

rack   The gait of a horse in which only one foot touches the ground at any one time, producing a four-beat gait. The legs move in lateral pairs but not quite in unison, so that each foot is lifted and put down alone.

rad   Another name or unit for "radiation energy absorbed" by food being processed with radiation. 1000 rad = 1 kilorad = 10 **gray**; 1,000,000 rad = 1 Mrad = 10 kGy; (the rad is being superseded by the **gray**).

radappertization   Sterilization by radiation processing. The resulting processed food can be stored at room temperature, in the same way as thermally sterilized foods (canned foods). Precooked food in hermetically sealed packaging is exposed to radiation at levels high enough to kill all organisms of food spoilage or public health significance. Doses used are typically greater than 1 Mrad.

radiant heat   Heat transmitted by radiation (such as that of the sun) as contrasted with that transmitted by **conduction** or **convection.**

radiation   The process of emitting radiant energy (heat) in the form of waves or particles. The sun transfers its energy by radiation. It is also an important method of heat loss from an animal to cooler objects and heat gain by the animal from warmer objects.

radiation sterilization   The use of radiation to cause a plant or animal to become sterile or incapable of reproduction. Also, the use of radiation to kill all forms of life (especially bacteria) in food, on equipment, etc.

radicidation   Radiation pasteurization intended to kill or render harmless all *disease-causing* organisms (except viruses and spore-forming bacteria) in food. Processing takes place at dose levels generally below 1 Mrad, and the processed foods usually must be stored under refrigeration (as in heat pasteurization).

radioactive dating   A technique for measuring the age of an object or sample of material by determining the ratios of various **radioisotopes,** or products of radioactive decay, it contains. For example, the ratio of carbon 14 to carbon 12 reveals the approximate age of bones, pieces of wood, or other archeological specimens that contain carbon extracted from the air at the time of their origin.

radioactive tracer   A small quantity of radioactive **isotope** used to follow biological, chemical, or other processes by detection, determination, or localization of the radioactivity.

radioactivity   The spontaneous decay or disintegration of an unstable atomic **nucleus,** usually accompanied by the emission of **ionizing radiation.**

radiograph   A record or photograph produced by x-rays or other rays on a photographic plate, commonly called an *x-ray picture.*

radioisotope   A radioactive isotope (the nucleus of such a species of atom). A nuclide. Such isotopes occur naturally but may be provided by bombardment of a common chemical element with high-velocity particles. A radioactive isotope transmutes into another element with emission of electromagnetic radiations. More than 1300 natural and artificial radioisotopes have been identified.

radiology The science that deals with the use of all forms of ionizing radiation in the diagnosis and therapy of disease.

radionuclide A radioactive nuclide.

radurization Radiation pasteurization designed to kill or inactivate *food-spoilage* organisms, thus extending the shelf life of a given food product. Processing takes place at dose levels generally below 1 Mrad, and the product usually must be stored under refrigeration, as in the case of **radicidized** food.

ram A male sheep. Also called a *buck*.

random mating A system of mating where every male has an equal chance of mating with every female.

rangy Designating an animal that is long, lean, leggy, and not too muscular.

rate Synonymous with *level, dosage, amount, quantity*, or *degree* measured in proportion to something else.

rate of passage The time taken by undigested residues from a given meal to reach the feces. A stained undigestible material is commonly used to estimate rate of passage.

ration The food allowed an animal for 24 h. A *balanced* ration provides all the nutrients required to nourish an animal for 24 h. See **balanced ration.**

rawhide The usual American name, which has spread to other English-speaking countries, for cattlehide that has been dehaired but is usually unfinished. Some rawhide is tanned with the hair left on. It is used principally for mechanical purposes, such as belt facings and pins, gaskets, pinions and gears, and also for trunk binding and luggage.

raw milk Fresh, untreated milk as it comes from the cow or another mammal.

raw wool Wool prior to removal of the grease.

razorback A type of hog with long legs and snout, sharp narrow back, and lean body; usually a half-wild mongrel breed (especially of the southern United States).

reactor An animal that reacts positively to a foreign substance; e.g., a tuberculous animal would be a reactor to **tuberculin.** See also **nuclear reactor.**

recessive gene A gene whose phenotypic expression is covered (masked) by its own dominant allele. For example, the blue-eyed gene *b* is recessive to the brown-eyed gene *B,* with *Bb* individuals having brown eyes. A *bb* individual would have blue eyes. Recessive genes appear to affect the phenotype only when present in a homozygous condition.

recombination A formation of genotypes and phenotypes that are new combinations of the parents in a given cross.

reconstituted milk The product that results from the recombining of milk fat and nonfat dry milk or dried whole milk with water in proportions to yield the constituent percentage occurring in milk.

red meat Meat that is red when raw. Red meat includes beef, veal, pork, mutton, and lamb.

redia A larval stage in the development of flukes. Redia of liver flukes of cattle, sheep, and goats occurs in snails.

reduction Chemically, the subtraction of oxygen from, or the addition of hydrogen to, a substance (or the loss of positive charges or the gain of negative charges). The atom or groups of atoms that lose electrons become oxidized.

reflectivity The ratio of the rate of reflection of radiant energy from a given surface to the rate of incidence of radiant energy on it.

reflex Action performed involuntarily in consequence of a nervous impulse transmitted from a receptor, or sense organ, to a nerve center.

regurgitation The casting up (backward flow) of undigested food from the stomach to the mouth, as by ruminants.

relative humidity (RH) The ratio of the weight of water vapor contained in a given volume of air to the weight that the same volume of air would contain when saturated. The quantity of water vapor that air can hold when saturated increases with temperature. The RH is expressed as a percentage. For example, if a sample of air at a given temperature contains 30 percent of the water vapor that it is possible for it to contain at that temperature, it is 30 percent saturated and therefore has a relative humidity of 30 percent.

reradiation The radiation emitted by the body as a result of its absorbing radiation incident on it.

research reactor A reactor primarily designed to supply neutrons or other **ionizing radiation** for experimental purposes.

reservoir host An animal that harbors the same species of parasite as humans. Also, an animal that becomes infected and serves as a source from which other animals can be infected.

respiratory quotient (RQ) The RQ is used to indicate the *type* of food being metabolized. This is possible because carbohydrates, fats, and proteins differ in the relative amounts of oxygen and carbon contained in their molecules. Also, the relative volumes of oxygen consumed and carbon dioxide produced during metabolism of each type of food vary. Respiratory quotient is calculated as follows:

$$RQ = \frac{\text{volume } CO_2 \text{ produced}}{\text{volume } O_2 \text{ consumed}}$$

retained placenta A placenta that was not expelled at parturition.

reticuloendothelial system A widely spread network of body cells concerned with blood cell formation, bile formation, and engulfing or trapping of foreign materials, which includes cells of bone marrow, lymph, spleen, and liver.

reticulum The second division of the stomach of a ruminant animal. Also called *honeycomb*.

retrogression Degeneration, deterioration, or a backward movement.

reversion Appearance of a trait in an individual that was possessed by remote ancestors but not by recent ones.

rickettsiae Intracellular parasites, i.e., ones that multiply inside the living cells of other larger organisms. In size they are intermediate between bacteria and viruses.

ride To mount and travel on a horse. To mount a cow, as another cow, indicative of estrus (heat).

ridgling Any male animal whose testicles fail to descend normally into the scrotum. Also called *cyptorchid*.

rigling A male sheep or horse that has only one testicle in the scrotum. See **ridgling.**

rigor mortis The stiffness of body muscles that is observed shortly after the death of an animal. It is caused by an accumulation of metabolic products, especially lactic acid, in the muscles.

ring test A test for brucellosis performed by mixing stained *Brucella* bacteria with **raw milk.** If **antibodies** to *Brucella* are present, the stained cells agglutinate (clump) and rise to the surface with the cream to form a blue ring.

RNA Ribonucleic acid.

roan   Designating the red-white-color phase of Shorthorn cattle. Red or black coat color of a horse intermingled with white; may be red or strawberry roan, blue roan, or chestnut roan, depending on the intermingling of the background colors.

roaster   A young chicken (meat type) weighing more than 3.5 lb (and usually 4 to 6 months old).

roasting pig   A pig weighing from 15 to 50 lb.

rodent   A classification of mammals, mostly vegetarians, characterized by their single pair of chisel-shaped, upper incisors (rabbits, rats, mice, squirrels).

rodenticide   Any poison that is lethal to rodents.

roe   The eggs or testes of fish. There are two types, the female eggs (hard roe) and the male testes (soft roe). They are considered a delicacy by many people.

roost   A resting or lodging place for fowls.

rooster (cock)   An adult male chicken.

roughage   Consists of pasture, silage, hay, or other dry fodder. It may be of good or poor quality. Roughages are usually high in crude fiber (more than 18 percent) and relatively low in **NFE** (approximately 40 percent).

rugged   Refers to a large, strong animal.

rumen   The first stomach compartment of a ruminant; also called *paunch*. The rumen is a large nutrient-producing fermentation vat that contains an amount of feed and water equal to approximately one-seventh of the ruminant's body weight.

rumen flora   The microorganisms of the rumen.

ruminant   One of the order of animals that has a stomach with four complete cavities—rumen, reticulum, omasum, abomasum—through which food passes in digestion. These animals chew their cud; they include cattle, sheep, goats, deer, antelopes, elk, and camels.

rumination   The casting up of food (cud) to be chewed a second time, as in cattle. A chewing of the cud, as by ruminants.

running horse   Any Thoroughbred.

run on   To graze or pasture on, as for cattle to run on the range.

runt   A term commonly used to denote a piglet of small size in relation to its littermates. Runts usually result from a shortage of milk in one or more teats of the sow.

rupture   The forcible tearing or breaking of a body part. See **hernia.**

rustle   To hunt for food, especially with reference to domestic animals inadequately fed by their owner. To steal livestock.

sagittal   Anteroposterior plane or section parallel to the long axis of the body.

salpingitis   An inflammation of a fallopian tube (oviduct).

saprophyte   Any vegetative organism, such as a bacterium, living on dead or decaying organic matter.

satiety   Full satisfaction of desire; may refer to satisfaction of sexual arousal, appetite, etc.

saturated fat   A completely hydrogenated fat; that is, each carbon atom is associated with the maximum number of hydrogens.

scale   The size of an animal.

schistosomiasis   Infestation with a schistosome, or blood fluke.

scintillation counter   An instrument that detects and measures ionizing radiation by counting the light flashes (scintillations) caused by radiation impinging on certain materials (e.g., phosphorus).

sclera   The tough, white, supporting covering of the eyeball, which encompasses all the eyeball except the cornea.

scours   A persistent diarrhea in animals.

scrub   An animal inferior in either breeding or individuality.

scurs   Small, rounded portions of horn tissue attached to skin at the horn pits of polled animals; also called *buttons.*

SDA (specific dynamic action)   The increased production of heat by the body as a result of a stimulus to metabolic activity caused by ingesting food.

sebum   The thick, semifluid substance composed of fat and epithelial debris secreted by the sebaceous glands.

secondary infection   Infection following an infection already established by other organisms.

second filial generation   See **F_2 generation.**

second meiotic division   The second of two divisions occurring in reductional cell division and resulting in the production of two cells, each of which is haploid, the chromosomes occurring singly (nonpaired).

seed tick   The newly hatched six-legged larva of a tick, especially of the cattle tick *Boophilis annulatus,* a one-host tick in which the larva, the nymph, and the adult are all found on cattle. Newly hatched larvae are found on the ground or on grass, weeds, and other objects in fields where infested cattle have grazed.

segregation of genes   This refers to the occurrence of genes in pairs in body cells, for example, *Aa.* However, when such an individual produces gametes, only one of these genes, either *A* or *a,* not both, goes into a single sex cell. Thus, although they are together in body cells, they segregate, or separate from each other, when gametes are formed.

selection   The causing or allowing of certain individuals in a population to produce the next generation. *Artificial selection* is that practiced by humans; *natural selection* is that practiced by nature.

self-feeding   Any feeding device by means of which animals can eat at will. See **ad libitum.**

senescence   The process or condition of growing old. Aging. (Chapter 8.)

sensible   Perceptible; as sensible heat loss (water) or weight loss (liquids or solids).

sensory   Pertaining to sensation. The eyes and ears are *sensory organs. Sensory nerves* convey impulses from the sense organs to a nerve center. Thus some nerves are *sensory* and pick up sensations from sense organs and carry them to main cords and the brain, whereas others are *motor* and carry impulses from the brain and main nerves to the muscles, which respond to the stimulation.

septicemia   Blood poisoning, which results from the presence of toxins or poisons of microorganisms in the blood.

serological   Pertaining to the use of blood serum of animals in various tests, which aids in detecting and treating certain diseases.

serotype   The type of microorganism as determined by the kind and combination of constituent antigens associated with the cell.

serum   The clear portion of animal fluids, separated from its cellular elements. Blood serum is the clear, pale-yellow, watery portion of blood that separates from the clot when blood coagulates.

serum therapy   The treatment of clinical cases of disease with serum of immunized animals.

service   A term used in animal breeding, denoting the mating of a male to a female. Also called *serving,* or *covering.* (See Appendix A.)

setting hen   A broody hen in the act of incubating eggs.

settled   A term commonly used to indicate that the animal has become pregnant.

sex chromosomes   One pair of chromosomes in an individual that determines the sex of that individual. In mammals, the female is XX and the male is XY. The X chromosome is considerably longer and carries more genes than the Y chromosome.

sex-influenced traits   Such traits are due to genes carried on autosomes; however, the gene is dominant in males and recessive in females. For example, the gene for baldness, *Ba,* in humans is a sex-influenced gene. Its allele is *Bn,* for nonbaldness phenotype.

Genotype	Men	Women
BaBa	Bald	Bald
BaBn	Bald	Not bald
BnBn	Not bald	Not bald

sex-limited traits   The appearance of such traits is limited to only one sex, for example, egg laying in hens and lactation in cows. Nevertheless, males of these species possess genes for these traits, even though they are not expressed phenotypically.

sex linkage   Refers to genes carried on the nonhomologous portion of the X chromosome. For recessive sex-linked genes, two are required to express the trait in females and one in males. A gene carried on the nonhomologous portion of the Y chromosome is always transmitted from father to son. It is referred to as *holandric inheritance.* Sex-linked genes are alleles, then, that have their loci on the sex chromosomes, usually only on the X chromosomes.

shear   To cut wool or hair from sheep, goats, etc.

shelf life   The time after processing during which a product remains suitable for human consumption, especially the time a food remains palatable.

shoat (shote)   A young pig of either sex less than 12 months old. (See Appendix A.)

shy breeder   A male or female of any domesticated livestock that has a low reproductive efficiency.

sibling   In genetics, a brother or sister.

sickle-hocked   Designating a horse, cow, or sheep having a crooked hock, which causes the lower part of the leg to be bent forward out of a normal perpendicular straight line.

silage (ensilage)   Prepared by chopping green **forage,** such as grass or clover, or **fodder,** such as field corn or sorghum, and blowing it into an airtight chamber (silo), where it is compressed so that air is excluded and it undergoes an acid fermentation (produces lactic and acetic acids) that retards spoiling. It usually contains 65 to 70 percent moisture.

silo   A pit, trench, aboveground horizontal container, or vertical cylindrical structure of relatively airtight construction into which chopped green crops, such as corn, grass, legumes, or small grain and other livestock feeds are placed and allowed to partially ferment into silage. See **silage.**

sinistral   Of or pertaining to the left side; left; or left-handed.

sire   The male parent. To father or to beget.

sire indexes   Various means of calculating the abilities of bulls to transmit economically important production traits. See also **USDA sire summary.**

sire summary   See **USDA sire summary.**

skewbald   A horse of any color except black, with white spots.

slip To abort. An incompletely castrated male.

slow gait One of the several forward movements, or gaits, of horses, faster than a walk but slower than a canter. There are three slow gaits: the running walk, the fox trot, and the slow pace.

slunk The skin of an unborn or prematurely born calf.

smooth mouth The mouth of a horse whose teeth have lost their natural cusps and have become smooth by use and wear, generally indicating that the horse is 10 or more years of age.

SMR See **standard metabolic rate.**

SNF Solids-not-fat of milk (proteins, lactose, and minerals). Same as *NFS* and *NMS.*

social insect Any insect that lives with others of its kind in a somewhat organized colony, as ants, bees, and wasps.

soilage Freshly cut green forage fed to animals in confinement. Also called *green chop.*

soiling A term previously used for the green chopping of forages.

soluble Designating a substance that is capable of being dissolved in another.

somatic Refers to body tissues; having two sets of chromosomes.

sorrel A coat color of horses. It includes the red shades of chestnut or yellowish brown.

sow Mature female swine.

sowbelly Salt pork; unsmoked fat bacon.

space spray Insect spray used in insect control in open spaces, e.g., in a dairy barn or hog house.

span (spann) A pair of animals usually harnessed together as a team.

spay To surgically remove the ovaries of a female.

species A group of animals having several common characteristics that differentiate them from others.

specific dynamic action See **SDA.**

sperm (spermatozoon) A mature male germ cell.

spermatogenesis The formation and development of spermatozoa.

SPF Specific-pathogen-free.

sphincter A ring-shaped muscle that closes an opening, e.g., the sphincter muscles in the lower end of a cow's teat.

spinnbarkeit The formation of a thread by cervical mucus when blown onto a glass slide and drawn out by a cover glass; the time at which it can be drawn to the maximum length usually precedes or coincides with the time of ovulation in women.

splayfooted See **toe out.**

split hide The outer (hair or grain) layer of a hide from which the under, or flesh, side has been split to give it a reasonably uniform thickness.

spontaneous Instinctive (performed apparently without the exercise of reason) and occurring without external influence.

spore From the Greek *spores,* meaning "seed." A single cell that becomes free and is capable of developing into a new plant or animal. The reproductive element of some lower organisms. Does not contain a preformed embryo, as do seeds.

springer A term commonly associated with female cattle showing signs of advanced pregnancy.

stable A building used for the feeding and lodging of horses and other livestock.

stable isotope An **isotope** that does not undergo radioactive decay.

stag A male animal castrated after the secondary sex characteristics have developed sufficiently to give it the appearance of a mature male.

stale A period when an animal does not work, lactate, etc., at normal standards, as opposed to **bloom.**

stallion A mature male horse, not castrated. (See Appendix A.)

stance Position, or posture, adopted when an animal is stationary.

standard metabolic rate (SMR) Reflects an animal's basic maintenance energy requirement. It is useful and important in studies of thermal physiology and productive efficiency to have such a reference metabolic rate. Because metabolic rate increases during thermal stress, an animal's reference metabolic rate should be measured in the thermoneutral zone of effective environmental temperature. Moreover, because metabolic rate increases postfeeding (due to the **heat increment of feeding**), the reference value should be determined sometime after the animal has absorbed its last meal. Additionally, because physical activity increases metabolic rate, to be meaningful the reference value must reflect metabolic rate when the animal is resting. Thus, the SMR takes these three conditions into account and is said to occur in a fasting, resting animal held in thermoneutral surroundings. Standard metabolic rate is based on the 0.75 power of body weight, the value commonly called **metabolic body size.** By means of SMR, comparisons can be made among animals of different sizes and species. In human physiology, SMR is called **basal metabolic rate (BMR).** In animal science literature, standard metabolic rate is synonymous with *fasting metabolic rate* and with *resting metabolic rate.*

starch equivalent (SE) A net-energy system of feed evaluation that is used extensively in Germany and other European countries. One kilogram of SE is equivalent to 2356 kcal net energy for fattening (NE_f). The system is based on work by Kellner and his successors. SE values are calculated based on digestible nutrients and crude fiber of the diet.

starvation The deprivation of an animal of any or all the food elements necessary to its nutrition.

steer A male bovine castrated before the development of secondary sex characteristics. (See Appendix A.)

sterilization The process of freeing a substance or an article from *all* living organisms.

sterilize To remove or kill all living organisms. To render an animal infertile.

sterol Any of a group of high-molecular-weight alcohols, as ergosterol and cholesterol.

stillborn Born lifeless; dead at birth.

stock cattle Usually, young steers or cows that are light and thin and lack **finish.**

stool Fecal material; evacuation from the digestive tract.

stover Fodder; mature, cured stalks of grain from which the seeds have been removed, such as stalks of corn without ears.

straggler An animal that strays or wanders from a herd or flock.

strain A group of animals within a breed differing in one or more characteristics from other members of the breed.

stress The sum of all nonspecific biological phenomena caused by adverse conditions or influences. It includes physical, chemical, and/or emotional factors to which an individual fails to make satisfactory adaptation and that cause physiological tensions that may contribute to disease.

stride The distance from one footprint of a horse to the print of the same foot when it next comes fully to the ground.

strobilation An asexual form of reproduction in which segments of the body separate to form new individuals, as in tapeworms.

stud   A unit of selected animals kept for breeding purposes, as of bulls and horses. Abbreviation for *stud horse:* a stallion.

sty   A pen in which swine are fed and housed.

subclinical   A disease condition without clinical manifestations.

subcutaneous   Situated or occurring beneath the skin.

sublimation   The process of sublimating or subliming; the direct transition from solid to vapor (bypassing the liquid form). The process of vaporizing and condensing a solid substance without melting it.

substrate   A substance on which cells or organisms may live and be nourished.

succulence   A condition of plants characterized by juiciness, freshness, and tenderness, making them appetizing to animals.

suckle   To nurse at the breast or mammary glands.

suede finish   A finish produced by running the surface of leather on an abrasive to separate the fibers in order to give the leather a velvetlike nap. The term denotes a finish, not a type of leather.

superfetation   Second impregnation of a female that is already pregnant.

supplement   Refers to the addition of minerals, vitamins, or other minor ingredients (bulkwise) of a diet.

supra-   A prefix meaning on, above, over, or beyond.

swarm   The simultaneous emergence or assembly in one location of large numbers of insects (especially bees), often to establish a new colony.

sweet butter   Unsalted butter.

swirl   Hair that grows in a whorl on an animal.

switch   The brush of hair on the end of a bovine tail.

symbiosis   The living together in intimate association of two dissimilar organisms, with a resulting mutual benefit.

synapsis   The pairing of a homologous set of chromosomes during the first meiotic division. The chromosomes occur as paired chromatids joined at the centromere.

syndrome   A group of signs and symptoms that occur together and characterize a disease; a disturbance or abnormality.

syngamy   Fusion of identical gametes.

synovia (synovial fluid)   A viscid fluid containing synovin, or mucin, and a small proportion of mineral salts. It is secreted by the synovial membrane and resembles the white of an egg. It is contained in joint cavities, bursae, and tendon sheaths.

tachycardia   Excessive rapidity in the action of the heart (pulse rate above 100/min in humans).

tack   Riding equipment, such as the bridle and saddle. Also refers to equipment used in the fitting and showing of animals.

tactile   Pertaining to the touch.

tag   See **tracer isotope.**

take   To accept a male in coitus. To result in a mild infection after vaccination.

tallow   The fat extracted from adipose tissue of cattle and sheep.

tanbark trail   A term commonly associated with those who exhibit animals in competition at fairs and shows. *Tanbark* is the bark of several trees (oak, chestnut, etc.), used as a ground covering in circus lots, racetracks, livestock pavilions, etc.

tankage   A **protein supplement** used as an animal feed. It consists of ground meat and bone by-products of animals that have been slaughtered.

tanning   The processing of perishable rawhides and skins into the permanent and durable form of leather by the use of tanning materials.

TDN   See **total digestible nutrients.**

teart   Molybdenosis of farm animals caused by feeding on vegetation grown on soil that contains high levels of molybdenum.

tease   To stimulate an animal to accept coitus.

teg   A sheep 2 years of age.

telophase   The fourth stage of mitosis, characterized by elongation of chromosomes, disappearance of spindle fibers, and reorganization of nuclear membrane.

temperament   Disposition of an animal.

tend   To care for, as a flock of sheep.

tendon   The strong tissue terminating a muscle and attached to a bone, affording leverage.

term   The gestation period.

test cross   Mating involving a recessive phenotype; used to determine heterozygosity of a stock.

tetany   A condition in an animal in which there are localized, spasmodic muscular contractions.

tether   To tie an animal with a rope or chain to allow grazing but prevent straying.

tetrad   A unit of four chromatids formed as a result of synapsis of homologous chromosomes, each of which consists of a pair of identical chromatids joined at the centromere.

therapy   The treatment of disease. Curative.

therm   See **megacalorie.**

thermal elements   Include temperature, humidity, air movement, and radiant heat.

thermocouple   A device consisting essentially of two conductors made of different metals, joined at both ends, producing a loop in which an electric current will flow when there is a difference in temperature between the two junctions.

thermogenesis   The chemical production of heat in the body.

thermolysis   The loss or dissipation of body heat.

thermoneutrality   The state of thermal balance between an organism and its environment so that the body thermoregulatory mechanisms are inactive. The thermoneutral zone is also referred to as the *comfort zone.*

thermoneutral zone   The relatively narrow zone of effective environmental temperature in which heat production at the animal's minimal or thermoneutral rate is offset by net heat loss to the environment without the aid of special heat-conserving or heat-dissipating mechanisms. Thus, the animal is under neither cold nor heat stress. See also **comfort zone.**

thorax   The chest.

Thoroughbred   The name of the English breed of running horses.

three-way cross   A system of rotation breeding involving three males of different breeds.

threshold   The level or point at which a physiological effect becomes evident as a result of stimulation.

throw   To cause an animal, as a horse or cow, to fall to the ground before branding, treating, etc. To abort an embryo or fetus.

thumps   An animal ailment resembling hiccups in humans that is seen, for example, in anemic baby pigs.

thymus   Glandlike organ in the upper part of the chest that reaches its maximum development during late childhood in humans. It is probably associated with immunity.

thyroid   Gland in the neck that helps to regulate many processes of growth and development.

tick   Any of the various bloodsucking arachnids that fasten themselves to warm-blooded animals. Some are important **vectors** of diseases.

titer   The quantity of a substance required to produce a reaction with a given volume of another substance, or the amount of one substance required to correspond to a given amount of another substance. *Agglutination titer* is the highest dilution of a serum that causes clumping of bacteria.

toe out   To walk with the feet pointed outward. Also called *splayfooted* or *slew-footed*.

tom   A male turkey. (See Appendix A.)

tonicity   The state of tension or partial contraction of muscle fibers while at rest; normal condition of muscle tone.

total digestible nutrients (TDN)   A standard evaluation of the usefulness of a particular feed for farm animals that includes all the digestible organic nutrients: protein, fiber, nitrogen-free extract, and fat.

toxemia   Generalized blood poisoning, especially a form in which the toxins produced by pathogenic bacteria enter the bloodstream from a local lesion and are distributed throughout the body.

toxins   Poisons produced by certain microorganisms. They are products of cell metabolism. The symptoms of diseases caused by bacteria, such as diphtheria and tetanus, are due to toxins.

toxoid   A detoxified toxin. It retains the ability to stimulate formation of antitoxin in an animal's body. The discovery that toxin treated with formalin loses its toxicity is the basis for preventive immunization against such diseases as diphtheria and tetanus.

tracer isotope   An isotope of an element, a small amount of which may be incorporated into a sample of material (the carrier) to follow (trace) the course of that element through a chemical, biological, or physical process and thus also follow the larger sample. The tracer may be radioactive, in which case observations are made by measuring the radioactivity. If the tracer is stable, mass spectrometers, density measurements, or **neutron activation analysis** may be employed to determine isotopic composition. Tracers are also called *labels* or *tags,* and materials are said to be labeled or tagged when radioactive tracers are incorporated in them.

trachea   The windpipe; in mammals it extends from the throat to the bronchi.

transduction   The transfer of genetic material from one cell to another when mediated by a bacteriophage.

transitory   Brief; momentary; lasting only a short time; fleeing; transient.

translocation   The attachment of a fragment of one chromosome to another that is not homologous to it.

translucent   Transmitting light, but diffusing it so that objects beyond are not clearly distinguished.

Trematoda   A class of the Platyhelminthes, which includes the flukes.

trematode   Any parasitic animal organism belonging to the class Trematoda.

tremor   An involuntary trembling or quivering.

trihybrid   An individual that is heterozygous for three pairs of alleles, such as *AaBbCc.*

trimester   A period of 3 months.

tripe   Beef consisting of the walls of the rumen and reticulum, used as food for people.

trophoblast   The enveloping layer of cells of the early embryo that will attach the ovum to the uterine wall and supply nutrition to the embryo.

tropism   The tendency of an organism (plant or animal) to react (turn or move) in a definite way in response to external stimuli.

trots A diarrheal, or abnormally loose, condition of the bowels.

true albino Solid white animal with pink eyes. The homozygous albino genes in horses may be lethal.

tuberculin A biological agent derived from the growth and further processing of the tubercle bacilli that is used for detection or diagnosis of tuberculosis in animals and humans.

tup A ram.

tush A tooth located between the incisors and molars; the eyetooth; tusk.

twitch To tightly squeeze the skin on the end of a horse's nose or its underlip by means of a small rope that is twisted.

type The physical conformation of an animal.

type classification A program sponsored by breed associations whereby a registered animal's conformation may be compared with the "ideal," or "true," type of animal of that breed by an official inspector (classifier).

udder The encased group of mammary glands provided with teats or nipples, as in a cow, ewe, mare, or sow. Also called *bag*.

UE See **urinary energy.**

ulceration Development of a condition whereby substance is lost from a cutaneous or mucous surface, causing gradual disintegration and necrosis of the tissues.

ungulate Referring to a hooved quadruped, as a cow.

unilateral That which affects only one side.

uniparous Producing only one egg or one offspring at a time.

unsaturated fat A fat having one or more double bonds; not completely hydrogenated.

unsex To castrate a male or female.

unthriftiness Lack of vigor, poor growth or development; the quality or state of being unthrifty in animals.

urea A nonprotein, organic, nitrogenous compound. It is made synthetically by combining ammonia and carbon dioxide.

uremia An accumulation of urinary constituents in the blood.

urinary energy (UE) The food energy lost through the urine.

urogenital (genitourinary) Pertaining to the urinary and genital tracts (including the kidneys and sex organs).

uropygial gland The preen gland (used by birds to waterproof their feathers). See also **oil gland.**

USDA United States Department of Agriculture.

USDA sire summary A summary of official milk production records of daughters of sires to aid in selection of the best genetic material available for breeding dairy cattle.

USDHHS United States Department of Health and Human Services.

USP United States Pharmacopeia. A unit of measurement or potency of biologicals that usually coincides with an international unit. See **IU.**

USPHS United States Public Health Service.

vaccination From the Latin *vacca,* meaning "cow." Artificial immunization. To inoculate with a mildly toxic preparation of bacteria or a virus of a specific disease to prevent or lessen the effects of that disease. Originally done with cow serum.

vaccine A suspension of attenuated or killed microorganisms (bacteria, viruses, or rickettsiae) administered for the prevention, amelioration (improvement), or treatment of infectious diseases.

vaporization The conversion of a solid or liquid into a vapor without chemical change.

variance "The clay of the breeder." Variance is a statistic that describes the variation that can be seen in a trait. Without variation no genetic progress is possible, since *genetically superior* animals would not be distinguishable from *genetically inferior* ones.

vascular   Concerning blood vessels.

vasectomy   The surgical removal of part or all of the vas deferens. This renders a male sterile without affecting his libido.

vasoconstriction   Constriction of blood vessels.

vasodilation   The dilation of blood vessels resulting from stimulation by a nerve or drug.

vealer   Calves fed for early slaughter (usually less than 3 months old).

vector   From the Latin *vector,* meaning "carrier." An organism, such as a mosquito or tick, that transmits microorganisms that cause disease.

venison   The edible flesh of deer.

venom   Poisonous secretion of bees, scorpions, snakes, and certain other animals.

ventilation rate   The volume of air exhaled per unit time.

ventral   Denoting a position toward the abdomen or belly (lower) surface. It means the same as *anterior* in human anatomy.

ventricular fibrillation   Very rapid uncoordinated contractions of the ventricles of the heart, resulting in the loss of synchronization between heartbeat and pulse beat. Ventricular fibrillation often results from a severe electrical shock.

vermicide   Any substance that kills internal parasitic worms.

VFA (volatile fatty acids)   Commonly used in reference to acetic, propionic, and butyric acids produced in the rumen of cattle, goats, and sheep, in the cecum of sheep, the cecum and colon of swine, the colon of the horse, and the cecum of the rabbit.

viability   Ability to live.

viremia   An infection of the bloodstream caused by a virus.

virosis   A disease caused by a **virus.**

virucide   A chemical or physical agent that kills or inactivates viruses; a disinfectant.

virulence   The degree of pathogenicity (ability to produce disease) of a microorganism as indicated by case fatality rates and/or its ability to invade the tissues of a host.

virulent   Poisonous or harmful; deadly; of a microorganism, able to cause a disease by breaking down the protective mechanisms of a host. Fully active organisms.

virus   One of a group of minute infectious agents. They are characterized by a lack of independent metabolism and by the ability to replicate only within living host cells. They include any of a group of disease-producing agents composed of protein and nucleic acid. Viruses are filterable and cause such diseases in people as rabies, poliomyelitis, chicken pox, and the common cold.

viscera   The internal organs of the body, particularly in the chest and abdominal cavities, such as the heart, lungs, liver, intestines, and kidneys.

vitamins   Exogenous organic catalysts (or essential components of catalysts) that perform specific and necessary functions in relatively small concentrations in an animal.

viviparous   Producing living young (as opposed to eggs) from within the body in the manner of nearly all mammals, many reptiles, and a few fishes.

void   To evacuate feces and/or urine.

volatile fatty acids   See **VFA.**

vomiting   The forcible expulsion of the contents of the stomach through the mouth.

walking horse   Any horse trained to do the running walk, fox trot, and canter.

wax gland   A wax-secreting gland of the worker bee.

weanling   A recently weaned animal.

wether   A male sheep or goat castrated before sexual maturity. (See Appendix A.)

wheal   A flat, usually circular, hard elevation of the skin, commonly accompanied by burning or itching. Its formation follows an irritation or other means of increasing the permeability of the vascular walls of the skin.

whelp   To give birth to, as by a female dog. (See Appendix A.)

whey   The water and solids of milk that remain after the curd is removed (e.g., in the manufacture of cheese). It contains about 93.5 percent water and 6.5 percent lactose, protein, minerals, enzymes, water-soluble vitamins, and traces of fat.

whinny   The gentle, soft cry of a horse.

WHO   World Health Organization. An agency of the United Nations founded in 1948. It seeks to promote worldwide health and prevent outbreak of disease. It assists countries in strengthening public health services. It plans and coordinates international efforts to solve health problems, with special attention to malaria, tuberculosis, and venereal, virus, and parasitic diseases. It works with member countries and other health organizations to collect information on epidemics; to develop international quarantine regulations; and to standardize medical drugs, vaccines, and treatment. More than 100 countries belong to WHO, which is headquartered in Geneva, Switzerland.

whole-body counter   A device used to identify and measure the radiation in the body of humans and animals; it uses heavy shielding (to keep out background radiation), ultrasensitive scintillation detectors, and electronic equipment.

whorl   A swirl, or cowlick, in an animal's hair.

with calf   Designating a cow that is pregnant.

wool   The soft and curly hair obtained from sheep.

woolskins   Sheepskins tanned with the wool on.

work   A term commonly associated with the use of horses to round up, cut, etc., cattle.

wriggler   The larva of a mosquito.

X   Designates the chromosome set for sex determination. Chromosomes occur in pairs, except for the sex chromosomes. There are two types of sex chromosomes, the X and Y. Males are XY and females are XX. Since the female can produce only ova that are X, the male sperm determines the sex of the individual at conception. The male has two kinds of sperm, X-carrying and Y-carrying. Union of the X sperm with the X ovum produces XX, a female. Union of the Y sperm with the X ovum produces XY, a male.

x-rays   Radiation produced when electrons in a vacuum tube are projected at very high tension and velocity to strike a metallic target. These are electromagnetic waves, but their wavelength is only about one-thousandth of that of visible light. X-rays are sometimes called roentgen rays, after their discoverer, Wilhelm Roentgen. See **radiograph.**

Y chromosome   The differential sex chromosome carried by one-half of the male gametes in humans and some other male-heterogametic species in which the homologue of the X chromosome has been retained. See also **chromosome** and **X.**

yean   To give birth to young, especially by goats and sheep.

yeanling   A young goat or sheep.

yearling   Refers to a male or female farm animal (especially cattle and horses) during the first year of its life.

yeld mare   A dry (nonlactating) mare or a mare that has not raised a foal during a particular season.

yogurt   Fermented milk, low-fat milk, or skim milk, sometimes **protein-fortified.** Milk-solid content is commonly 15 percent. Most yogurt is high in protein and low in calories.

Zebu   A strain of cattle originating in India; widely domesticated throughout India,

China, and East Africa, used as beasts of burden and meat animals and for their milk. The Zebu has a large hump over the shoulders, pendulous ears, and a large dewlap. Also called *Brahman*.

zo- The prefix *zo-* implies *animal*.

zoonoses (plural of zoonosis) Those diseases and infections that are naturally transmitted between vertebrate animals and humans.

zygote A diploid cell produced by the union of haploid male and female gametes.

INDEX*

Abomasum, relative size of, 464*f*

Absorption, 460

of feed nutrients, 470

Acaricides, 618

Accessory sex glands, 261

Acclimatization, 403

Accretionary growth, 230

Achromotrichia, 482*t*

Acromegaly, 239

ACTH (adrenocorticotropic hormone), 202, 280

Actin, 193

Actinomycosis, 571

Activation analysis, application of, in research, 721–723

Adaptation, 397

of cattle (regional), 401–402

to environment, 403–405

natural, 404

physical environmental factors affecting, 403–404

Adaptive physiology, comparative, of humans and animals, 405

Addisonian pernicious anemia, 507

Additive (*see* Gene action, additive)

ADG (*see* Average daily gain)

Adipoblast, 232

Adipocyte, 232

ADP (adenosine diphosphate), 194

Adrenal glands, 217*t*

Adrenal hormones:

and growth, 241–242

effect of, on lactation, 343

Adrenalin, 219

effect of: on milk letdown, 354

on muscles, 193

and temperature changes, 409

(*See also* Epinephrine)

Adrenocorticotropin, 216*t*

Advanced Animal Breeder, 320

Aedes aegypti, 550, 571, 613

Aflatoxin, 586*n*, 587

*Page numbers followed by *f* refer to figures; page numbers followed by *t* refer to tables; page numbers followed by *n* refer to footnotes. See also the Glossary, pp. 771–834.

African horse sickness, 590
African sleeping sickness, 572, 602t, 607t
African swine fever, 590
African trypanosomiasis, 572
Afterbirth, placental membranes, 277f
Agammaglobulinemia, 544
Age:
  effect of: on egg laying, 387–388
    on egg production, 248t
    on food conversion, 474–475
    on lactation, 357
    on litter size, 248t
    on milk composition, 363
    on social dominance, 655
    on weaning weight of calves, 248t
  and homeothermy, 408
  of horses, 686, 687–688f
  of puberty in farm mammals, 268t
Agglomeration, 100–101
Agglutination, 544–545
Agglutination test, 558–559
Aging, 247–253
  theories of, 251–252
  (See also Age)
Agonistic (combat) behavior of animals,
  642–644
Agricultural education and research, 5–6
Agricultural production, regional
  distribution of, 7f
Agricultural research, 5–6
  and animal production, 700, 757
  financing of, 741–743, 744f
  by governmental agencies, 741–742,
    744f
  by industry, 743, 744f
  and the United States economy, 757
Agricultural Research Service (see
  USDA, Agricultural Research
  Service)
Agriculture and the balance of nature,
  38–40
AI (see Artificial insemination)
Air conditioning, effect of, on milk
  production, 424
Albinos, 663, 667t
Albumen (see Eggs, albumen of)
Alkali disease, 522t, 532
Allantoin, 210f

Alleles, 137
Allelomorphs, 51, 135
Allen's rule, 410–411
Alopecia, 482t
Altitude, adaptation to, 484
Altricial birds, 372t, 374
Alveoli:
  of lungs, 205f
  of mammary glands, 332, 334f
American bison, 641–642, 647
American Veterinary Medical
  Association, 320
Amino acids, 473
  balance of, in milk, 69f
  as determined by GC, 729, 730f
  essential, 67–69, 441–442, 446f
  nonessential, 441–442, 446f
  in protein synthesis, 125t
  of selected foods, 68t
  sulfur-containing, 519–520
Amnion, 276
Amniotic fluid, 276
Amoebic dysentery, 582
Ampule, 317
Ampulla, 260
Amylase, 467–468t
Amylopsin, 467, 468t
Anaerobic digester, 748f
Anaplasmosis, 572, 601t, 602t, 612
Anatomy:
  of avian male reproductive system,
    376f
  of circulatory system, 195–202
  comparative, 184
  of digestive systems, 460, 461f
  of farm animals, 184–211
  of female reproductive tract, 261–265
  gross, 184
  of male reproductive tract, 256–261
  microscopic, 184
Ancestors, relationships to, 166t
Ancylostomiasis (see Hookworm)
Androgens, 217t, 219, 241–242
Anemia, 199, 481–482t, 504, 506f,
    521–522t, 523, 571, 599, 692t, 713
  in horses, 691, 692t
  pernicious, 482t, 507
  sickle-cell, 140, 523

Aneuploidy, 112
Animal and Plant Health Inspection
   Service (*see* USDA, Animal and
   Plant Health Inspection Service)
Animal agriculture:
   as an employer, 34
   and food production, 34
   history and development of, 1–2
   and humanity, 1–41
   as a source of recreation, 34–35
   and United States economy, 32–35
   and world economy, 6
Animal behavior, 630–657
   agonistic (combat), 642–644
   defined, 656
   eliminative, 641
   gregarious, 652–654
   and heredity, 632–633, 635
   hypothalamus in, role of, 636
   ingestive, 639
   instinctive, 633, 635, 656
   investigative, 651–652
   and malnutrition, 635
   mother-young, 645–651
   at parturition, 646–651
   and population density, 655–656
   in response to stimuli, 630
   sexual, 644–645
   shelter-seeking, 641–642
   social dominance, 654–655
   types of, 639–654
Animal breeding, laws of probability in,
   136–137
Animal communication, methods of,
   637–639
Animal diseases:
   and economic losses to farmers,
      537–538, 593
   and the health of humans, 536–593
   research on, 735–741
   transmissible to humans, 546–581
Animal fats:
   and atherosclerosis, 85–88
   and consumption trends, 86*f*
   and heart disease, 87–88
Animal health:
   governmental safeguards of, 588–590
   and the health of humans, 536–593

Animal health (*Cont.*):
   importance of, 588–590
Animal Import Center, 591
Animal intelligence, 635–636
Animal kingdom, important phyla of,
   599*t*
Animal learning, 633–635
Animal production:
   energy and efficiency of, 28–30
   world trends in, 11–20
Animal products, 64–107
   cholesterol in, 70*t*
   comparative nutritional contributions
      of, 66–78
   composition of, 67*t*
   consumption trends of, 80–84
   dietary contributions of, 80*f*
   economy of, 35–36, 78, 80
   export of, 37–38
   fatty acids in, 70*t*
   as a food buy, 66
   fortification of, 90–92
   future of, 101–105
   and growth in humans, 64–107
   history of availability of, 64–66
   and humans, 64–107
   labeling of, 90–91
   legal aspects of fortifying, 90
   lyophilization of, 101
   mineral content of, 76*t*
   new, 103–105
   as nutrient source, 81*t*
   pesticide residues in, 106–107, 619
   preservation of, 92–95
   synthetic, 105–106
   vitamins of, 78, 79*t*
Animal protein, 67–69
   availability of, 20–28
   nutritional merits of, 20–21
Animal quarantine, 590–591
Animal research, 698–760
   in animal diseases, 735–741
   associations and societies, 711
   atoms in, use of, 712–726
   in biotechnology, 749–753
   computers in, use of, 708–709
   contributions of, 699–700, 756–760
   control in, 706, 707*f*

Animal research *(Cont.)*:
  defined, 699
  disciplines of, 699
  and efficiency of production, 699
  in electronic technology, 753–755
  in endocrinology, 726–727
  financing of, 741–743
  fistulated animals in, use of, 734,
      735–736*f*
  gas chromatography (GC) in, use of,
      727–729
  goals of, 702
  high-pressure liquid chromatography
      (HPLC), 729–730, 731*f*
  in hybridoma technology, 752–753
  kinds of, 701–702
  library in, use of, 709–710
  low-level radiation laboratory in, use
      of, 713–718
  nonradioactive isotopes in, use of, 723,
      725–726
  nuclear reactor in, use of, 720–723
  opportunities in, 756
  organizations for, 710–711
  pigs in, use of, 730–731, 732–733*f,* 734
  in recombinant DNA, 750–752
  schematic pathway of, 705*f*
  and the scientific method, 703–706
  statistics in, use of, 706–708
  trends affecting, 743–755
  in waste management, 747–748
Animal science, societies and journals
    related to, 710–711
Animal scientist, identification of, 756
Animal size, effect of, on lactation,
    357–358
Animal wastes, utilization of, 30–32
Animals:
  behavioral responses of, 631–636
  classification of, 598
  in competition with humans for food,
      25–28
  composition of, 434, 447
  domestication of, 1–2, 3*t*
  ways they serve humans, 2
Anions, 513
Annelida, 599*t*
Anorexia, 514*t,* 516, 522*t*

Anterior pituitary gland, hormones of,
    216*t*
Anthrax, 565–566, 612
Antibiotics, 456–457, 605
  and feed digestibility, 472
  feeding to swine and poultry, 472–473
  use of, in semen, 314
  and vitamin K synthesis, 490
Antibodies, 456–457, 539–543, 734
  monoclonal, 543
Anticodons, 127
Antigen-antibody reactions, 544–546
Antigens, 540–542
Antioxidants, 487
Antiserum, 540
Antitoxins, 541, 546
Aortic rupture in poultry, 525*f*
APHIS *(see* USDA, Animal and Plant
    Health Inspection Service)
Appetite, factors affecting, 466
Appetitive behavior in animals, 645
Applied research, 701
Aquaculture, 19–20
Arable land per capita, 164
Arboviruses, 551*t*
Arsenic, 534, 722
Arteries, 196
Arterioles, 198
Arteriovenous differences, 345–347
Arthropod-borne diseases transmitted
    from animals to humans, 551*t*
Arthropod control, 617–627
Arthropoda, 599*t,* 602*t*
Arthropods:
  affecting humans and/or domestic
      animals, 551*t,* 608–617
  as vectors of disease, 607*t*
Artificial insemination, 295–327
  in bees, 321
  defined, 295
  and estrous synchronization, 324
  future of, 324–327
  and genetic effect of PDM, 170*f*
  and genetic engineering, 325
  history of, 295–298
  of horses, 671
  in humans, 321–324
  limitations of, 301–302

Artificial insemination *(Cont.)*:
  merits of, 300–301
  in poultry, 320–321
  in progeny testing, 325
  regulations in cattle, 318–319*t*
  species in which used, 298
  timing of, in cows, 318
  use of: in the United States, 298–299*t*
    in the world, 299*t*
Artificial vagina (AV), 305*f*, 306
Ascaridiasis, 602*t*
Ascarids of horses, 691
Ascaris, 580
Ascorbic acid, 492–494
  (See also Vitamin C)
Ash, 449
  and growth rate of young, 77*t*
Asiatic cholera, 607*t*
Assay techniques for vitamins:
  biological, 509
  chemical, 509–510
  microbiological, 509
Ataxia, 523
Atherosclerosis, 84–88, 433, 734
Atoms in animal research, 712–726
ATP, (adenosine triphosphate), 194
Atrophic rhinitis, 516
Autonomic nerves, 208
Autosomes, 111
AV (artificial vagina), 305*f*, 306
Average daily gain (ADG), 475
  calculation of, 152*t*
Avian digestion, 469–470
Avian digestive system, 462*f*
Avian distemper, 549–550
Avian pneumoencephalitis, 549–550
Avidin, 504
Axon, 206–207
Azoturia, 692*t*

B vitamins:
  and formation of red blood cells,
    491–492
  and release of energy from food,
    491–492
Bacillus anthracis, 565
Backcross, 180

Backfat of swine, 53*f*
Bacterial infections, 581–582
  transmissible from animals to humans,
    556–571
  transmissible from humans to animals,
    581–582
Bactericidal reaction, 545
Bactofugation, 93
Baldness and heredity, 129
Barrow, 258
Bartonellosis, 569
Basal heat production, 418–419
Basic research, 701
Basophils, 202
Bats, navigating of, by sound, 637
BCT (brucellosis card test), 559–560
Bedbugs, 600*t*, 617
Bee milk, 72
Bee venom, 72
Beebread, 72
Beef:
  cholesterol in, 70*t*
  composition of, 67*t*
  minerals of, 76*t*
  vitamins of, 79*t*
Beef breeds and artificial insemination,
  319*t*
Beef cattle, breeds of, 48*t*, 52–53,
  55–56*f*
Beef industry of the United States, 66
Beef steer:
  defined, 258
  external parts of, 187*f*
Beef tapeworm, 575, 576*f*
Bees, 600*t*
  (See also Honeybees)
Beeswax, 602
Beetles, 600*t*
Behavior:
  animal *(see* Animal behavior)
  maternal, 645–651
Behavioral responses of animals, 632–636
  external (learning experiences), 632
  internal (heredity), 632
Bell-shaped curve in phenotypic
  variations, 151*f*
Bergmann rule, 410–411
Beriberi, 478, 481*t*, 494–496

Best linear unbiased prediction (BLUP), 171–172
Bile, 468t
Binomen, 598
Bioassay of crops, use of insects in, 604, 605f
Biological assay of vitamins, 509
Biological control:
    of insects, 604, 620–626
    of rabbit population, 622
Biological half-life, 720
Biological value, 25, 68, 443
Biological year, 384
Biotechnology in animal research, 749–753
Biotin, 504–506
    clinical deficiency symptoms, 482t, 504–506
    common sources of, 482t, 506
    functions of, 482t, 504–506
Bird eggs, 371
    development of, 375–379
Birds:
    altricial, 372t, 374
    communication of, by sound, 637
    precocial, 372t, 374
Birth rate, effect of heterosis on, 179t
Birth size, maternal influence on, 235
Black death, 568
Blackbodies, 425
Blastoderm, 372, 374
Blastodisk, 372
Blastomeres, 134
Blind staggers, 522t, 532
Bloat, 437, 466
Blood:
    clotting, role of vitamin K in, 488, 490–491f
    compatability of types of, 200f
    composition of, 198–199
    plasma, 198
    portal system, 197
    serum, 198
    types of, 199–202
    typing of farm animals, 201–202, 320n
Blood flukes, 577
Blood platelets, 202
Blood spavin of horses, 682

Blood spots of eggs, 391
Blood-typing, 201–202, 320n
Bloodworms of horses, 691
Bloom of an egg, 373
Bluetongue, 616
BLUP (best linear unbiased prediction), 171–172
Body condition, effect of, on lactation, 358–359
Body conformation of horses, 668, 669–670f
Body weight and feed conversion, 475t
Boll weevil, 624–625
Bolus, 466
Bomb calorimeter, 452, 453f
Bone:
    growth of, 190
    parts of, 190f
Bone marrow, 190
Bone meal, 516
Botflies, 611–612, 690
Bots, 602t
Botulism, 582
Bovine tuberculosis, 593
Bovitec, 456
Bowman's capsule, 209, 210f
Brachiopoda, 599t
Brahmans, 46, 426f
    heat tolerance of, 404, 410, 411f, 419f
Brain:
    cholesterol in, 70t
    gross anatomy of, 207f
    size and lactose, 74t
Bran, 494
Breast-feeding, 73, 339, 351
    and conception control, 339
Breed associations, 45–46
Breeding mares, 672
Breeds:
    of beef cattle, 48t, 52–53, 55–56f
    of dairy cattle, 48t, 53, 57f
    defined, 43
    development of, 43–52
    dual-purpose, 47
    of fowl, 50t, 59, 61–62f
    genetic differences in, 50–52
    of horses, 49t, 57, 60f, 661t
    of sheep, 49t, 53, 58f

Breeds *(Cont.)*:
  of swine, 48*t*, 52, 54*f*
  triple-purpose, 47
Bronchi, 205
Broodiness, 650–651
  how delayed and induced, 386
Browning reaction, 100
BRT, (brucellosis ring test), 559
Brucella:
  abortus, 558–560, 562
  melitensis, 558–560, 562
  routes of infection, 561*f*
  suis, 558–560, 562
Brucellosis, 558–563
  card test, 559–560
  in cattle, swine, goats and sheep, 560
  diagnosis of, 544
  in humans, 561–562, 564*f*
  milk ring test, 559
  occurrence of, 558
  prophylaxis and control of, 562–563
  state classifications for cattle, 560*t*,
    562*f*
  tests for, 558–560
  therapy of, 563
Bryozoa, 599*t*
Bubonic plague, 568–569, 607*t*
Buffalo chips, 641
Bull stud, 301*n*
Buller syndrome, 644
Butter:
  cholesterol in, 70*t*
  composition of, 67*t*
  consumption of, 86*f*

Caffeine, libido and, 320*n*
Calcitonin, 217*t*
Calcium, 513–516
  and blood clotting, 490*f*, 514*t*, 515
  blood levels of, 221
  clinical deficiency symptoms of, 514*t*,
    516–517
  in forages, 513, 516
  functions of, 513, 514*t*, 516
  in grains, 516
  in muscle contraction, 195
  sources of, 514*t*, 516

Calcium/phosphorus ratio, 75, 513,
    515–516
  in feeding horses, 675
Calories, 250
  of carbohydrates, 452
  defined, 451
  expended during lifetime, 250
  of fats, 452
  of protein, 452
Campylobacter fetus, 569
Cancer of the eye, 427
Cancer cells, division of, 231
Canine distemper, 541
Cannibalism in animals, 656
Cannula, 734, 735*f*
Canter, 662
Capacitation of spermatozoa, 274
Capillaries, 196
Caponize, 655
Capybara as source of protein, 25
Carbohydrates, 435, 437
  of animal products, 67*t*, 70–75
  caloric value of, 452
  classification of, 436*t*, 446*f*
  consumption trends of, 80*t*
  digestion of, 467, 468*t*
Carbon monoxide, 206
Carboxyhemoglobin, 206
Carboxypeptidase, 468*t*, 469
Carcass, amount of heterosis for, 179*t*
Carnitine, 88–89
Carnivores, 459, 461*f*
Carotenes, 439
Carrier, 538
Carrion, 538–539
Carrion's disease, 569
Cartilage, 190
Castration, 257–258
  and animal behavior, 643
  chemical, 223
  effect of: on growth, 241–242
    on social dominance, 655
  of horses, 674
Cat ascarid, 579
Cat hookworm, 579–580
Cations, 513
Cattle:
  breeds of, 48*t*, 52–53, 55–57*f*,

Cattle *(Cont.)*:
  domestication of, 3*t*
  leading countries in production of, 15*f*
Cattle grubs, 507
Cattle tick fever, 613–614
Cattle warble, 611
Caudal, defined, 187, 333
Cecum, 461*f,* 463*t,* 468*t*
Cell division, 113–119
  of cancerous cells, 231
  of normal cells, 231
Cells:
  illustration of, 111*f*
  types of, in growth, 230–231
Cells of Leydig, 266
Cellulase, 468*t*
Cellulose, 436–437, 446*f,* 464, 467–468*t,*
    472, 746, 758
Centaur, 658
Cercariae, 577
Cerebellum, 207
Cerebrum, 207
Certified Semen Services, Inc., 320
Cervix, 262, 262*f,* 263*f,* 263–264
Cesarean operation, 280–281, 656
  in sows, 732–733*f*
Chalaza, 374
Chastek paralysis, 481*t,* 496
CHD (coronary heart disease), 84–88
Cheddar cheese:
  cholesterol in, 70*t*
  composition of, 67*t*
  minerals of, 76*t*
  vitamins of, 79*t*
Chédiak-Higashi syndrome, 544
Chemical assay of vitamins, 509–510
Chemical preservation of animal
    products, 94
Chemicals:
  and animal communication, 638
  in insect control, 618
Chicken:
  cholesterol in, 70*t*
  composition of, 67*t*
  minerals of, 76*t*
  vitamins of, 79*t*
Chicken pox, 581
Chickens:
  breeds of, 59, 61–62*f*

Chickens *(Cont.)*:
  classifying baby chicks, 375
  digestive system of, 204*f*
  domestication of, 3*t*
  external parts of, 189*f*
  skeleton of, 192*f*
Chiggers, 555, 607*t,* 615
  as vectors of disease, 551*t*
Chlordane, 618
Chlorine, 517–519
  clinical deficiency symptoms of, 514*t*
  functions of, 514*t,* 519
  sources of, 514*t*
Chlorophyll, 435, 440
Cholera, 567–568
Cholesterol, 219, 220*f,* 439, 446*f*
  in animal products, 70*t*
  effect of diet on amount in eggs,
      391
  and heart disease, 84–88
  synthesis of, 84–85
Choline, 482*t,* 507–508
  clinical deficiency symptoms of, 482*t,*
      507–508
  common sources of, 482*t,* 507–508
  functions of, 482*t,* 507–508
Chondrocytes, 232
Chordata, 599*t*
Chorion, 276
Chromosomes, 110–112, 291
  abnormalities of, 112–113, 114*f*
  of chickens, 113
  lethal, 112–113
  nonhomologous, 114*f*
  number of, in animals, 112*t*
  sex, 111, 142–143*f*
Chymotrypsin, 468*t,* 469
Circulatory system, 195–202
Cirrhosis, 482*t*
Classification:
  of digestive systems, 461*f*
  of nutrients, 446*f*
Claude Bernard rule, 412
Claudication, 680
Climate, 403
  effect of, on production, 428
  and the progress of people, 397
Climatic chamber, 428–430*f*
Cloaca, 209

Clone, 283
Cloning, 290
  of embryos, 134
Clonorchiasis (Oriental liver fluke
    disease), 578
Clutch, 385–387
Clutch size of selected birds, 387t
Coat color:
  changes with environmental
      temperatures, 426
  of horses, 663, 665–668
Cobalamin, 491, 507
Cobalt, 527–528
  clinical deficiency symptoms of, 522t,
      527–528f
  functions of, 522t, 527–528
  sources of, 522t
  and vitamin B_{12}, 507
Coccidiomycosis, 573
Cockroaches, 600t
Codon, 127
Coefficient of digestibility, 70
Coelenterata, 599t
Coenzymes, 481
Cold habituation, 407–408
Cold sterilization, 94
Cold sweat, 421
Cold tolerance of animals, 405
Colic in horses, 675, 683, 690, 692t
Collagen, effect of age on, 250
Collateral relatives, 165–166t, 668
  relationships of, 166t
Colorado tick fever, 551t, 607t
Colostrum, 336, 346
  composition of, 362t
    effect of feed on, 361n
  immunological aspects of, 364–
      366
  importance of, 364–366
Combat behavior, 642–644
Comfort zone:
  of cattle, 404–405
  of poikilotherms, 417–418
  (See also Thermoneutral zone)
Compaction, 437
Companion animals, 745–746, 758
Compensatory gains, 243
Compensatory growth, 230
Complement, 545–546

Complement fixation, principles of, 545f
Complement fixation reaction, 545–546
Complementary milk, 353
Computers:
  and animal research, 708–709
  and medicine, 709
Concentrate feeds for horses, 675–676
Concentrates, crude fiber of, 448
Conception rate, effect of heterosis on,
    179t
Conditioned reflex, 634
Conditioning:
  classical (associate learning), 634
  operant (trial and error), 634
Conduction and heat loss, 420
Confinement operations and waste
    management, 747
Conformation of horses, 668, 669–670f
Congenital malformation and toxic
    plants, 584f
Consumption trends:
  of dairy products, 83t
  of fats, 83t
  of fish, 83t
  of meat, 83t
  of oils, 83t
  of poultry products, 83t
Contagious bovine pleuropneumonia,
    590
Convection and heat loss, 420–421
Cooling in nonsweating species, 422–425
Cooling mechanisms, 407
Copper, 523–524
  clinical deficiency symptoms of, 522t,
      523–524f, 525f
  functions of, 522t, 523–524
  sources of, 522t, 523
Coronary heart disease (CHD), 84–88
Corpus albicans, 264–265
Corpus hemorrhagicum, 264
Corpus luteum, 264–266, 269, 282–283,
    324, 326, 338f
Correlations, genetic, 174
Corticotropin-releasing hormones
    (CRH), 216t
Cortisol, 220f
Cosmic radiation, 427
Cottage cheese, composition of, 67t
Cotyledons, 276, 277f

Coumarin, 586
Courtship among animals, 645
Cow Performance Index, 171
Cow ponies, 632, 659, 668, 685–686
Cow trap, 653*f*
Cowper's glands, 261
Cowpox, 552
Cows:
  brain of, 218*f*
  circulatory system of, 196*f*
  drying off, methods used in, 355
  external parts of, 185*f*
  as filter of radionuclides, 89–90
  insemination of, 318–319
  internal parts of, 203*f*
  muscles of, 193*f*
  reproductive organs of, 262*f*, 263*f*
  skeleton of, 191*f*
Coxiella burnetti, 556
Cranial, defined, 187, 333
Crazy-chick disease, 488
Creaming of milk, 362
Creeping eruption, 611
Cretinism, 241
Cribbing of horses, 683
Crisscross, 180
Crop, 460, 462*f*, 467–468
Crop milk, 340, 378
Crossbreeding, 178–180, 632
  and animal behavior, 632
  and growth, 246
Crossbreeding effect, type of gene action
    indicated by, 163*t*
Crossing, 163
Crossing effect, type of gene action
    indicated by, 163*t*
Crossing over, 117, 136
Crude fat, 448
Crude fiber, 448–449
Crude protein, calculation of, 448
Cryptorchid, 256–257
Cryptosporidiosis, 572–573
Ctenophora, 599*t*
Cultures, killed, 541
Cumulus oophorus, 274–275
Curled-toe paralysis, 481*t*, 498*f*
Cutaneous larva migrans, 579
Cutting horse, 632, 633*f*

Cyanide, 583
Cyanocobalamin (B_{12}), 482*t*, 507
  clinical deficiency symptoms of, 482*t*,
    507
  common sources of, 482*t*, 507
  functions of, 482*t*, 507
Cystic follicles, 272
Cytochrome C, 521
Cytokinesis, 115

Dairy breeds, composition of milks of,
    360, 361*t*
Dairy cattle, breeds of, 48*t*, 53, 57*f*
Dairy cows (*see* Cows)
Dairy products, consumption of, 83*t*
DDT (dichlorodiphenyltrichloroethane),
    106–107, 572, 617–619, 621, 627
Deerflies, 612
Dehydration, 93–94
Dehydration exhaustion, 415
7-Dehydrocholesterol, 91, 485
Delayed implantation, 276
Deletion, 113
Dementia, 481*t*, 501
Demodectic mange, 614–615
Dendrites, 206
Dengue, 550
Dermacentor andersoni, 554
Dermatitis, 439*t*, 482*t*, 498*f*, 505*f*
Dermis, 187
DES (diethylstilbestrol), 219, 456
Designer genes, 132–134
Dexamethasone, 292
DHIA (Dairy Herd Improvement
    Association), 454
  sire summaries, 170–171
Diaphysis, 191
Dicalcium phosphate, 516
Dicoumarin, 586
Dicoumarol, 490
Dieldrin, 618
Diestrus, 270
Diet composition, 472–473
Dietary nutrients furnished by major
    food groups, 81*t*
Diethylstilbestrol (DES), 219, 456
Differentiation, 226, 233

Digestibility:
  effects of diet composition on, 472–473
  factors affecting, 470–473
  of feeds, 449–451, 470–473
Digestible energy, 452–453, 455f
Digestion:
  avian, 469–470
  in body regions, 468t
  of carbohydrates, 467–468t
  of cellulose, 468t, 472–473
  effect of heating on, 472
  factors of, 463
  of fats, 468–469
  in monogastric animals, 463–464
  physiology of, 459–476
  in polygastric animals, 464–466
  process of, 463–466, 468t
  of proteins, 468–469
Digestion coefficient, 449–451
Digestion stall, 450f
Digestive systems, 203, 460–463
  anatomy of, 460–463
  avian, 460, 462f
  capacity of, 460–463t
  classification of, 461f
    of monogastric animals, 461f
    with functional cecum, 461f
  of polygastric animals, 461f
  types and capacities of, 460–463
Digestive tract, enzymes of, 467, 468t
Diluters (semen), 312–313
Diploid, 111
Disaccharides, 435–436, 446f
Disease:
  body defenses against, 539
  defined, 538
  diagnosis of, 546
  genetic resistance to, 540, 544
  infectious, 538
  and milk composition, 364
  modes of spreading, 538–539
  noninfectious, 538
  research on, 735–741
  symptoms in horses, 689–690
  testing for, 546
  types of, 538
Disease protection, recent application of
  technology to, 541–543

Disease resistance:
  acquired, 540–541
  natural, 540
Diseases:
  of animals transmissible to humans,
    546–581
  associated with selected parasites, 602t
  blood-typing of, 544
  epizootic, 590
  of humans associated with selected
    arthropods, 607t
  nutritional, of horses, 690, 692–695t
  protecting United States livestock from
    foreign, 590–592
  transmitted by arthropods, 607t
Diseases (bacterial) of animals
    transmissible to humans, 556–571
  actinomycosis, 571
  anthrax, 565–566
  bartonellosis, 569
  brucellosis, 558–563
  cholera, 567–568
  endemic relapsing fever, 569
  erysipelas, 570
  glanders, 570
  leptospirosis, 563–565
  listeriosis, 570
  paratyphoid fever, 567
  plague, 568–569
  salmonellosis, 566–567
  tuberculosis, 556–558
  tularemia, 568
  typhoid fever, 567
  vibriosis, 569–570
Diseases (fungal):
  actinomycosis, 573
  histoplasmosis, 573
Diseases (helminthic or worm)
    transmissible to humans by animals,
    573–581
  clonorchiasis (Oriental liver fluke
    disease), 578
  elephantiasis, 581
  enchinococcosis (hydatid disease),
    577
  fascioliasis (sheep liver fluke disease),
    577–578
  hookworm (ancylostomiasis), 578–580

Diseases (helminthic or worm) *(Cont.)*:
  paragonimiasis (Oriental lung fluke disease), 578
  roundworms, 580–581
  taeniasis, 574–577
  trematodiasis, 577
  trichinosis, 573–574
Diseases (protozoal) transmissible to humans by animals:
  African trypanosomiasis, 572
  anaplasmosis, 572
  cryptosporidiosis, 572–573
  malaria, 571–572
Diseases (rickettsial infections), 554–556
Diseases transmissible by animals as passive carriers, 582–583
  botulism, 582
  gas gangrene, 582
  tetanus, 582
Diseases (viral) of animals transmissible to humans, 546–554
  contagious ecthyma, 554
  dengue, 550
  encephalomyelitis, 550–551
  foot-and-mouth, 552–553
  Newcastle, 549–550
  psittacosis (ornithosis), 549
  rabies, 547–549
  smallpox (variola), 522
  vesicular stomatitis, 553
  yellow fever, 550
Distemper, 581
DNA (deoxyribonucleic acid), 120–132, 144, 199, 211, 242, 337, 515, 529, 546, 553, 636, 720, 729, 730, 749–752
DNA polymorphisms, 752*f*
Dog fly, 609
Dog hookworm, life cycle of, 579*f*
Dogs, domestication of, 3*t*
Domestication of animals, 1–2, 3*t*, 4–5
Dominance, 138–140
  and coat color of horses, 665
  complete, 154
  incomplete, 154
  lack of, 140
  partial, 140
Dominant, defined, 138

Dormancy, 407
Dorsal, 185–186
Double-yolked eggs, 374, 393
Down's syndrome, 112
Drosophilia, 604
Drugs, effect of: on egg laying, 389
  on milk composition, 364
Dry matter, calculation of, 447–448
Dry period, effect of, on lactation, 358
Drying off cows, methods used in, 355
Ducks, domestication of, 3*t*
Duplication, 113, 114*f*
Dystocia, 648

East Coast fever, 590
EATA (estimated average transmitting ability), 171
Echinococcosis, 577
Echinodermata, 599*t*
Ecology, 430
  defined, 400
  and environmental physiology, 397–431
Ecthyma, contagious, of goats and sheep, 554
Ectoparasites, 537
Edema, 334, 336*f*
Efficiency of feed conversion, factors affecting, 473–476
Efficiency of gain, heritability estimate of, 158*t*
Egg color, 370
  heritability estimate of, 158*t*
Egg hatchability:
  and manganese, 531
  and zinc, 530
Egg laying, 382–383
  effect of: artificial light on, 385, 398–399*f*
    light on, 384–386, 398, 399*f*
    nutrition on, 389
    persistency in, 384
    season on, 384–386
  and evaporative cooling, 424–425
  factors affecting, 383–389
  hormonal control of, 379–382

Egg laying *(Cont.)*:
  intensity, effect of clutch length on, 386–387
  physiology of, 369–394
  by selected birds, 387*t*
Egg production:
  effect of hen's age on, 248*t*
  per hen in the United States, 383
  heritability estimate of, 158*t*
Egg products, 103–105
Egg shape, 370–371
  heritability estimate of, 158*t*
Egg size:
  effect of: age on, 392
    bird size on, 392–393
    nutrition on, 393
    ovum on, 393
  factors affecting, 391–393
  heritability estimate of, 158*t*
Egg weight, 391*n*
  of selected birds, 372*t*
Egg white *(see* Eggs, albumen of)
Egg-white injury, 505
Egg yolk, 374
  double, 374, 393
  factors affecting color of, 390–391
  percent, 372*t*
  use of, in extending semen, 312–313
Eggs:
  albumen of, 374, 394
    percent, 372*t*
  of birds, 371
  cholesterol in, 70*t*
  components of, 373*f*
  composition of, 67*t*, 372*t*
    factors affecting, 390–391
  double-yolked, 374, 393
  formation of, in hen, 378*f*
  frozen scrambled egg mixes, 103–104
  hard-cooked–peeled, 104–105
  immunological and medical aspects of, 393–394
  kinds of, 371
  leading countries in production of, 17*f*
  minerals of, 76*t*
  nature's purpose of, 369–370
  of other species, 371

Eggs *(Cont.)*:
  pasteurization of, 99
  safety of, 98–99
  shell of: percent, 372*t*
    thickness, 373–374
  soft-shelled, 373*n*
  structure of, 372–374
  vitamins of, 79*t*
  within eggs, 382, 383*f*
  yolk of *(see* Egg yolk)
Elastin formation and copper deficiency, 525*f*
Electroejaculation, 306–307
Electron microscope, 736, 739*f*
Electron microscopy, 345–346
Electronic identification of animals, 753–755
Elephantiasis, 581
Eliminative behavior of animals, 641
Emaciation, 522*t*, 528, 533
Embryo transfer (ET), 284–289, 757–758
  procedures used in, 288*f*
  as a research tool, 289
  to test for Mendelian recessive alleles, 286
  and twinning, 286
  use of, in genetic testing, 286–287
Embryology, 184
Embryonic development:
  and dietary zinc, 530*f*
  in poultry, 377–379
Embryos:
  cloning of, 134–135
  freezing of, 289–290
  microsurgery with, 290
  sexing of, 291
Encephalitis, 601*t*
Encephalomalacia, 480*t*, 488
Encephalomyelitis, 550–551
  types of, 551*t*
Enchinococcus (hydatid disease), 577
Endemic relapsing fever, 569
Endemic typhus fever, 607*t*
Endocrine glands, 213–215
  of cows, 214*f*
  secretions of, 214–218
Endocrinology, 213–225, 726
  research in, 726–727

Endoskeleton, 190
Energy:
    of foods, 451–455
    required to double birth weight, 244t
    used by animals, 454–455
Enriched, 494n
Enrichment, 90n
Entomology, defined, 597
Envenomization, 600
Environment, effect of, on thyroxine
    secretion, 403
Environmental physiology, 397–431
    defined, 400
Environmental temperature, effect of, on
    lactation, 359
Enzymes:
    defined, 467
    of digestive tract, 465, 468t
Eosinophils, 202
EPA (Environmental Protection
    Agency), 617, 619
Epidemic typhus fever, 607t
Epidermis, 187
Epididymis, 259–260, 260f
Epiglottis, 204
Epinephrine, 194, 219f
    effect of, on milk letdown, 354
    and temperature changes, 409
Epiphysial groove, 240
Epiphysis, 190
Epistasis, 141, 161–162
Equine encephalomyelitis, 607t
Equine gonadotropin, 672
Equine piroplasmosis, 601t
Erepsin, 468t, 470
Ergocalciferol, 485
Ergosterol, 439, 446f, 485
Ergot, 588
Eructation, 466
Erysipelas, 570
Erysipeloid, 570
Erythema, 599
Erythroblastosis fetalis, 200–201
Erythrocytes in farm animals, 198t
Erythropoiesis, 506, 522t
Escherichia coli and genetic engineering,
    131
Esophageal groove, 464f

Essential amino acids, 441–442, 446f, 465
Essential fatty acids, 438
Essential oils, 440
Estimated average transmitting ability
    (EATA), 171
Estimated net energy, 454
Estivation of insects, 408
Estradiol, 220f
Estrogen, 224, 280, 338f, 341f
    effect of, on growth of mammary
        gland, 335–336, 338f, 341f
    secretion of, by beef and dairy cows,
        341n
    secretion variations, 341n
Estrogen compounds, 224
    and fattening, 224
    and growth, 224
Estrogens, 217t, 219
    effects of, on epiphysial cartilage, 242
    functions of, 269
Estrous cycle, 269–273
    abnormalities of, 272–273
    length of, 270t
Estrous detection, 304
Estrous period, length of, 270t
Estrus (heat), 269–273, 645
    continuous, 271
    detection in cattle, 326
    effect of, on lactation, 343, 358
    in mares, 671
    synchronization of, 222–223, 281–282,
        745
    and artificial insemination, 324
ET (see Embryo transfer)
Ethanol production, 748f, 749f
Ether extract, 448
Euploidy, 113
Evaporation and heat loss, 421–422
Evaporative cooling:
    for cows, 423f
    and egg laying, 424–425
    factors affecting, 421–422
    and milk production, 423–424
Exercise and milk composition, 364
Exocrine glands, 214, 330
Exoskeleton, 190
Exotic breeds, 47
Exploratory behavior, 651–652

Exsanguination, 614
Exudative diathesis, 480t
Eyes:
  of horses, 662
  of houseflies, 608

F body, 291
F$_1$ generation, 136
  and gene frequency, 153
F$_2$ generation, 136
  and gene frequency, 153
Face flies, 621
Famine, historical examples of, 4
FAO (Food and Agriculture
    Organization of the United Nations),
    8, 15n, 17n, 24f, 25, 28f, 32, 593
Farm exports, 37–38
Farm fuel system, schematic of, 749f
Farm products as part of food prices,
    35f
Farm purchases, 34
Fascioliasis (sheep liver fluke disease),
    577–578
Fat:
  of animal products, 67t, 69–70
  brown, 232
  consumption of, 86f
  consumption trends, 80t, 82f, 83t
  deficiency of, 439f
  effect of, on absorption, 439
  of feeds, 448
  on pork and beef carcasses, 81n, 82f
  synthesis in milk, 346
  white, 232
Fat-depressing diets, 363n
Fats, 437–440, 446f, 468t
  caloric value of, 452
  as dietary essentials, 69–70
  digestion of, 468–469
  effect of, on appetite, 438
  saturated, 70
  unsaturated, 70
Fats and oils, consumption of, 83t
Fatty acids, 438, 446f
  essential, 69–70
  saturated, 105
  unsaturated, 105

FCM (fat corrected milk), 474
FDA (see Food and Drug
    Administration)
Feather meal, 31
Feather pigmentation and dietary copper,
    524f
Feathers, 188–190
Febrile reaction, 539
Federal financing of research, 741–
    744
Federal Meat Inspection Act, 588
Feed:
  digestibility of, 449–451
  effect of: on egg composition, 391
    on lactation, 359
    on milk composition, 363
    on milk fat, 346
    on milk secretion in sows, 359n
  heating of, 472
  physical form of, 471–472
Feed additives, 456–457
Feed consumption, effect of hot weather
    on, 413–414
Feed conversion, 20, 470, 758
  of beef cattle, 21t
  of dairy cows, 21t
  efficiency of, 473–476
  factors affecting, 474–476
  of homeotherms, 414–415
  of poikilotherms, 414–415
  of poultry, 21t
  of sheep, 21t
  of swine, 21t
Feed digestibility:
  factors affecting, 470–473
  and ration composition, 472–473
Feed efficiency, 520, 714, 728, 745
  effect of heterosis on, 179t
Feed industry, 34
Feedback mechanism, 221
Fertility:
  heritability estimate of, 158t
  and population density, 655–656
Fertilization, 274–275
  in vitro, 283–284
Fetal membranes:
  of calves, 277f
  of pigs, 277f, 278f

Fever, 415–416, 599
  causes of, 415
Fibroblasts, 231
Filariae, 600
Filariasis, 601t, 607t
Filled milk, 105
Fish:
  composition of, 67t
  consumption of, 83t
  as convertors of feed into food, 745
  world catch of, 16–19, 19f
Fish tapeworm, 576
Fistulas, 734, 735f
  animals with, 735–736f
  defined, 734
  of horses, 682
Fleas, 576, 597, 600t, 604, 607t, 617
  as vectors of disease, 555t
Fleeceworms, 610
Flies, 600t, 601t, 607t, 608–612
  botflies, 611–612
  deerflies, 612
  dog, 609
  face, 612
  gadflies, 612
  heel, 610–611
  horn, 609, 621
  horseflies, 612
  houseflies (see Houseflies)
  irradiation of, 622–625
  sandflies, 616
  screwworm, 610
  stable, 609–610
  in transmission of African sleeping
    sickness, 572
  warble, 611
Flukes, 600
Flumethasone, 292
Fluorine, 533–534
  clinical deficiency (or excess) symptoms
    of, 522t, 533f
  functions of, 522t, 533–534
  sources of, 522t, 533–534
Fluorosis, 533
FMD (see Foot-and-mouth disease)
Foal heat, 272
Foals:
  care and management of, 673–674
  time to double birth weight, 77t

Foley catheter, 287–288, 289f
Folic acid, 482t, 506
  clinical deficiency symptoms of, 482t,
    506f
  common sources of, 482t, 506
  functions of, 482t, 506
Follicle-stimulating hormone (see FSH)
Follicles, 264–265
Food:
  absorption of, 470
  consumption trends, 80t
  costs by country, 36f
  mastication of, 466–467
  means of ingesting by animals, 639–641
  prehension of, 466
  projected consumption trends, 24f
  radionuclides of, 89
  utilization of, by cows, 455f
Food and Agriculture Organization (see
    FAO)
Food and Drug Administration (FDA),
    224, 532, 588–589, 619
Food for Peace (Public Law 480), 37
Food and Safety Inspection Service of
    USDA, 599
Food conversion, 473–476
  effect of age on, 474–475
  effect of average daily gain on, 475
  effect of inheritance on, 474
  effect of level of feeding on, 475
  effect of weight on, 474–475
Food mastication in farm animals, 466–
    467
Food nutrients, absorption of, 470
Food prehension in farm animals, 466
Food production:
  increases in, by world regions, 12f
  in the United States, 34
Food protein research, 748–749
Foodstuffs, analysis of, 447–449
Foot-and-mouth disease, 552–553, 590
Forages, 640
  for horses, 676
  preferences of, by animals, 640
Foreign diseases, protecting United
    States livestock from, 590–592
Formamidine compounds in insect
    control, 619
Fortification of animal products, 90–92

Founder of horses, 678, 682, 692*t*
Fowl, breeds of, 50*t*, 59, 61–62*f*
Fowl plague, 590
Fox trot, 662
Freemartin, 279
Freeze-drying of animal products, 101
Freezing of embryos, 289–290
FSH (Follicle-stimulating hormone), 216*t*, 266, 268–269, 273, 338*f*, 672
  role of: in egg laying, 379–381, 384
    in mammary gland growth, 338*f*
Full-feed and digestibility, 471
Full-siblings, relationship of, 166*t*
Fundamental research, 701
Fungal infections, 573
Fungicides, 618

Gadfly, 612
Gaits of horses, 661–662
Galactin, 340
Galactophores, 331–332
Galactopoiesis, 340, 359
Gametes, 111–112
Gamma globulin, 540, 544
Ganglion, 207
Gas chromotography (GC), 727–729, 730*f*
  components of unit, 731*f*
Gas gangrene, 582
Geese, domestication of, 3*t*
Gelding, 258
Gene action:
  additive, 141–142, 162–163, 178
  nonadditive, 138–141, 158–162, 178
  testing for additive effect, 162–163
  testing for nonadditive effect, 162–163
Gene frequency:
  factors affecting, 153–156
  and genetic drift, 156
  and mutations, 155
  in a population, 153–155
  and selection, 156
Gene transfer, 132
Generation interval, 173, 653
Genes, 119–129, 144, 633
  and coat color of horses, 663–665, 666–667*t*
  control, 128–129

Genes *(Cont.)*:
  and embryonical development, 129–130
  functions of, 124–126
  lethal, 130
  linkage of, 136
  operator, 129
  operon, 129
  regulator, 129
  segregation and recombination of, 135–136
  structural, 129
Genetic code, 127–128
Genetic correlations, 174
  for milk components, 361*t*
Genetic drift and gene frequency, 156
Genetic engineering, 130–135, 144–145, 758
  and artificial insemination, 325
  and growth hormone, 132
  and plant breeding, 133
  and production of growth hormone, 132
  and production of insulin, 131
  and production of vaccines, 131–132
Genetic progress:
  annual, 173
  for one generation, 172–173
  prediction of, 172–173
  for several years, 173
Genetic resistance to disease, 544
Genetics:
  and heat tolerance, 402
  principles of, 110–145
Genotypes, 136, 138–139, 154–155
  and blood, 200–201
  of horses, 665–667*t*
  mutations, 136–138
Germ-free pigs, 732–733*f*, 734
Germ oil, 480*t*
Gestation, 275
Gestation period of mares, 672–673
GH (*see* Growth hormone)
Gizzard, 460, 462*f*, 467–469
Glanders, 570
Gloger rule, 412
Glomerulus, 208
Glucagon, 217*t*

Glucocorticoids, 217*t,* 242, 280
Glycerol, 437
Glycogen, 436*t*
Gnats, 600*t,* 616
Goat fever, 561
Goats:
  domestication of, 3*t*
  leading countries in production of, 15*f*
Goiter, 525, 526–527*f,* 692*t*
Golgi apparatus, 348, 349*f,* 350
Gonadotropic hormones, 216*t*
Gonadotropin-releasing hormone
    (GnRH), 216*t*
Gonadotropins, 217*t,* 233, 266
  and twinning, 223
Goose-stepping, 481*t,* 499, 500*f*
Gossypol, 390
Gout, 210
Governmental safeguards for animal and
    human health, 588–590
Grass staggers, 517
Grass tetany, 517
Grazing by farm animals, 639–641
Gregarious behavior among animals,
    652–654
Grinding:
  of grain, 471
  of hay, 471
Gross energy, 452, 455*f*
Ground limestone, 516
Growth:
  accretionary, 230
  adolescent spurt, 229
  from birth to weaning, 245
  of body, heart and brain, 228*f*
  of bone, 232
  the cell as the unit of, 230–231
  compensatory, 230
  curve in cattle and swine, 238*f*
  and development of humans, 229
  effect of: limited feeding on, 243*t*
    undernutrition on, 243
  factors limiting, 237
  of fat, 232
  fetal and postnatal, 228*f*
  genetic control of, 246
  glucocorticoids, 242
  and heredity, 244

Growth *(Cont.):*
  hormonal control of, 238–242
  in humans, 229
  insulin, 242
  of muscle, 231*f,* 232
  and other traits, 246–247
  and ovulation rate, 246
  periods of, 232–238
  the phenomenon of, 229
  physiology of, 226–253
  postnatal, 235–238, 236*f*
  postweaning, 245–246
  prenatal, 233–235, 244–245
Growth-accelerating force, 235–237
Growth curves of unicellular and
    multicellular plants and animals, 236*f*
Growth hormone (GH or somatotropin),
    216*t,* 239–240
  concentration of, in pituitary gland,
    239*t*
  effect of, on lactation, 343
  functions of, 240
  and genetic engineering, 132
Growth hormone–inhibiting hormone,
    216*t*
Growth hormone–releasing hormone
    (GHRH), 216*t*
Growth-retarding force, 235–236
Guinea pigs, inbred strains of, 176*f*
Gut closure and antibody absorption, 365
Gyplure, 624

Habituation, 634
Hair, 188
Hair color, effect of environment on,
    426*f*
Hair follicle, 189*f*
Half-life of radioactive materials, 718–723
Half-siblings, relationship of, 166*t*
Hand, in horses, 661
Haploid, 111–112
Hardy-Weinberg law, 155
Harvest mite, 615
Hatch Act, 5, 698
Hatching of eggs, 379
Hatching date:
  and egg size, 392*t*

Hatching date *(Cont.)*:
  as related to age at sexual maturity,
    392*t*
Health:
  of animals and humans, 536–593
  defined, 538
  governmental safeguards for, 588–590
  role of vitamins in, 478–511
Heart, 195, 197*f*, 198
  diagram of, 197*f*
  disease and animal fats, 85–88
Heat (*see* Estrus)
Heat conservation and body size, 410
Heat cramps, 416
Heat dissipation, 420–428
  and body size, 410–411
Heat gain by radiation, 425
Heat increment, 409, 412–413, 416,
    452–455*f*
Heat loss, 409
  factors enhancing, 418*f*
  factors increasing, 418*f*
  by radiation, 425
  in swine, 420*f*
Heat production:
  of carbohydrates, 413–414
  effect of feed on, 412–413
  factors increasing, 418*f*
  of fats, 413–414
  in homeotherms, 417*f*
  of protein, 413–414
  variation among animals, 419
Heat-regulating mechanisms, 409–412
Heat regulation, chemical, 409
Heat stroke, 415
Heat tolerance:
  of animals, 405, 419*f*
  effect of genetics on, 402
Heating, effect of, on trypsin-inhibiting
    factor, 472
Heating diets, 472
Heaves of horses, 692*t*
Heel flies, 610–611
Helminthic (worm) infections. 573–581
Hemicellulose, 746
Hemoglobin, 199, 522*t*, 523
  sickle-cell, 199
Hemorrhagic sweet clover disease, 490

Hens, as filter of radionuclides, 89
Hepatoflavin, 496
Herbicides, 618
Herbivores, 442, 459–462
Herdbook, 45
Herdmate comparisons, 168
Hereditary patterns of animal behavior,
    632–633
Heredity:
  and environment, 400–402
  and growth, 244
  and social dominance, 655
  and social rank, 655
Heritability:
  and efficiency of feed conversion, 474
  of egg weight, 392
  of milk composition, 360–361
  of milk ejection, 353
  of peak flow of milk, 353
Heritability estimates, 157*t*, 158*t*, 709
  and genetic progress, 158*t*
  type of gene action indicated by, 163*t*
Hessian fly, 616
Heterosis, 47, 52, 178–179, 758
  effects of, on various traits, 179*t*
Heterospermic insemination, 316
Heterozygotes, 140
Heterozygous, 44, 138–141, 159–161
Hexapoda, 597
HGH (human growth hormone), 239
Hibernating species, 407
High-pressure liquid chromatography
    (HPLC), 729–730, 731*f*
Histoplasmosis, 573
Hogs:
  external parts of, 186*f*
  leading countries in production of, 15*f*
Holandric inheritance, 143
Homeostasis, 406–408
Homeothermic mechanism, 406
Homeotherms, 406–408, 612
  feed efficiency of, 414–415
Homeothermy, 406–409
  and age, 408
Homing behavior in animals, 639
Homing pigeons, 639
Homospermic insemination, 316
Homozygous, 44, 138–141, 159–161

Homozygous dominant, matings required
   to test for, 160t
Honey:
   composition of, 67t
   how made, 71–72
   kinds of, 71
   minerals of, 76t
   providing safe, 95
   specific gravity of, 71
   sugars of, 71
   uses of, 71
   vitamins of, 79t
   world production of, 73
Honeybees:
   communication between, means of,
      639
   drone, 72–73
   queen, 72–73
   sting of, 72
   worker, 72–73
Hookworm (ancylostomiasis), 578–580,
   602t
   of cats and dogs, 579–580
   of humans, 578–579
Hormone(s), 214–224, 633, 712
   and animal behavior, 633, 636
   balance of, 224–225
   and cancer, 224
   chemical nature of, 218–219
   effect of, on growth and fattening,
      224
   and egg laying, 379–382
   feedback mechanisms of, 221
   functions of, 216t, 219
   growth (see Growth hormone)
   interrelationships of avian females,
      380f
   juvenile, 625
   luteinizing (see LH)
   mechanism of action, 220–221
   and public health, 224
   regulation of their secretion, 221
   releasing, 216t
   role of, in reproduction, 223
   steroid, 219, 220f
   of thyroid gland, 241
Horn flies, 609, 621
   raising of, 623f

Horse botfly, 611–612
Horseflies, 612
Horses, 658–696
   anemia in, 691, 692t
   artificial insemination of, 296, 671
   bad habits of, 683–684
   birthday of, official, 678
   blemishes and unsoundnesses of,
      680–683
   body conformation of, 668–670
   brain of, weight of, 74t
   breeds of, 49t, 57, 59, 60f, 661t
   care and management of, 668, 671,
      674
   care of feet of, 684–686
   castration of, 258, 674
   coat color of, 663–665, 666–667t
   colic in, 675, 683, 690, 692t
   common defects and unsoundnesses of,
      679–684
   conformation of, 668, 669–670f
   cutting, 632, 633f
   determining age of, 686, 687–688f
   disease and parasite control in,
      689–691
   disease symptoms of, 689–690
   domestication of, 3t, 658
   draft, 659, 661t
   external parts of, 188f
   eyes of, 662
   feeding of, 675–678
   feet of: care of, 684–686
      parts of, 686f
   founder in, 678, 682, 692t
   gaits of, 661–662
   genotypes of, 666–667t
   gestation period of, 662
   gregarious behavior of, 653–654
   harness, 661t
   heavy, 661
   historical aspects of, 57, 59, 658–659
   leading countries in production of, 15f
   light, 669–670t, 699n
   markings of, 664f
   measuring height of, 661
   and medicine, 659–660
   moon blindness in, 677, 692t
   movement defects of, 680

Horses *(Cont.)*:
night blindness of, 694*t*
normal body functions of, 689
nutritional diseases of, 690
osteomalacia of, 694*t*
ponies, 659, 661*t*
and recreational use, 34–35
rickets in, 694*t*
saddle, 661*t*
selection of, 665–668
sizes and types of, 660–661
skeletal system of, 660*f*
smooth-mouthed, 686
as source of power, 29, 696
stable vices of, 683–684
training and grooming of, 678–679
trimming feet of, 684–686
uses of, 658–660, 691, 696
washing of, 679
wild, 661*t*
Host resistance, 626
Houseflies, 608–609, 620–621
life cycle of, 609
HPLC (*see* High-pressure liquid
chromatography)
Human botfly, 611
Human diseases transmissible to animals,
581–582
Human health and vitamins (*see*
Vitamins; *and specific vitamin*)
Hyaluronidase, 274–275
Hybrid vigor, 163, 178–179, 246
calculation of, 178
Hybridoma, 753
Hybridoma technology, 752–753
Hydatid disease, 577
Hydrocyanic acid, 583
Hydrophobia, 547–549
Hyperplasia, 230
Hyperthyroidism, 241
Hypertrophy, 230, 235
Hypophysectomy, 215, 240
effects of, 215
Hypophysis, 215*f*
Hypoprothrombinemia, 490
Hypothalmic releasing factors, 215, 216*t*
Hypothalamus:
in animal motivation, 636

Hypothalamus *(Cont.)*:
diagram of, 215*f*, 218*f*
and heat regulation, 409
hormones of, 216*t*
Hypothermia, experimental, 408
Hypothesis in animal research, 704
Hypothyroidism, 241

Ice cream:
composition of, 67*t*
minerals of, 76*t*
vitamins of, 79*t*
ICU (International Chick Unit) of
vitamin D_3, 510
Imitation milk, 105
Immune reaction, 539
Immune serum, 540
Immunity, 540–546
acquired, 540–541
defined, 540
differences between active and passive,
542*t*
natural, 540
species, 540
transfer of passive, from mother to
offspring, 365*t*
types of, 542*t*
Implantation, 275
Imprinting, 635
Inbreeding, 163, 174–177, 474
amount of, in relatives, 166*t*
and growth, 246
Inbreeding effect, type of gene action
indicated by, 163*t*
Incubation period:
length of, in poultry, 372*t*, 379
of selected birds, 372*t*
Induced parturition, 292
Inflammation, 539
Inflammatory reaction, four signs of, 539
Inflection, 236*f*, 237
Influenza, 581
Infundibulum, 264
Ingestive behavior of animals, 639–641
Inheritance:
cell theory of, 110–112
effect of, on food conversion, 474

Inheritance *(Cont.)*:
  and milk composition, 360–361
  modes of, in horses, 665–668*t*
  qualitative, 150–153
  quantitative, 150–153
  sex-influenced, 143–144
  sex-linked, 142–143
Inositol, 482*t*, 508–509
  clinical deficiency symptoms of, 482*t*,
    508–509
  common sources of, 482*t*, 509
  functions of, 482*t*, 508–509
Insect growth regulator (IGRs), 619
Insecta, 600*t*, 601*t*
Insecticides, 618
Insects, 596–628
  aesthetic value of, 604
  biological control of, 620–627
  body parts of, 597
  contributions of: to humans, 601–605
    to medicine, 604–605
  control of, 617–627
  in control of weeds, 604
  damage of, to crops, 606
  destructiveness of, 605–608
  economic significance of, 597
  as food for poultry, 602–603
  growth regulators, 619
  harmful effects of, 605–608
  instinct in, 627, 633–634
  as parasites, 606–607
    and/or predators, 603–604
  and pollination, 603
  reproductive capability of, 620
  resistance of, to insecticides, 621
  secretions of, 602
  sex attractants of, 625–626
  speed of, 596
  as test animals, 605*f*
  use of: in dyes, 602–603
    in research, 604
  value of, to soil, 604
  variation among, 596
  wings of, number of, 597
Insensible perspiration, 421–422
Insight learning (reasoning), 635
Instars, 597
Instinct, 633, 635, 656

Insulin, 217*t*, 242
  and genetic engineering, 131
  role of: in lactation, 344
    in mammary gland growth, 344
Integrated pest management, 628
Intelligence of selected animals, 635–636
International Code of Zoological
  Nomenclature, 598
Interspecies rearing, 650
Interstitial cells, 266
Intrauterine migration, 279
Intussusception, 230
Inversion, 113, 114*f*
Invertase, 467–468*t*
Investigative (exploratory) behavior of
  animals, 651–652
In vitro fertilization, 283–284
Involution of mammary gland, 340, 355–
  356
Iodinated casein, 456
Iodine, 524–526
  clinical deficiency symptoms of, 522*t*,
    525–527
  functions of, 522*t*, 524–526
  in secretion of thyroxine, 524
  sources of, 522*t*, 526
  supplementing diets with, 526*n*
Iodine value, 438
Ionizing radiation, 712
IPM (integrated pest management), 628
Iron, 521–523
  clinical deficiency symptoms of, 521–
    523
  functions of, 521–523
  of milk, 521
  sources of, 522*t*, 523
Iron absorption, effect of ascorbic acid
  on, 493*f*
Irradiation, 712
  of flies, 622
  ionizing, 712
Isotonic, 347
Iteration, 169
IU (International Unit), 510

Jerky, 93–94
Joint ill of horses, 673

Joints, types of, 192
Juvenile hormone, 625–626

Karyotype, 291
  of cattle, 115f
Kidney, 208
  as a secreting unit, 210f
Kilocalorie, 451
Korean hemorrhagic fever, 551t
Kwashiorkor, 68

Labeling of animal products, 90–91
Lactase, 467–468t
Lactation, 329–367
  artificial initiation of, 339–340
  effect of: age on, 357
    animal size on, 357–358
    feed on, 359
    frequency of milking on, 356
  endocrine control of, 341f
  factors affecting, 356–360
  hormonal control of, 340–344
  initiated by suckling stimulus,
    359–360
  peak of, 355, 357f
  persistency of, 356, 357f
  physiology of, 329–367
Lactochrome, 496
Lactoflavin, 496
Lactogen, 340, 341f
Lactogenesis, 340
Lactose, 435, 436t, 446f, 468t
  antipellagric properties of, 74
  and brain size, 73–74
  characteristics of, 73–75
  hygienic values of, 74–75
  in milks of mammals, 360t
  and mineral absorption, 74
  sweetness of, 75
  synthesis in milk, 346
Lamb:
  cholesterol in, 70t
  composition of, 67t
  minerals of, 76t
  vitamins of, 79t

Lambs:
  nutritional deficiencies in, 518f, 520f,
    527f, 528f, 532f
  skeletal muscular dystrophy in, 488
  swayback disease in, 523
Lameness of horses, 680
Laminitis, 678, 682, 692–693t
Land, world use of, 28f
Land classification, 18f
Land-grant colleges, 5
Larva migrans, 611
Larynx, 204, 206
Lasalocid, 456
Laws of probability and animal breeding,
  136–137
LD_{50}, 583
Learning in animals:
  associative, 634
  by conditioning, 634
  by habituation, 634
  by imprinting, 635
  insight (reasoning), 635
  operant conditioning (trial and error),
    634
  by reasoning, 635
  responses to, 633–635
  by trial and error, 634
Le Châtelier's second principle, 101
Leptospirosis, 563–565
  animals susceptible to, 563–565
  vaccination against, 565
Letdown of milk, 350–355
  early theories of, 350–351
Leukocytes, 202, 539, 544
  in farm animals, 198t
Leukopenia, 202
LH (luteinizing hormone), 216t, 221, 266,
  269, 271, 338f
  role of: in egg laying, 379–381, 384
    in mammary gland growth, 338f
Libido, 266, 306
Lice, 600t, 601t, 607t, 616
  biting, 616
  bloodsucking, 616
  control of, 616
  as vectors of disease, 555t
Life span:
  of birds, 388

Life span *(Cont.)*:
  effect of diet on, 250
  in relation to length of growth period, 249f
Ligaments, 192
Light:
  effect of: on reproduction, 398
    on semen, 308
    on sexual maturity in poultry, 385f
  and insect control, 625
Light photons, 717–718
Lignin, 446f, 746
Linebreeding, 177
Linecross, 162
Linkage of genes, 136
Lipase, 468t, 469
Lipids, 437–440, 446f, 452
  classification of, 446f
Lipizzans, 661t, 663
Lipoproteins, 88
Lipotropic factor, 508
Liquid nitrogen, 315
Listeriosis, 570
Litter size, effect of dam's age on, 248t
Littermates, 165
Liver:
  cholesterol in, 70t
  minerals of, 76t
  vitamins of, 79t
Liver fluke disease, 577–578
Liver flukes, 602t
Livestock:
  breeds of, 43–63
  number of, in the United States, 33
  and United States farm income, 33
  world number of, 13–16
Livestock production, leading countries in, 15f
Livestock products, worldwide use, 14f
Living insecticides, 621, 626–627
Lobule-alveolar system, 335–337
Longevity as related to velocity of growth, 248
Low-level radiation research laboratory, 713–718
Lumpy jaw, 571
Lung, mammalian, diagram of, 205f
Lung flukes, 602t

Luteinizing hormone (*see* LH)
Luteotropin, 340
Lymph vessels, 196
Lymphocytes, 202
Lyophilization of animal products, 101
Lysine, deficiency of, 442f
Lysozyme, 539n
Lytic reaction, 545

Macroelements, 513–520
Macrophages, 539
Magnesium, 517
  clinical deficiency symptoms of, 514t, 517, 518f
  functions of, 514t, 517
  in muscle contraction, 195
  sources of, 514t, 517
Maillard reaction, 100
Maintenance energy, 455
Malaria, 571–572, 601, 602t, 607t
Malnutrition, 13n, 65
  and animal learning, 635
Malpighian bodies, 209
Malta fever, 561–563
Maltase, 467–468t
Maltose, 435–436t, 446f, 468t
Mammary glands:
  anatomy of, 330–332
  defined, 330
  endocrine control of, 335, 338f, 341f
  growth of, 334–340
    measurement of, 337
  of higher mammals, 329–332
  internal structure of, 331–332
  involution of, 355–356
  lymph flow in, 333–334
  number of, in selected species, 329–330
  perfusion of, 345
  pressure changes in, 347, 348f
  retrogression of, 339
  of sows functional at weaning, 649t
Mammotropin, 340
Manganese, 530–531
  clinical deficiency symptoms of, 522t, 530, 531f
  functions of, 522t, 530–531

Manganese *(Cont.)*:
  and hatchability of eggs, 522*t*
  sources of, 522*t*
Marek's disease, 542, 745*n*
Mares:
  care and management of, 671–673
  milk composition of, 77*t*
  parturition in, 672–673
Margarine, consumption of, 86*f*
Markings of horses, 663–665
Mass spectrometer, 725*f*
Mastication of food, 466–467
Mastitis, 537, 754, 755*f*
  and milk composition, 347
Maternal behavior, 645–651
  of American bison, 647
Mating:
  in farm animals, 150–181
  habits among animals, 644–645,
    652–654
Mating systems for livestock
    improvement, 174–180
Maturity:
  at birth, 234*f*
  physiological, 251*f*
MCC (modified contemporary
    comparisons), 168
Meat(s):
  canning of, 96
  consumption of, 83*t*
  curing of, 96
  inspection of, 97
  preservatives for, 94
  processed, 103
  providing safe, 95–98
  smoking of, 96
  synthetic, 105, 106*f*
Meat analogues, 105
Meat inspection of USDA and states,
    589
Meat production:
  major sources of, 26*f*
  world production trends, 27*f*
  worldwide, 14*t*
Meat products restructured, 103
Meat tenderizers, 96–97
Mediterranean fever, 561–563
Medulla oblongata, 207–208

Megacalorie, 451
Megakaryocytes, 202
Meiosis, 113–114, 116–119
Meiotic division, 118–119*f*
Melanin, 401
Memory, long- and short-term, 636
Menadione, 480*t*
Mendel's law of segregation and
    recombination of genes, 135
Menopause, 270–271
Menstruation, 270–271
Mesocrine gland, 330
Metabolic rate, 342, 415
  and body size, 247*f*
  effect of fever on, 415
Metabolic water, 434
Metabolizable energy (ME), 453, 455*f*
Metacercariae, 578
Metamorphosis, 597
Metestrus, 270
Methane of rumen, 466
Methane production from livestock
    manure, 748*f*, 749*f*
Methoxychlor, 618
Methylene blue test of semen quality,
    311
Microbe engineering, 131–132
Microbiological assay of vitamins, 509
Microelements, 521–534
Microorganisms:
  attenuated, 541
  inhibiting growth of, 93
Microsurgery with embryos, 290
Migration and gene frequency, 155
Milk:
  antipellagric properties, 75
  antirachitic properties, 74
  Ca/P ratio of, 75–77
  cholesterol in, 70*t*
  composition of, 67*t*, 362*t*
    and growth rate, 76, 77*t*
  creaming of, 362
  defined, 360
  drying of, 93–94, 100
  early theory of secretion of, 344
  filled, 105
  fortification of, 90–92
  imitation, 105

Milk *(Cont.)*:
  leading countries in production of,
      17*f*
  minerals of, 76*t*
  multivitamin-mineral, 90
  progesterone and pregnancy testing,
      326
  residual, 353
  secretion of, 347–350
  synthesis of, 344–347
  synthetic, 105
  in treatment of ulcers, 75
  use of, in extending semen, 313
  vitamin A and D potency of, 90
  vitamins of, 79*t*
Milk composition:
  of dairy breeds, 361*t*
  factors affecting, 360–364
  heritability estimate of, 158*t*
  milk ejection, peak flow of, 353
  of selected mammals, 360*t*
Milk fat:
  dietary effects on, 363*n*
  in milks of mammals, 360*t*
Milk letdown, 350–355, 634
  factors favoring, 353–354
  factors inhibiting, 354
Milk precursors:
  determination of, 345–347
  of fat, 346
  of lactose, 346
  of minerals, 346
  of protein, 345–346
  of vitamins, 346–347
  of water, 347
Milk production:
  effect of heterosis on, 179*t*
  and evaporative cooling, 423–424
  heritability estimate of, 158*t*
  worldwide, 14*t*
Milk products, providing safe, 99–100
Milk ring test, 559
Milk secretion, when most rapid,
      347–348
Milk sickness in humans, 584
Milk toning, 92
Milk veins, 333
Milking, completeness of, 362

Milking frequency, effect of, on
      lactation, 356–357
Milking machine:
  for cows, 331*f*, 755*f*
  for sows, 354*f*
Mineral absorption and vitamin D,
      485–486
Mineral interrelationships, 512
Mineral nutrition, 512–535
Mineral research, use of activation
      analysis in, 721–722
Mineralocorticoids, 217*t*
Minerals, 444–445, 446*f*
  amount of, in milks of mammals, 77,
      360*t*
  of animal products, 67*t*, 75–78, 80
  classification of, 446*t*
  consumption trends of, 80*t*
  contributions of, to humans and
      animals, 512–535
  (See also specific mineral)
  daily allowance of, for humans, 76*t*
  deficiency symptoms of, 445
  essential, 445
  in feeding horses, 677, 692–694*t*
  of feeds, 449
  functions of, 444–445, 514*t*, 522*t*
  major sources of, 514*t*, 522*t*
  secreted in milk, 346
Miniature pigs, 731
Miracida, 577
Mites, 537, 600*t*, 601*t*, 607*t*, 614–615
  chorioptic, 615
  demodectic, 615
  fowl, 614
  mange, 614
  psoropic, 614
  sarcoptic, 615
  scab, 614
  as vectors of disease, 537, 555*t*
Mitochondria, 111*f*, 133
Mitosis, 113–116, 117*f*
Modified contemporary comparisons,
      168
Mollusca, 599*t*
Molt, 190
Molting, 380–381, 388
  how induced, 388

Molting *(Cont.)*:
  of insects, 597
Molybdenum, 532–533
  clinical deficiency (or excess) symptoms
    of, 522*t*, 523
  functions of, 522*t*, 533
  sources of, 522*t*, 532–533
Monensin, 456
Monoclonal antibodies, 543, 752–753,
  758
Monocytes, 202
Monoestrus, 271
Monogastric animals, 442, 461*f*,
  463–464
Monorchid, 256–257
Monosaccharides, 435–436, 446*f*
Moon blindness of horses, 677, 692*t*
Morbidity losses from diseases, 537–538
Morphogenesis, 226, 234
Morrill Act, 5
Mortality losses from disease, 537–538
Mosaic code, 574
Mosquitoes, 550, 571–572, 600*t*, 601*t*,
  602*t*, 607*t*, 613
  in transmission of malaria, 571–572,
    602
  as vectors of disease, 551*t*
Mother-young behavior of animals,
  645–651
Motivation of animals, 636
Mucin, 468*t*
Mule-foot, 190
Murine typhus, 554
Muscle contraction, physiology of,
  194–195
Muscles:
  smooth, 193
  striated, 193
  types of, 193–194
Muscular system, 193–195
Mutation and gene frequency, 155–156
Mutations, 137–138, 199
Mycotoxins, 586–588, 741
Myelitis, 607*t*
Myiasis, 600, 602*t*, 606
Myoblast, 231–232
Myoepithelial cells, 352*f*
Myosin, 193

Myotube, 231*f*, 232
Myxomatosis virus, 622

NAAB (National Association of Animal
  Breeders, Inc.), 320
Nasal fly, 611–612
National Animal Disease Center, 735,
  737*f*
National Association of Animal
  Breeders, Inc., 320
National Institutes of Health, 590
National Research Council (NRC), 455*n*,
  510*n*, 534*n*, 675*n*, 710
Natural foods, 92, 107
Nature as designer of milk, 75, 77–78
Navel ill of horses, 673
Navigation behavior in animals, 639
Negri bodies, 548
Nemathelminthes, 599*t*, 602*t*
Nematodes, 600
Nemertinea, 599*t*
Neomycin, use of, in semen, 314
Nervous system, 206–208
Net energy (NE), 453, 455*f*
Neurohormonal regulation, 215, 221
Neurons, 206–207
Neutron activation analysis, 721–723
Neutrophils, 202
Newcastle disease, 549–550, 590
Newton's law of cooling, 410
NFE (nitrogen-free extract), 447–449
Niacin (*see* Nicotinic acid)
Nickel, 534
  and rumen activity, 534
Nicking, 163
Nicotinic acid (niacin), 481*t*, 499–502
  clinical deficiency symptoms of, 481*t*,
    499–502, 503*f*
  common sources of, 481*t*, 502
  functions of, 481*t*, 499–502
Nidation, 275
Night blindness, 480*t*, 483
  of horses, 694*t*
Nitrates, 206, 583
Nitrogen-free extract (NFE), 447–449
Nitrogen metabolism, urinary end
  products of, 210*f*

Nitrogenous compounds, classification of, 446f
Nonadditive (see Gene action, nonadditive)
Nondisjunction, 112
Nonfat milk solids, fortifying with, 92
Nonprotein nitrogen (see NPN)
Nonradioactive isotopes, 723, 725
Nonthermal sweating, 421–422
Noradrenalin, effect of, on muscles, 193
Norepinephrine, 217t
  effect of, on muscles, 194
Norgestomet, 282
NPN (nonprotein nitrogen), 443–444, 465, 747, 758
NRC (see National Research Council)
Nuclear fusion, 135
Nuclear reactor, 720–723
Nuclear transfer, 134–135
Nucleic acid metabolism, urinary end products of, 210f
Null hypothesis, 707–708
Nursing intervals in piglets, 354f, 355, 631, 648f, 649–650
Nutrients, classification of, 446f
Nutrition:
  and behavior of humans and animals, 433
  defined, 432
  and egg laying, 389
  and egg size, 393
  environmental considerations of, 412–415
  and growth, 242–244
  of horses, 675–678
  principles of, 432–457
  role of minerals in, 512–535
  role of vitamins in, 478–511
Nutrition research, use of GC and HPLC in, 727–730, 731f
Nutritional diseases of horses, 690, 692–695t
Nutritional labeling, 90–91
Nutritional myopathy, 532
Nutritional wisdom of animals, 640
Nutritive ratio, 451

Nutritive values of beef, ham and lamb, 82f
Nymphomaniac, 272

Oils, 437–440
  consumption of, 83t, 86f
Oligospermia, 284
Omasum, relative size of, 464f
Omnivores, 442, 459–461
Oogenesis, 121f
Operant conditioning, 634
Operon, 129
Opisthotonos, 514t, 517
Oral contraceptives, 223
Oranges, vitamins of, 79t
Organogenesis, 234
Oriental liver fluke disease, 578
Oriental lung fluke disease, 578
Orientation behavior in animals, 639
Ornithosis, 549
Oscilloscope monitors, 740f
Osmotic pressure in mammary glands, 347
Osteoblasts, 232
Osteoclasts, 232
Osteomalacia, 480t, 487, 514t, 516, 522t, 694t
Outbreeding, 178–180
Ova transfer, 276f
Ovary, 264, 265f
  of avian female, 375–379
  endocrine control of, 268–269
  follicles of, 217t
  hormones of, 217t
  of sow, 265f
Overdifferentiation, 252
Overdominance, 140–141, 159–161
Overgrazing, 586
Oviduct, 262–263f
Oviposition, 382–383
Ovoflavin, 496
Ovulation, 223, 270t, 273–274
  in birds, 383
  synchronization of, 281–282
Ovum of sow, 275f
Ox warble, 611
Oxygen debt, 195

Oxyhemoglobin, 199
Oxytocin, 216*t*, 218, 280, 344, 382
    effect of, on semen output, 302
    and egg expulsion, 381
    and mastitis therapy, 353
    role of, in milk letdown, 351–355
Oxytocinase, 351–352
Oyster shells, 516

Pacemakers, 197
Palomino, 663
Pancreas, 469
    hormones of, 217*t*
Pantothenic acid, 412–413, 481*t*, 499
    clinical deficiency symptoms of, 481*t*,
        499
    common sources of, 481*t*, 499
    functions of, 481*t*, 499
Papanicolaou stain, 323
Papillae, 465*f*
    effect of diet on development of, in
        ruminants, 465*f*
Pappataci fever, 607*t*
Para-aminobenzoic acid (PABA), 482*t*,
    509
    clinical deficiency symptoms of, 482*t*,
        509
    common sources of, 482*t*, 509
    functions of, 482*t*
Paracaseinate, 469
Paragonimiasis (Oriental lung fluke
    disease), 578
Parakeratosis, 522*t*, 529*f*
Parasites:
    defined, 598
    external, 601*t*
    and the health of humans and animals,
        602*t*
    of horses, 690–691
    internal, 602*t*
    of significance to humans and animals,
        596–628
Parasitology, defined, 598
Parathyroid gland, 217*t*, 221
Parathyroid hormone, 217*t*, 221
    effect of, on lactation, 343
Paratyphoid fever, 567, 607*t*

Parthenogenesis, 72–73, 275, 604, 615
Parturient paresis, 334, 343
Parturition, 280–281
    and animal behavior, 646–651
    inducing, 292
    in mares, 672–673
Pasteurellosis, 593
Pasteurization, 712
    of deboned poultry meat, 98
    of eggs, 99
    of milk, 99–100
Pasture for horses, 676–677
Patency, 331
Pathogens:
    modes of entry of, 539
    resistance to, 540–541
PD (*see* Predicted difference)
PDM (predicted difference for milk),
    170
Peak flow of milk, 353
Peat scours, 533
Peck order, 266, 654
Pectin, 746
Pedigree, 44, 45*f*
Pedigree-Record Association, 45–46
Pellagra, 481*t*, 499–502, 503*f*
Pelleted diet, effect of, on papillae,
    465*f*
Pelleting of diets, 471
Penetrance, 159
Penicillin:
    production of, 712
    use of, in semen, 311
Pepsin, 468*t*, 469
Peptidase, 468*t*
Peptide linkage, 441
PER (protein efficiency ratio), 91
Perfusion of mammary gland, 345
Periosteum, 191
Peripheral nerves, 208
Pernicious anemia, 482*t*, 507
Perosis, 482*t*, 508*f*, 522*t*, 531
Pesticide residues as determined by GC,
    729
Pesticides, 722
    in animal products, 106–107, 619
Pet animals, number of, in the United
    States, 745–746

Pet foods, money spent for, 22n
PG (see Prostaglandins)
pH as used in preservation of animal
    products, 94–95
Pharynx, 204
Phenotype, 112, 138–141
Phenotypic variation, 150–151
    causes of, 156–157
Pheromones, 638
Phospholipids, 440, 446f
Phosphorus, 513–516
    clinical deficiency symptoms of, 514t,
        516f
    in forages, 516
    functions of, 513, 514t, 515–516
    in grains, 516
    phytin, 513
    sources of, 514t, 516
Photoperiodism, 377
Photosensitization, 427
Photosynthesis, 435, 446f, 459
Phyla, five most important, in animal
    science, 599–601
Physical form (feed) and digestibility,
    471
Physiological maturity, 251f
Physiology:
    adaptive, of humans and animals, 405
    of circulatory system, 195–198
    defined, 184–185
    of digestion, 459–476
    of egg laying, 369–394
    of farm animals, 184–211
    of growth, 226–253
    of lactation, 329–367
    of reproduction (see Reproduction,
        physiology of)
    of senescence, 226–253
Phytin, 513
Pica, 514t, 516
Pigeon milk, 220
Pigs:
    artificial rearing of, 366f
    domestication of, 3t
    external parts of, 186f
    as research animals, 730–731, 732–733f,
        734
Pinkeye, 612

Piroplasmosis, 590
Pitocinase, 351–352
Pituitary gland, 280
    anterior, 215–216, 218f, 341f, 524,
        636–637
    posterior, 215, 218f, 341f, 382
Placebo, 75
Placenta, 281
    hormones of, 217t
Placental lactogen, effect of, on
        lactation, 344
Placental membranes, 277f
Plague, 568
Plant protein, 67–69
Plants:
    composition of, 434–447
    toxic to animals, 583–586
Plasmodium and malaria, 572
Platyhelminthes, 599t, 602t
Play behavior in animals, 644
Pleitrophy, 174
Plus genes, 158
PMS (pregnant mare serum),
        659–660
Pneumonic plague, 568
Poikilothermic species, 408
Poikilotherms, 250–251, 556
    feed efficiency of, 414–415
Points of horses, 663
Poisons for insects, 618–619
Pollination by insects, 603
Polydactyly, 286–287
Polyestrus, 271
Polygastric animals, 461f, 462–466
Polymyxin, use of, in semen, 314
Polyneuritis, 481t, 496–497f
Polynucleotidase, 468t
Polyploidy, 113, 275
Polypnea, 405
Polysaccharides, 446f, 468t
Polyspermy, 275–280
Pons, 207
Population:
    doubling time of, 9
    and food production, increases in,
        12f
    regional distribution of, 7f
    relation of, to animal increase, 9t

Population *(Cont.)*:
world projections, 8*t*
world trends of, 6–11
Population density, factors affecting, in
animals, 655–656
Porifera, 599*t*
Pork:
cholesterol in, 70*t*
composition of, 67*t*
minerals of, 76*t*
vitamins of, 79*t*
Pork tapeworm, 574, 575*f*
Portal system, 197
Potassium, 517–519
clinical deficiency symptoms of, 514*t*,
518*f*
functions of, 514*t*, 519
sources of, 514*t*
Potassium 40 (^{40}K), 714–715
Potassium nitrate, 583
Poultry *(see* Chickens; Turkeys)
Poultry litter, use of, in farm animal
diets, 32
Poultry meat, leading countries in
production of, 17*f*
Poultry products:
consumption of, 83*t*
providing safe, 98–99
Pox diseases of humans and animals,
552–554
Precipitin reaction, 545
Precocial birds, 372*t*, 374
Predicted difference, 168–171
genetic effect of, 170*f*
Predicted difference for milk
(PDM), 170
Preejaculation preparation of bulls, 304
Pregnancy, 275–280
effect of, on lactation, 357
and milk composition, 364
tests for, in cows, 326
Pregnant mare serum (PMS), 659–660
Prehension of food in farm animals,
466
Prepotent, 175
Preservation of animal products, 92–95
Probability in animal breeding, 136–137
Proboscis, 608

Proestrus, 270
Professional societies of animal science,
710–711
Progeny testing and artificial
insemination, 325
Progeny tests, 159
Progesterone, 217*t*, 220*f*, 224, 280, 338*f*,
341*f*
effect of, on growth of mammary
gland, 335–336, 338*f*, 341*f*
functions of, 269
Progestins, 223
Prolactin (Prl), 216*t*, 219, 341*f*, 340–342
role of: in avian females, 381
in lactation, 340–342
Prolactin-inhibiting hormone (PIH), 216*t*
Prolactin-releasing hormone (PRH), 216*t*
Prostaglandins (PG), 221–222, 280*t*, 292,
324
chemical structure of, 222*f*
Prostate gland, 261
Protamone, 456
Protected proteins, 443–444
Protein(s), 440–444, 446*f*
of animal products, 67*t*, 67–70
calculation of crude, 448
caloric value of, 452
complete, 68
consumption trends of, 80*t*,
digestion of, 468–469
functions of, 440
and growth rate of young, 77*t*
from major food groups, 23*f*
in milks of mammals, 360*t*
in plants, 68
protected, 443–444
quality of, 442–444
synthesis in milk, 345–346
Protein/calorie ratios, 81
Protein efficiency ratio (PER), 91
Protein-free diet fed ruminants, 750*f*
Protoplasm produced by insects, 627
Protozoa, 599*t*, 602*t*
Protozoal infections, 571–573, 582
of humans transmissible to animals,
582
Proventriculus, 460, 462*f*, 467–468
Proximate analysis, 449

Pseudorabies, 593
Psittacosis, 549
Psychrometric chamber, 428–430f
Puberty, 335, 337n
    age of, in farm mammals, 268t
Public health, 536–538
Pulmonary system, 198
Pure Food and Drug Law, 588
Purebred, 44
Purified diet, 521, 750f
Pyrethrum, 618
Pyrexia, 415–416
Pyridoxal, 502
Pyridoxamine, 502
Pyridoxine (B₆), 482t, 502, 504
    clinical deficiency symptoms of, 482t,
        502, 504f
    common sources of, 482t, 504
    functions of, 482t, 502, 504

Q fever, 556
Quantitative traits, statistical evaluation
    of, 151–153
Quarantine, 590

Rabbits, use of virus to control
    population of, 622
Rabies, 547–549
    in selected wildlife hosts, 548f
    transmission of, 547
    vaccination of cats and dogs against,
        549
    vectors of, 547
Rack, 662
Radiation:
    and heat loss, 425–428
    in meat preservation, 94
Radiation sterilization, 622
Radioactive half-life, 718–723
Radioactive isotopes, 345–346
    and mineral nutrition, 513
Radioactive tracers, 712
Radiocarbon dating, 2n
Radiograph of sow uterus, 279f
Radioisotopes, 712, 719f
Ralgro, 456

Random sample, 707
Rate of passage:
    as affected by physical form, 471
    effect of, on digestibility of foods, 471
Reactions:
    antigen-antibody, 544–546
    bactericidal, 545
    complement fixation, 545–546
    febrile, 539
    immune, 539
    inflammatory, 539
    lytic, 545
    precipitin, 545
    toxin-antitoxin, 546
Recessive, defined, 139–141
Recessive gene, calculating frequency of,
    154–155
Recessiveness, 138–140
Recombinant DNA, 750–752
Records of relatives, 165–171
Rectal temperatures of selected animals,
    406
Red bug, 615
Red meat, leading countries in
    production of, 17f
Red water fever, 613–614
Reflexes, 63
Regeneration, 229
Registration certificate, 46f
Regression of mammary glands,
    355–356
Reinforcement and animal learning,
    634–635
Relapsing fever, 601t, 607t
Relaxin, 217t, 280
Releasing hormones, 216t
Releasing stimuli, 637
Renal cortex, 208
Renal medulla, 208
Rennin, 468t, 469
Repeatability, 169
Reproduction:
    effects of temperature on, 273
    physiology of, 255–292
        in avian females, 375–379
        in avian males, 375–376f
        in females, 268–281
        in males, 265–268

Reproductive efficiency, 755n
  effect of hot weather on, 423
  and population density, 655–656
Reproductive organs of cows and sows,
    262f, 263f
Reproductive physiology, recent findings
    of, 281–292
Reproductive physiology research and
    artificial insemination, 326–327
Reproductive tract:
  of boars, 257f
  of bulls, 257f
  of cows, 262f
  of females, 261–265
Research:
  agricultural (see Agricultural research)
  basic, 701
  in biotechnology, 749–753
  control, 706–707f
  in electronic technology, 753–755
  in food protein, 748–749
  opportunities in, 756
  publication of, 706
  in retrospect and prospect, 698–760
    (See also Animal research)
  use of computers in, 708–709
  use of statistics in, 706–708
  and waste management, 747–748
Research organizations, 710–711
Residual milk, 353
Respiratory center, 206
Respiratory system, 203–206
Rete testes, 259
Reticulorumen, 365f
Reticulum, relative size of, 464f
Rh factor, 200–202, 322
Rhesus monkey, 200
Riboflavin, (B₂), 481t, 496, 498
  clinical deficiency symptoms of, 481t,
    498
  common sources of, 481t, 498
  functions of, 481t, 496, 498
Rickets, 480t, 484n, 486f, 515t, 523, 694t
Rickettsial infections:
  Q fever, 556
  rickettsial pox, 607t
  Rocky Mountain spotted fever,
    554–555, 601t,

Rickettsial infections (Cont.):
  transmissible to humans by animals,
    554–556
  trench fever, 555, 601t
  tsutsugamushi, 555–556, 607t
  typhus fever, 554
Rift Valley fever, 590
Rinderpest, 590
Ringworm, 573
  and vitamin A, 484f
RNA (ribonucleic acid), 121, 123f,
    125–127, 128f, 144, 221, 242, 515,
    529, 546, 636, 720, 729, 730
Rocky Mountain spotted fever, 554–555,
    601t
Rodenticides, 618
Roughages, digestion of, 473
Roundworms, 580–581
  of horses, 691
  of humans, 580
  of swine, 580f
Royal jelly, 72, 481t
Rumen:
  development of, 461–463
  digestion in, 468t
  epithelium of, 465f
  methane of, 466
  microorganisms and, 464
  when functional, 462–463
Rumensin, 456
Ruminants, 442, 473, 476, 639–640, 746,
    758
  in competition with humans for food,
    25–28
  defined, 21
  role of, 22f
  stomach compartments of, 461f,
    462–463, 464f
  synthesis of amino acids, 441–444
  synthesis of B-complex vitamins,
    442
  synthesis of vitamin K, 442
  used to recycle wastes, 747–748,
    750f
  utilization of NPN, 443
Ruminate, 466
Rumination, 639–640
Runt pigs, 649–650

SAES (*see* State agricultural experiment stations)
St. Louis encephalomyelitis, 551*t*
Salmon, 639
  cholesterol in, 70*t*
  minerals in, 76*t*
  vitamins in, 79*t*
Salmonellosis, 566–567
Salt, 519
Sand flies, 616
Sarcolemma, 97
Sarcoptic mange, 614–615
Saturated fatty acids, 70, 438
Scar tissue, 229
Scarlet fever, 581
Scent posts, 644
Schistomiasis, 602*t*
Schistosomes, 577
Science, defined, 703
Scientific journals, 710
Scientific method, the, 703–706
Screwworm, 602*t*
Screwworm flies, 610
  sterilization of males, 610, 622–624, 712–713
Scrotum as heat-regulating mechanism, 256
Scrub typhus, 555–556
Scurs, 143
Scurvy, 471*t,* 478, 492–494
SDA (specific dynamic action), heat increment of feeding, 797
Season, effect of, on milk composition, 362
Sebaceous glands, 187
Selected at random, 707
Selecting farm animals, 150–181
Selection, 157–158, 653
  for additive gene action, 162–163
  and animal behavior, 632–633
  artificial, 158
  for dominance and recessiveness, 159
  effect of population density on, 655–656
  for epistasis, 161–162
  and gene frequency, 155–156
  of horses, 665, 668

Selection (*Cont.*):
  on individuality, 164–165
  and insect resistance to insecticides, 621–622
  for kinds of gene action, 158–164
  natural, 158
  for overdominance, 159–161
  on records of relatives, 165–171
  of superior breeding stock 164–172
Selection differentials, 172–173
  examples of, 173*t*
Selenium, 531–532
  clinical deficiency (or excess) symptoms of, 522*t,* 531, 532*f*
  functions of, 522*t,* 531–532
  in relation to vitamin E, 487–488
  sources of, 522*t,* 532
Semen:
  amount produced, 266–267
  appearance of, 307
  cold shock of, 310
  collection of, 302–307
  coloring of, 313–314
  custom freezing of, 298*t,* 318
  dead and alive cells, 310*f,* 311
  effect of light on, 308
  enumeration of, 307
  evaluation of, 307–312
  exports of, 300*n*
  extension of, 312–314
  freezing, in straws, 317
  freezing and thawing of human, 323
  fresh liquid, 314–315
  frozen, 315–318
  incubation test of, 311
  lyophilization of, 317–318
  morphology of, 309
  motility of, 308
  output and frequency of ejaculation, 302–303
  pelleting of, 317
  pH of, 311
  production of, in selected mature males, 304*t*
  reaction rates of, 311
  shell-freezing, of, 317
  staining of, 310

Semen *(Cont.)*:
  storage of, 314–318
  volume per ejaculate, 304*t*
Semen extenders:
  characteristics of, 312
  kinds of, 312–313
Semen storage, 314–318
Seminal vesicles, 261
Seminiferous tubules, 258*f*, 259
Senescence, 247–253, 388, 720
  physiology of, 226–253
Septicemia, 567
Seven-year itch, 615
Sex:
  chromosomes, 142–143
  determination of, 113
Sex attractants, 624–625
  of insects, 625–626
Sex control, 291
Sex hormones:
  and sex characteristics of birds,
    381–382
  and vitamin K, 490
Sex-influenced inheritance, 143–144
Sex-limited traits, 144
Sex-linked inheritance, 142
Sex reversal in poultry, 377, 633
Sexing embryos, 291
Sexual behavior of animals, 644–645
Sexual maturity:
  age at heritability estimate of, 158*t*
  factors affecting, in poultry, 384
Sexual preparation and semen collection,
    304
Shade:
  cooling effect of, 423*f*
  in nonsweating species, 422–425
Sheep:
  breeds of, 49*t*, 53, 57, 58*f*
  domestication of, 3*t*
  external parts of, 186*f*
  leading countries in production of, 15*f*
  scabies, 614*n*
  (See also Lambs)
Sheep botfly, 611–612
Sheep gadfly, 611–612
Sheep liver fluke disease, 577–578
Shelf life of poultry meats, 97–98

Shell membranes of eggs, 373–374
Shelter seeking among animals,
    641–642
Shivering, effect of, on heat production,
    419
Shorthorns, 46, 426*f*
  heat tolerance of, 410, 411*f*
Sickle-cell anemia, 140, 523
Sigmoid flexure, 260–261
Sigmoid growth curve, 235, 236*f*
Silkworms, 602
Single cross, 179
Sire indices, 168
Sire summary, 169–171
Skeletal muscular dystrophy, 480*t*,
    488
Skeletal system, 190–192
Skin pigmentation and vitamin D, 485
Skin temperature of humans, 416
Slaframine, 588
Slipped tendons, 531*f*
Smallpox, 552
Smith-Hughes Act, 6
Smith-Lever Act, 5
SMR *(see* Standard metabolic rate)
Social dominance of animals, 266, 654
Social pathology, effect of population
    density on, in rats, 655–656
Social rank of animals, 654–655
Sodium, 517–519
  clinical deficiency symptoms, 514*t*
  functions of, 514*t*
  sources of, 514*t*
Sodium chloride, 519
  clinical deficiency symptoms of, 514*t*
  functions of, 514*t*
  sources of, 514*t*
Somatomedins, 240
Somatostatin, 216*t*
Somatotropin (STH), 239, 343
  (See also Growth hormone)
Sonoray, 167*f*
Sore mouth disease, 554
Sound:
  in animal communication, 637–638
  and insect control, 625
Sows:
  effect of feed on milk of, 359*n*

Sows *(Cont.)*:
  ovary of, 265*f*
  ovum, 275*f*
  reproductive organs of, 263*f*
  teat location of, functional at weaning,
    649*t*
Soybeans, trypsin-inhibiting factor of,
    472
Spanish fly, 602
Species immunity, 540
Sperm capacitation, 274
Sperm-typing for sex preselection,
    325–326
Spermatogenesis, 120*f*, 309
  effect of temperature on, 428
Spermatozoa:
  abnormal, 309*f*
  cold shock of, 310
  concentration of, 304*t*
  enumeration of, 307–308
  fertile life of, 304*t*
  life of, in female tract, 304*t*
  live and dead, 267*f*
  microphotograph of, 259*f*
  morphology of, 308–309
  motility of, 308
  normal and abnormal, 309*f*
  numbers of, per ejaculate, 304*t*
  reaction rates of, 311
  staining of, 310
Spiders, 600*t*
Spinal cord, 208
Spinnbarkeit, 323–324
Splints of horses, 682
Spontaneous ovulation, 274
Spotted fever, 207*t*
Stable flies, 621
Stadia, 597
Stags, 258, 644
Stallions:
  care and management of, 671
  semen of, 671
Standard metabolic rate, 413,
    418–419
Starch, 436, 468*t*
Starvation, defined, 13*n*
State agricultural experiment stations
    (SAES), 700, 711, 743*f*, 746

Statistical evaluation of quantitative
    traits, 151–153
Statistically significant differences, 708
Statistics in research, 706–708
Steapsin, 468*t*, 469
Steely wool and copper deficiency,
    523–524
Steer (*see* Beef steer)
Sterilization, 712
  of male screwworm flies, 610, 622–624,
    712–713
Steroid hormones, 220*f*
Sterols, 439, 446*f*, 485
Stiff lamb disease, 488
Stigma, 375
Stillage residue as protein for livestock,
    749*f*
Stomach compartments of ruminants,
    460–464
Stomoxys calcitrans, 499
Strain 19 vaccine, 562
Streptomycin, use of, in semen, 314
Stress:
  effect of, on animal productivity,
    405
  and enzyme activity, 412–413
  measurement of, 405–406
Sucking stimulus, 359–360
Sucrase, 467–468*t*
Sucrose, 435–436*t*, 446*f*, 468*t*
Sulfur, 519–520
  clinical deficiency symptoms of, 514*t*,
    520*f*
  functions of, 514*t*, 519–520
  sources of, 514*t*, 519
Summit metabolism, 417
Sunburn, 427
  in farm mammals, 427
Superovulation, 274, 282–283
Surrogate mother, 283, 285*f*, 288
Swayback disease in lambs, 523
Sweating:
  cold, 421
  thermal, 421–422
Sweeney of horses, 682
Sweet clover and dicoumarin, 586
Sweet clover disease, hemorrhagic,
    490

Swine:
  in animal research, 730–731, 732–733f, 734
  backfat of, 53f
  breeds of, 48t, 52, 54f
  domestication of, 3t
  heat loss in, 420f
Swine roundworm, life cycle of, 580f
Symbiotic relationships of animals and humans, 398
Synapse, 206
Synchro-mate B, 282
Synchronization:
  of estrus, 223, 281–282
  of ovulation, 281–282
Syndactyly, 286–287
Synovial fluid, 192

Taeniasis, 574–575
Tapeworm, 602t
  of dogs, 601t
  of humans, 574–577
Taxonomist, 598
Taxonomy, 598
TDN (total digestible nutrients), 451–452
Teart, 522t, 533
Teat location of sows, functional at weaning, 649t
Teeth:
  effect of fluorine on, 533f
  of horses, 687–688f
Temperature:
  effect of: on feed digestibility, 470–471
    on reproduction, 273
  rectal, of selected animals, 406
Temperature-regulating mechanisms, body, 409–413
Temperature regulation, 409–412
Tenderizers, meat, 96–97
Termites, 600t, 620
Test-tube babies, 283–284
Testes:
  of avian male reproductive system, 375–376f
  cross section of, 258f

Testes (Cont.):
  hormones of, 217t
Testicle of a boar, 260f
Testicular development and semen output, 303–304
Testosterone, 220f, 223, 266
  and animal behavior, 643
Tests of significance, 707–708
Tetanus, 582
Texas armadillo, 233n
Texas cattle fever, 613–614
Texas fever, 613–614
Thalidomide, 130
Therm, 451
Thermal sweating, 421–422
Thermogenesis, 409, 417
Thermogenesis, 409, 417
Thermolysis, 409
Thermoneutral zone, 402, 404–405, 416–418
  of homeotherms, 417–418
  of poikilotherms, 417–418
Thermoneutrality, 416–417f
Thiaminase, 495, 583
Thiamine (B_1), 494–497
  clinical deficiency symptoms of, 481t, 494–497
  common sources of, 481t, 496
  functions of, 481t, 494–497
  synthesis of, 496
Thiouracil, 457
Thoroughbreds, 661, 668
Thoroughpin, 683
Three-breed cross, 179
Thrush, 683
Thumps, 521
Thyroid gland, hormones of, 216–217t
Thyroidectomy, effect of, on growth, 342f
Thyroprotein, 456
Thyrotropin or thyroid-stimulating hormone (TSH), 216t, 241, 524–525
Thyrotropin-releasing hormone (TRH), 216t
Thyroxine, 216t, 219f, 241, 726f
  effect of: on growth, 342f
    on lactation, 342–343

Thyroxine *(Cont.)*:
  and egg laying, 380–381
  measuring secretion of, in cow, 726*f*
  seasonal changes in secretion of, 343
  and temperature changes, 409
Ticks, 600*t*, 601*t*, 607*t*, 613–614
  life cycle of, 613
  nonscutate, 614
  scutate, 614
  as vectors of disease, 551*t*, 555*t*
Tissue culture, 546–547
Tocopherol, 487
α-Tocopherol deficiency, 489*f*
Total digestible nutrients (TDN), 451–452
Toxaphene, 618
Toxic plants, 583–586
Toxin-antitoxin, reaction, 546
Trace elements, 521–534
Trachea, 204, 205*f*
Traits:
  economic, 150
  qualitative, 150–153
  quantitative, 150–153
Tranquilization of bulls in semen collection, 307
Translocation, 113
Trematodiasis, 577
Trembles, 584
Trench fever, 555, 601*t*
Trial and error, 634
Trichinella spiralis, 573–574
  life cycle of, 575*f*
Trichinosis, 95–96, 573–574, 575*f*, 602*t*
Triglycerides, 88
Triiodothyronine, 216*t*
Trisaccharides, 436*t*
Trypansomiasis, 590, 602*t*
Trypsin, 468*t*, 469
Trypsin-inhibiting factor, 472
Tsetse fly in transmission of African sleeping sickness, 572
TSH *(see* Thyrotropin or thyroid-stimulating hormone)
Tsutsugamushi disease, 555–556, 607*t*
Tuberculosis, 556–558
  in mammals other than cattle, 557–558

Tuberculosis *(Cont.)*:
  methods of controlling infection from cattle to people, 557
Tularemia, 601*t*, 607*t*, 612
  diagnosis of, 544
Turkey meat:
  cholesterol in, 70*t*
  composition of, 67*t*
  minerals of, 76*t*
  vitamins of, 79*t*
Turkeys:
  breeds of, 59, 61
  domestication of, 3*t*
  external parts of, 189*f*
Twinning and hormones, 223
  with ET, 286
Twins, relationships of, 166*t*
Typhoid fever, 567, 607*t*
  diagnosis of, 544
Typhus, 601*t*
Typhus fever, 554

Udder:
  blood circulation in, 332–333
  high-quality, 330, 363
  internal structure of, 331–332
  pendulous, 331
  supporting structures of, 330–331
Ultraviolet (UV) light, 401
Undulant fever, 561–563
United States, the, as an exporter and importer of food, 699*n*
U.S. Department of Agriculture *(see* USDA)
USDA (U.S. Department of Agriculture), 5, 537, 599, 700, 710–711, 737*f*, 739, 742, 747, 756
  Agricultural Research Service (ARS), 711, 742*f*, 743, 744*f*, 750
  Animal and Plant Health Inspection Service (APHIS), 560*t*, 589, 591, 739, 741
  Animal Science Institute, Sire Summary of, 170–171
  biological control of insects facility, 623*f*
  Bureau of Animal Industry, 590

USDA (U.S. Department of Agriculture) *(Cont.)*:
Labeling and Registration Section of, 589
meat inspection by, 588–589
National Animal Disease Center (NADC), 735, 737*f*, 739
Parasitology Research Center, 739
U.S. Meat Animal Research Center, 711, 739
Veterinary Services Division, 589–590, 592*f*, 739, 741
U.S. Department of Health and Human Services, 589–590
sister institutes of, 590
U.S. Food and Drug Administration (FDA), 224, 532, 588–589, 619
U.S. Meat Animal Research Center, 711, 739
USP (U.S. Pharmocopeia units), 510
U.S. Public Health Service (USPHS), 589–590
U.S. Recommended Dietary Allowances (USRDA), 76*t*, 79*t*, 90–91
Unsaturated fatty acids, 70, 438
Urea, 210*f*, 443, 446*f*, 472–473, 519
Urease, 534
Urethra, 209, 260*f*, 263*f*
Uric acid, 210*f*
Urinary system, 208–211
parts of, 209*f*
Urine, components of, 209, 210
Uterine horns, 264

Vaccination, 552
Vaccine, 541
Vagina, 262*f*, 263*f*
artificial, 305*f*, 306
Vanadium, 534
Van't Hoff-Arrhenius equation, 311
Vaporization, 421–422
of moisture, 407
Variation in animal breeding, 151
Variola, 552
Vas deferens, 260*f*
Vasectomy, 260, 322
Vasopressin, 216*t*, 218

Veins, 196
Venezuelan equine encephalomyelitis, 551*t*
Ventral, defined, 187
Ventricular fibrillation, 408
Ventriculus, 460, 462*f*, 467–468
Venules, 198
Verdoflavin, 496
Vesicular stomatitis, 553
Vestibule, 263
Veterinary Services Division (VSD) of USDA, 589–590, 592*f*, 739, 741
VFA (volatile fatty acids), 464, 470
production of, effect of temperature on, 428
as source of energy, 464–465
Vibrio fetus, 569
Vibriosis, 569–570
Vigor tolerance, 626
Villi, 470
Viral infections, 581
transmissible from animals to humans, 546–554
transmissible from humans to animals, 581
Viral vaccine, use of, in controlling cancer, 541–543
Virus, means of reproducing, 130
Visceral larva migrans, 579–580
Visual displays in animal communication, 638–639
Vitamin A, 479–480, 483–484
clinical deficiency symptoms of, 480*t*, 483
common sources of, 480*t*, 483–484
expressed quantitatively, 510
functions of, 479, 480*t*, 483
and ringworm, 484*f*
seasonal variation of, in milk, 363
Vitamin B₂ *(see* Riboflavin)
Vitamin B₁₂:
and cobalt, 527–528
(See also Cyanocobalamin)
Vitamin C, 492–494
and absorption of iron, 493*f*
clinical deficiency symptoms of, 481*t*, 493–494
common sources of, 481*t*, 494

Vitamin C *(Cont.)*:
functions of, 481*t*, 492–494
Vitamin D, 439, 484–487, 510
in bone development, 484–485
clinical deficiency symptoms of, 480*t*,
486–487
common sources of, 480*t*, 487
danger of excess, 91
effect of excess, 486
and egg size, 393
expressed quantitatively, 510
functions of, 480*t*, 484–486
and parathyroid hormone, 343
and parturient paresis, 343
Vitamin E, 487–488
clinical deficiency symptoms of, 480*t*,
487–488, 489*f*
common sources of, 480*t*, 488
functions of, 480*t*, 487
and reproduction, 487
and selenium, 487
Vitamin K, 490–491
clinical deficiency symptoms of, 480*t*,
490–491
common sources of, 480*t*, 491
functions of, 480*t*, 488, 490
role of, in blood clotting, 490*f*
and sex hormones, 490
synthesis of, 442, 488, 490
Vitamins, 445–447, 478–511
of animal products, 78–80
assay techniques for, 509–510
classification of, 446*f*
consumption trends of, 80*t*
daily allowance for, in humans, 79*t*
defined, 478–479
fat-soluble, 445–447, 479–491
fortifying milk with, 91
for horses, 677, 692–694*t*
nomenclature of, 479
providing, for farm animals, 510
secreted in milk, 346
storage loss in milk, 100
term first introduced, 478–479
water-soluble, 445–447, 491–509
Voice box, 204
Volatile fatty acids *(see* VFA)
Vulva, 262*f*, 263*f*

Warble fly, 611
Warbles, 602*t*
Warfarin, 586
Warm-blooded species, 406–407
Wasps, 621
Wassermann test, 394, 546
Waste-management research, 747
Wastes, recycling of, 747–748
Water:
of animals, 434
for horses, 677–678
metabolic, 434
of plants, 434
Waxes, 440, 446*f*
Weaning weight:
effect of: dam's age on, 248*t*
heterosis on, 179*t*
heritability estimate of, 158*t*
Weight effect on food conversion,
474–475
Wernicke's disease, 496
Western encephalitis, 551*t*
Wether, 258
Whey, 31
White blood cells *(see* Leukocytes)
White muscle disease, 488–489, 722
WHO (World Health Organization), 538
Whole-body counter, 717–718
Whole-body counting, 713–718
Wilson rule, 411–412
Witch's milk, 337
Wool, 188, 190
leading countries in production of, 17*f*
World Health Organization (WHO),
538
World land resources, classification of,
18*f*
Worm infections, 573–581

Xerophthalmia, 479–480, 483

Yellow fever, 550, 551*t*, 607*t*
Yogurt, composition of, 67*t*
Yolk of eggs *(see* Egg yolk)
Young horses, care and management of,
674

Zearalenone, 587
Zebu cattle, 404, 411, 419
Zeranol, 456
Zinc, 528–530
    clinical deficiency symptoms of, 522*t*, 528, 529–530*f*
    effect of, on egg laying, 388
    functions of, 522*t*, 528–530

Zinc *(Cont.)*:
    and hatchability of eggs, 530
    sources of, 522*t*, 528, 530
Zona pellucida, 275*f*
Zone cooling, effect of, on milk production, 424
Zoo animals, 590–591, 592*f*
Zoonoses, 593